Principles of
ELECTRODYNAMICS

B. Chakraborty

CELEBRATING 50 YEARS 1960-2010

Since 1960

[*Meant for the* **Postgraduate** *students of* **Physics, Applied Mathematics** & **Engineering**]

Principles of ELECTRODYNAMICS

Bishwanath Chakraborty

Professor (Retired), Mathematics Department
Jadavpur University, Kolkata

BOOKS AND ALLIED (P) Ltd.

No. 1-E(1), 'Shubham Plaza' (1st Floor)
83/1 Beliaghata Main Road, Kolkata 700010 (India)
Tel : (033) 6535-3844 • (033) 6548-5530
e-mail : books1960@ymail.com • booksandallied1960@gmail.com

Principles of
ELECTRODYNAMICS

First Published : April, 2002
Reprinted : 2010
Reprinted : 2016

Publisher : Arunabha Sen
BOOKS & ALLIED (P) LTD.
8/1 Chintamoni Das Lane, Kolkata 700009

Printer : EUREKA
91/A, Baithakkhana Road, Kolkata 700009

Typesetter : BOOKS & ALLIED (P) LTD.
8/1 Chintamoni Das Lane, Kolkata 700009

Price : ₹575.00

Dedicated to the Guide

SHANKAR BRAHMACHARI

ACKNOWLEDGEMENT

I am indebted Prof. C. Kooy of the Department of Electrical Technology of Eindhoven University of Technology (Netherlands) for helpful discussions and information on the literature on the electrodynamics of rotation and other accelerations, during a visit there; and to a lesser extent, for the same, to Prof. F. Verheest of Royal University of Gent(Belgium) at that time.

My colleague Prof. Nemai C. Das once arranged a seminar talk for me on the subject of electrodynamics of rotation. After this talk began to make plans for developing the relevant notes further for publication. So, I remain thankful to Prof. Das. I am indebted to my colleague Prof. Arun K. Ghosh for taking measures which were helpful for may work on the book. I thank my colleague Prof. B.R. Das for his help and cooperation while he was the head of the department of Mathematics. My typist friend Mr. Dilip Mukherjee has made my task of preparing the script much easier than what it would have been. So, I am thankful to him for this favour. I am also thankful to Mr. Dibyendu Ghosh and Dr. Rajarshi Bhattacharjee for helpful suggestions relating to my work on the book. I am obliged to Mr. Somnath Nandi for his help in the matter of publication of the book. Prof. Dipak Ghosh, Dean of Faculty of Science, and my friends Prof. Pradip Niyogi, Prof. Dilip Das, Prof. Subenoy Chakraborty, Dr. Narayan Chakraborty, Prof. Pulin Bhattacharya and Prof. Baidynath Basu are thanked for encouragement and normal support.

I am indebted to my wife Manjula and daughter Madhurima for full cooperation and cheerful adjustment with my various academic demands, including, very much, that for the work on this book.

Bishwanath Chakraborty

PREFACE

Once I thought of solving a problem of wave precession of electrodynamics by transforming the Maxwell-Lorentz equations to a rotating frame. But, I found that even the best books on electrodynamics do not deal with the Maxwell equations for rotation, even for small rotational velocities. To know the reasons I began search of relevant literature for study. The outcome is this book. I contains several chapters on electrodynamics of rotation and other accelerated motions and on the related physics.

Greek symbols α, β, γ when used as subscripts or superscripts run from 1 to 4; similar Latin scripts like i, j, k, ..., run from 1 to 3. Einstein summation convention for repeated indices has been adopted. When the Latin scripts denote generalized coordinates they run for the first n positive integers. Then, in place of the Einstein summation convention, the symbol S stands for sum. Summation is carried out over a contravariant index and a covariant index. An index repeated on the same side of a relation means summation over that index. Free and non-repeated indices on one side of a tensor relation must be balanced by the same indices on the other side. Contravariant components of a vector are denoted by superscripts. CGS Gaussian units have been used.

May 15, 2001

Bishwanath Chakraborty

147/1, Prince Gholam Hussain Shah Road

Flat 2A, Kolkata 700032

ABBREVIATIONS

Lorentz transformation	...	LT
Special theory of relativity	...	STR
General theory of relativity	...	GTR
Galilean transformation	...	GT
Electromagnetics or Electromagnetism	...	EM
Laboratory frame	...	LF
Electromotive force	...	emf
Physics in rotating frames	...	PRE
Electrodynamics in rotating frames	...	ERF
Euler-Lagrange	...	EL
Clockwise	...	CW
Counter clockwise	...	CCW
Classical electrodynamics	...	CE
Galilean Newtonian transformation	...	GNT
General relativity and gravitation	...	GRG
Magnetohydrodynamics	...	MHD
Left circularly polarized	...	LCP
Right circularly polarized	...	RCP
Nuclear magnetic resonance	...	NMR
Finite Larmor Radius	...	FLR
Chew, Goldberger,Low	...	CGL

Contents

Appendices

Chapter 1

Fundamentals of Electrodynamics at a Glance

1.1 Introduction

The fundamentals of electrodynamics are discussed in this chapter. Non-validity in non-inertial frames of reference, of the existing theory calls for a critical discussion on foundations of electrodynamics which has been provided. The field equations of non-inertial electrodynamics with special reference to rotation are considered in terms of the Riemann tensors, in the covariant form in the four space of position co-ordinates and time and also in terms of the familiar 3-vectors in position space, and time of special theory of relativity. Several versions of covariant formulation of the constitutive relations are possible for non-inertial motion. These, using conventional vectors for small rotational velocities, have been discussed. A sequential projection of the fundamentals of the electromagnetic (EM) theory have been attempted for understanding in various aspects of electrodynamics.

Difficulties of non-uniqueness and ambiguities are noticed in the formulation of the theory for electrodynamics in rotating and other accelerating frames of references, and for application of this theory. Knowledge of theory of relativity, both special and general, and other modern activities have created scopes of development of the sophisticated aspects of the subject, which have been discussed. The subsequent chapters provide an extended version of the subject matter of this chapter.

1.2 Electrodynamics and Electrodynamical Laws

1.2.1 Electrodynamics

In electricity and magnetism the primary sources for fields are charges and currents and secondary sources are polarizable and magnetizable matter. Charges in motion give rise to electric currents. Matter is polarized by the Coulomb field and magnetized by electric current.

Considering only electric and magnetic properties, all bodies are either dielectric materials (example : glass, water, crystals) or insulators (example : wood, porcelain) or conductors (example : copper, gold, solver, aluminium) or even mixtures having properties of all these types (example : plasmas, alloys). Bodies, through which electricity does not pass, are called *insulators.* Polarization for the dielectric behaviour occurs when charges have been displaced slightly from prefield positions to new positions by an applied field. Insulators are polarized very little. Dielectric materials are polarized to an extent sufficient for use. In electrical conductors charges are free to move under the action of applied electric field through distances large compared to atomic dimensions. This movement occurs even when electric field is small.

Dielectrics are composed of aggregates of dipoles which are systems of two equal but opposite charges at small distances apart, and bound together by the Coulomb force and other internal forces. Two mechanisms polarize matter. One is by the externally applied field which induces a dipole moment in molecules that have no dipole moment in the absence of an applied field; the other is that, an externally applied field reorients molecules that have permanent dipole moment. Negative charges, owing to their small masses, are displaced more by applied fields than ions which are heavy. Since magnetic poles do not exist in volume large or small, only magnetic dipoles occur; magnetic monopole density and current do not appear in the field equations, though they are responsible for magnetization of matter. Matter is also magnetized by currents of electric charges. The current of a moving electron produces a magnetic dipole both by its own field and by an externally applied field.

The dielectric polarization of matter depends on the translational displacement and the magnetic moment depends on the rotational displacement, both induced by applied fields. So, these are complementary effects because rotations and translations are complementary displacements. Electric and magnetic fields are different aspects of resultant electromagnetic fields. These fields exist at all points in the vicinity of their sources, and are detected by charges if placed at these points. Fields of radiation of moving charges also exist away from their sources and other materials. A simultaneous determination is necessary of motion of charges and the EM field in which they move. Only a strong external field can be considered as predetermined. Electric and magnetic fields exert forces on charges at points in the vicinity of their sources (charges and currents). EM fields exist even in isolation from matter and their sources. So, these sustain themselves independently of any matter, in matter-free space. Electromagnetic fields interact with charged particles in motion. So, fields interact with matter in which charges are distributed continuously. At high velocities, effects by particles are considered through notions about relativity. Also, fields generated by accelerated charges form EM radiation. The field transmits the results of interaction between the charges with the light velocity $c(= 3 \times 10^{10}$ cm/sec). So, the force acting on a given charge depends on the past state of motion of all other charges. The field of the accelerated motion of charges sends EM radiation to the outside space, at the expense of energy of the charges.

Electromagnetic energy is exchangeable with energy of particles, including energy of interaction with charged particles or with energy of a continuous material body. The action at a distance EM theory treats the interaction of particles without fields, and can represent the particles by classical fields. Fields form continuous distributions. Non-static distributions of fields have continuous momentum which can be exchanged with material momentum. The momentum containing those of fields also have their own angular momentum, relative to any chosen origin. The field angular momentum can be resolved into orbital and intrinsic spin components. The electromagnetic field energy $\mathcal{E}$ has also its equivalent mass according to the Einstein relation $\mathcal{E} = Mc^2$. This equivalent mass can be subjected to gravitation. Whenever the field energies are changed by interaction, their masses also undergo change. The uncharged masses are very closely packed assemblies of masses of charges of both signs, such that outside an assembly their net Coulomb field vanishes.

1.2.2 Electric charge and its mass

Classically, an atom is a point nucleus of positive charge (an ion) with a number of circulating point charges of negative sign (electrons) in vacuum. Theoretically, the point charges are without any structure. A charge, positive or negative, has mass and dimension. Classically, the electron has the smallest unit of electric (negative) charge, of magnitude 4.8×10^{-10} esu and mass 9.2×10^{-28} gm. So, the electron charge is the natural unit of electricity. The smallest unit of positive charge is the charge of the hydrogen nucleus, having the magnitude of that of an electron; its mass is about 2000 times heavier. Two charges of opposite sign attract each other according to Coulomb's law of inverse square of the distance. Two charges of the same sign have a similar law of repulsion. Two masses attract each other according to the Newtonian law of inverses square of distance of their separation. Electrons, being much lighter than the ions, are relatively mobile; so, ions generally provide a static background of positive charges for the maintenance of overall charge neutrality of electrified media. Resemblence of law of Newton and Coulomb led to futile attempts to combine electrodynamics and gravitation.

When very large number of point like particles of charges of both sign are closely packed in finite volumes, electric charge can be regarded as granular in nature, and as a continuous fluid, or as mixture of fluids of charges of some species. The charges are confined to materials which are conductors or dielectrics, where their individualities are lost. But, some charges move faster or slower than other charges, as in electron streams, where their individuality is not lost. The laws of mechanics and the laws of electromagnetics describe the motion of charges. The charge and mass of an electron are determined precisely. It has wave nature, spin and magnetic moment, Klemperer (1959, section 1.1 p. 27). In dielectrics the electrons are bound to ions; in conductors they are more or less mobile and in magnetic materials, due to their spin electrons are aligned to the magnetic field. These aspects of charge motion can be studied with the help of particle dynamics, or statistical mechanics of charged particles, Balescu (1963).

1.2.3 Distinction between the kinematics of mass and charge

Charge and associated mass are normally identified as riding on one and the same carrier, say a particle, Post (1986). Since a moving mass cannot exceed the velocity of light in vacuum, the charge also cannot exceed the velocity of light. It is implied that the charge and the mass are at the same point of the 4-space of space-time continuum, and the Newtonian concept of mass point emerges as a charge point. The charge and mass then become the two attributes of a point particle.

Since the EM laws depend on charge and not on mass, charges can be regarded as a primary thing. The mass-energy theorem of relativity theory indicates that mass is the energy manifestation of electric charge, and its associated electric and magnetic fields. In this sense the kinematics of charge is distinct from the kinematics of mass. It is not necessary to impose the restriction that like mass, charge motion also should have the relativistic upper bound, Post (1986).

These and other arguments of quantum mechanical basis indicate the possibility of a separate kinematics for charge and mass. Also the electron is regarded as a ring current involving the elementary charge e, circulating azimuthally as a tabular sheath of distributed charge. Its velocity of circulation may be taken equal to that of light, which would asymptotically promote a balance of counteracting Coulomb and Ampere forces in the charge sheath. This system of circulating charge is non-dissipative and is presumably stable.

Quantum mechanically, each electron is a cloud of charges with the total equal to that of the point electron of the classical theory, but extending over a region of space comparable with the size of the atom. This charge "cloud" represents, for each electron, the spatial probability distribution of its position, and quantum mechanical expected values replace the classical quantities of the microscopic theory. Actually, the atoms must be described quantum mechanically, and the theory of their behaviour developed closely parallel to the classical theory. Critical review is available of the concepts associated with the existence of the electron and its motion, Caldirola (1979).

1.2.4 Electric charge and current

Electric current is due to charge of electricity in motion. It is measured in units of Coulomb per second or Amperes.

The current element $Id\mathbf{r}$ along a filament at a point $P(r)$ on it, where I is the electric current flowing along the filament, equals the current element qv, where q is the charge at P on a small volume and v is its velocity. So

$$Id\mathbf{r} = q\mathbf{v} = \rho^c dV\mathbf{v},$$

where ρ^c is the charge density and dV is the three dimensional volume element at P. write this relation as

$$Id\mathbf{r} = \mathbf{j}dV,$$

where $\mathbf{j}$ is a current vector. Then we can write

$$Id\mathbf{r} = (\mathbf{j}\cdot d\mathbf{\Sigma})d\mathbf{r},$$

where $(\mathbf{j}\cdot d\mathbf{\Sigma})$ is the flux of the current vector $\mathbf{j}$ of electricity through the elementary cross sectional area $d\mathbf{\Sigma}$ of the filament at P. Evidently, the volume element is $dV = (d\mathbf{\Sigma}\cdot d\mathbf{r})$.

If $A = B$ is a field equation, then with the help of the relation $Id\mathbf{r} = \mathbf{j}dV$ we get an integral relation between a line integral in a closed circuit Γ, and a volume integral :

$$\oint_\Gamma AId\mathbf{r} = \iiint_V B\mathbf{j}dV,$$

without the help of theorems of Gauss and Stokes. Such relations are useful in many problems of electrodynamics.

If q is the charge of a particle of mass m and $\mathbf{r}$ is its position vector at time t, and $\mathbf{v}(=\dot{r})$ is its velocity, then the electric current of this charge is $q\mathbf{v}$. For N discrete charges, the electric current is the sum $\sum_i (q_i\dot{\mathbf{r}}_i)$, where q_i is the ith charge and $\mathbf{r}_i$ is its position vector at time t. If V is the volume containing these charges, then the density of charge, ρ^c, and the corresponding current $\mathbf{j}$, are $\left(\sum_i q_i\right)/N$ and $\sum_i (q_i\mathbf{v}_i)/N$ respectively. Using concepts of delta function $\delta(x)$ of Dirac, Appendix 1, for the charge density, we have

$$\rho^c = \frac{1}{V}\left(\sum_i q_i\delta(\mathbf{r}_i - \mathbf{r}(t)\right), \quad \mathbf{j} = \frac{1}{V}\sum_i q_i\dot{\mathbf{r}}_i\delta(r - r_i(t)).$$

When a number of species of charged particles is present in a volume V, and for the of the ith species $\mathbf{v}_i, m_i$ and q_i are the velocity, mass and charge respectively, then the density of charge, mass and current, respectively, are $(N_i m_i)$, $(N_i q_i)$, $(N_i q_i \mathbf{v}_i)$, of the ith species; and the average density of mass, charge and current in V are

$$\frac{1}{V}\sum_i^N (N_i m_i), \quad \frac{1}{V}\sum_i (N_i q_i), \quad \frac{1}{V}\sum_i (N_i q_i \mathbf{v}_i)$$

respectively, where N is the number of species present in V. For a continuous distribution of all species of particles, these sums are replaced by the definite integrals

$$\frac{1}{V}\sum_i \iiint_V (N_i m_i) dV, \quad \frac{1}{V}\sum_i \iiint_V (N_i q_i) dV, \quad \frac{1}{V}\sum_i \iiint_V (N_i q_i \mathbf{v}_i) dV.$$

In a fully ionized hydrogen plasma $N = 2$ and there are only free electrons, each of charge $-e$ and fully stripped ions, each of charge e.

If l is the distance separating an electron from an ion, the dipole moment of this system is le. For a continuous distribution, if the position vector of ions is $\mathbf{R}$ and ρ^c is the density of electrons at $\mathbf{R} + \mathbf{r}$, the dipole moment is $\rho^c \mathbf{r}$, and the total dipole moment in a volume V is the integral $\iiint_V \rho^c \mathbf{r} dV$. The magnetic moment of a charge q at the point $P(\mathbf{r})$, moving with velocity $\dot{\mathbf{r}}$ is

$$\mathbf{m} = \frac{q}{2c}(\mathbf{r} \times \dot{\mathbf{r}}).$$

For a continuous distribution of currents in space, if $\mathbf{j}$ is the current density at P, then the magnetic moment per unit volume at P, is

$$\mathbf{M} = \frac{1}{2c}(\mathbf{r} \times \mathbf{j}).$$

1.2.5 Electrodynamical laws

Oersted, Biot-Savart and Ampere's law for the magneto electric induction is : Wires carrying electric current produce magnetic flux and force acts between two such wires. This is evident from deflection of permanent magnetic dipoles situated nearby. So the electric currents are the sources of magnetic flux. These basic laws relate properly the magnetic induction $\mathbf{B}$ to the currents $\mathbf{j}$. The law of force between two currents was found to be inversely proportional to the square of the distance separating the current elements in the two circuits.

This law of field induction was generalized by Maxwell through an involvement of the displacement current in this process. The displacement current is effective in rarefied plasmas and for very high frequency waves (radio frequency and still higher frequencies but less than the X-ray frequencies for study of which physics of quantum mechanics is necessary). For supply of electric power through conducting wires for human use the displacement current is negligible, because the field frequency (about 60 per second) is not so high for this purpose. So, magnetostatic laws are valid for currents for household purposes.

If, at a point P, $d\mathbf{l}$ is an element of length along the direction of a filament which carries a current I and $\mathbf{r}$ is the position vector drawn from P to an observation point Q, then experiments of Oersted, Biot and Savart for the magnetic flux differential $d\mathbf{B}$ at the observation point Q give

$$d\mathbf{B} = I(d\mathbf{l} \times \mathbf{r})/cr^3.$$

Like Coulomb's law of electrostatics, it is a law of inverse square distance. This $d\mathbf{B}$ field exists as one infinitesimal element of a sum over a continuous set. So, the sum gives the magnetic induction of a loop or circuit of current. Actually, currents of electricity are charges in motion, and this B is for one charge only. This current is such that as the element of charge dq goes to zero the number of sub-divisions of the N charges goes to infinity. Then the sum of the exact fields including acceleration fields gives the B field from integration over the current. So,

$$I d\mathbf{l} = \mathbf{v} dq,$$

where q is the charge and $\mathbf{v}$ is its velocity. The magnetic flux density for such a charge in motion is

$$\mathbf{B} = (q/c)(\mathbf{v} \times \mathbf{r})/r^3.$$

For steady state magnetic induction fields useful results obtain from integration of $d\mathbf{B}$ over the circuits.

Theory of electromagnetics is based on two main sets of laws, which are discussed in the next section. One set gives the universally valid Maxwell equations which do not contain any material constants. The other set of laws, complementary to these, called the *constitutive relations* (or the *state relations*), depend on the nature of the interaction of the applied field with matter. So, these laws depend on the nature and intensity of the applied field and on the constitution and motion of the material under examination.

Newton's third law (Forces of action and reaction are equal and opposite) provides both physical significance and the real content to the concept of force, and is meaningful only when both forces of action and reaction act at the same 4-point (the concept of absolute simultaneity). The laws of electrodynamics were originally developed for weak fields acting in media having simple structure in the laboratory rest frames. For modern applications generalization becomes often necessary of these laws.

The problems from observations of rotating and stationary observers, relative to rotating conductors, and other difficulties cannot be explained by the traditional renderings of the basic laws of electrodynamics. So, for the right theory, either the Maxwell equations are to be regarded as valid for accelerated frames, and the constitutive relations are to be changed, or the Maxwell equations are to be modified without changing the constitutive relations or both are to be modified. The current tendency is in favour of changing the constitutive relations without changing the Maxwell equations. These problems have lead to questions on the justification of the present practice of depending, on the field vectors which are local functions of the space-time continuum. It is found that two classes of field quantities are involved in experimental measurements and theoretical investigations. One is the observable quantities, called the *primary quantities* or the *domain quantities* :

$$\text{emf} = \int_{\Gamma} (E \cdot dl)$$

$$\text{Field flux} = \iint_{\Sigma} (B \cdot d\Sigma)$$

$$\text{Current flux} = \iint_{\Sigma} (j \cdot d\Sigma),$$

where Γ is the closed contour of a curve and Σ is the surface of a material, dl is a line element and $d\Sigma$ is a surface element. Evidently, these are integrals over well defined domains. Faraday's law of EM induction deals with relations between the observable primary quantities, like the emf and the rate of change of magnetic flux through a surface, which are integrated quantities. These are valid over specific domains, so these are global laws. The other electromagnetic quantities are the field variables. These are distinct from the domain quantities, but follow from these, and obtain their properties from these. The field variables are local point functions of space co-ordinates and time, and are derived quantities. Examples are the field vectors $\mathbf{E}, \mathbf{B}, \mathbf{H}, \mathbf{D}, \mathbf{J}$. From the generally valid global laws, involving the domain quantities, the local laws (the partial differential equations of electrodynamics of Maxwell), are derived under some assumptions.

The EM laws of induction and their applications depend on inertial frame observations involving the domain quantities. In the purely spatial domain Faraday originally imagined, for the magnetic flux, the tubes of force which close on themselves, which are integrated quantities and which define the laws of EM induction. A primary role should, therefore, be given to the observable quantities, with respect to the frame of reference.

The EM field concepts of the macroscopic continuum physics are distinct from the particle interaction concepts of the Lorentz theory in classical electrodynamics. Both field concepts and the concepts of particle motion in vacuum, in presence of these fields, are simultaneously necessary in electrodynamics in non-inertial frames.

A proper sequential presentation of electrodynamics, including the related constitutive relations, is desirable for finding its complete evolution in accelerated frames. A systematic summarization of the available information should be able to expose the problems and the ideas working for the emergence of an improved EM theory.

1.2.6 Basic postulates of Einstein for the special theory of relativity

Einstein accepted the experimental conclusions of Michelson that the velocity of light (electromagnetic waves) does not depend on the velocity of the observer and the source. So, light and other EM waves spread out in straight lines in all directions with the constant velocity c (the wave phase velocity) which is independent of the motion of the observer and the source of the field. So, the real problem is to obtain a space-time relation (the Lorentz transformation) that gives this constancy of the velocity of light; and to see whether determination is possible of a system at absolute rest, and of the same scale of time for all systems. The postulates of Einstein are : (i) No means exist of determining a system at absolute rest and no time scale exists which is the same for all systems. Only the relative velocity can be measured. (ii) Vacuum velocity of light, c, is independent of the direction of its propagation and motion of both the source and the observer. This velocity is independent of its direction of propagation (the principle of isotropy of space) and motion of both the source and the observer. It is an absolute constant, $c = 3 \times 10^{10}$ cm/sec; and no material velocity can reach or exceed this value.

1.3 Basic Laws of Electrodynamics

Electrodynamics is based on laws of universal validity which have been formulated slowly and gradually with the help of experiments, conducted by many persuers, well-known and little known, in the last few centuries of this era. The basic field variables of classical electrodynamics are **E** (the electric field intensity vector), **D** (the electric displacement vector), **H** (the magnetic field intensity vector) and **B** (the magnetic induction, or the magnetic flux density vector). In CGS Gaussian units they have the same dimension. The distinction between **E** and **D**, and between **B** and **H** should be evident from the field equations of electrodynamics. The accepted convention for using these are that **E** and **B** shall appear in the Lorentz force law and **D** and **H** shall be related to the charge current density in the Ampere-Maxwell equations.

1.3.1 Laws of Gauss for electricity and magnetism

The first law, the law of Gauss for electricity, states that the normal flux of **D** from a closed surface Σ, in CGS units, is 4π times the total charge inside $\mathbf{\Sigma}$. It is a consequence of the experimental law of Coulomb which states that the force between two fixed charges is proportional to the square of the inverse distance between these. The second law is for magnetism, and is similar. It states that B is solenoidal (divergence free), that is, in other words, the total normal flux of magnetic induction from a closed surface vanishes because magnets have no monopoles. The integral forms of these two laws are

$$\iint_{\Sigma} (D\cdot d\Sigma) = 4\pi\rho^c \tag{1.3.1}$$

$$\iint_{\Sigma} (B\cdot d\Sigma) = 0, \tag{1.3.2}$$

where Σ is the surface of a material of volume V, $d\Sigma$ is a surface element on $\mathbf{\Sigma}$ and ρ^c is the net charge per unit volume at the integration point inside V.

1.3.2 Faraday's law of electromagnetic induction

The third law, known as *Faraday's law of electromagnetic induction,* states that a steady current flows in a circuit consisting of a loop of conducting wire, if a magnet is moved across it or the wire circuit is moved to cut the magnetic lines of force. This current lasts so long as there is motion between the area covered by the circuit loop and the lines of force of the magnet; the induced electromotive force[1] around the circuit is proportional to the time rate of change of magnetic flux through the circuit; and the induced current is opposed. Briefly, the law reads : The total emf induced in a closed circuit is equal to the time rate of decrease of the total magnetic flux linking the circuit. So, in CGS units, the law is

$$\oint_{\Gamma} (E\cdot dl) = -\frac{1}{c}\frac{d}{dt}\iint_{\Sigma} (B\cdot d\Sigma), \tag{1.3.3}$$

where Γ is a closed curve on Σ, dl is a line element in Γ, and c is the vacuum speed of light. The integral for **E** represents the work done by the electric field in moving a unit positive charge

---

[1] In the expressions for the electromotive force $\mathbf{E}\left(=\oint_{\Gamma}(E\cdot dl)\right)$ of equation (1.3.3) and the magnomotive force $\left(\oint_{\Gamma}(H\cdot dl)\right)$ of equation (1.3.5) the word 'force' is used in its primitive sense of energy.

once round the contour Γ; the surface integral depends only on the field and the contour. For stationary circuits this law gives

$$\oint_{\Gamma} (E \cdot dl) = -\frac{1}{c} \iint_{\Sigma} \left(\frac{\partial B}{\partial t} \cdot d\Sigma \right) \tag{1.3.4}$$

1.3.3 The Ampere-Maxwell law of magneto-electric induction

The fourth law, associated with the names of Oersted, Biot-Savart, Ampere, Faraday and Maxwell, states that the line integral of the magnetic field intensity H around a closed contour (the magnetomotive force) equals the current enclosed. In other words, the law states that the number of electric lines, which traverse an arbitrary surface is accompanied by the magnetomotive force and is equal to it both in magnitude and direction. In a material body, it gives the connection between motion of magnets, currents in circuits and dielectric behaviour and continuity of charges. Mathematically, the law reads,

$$\oint_{\Gamma} (H \cdot dl) = \frac{1}{c} \iint_{\Sigma} (K \cdot d\Sigma), \tag{1.3.5}$$

where K is the electric current flux. This current is the sum

$$4\pi \iint_{\Sigma} (j \cdot d\Sigma) + (d/dt) \iint_{\Sigma} (D \cdot d\Sigma), \tag{1.3.6}$$

where j includes both conduction and convection currents of electricity; the second term is the rate of change of the flux of D across the surface Σ and is named the displacement current by Maxwell. If I is the volume current inside Σ, then

$$I = \iint_{\Sigma} (j \cdot d\Sigma) \tag{1.3.7}$$

and (1.3.5) reads

$$\oint_{\Gamma} (H \cdot dl) = \frac{4\pi}{c} I + \frac{1}{c} \frac{d}{dt} \iint_{\Sigma} (D \cdot d\Sigma). \tag{1.3.8}$$

So, another version of the law is that the magnetomotive force, acting over a closed loop, equals the sum of the current through the loop and the rate of change of flux of the electric displacement vector through surface Σ containing the loop. For stationary circuits, therefore,

$$\oint_{\Gamma} (H \cdot dl) = \frac{4\pi}{c} \iint_{\Sigma} (j \cdot d\Sigma) + \frac{1}{c} \iint_{\Sigma} \left(\frac{\partial D}{\partial t} \cdot d\Sigma \right) \tag{1.3.9}$$

$\partial D / \partial t$ is the displacement current per unit area at a point on Σ. Maxwell introduced it to account for the production of the magnetic field in empty space. The induction laws in the form (1.3.4) and (1.3.9) give the partial differential equations of electrodynamics in vector form, in stationary media. The integral laws (1.3.3) and (1.3.8) are valid for moving media, as well.

1.4 Differential Equations of Maxwell

1.4.1 In material media

In stationary circuits, in inertial frames of reference, the differential equations for the four electromagnetic laws, obtained from (1.3.1), (1.3.2), (1.3.3), (1.3.6), with the help of Green's theorem and Stoke's theorem (Appendix 2), and assuming the validity of the laws in arbitrarily small volumes and surfaces, in CGS Gaussian units, are

$$(\nabla \cdot D) = 4\pi\rho^c, \tag{1.4.1}$$

$$\nabla \cdot B = 0, \tag{1.4.2}$$

$$\nabla \times E = -\frac{1}{c}\frac{\partial B}{\partial t}, \tag{1.4.3}$$

$$\nabla \times H = \frac{1}{c}\frac{\partial D}{\partial t} + \frac{4\pi}{c}j \tag{1.4.4}$$

$$(\nabla \cdot j) + \frac{\partial \rho^c}{\partial t} = 0. \tag{1.4.5}$$

Taking divergence of (1.4.4) and using (1.4.1) gives the continuity condition (1.4.5) for the current of electricity.

For finding the standard form of equations of electrostatics and magnetostatic we emphasize that the charge density ρ^c of (1.4.1) is the density ρ_f of free charges and the current density j is current j_f of this free charge density. So these two equations may be written as

$$(\nabla \cdot D) = 4\pi\rho_f, \quad \nabla \times H = \frac{1}{c}\frac{\partial D}{\partial t} + \frac{4\pi}{c}j_f. \tag{1.4.6}$$

Then we write

$$\rho^c = \rho_f + \rho_p = \rho_f - (\nabla \cdot p), \quad j = j_f + j_b + j_p = j_f + \nabla \times M + \frac{\partial p}{\partial t}, \tag{1.4.7}$$

where ρ_p is the polarization charge density, and M is the magnetization density, j_b is the current density of the bound charges responsible for generation of M and j_p is the polarization current; so it equals $\partial p/\partial t$. Using (1.4.7) in (1.4.6) and the state relations $D = E + 4\pi P$, and $B = H + 4\pi M$, we obtain

$$(\nabla \cdot E) = 4\pi\rho^c, \quad \nabla \times B = \frac{1}{c}\frac{\partial E}{\partial t} + \frac{4\pi}{c}j. \tag{1.4.8}$$

The standard form of Ampere's law

$$\nabla \times B = \frac{4\pi}{c}j, \tag{1.4.9}$$

is free from the displacement current $\partial E/\partial t$. The magnetization M obtains from the current loops which penetrate only once the surface Σ of the material. Orbits which penetrate twice contribute nothing to the net current. If dl is the line element of a closed circuit Γ in Σ, and N is the number of current loops per unit volume, each containing the average current I, then

$$M = N\mu = NI\Sigma, \tag{1.4.10}$$

where μ is the magnetic moment vector per loop.

1.4.2 Electrodynamics of isolated charges in vacuum

For motion of isolated charges in vacuum, there is no mechanism of material polarization and magnetization, so we take $\varepsilon = 1$ and $\mu = 1$. Then the state relations $B = \mu H$ and $D = \varepsilon E$, simplify to the Gaussian field identification criteria

$$B = H, \quad D = E \tag{1.4.11}$$

and the Maxwell equations become

$$\nabla \times E = -\frac{1}{c}\frac{\partial H}{\partial t}, \tag{1.4.12}$$

$$\nabla \cdot H = 0, \tag{1.4.13}$$

$$\nabla \times H = \frac{1}{c}\frac{\partial E}{\partial t} + \frac{4\pi}{c}j, \tag{1.4.14}$$

$$\nabla \cdot E = 4\pi\rho^c. \tag{1.4.15}$$

Since

$$j = \rho^c u, \tag{1.4.16}$$

the charge conservation condition (1.4.5) becomes

$$\frac{\partial \rho^c}{\partial t} + \nabla \cdot (\rho^c u) = 0. \tag{1.4.17}$$

An alternative form of this relation is

$$\frac{d}{dt}(\rho^c dV) = 0, \tag{1.4.18}$$

where dV is the volume element occupied by the charge element $de(= \rho^c dV)$. When the terms of (1.4.17) are integrated over the whole volume V of a body, the total charge q remains constant in time. For a particular particle of volume V in a system, the volume integral

$$q = \iiint_V \rho^c(r,t)dV \tag{1.4.19}$$

gives the charge q. Integration of (1.4.5) over a volume V gives

$$\frac{\partial}{\partial t}\left(\iiint_V \rho^c dV\right) = -\iint_\Sigma (j \cdot d\Sigma).$$

If Σ is large enough to contain all matter, including all currents in the Universe, then, over such a surface, the integral on the right hand side vanishes and the conservation condition becomes

$$\iiint_V \rho^c dV = \text{constant},$$

where V denotes the whole volume of the universe. This conservation condition shows that the total charge in the universe is conserved.

1.5 Electrodynamics Chronicle

Coulomb gave the inverse square law of force between two charges. Oersted noticed that an electric current deflects a magnetic needle. A magnetostatic field is produced by a steady current by the *law of Oersted, Biot-Savart and Ampere.*

$$B(r) = \frac{I}{c} \oint_{\Gamma} \frac{1}{r^2} (d\mathbf{l} \times \mathbf{r}),$$

where dl is a line element along the wire current I, integration is along this direction of flow and $\mathbf{r}$ is the position vector joining the field point P to the integration point in the wire.

Stationary charges produce electrostatic fields. The field of *non-fluctuating charge distributions* is also practically electrostatic. Steady currents produce magnetostatic field. Slowly fluctuating fields are approximately static fields of magnetism and electricity. The city power supply, to residents, fluctuates at the rate of 60 per second, so approximation of magnetostatics (*Ampere's law*) is valid for this electric power.

The *displacement current* $\partial D/\partial t$ was incorporated by Maxwell into *Ampere's law,* (1.4.6), to make electrodynamics theoretically complete. The *law of Oersted, Biot-Savart and Maxwell is the law of magneto-electric induction,* because it involves the *magnetomotive force* $\oint_{\Gamma}(H \cdot dl)$ and the time rate of variation of the *electric displacement vectors* D. *Faraday* discovered that a moving magnet generates an electric current. The subjects under the headings of Electricity and Magnetism, have been guided by the four universal laws into the single subject called *Electromagnetics* or *Electrodynamics,* or *simply Electricity and Magnetism,* as a result of discoveries of the theories of relativity so that the two fields (the electric field and the magnetic field) lost their indeviduality. The efforts of Maxwell, Lorentz, Einstein, Minkowski in this regard have brought the subject into its present vast form.

The concept of photon (light radiation consisting of discrete amount of energy at points in the form of particles or light quanta), given by Einstein is important in physical sciences. Another important work is of the two volumes, Electrodynamics of continuous media and the Classical theory of fields by Landau and Lifshitz. The potentiality of these masterpieces on electrodynamics for guiding the activity on this subject has not exhausted.

Einstein's general theory of relativity has opened many vistas, including that for the study of electrodynamics of rotating and other accelerating frames of reference. This theory has also created problems for full development of electrodynamics. The question of which of the basic laws and assumptions of EM theory are to be modified, to make the subject free from difficulties, for acceleration, remains unanswered. There is tremendous expectation of service to human civilization from electricity and magnetism. The expectations are realized through experiments which give rise to newer expectations.

1.6 The Constitutive Relations

The Maxwell equations of electrodynamics are not sufficient in number to make of these a closed system[2] of equations, which is necessary for analytical study of problems. For this purpose finding relations, called the *constitutive relations* or the *state relations* is necessary, the number

[2] A set of equations form a closed system when the number of independent equations equals the number of unknown quantities, so that, in principle the equations can be solved.

of which together with the number of independent Maxwell equations should be sufficient to form a closed system, or a *complete system* of equations for study of electrodynamics. The *constitutive relations* are guessed empirically for convenience of experiments. So these depend on the parameters defining the applied field of force (the field amplitude, frequency, etc. for wave fields) and the material used for investigation (its electrical conductivity σ, the dielectric constant ε, magnetic permeability μ). But the Maxwell equations (section 1.3), being fundamental laws, are universally valid relations. So, these do not depend on the parameters of the medium and of the applied force. These universally valid relations bear only some universal constants like the vacuum speed of light $c(= 3 \times 10^{10}$ cm/sec) and $\pi(= 22/7)$. So, the Maxwell equations are *generally true* and the constitutive relations are *specifically true.*

Evidently, the *microscopic equations of electrodynamics* contain only universal constants like the mass and charge of ionized particles and the vacuum speed of light. But the phenomenologically based macroscopic equations of Maxwell contain material parameters ε, μ, σ, etc. In the macroscopic approach, therefore, electrodynamics is linked with mechanics of deformation which is suitable for elastic changes. In the macroscopic approach, electrodynamics is combined with analytical mechanics of charges in position space, or statistical mechanics of *charges in phase space.*

Equation (1.4.2) is implied in equation (1.4.3) because the divergence of (1.4.3) gives $(\partial/\partial t)(\nabla \cdot B) = 0$. So, $(\nabla \cdot B) = 0$ for all time, if $\nabla \cdot B = 0$ at any time, and all the four equations are not linearly independent. Equation (1.4.5) and the divergence of the Ampere-Maxwell equation give $\partial((\nabla \cdot D) - 4\pi\rho^c)/\partial t = 0$. So, if $\nabla \cdot D = 4\pi\rho^c$ at any instant, this relation will remain valid for all time.

Evidently, therefore, equations (1.4.1)–(1.4.4) and the continuity condition (1.4.5) taken together give only seven linearly independent differential equations for the sixteen field variables E, H, D, B, j and ρ^c. Hence, the Maxwell equations are not a closed (or, complete) system of equations. For their solution, some more linearly independent relations are necessary. These are the experimentally determined constitutive relations (or the state relations). For weak fields and large scale phenomena in stationary materials having simple internal structures, these relations are

$$j = \sigma E \tag{1.6.1}$$

$$B = \mu E \tag{1.6.2}$$

$$D = \varepsilon E, \tag{1.6.3}$$

where the constants σ (electrical conductivity), μ (magnetic permeability) and ε (electrical permittivity) depend on the nature of the material and the parameters of the applied fields. Equation (1.6.1) is Ohm's law. For a conductor which moves with velocity $\mathbf{v}$, Ohm's law reads

$$\mathbf{j} = \sigma \mathbf{E} + \frac{\sigma}{c}(\mathbf{v} \times \mathbf{B}). \tag{1.6.4}$$

The universally valid Maxwell equations (1.4.1)–(1.4.5) are free from the material constants.

The general forms of the constitutive relations, involving the vectors, D, E, B, H, follow from the relativistically valid Minkowski formulation of electrodynamics in moving media, (Kong (1975), Chapter 2 and 7). These are be written in terms of the constitutive parameters $\lambda, \xi, \eta, \zeta$,

$$D = \lambda : E + \xi : B \tag{1.6.5}$$

$$H = \eta : E + \zeta : B. \tag{1.6.6}$$

The components of $\lambda, \xi, \eta, \zeta$ are elements of 3×3 matrices. These depend on the nature of the material under consideration. These are dimensionless real or complex constants or functions of the position co-ordinates and time. For weak fields these parameters are constants, and for materials of simple structures, the matrices $\lambda, \xi, \eta, \zeta$ are proportional to unit matrices.

Equations (1.6.5) and (1.6.6), in matrix form, are

$$\begin{bmatrix} D \\ H \end{bmatrix} = \begin{bmatrix} D_1 \\ D_2 \\ D_3 \\ H_1 \\ H_2 \\ H_3 \end{bmatrix} = \nu : \begin{bmatrix} E \\ B \end{bmatrix} = \begin{bmatrix} \nu_{11} & \nu_{12} & \nu_{13} & \nu_{14} & \nu_{15} & \nu_{16} \\ \nu_{21} & \nu_{22} & \nu_{23} & \nu_{24} & \nu_{25} & \nu_{26} \\ \nu_{31} & \nu_{32} & \nu_{33} & \nu_{34} & \nu_{35} & \nu_{36} \\ \nu_{41} & \nu_{42} & \nu_{43} & \nu_{44} & \nu_{45} & \nu_{46} \\ \nu_{51} & \nu_{52} & \nu_{53} & \nu_{54} & \nu_{55} & \nu_{56} \\ \nu_{61} & \nu_{62} & \nu_{63} & \nu_{64} & \nu_{65} & \nu_{66} \end{bmatrix} : \begin{bmatrix} E_1 \\ E_2 \\ E_3 \\ B_1 \\ B_2 \\ B_3 \end{bmatrix}, \tag{1.6.7}$$

where

$$\nu = \begin{bmatrix} \lambda & \xi \\ \eta & \zeta \end{bmatrix}. \tag{1.6.8}$$

This λ is a 6×6 constitutive matrix.

Equations (1.6.5) and (1.6.6) relate D and H with E and B. Alternative equivalent constitutive relations, in which D and B are related to E and H, are

$$D = \varepsilon : E + \alpha : H \tag{1.6.9}$$

$$B = \delta : E + \mu : H \tag{1.6.10}$$

Their matrix form is

$$\begin{bmatrix} D \\ B \end{bmatrix} = P \begin{bmatrix} E \\ H \end{bmatrix} = \begin{bmatrix} \varepsilon & \alpha \\ \delta & \mu \end{bmatrix} : \begin{bmatrix} E \\ H \end{bmatrix}. \tag{1.6.11}$$

Comparison with (1.6.5) and (1.6.6) shows that

$$P = \begin{bmatrix} \lambda - \xi : (\eta \cdot \zeta)^{-1} & \xi : \zeta^{-1} \\ -\eta : \zeta^{-1} & \zeta^{-1} \end{bmatrix}, \tag{1.6.12}$$

where ζ^{-1} is the matrix inverse to ζ, etc., and all the constitutive matrices, for physical reasons, are non-singular. The fact is that those 3×3 matrices in (1.6.5) and (1.6.6), which give the 3×3 *dielectric tensor* ε_{ij} and the 3×3 *magnetic permeability tensor* μ_{ij} are essentially non-singular matrices. So, matrices of inversion of all these matrices exist. Some of these matrices are even purely null 3×3 matrices for a medium of very simple structure.

Constitutive relations for *anisotropic crystals* are

$$D_i = \varepsilon_{ij} E_j, \quad j = 0, \quad B = \mu H, \tag{1.6.13}$$

where the Einstein summation convention (repeated index means summation over that index) has been adopted. Strong fields generate non-linear interaction in matter. The relevant non-linear constitutive relation is

$$D_i = \varepsilon_{ij} E_j + \varepsilon_{ijk} E_j E_k + \varepsilon_{ijkl} E_j E_k E_k + \cdots .. \tag{1.6.14}$$

To explain some aspects of *optical rotation* of elliptically or circularly polarized waves, it becomes useful to assume that the linear relation between D and E also contains a linear combination of the space derivatives of E, so

$$D_i = \varepsilon_{ii} E_j + \gamma_{ijk} \frac{\partial E_j}{\partial \chi_k}, \tag{1.6.15}$$

where γ_{ijk} is a third order tensor. For *optically active isotropic substances,* we put $\gamma_{ijk} = \gamma\varepsilon_{ijk}$ (where ε_{ijk} is the alternative unit tensor, Appendix 4) and obtain

$$D_i = \varepsilon E_i + \gamma \left(\frac{\partial E_j}{\partial \chi_k} - \frac{\partial E_k}{\partial \chi_j} \right). \tag{1.6.16}$$

The corresponding vector constitutive relation is

$$D = \varepsilon E + \gamma \nabla \times E. \tag{1.6.17}$$

Also, the integral equations are valid over finite space domains at a time; whereas, the differential equations are valid over all space-time points in a region.

Equations (1.4.2) and (1.4.3) are the *homogeneous* (or, *source free,* or *kinematical*) Maxwell equations and (1.4.1) and (1.4.4) are the *inhomogeneous* (or *source-full,* or *dynamical*) Maxwell equations. Also some authors call the *curl equations* (1.4.3) and (1.4.4) as the Maxwell equations and (1.4.1) and (1.4.2) as the *divergence equations.* The *homogeneous Maxwell equations* are independent of the characteristics of the medium in which the circuit is located.

This *phenomenological theory* for the constitutive relations is free from the Lorentz force $\rho^c E + (j \times B)/c$. Actually, the effect of the electromagnetic force is here indirectly included in *Ohm's law* (1.6.1) or (1.5.4) which empirically accounts for the motion of charges in the presence of electric and magnetic fields and for dissipation of electromagnetic power into heat.

The *non-linear constitutive relation* (1.6.14) has been formulated for use in *non-linear optics* (also called *quantum electronics* or *laser optics*) because of its importance in lasers (Light Amplification by Stimulated Emission of Radiation) and masers (Microwave Amplification by Stimulated Emission of Radiation). Laser fields have potentials of the order of 10^5 volts/cm or more. These being of the order of quantum mechanical forces binding charges inside molecules, distort the molecular and atomic structures of material bodies. Action of these fields in matter is classically explained in terms of non-linear interaction of waves with matter with the help of the *non-linear constitutive relation* (1.6.14).

Laser is a source of light with a well-defined amplitude and phase, like a traditional radio transmitter. Spontaneous emission of energy from a laser source is a consequence of the quantum nature of light. Very high pressure of laser light generates forces so strong that mechanical phenomena overtake optical effects, so that the two actions mix up in laser affected matter, and thermal effects become important. Laser light has a high degree of spatial and temporal coherence. So, it is a source of very intense monochromatic EM waves. Lasers are not continuous suppliers of such light, pulses of short duration are emitted by these sources.

At a laser spot the field intensity is about 10^{15} watts/cm^2. So the oscillating electric and magnetic fields have peak value of about 9×10^{18} volts/cm and 3×10^6 gauss respectively, the radiation pressure is about 3×10^5 atmosphere and the field induced quiver velocity is that of 16 kcv electrons. The internal field holding the electrons in orbits about their inner positively charged cores are, for hydrogen atom is 5×10^9 volts/cm; for condensed matter it is smaller; for some semiconductors, it is about 10^7 volts/cm. These fields are comparable to the pressure of fields of focussed intense laser beams. Their optical effects are given by the non-linear state relation (1.6.14). A consequence of the non-linearity is the appearance of higher harmonics at twice and thrice of the fundamental frequency in non-linear crystals which have been observed experimentally.

1.7 Analysis of the Equations of Electrodynamics

The four Maxwell equations have been written below as a set of differential equations or integral equations.

Name of the law	Differential equations	Integral equations
Faraday's law for electromagnetic induction	$\nabla \times E = -\frac{1}{c}\frac{\partial B}{\partial t}$	$\oint_\Gamma (E\cdot dl) = -\frac{1}{c}\frac{\partial}{\partial t}\iint_\Sigma (B\cdot d\Sigma)$
The Ampere-Maxwell law for magneto-electric induction	$\nabla \times H = \frac{1}{c}\frac{\partial D}{\partial t} + \frac{4\pi}{c} j$	$\oint_\Gamma (H\cdot dl) = \frac{4\pi}{c} I + \frac{1}{c}\frac{\partial}{\partial l}(D\cdot d\Sigma)$
Gauss's law for electricity	$\nabla\cdot D = 4\pi\rho^c$	$\iint_\Sigma (D\cdot d\Sigma) = 4\pi \iiint_V \rho^c dV$
Gauss's law for magnetism	$\nabla\cdot B = 0$	$\iint_\Sigma (B\cdot d\Sigma) = 0$

The integral relations are formulated on the basis of experimental findings. The set of differential equations follows from these for phenomena in stationary media. So, the integral set is more generally valid than the set of differential equations.

1.8 The Lorentz Force

The Lorentz force, the force of electromagnetic field, is an experimentally determined quantity. For an electron of charge q, velocity $\mathbf{v}$, this force is $qE + q(\mathbf{v} \times H)/c$. The first term, qE, is the force exerted by the electric field on the charge q; it acts along the direction of E. The second part, $q(\mathbf{v} \times H)/c$, is due to the field of H, and is perpendicular to both H and the instantaneous direction of $\mathbf{v}$. The Lorentz force is the only theoretical means available as a force of electromagnetic fields, which acts on materials for producing physical effects. It gives an operational definition of fields of E and H, measuring the acceleration of a charged particle. In a continuous medium, if ρ^c is the net charge density, and $\mathbf{j}$ is the electric current vector, then the Lorentz force per unit volume of charge and mass at a field point P, position vector $\mathbf{r}$, at time t, is

$$F = E + \frac{1}{c}(j \times B). \tag{1.8.1}$$

If $\mathbf{v}$ is the average velocity of the charges, then $j = \rho^c\mathbf{v}$, and

$$F = \rho^c E + \frac{\rho^c}{c}(\mathbf{v} \times B). \tag{1.8.2}$$

The law of motion for the ith particle, in a system of n particles, is

$$\frac{dp_i}{dt} = F_i, \tag{1.8.3}$$

where $\mathbf{p}_i$ is the linear momentum ($= m_i \mathbf{v}_i$) and F_i is the force acting on the particle mass m_i. For relativistic motion

$$\mathbf{p}_i = m_{0i}\mathbf{v}_i/(1 - v_i^2/c^2)^{1/2}, \tag{1.8.4}$$

where m_{0_i} is its rest mass. For a single particle, the equation of motion is

$$\frac{dp}{dt} = qE + \frac{q}{c}(\mathbf{v} \times H) \tag{1.8.5}$$

because the Lorentz force per particle of charge q is

$$F = qE + \frac{q}{c}(\mathbf{v} \times H). \tag{1.8.6}$$

1.9 Field Potentials

A vector A is defined in terms of the solenoidal magnetic induction vector B by the relation

$$B = \nabla \times A. \tag{1.9.1}$$

The equation (1.4.3) then gives

$$\nabla \times \left(E + \frac{1}{c}\frac{\partial A}{\partial t} \right) = 0 \tag{1.9.2}$$

$$\text{or,} \quad E + \frac{1}{c}\frac{\partial A}{\partial t} = \nabla\phi, \tag{1.9.3}$$

where ϕ is a scalar function of space co-ordinates and time. For electrostatic fields $E = \nabla\phi$.

Putting the values of E and H from (1.9.1) and (1.9.3) in the Maxwell equation (1.4.4) we obtain

$$\nabla \times \left(\nabla \times A) + \frac{\mathbf{E}\mu}{c^2}\frac{\partial^2 A}{\partial t^2} \right) A - \nabla \left(\nabla \cdot A - \frac{\mathbf{E}\mu}{c^2}\frac{\partial \phi}{\partial t} \right) = -\frac{4\pi\mu}{c} j. \tag{1.9.4}$$

Writing $c_1^2 = c^2/\mathbf{E}\mu$, and using the vector formula (A3.11) this equation becomes

$$\left(\nabla^2 - \frac{1}{c_1^2}\frac{\partial^2}{\partial t^2} \right) A - \nabla \left(\nabla \cdot A - \frac{\mathbf{E}\mu}{c^2}\frac{\partial \phi}{\partial t} \right) = -\frac{4\pi\mu}{c} j. \tag{1.9.5}$$

1.9.1 The Lorentz gauge

In the Lorentz gauge, A is defined uniquely by the condition

$$(\nabla \cdot A) - \frac{\varepsilon\mu}{c}\frac{\partial \phi}{\partial t} = 0. \tag{1.9.6}$$

This A is called the *vector potential* for the Lorentz gauge. Putting (1.9.6) in (1.9.4) we obtain the wave equation

$$\nabla^2 A - \frac{1}{c_1^2}\frac{\partial^2 A}{\partial t^2} = -\frac{4\pi\mu}{c} j. \tag{1.9.7}$$

Taking divergence of (1.9.3), putting in that result the condition (1.9.6) to eliminate A, the equation for ϕ is obtained :

$$\nabla^2\phi - \frac{1}{c_1^2}\frac{\partial^2\phi}{\partial t^2} = \frac{4\pi}{c}\rho^c. \qquad (1.9.8)$$

Solution of these inhomogeneous wave equations for A and ϕ depends on the sources, which are the current J and charge density ρ^c. Besides, the characteristic source-free wave solutions for A and ϕ (one forward going and one backward going) are obtained for $j = 0$ and $\rho^c = 0$. The source dependent solutions for A and ϕ in the Lorentz gauge are called the *radiation solutions.*

With the help of (1.9.1) and (1.9.3) the Lorentz force F of (1.8.2) is expressed as

$$E = \rho^c\nabla\phi - \frac{\rho^c}{c}\frac{\partial A}{\partial t} + \rho^c(\mathbf{v}\times(\nabla\times A)). \qquad (1.9.9)$$

Using (A3.10) it becomes

$$E = \rho^c\nabla\phi - \frac{\rho^c}{c}\frac{\partial B}{\partial t} + \frac{\rho^c}{c}\nabla(A\cdot v) - \frac{\rho^c}{c}(v\cdot\nabla)A - \frac{\rho^c}{c}(A\cdot\nabla)v - \frac{\rho^c}{c}A\times(\nabla\times v). \qquad (1.9.10)$$

1.9.2 The Coulomb gauge

In the Coulomb gauge (also called the *radiation gauge*, or the *transverse gauge*)

$$(\nabla\cdot A) = 0, \qquad (1.9.11)$$

so that, A is source free, and the scalar potential ϕ satisfies the Poisson equation

$$\nabla^2\phi = 4\pi\rho_0^c, \qquad (1.9.12)$$

where in ρ_0^c the subscript 0 is inserted to distinguish this ρ^c with the ρ^c of the Lorentz gauge. It represents the instantaneous response of ϕ to fluctuations in the charge density without linking that with the EM radiation field. This apparent *non-local behaviour* of ϕ is allowed because ϕ is not an observed physical quantity. The E field propagates at the wavephase velocity v_ρ (which in matter-free space is c) and its interaction with the material dipole is local. So, equation (1.9.5) gives

$$\nabla^2 A - \frac{1}{c_1^2}\frac{\partial^2 A}{\partial t^2} = -\frac{4\pi}{c}j - \frac{1}{c^2}\frac{\partial}{\partial t}(\nabla^2\phi). \qquad (1.9.13)$$

It is the wave equation for A with the source on the right hand side. The ϕ source is not important for the A wave.

The solution of the *Poisson equation* for ϕ is

$$\phi(r) = \iiint\limits_V \frac{\rho^c(r')}{|r-r'|}dV'. \qquad (1.9.14)$$

It is proportional to $1/r$ at distances large compared to the charge size. The deviations of ϕ from spherical symmetry vanish asymptotically. The consequent time variation of charge density affecting A through the last term in the wave equation (1.9.13) affects only the component of A parallel to $\mathbf{r}$. Since the A waves are purely transverse in nature, relative to the direction of $\mathbf{r}$, we can write

$$A = A_\perp + A_\parallel \tag{1.9.15}$$

and separate the equation for A into the two equations

$$\left(\nabla^2 - \frac{1}{c^2}\frac{\partial^2}{\partial t^2}\right) A_\perp = -\frac{4\pi}{c} j \tag{1.9.16}$$

$$\left(\nabla^2 - \frac{1}{c^2}\frac{\partial^2}{\partial t^2}\right) A_\parallel = -\frac{1}{c^2}\frac{\partial}{\partial t}(\nabla^2 \phi) \tag{1.9.17}$$

for $A_\perp$ and $A_\parallel$ for convenience in the study of radiation from distant sources. For the EM radiation problems only the wave equation for A is important. The time varying component $A_\parallel$ of A is not involved in the $A_\perp$ wave propagating from or into the charge region. The transverse nature of the A wave follows from the Coulomb gauge condition

$$\nabla \cdot A = 0. \tag{1.9.18}$$

The radiation electric field is therefore given by

$$E = -\frac{1}{c}\frac{\partial A_\perp}{\partial t}. \tag{1.9.19}$$

The Coulomb gauge is useful in the study of optical radiation and in quantum electrodynamics.

1.10 Dielectric Polarization and Magnetic Moment

The dielectric polarization from electrification of matter depends on the translational displacement. It is the time integral of the linear momentum per unit mass. The magnetic moment depends on the rotational displacement. So, polarization is related to linear momentum and magnetization to angular momentum. The linear momentum and angular momentum of particle mechanics are the mechanical equivalents of the dielectric polarization and magnetic moment of charged particles of electrodynamics of media. Polarization, linear velocity, linear momentum, translational displacement are polar vectors.[3] And angular momentum, rotational velocity, magnetic moment are axial vectors.[4] Therefore, any material displacement induced by a force field, consists of the sum of the translational displacement and the rotational displacement. They form the set of elements of the displacement in three dimensional position space, and so are mutually complementary.

Both angular momentum and magnetic moment are excited by the action of the restoring force of the electric field on the charged particles at all points of their motional paths. The magnetic moments vanish when and where the material displacement is parallel to the electric current which means no bending of the motional path of the charge, then or there.

[3]The sign of a polar vector changes when the signs of all its co-ordinates change, which happens under a reflection (inversion) of the co-ordinate system.

[4]The components of a vector that can be written as the cross product of polar vectors do not change sign under inversion. Such vectors are called *axial vectors*, or *pseudovectors*. Under rotation, the components of a pseudovector transform according to the vector law, Sard (1970), p. 36, Jackson (1975), p. 247.

Macroscopically considering, an EM field polarizes and magnetizes a medium by exciting electric and magnetic multipole moments. The influence of the multipole moments on the EM fields are taken into account in the constitutive relations. These supplementary relations depend on the parameters of the medium and the force field applied. These are necessary because the number of independent Maxwell equations is not sufficient to form a closed system of equations. These relations determine the dielectric and magnetic permeability tensors, excited by the applied field, and are formulated completely independent of the derivation of the macroscopic Maxwell equations from the microscopic analytical formulation by Lorentz. The electric and magnetic moments may be induced by fields exerted on the atoms, these may be permanent magnetic moments (as in magnetic materials; example, ferromagnets).

The dominant molecular electric and magnetic multipoles are the electric dipole and the magnetic dipole. Only averages of these multipoles, taken over many molecules, are considered. In the absence of applied fields, the multipole moments vanish.

1.10.1 Polarization physics

Charges in a body are displaced from their positions in the field free state by applied fields and fields of other charges, and by collisions with the uncharged particles. If q_α is the charge of a particle, N_α is the number density and Σ_α is the field induced displacement of the αth species, then the polarization density is

$$\mathbf{P} = \sum_\alpha (N_\alpha q_\alpha \mathcal{E}_{\boldsymbol{\alpha}}) \tag{1.10.1}$$

and the Ampere-Maxwell law reads

$$\nabla \times H = \frac{1}{c}\frac{\partial E}{\partial t} + \frac{4\pi}{c}\frac{\partial P}{\partial t} + \frac{4\pi}{c} j = \frac{1}{c}\frac{\partial E}{\partial t} + \frac{4\pi}{c} j_t, \tag{1.10.2}$$

where

$$\mathrm{j}_t = j + \frac{\partial}{\partial t}\sum_\alpha (N_\alpha q_\alpha \mathcal{E}_{\boldsymbol{\alpha}}). \tag{1.10.3}$$

So j_t is the sum of the polarization current and other currents in a material.

The dielectric polarization P satisfies the relations

$$D = E + 4\pi P = \varepsilon : E, \quad P = \eta : E, \quad (\varepsilon - 1) : E = 4\pi P. \tag{1.10.4}$$

These constitutive relations link D with E and P; η is the electric susceptibility tensor.

1.10.2 Magnetic moment physics

The analytical formula for the magnetic moment M per unit volume, in CGS Gaussian units is

$$M = \frac{1}{2c}(\xi \times j), \tag{1.10.5}$$

where c is the vacuum velocity of light. It vanishes when (or where) the material displacement is parallel to the electric current, which means no bending of the path of the charge, $\boldsymbol{\xi}$ and $\mathbf{j}$ are determined from the equations of motion, or from the Maxwell equations. The direction of M is recovered on reversal of rotations of the field E of transverse waves. This M is a point function of the space-time continuum.

The magnetic permeability elements μ_{ij} are obtained from the constitutive relations

$$B = H + 4\pi M = \mu : H, \quad M = \chi : H. \tag{1.10.6}$$

The fields B, H, M being solenoidal

$$\nabla \cdot B = 0, \quad \nabla \cdot H = 0, \quad \nabla \cdot M = 0 \tag{1.10.7}$$

The elements of χ, η, ε, μ are dimensionless material parameters and form 3×3 matrices. These and $I - 4\pi\chi$ and $I - 4\pi\eta$ are non-singular matrices, where I is the 3×3 unit diagonal matrix. In vacuum $\boldsymbol{\mu} = \boldsymbol{\varepsilon} = \mathbf{I}$ and $\boldsymbol{\chi}$ and $\boldsymbol{\eta}$ are null matrices. $\boldsymbol{\mu}$ and $\boldsymbol{\chi}$ are scalar quantities and for paramagnetic substances $\mu > 1$; for diamagnetic substances $\mu < 1$. For ferromagnetic substances μ is a tensor having elements possessing large numerical values ($= 10^6$). These parameters depend on temperature, density, chemical constitution and frequency and field strength of the applied wave fields.

1.11 Electrodynamics Including Polarization and Magnetization

In anisotropic, non-linear dielectrics and magnetized media, the phenomenological, vector form, partial differential equations of Maxwell for electrodynamics are :

(1) The Faraday law of EM induction,

$$\nabla \times E = -\frac{1}{c}\frac{\partial}{\partial t}(\mu : H) = -\frac{1}{c}\frac{\partial}{\partial t}(\mu^l : H) - \frac{4\pi}{c}\frac{\partial M^N}{\partial t} = \frac{4\pi}{c}\left(j^l_M + j^N_M\right). \tag{1.11.1}$$

(2) The Ampere-Maxwell law of magneto electric induction, correct up to the order of the quadrupole moment Q (section 3.12),

$$\nabla \times H = -\frac{1}{c}\frac{\partial}{\partial t}(\varepsilon : E) + \frac{4\pi}{c}j + \frac{4\pi}{c}(\nabla : \dot{Q}). \tag{1.11.2}$$

It can be written as

$$\nabla \times H = \frac{1}{c}\frac{\partial}{\partial t}(\varepsilon^l : E) + \frac{4\pi}{c}\frac{\partial}{\partial t}P^N + \frac{4\pi}{c}j_c + \frac{4\pi}{c}(\nabla : \dot{Q}). \tag{1.11.3}$$

(3) Gauss' law for the normal flux of the electric displacement vector,

$$\nabla \cdot D = 4\pi\rho^c = \nabla \cdot (\varepsilon^l : E) + 4\pi(\nabla \cdot P^N), \tag{1.11.4}$$

where ρ^c is the charge density.

(4) Gauss's law for the normal flux of the magnetic induction,

$$\nabla \cdot B = 0 = \nabla \cdot (\boldsymbol{\mu}^l : H) + 4\pi(\nabla \cdot M^N), \tag{1.11.5}$$

where $\boldsymbol{\mu}$ is the magnetic permeability tensor of second order (so that its components can be expressed as the elements of a 3×3 matrix) and

$$\boldsymbol{\mu} = \boldsymbol{\mu}^1 + \boldsymbol{\mu}^N, \tag{1.11.6}$$

so that $\boldsymbol{\mu}$ can be resolved into the sum of a linearized approximation $\boldsymbol{\mu}^1$ and a non-linear part $\boldsymbol{\mu}^N$, M^N is the non-linear part of the magnetization, $\mathbf{j}_M$ is the magnetization current,

$$\mathbf{j}_M = \mathbf{j}_M^l + \mathbf{j}_M^N, \quad (\boldsymbol{\mu}^N : H) = 4\pi M^N, \quad \mathbf{j}_H^l = \frac{1}{4\pi}\frac{\partial}{\partial t}(\boldsymbol{\mu}^l : H), \quad \mathbf{j}_M^N = \frac{\partial}{\partial t}M^N \tag{1.11.7}$$

j_M^l and j_M^N are the magnetization currents of the linearized approximation and the non-linear residue respectively,

$$\boldsymbol{\varepsilon} = \boldsymbol{\varepsilon}^l + \boldsymbol{\varepsilon}^N, \quad (\boldsymbol{\varepsilon}^l : E) = 4\pi P^N. \tag{1.11.8}$$

The conduction current $\mathbf{j}_c$ in a body moving with velocity $\mathbf{v}$, obeying Ohm's law, is given by

$$\mathbf{j}_c = \sigma E + \frac{\sigma}{c}(\mathbf{v} \times B) \tag{1.11.9}$$

$\boldsymbol{\varepsilon}^l$ is the dielectric tensor of the linearized approximation and $\boldsymbol{\varepsilon}^N$ is the non-linear part of it. The current density $\mathbf{j}$ is expressed in terms of a multipole series and the conduction current, so that

$$\mathbf{j} = \mathbf{j}_c + \frac{\partial P}{\partial t} + (\nabla \times M) - \frac{\partial}{\partial t}(\boldsymbol{\nabla} : \mathbf{Q}) \tag{1.11.10}$$

$\boldsymbol{\mu}^l$ and $\boldsymbol{\varepsilon}^l$ are independent of the amplitudes of the field of perturbation.

In general $\boldsymbol{\varepsilon}$ is a function of $\mathbf{E}$ and $\boldsymbol{\mu}$ is a function of $\mathbf{H}$, and we can write

$$\boldsymbol{\varepsilon}(\mathbf{E}) = \boldsymbol{\varepsilon}^l + \boldsymbol{\varepsilon}^N(\mathbf{E}), \quad \boldsymbol{\mu}(\mathbf{H}) = \boldsymbol{\mu}^l + \boldsymbol{\mu}^N(\mathbf{H}), \tag{1.11.11}$$

$$\mathbf{D} = \boldsymbol{\varepsilon}(\mathcal{E}) : \mathcal{E}, \quad \mathbf{B} = \boldsymbol{\mu}(\mathbf{H}) : \mathbf{H}. \tag{1.11.12}$$

For dielectrics there is no space charge separation; so that for these $\rho^c = 0$. The theories of determination of $\boldsymbol{\varepsilon}$ and $\boldsymbol{\mu}$, from the constitutive relations, are independent of the derivation of these macroscopic Maxwell equations from the Lorentz theory. Moreover, for the non-linear behaviour of the medium we have the useful relations

$$D = \boldsymbol{\varepsilon}E = E + 4\pi P^l + 4\pi P^N \tag{1.11.13}$$

$$B = \boldsymbol{\mu}H = H + 4\pi M^l + 4\pi M^N. \tag{1.11.14}$$

1.12 Distinction between the Laws for Displacements and EM Fields

The Lorentz transformations for the displacement 4-vector can be written as

$$d\mathbf{r}_\perp = d\mathbf{r}_\perp^0, \quad dr_\parallel = \gamma(dr_\parallel^0 - \beta cdt^0), \quad cdt = \gamma(cdt^0 - \beta dr_\parallel^0). \tag{1.12.1}$$

For EM fields the consequent laws are

$$H_\perp = \gamma[H_\perp^0 - (\beta \times D)], \quad H_\parallel = H_\parallel^0, \tag{1.12.2}$$

$$D_\perp = \gamma[D_\perp^0 + (\beta \times H^0)], \quad D_\parallel = D_\parallel^0, \tag{1.12.3}$$

$$E_\perp = \gamma[E_\perp^0 + (\beta \times B^0)], \quad E_\parallel = E_\parallel^0, \tag{1.12.4}$$

$$\mathbf{B}_\perp = \gamma[\mathbf{B}_\perp^0 - (\beta \times \mathbf{E}_\perp^0)], \quad B_\parallel = B_\parallel^0. \tag{1.12.5}$$

These relations show that the parallel components of the field vectors are unaffected and only the perpendicular components are affected. This is unlike the transformation law for the components of the displacement vector of (1.12.1). This distinction between space displacements and fields follows Faraday's law of EM induction : *When an electrically conducting wire is moved in one direction, the magnetic field perpendicular to that is affected, and the parallel magnetic field remains unchanged.* This correspondence gives rise to the question : Is there any such correspondence between velocity displacements (which are proportional to accelerations) and fields in accelerated frames? In other words, the question raised is : Is it meaningful to assume that for accelerated motion, the field evolution is perpendicular to the velocity displacement, as here?

1.13 Two Categories of Field Equations and Field Expressions

The basic equations of electrodynamics are of the following two kinds :

Electrodynamics

(1) **Universally valid Maxwell equations**

These do not contain any material constant, but these do not form a closed system of equations, because the number of these is not sufficient for solving problems.

(2) **Constitutive relations (or the state relations)**

These are complementary to the Maxwell equations; their number together with the number of Maxwell equation is sufficient to make a closed system of equations of electrodynamics. These relations depend on the material constitution and parameters of applied fields.

The universally valid Maxwell equation are of the following two kinds :

The Maxwell equations

$$\nabla \times E = -\frac{1}{c}\frac{\partial B}{\partial t}, \quad \nabla \cdot B = 0.$$

The homogeneous pair or the source-free pair.

$$\nabla \times H = \frac{1}{c}\frac{\partial D}{\partial t} + \frac{4\pi}{c}j, \quad \nabla \cdot D = 4\pi\rho.$$

The inhomogeneous pair or the source full pair.

The field quantities are of the following two kinds :

Field quantities

(1) **Observable quantities,** also called *primary quantities* or the *domain quantities*, example :

$$\text{EMf} = \oint_\Gamma (E \cdot dl)$$

$$\text{Field flux} = \iint_\Sigma (B \cdot d\Sigma)$$

$$\text{Current flux} = \iint_\Sigma (j \cdot d\Sigma)$$

(2) **Derived quantities** or the locally valid point functions; example :

E, H, D, B, J, ρ^c. These follow from the domain quantities of Flux, and EMF, and obtain their properties as well from these quantities, and are functions of space and time.

Evidently, force vectors are associated with line integrals, and flux vectors, which are distinct from these, are associated with surface integrals. The EM laws and their applications depend entirely on the inertial frame observations involving the domain quantities. So, full importance should be given to the integrals of fields, because only these are involved in the experimentally based formulation of basic laws of electrodynamics.

1.14 The Minkowski Theory of Macroscopic Electrodynamics in Moving Media

Einstein (1905) considered electrodynamical problems of isolated charges like electrons only. For ponderable bodies Minkowski (1908) then solved completely the macroscopic problem of electrodynamics of moving media, and established extension of the Maxwell theory, from media at rest to those in translation at a uniform rate, by uniting space and time to form a single 4-space continuum.

Minkowski postulated that the laws of transformation for all bodies follow from the LT. So, the Maxwell equations, together with the conservation conditions are Lorentz covariant, that is they have the same form in all inertial frames. Minkowski (1908) noted that (1) all material media are sources in the Maxwell equations; (2) only two EM vectors, E and B, or D and H are the fundamental quantities; (3) the constitutive relations are necessary for the complete formulation of EM problems, and to specify the EM properties of the medium a priori. Parameters like σ, μ, ε specify phenomenologically the EM laws, purely on the basis of experiments. The Maxwell-Minkowski equations are valid for a medium moving with uniform velocity $\mathbf{v}$ with respect to a reference system, K° say, which is the laboratory system. The constitutive relations are traditionally introduced in terms of two material vectors $(i)P$, the polarization density, or the electric dipole moment per unit volume, and (2) M, the magnetization density (or the magnetic moment per unit volume). Their connections with D, E, H and B are $\mathbf{D} = \varepsilon\mathbf{E} = \mathbf{E} + 4\pi\mathbf{P}$, $\mathbf{B} = \mu\mathbf{H} = \mathbf{H} + 4\pi\mathbf{M}$. The constitutive relations in moving media are determined on the assumption that for a medium at rest these are known.

If terms of the order $(v/c)^2$ are neglected, the Minkowski relations, consistent with the constitutive relations and the simple Ohm's law, Tai (1964), are

$$\mathbf{D} = \varepsilon\mathbf{E} + (\varepsilon\mu - \varepsilon_0\mu_0)(\mathbf{v} \times \mathbf{H})/c, \quad \mathbf{B} = \mu\mathbf{H} - (\varepsilon\mu - \varepsilon_0\mu_0)(\mathbf{v} \times \mathbf{E})/c \tag{1.14.1}$$

$$\mathbf{j} = \rho^c\mathbf{v} + \sigma\mathbf{E} + \frac{\sigma}{c}(\mathbf{v} \times \mathbf{B}), \tag{1.14.2}$$

where ε_0 and μ_0 are the vacuum values of the dielectric constant and magnetic permeability. So, in CGS Gaussian units both equal unity. When $\sigma = 0$, $\varepsilon\mu = \varepsilon_0\mu_0$ and $\rho^c = 0$ these reduce to the vacuum relations $\mathbf{D} = \mathbf{E}$, $\mathbf{B} = \mathbf{H}$. These ideas of Minkowski led to the tensor formulation of macroscopic electrodynamics, including the state relations and Ohm's law. All the other formulations of the EM theory, of moving media, depend on the Minkowski formulation, can be directly deduced from it by appropriate linear transformation relations, Penfield and Haus (1967), Chapter 7, Shiozawa (1973).

Relations of transformation of fields, based on the Minkowski theory for moving media, correct up to the first order of the relative velocity $\mathbf{v}$, are $\mathbf{E} = \mathbf{E}^0 + (\boldsymbol{\beta} \times \mathbf{B}^0)$, $\mathbf{H} = \mathbf{H}^0 + (\boldsymbol{\beta} \times \mathbf{D}^0)$, $\mathbf{D} = \mathbf{D}^0 + (\boldsymbol{\beta} \times \mathbf{H}^0)$, $\mathbf{B} = \mathbf{B}^0 + (\boldsymbol{\beta} \times \mathbf{E}^0)$, $\mathbf{j} = \mathbf{j}^0 - \rho^{0c}\mathbf{v}$, $\rho^c = \rho^{0c}$, $\mathbf{A} = \mathbf{A}^0$, $\phi = \phi^0 + (\boldsymbol{\beta} \cdot \mathbf{A}^0)$, the Lorentz force $\mathbf{F} = \mathbf{E} + (\mathbf{j} \times \mathbf{B})/c = \mathbf{E}^0 + (\mathbf{j}^0 \times \mathbf{B}^0)/c$.

The general results of field transformation by LT are

$$\mathbf{E} = \gamma[\mathbf{E}^0 + (\boldsymbol{\beta} \times \mathbf{B}^0)] - \gamma^2\boldsymbol{\beta}(\boldsymbol{\beta} \cdot \mathbf{E}^0)/(\gamma + 1), \tag{1.14.3}$$

$$\mathbf{D} = \gamma[\mathbf{D}^0 + (\boldsymbol{\beta} \times \mathbf{H}^0)] - \gamma^2\boldsymbol{\beta}(\boldsymbol{\beta} \cdot \mathbf{D}^0)/(\gamma + 1), \tag{1.14.4}$$

$$\mathbf{B} = \gamma[\mathbf{B}^0 - (\boldsymbol{\beta} \times \mathbf{E}^0)] - \gamma^2\boldsymbol{\beta}(\boldsymbol{\beta} \cdot \mathbf{B}^0)/(\gamma + 1), \tag{1.14.5}$$

$$\mathbf{H} = \gamma[\mathbf{H}^0 - (\boldsymbol{\beta} \times \mathbf{D}^0)] - \gamma^2\boldsymbol{\beta}(\boldsymbol{\beta} \cdot \mathbf{H}^0)/(\gamma + 1). \tag{1.14.6}$$

For the reverse transformation results, obviously, change the sign of the vector β. Hence a purely electric or magnetic field, in one inertial system, gives rise to a mixture of these in another such system, and these fields have no independent existence in an inertial frame. Being thus completely inter related, the fields are regarded as the elements of a second order tensor $F^{\mu\gamma}$ (or $F_{\mu\gamma}$) in place of the apparently distinguishable field vectors E and B, which, as such, do not give the correct picture of the situation.

1.15 Tensor Description of Classical Electrodynamics

The homogeneous pair of Maxwell equations becomes

$$\frac{\partial}{\partial x^\lambda}F_{\mu\nu} + \frac{\partial}{\partial x^\mu}F_{\nu\lambda} + \frac{\partial}{\partial x^\nu}F_{\lambda\mu} = 0, \quad (\lambda, \mu, \nu) = (1, 2, 3, 4), \tag{1.15.1}$$

where

$$F_{\mu\nu} = \frac{\partial A_\nu}{\partial x_\mu} - \frac{\partial A_\mu}{\partial x_\nu}, \quad A_\mu \equiv (A, \phi) \quad \text{or,} \quad A^\mu \equiv (A, -\phi). \tag{1.15.2}$$

$$dx^\mu \equiv (d\mathbf{r}, cdt), \quad \text{or}, \quad dx_\mu \equiv (d\mathbf{r}, -cdt). \tag{1.15.3}$$

$$F_{\mu\nu} = \begin{bmatrix} 0 & B_Z & -B_Y & E_X \\ -B_Z & 0 & B_X & E_Y \\ B_Y & -B_X & 0 & E_Z \\ -E_X & -E_Y & -E_Z & 0 \end{bmatrix}. \tag{1.15.4}$$

The inhomogeneous pair of Maxwell equations becomes

$$\frac{\partial \mathcal{H}^{\mu\nu}}{\partial x^\nu} = \frac{4\pi}{c} j^\mu, \quad (\mu, \nu) = (1, 2, 3, 4). \tag{1.15.5}$$

$$\mathcal{H}_{\mu\nu} = \begin{bmatrix} 0 & H_X & -H_Y & -D_X \\ -H_Z & 0 & H_X & -D_Y \\ H_Y & -H_X & 0 & -D_Z \\ D_X & D_Y & D_Z & 0 \end{bmatrix}. \tag{1.15.6}$$

$$j^\mu = (j, c\rho^c), \quad j_\mu = (j, -c\rho^c).$$

Ohm's law : (1) In stationary media

$$j^\mu = \frac{\sigma}{c} F^{\mu\nu} V_\nu, \quad V_\nu \equiv (0, 0, 0, c). \tag{1.15.7}$$

(2) In moving media

$$j^\mu + (j^\nu V_\nu) V^\mu / c^2 = \sigma F^{\mu\nu} V_\nu / c, \tag{1.15.8}$$

where $V^\mu (= dx^\mu/dT)$ is the Minkowskian 4-velocity. Equation of particle motion due to the Lorentz force :

$$\frac{d}{dT} p^\mu = \frac{q}{c} F^{\mu\nu} V_\nu, \quad \text{or}, \quad \frac{d}{dT} P^\mu = \frac{q}{c} \left(\frac{\partial}{\partial x_\mu} A^\nu \right) V^\nu, \tag{1.15.9}$$

whee the canonical moment $P^\mu = p^\mu + qA^\mu/c$. Another form of this equation, which is suitable for generalization to rotating and other accelerated frames, is

$$mc \frac{d}{\sqrt{-ds^2}} W^\mu = -\frac{q}{c} \eta_{\nu\xi} F^{\mu\xi} W^\nu, \tag{1.15.10}$$

where $W^\mu = dx^\mu/\sqrt{-ds^2}$, and m is the rest mass.

In vacuum, in presence of free charges, the constitutive relations are $D = E$, $B = H$ and the tensor form of the inhomogeneous pair of the Maxwell equations, is

$$\frac{\partial}{\partial x^\nu} F^{\mu\nu} = \frac{4\pi}{c} j^\mu.$$

This Gaussian field identification is not valid in rotating and other accelerated frames of references, because even vacuum becomes anisotropic by rotation.

1.16 Covariant Description of Electrodynamics

Equation of motion of a charge q :

$$mc\left[\frac{d}{\sqrt{-ds^2}}W^{\mu}+\Gamma^{\mu}_{\nu\eta}W^{\nu}W^{\eta}\right]=\frac{q}{c}q_{\nu}g_{\nu\xi}F^{\mu\xi}W^{\nu},$$

where the Christofell symbol

$$\Gamma^{\mu}_{\nu\eta}=\frac{1}{2}g^{\mu\alpha}\left(\frac{\partial g_{\alpha\nu}}{\partial x^{\eta}}+\frac{\partial g_{\alpha\eta}}{\partial x^{\nu}}-\frac{\partial g_{\nu\eta}}{\partial x^{\alpha}}\right)$$

The continuity condition :

$$\frac{1}{\sqrt{-g}}\frac{\partial}{\partial x^{\nu}}\left(\sqrt{-g}j^{\mu}\right)=0,$$

where j^{μ} is the 4-current and

$$g=\det g_{\mu\gamma}.$$

The Maxwell equations in covariant tensor form :

$$F_{\lambda\mu;\nu}+F_{\mu\nu;\lambda}+F_{\nu\lambda;\mu}=0,$$

where the semi-colon denotes the covariant derivative with respect to the co-ordinate whose index lies after the semi-colon, and

$$F_{\mu\nu}=\begin{array}{c}\rightarrow\nu\\ \downarrow\mu\end{array}\begin{bmatrix}0 & B_3 & -B_2 & E_1\\ -B_3 & 0 & B_1 & E_2\\ B_2 & -B_2 & 0 & E_3\\ -E_1 & -E_2 & -E_3 & 0\end{bmatrix}.$$

$$\mathcal{H}^{\mu\nu}=\begin{array}{c}\rightarrow\nu\\ \downarrow\mu\end{array}\begin{bmatrix}0 & H_3 & -H_2 & -D_1\\ -H_3 & 0 & H_1 & -D_2\\ H_2 & -H_1 & 0 & -D_3\\ D_1 & D_2 & D_3 & 0\end{bmatrix}.$$

These Maxwell equations can also be written, avoiding the semi-colon for the covariant derivative as

$$\frac{\partial}{\partial x^{\nu}}F_{\lambda\mu}+\frac{\partial}{\partial x^{\lambda}}F_{\mu\nu}+\frac{\partial}{\partial x^{\mu}}F_{\nu\lambda}=0.$$

$$\frac{1}{\sqrt{-g}}\frac{\partial}{\partial x^{\nu}}\left(\sqrt{-g}\mathcal{H}^{\mu\nu}\right)=\frac{4\pi}{c}j^{\mu}.$$

Vector equations of Maxwell in accelerated frames :

$$\operatorname{curl} E = -\frac{1}{c\sqrt{\tilde{\gamma}}}\frac{\partial}{\partial t}\left(\sqrt{\tilde{\gamma}}B\right),$$

$$\operatorname{div} B = \frac{1}{\sqrt{\tilde{\gamma}}}\frac{\partial}{\partial x^i}\left(\sqrt{\tilde{\gamma}}B^i\right) = 0,$$

$$\operatorname{curl} H = -\frac{1}{c\sqrt{\tilde{\gamma}}}\frac{\partial}{\partial t}\left(\sqrt{\tilde{\gamma}}D\right) + \frac{4\pi}{c}\rho^c u,$$

$$\operatorname{div} D = -\frac{1}{\sqrt{\tilde{\gamma}}}\frac{\partial}{\partial x^i}\left(\sqrt{\tilde{\gamma}}D^i\right) = 4\pi\rho^c.$$

1.17 Constitutive Relations

(1) Shiozawa (1974) :

$$\mathcal{H}^{\mu\nu}V_\nu = \varepsilon F^{\mu\nu}V\nu,$$
$$^*F^{\lambda\mu}V_\mu = \mu_0\mathcal{H}^{\lambda\eta}V_\eta,$$

where $^*F^{\alpha\beta}(= -e^{\alpha\beta\gamma})\delta F^{\alpha\beta}/2$, $e^{\alpha\beta\gamma\delta}$ is the contravariant fourth order Levi Civita symbol) is the dual of $F^{\alpha\beta}$, and

$$^*F^{\alpha\beta} = \begin{bmatrix} 0 & E_Z & -E_Y & B_X \\ -E_Z & 0 & E_X & B_Y \\ E_Y & -E_X & 0 & B_Z \\ -B_X & -B_Y & -B_Z & 0 \end{bmatrix}.$$

(2) Tamm (1924) and Mandel'sshtam (1924) :

$$\mathcal{H}^{\mu\nu} = g^{\mu\alpha}g^{\nu\beta}\xi^\gamma_\alpha\xi^\delta_\beta F_{\gamma\delta}$$

$$\text{or,}\quad g_{\alpha\lambda}g_{\beta\nu}\mathcal{H}^{\alpha\beta} = \xi^\gamma_\lambda\xi^\delta_\nu F_{\gamma\delta},$$

where the non-zero components of ξ^λ_μ are the diagonal elements $\xi^1_1 = \xi^2_2 = \xi^3_3 = 1/\sqrt{\mu_0}$, $\xi^4_4 = \varepsilon\sqrt{\mu_0}$. Here, the coefficient μ_0 is the magnetic permeability and ε is the dielectric constant of a medium.

(3) Heer (1964) and Khromykh (1969) :

For stationary media,

$$\mathcal{H}_{i4} = \varepsilon F_{i4}, \quad F^{ij} = \mu_0\mathcal{H}^{ij}.$$

For moving media

$$g_{\alpha\xi}g_{\beta\eta}\mathcal{H}^{\xi\eta}V^\beta = \varepsilon F_{\alpha\delta}V^\delta,$$
$$\varepsilon_{\alpha\beta\gamma\delta}g^{\beta\xi}g^{\gamma\eta}F_{\xi\eta}V^\delta = \mu_0 e_{\alpha\beta\gamma\delta}\mathcal{H}^{\beta\gamma}V^\delta,$$

where $e_{\alpha\beta\gamma\delta}$ is the covariant fourth order Levi Civita symbol.

(4) Post (1967), Post and Yildiz (1965), Yildiz and Tang (1965) (PYT) :

$$\mathcal{H}^{\alpha\beta} = \frac{1}{2}\chi^{\alpha\beta\gamma\delta}F_{\gamma\delta},$$

where $\chi^{\alpha\beta\gamma\delta}$ is the general constitutive tensor of fourth order, which is specified in an inertial frame, for linear isotropic media, by the conditions

$$\chi^{4^0 1^0 4^0 1^0} = \chi^{4^0 2^0 4^0 2^0} = \chi^{4^0 3^0 4^0 3^0} = -\varepsilon$$
$$\chi^{2^0 3^0 2^0 3^0} = \chi^{3^0 1^0 3^0 1^0} = \chi^{2^0 1^0 2^0 1^0} = 1/\mu_0.$$

We can write these in short as

$$\chi^{4^0 i^0 4^0 i^0} = -\varepsilon, \quad \chi^{i^0 j^0 i^0 j^0} = 1/\mu_0,$$

where $(i, j) = (1, 2, 3)$.

1.18 Field Equations for Small Rotational Velocities

Vector relations, correct up to the first order of rotation velocity, are obtained from tensor formulation of transformation laws:

$$\nabla = \nabla', \quad t = t', \quad \rho^c = \rho^{c'}, \quad j = j' + \rho^{c'}\mathbf{v}, \quad \mathbf{v} = (\Omega \times \Omega),$$

$$\frac{\partial B}{\partial t} = \frac{\partial B'}{\partial t} + \nabla \times (\mathbf{v} \times B), \quad \frac{\partial D}{\partial t} = \frac{\partial D'}{\partial t} + \nabla \times (\mathbf{v} \times D') - 4\pi\rho^{c'}\mathbf{v},$$

$$B = B', \quad E = E' - (\mathbf{v} \times H')/c, \quad D = D', \quad H = H' + (\mathbf{v} \times D')/c,$$

$$A = A', \quad \phi = \phi' - (\boldsymbol{\beta}\cdot\mathbf{B}').$$

Then the form invariant, vector Maxwell equations of electrodynamics, for this rotation, are

$$\nabla'\cdot B' = 0, \quad \nabla' \times E' = -\frac{1}{c}\frac{\partial B'}{\partial t'},$$

$$\nabla'\cdot D' = 4\pi\rho^{c'}, \quad \nabla' \times H' = \frac{1}{c}\frac{\partial D'}{\partial t'} + \frac{4\pi}{c}j'.$$

These are to be supplemented by the admissible set of the constitutive relations, the choice of which should be decided experimentally. On using rotating frames, these state relations simplify to :

(1) Shiozawa (1974) :

$$D = \varepsilon E - \varepsilon(\beta \times B), \quad B = \mu_0 H + \mu(\beta \times D).$$

(2) Tamm (1924) and Mandel'stham (1924) :

$$D = \varepsilon E - (\beta \times B)/\mu_0, \quad B = \mu_0 H + \mu_0(\beta \times D).$$

(3) Heer (1964) and Khoromykh (1969) :

$$D = \varepsilon E - (\beta \times H), \quad B = \mu_0 H + (\beta \times E).$$

(4) Post (1967), Post and Yildiz (1965), Yildiz and Tang (1966) (PYT) :

$$D = \varepsilon E - \varepsilon(\beta \times D), \quad B = \mu_0 H + \mu_0(\beta \times D).$$

For charges in vacuum, we put $\varepsilon = 1$, $\mu_0 = 1$, then the constitutive relations in (1), (2) and (4), are $D = E - (\beta \times B)$, $B = H + (\beta \times D)$; and in (3) these are $D = E - (\beta \times H)$, $B = H + (\beta \times E)$. So, the rotation induced anisotropy of vacuum is not same for the different covariant formulations of the constitutive relations.

1.19 Magnetic Field

Electrodynamics has influenced the development of the unified field theory, which is the unification of various natural forces into a single theory. The resemblance between the law of Coulomb on interaction between charges, and the Newtonian law of gravitational attraction of masses raised the hope of combining electrodynamics and gravitation. But gravitation always produces attraction between masses, and there is nothing like negative and positive masses and repulsive gravitation, or quasistatic electric lines of force. However, there are charges of both positive and negative sign; so that there is attraction between charges of opposite sign and repulsion between changes of same sign. This connection between two charges by field lines is not possible between two masses.

All celestial bodies, on account of their rotation, are electrically polarized when they are observed from a system at rest, though these are electrically neutral, or nearly so on an astronomical scale. On the atomic scale the EM attractions and repulsions far exceed gravitational attraction. So, it is the other way round, and the EM theory and that of gravitation can be studied separately.

On large spatial scales the universe is homogeneous and isotropic. Cosmic or global magnetic fields though situated very remotely with respect to stations of observation in earth, violate the isotropy but not the homogeneity of the Universe, Zeldovich et al (1983). In neutral plasma of earth, sun and other astrophysical bodies, temperature is high, so thermal energy is larger than the magnetic energy, and the surface of constant pressure and a constant density are not coincident.

The expression for the magnetic moment field follows from the basic law in the names of Oersted, Ampere, Biot-Savart in electrodynamics, which states that a current through a closed loop produces an axial magnetic field. The magnetic moment of an atom is generated by the electron spin, by the electron angular momentum about its nucleus and by applied fields. The magnetic moment is a primary source of magnetization in matter, because no seed magnetic field is necessary for its evolution. Any charge in the direction of a motion is a consequence of disturbance in matter which is not symmetric in all directions. Asymmetry in motion of charges and electrically conducting fluids generate the moment field because it implies a gain or loss of magnetic moment. So, all magnetization is generated by this process, whatever be the apparent reason for its excitation.

Hence the assumption of the existence of an all space pervading primeval magnetic field—the cosmic magnetic field, Parkar (1979, p. 2) or the magnetic universe, Piddington (1969, Preface), seems to be not essential. And, for the secondary sources which require for their evolution the existence of a seed magnetic field, the seed field is the magnetic moment field from thermal noise fields. This field may therefore be the seed field in the dynamo model (a secondary source) for magnetic fields of the earth, sun and other astrophysical objects. So, the evolution of all other intuitively derived macroscopic primary sources of field generation should be determined by distributions of the magnetic moment.

MHD dynamo theories for magnetic field generation in electrically conducting fluids are applied to explain the magnetic fields of planets and stars. These theories ensure that kinetic energy of motion is converted to magnetic energy, provides maintenance for an indefinite period of time of the initially existing seed magnetic field lines of force, and for their amplification. These theories should not violate Cowling's (1934) theorem which shows the impossibility of maintenance of an axisymmetric magnetic field by flows of electrically conducting fluids.

Electric currents maintain earth's magnetic field in its core, where the magnetic viscosity is large ($= 3 \times 10^4$ cm^2/sec). A characteristic time for the Ohmic resistance of the field in the outer liquid metal core of the earth, which forms the spherical layers between the radii $0.19 \times R_e$ and 0.51 R_e (where R_e is the radius of earth) is about 10^4 years. On the other hand earth's magnetic field has existed for not less than 10^4 years. There is also the random reversal of direction of the geomagnetic dipole field in times of the order of 10^5 years, followed by quicker deflections in about 10^3 to 10^4 years, of the dipole field relative to the axis of rotation. The dipole axis makes a small angle (of the order of 11^0) with the rotation axis in a time of the order of 10^4 years.

Pulsar magnetic fields are of the order of 10^{12} gauss. Such high fields are produced by the compression of magnetic fields of stars from which the pulsars emerge. A galaxy can provide the initial field for its planets.

Physics of magnetism is vital in the programmes of Magnetic Confinement Fusion (MCF) of tokamaks, other torus devices, the mirror machines and pinch devices and in Inertial Confinement Fusion (ICF) involving the outputs from laser irradiation of target materials. The MHD power generation schemes require magnetic field of superconducting magnets which are of the order of few teslas, for their running. Magnetic birefringence, including the Zeeman effect of splitting of waves and Faraday rotation, Whistler modes, the freezing effect including that responsible for sun-spots and solar prominences, the Hall effect, Ferraro's law of isorotation, gyration of free charges and of bound charges for the Larmour precession effect, the pinch effect, magnetic bottling, magnetic tray and the magnetic mirror effect, generation of some drifts, currents and forces and the consequences of the finite Larmour radius (FLR) expansion, are some of the interesting manifestations of magnetic field of MHD and plasma physics. Alfven waves blossom out from the stresses and strains created by the magnetic field in matter. These waves are carriers of energy, electrically conducting mobile matter and heat and act as a refrigerator on matter.

Magnetism in matter, behaviour of magnetic materials, magnetism in super conductivity, and the impact of NMR (Nuclear Magnetic Resonance) in high energy physics are important subjects of quantum physics.

Magnetic effects on rocks, fossils and archeological findings are important topics of geophysics. Medical science makes use of magnetic resonance imaging (MRI). Magnetic healers of diseases also exist, whose success depends on the correctness of the application of the influence of magnetic stresses and strains on human body.

Without magnetism the physical world would be devoid of much of its interesting manifestations. The magnetic field-free universe would have a sustained serenity due to the preservation, from thermal degradation, of its calmness. So, the existence of the cosmic magnetic field is responsible for the wonderful cosmic manifestations. Possibilities of new revelations are not over in the magnetic field centered physics. The story is endless of the many facets of this field and the output of their interaction with matter.

1.20 Remark on Problems of Electrodynamics in Arbitrary Frames of Reference

The EM laws of induction and their application depend entirely on *inertial frame observations* involving the domain quantities. So, for a better understanding of the problems of electrodynamics a primary role should be given to the integrals as the observable quantities, and not to the field quantities like E, H, D etc. which are point functions of space time.

Failure to explain observations of rotating and stationary observers, relative to rotating conductors and other effects, is an indication that the familiar, experimentally based, basic laws of electrodynamics and the material laws defining the *constitutive relations* should be regarded as valid only for *inertial frames of reference.* For application to accelerated frames, therefore, the existing laws are incomplete, and require rectification. To determine the *universally valid forms* of these *laws* either the Maxwell equations are to be accepted as valid for accelerated frames, and the constitutive relations are to be changed, or the Maxwell equations are to be modified without changing the constitutive relations, or both are to be modified for that purpose.

In the relativistic theory of rotation, Ise and Uretsky (1958), the *transformation equations* (to *non-inertial* frames) are non-physical, because they have in themselves no information on the rate of a clock or the length of a meter stick, at some arbitrary points of the rotating system. Such quantities as these follow from the principle of covariance, which determines only the fore shortening and time dilation effects of special theory of relativity, if naively applied. Trocheries (1949) remarks "It is impossible in general relativity, to define, in a unique way, what is a frame of reference in uniform rotation about an axis. Various definitions have been proposed in each of which a particular property of *classical rotation* is retained". Post (1967) mentioned that a physically meaningful transformation for rotation is not obtained from the principle of space time covariance; and the space time theory of gravitation does not establish physically correct transformations. The principle of general space-time covariance, mathematically, accommodates within it both realistic and unrealistic transformations. Galilean or Lorentz transformation, Galilean or other forms of non-Euclidean rotation, are all equally permissible from a mathematical point of view because all of them are sub-groups in the set of general space time transformations implied by the principle of general covariance. The known transformations for rotation, both Newtonian and non-Newtonian, do not determine uniquely the electromagnetic field of rotating bodies as a Lorentz transformation does for translating bodies, Trocheris (1949). For instance, knowing the field of a magnet, at rest, one can derive the field it produces, when it is in uniform translation, with the help of a Lorentz transformation. For rotation there is no such principle. Knowing the results of translation, one cannot tell what a body becomes when it is set in rotation; the centrifugal force distorts its elastic properties.

The Newtonian attraction is a real force, whereas the centrifugal force of gravitation is a fictitious force. The gravitational force of matter and rotation (the Coriolis force and the centrifugal force) are manifestations of curvature of space, due to matter. The world point, representing a particle, moves along a geodesic, and the forces are equivalent to a change in the metric of space-time, $g_{\mu\gamma}$, which removes the non-permanent gravitational fields from the field equations in tensor form.

Non-uniqueness due to several non-identical covariant formulations of the constitutive relations is a problem of electrodynamics. These give different vector constitutive relations even for small rotational velocities. The correct choice should depend on experimental indications, which are not yet available. In inertial frames all these relations give the same correct state relations.

Spinning electrons in orbital motion in atoms, motion of the planets round the sun, and of astrophysical bodies, are translations together with rotations. Also, translations are the limits of rotation, about axes at infinity. Constant velocity is inertial motion. Rotation is accelerated (non-inertial) motion. Rotation generates the Coriolis force and the centrifugal force; the effective gyro-frequency is the sum of the usual gyro-frequency, plus twice the rotation frequency, Èngles and Verheest (1975). The dispersion law, for wave propagation in rotating plasmas, is also modified by the Coriolis force.

The earth is a rotating, orbiting, electrically conducting and magnetized sphere. It has, in the interior, electrically conducting fluid, and it is surrounded by the ionosphere, which is also rotating with the earth at the same rate. Other planatory, and Stellar and other astrophysical bodies have also their own specific rotation rate, magnetic field and electrical properties.

A co-ordinate frame can be found near a field point, in which a magnetic field can exist without an electric field to an observer. In another co-ordinate frame the reversed observation is also possible, namely an electric field can exist without a magnetic field. However, in the first approximation, magnetic fields are found to be independent of the co-ordinate system, but not the electric field. So, these field determinations are not analogous; in other words, conclusions drawn from determination of the magnetic field for zero electric field cannot be guessed from those of the electric field for zero magnetic field. Another important conclusion is that all celestial magnetized bodies, due to their rotation, are electrically polarized when seen from a system at rest, though these are electrically neutral or nearly so on the astronomical scale. On the atomic scale EM attractions and repulsions far exceed gravitational attraction; so it is the other way round, and the electromagnetic theory and that of gravitation can be studied separately.

A rotation dependent space charge effect, Schiff (1939), appears in the Minkowski theory for moving media, and so in the non-rigorous, but largely used, theory for rotating media. This spurious Schiff charge (Chapter 13) and current of rotation are avoided by using covariant tensor form of ERF for formulation of the EM theory in the familiar vector form for rotation. However, the non-covariant treatment gives results which are sufficiently correct for experiments. So, the huge extra mathematical labour for finding a more rigorous theory by involving the principle of space-time covariance and Riemannian tensors, is practically worthless.

The Gaussian field identifications $D = E$, $B = H$ have the implication that there is no physical distinction between the electric displacement vector D and the electric field vector E, and between the magnetic induction vector B and the magnetic intensity H. Actually, these define the properties of free space only, as seen from inertial frames of reference because material polarization mechanism, responsible for non-validity of these relations, are absent in vacuum. For ERF these EM field identifications in free space have not been accepted because the gravitional field of rotation makes even matter-free space (vacuum) anisotropic.

The 4-space can be defined both as a contravariant 4-vector, and as a covariant 4-vector. If force is defined as the product of mass and 4-acceleration, then it is a contravariant 4-vector. If $\delta\omega$ is the invariant infinitesimal work done due to the infinitesimal displacement δx^μ, which is a contravariance 4-vector, then $\delta\omega = \delta x^\mu F_\mu$. So, the 4-force, thus defined, is a covariant four vector. The Lorentz force is a tensorial force field if it is regarded as a contravariant vector density. The sum of inertial and gravitational forces may be tensorial. Hence the elementary definition of physical quantities must be reconsidered to decide whether a physical vector is contravariant or covariant. The consequences of these two formulations are not always identical for non-inertial motion.

When a spherical conductor is in rotation about a diameter, a magnetic field is seen to exist outside, and in the absence of rotation no field is observed outside the conductor. Moreover, detection, by experiments, of surface charge in the plates of rotating cylindrical condensers and emf of these plates have been reported as effects of rotation. These and some other related experimental findings could not be explained by the familiar theory of classical electrodynamics when naively extended to rotating frames.

Rotating objects are usually studied by the Maxwell equations and the constitutive relations of Minkowski, both referred to an inertial frame and the equation of motion, containing only the Coriolis force. For convenience of calculation, the centrifugal force is also neglected. Only the homogeneous pair of Maxwell equations, that carry over to rotating co-ordinates without any change of form, are used if necessary. Rotating co-ordinates are avoided for the other equations of Maxwell, if their use becomes necessary, Webster and Whitten (1973) and Schiff (1939). So, a rotating object is practically studied by the Maxwell equations, referred to an inertial frame, Gron (1984) and the equation of motion containing the Coriolis force. Working rules have been framed (Van Bladel (1973, 1976), Mo (1969), Videc and Kooy (1982)), for application to technical problems of rotating objects, which do not follow from the inertial frame theory of electrodynamics.

The field equations of Maxwell, in a rotating frame, even for small rotational velocities at a constant frequency Ω, do not exist even in the best text books on *Classical Electrodynamics,* which are by Sommerfeld (1952), Jackson (1975), Landau and Lifshitz (1960, 1975), Panofsky and Phillips (1964). There is no name like electrodynamics in Rotating Frame, or topics related to it in the list of subjects, numbering several thousands, under physics and astronomy classification scheme of Journals of American Physical Society. So far, there has not been any serious review of topics which come under the preview of this subject.

Chapter 2

Special Theory of Relativity

2.1 The Problems of Absolute System and Absolute Rest

2.1.1 Postulate of absolute system

The equations of classical dynamics and electrodynamics assume the existence of a system of axes at absolute rest and absolute universal time which is identical for all observers whatsoever. The absolute system, based on these concepts, is the Galilean system.[1]

2.1.2 Postulate of absolute rest

The Newtonian equations of motion are valid in a fixed co-ordinate system in absolute static space, and the equations (2.3.6)–(2.3.8) of Lorentz for electrodynamics are referred to co-ordinate axes which are fixed in ether, which is at absolute rest. These implied assumptions raise the question : What is the system of reference which we can consider to be at absolute rest? For answer of this problem consider the Newtonian equation of motion

$$m\ddot{r} = F, \tag{2.1.1}$$

which is valid if the system of reference is at rest. If, however, the frame of reference rotates about the axis of z with constant angular velocity Ω, the equation becomes invalid. The equation of motion will not simply be product of mass and acceleration = force, but

$$m\ddot{r} = F - m(\Omega \times (\Omega \times r)) - 2m(\Omega \times v), \tag{2.1.2}$$

where the centrifugal force $-m(\Omega \times (\Omega \times r))$ and the Coriolis force $-2m(\Omega \times v)$, have been added, because the force F is independent of the system of reference. Therefore, in the rotating system, by measuring the force F the fixed axes of co-ordinates about which the given system is rotating, are determined.

[1] This Chapter is largely based on Ray (1957). The section 2.35 is based on Ghosh (1957).

2.1.3 Postulate of universal time

Newton assumed the existence of absolute universal time to be measured by a clock which is a periodic system. To measure the time in any other system of reference the clock is simply transferred to the new system of reference, such that its periodic motion is not affected, and consequently the time measured by it is not affected.

2.2 The Notion of Simultaneity

Consider two frames of reference, Fig. 2.2.1. The lines AB and $A'B'(AB = A'B')$ are placed side by side and moving with respect to each other with velocity u. To be definite suppose $A'B'$ be considered at rest with respect to which AB moves. If two light signals be given when A and B come just opposite to A' and B', they should be simultaneous events according to the ordinary notion of simultaneous events. In the time T taken by light from A' to reach C' or from B' to reach C' (C' being the mid point of $A'B'$), the point C (the mid point of AB) has moved to C_1, so that the light signal A reaches C which is at C_1 at time $T + \delta T$. The signal from B reaches C at $T - \delta T$. To a man at the middle point C of AB, the events (that is signals) will not appear to be simultaneous, whereas to a man at C' which is considered at rest the event will be simultaneous. Thus, what is simultaneous to C' is not simultaneous to C. So, unless we are definite that $A'B'$ is at absolute rest the word simultaneity loses its significance without the frame of reference for the events.

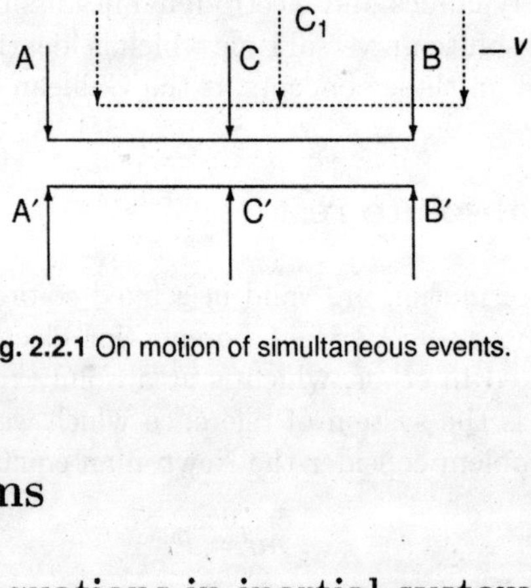

Fig. 2.2.1 On motion of simultaneous events.

2.3 Inertial Systems

2.3.1 The dynamical equations in inertial systems

If the absolute inertial system of reference existed and had it been possible to determine it, it would have given us the absolute motion of the earth through ether. The rotation of earth is determined by noting the deviation of Focault's pendulum from the ordinary pendulum. So, we might expect to obtain the motion of the earth by observing the deviation of motion of bodies upon earth from theoretical results obtained from Newton's equations. Unfortunately, there is something inherent in the form of the dynamical equations of motion that is responsible for the experimental failure to detect the class of uniform motions of translation (with respect to the hypothetical inertial system).

Consider the equation of motion

$$m\ddot{r} = F. \tag{2.3.1}$$

Let x, y, z, t be the space-time co-ordinates of one system and x_1, y_1, z_1, t_1 be the space-time co-ordinates with respect to another. The Galelian transformation relations

$$r_1 = r - ut, \quad t_1 = t \text{ (universal time)}, \tag{2.3.2}$$

where the last relation represents the postulate of absolute time, give

$$\dot{r}_1 = \dot{r} - u, \quad \ddot{r}_1 = \ddot{r}. \tag{2.3.3}$$

These show that the acceleration is the same for the two systems which move with respect to each other with uniform velocity of translation. Hence, the force being independent of the choice of axes, the equation

$$\text{Mass} \times \text{acceleration} = \text{Force}, \tag{2.3.4}$$

$$\text{or,} \quad m\ddot{r}_1 = F \tag{2.3.5}$$

is same in both the systems of axes, and the equations of dynamics are invariants of the Galelian transformation. Thus, every particular set of Galelian co-ordinate axes is a system at rest like any other set of axes of the system, so that there is no preferred natural inertial system, and the term absolute rest loses its significance in dynamics.

The equations (2.3.1) do not involve the velocity $\dot{r}$, the components of which are the first derivatives of the co-ordinates with respect to time t. And, it is impossible to detect uniform motion of translation from (2.3.1); consequently, the inertial system cannot be determined by experiments in classical dynamics.

2.3.2 From electrodynamical experiments

The equations of electrodynamics were also considered for the determination of the inertial system, because these contain the velocities $\dot{x}, \dot{y}, \dot{z}$ or first order derivatives with respect to time. The Lorentz equations of electrodynamics, based on the postulate of universal time and co-ordinates referred to ether (which is at absolute rest), are

$$\nabla \times H = \frac{1}{c}\frac{\partial E}{\partial t} + \frac{4\pi}{c}\rho v, \tag{2.3.6}$$

$$\nabla \times E = -\frac{1}{c}\frac{\partial H}{\partial t}, \quad \nabla \cdot H = 0, \tag{2.3.7}$$

$$\nabla \cdot E = 4\pi\rho. \tag{2.3.8}$$

These contain v, the velocity, and the derivative operator $\partial/\partial t$ with respect to universal time. The Galelian principle of invariance will not hold for these equations. The simple case of $\rho = 0$ ensures no macroscopic space charge separation, at the point of measurement, without removing v and $\partial/\partial t$. Then, since the light waves of E and H travel with the constant velocity c, referred to ether at rest, determination of the inertial system by measuring the velocity of light from the moving earth, and detection of the absolute motion of the earth through ether seem to be possible by electrodynamics including optical experiments performed in the earth laboratory, noted Maxwell.

Michelson, in 1881 and Michelson and Morley in 1887 performed experiments on this suggestion of Maxwell, but failed to detect the absolute motion of the earth. All other subsequent experiments on this issue also failed to determine the absolute motion of the earth (and therefore the inertial system of the earth) even by observing electrodynamical phenomena.

2.4 Experiment of Michelson and Morley

The light source, Fig. 2.4.1, is L, from which light is split up into two beams PA and PB by partial reflection at a half silvered glass plate P. These two portions are reflected back by the mirrors A and B and brought into interference at the screen s. The earth moves in the direction

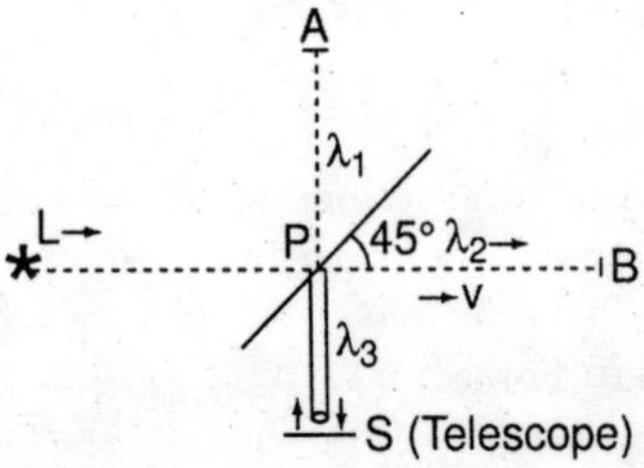

Fig. 2.4.1 Experiment of Michelson and Morley.

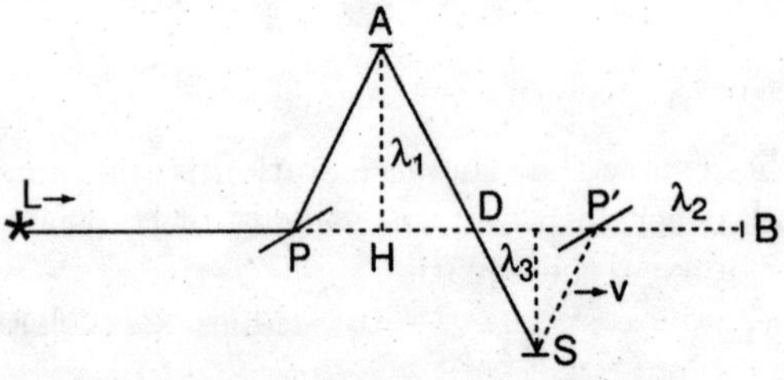

Fig. 2.4.2 It shows the optical arrangement in the experiment of Michelson and Morley.

of PB with velocity v, and the whole apparatus could be rotated into any position desired. The times $T_1', T_1'', T_1''', T_2', T_2'', T_3'''$ for light to travel the distances PA, AD, DS, PB, BP' and $P'S$ respectively, fig. 2.4.2, are given by

$$(cT_1')^2 = PA^2 = PH^2 + \lambda_1^2 = \lambda_1^2 + (vT_1')^2, \tag{2.4.1}$$

$$(cT_1'')^2 = AD^2 = PH^2 + \lambda_1^2 = \lambda_1^2 + (vT_1'')^2. \tag{2.4.2}$$

And, since $AD = PA$, we find that

$$T_1'' = T_1', \tag{2.4.3}$$

$$(cT_1''')^2 = DS^2 = \lambda_3^2 + (vT_1''')^2, \tag{2.4.4}$$

$$T_1' = \frac{\lambda_1}{\sqrt{c^2 - v^2}}, \quad T_1'' = \frac{\lambda_1}{\sqrt{c^2 - v^2}}, \quad T_1''' = \frac{\lambda_3}{\sqrt{c^2 - v^2}}. \tag{2.4.5}$$

Hence the total time taken by light to reach s after reflection at the point A, is

$$T_1 = T_1' + T_1'' + T_1''' = \frac{2\lambda_1}{\sqrt{c^2 - v^2}} + \frac{\lambda_2}{\sqrt{c^2 - v^2}}. \tag{2.4.6}$$

$$cT_2' = PB = vT_2' + \lambda_2, \quad \text{since } \lambda_2 = P'B; \tag{2.4.7}$$

$$cT_2'' = BP' = \lambda_2 - vT_2''', \tag{2.4.8}$$

$$(cT_2''')^2 = P'S^2 = \lambda_2^2 + (VT_2''')^2, \tag{2.4.9}$$

$$\begin{aligned} T_2 = T_2' + T_2'' + T_2''' &= \frac{\lambda_2}{c - v} + \frac{\lambda_2}{c + v} + \frac{\lambda_3}{\sqrt{c^2 - v^2}} \\ &= \frac{2\lambda_2 c}{c^2 - v^2} + \frac{\lambda_3}{\sqrt{c^2 - v^2}}, \end{aligned} \tag{2.4.10}$$

$$\Delta T = T_2 - T_1 = \frac{2\lambda_2 c}{c^2 - v^2} - \frac{2\lambda_1}{\sqrt{c^2 - v^2}}. \tag{2.4.11}$$

Since $\lambda_2 = \lambda_1 = \lambda$, putting $\beta = v/c$, we find that

$$\Delta T = \frac{2\lambda}{c}(1 + \beta^2 + \cdots) - \frac{2\lambda}{c}\left(1 + \frac{1}{2}\beta^2 + \cdots\right). \tag{2.4.12}$$

Approximately, correct up to the second order in v/c, then

$$\Delta T = \frac{\lambda}{c}\beta^2. \tag{2.4.13}$$

The time difference being a phase difference of the two split up beams is indicated at S by a shift of interference fringes. So, by measuring the phase shift, it would be possible to detect the time differences. When the apparatus is rotated through a right angle, B takes position of A and A takes the position of B, and fringes are shifted in the opposite direction, since in the first case $T_1 < T_2$ and in the second case the corresponding time $\bar{T}_1 > \bar{T}_2$. Very careful observations, and other more accurate subsequent experiments did not yield at all the expected shift of fringes.

The assumption that c is independent of the velocity of the source, is in accordance with the electromagnetic theory of light. Since ether is conceived to be at absolute rest, to explain the Michelson experiment, it has to be assumed to be carried along with the earth, which means ether is only at relative rest. It also fails to explain the aberration of stars due to motion of the earth, to explain which the ether must be assumed to be at rest.

The Michelson interferometer shows fringes even when the earth is not moving with respect to the source, that is, if the source be not star light but a source of light on the surface of the earth. The motion of the earth only causes the fringes to shift in one direction. The object of rotating the apparatus is to cause the shift of the fringes to travel in the direction opposite to the original to an amount twice the original. The zero shift of Michelson means that the time difference $T_2 - T_1 = 0$. This would indicate $c(c^2 - v^2)^{1/2} = c^2 - v^2$ or $v^2(v^2 - c^2) = 0$. Since v cannot equal the large value c, $v = 0$, which is certainly not true. We can only say that the velocity of earth affecting c is zero. This would be possible if the ether be carried along with the motion of the earth.

Michelson's experiment proves that the velocity of light is neither affected by the velocity of source nor by the velocity of the observer. It is independent of both the motion.

2.5 Experiments of Bradley and Fizeau

2.5.1 Bradley's Experiment

The idea of ether drag cannot be reconciled with the experimental findings of Bradley. A telescope at absolute rest in ether, when pointed towards a fixed star, will have its image at the focus M of Fig. 2.5.1(a) of the objective lens. If now the telescope is moved at a right angle to the direction of the fixed star, the image will not coincide with M, the focus, as the telescope has moved through the distance d during the time the light travels through the distance F. Then

$$\frac{F}{d} = \frac{c}{v}, \tag{2.5.1}$$

only if the ether remains at rest, and unaffected by the moving telescope. The telescope has to be tilted through the angle

$$\delta = \tan^{-1}\frac{d}{F} = \tan^{-1}\frac{v}{c} \tag{2.5.2}$$

for an image at the focus. Bradley observed the fixed stars at an interval of six month, that is at two points of the orbit of the earth diametrically opposite to each other. The telescope had to be rotated through an angle which gave exactly, as far as experimental errors are allowed, the result (2.5.2).

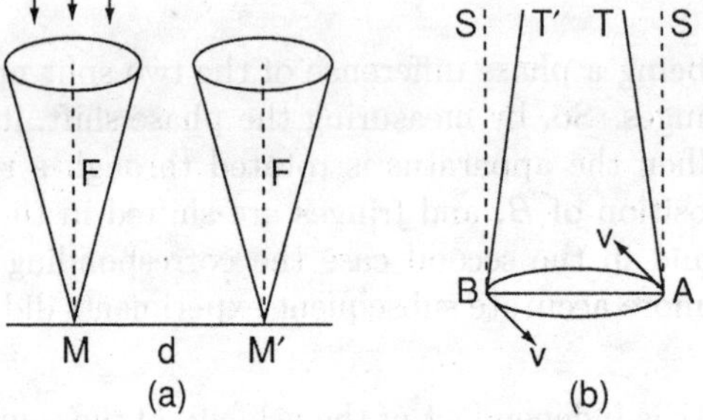

Fig. 2.5.1 Bradley's Experiment.

2.5.2 Fizeau's Experiment

The refraction of light through a material medium depends on the ratio of velocity of light in ether to that in it. The amount of refraction may be affected if the medium moves. In 1818 Arago carried out an experiment but found no change in the direction of the refracted ray by the movement of the medium. Arago concluded that ether was dragged along with the earth so that the velocity in the medium is not affected by the movement of the earth. Fresnel's explanation is that the ether is at rest, but the velocity of light in the medium is affected by the motion of the medium through the ether. So, the direction of a ray of light would not change when it is passed through a medium (say water) moving with velocity v_0 relative to an observer if his measured phase velocity of light v is given by

$$v = \frac{c}{n} \pm \left(1 - \frac{1}{n^2}\right) v_0 = \frac{c}{n} \pm \mu v_0, \tag{2.5.3}$$

where n is the refractive index, and Fresnel' coefficient of ether drag $\mu = 1 - 1/n^2$. But the Galilean principle gives

$$v = \frac{c}{n} \pm v_0. \tag{2.5.4}$$

Experimentally Fizeau showed that the velocity v relative to the observer must be multiplied by the factor μ of Fresnel's ether drag.

2.6 Zeeman's Water-tube Experiment

In this arrangement, Fig. 2.6.1, light from source L is incident on a half silvered glass plate and divided into two partial beams (1) $S_1P_1P_2S_2$ to G and (2) S_2 to P_2 to P_1S_1G, which is just the opposite path. The two beams unite at G. *Interference fringes* are observed by a telescope T, on the way of rays through tubes through which water flows. For water $n = 4/3, \mu = 7/16$, which is nearly equal to 1/2. This was in good arrangement with the observed value.

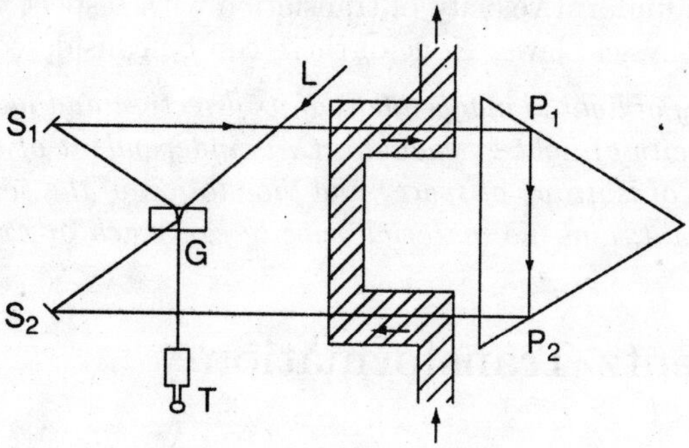

Fig. 2.6.1 Zeeman's experiment.

2.7 The Contraction Hypothesis of Fitzerald and Lorentz

After the experimental conclusions that ether is at rest, which gave the zero result of Michelson's experiment, for explanation Fitzerald and Lorentz assumed that material bodies contract slightly along the direction of motion. For, in this case λ for PB (Fig. 2.4.1) and λ for PA differ from those of equation (2.4.11). So, we have

$$T_2 - T_1 = \frac{2}{c}\left[-\frac{\lambda_1}{\sqrt{1-\beta^2}} + \frac{\lambda_2}{1-\beta^2}\right], \tag{2.7.1}$$

which is zero if

$$\lambda_2 = \lambda_1(1-\beta^2)^{1/2}. \tag{2.7.2}$$

This gives the length contraction formula

$$l = l_0\sqrt{1-\beta^2} \tag{2.7.3}$$

of Fitzerald and Lorentz.

2.8 The Postulates of Einstein

Einstein took for granted that uniform rectilinear motion cannot be detected by means of any process (mechanical or electrodynamical) performed within the system. And, on the problem of detection of the uniform rectilinear motion through ether, he referred to Michelson's conclusion that the velocity of the light does not depend on the velocity of observer, or the velocity of the source. So, light spreads out in straight lines in all directions with constant velocity independent of motion of the observer or of the source. And, according to Einstein, *The real problem is to obtain a space-time relation that gives this constancy of the velocity of light.* Einstein's postulates are :

(a) *On determination of a system at absolute rest and of the same scale for all systems. No means exist of determining a system at absolute rest and no time scale exists which is the same for all systems. We can only measure relative velocity.* In other words, all systems of reference, moving with uniform velocity of translation with respect to each other, are perfectly equivalent as regards physical laws. None of these can be considered privileged.

(b) *Vacuum velocity of light is independent of its direction and motion of both the source and the observer. The velocity of light in vacuum, C, is independent of its direction of propagation (which is the principle of isotropy of space) and the motion of the source and the observer both. It is an absolute constant; and, no material velocity can reach or exceed this value.*

2.9 The Lorentz Transformation

Two distances on different bodies are equal (as we say) if the two end points coincide when placed side by side at the same instant, that is, simultaneously. For relative velocity v, as in Fig. 2.2.1, the instant when A and A' coincide will appear earlier to C than to C' and the instant when B and B' coincide will appear later to C than to C'. To an observer C in the moving system the length of the beam will appear greater than $A'B'$. The metric of space therefore will differ for different systems, and simultaneity (section 2.2) is a relative notion, depending on the motion of a system.

To show that time differs for different systems moving relative to each other, consider two clocks B and B' on the moving bodies, which agree when they coincide. They will be at different places after a while and their hands cannot assume equivalent positions simultaneously for the two observers.

Consider relations of transformation between two sets of co-ordinates referred to two systems of axes moving with uniform relative velocity v of translation, the two space-time co-ordinate systems are $K(x, y, z, t)$ and $K'(x', y', z', t')$. Their x-axis is along the direction of translation; OY, OZ and $O'Y', O'Z'$ are parallel to each other, and the systems K and K' of reference coincide initially. So

$$x = x', \ y = y', \ z = z' \ \text{ at } \ t = t' = 0. \tag{2.9.1}$$

The X co-ordinate of O' with respect to O is given by

$$x - vt = 0. \tag{2.9.2}$$

So, co-ordinates of O' with respect to O are $(vt, 0, 0)$. Light wave from the point O, at time $t = 0$, will, at time t, lie on the sphere

$$x^2 + y^2 + z^2 - c^2t^2 = 0. \tag{2.9.3}$$

For the observer at O' also this light wave has a velocity c according to the postulates of Einstein. So that, at the time t' in the K' system, corresponding to the time t of the K system, the wave will lie on the sphere

$$x'^2 + y'^2 + z'^2 - c^2t'^2 = 0. \tag{2.9.4}$$

As the Galelian transformation

$$x' = x - vt, \ y' = y, \ z' = z, \ t' = t, \tag{2.9.5}$$

transforms (2.9.3) into

$$x'^2 + y'^2 + z'^2 - c^2t'^2 + 2vx't' + v^2t'^2 = 0, \tag{2.9.6}$$

it fails to transform (2.9.3) into (2.9.4). The correct transformation relations between x, y, z, t and x', y', z', t' must be linear; for light will appear to move in straight lines in both the systems. So, a straight line in the first system is transformed into a straight line in the second system. The axes OX and $O'X'$ coincide initially, $y' = z' = 0$ whenever $y = z = 0$ and the plane $X'OY'$ coincides with the plane XOY so that $z' = 0$ when $z = 0$. We can suppose

$$Z' = DZ, \quad Y' = BY, \quad x' = A(x - vt) \tag{2.9.7}$$

since $x' = 0$ when $x = vt$. The forms (2.9.3) and (2.9.4) show that the transformation must be homogeneous; so t' cannot contain a term free from x, y, z, t. This is also clear from the fact that $t' = 0$ at $t = 0$. Therefore, we can write

$$t' = P(t + \alpha y + \beta' z + \lambda x). \tag{2.9.8}$$

The quantities $A, B, D, P, \alpha, \beta', \lambda$ are independent of the variables x, y, z, t, since whatever be the values of z, x and t, the relation $y' = BY$ (say) would remain unchanged. Since the product terms vanish, $\alpha = \beta' = 0$, so that

$$t' = P(t + \lambda x). \tag{2.9.9}$$

To transform (2.9.3) into (2.9.4) we write

$$x'^2 + y'^2 + z'^2 - c^2t'^2 = a(x^2 + y^2 + z^2 - c^2t^2), \tag{2.9.10}$$

$$\begin{aligned} x'^2 + y'^2 + z'^2 - c^2t'^2 &= A^2(x - vt)^2 + B^2y^2 + D^2z^2 - P^2c^2(t + \lambda x)^2 \\ &= B^2\left[\frac{A^2}{B^2}(x - vt)^2 + Y^2 + \frac{D^2}{B^2}z^2 - \frac{P^2}{B^2}C^2(t + \lambda x)^2\right], \end{aligned} \tag{2.9.11}$$

where a is a constant. The right hand side of (2.9.11) must be identical with $a(x^2 + y^2 + z^2 - c^2t^2)$; so

$$D = B = 1. \tag{2.9.12}$$

Hence writing

$$A^2 = \xi, \quad P^2 = \eta \tag{2.9.13}$$

in (2.9.11) we obtain

$$\xi(x - vt)^2 + y^2 + z^2 - c^2\eta(t + \lambda x)^2 = x^2 + y^2 + z^2 - c^2t^2, \tag{2.9.14}$$

then

$$\xi - c^2\lambda^2\eta = 1, \quad \xi v + \eta\lambda c^2 = 0, \quad \xi v^2 - c^2\eta = -c^2. \tag{2.9.15}$$

From the first two relations of (2.9.15) eliminating η and from the second and third relations, eliminating ξ we get

$$\xi = \frac{1}{1 + \lambda v}, \tag{2.9.16}$$

$$\eta = \frac{1}{1 + \lambda v} = \xi, \tag{2.9.17}$$

$$\lambda = -\frac{v}{c^2}, \tag{2.9.18}$$

$$\xi = \eta = \frac{1}{1-\beta^2} = \gamma^2, \tag{2.9.19}$$

$$\beta = \frac{v}{c}, \quad \gamma = \frac{1}{\sqrt{1-\beta^2}}. \tag{2.9.20}$$

The transformation relations simplify to the Lorentz transformation

$$x' = \gamma(x - vt), \;\; y' = y, \;\; z' = z, \;\; t' = \gamma(t - \beta x/c), \tag{2.9.21}$$

where γ is the relativistic factor. The reverse LT relations,

$$x = \gamma(x' + vt'), \tag{2.9.22}$$

$$t = \left(t' + \frac{\beta}{c}x'\right), \tag{2.9.23}$$

are obtained by changing the sign of v.

2.10 Length Contraction and Time Dilatation

2.10.1 Length contraction

The length of a rod changes due to motion. When the rod is at right angle to the direction of motion, this change is different from the change if the rod is held in the direction of motion. If lx, ly, lz are the components of the length of the rod parallel to the co-ordinate axes, then the length perpendicular to the direction of motion remains unaltered, because

$$y_2' - y_1' = y_2 - y_1 = l_y, \quad z_2' - z_1' = z_2 - z_1 = l_z.$$

Let (x_1, y_1, z_1) and (x_2, y_2, z_2) be the extremities of the rod of length l placed in the rest system k. Then

$$x_2 = \gamma(x_2' + vt'), \quad x_1 = \gamma(x_1' + vt'), \tag{2.10.1}$$

$$l = x_2 - x_1 = \gamma(x_2' - x_1') = \gamma l', \quad l' = \frac{1}{\gamma} l < l. \tag{2.10.2}$$

So, along the direction of motion, the rod contracts in the ratio $\sqrt{1-\beta^2} : 1$. This explains the Fitzerald and Lorentz contractions and the experiments of Michelson and Moreley. Length, therefore, depends on the velocity, it has no absolute value. It is a relative quantity. Euclidean geometry applies only to figures at rest relative to the observer.

2.10.2 Time dilatation

Suppose at a point $x = 0$, $T = t_2 - t_1$. If $T' = t_2' - t_1'$ be the corresponding interval as observed from the moving system k', then

$$T' = t_2' - t_1' = \gamma(t_2 - t_1) = \gamma T. \tag{2.10.3}$$

So, the duration between two instants at any point of a system appears to increase when observed from a moving system in the ratio $1 : (1-\beta^2)^{1/2}$. Every clock becomes slow in a moving system and material bodies have no rigidity. A body appears shorter in the direction of motion of the system. Velocity being the ratio of distance and time, the constancy of the velocity of light with respect to moving observers becomes evident.

The velocity v is always less than c, for otherwise t' would be imaginary, and simultaneity losses its significance. For if two events be simultaneous and x_1 and x_2 are the x co-ordinates of their positions in the k system, then

$$t_1' = \gamma\left(t - \frac{\beta}{c}x_1\right), \quad t_2' = \gamma\left(t - \frac{\beta}{c}x_2\right). \tag{2.10.4}$$

So, $t_2' \neq t_1'$ although t is the same at x_1 and x_2. Similarly, when simultaneous at different points in K', the two events will not be simultaneous with respect to K.

2.11 Applications

2.11.1 Explanation of the experimental findings of Fizeau

Taking the k' axes on the moving water and the K axes fixed with the observer, the velocity of light in the K frame

$$\begin{aligned} &= \frac{\frac{c}{n} \pm v_0}{1 \pm \frac{v_0}{nc}} = (v \pm v_0)\left(1 \mp \frac{vv_0}{c^2}\right) \\ &= v \pm v_0\left(1 - \frac{v^2}{c^2}\right) - v \cdot \frac{v_0^2}{c^2} \approx v \pm v_0\left(1 - \frac{1}{n^2}\right), \end{aligned} \tag{2.11.1}$$

where $v = c/n$ and vv_0^2/c^2 has been neglected. Hence Fizeau's formula follows from the addition theorem for velocities, section 7.2(c).

2.11.2 Doppler effect

Electromagnetic fields satisfy the wave equation

$$\nabla^2\phi = \frac{1}{c^2}\frac{\partial^2\phi}{\partial t^2}, \tag{2.11.2}$$

which has the wave solution

$$\phi = Ae^{i\psi} = Ae^{2\pi i\nu\left(t - \frac{lx+my+nz}{c}\right)}. \tag{2.11.3}$$

The wave seems to come from a certain source outside the earth moving with velocity v say. So, in the moving K' system of the earth, with co-ordinates x', y', z', t',

$$\begin{aligned} \Psi' &= 2\pi\nu\left[\gamma\left(t' + \frac{\beta}{c}x'\right) - \frac{l}{c}\gamma(x' + vt') - \frac{m}{c}\gamma' - \frac{n}{c}z'\right] \\ &= 2\pi\nu\gamma\left[(1 - l\beta) + \frac{x'}{c}(\beta - l) - \frac{m\gamma' + nz'}{\gamma c}\right] \\ &= 2\pi\nu\gamma(1 - l\beta)\left[t' - \frac{x'}{c}\frac{(l-\beta)}{1 - l\beta} - \frac{m\gamma' + nz'}{\gamma c(1 - l\beta)}\right], \end{aligned} \tag{2.11.4}$$

where Ψ' is the transformation of the wave phase to the K' frame. This is a plane wave with the Doppler shifted frequency $\nu' = \nu\gamma(1 - l\beta)$, and the Doppler shifted direction of wave propagation having direction cosines

$$l' = \frac{l - \beta}{1 - l\beta}, \quad m' = \frac{m}{\gamma(1 - l\beta)}, \quad n' = \frac{n}{\gamma(1 - l\beta)}. \tag{2.11.5}$$

For propagation parallel to the direction of translation, $l = 1, m = n = 0$, so

$$\nu' = \nu\gamma(1 - \beta) = \gamma\sqrt{1 - \beta}/\sqrt{1 + \beta} \approx \nu(1 - \beta) \quad \text{and} \quad l' = 1.$$

If the wave normal is perpendicular to the observer, then $l = 0$ and

$$\nu' = \frac{\nu}{\sqrt{1 - \beta^2}} \approx \nu\left(1 + \frac{\beta^2}{2}\right). \tag{2.11.6}$$

It is the transverse Doppler effect, and is of second order of smallness.

2.11.3 Aberration

The direction of the normal changes with the change of ν, causing the aberration effects. The angle between the old and the new directions is given by

$$\cos\delta = ll' + mm' + nn' = \frac{l(l - \beta)}{l - l\beta} + \frac{(m^2 + n^2)}{\gamma(l - l\beta)}. \tag{2.11.7}$$

When $l = 0$ the ray is perpendicular to the direction of motion, $\cos\delta = 1/\gamma$ and $\tan\delta \approx \beta$.

2.12 The Invariant Character of some Equations

The invariance of the D'Alembartian wave equation and the elctromagnetic field equations prove the validity of these equations for all uniformly moving systems of co-ordinates and indicates the impossibility of detection of absolute motion through ether.

2.12.1 The homogeneous wave equations

The transformations of the differential operators are

$$\frac{\partial}{\partial x'} = \frac{\partial}{\partial x}\frac{\partial}{\partial x'} + \frac{\partial}{\partial t}\frac{\partial}{\partial x'} = \gamma\left(\frac{\partial}{\partial x} + \frac{\beta}{c}\frac{\partial}{\partial t}\right). \tag{2.12.1}$$

$$\frac{\partial^2}{\partial x'^2} = \gamma^2\left(\frac{\partial^2}{\partial x^2} + \frac{2\beta}{c}\frac{\partial^2}{\partial t\partial x} + \frac{\beta^2}{c^2}\frac{\partial^2}{\partial t^2}\right). \tag{2.12.2}$$

$$\frac{\partial}{\partial y'} = \frac{\partial}{\partial y}, \quad \frac{\partial^2}{\partial y'^2} = \frac{\partial^2}{\partial y^2}, \quad \frac{\partial}{\partial z'} = \frac{\partial}{\partial z}, \quad \frac{\partial^2}{\partial z'^2} = \frac{\partial^2}{\partial z^2}. \tag{2.12.3}$$

$$\frac{\partial}{\partial t'} = \gamma\left(v\frac{\partial}{\partial x} + \gamma\frac{\partial}{\partial t}\right), \quad \frac{\partial^2}{\partial t'^2} = \gamma^2\left(v\frac{\partial^2}{\partial x^2} + 2v\frac{\partial^2}{\partial x\partial t} + \frac{\partial^2}{\partial t^2}\right). \tag{2.12.4}$$

$$\Box'^2\phi = \frac{\partial^2\phi}{\partial x'^2} + \frac{\partial^2\phi}{\partial y'^2} + \frac{\partial^2\phi}{\partial z'^2} - \frac{1}{c^2}\frac{\partial^2\phi}{\partial t'^2} = 0 \tag{2.12.5}$$

$$= \gamma^2\left(\frac{\partial^2}{\partial x^2} + 2\frac{\beta}{c}\frac{\partial^2}{\partial x\partial t} + \frac{\beta^2}{c^2}\frac{\partial^2}{\partial t^2}\right)\phi + \frac{\partial^2\phi}{\partial y^2} + \frac{\partial^2\phi}{\partial z^2}$$

$$-\gamma^2\left(\beta^2\frac{\partial^2}{\partial x^2} + \frac{2\beta}{c}\frac{\partial^2}{\partial x\partial t} + \frac{1}{c^2}\frac{\partial^2}{\partial t^2}\right)\phi \tag{2.12.6}$$

$$= \gamma^2(1-\beta^2)\frac{\partial^2\phi}{\partial x^2} + \frac{\partial^2\phi}{\partial y^2} + \frac{\partial^2\phi}{\partial z^2} + \frac{\gamma^2}{c^2}(\beta^2-1)\frac{\partial^2\phi}{\partial t^2} \tag{2.12.7}$$

$$= \left(\nabla^2 - \frac{1}{c^2}\frac{\partial^2}{\partial t^2}\right)\phi = \Box^2\phi = 0, \tag{2.12.8}$$

where $\Box^2$ is the D'Alembartian differential operator, $\Box^2 = \nabla^2 - \frac{1}{c^2}\frac{\partial^2}{\partial t^2}$.

2.12.2 The Maxwell equations in vacuum

The field E_0 of the rest system, in the classical theory, becomes $E+(\beta \times H)$ in a system moving with velocity v. So, the transformation behaviour of E and H, and the invariance study of their guiding equations are important. In vacuum the charge density $\rho^c = 0$, and the Maxwell equations are

$$\nabla \times H = \frac{1}{c}\frac{\partial E}{\partial t}, \quad \nabla \times E = \frac{1}{c}\frac{\partial H}{\partial t}, \tag{2.12.9}$$

$$\nabla \cdot E = 0, \quad \nabla \cdot H = 0. \tag{2.12.10}$$

The x component of the first equation of (2.12.9) reduces to

$$\frac{\partial H_z}{\partial y'} - \frac{\partial H_y}{\partial z'} = \frac{\gamma}{c} - \frac{\partial E_x}{\partial t'} - \beta\gamma\frac{\partial E_x}{\partial x'}; \tag{2.12.11}$$

and since

$$\nabla \cdot E = 0 = \gamma\frac{\partial E_x}{\partial x'} - \frac{\beta}{c}\gamma\frac{\partial E_x}{\partial t'} + \frac{\partial E_y}{\partial y'} + \frac{\partial E_z}{\partial z'} \tag{2.12.12}$$

we find that

$$\frac{\partial H_z}{\partial y'} - \frac{\partial H_y}{\partial z'} = \frac{\gamma}{c}(1-\beta^2)\frac{\partial E_x}{\partial t'} + \beta\left(\frac{\partial E_y}{\partial y'} + \frac{\partial E_z}{\partial z'}\right), \tag{2.12.13}$$

$$\frac{\partial}{\partial y'}(H_z - \beta E_y) - \frac{\partial}{\partial z'}(H_y + \beta E_z) = \frac{1}{c\gamma}\frac{\partial E_x}{\partial t'}. \tag{2.12.14}$$

On multiplication of both sides by γ this equation becomes

$$\frac{\partial}{\partial y'}[\gamma(H_z - \beta E_y)] - \frac{\partial}{\partial z'}[\gamma(H_y + \beta E_z)] = \frac{1}{c}\frac{\partial E_x}{\partial t'}. \tag{2.12.15}$$

Similarly transformation of the y and z components gives

$$\frac{\partial H_x}{\partial z'} - \frac{\partial}{\partial x'}[\gamma(H_z - \beta E_y)] = \frac{1}{c}\frac{\partial}{\partial t'}[\gamma(E_y - \beta H_z)]; \tag{2.12.16}$$

and the components of the second equation of (2.12.9) give

$$\frac{\partial}{\partial y'}[\gamma(E_z+\beta H_y)] - \frac{\partial}{\partial z'}[\gamma(E_y-\beta H_z)] = -\frac{1}{c}\frac{\partial H_x}{\partial t'}, \tag{2.12.17}$$

$$\frac{\partial E_x}{\partial z'} - \frac{\partial}{\partial x'}[\gamma(E_z+\beta H_y)] = -\frac{1}{c}\frac{\partial}{\partial t'}[\gamma(H_y+\beta E_z)], \tag{2.12.18}$$

$$\frac{\partial}{\partial x'}[\gamma(E_y-\beta H_z)] - \frac{\partial E_x}{\partial y'} = -\frac{1}{c}\frac{\partial}{\partial t'}[\gamma(H_z-\beta E_y)]. \tag{2.12.19}$$

Eliminating $\partial E_x/\partial t'$ from (2.12.11) and (1.12.12) we obtain

$$\frac{\partial E_x}{\partial x'} + \frac{\partial}{\partial y'}[\gamma(E_y-\beta H_z)] + \frac{\partial}{\partial z'}[\gamma(E_z+\beta H_y)] = 0. \tag{2.12.20}$$

Similarly eliminating $\partial H_x/\partial t'$ between the x component of the second equation of both (2.12.9) and (2.12.10), we have

$$\frac{\partial H_x}{\partial x'} + \frac{\partial}{\partial y'}[\gamma(H_y+\beta E_z)] + \frac{\partial}{\partial z'}[\gamma(H_z-\beta E_y)] = 0. \tag{2.12.21}$$

Therefore, if we write

$$E'_x = E_x, \quad E'_y = \gamma(E_y-\beta H_z), \quad E'_z = \gamma(E_z+\beta H_y), \tag{2.12.22}$$

$$H'_x = H_x, \quad H'_y = \gamma(H_y+\beta E_z), \quad H'_z = \gamma(H_z-\beta E_y), \tag{2.12.23}$$

we obtain the relations of transformation of the field vectors E and H for the moving system.

Then the field equations reduce to

$$\nabla' \times H' = \frac{1}{c}\frac{\partial E'}{\partial t'}, \quad \nabla' \times E' = -\frac{1}{c}\frac{\partial H'}{\partial t'}, \tag{2.12.24}$$

$$\nabla' \cdot E' = 0, \quad \nabla' \cdot H' = 0. \tag{2.12.25}$$

This is of the form of (2.12.9) and (2.12.10), and establishes the form invariance of electrodynamical field equations under LTs. The formulas (2.12.22) and (2.12.23) may be put in the vector form

$$E' = \gamma[E + (\beta \times H)] - \frac{1}{\beta^2}(\gamma-1)\beta(\beta\cdot E). \tag{2.12.26}$$

$$H' = \gamma[H - (\beta \times E)] - \frac{1}{\beta^2}(\gamma-1)\beta(\beta\cdot H). \tag{2.12.27}$$

The reverse transformation obtains by replacing v by $-v$ and interchanging E, H with E', H'. If $v \ll c$, $\gamma \to 1$, then

$$E' = E + (\beta \times H), \quad H' = H - (\beta \times E). \tag{2.12.28}$$

2.13 The Four Dimensional Continuum of the Minkowski World

The time t, together with the three space co-ordinates form the 4-dimensional space-time continuum of the Minkowski world. The y and z co-ordinates are dropped for simplicity, in this geometrical study of light propagation in this world.

Consider the axes of the space co-ordinate OX and the time co-ordinate $u = ct$ in the $x - t$ plane for study of a two-dimensional Minkowski geometry. An event occurring at any point of space at any instant is represented in this world by a point called the *world point.* The world line is the locus of a point in the UX plane, which moves in position space with time t; the $U(= ct)$ co-ordinate changes with the space co-ordinate x, to describe the curve. A point fixed in the fundamental system k (the system which does not move in time) describes a straight world line parallel to the u axis. For a point moving with velocity q along any path

$$\tan\theta = \frac{\delta x}{\delta u} = \frac{\delta x}{c\delta t} = \frac{q}{c}. \tag{2.13.1}$$

Since $q \ll c$, the maximum value of $\tan\theta$ is $c/c = 1$. The path of light moving with the constant velocity c, is straight and bisects the UX axes. For other paths $\delta x/\delta q < 1$, and the world line for material particles is inclined at an angle less than $\pi/4$.

2.13.1 Relativity of simultaneity

Events represented by points on the x' axis are simultaneous to the observer in the k' system, but they occur at successive instants to an observer in the k system; the event A' of Fig. 2.13.1 occurs at time $u/c(= AA'/C)$ latter than the event 0.

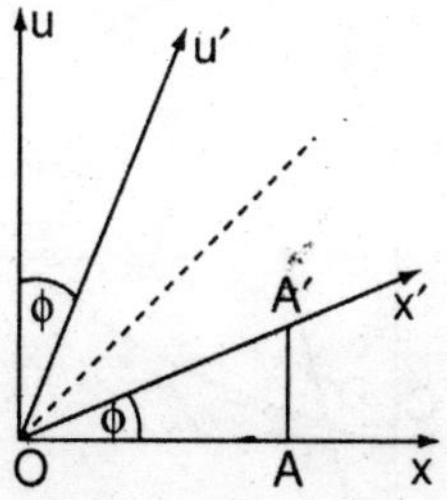

Fig. 2.13.1 The straight worldline of a uniformly moving particle.

2.13.2 Representation of events

To determine the unit (that is, the scale), draw the hyperbolas, of Fig. 2.13.2,

$$x^2 - u^2 = +1. \tag{2.13.2}$$

$$x^2 - u^2 = -1. \tag{2.13.3}$$

The u and x axes cut these hyperbolas at the two points $u = 1, x = 0$ and $u = 0, x = 1$. The points on the new axes are $OU'(x' = 0)$ and $OX'(u' = 0)$. The u' axis intersects the hyperbola (2.13.3) at Q and the x' axis intersects (2.13.2) at R. At Q, $x = \gamma\beta$, $u = \gamma$, and $x' = 0$, $u' = 1$. Similarly the conditions $u' = 0$, $x' = 1$, satisfying (2.13.2), means $x = \beta, u = \gamma\beta$.

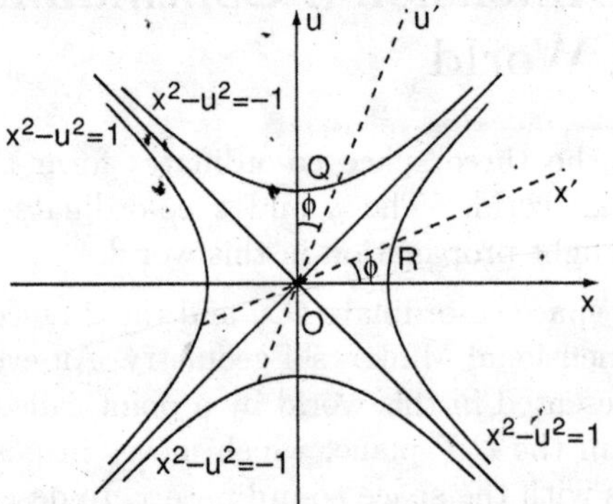

Fig. 2.13.2 Representation of events by the hyperbolas $x^2-u^2=\pm 1$.

2.13.3 Change of length

In Fig. 2.13.3 the unit lengths in the two systems are OA and OB' and the world line of O and A, the extremities of the unit vector, are along OX. To an observer in the k' system the simultaneous position for $u' = 0$ of these end points are O and A', the length of unit rod of the k system, placed along OX, appears to be shorter than OB' and the unit rod at rest in k'. On the other hand, the end points O and B' of a unit rod along OX' have the world lines $OC'D'$ and BB', so that to an observer in k at $u = 0$ its end points are at O and B. Hence OB', the length of unit rod in k', as observed from k, is less than OA (the unit rod in k) and it is shorter to this observer.

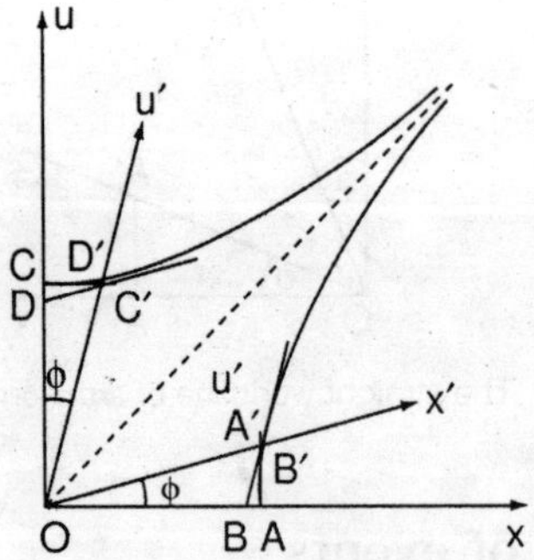

Fig. 2.13.3 Events and worldline.

2.13.4 Relativity of time

In the same way the unit interval of k', as observed from k is $OD(< OC$, where OC is the unit interval of time in k). Also, unit time of k as observed from k' is $OC'(< OD'$, where OD' is the unit time of k').

2.13.5 The regions of past and future events

Future events are marked in the region of the upper half of the x axis and the portion below stands for the past events, fig. 2.13.3. The portions on right and left do not stand for event, because the velocity of light there is greater than C.

2.14 Another procedure, using Imaginary Coefficients

It is convenient to take x, y, z and ict as the four co-ordinates or x_1, x_2, x_3, x_4, where $i = \sqrt{-1}$, $x_1 = x$, $x_2 = y$, $x_3 = z$ and $x_4 = ict$. Then equation (2.9.3) gives

$$x^2 + y^2 + z^2 - c^2t^2 \equiv x_1^2 + x_2^2 + x_3^2 + x_4^2 = d^2. \tag{2.14.1}$$

In this 4-dimensional world continuum of Minkowski d gives the distance reached by light at time t in the k system, and, any three lines perpendicular to the x_4 axis represent events that take place simultaneously. In the k' system the four co-ordinates are (x_1', x_2', x_3', x_4'), and the LT transform (2.14.1) into

$$x_1'^2 + x_2'^2 + x_3'^2 + x_4'^2 = d^2, \tag{2.14.2}$$

keeping the distance from the origin in 4-space constant. It is, therefore, a rotation with origin fixed in the world continuum. This particular LT is equivalent to a rotation in the x_1x_4 plane through an imaginary angle, determined by the relations

$$x_1 \cos\theta - x_4 \sin\theta = x_1', \quad x_1 \sin\theta + x_4 \cos\theta = x_4'. \tag{2.14.3}$$

The first relation is compared with the LT

$$x_1' = \gamma(x_1 - vt) = \gamma(x_1 - i\beta x_4) \tag{2.14.4}$$

to yield

$$\cos\theta = \gamma, \quad \sin\theta = i\gamma\beta. \tag{2.14.5}$$

The real angle $\phi(= \tan^{-1}\beta)$ is given by

$$\tan\theta = i\beta = i\tan\phi. \tag{2.14.6}$$

A rotation of the axes through the imaginary angle $\tan^{-1}(i\tan\phi)$ gives the LT.

The successive transformation of the co-ordinate axes from k to k' and from k' to k'', give the rotation of the world diagram successively through angles θ_1 and θ_2. The total shift and the final velocity,

$$\theta = \theta_1 + \theta_2. \tag{2.14.7}$$

$$\begin{aligned} V &= -ic\tan\theta = -ic\tan(\theta_1 + \theta_2) \\ &= -ic\frac{\tan\theta_1 + \tan\theta_2}{1 - \tan\theta_1\tan\theta_2} = \frac{v_1 + v_2}{1 + \beta_1\beta_2} \end{aligned} \tag{2.14.8}$$

give the addition theorem of Einstein for velocity.

2.15 General Lorentz Transformation

Since

$$x_1^2 + x_2^2 + x_3^2 + x_4^2 \to x_1'^2 + x_2'^2 + x_3'^2 + x_4'^2, \tag{2.15.1}$$

$$x_\mu x_\mu = x'_\mu x'_\mu = d^2, \quad \mu = (1, 2, 3, 4), \tag{2.15.2}$$

where the Einstein notation (repeated index means summation over that index) is adopted. The most general LT is represented by the scheme

$$\begin{bmatrix} x_1^1 \\ x_2^1 \\ x_3^1 \\ x_4^1 \end{bmatrix} = \begin{bmatrix} \alpha_{11} & \alpha_{12} & \alpha_{13} & \alpha_{14} \\ \alpha_{21} & \alpha_{22} & \alpha_{23} & \alpha_{24} \\ \alpha_{31} & \alpha_{32} & \alpha_{33} & \alpha_{34} \\ \alpha_{41} & \alpha_{42} & \alpha_{43} & \alpha_{44} \end{bmatrix} \quad (\text{columns } x_1, x_2, x_3, x_4) \tag{2.15.3}$$

when subjected to the following restrictions

$$\sum_\nu \alpha_{\mu\nu}^2 = \alpha_{\mu 1}^2 + \alpha_{\mu 2}^2 + \alpha_{\mu 3}^2 + \alpha_{\mu 4}^2 = 1, \quad (\mu, \nu) = (1, 2, 3, 4) \tag{2.15.4}$$

$$\begin{aligned} \alpha_{\mu\eta}\alpha_{\mu\nu} &= \alpha_{1\eta}\alpha_{1\nu} + \alpha_{2\eta}\alpha_{2\nu} + \alpha_{3\eta}\alpha_{3\nu} + \alpha_{4\eta}\alpha_{4\nu} \\ &= 0 \text{ if } \eta \neq \nu \\ &= 1 \text{ if } \eta = \nu, \end{aligned} \tag{2.15.5}$$

where $\alpha_{\mu\nu}$ are the elements of a 4×4 matrix. Thus, LTs are orthogonal transformations. Two successive transformations are given by the matrix product

$$\begin{bmatrix} \alpha_{11} & \alpha_{12} & \alpha_{13} & \alpha_{14} \\ \alpha_{21} & \alpha_{22} & \alpha_{23} & \alpha_{24} \\ \alpha_{31} & \alpha_{32} & \alpha_{33} & \alpha_{34} \\ \alpha_{41} & \alpha_{42} & \alpha_{43} & \alpha_{44} \end{bmatrix} \begin{bmatrix} \alpha_{11} & \alpha_{12} & \alpha_{13} & \alpha_{14} \\ \alpha_{21} & \alpha_{22} & \alpha_{23} & \alpha_{24} \\ \alpha_{31} & \alpha_{32} & \alpha_{33} & \alpha_{34} \\ \alpha_{41} & \alpha_{42} & \alpha_{43} & \alpha_{44} \end{bmatrix} = \begin{bmatrix} 1 & 0 & 0 & 0 \\ 0 & 1 & 0 & 0 \\ 0 & 0 & 1 & 0 \\ 0 & 0 & 0 & -1 \end{bmatrix} \tag{2.15.6}$$

with

$$(\det A)^2 = 1 \tag{2.15.7}$$

$$\det A = \pm 1, \tag{2.15.8}$$

where $A = [\alpha_{\mu\nu}]$. For identical transformations $A = +1$. For reasons of continuity, always take $\det A = +1$, and det LT $= +\ 1$. The LTs form a group and the general LT is characterized by (a) the orthogonality, (2.15.4) and (2.15.5); (b) the determinant $A = +1$; (c) all coefficients into which the index 4 appears only once are purely imaginary and every other coefficient is real, because the product $x_1 x_2 x_3$ is purely real and x_4 purely imaginary in the k system. In any other system

$$x'_1 = \alpha_{11}x_1 + \alpha_{12}x_2 + \alpha_{13}x_3 + \alpha_{14}x_4 \tag{2.15.9}$$

$$x'_4 = \alpha_{41}x_1 + \alpha_{42}x_2 + \alpha_{43}x_3 + \alpha_{44}x_4. \tag{2.15.10}$$

Hence the elements α_{11}, α_{12}, $\cdots$, not containing the index 4 are real and α_{14}, α_{24} etc., containing the index 4 only once, are imaginary. Also α_{44} is real. The particular LT of the special theory of relativity is

$$\begin{bmatrix} \gamma & 0 & 0 & i\beta\gamma \\ 0 & 1 & 0 & 0 \\ 0 & 0 & 1 & 0 \\ i\beta\gamma & 0 & 0 & \gamma \end{bmatrix} \tag{2.15.11}$$

2.16 Some useful Four Vectors

For motion of a point P we write the co-ordinates x_1, x_2, x_3, x_4 as functions of the proper time parameter T. Then the displacement 4-vector has the components dx_1, dx_2, dx_3, dx_4 and the Minkowski velocity 4-vector $\mathbf{V}$ is defined as $(dx_1/dT,\ dx_2/dT,\ dx_3/dT,\ dx_4/dT)$ because $dT\left(= dt\sqrt{1-\beta^2}\right)$ is an invariant. We find that

$$V = \left(\frac{dx_1}{dT}, \frac{dx_2}{dT}, \frac{dx_3}{dT}, \frac{dx_4}{dT}\right) = \gamma\left(\frac{dx_1}{dt}, \frac{dx_2}{dt}, \frac{dx_3}{dt}, \frac{dx_4}{dt}\right) \tag{2.16.1}$$

$$V^2 = \gamma^4(\dot{x}^2 + \dot{y}^2 + \dot{z}^2 - c^2) = \gamma^2(u^2 - c^2) = -c^2. \tag{2.16.2}$$

Hence V is time-like. When P is at rest in a moving system

$$V_1^0 = V_2^0 = V_3^0 = 0 \quad \text{and} \quad V_4^0 = ic. \tag{2.16.3}$$

From the 4-velocity $\mathbf{V}$ (V_1, V_2, V_3, V_4) we construct the infinitesimal vector $dV(dv_1, dv_2, dv_3, dv_4)$. Dividing dV by the scalar invariant infinitesimal dT we find the 4-acceleration

$$f = (f_1, f_2, f_3, f_4) = (dV_1/dT, dV_2/dT, dV_3/dT, dV_4/dT). \tag{2.16.4}$$

It follows that

$$(\mathbf{v}\cdot\mathbf{f}) = \left(V_1\frac{dV_1}{dT} + V_2\frac{dV_2}{dT} + V_3\frac{dV_3}{dT} + V_4\frac{dV_4}{dT}\right) = 0, \tag{2.16.5}$$

the 4-acceleration f is perpendicular to the 4-velocity $\mathbf{V}$ and is space-like; so

$$f_\mu f_\mu > 0. \tag{2.16.6}$$

2.17 Proper or Intrinsic Time

If P_1 and P_2 are two close points in 4-space, we can write

$$y_1 = x_1 + dx_1, \quad y_2 = x_2 + dx_2,$$

$$y_3 = x_3 + dx_3, \quad y_4 = x_4 + dx_4. \tag{2.17.1}$$

The four components of P_1P_2 being dx_1, dx_2, dx_3, dx_4, the integral

$$\int_0^S dS = \int_0^S \sqrt{dx_1^2 + dx_2^2 + dx_3^2 + dx_4^2} \tag{2.17.2}$$

of an element dS of the worldline is a definite portion of the world line which is independent of the choice of axes. Hence,

$$\frac{S}{ic} = \int_0^S \sqrt{dt^2 - \left(dx_1^2 + dx_2^2 + dx_3^2\right)/c^2}$$

$$= \int_0^S \left(1 - \frac{v^2}{c^2}\right)^{1/2} dt = \int_0^T dT = T \text{ (say)}. \tag{2.17.3}$$

This T has the dimensions of time, and is independent of the choice of co-ordinates, so it is an invariant of the LT, though t and v are not. If we take a system of axes k' fixed in a particle describing a worldline, the particle velocity with respect to the k' axes is zero. Then $dT = dt$ since $v = 0$ in k'; i.e., dT is a measure of the increase in time in the k' system. In place of t, the time in the laboratories frame for the particle motion, T, called the *proper time*, (or *intrinsic time*, or *local time*), is necessary as the independent variable, for 4-tensors.

2.18 The Field of a Uniformly Moving Electron

Suppose the charge e is placed at O' in the k' system. Then

$$(E'_x, E'_y, E'_z) = \frac{e(x', y', z')}{r'^{3/2}}, \quad H' = 0, \tag{2.18.1}$$

$$E_x = E'_x, \quad E_y = rE'_y, \quad E_z = rE'_z,$$

$$H_x = H'_x = 0, \quad H_y = -\gamma\beta E'_z, \quad H_z = \gamma\beta E'_y. \tag{2.18.2}$$

Hence

$$E_x = \frac{e\gamma(x - vt)}{[\gamma^2(x - vt)^2 + y^2 + z^2]^{3/2}},$$

$$(E_y, E_z) = \frac{e\gamma(y, z)}{[\gamma^2(x - vt)^2 + y^2 + z^2]^{3/2}} \tag{2.18.3}$$

$$H_x = 0, \quad (H_y, H_z) = \frac{e\beta\gamma(-z, y)}{[\gamma^2(x - vt)^2 + y^2 + z^2]^{3/2}}. \tag{2.18.4}$$

To an observer in the laboratory rest system k (with respect to which the velocity of the electron is v) the field varies in time. So, in addition to the electric field moving with velocity v it experiences a magnetic field which moves with the charge. Putting $\xi = x - vt$ we find that

$$E_x = \frac{e\gamma\xi}{(\gamma^2\xi^2 + y^2 + z^2)^{3/2}}, \quad (E_y, E_z) = \frac{e\gamma(y, z)}{[\gamma^2\xi^2 + y^2 + z^2]^{3/2}} \tag{2.18.5}$$

$$H_x = 0, \quad (H_y, H_z) = \frac{e\beta\gamma((-z, y)}{[\gamma^2\xi^2 + y^2 + z^2]^{3/2}}. \tag{2.18.6}$$

If $v \ll c$, then $\gamma = 1$ and we obtain

$$(E_x, E_y, E_z) = \frac{e(\xi, y, z)}{(\xi^2 + y^2 + z^2)^{3/2}} \tag{2.18.7}$$

$$H_x = \sqrt{H_y^2 + H_z^2} = \frac{e\beta\sqrt{y^2 + z^2}}{(\xi^2 + y^2 + z^2)^{3/2}} = \frac{e\beta}{r^2}\sin\theta. \tag{2.18.8}$$

This is a form of Biot-Savart's law. The equipotential surfaces are

$$\phi' = \frac{e}{(x'^2 + y'^2 + z'^2)^{1/2}} = \frac{e}{r} = \text{constant.} \tag{2.18.9}$$

Therefore, the equipotential surfaces $x'^2 + y'^2 + z'^2 =$ constant of the k' frame, in the k system are equivalent to the time dependent surface given by

$$\gamma^2(x - vt)^2 + y^2 + z^2 = \text{constant.} \tag{2.18.10}$$

2.19 A Straight Worldline Description of a Particle in Motion

The co-ordinate system k' moves with velocity v in the k system. The particle starts at the origin, so

$$x' = \gamma(x - \beta u), \quad u' = ct' = \gamma(u - \beta x) \tag{2.19.1}$$

$$x = \gamma(x' + \beta u'), \quad u = ct' = \gamma(u' + \beta x'). \tag{2.19.2}$$

Comparing with the transformation relations of the oblique axes

$$x = \frac{x'\sin(\omega - \alpha)}{\sin\omega} + \frac{u'\sin(\omega - \beta)}{\sin\omega}, \tag{2.19.3}$$

$$u = \frac{x'\sin\alpha}{\sin\omega} + \frac{u'\sin\beta}{\sin\omega}, \tag{2.19.4}$$

of Fig. 2.13.1, we find that these relations are those of a transformation formula in oblique co-ordinate axes. The equation for the u' axis on the x- plane,

$$x - \beta u = 0, \tag{2.19.5}$$

obtains by putting $x' = 0$ in (2.19.1) and the equation for the x' axis,

$$u - \beta x = 0 \tag{2.19.6}$$

obtains by putting $u' = 0$ in (2.19.2). The oblique axes are equally inclined to the original u and x axes, because equations (2.19.5) and (2.19.6) give

$$\frac{x}{u} = \beta = \tan\phi \tag{2.19.7}$$

$$\frac{u}{x} = \beta = \tan\phi \tag{2.19.8}$$

for inclinations of the u' axis to the u axis and the x' axis to the x axis.

2.20 Tensors and Invariants in Relativity Theory

Relativistically, all physical laws are independent of the co-ordinates of the observers. So, these laws must be invariants of the LT. The invariant character is evident when these laws are put in specific forms. The procedure is to put the equations of physics in the vector form, because vector relations are invariants of co-ordinate transformation in three dimensions. Four-vectors correspond to the polar vectors and six vectors, correspond to the axial vectors, both in three dimensional position space.

Four-vectors are spatial vectors or temporal vectors; in other words, they are space like or time like. A 4-vector, divided by a quantity Q is a 4-vector if Q is independent of the choice of co-ordinates, i.e., if Q is an invariant. Although the EM field equations are invariants, the 4-vectors themselves vary, as they must, with change of axes. Either the first three terms of a 4-vector are real and the fourth term is purely imaginary, or the first three terms are purely imaginary and the fourth term is real.

2.21 LT of Vectors and Tensors

The ordinary LT,

	x_1	x_2	x_3	x_4
x_1'	γ	0	0	$i\beta\gamma$
x_2'	0	1	0	0
x_3'	0	0	1	0
x_4'	$-i\beta\gamma$	0	0	γ

(2.21.1)

transforms the components A_1, A_2, A_3, A_4 of a 4-vector A in the k system into A' in the k' system, as

$$A_1' = \gamma(A_1 + i\beta A_4), \quad A_2' = A_2, \quad A_3' = A_3, \quad A_4' = \gamma(A_4 - i\beta A_1). \tag{2.21.2}$$

Following (2.12.3) and (2.12.5) the transformation of the components of the second order tensor $A_{\mu\nu}$ is given by

$$\begin{aligned}
A_{11}' &= \gamma^2[A_{11} + i\beta(A_{14} + A_{41}) - \beta^2 A_{44}], \\
A_{12}' &= \gamma^2[A_{12} + i\beta A_{42}], \quad A_{13}' = \gamma^2[A_{13} + i\beta A_{43}], \\
A_{14}' &= \gamma^2[A_{14} - i\beta(A_{11} - A_{44}) + \beta^2 A_{41}], \\
A_{21}' &= \gamma^2[A_{21} + i\beta A_{24}], \quad A_{22}' = A_{22}, \quad A_{33}' = A_{33}, \\
A_{23}' &= A_{23}, \quad A_{32}' = A_{32}, \quad A_{41}' = \gamma^2[-i\beta(A_{11} + i\beta A_{14} + A_{44}) + A_{41}], \\
A_{24}' &= \gamma(A_{24} - i\beta A_{24}), \quad A_{34}' = \gamma(A_{34} - i\beta A_{34}).
\end{aligned} \tag{2.21.3}$$

If $A_{\mu\nu}$ is an antisymmetric tensor, then

$$A_{11} = A_{22} = A_{33} = A_{44} = 0,$$

$$A_{12}' = \gamma(A_{12} - i\beta A_{24}), \quad A_{13}' = \gamma(A_{13} - i\beta A_{34}), \quad A_{14}' = \frac{1}{\gamma} A_{14},$$

$$A_{24}' = \gamma(A_{24} + i\beta A_{12}), \quad A_{34}' = \gamma(A_{34} + i\beta A_{13}).$$

2.22 Transformation of the Field Equations

The invariant character of the field equations of electrodynamics is evident when these are expressed in the 4-space continuum. In such forms the equations are more symmetrical than in the three dimensional form of classical physics. The relationship between the electric and magnetic field strengths or between energy, Poynting vector, etc., thus revealed, gives a deeper understanding of the electromagnetic process, then otherwise.

2.22.1 Invariance of the charge

The equation of continuity, in the k system, moving with the constant velocity v, for a charge ρ at rest in the k_0 frame, is

$$\frac{\partial \rho^c}{\partial t} + \text{div}(\rho^c v) = 0 \quad \text{or} \quad \frac{d}{dt}(\rho^c V) = 0, \tag{2.22.1}$$

where V is the volume occupied by the charge $e(= \rho^c V)$. Since this charge remains unaltered in the k system, the amount of charge remains unaltered by the LT, so $e_0 = e$. Hence,

$$\rho_0^c V_0 = \rho^c V = \rho_c' V' \tag{2.22.2}$$

$$\rho^c = \frac{\rho_0^c V_0}{V} = \frac{\rho_0^c \delta x_0 \delta y_0 \delta z_0}{\delta x \delta y \delta z} = \rho_0 \gamma \tag{2.22.3}$$

because $V = V^0/\gamma$ due to the relativistic length contraction, where V_0 is the volume occupied by the charge in the k_0 system. Hence the charge density increases in the ratio $\gamma : 1$.

2.22.2 The 4-current and the 4-potential

The equations for the vector potential A and the scalar potential ϕ, in the Lorentz gauge, are given by

$$\nabla^2 A - \frac{1}{c^2}\frac{\partial^2 A}{\partial t^2} = -4\pi\rho\beta, \quad \nabla^2 \phi - \frac{1}{c^2}\frac{\partial^2 \phi}{\partial t^2} = 4\pi\rho \tag{2.22.4}$$

$$\nabla \cdot A - \frac{1}{c}\frac{\partial \phi}{\partial t} = 0. \tag{2.22.5}$$

$$\Box^2 A_\mu = -\frac{4\pi}{c} S_\mu, \tag{2.22.6}$$

where

$$S_\mu = \rho(v_x, v_y, v_z, -ic), \quad A_4 = i\phi, \quad x_4 = ict \tag{2.22.7}$$

the D'Alembartian

$$\Box^2 = \frac{\partial^2}{\partial x_1^2} + \frac{\partial^2}{\partial x_2^2} + \frac{\partial^2}{\partial x_3^2} + \frac{\partial^2}{\partial x_4^2} \tag{2.22.8}$$

and equation (2.22.5) transforms to

$$\text{Div } A = \frac{\partial A_1}{\partial x_1} + \frac{\partial A_2}{\partial x_2} + \frac{\partial A_3}{\partial x_3} + \frac{\partial A_4}{\partial x_4} = \frac{\partial A_\mu}{\partial x^\mu} \quad (\mu = 1, 2, 3, 4) \tag{2.22.9}$$

A_μ transforms as S_μ, because the operator $\Box^2$ on the left hand side of (2.22.6) is a scalar invariant of LT; so, A_μ is a 4-vector, and the invariant equation for A_μ is (2.22.6).

2.23 Invariance of the Maxwell-Lorentz Equations

Since $H = \text{curl}\, A$ and $E = \nabla\phi - \partial A/c\partial t$, taking $A_\mu = (A - i\phi)$, we find that

$$H_x = H_1 = \frac{\partial A_3}{\partial x_2} - \frac{\partial A_3}{\partial x_3} = F_{23},$$

$$H_y = H_2 = \frac{\partial A_1}{\partial x_3} - \frac{\partial A_3}{\partial x_1} = F_{31},$$

$$H_z = H_3 = \frac{\partial A_2}{\partial x_1} - \frac{\partial A_1}{\partial x_2} = F_{12}, \tag{2.23.1}$$

$$E_x = E_1 = \frac{\partial \phi}{\partial x_1} - \frac{i\partial A_1}{\partial x_4} = i\left(\frac{\partial A_4}{\partial x_1} - \frac{\partial A_2}{\partial x_4}\right) = iF_{14},$$

$$E_y = E_2 = \frac{\partial \phi}{\partial x_2} - \frac{i\partial A_2}{\partial x_4} = i\left(\frac{\partial A_4}{\partial x_2} - \frac{\partial A_2}{\partial x_4}\right) = iF_{24},$$

$$E_z = E_3 = \frac{\partial \phi}{\partial x_3} - \frac{i\partial A_3}{\partial x_4} = i\left(\frac{\partial A_4}{\partial x_3} - \frac{\partial A_3}{\partial x_4}\right) = iF_{34}. \tag{2.23.2}$$

These equations can be represented as

$$F_{\mu\nu} = \frac{\partial A_\nu}{\partial x_\mu} - \frac{\partial A_\mu}{\partial x_\mu} \tag{2.23.3}$$

or as

$$F = \text{curl}\, A. \tag{2.23.4}$$

The field vectors H and E become the components of the antisymmetric tensor (six-vector)

$$F_{\mu\nu} = \begin{bmatrix} 0 & H_z & -H_y & -iE_x \\ -H_z & 0 & H_x & -iE_y \\ H_y & -H_x & 0 & -iE_z \\ iE_x & iE_y & iE_z & 0 \end{bmatrix}. \tag{2.23.5}$$

The electric and magnetic vectors do not form two separate 4-vectors, but, taken together, constitute a six-vector. They are not two separate vectors, but are dependent as they appear in the unified form as components of a tensor. That, in an LT, E and H appear coupled together is now natural to expect. Thus an electric charge, fixed in the k_0 system, is accompanied by a magnetic field to an observer in the k system and a purely magnetic field in k_0 has an electric field in k.

2.23.1 Transformation formulas for EM 6-vectors

The formulas (2.21.3), (2.23.1) and (2.23.2) give

$$E'_x = iF'_{14} = iA'_{14} = iA'_{14} = iF_{14} = E_x. \tag{2.23.6}$$

$$E'_y = iF'_{24} = \gamma(iF_{24} - \beta F_{12}) = \gamma(E_y - \beta H_z). \tag{2.23.7}$$

$$E'_z = iF'_{34} = \gamma(iF_{34} - \beta F_{13}) = \gamma(E_z + \beta H_y). \tag{2.23.8}$$

$$\left.\begin{aligned} H'_x &= F'_{23} = F_{23} = H_x, \\ H'_y &= F'_{31} = -F'_{13} = \gamma(-F_{13} + i\beta F_{34}) = \gamma(H_y + \beta E_z) \\ H_z &= F'_{12} = \gamma(F_{12} - i\beta F_{24}) = \gamma(H_z - \beta E_y) \end{aligned}\right\}. \tag{2.23.9}$$

If

$$\beta^2 \ll 1, \quad \text{then} \quad E' = E + (\beta \times H). \tag{2.23.10}$$

2.23.2 The tensor form of the Maxwell equations

The inhomogeneous pair of Maxwell equations

$$\nabla \times H = \frac{1}{c}\frac{\partial E}{\partial t} + \frac{4\pi}{c}\rho v, \quad \nabla \cdot E = 4\pi\rho, \tag{2.23.11}$$

equations (2.22.7) and (2.23.5) and the Einstein summation rule give

$$\frac{\partial F_{\mu\nu}}{\partial x_\nu} = \frac{4\pi}{c} S_\mu. \tag{2.23.12}$$

The homogeneous pair of Maxwell equations

$$\nabla \times E = -\frac{1}{c}\frac{\partial H}{\partial t}, \quad \nabla \cdot H = 0 \tag{2.23.13}$$

reduces to

$$\frac{\partial F_{\mu\nu}}{\partial x_\lambda} + \frac{\partial F_{\mu\lambda}}{\partial x_\mu} + \frac{\partial F_{\lambda\mu}}{\partial x_\nu} = 0, \tag{2.23.14}$$

where

$$\lambda\mu\nu = (2,3,4) \quad \text{or} \quad (3,4,1) \quad \text{or} \quad (1,2,3). \tag{2.23.15}$$

► **Example 2.23.1** *Prove that the scalar product of E and H is an invariant.*

Solution : In the k' system

$$\begin{aligned} (E' \cdot H') &= E'_x H'_x + E'_y H'_y + E'_z H'_z \\ &= E_x H_x + \frac{(E_y - \beta H_z)(H_y + \beta H_z)}{1-\beta^2} + \frac{(E_z + \beta H_y)(H_z - \beta E_y)}{1-\beta^2} \\ &= E_x H_x + E_y H_y + E_z H_z = (E \cdot H). \end{aligned}$$

► **Example 2.23.2** *Prove that $E^2 - H^2$ is an invariant.*

Solution : In the k' system

$$\begin{aligned} E'^2 - H'^2 &= E'^2_x + E'^2_y + E'^2_z - H'^2_x - H'^2_y - H'^2_z \\ &= E_x^2 + \gamma^2[(E_y - \beta H_z)^2 + (E_z + \beta H_z)^2] \\ &\qquad - H_x^2 - \gamma^2[(H_y + \beta E_z)^2 + (H_z - \beta E_y)^2] \\ &= E_x^2 - H_x^2 + (E_y^2 + E_z^2 - H_y^2 - H_z^2) = E^2 - H^2. \end{aligned}$$

2.24 The Lorentz Force

The first three components of the Lorentz force density,

$$F = \rho E + \frac{1}{c}\rho^c(v \times H), \tag{2.24.1}$$

can be written as

$$F_x = \frac{1}{c}(i\rho^c c)(-iE_x) + \frac{1}{c}\rho^c v_y H_z - \frac{1}{c}\rho^c v_z H_y = \frac{1}{c}F_{1\mu}S_\mu \tag{2.24.2}$$

$$F_y = \frac{1}{c}F_{2\mu}S_\mu, \quad F_z = \frac{1}{c}F_{3\mu}S_\mu. \tag{2.24.3}$$

These are the space components of a 4-vector, where the analogy suggests that

$$F_4 = \frac{1}{c}F_{4\mu}S_\mu = \frac{i}{c}\rho^c(E\cdot v) = \frac{1}{c}(E\cdot j). \tag{2.24.4}$$

F_4 is the rate of work done by the field on the charge per unit volume. It vanishes in the $\mathbf{K}_0$ system, in which the electron is at rest,

$$F_4^0 = 0. \tag{2.24.5}$$

In the $\mathbf{K}$ system, therefore,

$$F_\mu = \frac{1}{c}F_{\mu\nu}S_\nu.$$

$$F_\mu S_\mu = \frac{1}{c}F_{\mu\nu}S_\nu S_\mu = \frac{1}{c}F_{\mu\mu}S_\mu^2 + \frac{1}{c}(F_{\mu\nu} + F_{\nu\mu})S_\mu S_\nu = 0 + 0, \tag{2.24.6}$$

and because $F_{\mu\nu}$ is antisymmetric. In the $\mathbf{K}_0$ system $v = 0$, so $F = \rho^c E$ and $S_1 = 0 = S_2 = S_3$. Hence (2.24.6) is valid in all inertial systems, though the total force is not an invariant since it changes with change of velocity.

The total force acting on an electron in a slowly varying field,

$$= \iiint_V \left[\rho^c E + \frac{1}{c}\rho^c(v \times H)\right] dV = eE + \frac{e}{c}(v \times H), \tag{2.24.7}$$

cannot be considered for a 4-vector formalism, though F_μ of (2.24.2)–(2.24.4), is a 4-vector because the volume element dV is not a scalar invariant. Therefore, the total force changes from one inertial system to another. In the rest system $\mathbf{K}_0$, since $F_4^0 = 0$, the transformation formulas are

$$F_x = F_1 = \gamma F_1^0, \quad F_y = F_2 = F_2^0, \quad F_z = F_3 = F_3^0. \tag{2.24.8}$$

Also, from relativistic length contraction,

$$dV = \frac{1}{\gamma}dV^0. \tag{2.24.9}$$

The components of the total force F are

$$F_x = \iiint_V F_x dV = \iiint_{V^0} F_x^0 \gamma \frac{dV^0}{\gamma} = \iiint_{V^0} F_x^0 dV^0 = F_x^0 \tag{2.24.10}$$

$$F_y = \iiint_V F_y dV = \iiint_{V^0} F_y^0 \frac{dV^0}{\gamma} = \frac{1}{\gamma} F_y^0, \tag{2.24.11}$$

$$F_z = \iiint_V F_z dV = \iiint_{V^0} F_z^0 \frac{dV^0}{\gamma} = \frac{1}{\gamma} F_z^0, \tag{2.24.12}$$

$$F_4 = \iiint_V F_4 dV = 0 \text{ because } F_4^0 = 0. \tag{2.24.13}$$

2.25 The Energy Momentum Tensor

The 4-force is written as

$$F_\mu = \frac{1}{c} F_{\mu\nu} S_\nu = \frac{1}{4\pi} F_{\mu\nu} \frac{\partial F_{\nu\lambda}}{\partial x_\lambda} = \frac{1}{4\pi} \left[\frac{\partial}{\partial x_\lambda} (F_{\mu\nu} F_{\nu\lambda}) - F_{\nu\lambda} \frac{\partial F_{\mu\nu}}{\partial x_\lambda} \right]. \tag{2.25.1}$$

Using (2.23.14) we find that

$$-F_{\nu\lambda} \frac{\partial F_{\mu\nu}}{\partial x_\lambda} = F_{\nu\lambda} \left(\frac{\partial F_{\nu\lambda}}{\partial x_\mu} + \frac{\partial F_{\lambda\mu}}{\partial x_\gamma} \right) = \frac{1}{4} \frac{\partial}{\partial x_\nu} (F_{\mu\nu})^2 = \frac{1}{4} \frac{\partial}{\partial x_\lambda} (F_{\mu\nu})^2 \delta_{\nu\lambda}, \tag{2.25.2}$$

where $\delta_{\nu\lambda}$ is the Kronecker delta. Hence, we can write

$$F_\mu = \frac{\partial}{\partial x_\lambda} \left[\frac{1}{4\pi} (F_{\mu\nu} F_{\nu\lambda}) + \frac{1}{16\pi} (F_{\mu\nu})^2 \delta_{\nu\lambda} \right] = \frac{\partial T_{\mu\lambda}}{\partial x_\lambda}, \tag{2.25.3}$$

where

$$T_{\mu\nu} = \frac{1}{4\pi} F_{\mu\nu} F_{\nu\lambda} + \frac{1}{16\pi} (F_{\mu\nu})^2 \delta_{\nu\lambda} \tag{2.25.4}$$

$T_{\mu\nu}$ is the symmetric energy-momentum tensor.

2.26 The Maxwell Stress Elements

We observe that

$$\frac{1}{16\pi} (F_{\mu\lambda})^2 \delta_{\nu\lambda} = \frac{1}{8\pi} (H^2 - E^2) \delta_{\nu\lambda}. \tag{2.26.1}$$

For $\nu = \lambda = 1$, we obtain

$$\frac{1}{4\pi} F_{1\mu} F_{\mu 1} = \frac{1}{4\pi} (H_y^2 - H_z^2 + E_x^2). \tag{2.26.2}$$

For $(\nu, \lambda) = (1, 2, 3, 4)$ equation (2.25.4) gives

$$T_{xx} = T_{11} = \frac{1}{4\pi} \left[E_x^2 + H_x^2 - \frac{1}{2} (E^2 + H^2) \right] \tag{2.26.3}$$

$$T_{yy} = T_{22} = \frac{1}{4\pi} \left[E_y^2 + H_y^2 - \frac{1}{2} (E^2 + H^2) \right] \tag{2.26.4}$$

$$T_{zz} = T_{33} = \frac{1}{4\pi}\left[E_z^2 + H_z^2 - \frac{1}{2}(E^2 + H^2)\right] \tag{2.26.5}$$

$$T_{yz} = T_{zy} = T_{23} = T_{32} = \frac{1}{4\pi}(E_y E_z + H_y H_z) \tag{2.26.6}$$

$$T_{zx} = T_{xz} = T_{31} = T_{13} = \frac{1}{4\pi}(E_z E_x + H_z H_x) \tag{2.26.7}$$

$$T_{xy} = T_{yx} = T_{12} = T_{21} = \frac{1}{4\pi}(E_x E_y + H_x H_y), \tag{2.26.8}$$

$$S(\text{Poynting flux}) = \frac{c}{4\pi}(E \times H), \tag{2.26.9}$$

$$g(\text{Electromagnetic momentum per unit volume}) = \frac{(E \times H)}{4\pi}, \tag{2.26.10}$$

$$\xi(\text{Electromagnetic energy density}) = \frac{(E^2 + H^2)}{8\pi}. \tag{2.26.11}$$

The energy momentum tensor

$$T = \begin{bmatrix} T_{xx} & T_{xy} & T_{xz} & -icg_x \\ T_{yx} & T_{yy} & T_{yz} & -icg_y \\ T_{zx} & T_{zy} & T_{zz} & -icg_z \\ icg_x & icg_y & icg_z & \xi \end{bmatrix} \tag{2.26.12}$$

Integrating F_4 over volume V, we have

$$ic\iiint_V F_4 dV = ic\iiint_V \left(\frac{\partial T_{41}}{\partial x} + \frac{\partial T_{42}}{\partial y} + \frac{\partial T_{43}}{\partial z}\right) dV + ic\iiint_V \frac{\partial T_{44}}{\partial t} dV$$

$$= \iiint_V \left(\frac{\partial S_x}{\partial x} + \frac{\partial S_y}{\partial y} + \frac{\partial S_z}{\partial z}\right) dV + \frac{d}{dt}\iiint_V \xi dV. \tag{2.26.13}$$

Since $icF_4 = -(F \cdot v)$, applying Green's theorem this equation becomes

$$-\iiint_V (F \cdot v) dV - \iint_\Sigma S_n d\Sigma = \frac{\partial}{\partial t}\iiint_V \xi dV, \tag{2.26.14}$$

where $\partial/\partial t$ has been replaced by d/dt because t is independent of x, y, z. This is the energy principle : The rate at which work is done plus the influx of the Poynting vector **S** through the surface Σ equals the rate of increase of the total energy.

Integrating (2.25.4) over the volume V we obtain

$$\iiint_V F_x dV = \iiint_V \left(\frac{\partial T_{xx}}{\partial x} + \frac{\partial T_{xy}}{\partial y} + \frac{\partial T_{xz}}{\partial z}\right) dV + \frac{1}{ic}\frac{d}{dt}\iiint_V T_{14} dV$$

$$= \iint_\Sigma (T_x)_n d\Sigma - \frac{d}{dt}\iiint_V g_x dV, \tag{2.26.15}$$

$$\left(\frac{d}{dt}G_{\text{material}}\right)_x = \iint_\Sigma (T_x)_n d\Sigma - \frac{d}{dt}\iiint_V g_x dV, \tag{2.26.16}$$

$$\text{or} \quad \frac{d}{dt}(G_{\text{material}} + G_{elm}) = \iint_\Sigma (T_x)_n d\Sigma. \tag{2.26.17}$$

Hence g is the electromagnetic momentum density; symmetry of these shows that

$$S = gc^2. \tag{2.26.18}$$

2.27 The Equation of Motion

The dynamical equations of Newton and the Galelian addition theorem for velocities do not remain invariant under the universal time. Einstein supposed that the Newtonian equations are true only for velocities small compared to the velocity of light $C(= 3 \times 10^{10}$ cm/sec), made the Newtonian dynamical equations invariant under an LT, and gave them forms which reduce to the equations of Newton when the velocities become much less than C. The Newtonian equation

$$m_0 \frac{d^2 r_0}{dt_0^2} = F_0 \tag{2.27.1}$$

remains true in the rest system of the particle. Then the three-dimensional vectors of Newtonian motion are replaced by the 4-dimensional vectors of Minkowski, starting with the Newtonian equation of motion in the rest system.

2.28 The Procedures of Einstein and Minkowski for Laws of Motion

2.28.1 The procedure of Einstein

Equation (2.27.1) does not tell us how F is to be defined, so that it may depend on the velocity of the system. But, the transformation rule for electromagnetic forces being known, we consider the motion of a material particle of rest mass m_0 and charge e.

Let v be velocity of the particle in the **K** system along the x axis. In the rest system $\mathbf{K}_0$ the four co-ordinates of a point are ξ^0, η^0, ζ^0, T^0, and the equations of motion are

$$m_0 \frac{d^2\xi^0}{dT^{02}} = eE^0_\xi, \quad m_0 \frac{d^2\eta^0}{dT^{02}} = eE^0_\eta, \quad m_0 \frac{d^2\zeta^0}{dT^{02}} = eE^0_\zeta, \tag{2.28.1}$$

$$\frac{d\xi^0}{dT^0} = \frac{d\eta^0}{dT^0} = \frac{d\zeta^0}{dT^0} = 0. \tag{2.28.2}$$

If (ξ, η, ζ, T) be the four co-ordinates in the **K** system, then

$$\left.\begin{aligned} \xi^0 &= \gamma(\xi - vT), \quad \eta^0 = \eta, \quad \zeta^0 = \zeta, \\ T^0 &= \gamma(T - (\beta\xi/c)), \quad \gamma = (1 - \beta^2)^{-1/2} \end{aligned}\right\} \tag{2.28.3}$$

$$\frac{d\xi^0}{dT} = \gamma(\dot{\xi} - vT), \quad \frac{dT^0}{dT} = \gamma\left(1 - \frac{\beta}{c}\dot{\xi}\right), \tag{2.28.4}$$

$$\frac{d\xi^0}{dT^0} = \frac{\dot{\xi} - v}{1 - \beta\dot{\xi}/c}, \quad \frac{d\eta^0}{dT^0} = \frac{\dot{\eta}}{\gamma(1 - \beta\dot{\xi}/c)}, \quad \frac{d\zeta^0}{dT^0} = \frac{\dot{\zeta}}{\gamma(1 - \beta\dot{\xi}/c)}, \tag{2.28.5}$$

$$\frac{d}{dT}\left(\frac{d\xi^0}{dT^0}\right) = \frac{\ddot{\xi}(1 - \beta\dot{\xi}/c) + (\dot{\xi} - v)\ddot{\xi}(\beta/c)}{(1 - \beta\dot{\xi}/c)^2} = \frac{\ddot{\xi}}{\gamma^2(1 - \beta\dot{\xi}/c)^2} \tag{2.28.6}$$

$$\frac{d}{dT^0}\left(\frac{d\xi^0}{dT^0}\right) = \frac{\ddot{\xi}}{\gamma^2(1 - \beta\dot{\xi}/c)^3}, \tag{2.28.7}$$

$$\frac{d}{dT^0}\left(\frac{d\eta^0}{dT^0}\right) = \frac{\ddot{\eta} + (\dot{\eta}\ddot{\xi} - \ddot{\eta}\dot{\xi})(\beta/c)}{\gamma^2(1 - \beta\dot{\xi}/c)^3}, \tag{2.28.8}$$

$$\frac{d}{dT^0}\left(\frac{d\zeta^0}{dT^0}\right) = \frac{\ddot{\zeta} + (\dot{\zeta}\ddot{\xi} - \ddot{\zeta}\dot{\xi})(\beta/c)}{\gamma^2(1 - \beta\dot{\xi}/c)^3}. \tag{2.28.9}$$

Using (2.28.2) in (2.28.5), since

$$\dot{\xi} = v, \quad \dot{\eta} = 0 = \dot{\zeta} \tag{2.28.10}$$

equations (2.28.7) to (2.28.9) reduce to

$$\frac{d^2\xi^0}{dT^{02}} = \gamma^3\ddot{\xi}, \quad \frac{d^2\eta^0}{dT^{02}} = \gamma^2\ddot{\eta}, \quad \frac{d^2\zeta^0}{dT^{02}} = \gamma^2\ddot{\zeta}. \tag{2.28.11}$$

Hence the relations (2.28.1) become

$$\left.\begin{aligned} \gamma^3 m_0\ddot{\xi} &= eE^0_\xi = eE_\xi \\ \gamma^2 m_0\ddot{\eta} &= eE^0_\eta = e\gamma(E_\eta - \beta H_\zeta) \\ \gamma^2 m_0\ddot{\zeta} &= eE^0_\zeta = e\gamma(E_\zeta + \beta H_\eta) \end{aligned}\right\} \tag{2.28.12}$$

These equations simplify to

$$\left.\begin{aligned} m_0 r^3\ddot{\xi} &= eE_\xi \\ m_0 r\ddot{\eta} &= e(E_\eta - \beta H_\zeta) \\ m_0 r\ddot{\zeta} &= e(E_\zeta + \beta H_\eta) \end{aligned}\right\} \tag{2.28.13}$$

If we write

$$m_0\gamma^3 = m_l \quad \text{and} \quad m_0\gamma = m_t, \tag{2.28.14}$$

where m_l is the longitudinal mass and m_t is the transverse mass, these have the usual form. Since $\dot{\eta} = 0$ and $\dot{\zeta} = 0$, we find that

$$\frac{d}{dT}\left(\frac{m_0\dot{\xi}}{\sqrt{1 - \dot{\xi}^2/c^2}}\right) = \frac{d}{dT}\left(\frac{m_0 c\beta}{\sqrt{1 - \beta^2}}\right) = m_0\gamma\left[\frac{\ddot{\xi}}{\gamma} + \beta^2\gamma\ddot{\xi}\right] = m_0\gamma^3\ddot{\xi}, \tag{2.28.15}$$

$$\frac{d}{dT}\left(\frac{m_0\dot{\eta}}{\sqrt{1-\dot{\xi}^2/c^2}}\right) = m_0\gamma^3\left(\frac{\ddot{\eta}}{\gamma^2}+\frac{1}{c^2}\dot{\eta}\dot{\xi}\ddot{\xi}\right) = m_0\gamma\ddot{\eta}, \tag{2.28.16}$$

$$\frac{d}{dT}\left(\frac{m_0\dot{\zeta}}{\sqrt{1-\dot{\xi}^2/c^2}}\right) = m_0\gamma\ddot{\zeta}. \tag{2.28.17}$$

Thus the transformed equations are

$$\left.\begin{aligned} \frac{d}{dT}(m_0\gamma\dot{\xi}) &= eE_\xi,\\ \frac{d}{dT}(m_0\gamma\dot{\eta}) &= e(E_\eta - \beta H_\zeta),\\ \frac{d}{dT}(m_0\gamma\dot{\zeta}) &= e(E_\zeta + \beta H_\eta). \end{aligned}\right\} \tag{2.28.18}$$

Since these have the form

$$\frac{d}{dt}(m_0\gamma v) = eE + \frac{e}{c}(v \times H) = F \text{ (say)}. \tag{2.28.19}$$

Einstein gave the law of motion as

$$\frac{d}{dt}(m_0\gamma\mathbf{v}) = \mathbf{F} \tag{2.28.20}$$

in place of the Newtonian law

$$\frac{d}{dt}(m_0\mathbf{v}) = \mathbf{F}. \tag{2.28.21}$$

The relativistic momentum is $m_0\gamma\mathbf{v}$, and $m_0\gamma$ is the relativistic mass or the effective mass. The inertial mass is different in different inertial systems and also in different directions in the same inertial system. Along the direction of motion the effective mass is $m_0\gamma^3$ and perpendicular to the direction of motion the effective mass is $m_0\gamma$. For small velocities, for instance up to 30 km/sec, $m = m_0(1 + 1/2 \times 10^{-8})$. This discrepancy is beyond the range of ordinary measurement. So m is taken as the rest mass m_0 for practical purposes.

2.28.2 The 4-space procedure of Minkowski

The 4-velocity of Minkowski and the 4-acceleration vector components

$$V_1 = \frac{dx}{dT} = \gamma\dot{x}_1, \quad V_2 = \gamma\dot{x}_2, \quad V_3 = \gamma\dot{x}_3, \quad V_4 = ic\gamma, \tag{2.28.22}$$

$$\frac{d^2x_1}{dT^2} = \frac{dV_1}{dT}, \quad \frac{d^2x_2}{dT^2} = \frac{dV_2}{dT}, \quad \frac{d^2x_3}{dT^2} = \frac{dV_3}{dT}, \quad \frac{d^2x_4}{dT^2} = \frac{dV_4}{dT} = 0 \tag{2.28.23}$$

give the equations of motion

$$m_0\frac{dv_\mu}{dT} = F_\mu \quad (\mu = 1, 2, 3, 4), \tag{2.28.24}$$

where F_μ is a 4-force. To understand the significance of F_μ we change T to t and obtain

$$m_0\frac{dV_\mu}{dt} = \frac{1}{\gamma}F_\mu. \tag{2.28.25}$$

2.29 Inertia of Momentum and Energy

The kinetic energy

$$\xi = m_0\gamma c^2 \tag{2.29.1}$$

does not vanish when $v = 0$. The constant energy $\xi_0 = m_0c^2$ is the intrinsic energy or the inner energy. Every mass is associated with energy, and energy, given by ξ_0 corresponds to the mass m_0 (Einstein). If, therefore, a body gives up energy, it looses its mass. Hot bodies lose mass when cooled and a cold body gains mass when heated. Of course, the gain or loss is extremely small. So, a mass comes into existence at the cost of energy and can be annihilated by radiation. Thus, the laws of conservation of energy and matter are one and the same in principle.

Sun loses its mass at the rate of 2.5×10^9 kgm per min or 1.21×10^{15} kgm per year. This number, although very large, is small compared with the total mass of the sun. If the sun radiates the same constant energy it will take 11 billion years for its annihilation.

2.30 Hamilton's Form of Relativistic Equations

The x component of the equation of motion of an electron in an electromagnetic field, (2.28.19), when expressed in terms of the 4-potential A_μ, reads

$$\frac{d}{dt}(m_0V_1) = \frac{e}{c}\gamma(cE_x + (u \times H)_x) \tag{2.30.1}$$

$$= \frac{e}{c}(F_{11}V_1 + F_{12}V_2 + F_{13}V_3 + F_{14}V_4), \tag{2.30.2}$$

where

$$F_{23} = \frac{\partial A_z}{\partial y} - \frac{\partial A_y}{\partial z} = \frac{\partial A_3}{\partial x_2} - \frac{\partial A_2}{\partial x_3}, \text{ etc.} \tag{2.30.3}$$

$$iF_{14} = E_x = (\nabla\phi)_x - \frac{1}{c}\frac{\partial A_x}{\partial t} = -i\left(\frac{\partial A_4}{\partial x_1} - \frac{\partial A_1}{\partial x_4}\right), \text{ etc.} \tag{2.30.4}$$

Then (2.30.1) becomes

$$\frac{d}{dT}(m_0V_1) = \frac{e}{c}\left(\frac{\partial A_\mu}{\partial x_1} - \frac{\partial A_1}{\partial x_\mu}\right)V_\mu \tag{2.30.5}$$

A_μ is a function of x_ν and these co-ordinates depend on the proper (intrinsic) time T. We can, therefore, write

$$\frac{dA_\nu}{dT} = \frac{\partial A_\nu}{\partial x_\mu}\frac{\partial x_\mu}{\partial T} = \frac{\partial A_\nu}{\partial x_\mu}V_\mu \tag{2.30.6}$$

$$\frac{d}{dT}\left(m_0V_\nu + \frac{e}{c}A_\nu\right) = \frac{e}{c}\frac{\partial A_\mu}{\partial x_\nu}V_\mu. \tag{2.30.7}$$

To transform this equation further we introduce the Lagrangian function

$$\mathcal{L}(x_\mu, V_\mu) = \sum_\nu\left(\frac{1}{2}m_0V_\nu^2 + \frac{e}{c}V_\nu A_\nu\right) \tag{2.30.8}$$

Then

$$\frac{\partial \mathcal{L}}{\partial V_\nu} = m_0 V_\nu + \frac{e}{c} A_\nu, \tag{2.30.9}$$

$$\frac{\partial \mathcal{L}}{\partial x_\nu} = \frac{e}{c} \frac{\partial A_\mu}{\partial x_\nu} V_\nu \tag{2.30.10}$$

and equation (2.30.7) reduces to the Euler-Lagrangian equation

$$\frac{d}{dT}\left(\frac{\partial \mathcal{L}}{\partial V_\nu}\right) = \frac{\partial \mathcal{L}}{\partial x_\nu} \tag{2.30.11}$$

because

$$\frac{d}{dT}\left(\frac{\partial \mathcal{L}}{\partial V_\nu}\right) = \frac{d}{dT}\left(m_0 V_\nu + \frac{e}{c} A_\nu\right) = \frac{e}{c} \frac{\partial A_\mu}{\partial x_\nu} V_\mu = \frac{\partial \mathcal{L}}{\partial x_\nu}. \tag{2.30.12}$$

Let us now define

$$p_\nu = \frac{\partial \mathcal{L}}{\partial V_\nu} = m_0 V_\nu + \frac{e}{c} A_\nu \tag{2.30.13}$$

with $\nu = 1, 2, 3, 4$, solve these four equations for v_ν, and regard these V_ν as functions of $x_1, x_2, x_3, x_4, p_1, p_2, p_3, p_4$ in (2.30.8). Next we write

$$H^{-1}\left(\begin{array}{c|c} x_1 & p_1 \\ x_2 & p_2 \\ x_3 & p_3 \\ x_4 & p_4 \end{array}\right) = p_\mu V_\mu - \mathcal{L}\left(\begin{array}{c|c} x_1 & V_1 \\ x_2 & V_2 \\ x_3 & V_3 \\ x_4 & V_4 \end{array}\right), \tag{2.30.14}$$

and find that

$$\frac{\partial H^{-1}}{\partial x_\nu} = p_\mu \frac{\partial V_\mu}{\partial V_\nu} - \frac{\partial \mathcal{L}}{\partial x_\nu} - \frac{\partial \mathcal{L}}{\partial V_\mu} \frac{\partial V_\mu}{\partial x_\nu} = -\frac{\partial \mathcal{L}}{\partial x_\nu} = -\frac{\partial p_\nu}{\partial T} \tag{2.30.15}$$

$$\frac{\partial H^{-1}}{\partial p_\nu} = V_\nu + \frac{\partial V_\mu}{\partial p_\nu} p_\mu - \frac{\partial \mathcal{L}}{\partial V_\mu} \frac{\partial V_\mu}{\partial p_\nu} = V_\nu = \frac{dx_\nu}{dT}. \tag{2.30.16}$$

We thus have the equations of motion in the Hamiltonian form

$$\frac{dp_\nu}{dT} = -\frac{\partial H^{-1}}{\partial x_\nu}, \quad \frac{dx_\nu}{dT} = -\frac{\partial H^{-1}}{\partial p_\nu}, \tag{2.30.17}$$

where

$$\begin{aligned} H^{-1} &= \frac{p_\mu}{m_0}\left(p_\mu - \frac{e}{c} A_\mu\right) - \frac{1}{2m_0}\left(p_\mu - \frac{e}{c} A_\mu\right)^2 - \frac{e}{m_0 c} A_\mu \left(p_\mu - \frac{e}{c} A_\mu\right) \\ &= \frac{1}{2m_0}\left(p_\mu - \frac{e}{c} A_\mu\right)^2. \end{aligned} \tag{2.30.18}$$

2.31 Photoelectric Effect

When rays of high frequency (X-rays or ultra-violet rays or even visible light) impinge on a metal, electrons are emitted and the metal surface becomes charged positively if it is originally uncharged. This is the photoelectric effect. Lenard discovered that kinetic energy of emitted electrons is independent of the intensity of the incident ray. The intensity only increases the rate of emission. The kinetic energy depends on the observed light frequency. These experimental facts, when supplemented by the relation

$$\frac{1}{2}mv^2 = h\nu - P \tag{2.31.1}$$

between the electron kinetic energy and the incident light frequency (where h is the Planck constant), ν is the frequency (number of oscillations per units time) of the incident light and P depends on the nature of the matter representing the work necessary to draw the electron out of the metal, indicate that the energy acquired by the photoelectron is obtained from the incident light. The classical EM theory of light fails to give support to this presumption. Einstein assumed that light (radiation) consists of discrete amount of energy at points in the form of particles or light quanta, photons. These photons travel in straight lines with the velocity of light, and the amount of energy carried by a photon of light of frequency ν is $h\nu$. When a photon strikes a metal, it gives up its energy $h\nu$ to one of the electrons of the metal; this energy is converted to the kinetic energy of the electron. When no work is necessary to extract the electron out of the metal, we obtain the experimentally verified relation

$$h\nu = \frac{1}{2}mv^2. \tag{2.31.2}$$

Taking account of the work necessary to draw the electron from the metal, we obtain (2.31.1), which is more general than (2.31.2).

2.32 Photon and the Special Theory of Relativity

We write (2.28.24) as

$$\frac{d}{dT}(m_0\gamma\mathbf{v}) = \mathbf{F}\frac{dT}{dt}, \quad \frac{d}{dT}(m_0ic\gamma) = F_4\frac{dT}{dt}. \tag{2.32.1}$$

Then the three dimensional force

$$\mathbf{f} = \mathbf{F}\frac{dT}{dt} = \frac{1}{\gamma}\mathbf{F}. \tag{2.32.2}$$

The second equation of (2.32.1) is not independent of the first equation because

$$\left(\frac{dx}{dT}\right)^2 + \left(\frac{dy}{dT}\right)^2 + \left(\frac{dz}{dT}\right)^2 + \left(\frac{d}{dT}(ict)\right)^2 = -c^2. \tag{2.32.3}$$

Differentiation gives

$$\frac{dx}{dT}\frac{d^2x}{dT^2} + \frac{dy}{dT}\frac{d^2y}{dT^2} + \frac{dz}{dT}\frac{d^2z}{dT^2} + \frac{d}{dt}(ict)\frac{d^2}{dT^2}(ict) = 0. \tag{2.32.4}$$

Using (2.28.24) it is written as

$$F_x\frac{dx}{dT} + F_y\frac{dy}{dT} + F_z\frac{dz}{dT} + F_4\frac{d}{dT}(ict) = 0.$$

or as

$$-icF_4 = \left(f_x\frac{dx}{dt} + f_y\frac{dy}{dt} + f_z\frac{dz}{dt}\right)\frac{dt}{dT} = (f\cdot v)\frac{dt}{dT}. \tag{2.32.5}$$

Hence the conservation condition (the rate of change of momentum equals the rate at which work is done) is

$$\frac{d}{dt}(m_0\gamma c^2) = (f \cdot v). \tag{2.32.6}$$

Therefore, the energy of the particle,

$$\xi = m_0\gamma c^2, \tag{2.32.7}$$

consists of two parts : (1) the kinetic energy

$$m_0c^2(\gamma - 1) = \frac{1}{2}m_0v^2 \tag{2.32.8}$$

and (2) the rest energy m_0c^2 proportional to the rest mass. If the mass, momentum and rest energy are defined as

$$m = m_0\gamma, \tag{2.32.9}$$

$$g = mv, \tag{2.32.10}$$

$$\xi = mc^2, \tag{2.32.11}$$

then v can never exceed c. The limiting velocity $v = c$ is reached by a particle of zero rest mass. Though $m_0 = 0$ has no sense in Newtonian mechanics, it represents these particles (if, of course, they exist at all) which move with speed C. These are light particles, or photons, of the corpuscular theory of light, moving always with the constant velocity c. Then the diffraction of light was explained as the absorption and re-emission of light corpuscles by the atoms of the body by Einstein. The photons exist only when they move with the velocity c, possess no mass, and they vanish as soon as their velocity is less than c. When the photon is absorbed by a material atom, the light quantum does not stick to the atom like the light corpuscle of Newton, but ceases to exist as a particle. Its energy ξ and the momentum g, both are transferred to the absorbing atom. So, a mass and a momentum, given by

$$m = \frac{\xi}{c^2} = \frac{|g|}{c}, \tag{2.32.12}$$

$$|g| = \xi\left|\frac{v}{c^2}\right| = \frac{\xi}{c}, \tag{2.32.13}$$

can be assigned to the photon. In collision with another atom or electron the photon is either reflected, scattered or absorbed. These ideas of Einstein have been confirmed by the photoelectric effect, Compton effect and other experiments.

2.33 Compton Effect

When a beam of X-rays falls on a body the frequency of the scattered radiation remains unaltered in the direction of the incident ray, but becomes slower (i.e. less) in all other directions, depending on the angle of deviation. Debye and Compton explained this deviation assuming that a photon with energy $h\nu$ collides with a free electron of the body, to generate a radiation of increased wave length i.e., less frequency ν.

The energy and momentum of the photon were partly transferred by collision to the electron and the new photon of lower energy and similar momentum is formed, which moves in a direction inclined at angle θ to the original direction of the incident photon. If ξ be the energy of the photon, its mass $m = \xi/c^2$ and its momentum $mc = \xi/c$, where $\xi = h\nu$. Due to collisions the electron travels in the direction ϕ with velocity v, the energy of motion is supplied by the photon and the remainder of the energy travels in the direction of θ as a photon of energy $h\nu$ (Fig. 2.33.1). The relation between $\nu, \nu_0, v, \theta, \phi$ is determined by the conservation of relativistic energy and momentum, because v is large.

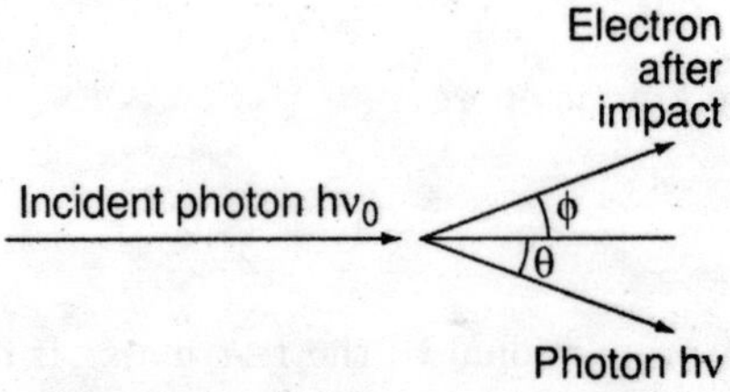

Fig. 2.33.1 Compton scattering effect.

Since energy of the incident photon equals the sum of the energy of the electron removed and the energy of the deviated photon,

$$h\nu_0 = m_0c^2(\gamma - 1) + h\nu. \tag{2.33.1}$$

Also, the conservation of momentum along the direction of incident radiation gives

$$\frac{h\gamma_0}{c} = m_0c\beta\gamma\cos\phi + \frac{h\nu}{c}\cos\theta. \tag{2.33.2}$$

Conservation of momentum, perpendicular to the direction of incident radiation, gives

$$0 = -m_0c\beta\gamma\sin\phi + \frac{h\gamma}{c}\sin\theta. \tag{2.33.3}$$

Eliminating ϕ between (2.33.2) and (2.33.3) we obtain

$$\frac{h^2\nu_0^2}{c^2} + \frac{h^2\nu^2}{c^2} - \frac{2h^2}{c^2}\nu\nu_0\cos\theta = m_0^2c^2\beta^2\gamma^2. \tag{2.33.4}$$

Also, from (2.33.1) we find that

$$m_0^2c^2\beta^2\gamma^2 = m_0^2c^2\left\{\frac{h(\nu_0-\nu)}{m_0c^2} + 1\right\}^2 - m_0^2c^2, \tag{2.33.5}$$

$$\nu = \frac{\nu_0}{1 + (2h\nu_0/m_0c^2)\sin^2(\theta/2)}. \tag{2.33.6}$$

Since $c = \lambda_0\nu_0 = \lambda\nu$, this equation reads

$$\frac{c}{\nu} = \frac{c}{\nu_0}\left(1 + \frac{2h\nu_0}{m_0c^2}\sin^2\frac{\theta}{2}\right), \tag{2.33.7}$$

$$\text{or}\quad \Delta\lambda = \lambda - \lambda_0 = \frac{2h}{m_0c}\sin^2\frac{\theta}{2}. \tag{2.33.8}$$

For perpendicular deviation

$$\Delta\lambda = \frac{h_0}{m_0^c} = 0.0242A^0, \tag{2.33.9}$$

which agrees with observed results.

2.34 Field of an Electron

Let a point charge e move along the x_3 axis with constant velocity v. Then the charge density is given by

$$\rho = \delta(x_1)\delta(x_3)\delta(x_3 - vt), \tag{2.34.1}$$

where $\delta(x)$ is the delta function of Dirac, Appendix 1. The wave equations for ϕ and A_3 are

$$\left.\begin{aligned}\left(\nabla^2\phi - \frac{1}{c^2}\frac{\partial^2\phi}{\partial t^2}\right) &= 4\pi e\delta(x_1)\delta(x_2)\delta(x_3 - vt)\\ \left(\nabla^2 A_3 - \frac{1}{c^2}\frac{\partial^2 A_3}{\partial t^2}\right) &= -4\pi e\beta\delta(x_1)\delta(x_2)\delta(x_3 - vt)\end{aligned}\right\}, \tag{2.34.2}$$

where $\beta = v/c$ and

$$A_1 = 0, \quad A_2 = 0, \quad A_4 = -\beta\phi, \tag{2.34.3}$$

$$E_1 = \frac{\partial\phi}{\partial x_1}, \quad E_2 = \frac{\partial\phi}{\partial x_2}, \quad E_3 = \frac{\partial\phi}{\partial x_3} - \frac{1}{c}\frac{\partial A_3}{\partial t} = (1-\beta^2)\frac{\partial\phi_3}{\partial x_3}$$

$$H_1 = \frac{\partial A_3}{\partial x_2} = -\beta E_2, \quad H_2 = \beta E_1, \quad H_3 = 0. \tag{2.34.4}$$

$$\frac{1}{c}\frac{\partial A_3}{\partial t} = -\beta\frac{\partial A_3}{\partial x_3}, \text{ etc.} \tag{2.34.5}$$

because the field components depend on the quantity $x_3 - vt$.

The Fourier integral representation of the three dimensional delta function is

$$\delta(x_1)\delta(x_2)\delta(x_3 - vt) = \frac{1}{8\pi^3}\iiint dk_1 dk_2 d\chi e^{i(k_1x_1 + k_2x_2 + \chi(x_3 - vt))}. \tag{2.34.6}$$

Application of the inverse of D'Alemberts' operator gives

$$\phi = \frac{e}{2\pi^2}\iiint e^{i(k_1x_1+k_2x_2+\chi(x_3-vt))}$$

$$\left[\frac{1}{k_1^2 + k_2^2 + \chi^2(1-\beta^2)} + \pi i\frac{\chi}{|\chi|}\delta(k_1^2 + k_2^2 + \chi^2(1-\beta^2)\right] dk_1 dk_2 dx. \tag{2.34.7}$$

For obtaining the advanced solution for $\phi, \pi i$ should be replaced by $-\pi i$ in the second term inside the square brackets. When $0 \leq \beta < 1$, the argument of the delta function has no zero, so we omit the delta function. Then the advanced and retarded potentials have the same solution. Substitution of

$$k_3 = \chi\sqrt{1-\beta^2}. \tag{2.34.8}$$

gives

$$\phi = \frac{e\gamma}{2\pi^2}\iiint dk \frac{e^{i(k_1x_1+k_2x_2+k_3\gamma(x_3-vt))}}{k_1^2 + k_2^2 + k_3^2},$$

where $\gamma = (1-\beta^2)^{-1/2}$. This integral is proportional to the Green function for the Laplace Poison equation. We write

$$k_1x_1 + k_2x_2 + \gamma k_3(x_3 - vt) = (k\cdot x) = |k||x|\cos\theta. \tag{2.34.9}$$

So x is along the k_3 direction. Then

$$\phi = \frac{e\gamma}{2\pi^2}\int_{|k|=0}^{\infty}\int_{\theta=0}^{\pi}\int_{\phi=0}^{2\pi}\frac{e^{i|k||x|\cos\theta}}{|k|^2}|k|^2 d|k|\sin\theta d\theta d\phi$$

$$= \frac{e\gamma}{\pi}\int_{|k|=0}^{\infty}\frac{2i\sin(|k||x|)}{i|x||k|}d|k| = \frac{2e\gamma}{\pi|x|}\frac{\pi}{2}$$

$$= \frac{e\gamma}{[x_1^2 + x_2^2 + \gamma^2(x_3 - vt)^2]^{1/2}} = \frac{e}{r}, \tag{2.34.10}$$

where

$$r^2 = (1-\beta^2)(x_1^2 + x_2^2) + (x_3 - vt)^2. \tag{2.34.11}$$

2.35 Cerenkov Radiation

Relativistically, the velocity of an electron cannot exceed the velocity of light c in vacuum ($\beta = v/c < 1$). But the electron velocity v, although remains less than c, can be greater than the phase velocity c' of light in the medium, where $c' = c/n < c$, n is the refractive index of the medium. So it is possible to have a material velocity v in the interval $c/n < v < c$. Then the electron leaves its field and radiates (Cerenköv radiation). This is an extremely weak visible radiation due to a fast electron moving in different liquids and solids with high refractive indices (for example, for cyclohexane, $n = 1.4367$). The electrons are generated in an external source or by the Compton effect and the photo electric effect of γ radiation of a radio active preparation. It is difficult to separate the new radiation from different kinds of luminiscences and to observe this radiation of small intensity.

For velocities above that of light the electron leaves its field behind, in the shape of a Mach cone. The field radiates in directions perpendicular to the surface of the cone, and because of the nature of dispersion, this radiation consists mainly of visible light. Due to radiation, the electron velocity decreases to that there of light velocity c/n. The radiated vector is polarized, so that the electric vector lies in the plane passing through the trajectory of the electron.

If a point electron moves in a medium with refractive index $n = \sqrt{\varepsilon}$ (where ε is the dielectric constant) with constant velocity v along the x_3 axis, the Maxwell Lorentz equations take the form

$$\nabla \times H = \frac{1}{c}\frac{\partial D}{\partial t} + \frac{4\pi}{c}j, \nabla\cdot D = 4\pi\rho^c, \tag{2.35.1}$$

where

$$H = \nabla \times A, \quad E = \nabla\phi - \frac{1}{c}\frac{\partial A}{\partial t}, \quad j_3 = \rho^c v, \quad j_1 = 0, \quad j_2 = 0 \tag{2.35.2}$$

$$\rho^c = e\delta(x_1)\delta(x_2)\delta(x_3 - vt), \quad D = \varepsilon E = n^2 E. \tag{2.35.3}$$

This constant refractive index n depends on the light frequency. Equation (2.35.3) has an operator character and it holds only for individual Fourier components i.e., for individual frequencies. To express this operator character we enclose n^2 in the first bracket, develop the fields in Fourier integrals, as

$$D = \int D(\omega)e^{-i\omega t}d\omega, \quad E = \int E(\omega)e^{-i\omega t}d\omega, \tag{2.35.4}$$

where ω is the frequency, and obtain, for the Fourier components, the formula

$$(n^2E) = \int n^2 E(\omega) e^{-i\omega t} d\omega \tag{2.35.5}$$

$$D(\omega) = n^2 E(\omega). \tag{2.35.6}$$

Division by an n^2 like operator is like the division by a linear operator, constructed by partial differentiation. Equations (2.35.3) and (2.35.6) give

$$E = \left(\frac{1}{n^2}D\right) = \int \frac{1}{n^2} D(\omega) e^{-i\omega t} d\omega. \tag{2.35.7}$$

We do not indicate explicitly the operator character of a quantity containing the refractive index n.

The Lorentz gauge condition and other field equations give

$$A_1 = 0, \quad A_2 = 0, \quad \frac{\partial A_3}{\partial x_3} - \frac{n^2}{c}\frac{\partial \phi}{\partial t} = 0; \quad A_3 = -\beta n^2 \phi,$$

$$\left(\nabla^2 - \frac{n^2}{c^2}\frac{\partial^2}{\partial t^2}\right)\phi = \frac{4\pi e}{n^2}\delta(x_1)\delta(x_2)\delta(x_3 - vt). \tag{2.35.8}$$

$$\left.\begin{aligned} E_1 &= \frac{\partial \phi}{\partial x_1}, \quad E_2 = \frac{\partial \phi}{\partial x_2}, \quad E_3 = -\gamma^2 \frac{\partial \phi}{\partial x_3}, \\ H_1 &= -\beta n^2 \frac{\partial \phi}{\partial x_2}, \quad H_2 = \beta n^2 \frac{\partial \phi}{\partial x_1}, \quad H_3 = 0. \end{aligned}\right\}, \tag{2.35.9}$$

where $\gamma = n^2\beta^2 - 1$. Using (2.35.8) and (2.34.6), for the retarded potential we obtain

$$\begin{aligned} \phi = \frac{e}{2\pi e^2} \int \frac{1}{n^2} e^{i(k_1x_1 + k_2x_2 + \chi(x_3 - vt))} \\ \times \left\{\frac{1}{k_1^2 + k_2^2 - \chi^2\gamma^2} + \pi i \frac{\chi}{|\chi|}\delta(k_1^2 + k_2^2 - \gamma^2\chi^2)\right\} dk_1 dk_2 d\chi. \end{aligned} \tag{2.35.10}$$

The electron motion being at velocity $v > c/n$, and $\gamma^2 (= \beta^2 n^2 - 1)$ being greater than zero, we cannot neglect the term of radiation proportional to the delta function. We use cylindrical co-ordinates $|k| \left(= \sqrt{k_1^2 + k_2^2}\right)$ and $\phi (= \tan^{-1} k_2/k_1)$. Integrating over the angle ϕ we obtain

$$\begin{aligned} \phi = \frac{e}{\pi} \int \frac{1}{n^2} e^{i\chi(x_3 - vt)} dx \int_0^\infty |k| j_0(|k|r) \\ \times \left[\frac{1}{|k|^2 - \chi^2\gamma^2} + \pi i \frac{x}{\omega}\delta(|k|^2 - \chi^2\gamma^2)\right], \end{aligned} \tag{2.35.11}$$

where $r = (x_1^2 + x_2^2)^{1/2}$, because

$$\int_0^{2\pi} e^{i|k|r\cos\psi} d\psi = 2\pi J_0(|k|r).$$

Using the formula

$$\int_0^\infty \frac{J_0(|k|r)}{|k|^2 - a^2}|k|d|k| = -\frac{\pi}{2}N_0(ar),$$

$$\int_0^\infty J_0(|k|r)\delta(|k|^2 - a^2)|k|d|k| = \frac{1}{2}J_0(ar)$$

we find that

$$\phi = -\frac{ie}{2}\int \frac{1}{n^2}e^{i\chi(x_3-vt)}\left[\frac{x}{|x|}J_0(r\gamma|x|) + iN_0(r\gamma|x|)\right]dx. \tag{2.35.12}$$

From the Poynting theorem of radiation of energy per unit time by the electron, the total energy u, in the volume enclosed by the surface Σ, is given by

$$-\frac{du}{dt} = \frac{c}{4\pi}\iint((E\times H)\cdot d\Sigma). \tag{2.35.13}$$

We integrate over the surface of a cylinder of radius r whose geometrical axis is the x_3 axis. The surface element being

$$d\Sigma = rd\phi dx_3\frac{r}{r} \text{ with } r_1 = x_1, r_2 = x_2, r_3 = 0, \tag{2.35.14}$$

we find that

$$((E\times H)\cdot d\Sigma) = \beta r d\phi dx_3\left(\gamma^2\frac{\partial\phi}{\partial x_3}\right)\left(n^2\frac{\partial\phi}{\partial r}\right), \tag{2.35.15}$$

$$\left.\begin{aligned}\left(\gamma^2\frac{\partial\phi}{\partial x_3}\right) &= -\frac{e}{2}\int\frac{\gamma^2}{n^2}e^{i\chi(x_3-vt)}[|x|J_0(r\gamma|x|) + i\chi N_0(r\gamma|x|)]dx\\ \left(n^2\frac{\partial\phi}{\partial r}\right) &= -\frac{ie}{2}\int\gamma e^{-i\chi'(x_3-vt)}\left[x'J_0'(r\gamma|x'| - i|x'|N_0'(r\gamma|x'|)\right]dx'\end{aligned}\right\}. \tag{2.35.16}$$

$$\frac{du}{dt} = \frac{1}{2}\pi\beta e^2 c\gamma\int_0^\infty\frac{\chi^2}{n^2}\gamma^3\left[N_0'(r\gamma\chi)J_0(r\gamma\chi) - J_0'(r\gamma\chi)N_0(r\gamma\chi)\right]dx. \tag{2.35.17}$$

This integrand is symmetric and so independent of the angle ϕ, and the relation

$$\int_{-\infty}^\infty e^{i(\chi-\chi')x_3}dx_3 = 2\pi\delta(x - x')$$

has been used. Also, while integrating with respect to χ from $-\infty$ to $+\infty$ the odd function has been omitted under the integral, and the integral over the even function is put equal to twice the integral from 0 to ∞. The Bessel theory formula

$$N_0'(\chi)J_0(\chi) - N_0(\chi)J_0'(\chi) = \frac{2}{\pi x} \tag{2.35.18}$$

gives

$$-\frac{du}{dt} = \frac{e^2}{2}u\int_0^{w_m} w\left(1 - \frac{1}{\beta^2 n^2}\right)dw, \tag{2.35.19}$$

where, instead of the variable χ, we have introduced the frequency $\omega(=\chi v)$. The refractive index decreases with the increase in ω and for X-rays it is less than one. The upper limit of the integral over ω is determined from the equation

$$\beta n(\omega_m) = 1. \tag{2.35.20}$$

For the equipotential surface when n = constant, we transform equation (2.35.12) into

$$\phi = -\frac{e}{m}\int_0^\infty [\sin x(x_3 - vt)J_0(r\gamma x) + \cos x(x_3 - vt)N_0(r\gamma x)]dx. \tag{2.35.21}$$

The Bessel theory relations

$$\int_0^\infty \sin(\chi\alpha)J_0(\chi\beta)dx = \int_0^\infty \cos(\chi\alpha)N_0(\chi\beta)d\chi = \begin{cases} \dfrac{1}{\sqrt{\alpha^2-\beta^2}}, & \alpha > \beta > 0 \\ 0, & \beta > \alpha > 0 \end{cases} \tag{2.35.22}$$

give

$$\phi = \begin{cases} \dfrac{2e}{n^2\sqrt{(vt-x_3)^2 - \gamma^2 r^2}}, & vt - x_3 > \gamma r = \gamma\sqrt{x_1^2 + x_2^2} \\ 0, & vt - x_3 < \gamma r. \end{cases} \tag{2.35.23}$$

In the direction of the Poynting vector at the instant $t = 0$ we find that

$$\phi = \begin{cases} \dfrac{2e}{n^2\sqrt{x_3^2 - \gamma^2 r^2}}, & -x_3 > \gamma r \\ 0, & -x_3 < \gamma r. \end{cases} \tag{2.35.24}$$

Therefore, the equivalent surfaces are the hyperboloids of revolution of Fig. 2.35.1 :

$$x_3^2 - \gamma^2 r^2 = \text{constant}. \tag{2.35.25}$$

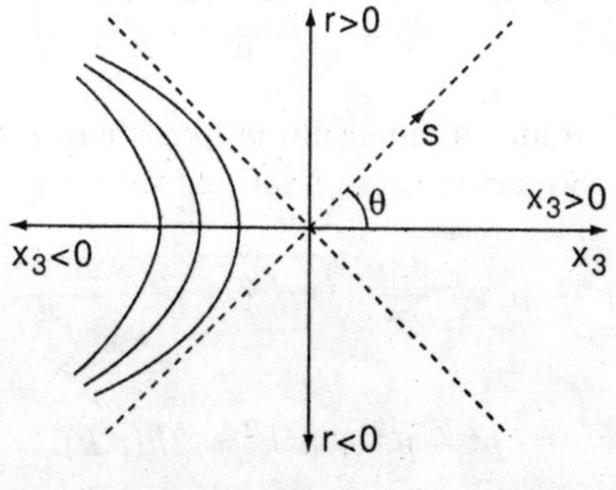

Fig. 2.35.1 The equipotential surfaces in cerenkov radiation.

Hence the potential and the field are non-vanishing only for $x_3 < 0$. The fields lie within a cone whose generators are described by the equation $-z = \gamma r$. With the help of the formula (2.35.9) the field is found to be

$$E = -\frac{2\gamma^2 eR}{n^2(R^2 - \beta^2 n^2 r^2)^{3/2}}, \quad H = \frac{1}{c}\beta n^2(v \times E), \tag{2.35.26}$$

where $R = (x_1, x_2, x_3)$ and v_0 is the unit vector in the direction of motion of the particle. The components of the Poynting vector **S** are

$$S_r = -\frac{c}{4\pi}\beta n^2 E_r E_3 > 0, \quad z < 0 \tag{2.35.27}$$

$$S_3 = \frac{c}{4\pi}\beta n^2 E_r^2 > 0, \tag{2.35.28}$$

where

$$E_r = \frac{r}{R}|E| = \sqrt{E_1^2 + E_2^2}.$$

It therefore follows that

$$|s| = \sqrt{s_r^2 + s_3^2} = \frac{c}{4\pi}\beta n^2 |E| E_r \tag{2.35.29}$$

$$\cos\theta = \frac{s_3}{|s|} = \frac{r}{R}. \tag{2.35.30}$$

The fields and the Poynting vector of (2.35.26) are infinite for

$$\cos\theta = \frac{1}{\beta n}. \tag{2.35.31}$$

Hence the entire radiation goes in the direction perpendicular to the generators of the cone which encloses the field, and in the directions defined by (2.35.31) the emitted light (electromagnetic) waves strengthen one another, while in other directions they weaken one another.

In the quantum mechanical treatment (not considered here) of Photon radiation, for every elementary electron, principles of conservation of energy and momentum hold good. The energy ξ of an electron is connected with its momentum p and the rest mass m_0 by the relation

$$\xi = c\sqrt{m_0^2 c^2 + p^2}.$$

If the momentum of the electron before and after the emission of a photon are denoted by p and p' respectively, then the principles of energy conservation and momentum conservation give

$$c\sqrt{m_0^2 c^2 + p^2} - h\nu = c\sqrt{m_0^2 c^2 + p'^2}, \quad p - hk = p',$$

where the photon energy is $h\nu$ and momentum is hk, where ν is the frequency and $|k|(= 1/\lambda)$ is the wave number of light. Squaring these relations we have

$$p^2 - p'^2 - \frac{2h\nu}{c}\sqrt{m_0^2 c^2 + p^2} + \frac{h^2\nu^2}{c^2} = 0$$

and

$$p^2 - p'^2 + h^2 k^2 = 2h(p.k).$$

Hence

$$(p.k) = (hk^2/2) + (\nu/c)\sqrt{m_0^2 c^2 + p^2} - (h\nu^2/2c^2),$$

$$\cos\theta = (p.k)/|p||k|,$$

$$\beta = v/c = c|p|/\varepsilon = |p|/\sqrt{m_0^2 c^2 + p^2},$$

where θ is the angle between the direction of motion of the electron and the direction in which the photon is emitted. Evidently,

$$\cos\theta = \frac{\nu\sqrt{m_0^2 c^2 + p^2}}{c|p||k|} + \frac{h|k|^2}{2|p||k|}\left(1 - \frac{\nu^2}{|k|^2 c^2}\right) = \frac{\nu}{|k|c\beta} + \frac{h|k|}{2|p|}\left(1 - \frac{\nu^2}{|k|^2 c^2}\right) \tag{2.35.32}$$

In vacuum $\nu = |k|c$ and $\cos\theta = 1/\beta > 1$ which is an inconsistent relation. Hence, an electron at rest in an inertial frame cannot radiate as in the classical case because then the principles of conservation of energy and momentum cannot hold good simultaneously. Instead of in a vacuum, in a dielectric, light propagates with the phase velocity $c' = c/n$, and $\nu = |k|c' = |k|c/n$; then the formula (2.35.32) becomes

$$\cos\theta = \frac{1}{\eta\beta} + \frac{\Lambda}{2\lambda}\left(1 - \frac{1}{n^2}\right),$$

where $\Lambda(= h/|p|)$ is the DeBroglie wave length of the emitted radiation $\lambda = 1/|k|$. In the Cerenkov case $\lambda \gg \Lambda$, so that the classical formula is obtained.

Chapter 3

Lorentz Theory of Electrodynamics

3.1 Introduction

Following the discovery of the classical atomistic theory of matter, Lorentz initiated work on obtaining the phenomenologically developed macroscopic Maxwell equations of electrodynamics in terms of the averages (mean values) of the analytically deduced fields of a system of microscopic charges in motion in vacuum. Lorentz wanted to explain analytically all EM phenomena in terms of the motion of electrons bound to atoms and their interaction with light (EM waves). The theory of Lorentz is meant for explaining the interaction between the sources (an assembly of elementary charges (electrons and ions) and electric currents in vacuum) via the EM field. The Maxwell equations determine the field of the Lorentz force produced by these sources. The two theories, essentially, are complementary to each other, and EM fields can be well understood by models of classical atomic theory of matter.

The macroscopic field equations of electrodynamics, including optics, in matter, obtain from these averages over volumes and times large enough compared with the atomic scales for their validity. In this theory the phenomenological parameters σ, ε, μ are used without enquiries into the internal structure of the material. And, all phenomena are reduced to the knowledge of the relations between B and H and between D and E, which must be known *experimentally a priori*. Also, macroscopic devices for experiments do not yield any information on individual particles.

The field equation of Lorentz involve only two field vectors E and H, and there is no complication from material parameters like ε, μ and σ. These quantities give rise to other field vectors like D and B and field relations called the *constitutive relations* in the phenomenological theory. So, the Lorentz theory is relatively simpler. Moreover, it explains the Zeeman effect, the dispersion of light, and the propagation of light in moving bodies, including the wave theory of Fresnel. Lorentz established the electromagnetic origin of the dielectric constant and magnetic permeability. His theory also lead to the *electron theory of metal* of Sommerfeld for explaining properties of electric and thermal conductivity, which was subsequently perfected by the ideas of quantum mechanics. Thus, the interaction between charges and electromagnetic fields has been conceptually established as the only fundamental interaction down to the atomic level. However, the local macroscopic description for fields in dielectrics using the concepts of polarization is much simpler than the description in terms of the microscopic charge motion in vacuum of the Lorentz theory.

The microscopic theory which depends on the particle picture has problems. One problem is regarding the repulsive Coulomb force between parts of the total charge which is responsible for its destabilization. If the charge is a small sphere, then owing to its symmetry, and the existence of the repulsive force between its parts, it would expand like a balloon and continue to do so to arbitrary large radius. The fact that macroscopic charged spheries do not explode in this manner is entirely due to interatomic and intermolecular forces which keep the shape of the sphere.

Macroscopically, material bodies are continuous pieces in specified domains in position space. So, continuum dynamics, in one sense, is a fundamental branch of mechanics. In that sense, particle dynamics deduced from it to treat small bodies is an idealization. However, speeds of both macroscopic particles and molecules in gases are such that under normal conditions the Newtonian laws of motion are a good approximation to the relativistic ones. This ceases at the next level of *subnuclear particles,* which are of concern in high speed collisions and are studied with the help of the relativistic laws of physics. So, in another sense, particle dynamics is more fundamental than the continuum dynamics, and deserves to be treated first. Study of subnuclear physics also shows that one should not enquire too closely as to what is met by a particle, or by a collision.

The theory of Lorentz has been establish without reservation. This theory is regarded as the foundation of electrodynamics by De Groote and Suttorp (1972). Sommerfeld (1956, p. 236) has suggested the name Electron Dynamics and Rohrlic (1965) the name Particle Electrodynamics. This theory is also called the *Maxwell-Lorentz theory of electrodynamics.* Including the phenomenological macroscopic theory which is complementary to the Lorentz theory, the name Macro electrodynamics and Micro electrodynamics has been by proposed Post (1967).

The ideas of Lorentz (1915) were further developed and perfected by Rosenfeld (1951), Mazur and Nijboer (1953), Mazur (1958), De Groot (1965), Russakoff (1970), De Groot and Suttorp (1972) and Robinson (1973). This topic has been briefly discussed in some text-books: Jackson (1975), Landau and Lifshitz (1960, 1975), Fano, Chu and Adler (1960), Portis (1978). In this chapter we provide a non-relativistic treatment of the Lorentz theory.

3.2 Basic Postulates

Electrodynamics is based on the laws of Coulomb, Gauss, Biot-Savart, Amperes, Ohm and Faraday. These laws were formulated purely from experimental observations. The basic postulates, of the Lorentz theory are : (1) There exists two kinds of electric charges : a positive charge and a negative charge; (2) Electric charge is conserved in any isolated system or in the whole of universe; (3) All charges are integral multiples of the electronic negative charge whose magnitude is 4.8×10^{-10} esu; (4) An electric charge in motion may be acted on by a force independent of its velocity and also by a force proportional to its velocity, and directed at right angles to it; (5) Magnetic monopoles do not exist; these poles occur only in pairs.

3.3 The Faraday-Maxwell Theory

Originally, the Maxwell equations were developed as a synthesis of experimental laws for large scale phenomena not depending on the details regarding the structure of matter. There is no guarantee that these equations are exact. But the conclusions have been found to be in excellent agreement with experiment. The validity of these equations, therefore, has never been questioned. The discrepancies between this correct and reliable theory and experiment are regarded as due to inadequacies of the constitutive relations, which are phenomena based supplementary relations, and guessed from experiments.

Concept of an elastic medium for matter in electrodynamics experiments proved helpful in many ways to the early studies. Charges were considered as forming macroscopic smooth distributions in space, that is, the discrete character of the elementary charge is disregarded. Faraday gave the idea of fields in all space, due to localized charges and currents and the resulting stresses and strains in matter.

All forces and energy sources of chemical and technical origin, interactions between nuclei and electrons, between the atoms and molecules of gases and of liquids, and between the atoms of a crystal or any solid are of electromagnetic nature. Many features of these interactions are explained by models of classical atomic theory of matter; however, quantum mechanics, though complicated, is appropriate.

The macroscopic theory, apart from furnishing a connected account of the electrical properties of matter, has brought uniformity in the whole science of theoretical physics. It has made optics a branch of electromagnetic. This electrodynamics is universely valid both theoretically, and experimentally, and has led to the ideas of Lorentz, Abraham and others for the classical electron theory. It is a correct, useful and reliable theory, and to the point in many fields of microscophycics, because everything of it is experiment based. So, there is no doubt in its validity. No other field theory in classical or modern physics, is so old, and has such an impact as the phenomenological theory of electricity and magnetism in modern life.

Nevertheless, this theory is not without some limitations. It has not attained the logical completeness that is desired in theoretical science. It has not explained important physical effects like Ohm's law, the parameters of dielectric permittivity, the magnetic permeability, and the electrical conductivity of materials. That the constants, ε, μ, σ vary in a rapidly changing field is ignored and consequently the phenomenon of dispersion of light has not been explained in the phenomenological theory of Maxwell, Faraday and Hertz. The constitutive relations for the electromagnetic properties of a medium are approximate, and phenomena based. These have always the scope of further improvement. So, further extension of the theory of phenomenological electrodynamics is necessary.

3.4 Concept of Electromagnetic Mass and Charge

The current of a charge q, confined in a sphere of radius a, moving with velocity $\mathbf{v}$, is $j(= q\mathbf{v})$. The Biot-Savart Law (equation (3.19.1)) gives the magnetic field

$$H = \frac{1}{cr^2} j \sin\theta = \frac{qv \sin\theta}{cr^2} \tag{3.4.1}$$

at the distance r. The magnetic energy of the field of the charge in the infinite space beyond its volume, in the sphere of radius a, is

$$W_H = \iiint\limits_V \frac{H^2}{8\pi} dV = \frac{q^2 v^2}{8\pi c^2} \int_{r=a}^{\infty} \int_{\theta=0}^{\pi} \int_{\phi=0}^{2\pi} \frac{1}{r^4} \sin^2\theta \cdot dV. \tag{3.4.2}$$

If we equate this energy with the kinetic energy $mv^2/2$ of the charge, then its mass m is given by

$$m = \frac{q^2}{4\pi c^2} \int_{r=a}^{\infty} \int_{\theta=0}^{\pi} \int_{\phi=0}^{2\pi} \frac{1}{r^4} \sin^2\theta r^2 \sin\theta dr d\theta d\phi = \frac{2q^2}{3ac^2}. \tag{3.4.3}$$

3.5 The Elementary Charge

The quantity of electricity, q, is defined by

$$q = \frac{1}{4\pi} \iiint\limits_V (\nabla \cdot D) dV = \frac{1}{4\pi} \iint\limits_\Sigma D_n d\Sigma. \tag{3.5.1}$$

Thus the charge q is only a short and convenient expression for this integral, and not a physical reality. This point of view is, however, not suitable to explain the effects of electrolysis in which the quantity of electricity is of fundamental importance.

3.6 Atomic Nature of Electric Charge

Each gramme equivalent of a substance transports 96494 Coulomb of electricity. So, electric charge is atomic in nature (i.e., discontinuous) and any charge is an integral multiple of a least amount e, below which no charge can exist. Since a gramme equivalent consists of N atoms, where N is about 10^{23} atoms of the substance, each atom carries a charge

$$= \frac{3 \times 96650 \times 10^9}{6 \times 10^{23}} = 4.77 \times 10^{-10} \text{ esu}.$$

Millikan and others measured e directly, and showed that this amount of charge is the smallest that can exist in nature, and that all other charges are integral multiples of this charge.

3.7 Measurement of the Charge

A particle was charged with different amounts of electricity and the charge was measured first by Millikan. All the charges were found to be integral multiples of an amount, which is the smallest elementary charge that can exist. In this experiment, the particle of mass m and charge e is suspended in a homogeneous electric field E. The field strength is so adjusted that the electric force is directed upwards and is just sufficient to balance the weight. From the principle of conservation of energy, the work done for moving the mass m upward through the distance $d(= mgd = e\phi$, where ϕ is the electric potential difference). Hence

$$e = \frac{mgd}{\phi}. \tag{3.7.1}$$

Thus charges are integral multiples of the minimum amount 4.7×10^{-10} esu. This charge is the same as the positive charge of a hydrogen nucleus. Existence of this amount of elementary charge is established also by several other methods. So, all charges are integral multiples of this elementary minimum charge.

3.8 Cathode Rays and the Electron

When a cathode is heated to a high temperature, in a highly evacuated glass tube, cathode rays are emitted due to the difference of the potential between the cathode K and the anode A. The rays behave exactly as ordinary electric current, and no chemical change of the cathode or anode takes place, however long the rays may be sent. So, (1) the current which flows through a conducting wire may be made to flow through vacuum; (2) the motion of electricity is not accompanied by any transport of matter; (3) the charge that moves is negative; (4) the rays are deviated by a transverse magnetic field. Evidently, therefore, the cathode rays are small charged particles moving with velocity v. The ratio e/m, called the *specific charge*, and the velocity are same for all particles of the cathode rays; specifically,

$$\frac{e}{m} = 1800 \times \frac{e}{m_H}, \tag{3.8.1}$$

where e/m_H is the specific charge of Hydrogen ion as obtained from experiments on electrolysis. The explanation that the particles may have a mass of the order of magnitude of ordinary hydrogen atom and a charge equal to 1800 times the unit charge occurring in electrolysis, is improbable from findings of the experiments of Millikan and others that the particles carrying charges are elementary in nature. The mass of each cathode ray particle is (1/1800) times the mass of the lightest atom, hydrogen. Moreover, the rays are independent of the materials of which the cathodes are made. So, cathode rays are not material substances and the name electron is attributed to these particles.

The equation of motion of a particle of mass m and charge e is

$$\frac{d\mathbf{v}}{dt} = -\frac{e}{mc}[c\mathbf{E} + (\mathbf{v} \times \mathbf{H})] = -\eta\mathbf{E} - \eta\left(\frac{\mathbf{v}}{c} \times \mathbf{H}\right),$$

where $\eta = e/m$. If $E = \nabla\phi$, then the scalar product of this equation with v on integration, gives the relation

$$m(v^2 - v_0^2)/2 = -e(\phi - \phi_0).$$

Hence, the change in kinetic energy of an electron equals the electrostatic potential difference. When $v_0 = 0$, $v = [2\eta(\phi_0 - \phi)]^{1/2}$.

3.8.1 Motion by a constant electric potential and a uniform magnetic field

In this case the electron moves initially with uniform velocity v_0 in the field of $\phi =$ constant, $E = \nabla\phi = 0$, a uniform magnetic filed H perpendicular to v. So, $dv/dt = -\eta(v \times H)$. The centrifugal acceleration $v^2/\rho = \eta vH/c$; the radius of curvature $\rho(= cv/\eta H)$ is perpendicular to v, and the particle remains in the plane perpendicular to H. So, the curvature increases as H increases and v decreases.

When the magnetic field is along the z axis, the equations of motion and their solution can be written as

$$\ddot{x} = \Omega\dot{y}, \quad \ddot{y} = -\Omega\dot{x}$$

$$m\ddot{\zeta} = -\frac{ie}{c}\dot{\zeta}H,$$

$$\zeta = \zeta_0 + (\dot{\zeta}_0/i\Omega)(1 - \exp(-i\Omega t)),$$

where $\zeta = x + iy, \Omega = eH/mc$. ζ_0 and $\dot{\zeta}_0$ being the initial values of ζ and $\dot{\zeta}$ respectively, we obtain

$$\left|\zeta = \zeta - \frac{1}{i\Omega}\dot{\zeta}_0\right| = \left|\frac{\dot{\zeta}_0}{\Omega}\right| = \text{constant}.$$

And since $|\dot{\zeta}_0| = |\ddot{x}_0 + i\dot{y}_0| = v$, the particle describes a circle in the xy-plane with radius

$$\rho = \left|\frac{\dot{\zeta}_0}{\Omega}\right| = \frac{vmc}{eH}.$$

3.8.2 Motion when H is not perpendicular to velocity

The two parts of v are v_1 and v_2, where v_1 is parallel to H and v_2 is perpendicular. Then

$$\dot{v}_1 = 0, \quad \dot{v}_2 = -\eta v_2 H/c,$$

the velocity parallel to H remains constant,

$$\frac{1}{\rho} = \eta\frac{H}{cv_2},$$

where ρ is the curvature of the path projected on the perpendicular plane, and the particle describes a spiral curve. When $v_i = 0$ we get

$$\frac{1}{\rho} = \eta\frac{H}{cv} = \frac{\eta H}{c\sqrt{2\eta(\phi - \phi_0)}} = \frac{1}{c}H\sqrt{\frac{\eta}{2(\phi - \phi_0)}}.$$

The curvature, therefore, varies as the square root of the potential difference. This effect is experimentally observed and $\eta(= 1.80 \times 10^7$ esu) is measured from the deflection of the path.

3.8.3 Motion by constant parallel fields

When both electric and magnetic fields are constant and parallel to the x-axis, we have

$$E_x = E,\ H_x = H,\ E_y = E_z = H_y = H_z = 0$$

and the equations of motion are

$$\ddot{x} = -\frac{e}{m}E, \quad \ddot{y} = -\dot{z}\Omega, \quad \ddot{z} = 0.$$

Since motion is in the xy-plane, the solution is

$$mx = \frac{1}{2}eEt^2, \quad my = \frac{e}{2c}Hvt^2, \quad z = vt,$$

where v_z is replaced by the symbols v and $x(= 0 = y$ at $t = 0)$.

The cathode rays are made to fall on a screen at a distance l from the slit through which they initially pass. The time to reach the screen is $t(= l/v)$, so

$$x = \frac{eEl^2}{2mv^2}, \quad y = \frac{eHl^2}{2mcv}, \quad z = 1, \quad \frac{y}{x} = \frac{Hv}{EC}, \quad \frac{y^2}{x} = \frac{eH^2l^2}{2mc^2E}.$$

Hence, if the particles have the same velocity v, but different e/m, they lie on a straight line passing through the origin. When e/m is the same for all particles, as in cathode rays, the particles lie on a parabola. Experiments have verified, for cathode rays of different velocities through a slit, the dimensions of the parabola $y^2/x = el^2H^2/2mc^2E$, and the value of e/m. So, these are rays of negatively charged particles.

For every rapidly moving cathode ray the electromagnetic mass m is relativistic, so

$$m = \left(= m_0/\sqrt{1 - v^2c^2}\right);$$

then the trajectory deviates from the parabola, as in β rays.

► **Example 3.8.1** *A charge q is uniformly distributed in the shape of a spherical shell of radius a. Inside and outside the shell three is a vacuum. Considering the shell as a sum of circular loops and assuming that the shell rotates with angular velocity Ω, show that the field at a large distance is that due to a magnetic moment $\mu = (qa^2/3e)$.*

Answer. The surface charge density is $q/4\pi a^2$ and the surface element $d\Sigma = 2\pi a \sin\theta \cdot a d\theta$. So, on $d\Sigma$, the charge element $dq = (q/4\pi a^2)(2\pi a^2 \sin\theta d\theta)$. Putting $dt = 2\pi/\Omega$ (the time period of rotation), the volume current

$$I = \frac{1}{c}\frac{dq}{dt} = \frac{1}{c}\frac{q}{4\pi a^2} 2\pi a^2 \sin\theta d\theta \cdot \frac{1}{2\pi/\Omega} = \frac{\Omega q}{2\pi c} \sin\theta d\theta.$$

Hence the magnetic moment element

$$d\mu = \frac{1}{c} I \cdot \pi (a \sin\theta)^2 = \frac{\pi a^2 \Omega q}{2\pi c} \sin^3\theta d\theta.$$

Therefore,

$$\mu = \int_0^\pi d\mu = \int_0^\pi \frac{\pi a^2 \Omega q}{2\pi c} \sin^3\theta d\theta = \frac{a^2 \Omega q}{3c}$$

and $\mu = (a^2q/3c)\Omega$, because, the magnetic moment is directed along the direction of rotation (the direction of cylindrical symmetry).

3.9 Concepts of Averaging in Electrodynamics

Macroscopic field quantities as defined are statistical averages of the corresponding analytically derived quantities, over a number of atoms or molecules contained in mass element which is large enough for the validity of the *principles of statistical averaging,* but which is still small for a macroscopic point of view. The spatial dimensions of volume elements of microscopic interest are large compared to the distance between atoms and molecules of the neighbourhood, and also small compared to macroscopic distances. Also, times of macroscopic interest should be much larger than those of microscopic evolution and averages are taken over large number of particles over domains which are small compared to the total system. For these reasons, macroscopically averaged fields are relatively smooth and are slowly varying functions of space and time co-ordinates (except at the boundaries) compared to the fields of microscopic theory. Macroscopic quantities are free from the influence of rapid changes in space and time. The corresponding microscopic quantities are capable of showing these changes.

The macroscopic field variables E (electric field intensity vector), D (electric displacement vector), H (magnetic field intensity vector), B (magnetic induction vector), j (electric current vector) and ρ^c (electric charge per unit volume) must be correlated in some way with appropriate averages of microscopic fields and charge over a large number of molecules. Averaging obtains the macroscopic fields and their governing equations from the microscopic fields and their equations. Derivation of these equations is carried out within a classical frame work, and not quantum mechanically, here. The paraméters ε, μ, σ depend on temperature, density, chemical constitution, wave frequency, field strength, etc., and predict the average behaviour of the particle and fields. Any special problem requires a specification of the EM macroscopic properties of the material as expressed in the constitutive relations.

For macroscopic description of electrodynamics, the lower limit of length is $l = 10^{-6}$ cm $= 10^2 A^0$. The volume l^3 cc has about 10^6 ions and electrons. Any region of macroscopic interest is of dimensions larger than 1 cm. The number of ions and electrons would be large enough in such regions to nullify by spatial averaging over these regions the very rapid fluctuations. But, electromagnetic waves of such wave lengths have the time period of oscillation $1/c = 3 \times 10^{-17}$ sec. This time being of the order of times of orbital motion of electrons (or of atomic and molecular motions) averaging over such small periods would not be appropriate. In other words, since the time variations of the microscopic fields are uncorrelated over distances of the order l cm, microscopic time fluctuations of a medium cannot be obtained from the equations for the spatially averaged field quantities. These equations give only the frequencies corresponding to oscillators driven at not so high external applied frequencies. Hence, use of time averaging would be improper for physical reasons; but spatial averaging is alright.

Both ions and electrons, for dimensions large compared with about 10^{-6} cm, exist, as point systems, in states of continuous rapid motion due to thermal agitation, and zero point vibration or orbital motion. The microscopic fields due to these motions vary extremely rapidly in space and time, in distances of the order of 10^{-8} cm or less, and in times of the order of 10^{-13} sec for nuclear (ionic) vibrations and of 10^{-17} sec for orbital motion of electrons, which is of the order of time periods of high frequency waves. Macroscopic averáges are valid over intervals in space and time much larger than these. Such averages are smooth, and valid for slowly varying fields of the phenomenological theory of Maxwell.

Field variables depend both on time and space co-ordinates, and on integrals of distribution in phase space. In statistical physics, specification of the averaging process is important, because it is not unique, because the independent variables are of several different category. The averaging procedures for determination of macroscopic field quantities, Russakoff (1970), are (1) spatial averaging, (2) time averaging and (3) ensemble averaging, Tolman (1938, Chapter III). The lower limits of intervals of length and time for macroscopic averages are ascertained on the basis of some facts of physical significance. For instance, the phenomena of reflection and refraction are described by a continuous dielectric constant ε, which is a consequence of the macroscopic theory. It fails for the X-ray diffraction because the X-ray wave lengths are small enough to ionize matter and to excite quantum effects.

Since the intervals for averaging purpose should be large compared to atomic volumes and times respectively, for validity of the classical theory, the ratio of the size of a molecule to the size of the averaging volume may be used as the expansion parameter in spatial averaging. Spatial averaging requires the minimum number of assumptions, takes the configuration of the system as given and requires no knowledge of the probability of the configuration or how the configuration will evolve in time. After carrying out the space averaging, time and ensemble averaging can be taken, because both these commute with time and space derivatives.

Denoting by $F(t)$ the average at the time t, of $F(t)$, the time averaging is given by

$$\bar{F}(t) = \frac{1}{T}\int_{-T/2}^{T/2} F(t+t')dt', \tag{3.9.1}$$

where integration is taken over some suitably chosen time span, T say, over the interval $-T/2 < t' < T/2$. The ensemble average of the field at a point is over some configuration (say, in the position space and momentum space) of the particle. Averaging over possible configurations is done according to some probability distributions (say, the law of Boltzmann). The space average of a function $F(r,t)$ is

$$\bar{F}(r,t) = \frac{\iiint\limits_V f(R)F(r+R)dV_R}{\iiint\limits_V f(R)dV_R}, \tag{3.9.2}$$

where $f(\mathbf{r})$ is a real, non-negative and non-zero spatial distribution function in the neighbourhood of $R = 0$ and dV_R is the three dimensional volume element at the integration point R. $f(\mathbf{r})$ should be continuous and smooth enough for a rapid convergence of its Taylor series expansion over distances of the order of atomic dimensions. *Commutativity of operation of differentiation* with respect to time and space co-ordinates is valid, because

$$\frac{\partial \bar{F}(r,t)}{\partial r_i} = \frac{\iiint\limits_V f(R)\dfrac{\partial F}{\partial r_i}(R+r)dV_R}{\iiint\limits_V f(R)dV_R} = \overline{\frac{\partial}{\partial r_i}F(r,t)}, \tag{3.9.3}$$

$$\frac{\partial \bar{F}(r,t)}{\partial t} = \overline{\frac{\partial \bar{F}(r,t)}{\partial t}}. \tag{3.9.4}$$

Two useful spatial distribution functions are :

(1) the Gaussian distribution function $f(r) = \exp(-r^2/l^2)$,

(2) the step function $f(r) = \begin{cases} 1, & r < 1 \\ 0, & r > 1. \end{cases}$

Classical mechanics determines behaviour of mechanical systems in time. But statistical methods are necessary to study the behaviour of a system, when the available knowledge is not enough for a complete specification of a precise state of it. These methods give the average behaviour of a collection of systems, called *statistical ensembles,* and distributed over a range of different possible states, but having the structure of the system concerned.

The state of each system in the ensemble, and the condition of the ensemble, as a whole, are studied quasi-geometrically in the phase space or the Γ space, which is a multidimensional Euclidean space. If the system has n degrees of freedom, then this space has $2n$ dimensions, with $2n$ rectangular co-ordinate axes, one for each of the position co-ordinates $q_1, q_2, \cdots, q_n$ and one for each of the momenta $p_1, p_2, \cdots, p_n$ whose values determine the state of the system. The system at a time is denoted by a point in the $2n$ dimensional Γ space, and the changes in it are given by changes in phase.

The motion of a phase point is an uniquely defined function of its instantaneous position. An ensemble is a "cloud" of the representative points for it in the Γ space, one for each system in the ensemble. The "streaming" motion of this cloud of the ensemble in time, as its points describe trajectories in the Γ space, follows the laws of classical mechanics.

A system of non-interacting N point particles in a volume V has $3N$ position co-ordinates $r_i (i = 1, 2, \cdots, N)$ and $3N$ momenta p_i. This system has $6N$ degrees of freedom. The $6N$ co-ordinates are denoted either by $q_1, q_2, \cdots, q_{3N}, p_1, p_2, \cdots, p_{3N}$, or by the vectors $r_1, r_2, \cdots, r_N, p_1, p_2, \cdots, p_N$, each having three components. The $3N$ co-ordinates of the momentum space are conjugate to the $3N$ co-ordinates of the position space (or the configuration space) that corresponds to the set of $3N$ position co-ordinates.

The ensemble, at any time, has the density f in the phase space of the representative points. This density is a function of the $6N$ co-ordinates of the Γ space and also of time t, since the motion of the phase points changes f in time at all points. The distribution function in the phase space, $f(q,p)$, is the probability per unit volume of finding the phase point for a system picked up at random from the ensemble in different regions of the phase space. It satisfies the normalization condition

$$\iint \cdots \iint f(q,p)\,dq\,dp = 1, \tag{3.9.5}$$

where

$$(q,p) = (q_1, q_2, \cdots, q_{3N}, p_1, p_2, \cdots, p_{3N})$$
$$dq\,dp = dq_1 dq_2 \cdots dq_{3N} dp_1 dp_2 \cdots dp_{3N}. \tag{3.9.6}$$

Knowing f as a function of the q's and p's, averages over all the systems in the ensemble, of any quantity, say X, which depends on the q's and p's and which belongs to the systems, are determined. The mean value of $X(q,p)$, a function of q and p, for all the systems in the ensemble at time t, is

$$\bar{X}(q,p) = \int \cdots \int X(q,p) f(q,p,t)\,dq\,dp. \tag{3.9.7}$$

Here the bar denotes mean over all the systems of the ensemble.

3.10 Macroscopic Averages

If $\psi(\mathbf{r},t)$ is a microscopic field and $\langle\psi(\mathbf{r},t)\rangle$ is its average over space and time, then

$$\langle\psi(\mathbf{r},t)\rangle = \frac{1}{\delta V \delta t} \iiint_{\delta V} dV_\xi \int_\delta dt_0 \psi(\mathbf{r}+\boldsymbol{\xi}, t+t_0), \tag{3.10.1}$$

where dV_ξ is the volume element in the integration point Q, whose position vector with respect to the observation point P is ξ. Hence, if q_j is the charge in the jth molecule and $\rho^c(\mathbf{r}, t)$ is the molecular charge density, we can write

$$\frac{1}{\delta V \delta t} \iiint_{\delta V} dV_\xi \int_{\delta t} dt_0 \rho^c(\mathbf{r} + \boldsymbol{\xi}_i, t + t_0) = \frac{1}{\delta V \delta t} \sum_j q_j, \tag{3.10.2}$$

where the sum is taken over all the molecules in δV in the time interval $(t, t + \delta t)$ and $\mathbf{r}$ is measured from the center of mass of the molecules. If $\rho_j^c(\mathbf{r}, t)$ is the molecular charge density, then

$$q_j(\mathbf{r}_j, t) = \iiint_{\text{mol}} \rho_j^c(\mathbf{r} + \mathbf{r}_j) dV_r. \tag{3.10.3}$$

Denoting the average number of molecules in unit volume at $\mathbf{r}$ at time t by $N(\mathbf{r}, t)$, the macroscopic charge density is

$$\rho^c(\mathbf{r}, t) = \sum_{j=1}^{N(\mathbf{r},t)} q_j \delta(\mathbf{r} - \mathbf{r}_j). \tag{3.10.4}$$

The average charge per unit volume at $\mathbf{r}$ at time t is

$$\langle \rho^c(\mathbf{r}, t) \rangle = \frac{1}{\delta V \delta t} \iiint_{\delta V} dV_\xi \int_{\delta t} dt_0 \rho^c(\mathbf{r} + \boldsymbol{\xi}, t + t_0) = N(\mathbf{r}, t) \langle q(\mathbf{r}, t) \rangle, \tag{3.10.5}$$

and the average moment of elementary charge in molecules in unit volume is

$$\langle P^c(\mathbf{r}, t) \rangle = \frac{1}{\delta V \delta t} \iiint_{\delta V} dV_\xi \int_{\delta t} dt_0 P(\mathbf{r} + \boldsymbol{\xi}, t + t_0) = N(\mathbf{r}, t) \langle p(\mathbf{r}, t) \rangle. \tag{3.10.6}$$

Here $P(\mathbf{r}, t)$ is the dipole moment of all the charges and $\langle p \rangle$ is the average dipole moment per unit volume. If p_j is the dipole moment of the jth charge, then

$$P(\mathbf{r}, t) = \sum_{j=1}^{N} p_j \delta(\mathbf{r} - \mathbf{r}_j), \tag{3.10.7}$$

where

$$p_j(\mathbf{r}_j, t) = \iiint_{\text{mol}} r \rho_j^c(\mathbf{r}_j + \mathbf{r}, t) dV_r. \tag{3.10.8}$$

Here dV_r is the volume element at $\mathbf{r}$.

3.11 Forces, Torques and Moments

3.11.1 Torques

Torques on a particle P, due to a force F, is

$$N = (\mathbf{r} \times F), \tag{3.11.1}$$

where $\mathbf{r}$ is the position vector of P, with respect to a given origin. The net torque, for a given system of particles, is

$$\mathbf{N} = \sum_i \left(\mathbf{r}_i \times \frac{d\mathbf{p}_i}{dt} \right) = \frac{d}{dt} \left(\sum_i (\mathbf{r}_i \times \mathbf{p}_i) \right) = \frac{d\mathbf{L}}{dt}, \tag{3.11.2}$$

where

$$\mathbf{L} = \sum_i (\mathbf{r}_i \times \mathbf{p}_i) \tag{3.11.3}$$

because $\dot{\mathbf{r}}_i$ and $\mathbf{p}_i$ are parallel. The torque, therefore, equals the rate of change of angular momentum of the system of particles (or, the moment of momentum of the system). So, the angular momentum is constant in the direction in which the torque vanishes. The pseudo vector[1] $\mathbf{L}$ depends on the choice of origin of the co-ordinate system; because, shifting the origin O to another position O' means the position vector $\mathbf{r}_i$ is replaced by $\mathbf{r}_i - \mathbf{OO'}$. The linear momentum p_i is a polar vector. Angular velocity and spin angular momentum are axial vectors. The angular momentum changes by the amount $\mathbf{OO'} \times \mathbf{p}_i$. In the rest frame of the system of particles, the angular momentum is a multiple of one-half of $\hbar$ (Planck's constant), a unique value.

3.11.2 Dipole field

The potential of a single charge (monopole) (which is the simplest case of a system having macroscopically a small space charge separation per unit volume at a point) is proportional to the inverse of the first power of the distance. A pair of equal and opposite charge $+q$ at $\mathbf{r}$ and $-q$ at $\mathbf{r} - 1$ form a doublet with moment

$$\nu = ql, \tag{3.11.4}$$

where l tends to zero in such a manner that ql remains finite, the doublet is called a *dipole* and ν is the dipole moment of charges. For a volume distribution of charges of density $\rho^c(r)$

$$\nu(r,t) = \iiint\limits_V R\rho^c(r+R)dV_B, \tag{3.11.5}$$

where dV_R is the volume element at the point $\mathbf{R}+\mathbf{r}$ and $\mathbf{r}$ is the position vector of the point at which ν is evaluated. Dipole formation is useful in the study of behaviours of a macroscopically neutral system of charges. The potential of the dipole is

$$\phi(r) = \frac{q}{r} - \frac{q}{|r-l|} = \frac{q}{r}\left[1 - \frac{1}{\left\{1 - \frac{2}{r^2}(r.l) + \frac{l^2}{r^2}\right\}^{1/2}}\right]$$

$$= -\frac{q}{r^3}(r.l) = -\frac{1}{r^3}(\nu.r) + \cdots, \tag{3.11.6}$$

where l is assumed to be an infinitesimal length; so l^2 and other higher powers of l are neglected. The electrostatic force for a dipole is

$$F = qE(r) - qE(r-l) = q(l\cdot\nabla)E = (\nu\cdot\nabla)E. \tag{3.11.7}$$

[1] Section 1.10.

Therefore, in an extended polarized body

$$F = \iiint\limits_V (\nu \cdot \nabla) E dV, \tag{3.11.8}$$

$$E = \nabla\phi = \frac{1}{r^3}\nabla(\boldsymbol{\nu}\cdot \mathbf{r}) + (\boldsymbol{\nu}\cdot \mathbf{r})\nabla\left(\frac{1}{r^2}\right) = \frac{1}{r^3}\nu - \frac{3}{r^3}(\boldsymbol{\nu}\cdot \mathbf{r})r. \tag{3.11.9}$$

So, the potential varies as the inverse square of the distance, and the electric field as the inverse cube of the distance, from the dipole. So, the dipole potential falls off more rapidly with distance than the potential of a monopole (single charge).

A constant electric field $\mathbf{E}$ can rotate a dipole ν about its center. The torque of the couple due to the force, $+qE$ on $+q$ at r and $-qE$ on $-q$ at $(r-1)$, is

$$\begin{aligned} N &= q(\mathbf{r}\times \mathbf{E}(\mathbf{r})) - q((\mathbf{r}-\mathbf{l})\times(\mathbf{r}-\mathbf{l})) \\ &= q(\mathbf{l}\times \mathbf{E}(\mathbf{r})) + q(\mathbf{r}\times(\mathbf{l}\cdot \boldsymbol{\nabla})\mathbf{E}(\mathbf{r}) \\ &= (\boldsymbol{\nu}\times \mathbf{E}) + (\mathbf{r}\times(\boldsymbol{\nu}\cdot \nabla)\mathbf{E}) \\ &= (\boldsymbol{\nu}\times \mathbf{E}) + (\mathbf{r}\times \mathbf{F}), \end{aligned} \tag{3.11.10}$$

where $\mathbf{F}$ is given by (3.11.7).

The work done to bring the electric field, W_E, is also its potential energy. The couple tending to increase the angle θ is

$$-\frac{dW_E}{d\theta} = -\nu E \sin\theta. \tag{3.11.11}$$

Hence if $W_E = 0$ when $\phi = \pi/2$, for the energy of a dipole of fixed moment $\boldsymbol{\nu}$, as a function of its orientation and position in the field, integration gives

$$W_E = -\nu E \cos\theta = -(\nu.E). \tag{3.11.12}$$

It is the energy in a curl free (electrostatic) field, because equation (3.11.7) can be written as

$$\mathbf{F} = \boldsymbol{\nabla}(\boldsymbol{\nu}.\mathbf{E}) - (\boldsymbol{\nu}\times(\boldsymbol{\nabla}\times \mathbf{E})). \tag{3.11.13}$$

If the field is not irrotational, $\nabla \times E \neq 0$, and the force expression (3.11.7) cannot be derived from the expression for a potential energy.

3.11.3 Field of a system of dipoles

The potential of a system of n dipoles, at a point $P(r)$, fig. 3.11.1, is

$$\phi(r) = \sum_{i=1}^{n}\left[\frac{q_i}{r_i} - \frac{q_i}{|\mathbf{r}_i - \mathbf{l}_i|}\right], \tag{3.11.14}$$

where q_i is the charge per particle of the ith species and l_i is the distance of separation of its negative charge to its positive charge. The moment of the ith dipole is

$$\nu_i = \lim_{l_i \to 0} [q_i r_i - q(r_i - l_i)] = \lim_{l_i \to 0} (q_i l_i). \tag{3.11.15}$$

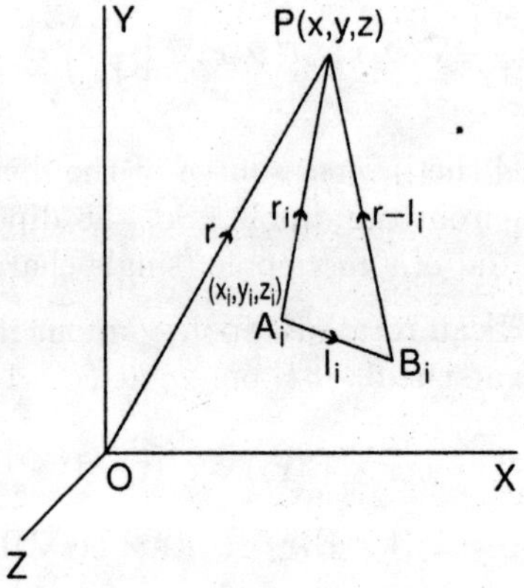

Fig. 3.11.1 Shows the field of the ith dipole A_iB_i at a point P in space.

The potential and the electrostatic force of this dipole system are

$$\phi(r) = -\sum_{i=1}^{n} (q_i(l_i - r_i))\frac{1}{r_i^3} = -\sum_{i=1}^{n} (\nu_i \cdot r_i)\frac{1}{r_i^3}, \tag{3.11.16}$$

$$F = \sum_i [q_i E(r) - q_i E(r_i - l_i)] = \sum_i (\nu_i \cdot \nabla) E(r). \tag{3.11.17}$$

The potential energy W of the system of two dipoles,

$$W = -(\boldsymbol{\nu}_1 \cdot \mathbf{E}(\boldsymbol{\nu}_2)) = -(\boldsymbol{\nu}_2 \cdot \mathbf{E}(\boldsymbol{\nu}_1)) = -\frac{(\boldsymbol{\nu}_1 \cdot \boldsymbol{\nu}_2)}{r^3} - \frac{3}{r^5}(\boldsymbol{\nu}_1 \cdot \mathbf{r})(\boldsymbol{\nu}_1 \cdot \mathbf{r}) \tag{3.11.18}$$

is the work done in bringing the dipole $\boldsymbol{\nu}_1$ (or $\boldsymbol{\nu}_2$) into its position in presence of the dipole $\boldsymbol{\nu}_2$ (or $\boldsymbol{\nu}_1$) in its position. Similarly, the potential energy of a system of n dipoles is

$$\begin{aligned} W &= -(\boldsymbol{\nu}_1.\mathbf{E})(\boldsymbol{\nu}_2, \boldsymbol{\nu}_3, \cdots, \boldsymbol{\nu}_n) = \boldsymbol{\nu}_1 \cdot \boldsymbol{\nabla}\phi(\boldsymbol{\nu}_2, \boldsymbol{\nu}_3, \cdots, \boldsymbol{\nu}_n) \\ &= -(\boldsymbol{\nu}_1 \cdot \boldsymbol{\nabla}) \left[\sum_{i=2}^{n} \left\{ (\boldsymbol{\nu}_i \cdot \mathbf{r}_i) \frac{1}{r_i^3} \right\} \right]. \end{aligned} \tag{3.11.19}$$

Hence

$$\phi(x, y, z) = \sum_{i=1}^{n} \left[\frac{q_i}{\sqrt{(x - x_i)^2 + (y - y_i)^2 + (z - z_i)^2}} - \frac{q_i}{\sqrt{(x - x_i - l_{ix})^2 + (y - y_i - l_{iy})^2 + (z - z_i - l_{iz})^2}} \right], \tag{3.11.20}$$

where the components of l_i are (l_{ix}, l_{iy}, l_{iz}).

3.12 Multipole Moments

A pair of equal and opposite dipoles makes an quadrupole. A pair of back-to-back quadrupoles makes an octuple. In that sense, a single charge is a monopole. Generalizing, potentials and fields of multipole moments exist for 2^n number of point charges, where $n = 0$ gives the field of a monopole, $n = 1$ gives that for a dipole, $n = 2$ for a quadrupole and $n = 3$ for an octuple. Next is the hexa decapole for which $n = 4$. Fields of quadrupoles and higher order poles are important in the study of fields of atomic nuclei. Dipole fields are important for dielectric materials. The monopole moment $(= ql^0 = q)$ is a single element. The dipole moment $\boldsymbol{\nu}$ has three independent components. Quadrupole moments have five independent components.

The dipole moment of the pair of charges q and $-q$, separated by the vector distance l, is the constant $\nu = ql$ as l goes as zero. The quadrupole moment of the pair of identical but opposite dipoles $\nu(= ql_1)$ and $-\nu(= -ql_1)$, separated by the vector distance l_2 is the constant product $Q = ql_1l_2$ as l_2 tends to zero. The octupole moment R is of the pair of quadrupoles $Q(= (ql_1l_2))$ and $-Q(= -ql_1l_2)$ separated by the vector distance l_3, is the constant product $R(= (ql_1l_2l_3))$ as l_3 tends to zero.

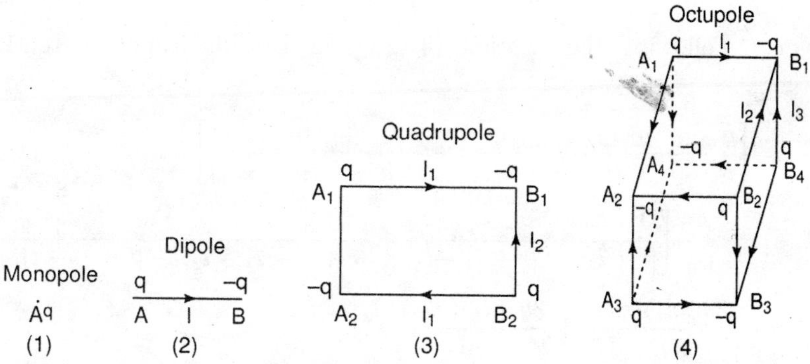

Fig. 3.12.1 These show a monopole (1), dipole (2), quadrupole (3) and an octupole (4).

Potential of the quadrupole moment of fig. 3.12.1 is

$$\phi(r) = \frac{q}{r} - \frac{q}{|r - l_1|} + \frac{q}{|r - l_1 + l_2|} - \frac{q}{|r - l_1 + l_2 + l_3|} = -\frac{q}{r^3}(r.l_1) + \frac{q((r + l_2).l_1)}{|r + l_2|^3}$$

$$= -\frac{q}{r^3}(r.l_1) + \frac{q(r.l_1)}{|r + l_2|^3} = -\frac{q}{r^3}(r.l_1) + \frac{q(r.l_1)}{r^3}\left(1 - 3\frac{(r.l_2)}{r^2}\right) + \cdots$$

$$= -\frac{3q(r.l_1)(r.l_2)}{r^5} + \cdots. \tag{3.12.1}$$

The largest term of the expression for the quadrupole potential, $\nabla(1/r^3)$, is inversely proportional to the fourth power of the distance. Evidently, therefore,

$$\boldsymbol{\nabla}\phi(r) = \frac{3q}{r^5}\mathbf{r} \times (\mathbf{l}_1 \times \mathbf{l}_2) + \frac{3qr}{r^5}(\mathbf{l}_1 - \mathbf{l}_2) + \frac{15q}{r^7}(\mathbf{l}_1 \cdot \mathbf{r})(\mathbf{l}_2 \cdot \mathbf{r})\mathbf{r}. \tag{3.12.2}$$

If Q^0_{lm} is the lmth element of the *quadrupole* moment of the charges in the jth molecule, then

$$Q^0_{lm}(\mathbf{r}, t) = \iiint_{\text{mol}} r_l r_m \rho^c_j(\mathbf{r}_j + \mathbf{r}, t)dV_r, \tag{3.12.3}$$

where r_l is the component of r along the lth co-ordinate direction. We define the total quadrupole moment Q_{lm} by

$$Q_{lm} = \sum_{j=1}^{N(\mathbf{r},t)} Q^0_{lm}\delta(\mathbf{r} - \mathbf{r}_j) \tag{3.12.4}$$

and its average by

$$\langle Q_{lm}\rangle = \frac{1}{\delta V \delta l}\iiint_{\delta V} dV_\Sigma \int_{\delta t} Q_{lm} dt_0 = N(\mathbf{r},t)\langle Q^0_{lm}(\mathbf{r},t)\rangle. \tag{3.12.5}$$

Similarly,

$$\iiint_{\text{mol}} r_x^\alpha r_y^\beta r_z^\gamma \rho^c(\mathbf{r}_j + \mathbf{r}, t) dV_r, \tag{3.12.6}$$

where $\alpha + \beta + \gamma = m$. The multipole expansion would be further discussed in section 3.17.

▶ **Example 3.12.1** *Find the potential of the octupole moment of fig* 3.12.1(4).

Answer. Following the relation (3.12.1) for the quadrupole potential the octupole potential is

$$\begin{aligned}
\phi(r) &= \left[\frac{q}{r} - \frac{q}{|r - l_1|} + \frac{q}{|r - l_1 + l_2|} - \frac{q}{|r - l_1 + l_2 + l_1|}\right] \\
&\quad - \left[\frac{q}{|r + l_3|} - \frac{q}{|r + l_3 - l_1|} + \frac{q}{|r + l_3 - l_1 + l_2|} - \frac{q}{|r + l_3 - l_1 + l_2 + l_1|}\right] \\
&= -\frac{3q(r.l_1)(r.l_2)}{r^5} + \frac{3q(r.l_1)(r.l_2)}{|r + l_3|^5} + \cdots \\
&= -\frac{3q(r.l_1)(r.l_2)}{r^5} + \frac{3q(r.l_1)(r.l_2)}{r^5\left(1 + 2\dfrac{(r.l_3)}{r^2}\right)^{5/2}} + \cdots \\
&= -\frac{3q(r.l_1)(r.l_2)}{r^5} - \frac{3q(r.l_1)(r.l_2)}{r^5}\left[1 - \frac{5(r.l_3)}{r^2}\right] + \cdots \\
&= \frac{15q(r.l_1)(r.l_2)(r.l_3)}{r^7}.
\end{aligned}$$

3.13 Liouville's Equation for a system of charged Particles in EM Fields

The development of a state of a system of charged particles out of the previous states and its evolution in future, is possible, Delcroix (1965, Chapter 7), using the microscopic equations which evolve with respect to the density in the multi-dimensional phase space. For a system, consisting of one kind of charged particles, we find the Liouville equation of evolution of the number density $F_n(\mathbf{r}_i, \mathbf{p}_i, t)$ in the phase space, approximately, by taking into account the electrical interactions, and neglecting the magnetic interaction between the particles, the finite speed of propagation leading to retarded potential solution of the fields, and the relativistic variation of mass with velocity.

The volume element in the phase space $(\mathbf{r}_i, \mathbf{v}_i)$ is dz_0. In the phase space $(\mathbf{R}_i, \mathbf{p}_i)$ it is dz_1. The canonical momentum

$$p_i = mv_i - \frac{e}{c}A_i, \tag{3.13.1}$$

where $\mathbf{v}_i$ is the velocity vector of the ith particle, $\mathbf{r}_i$ and $\mathbf{R}_i$ are its position vectors in different systems. Then, if J is the Jacobian of the transformation, following (3.14.10)–(3.14.12), we obtain

$$f_N dZ_0 = F_N dZ_1 = F_N J dZ_0 \tag{3.13.2}$$

$$R = r_i \tag{3.13.3}$$

and arrange the components of $\mathbf{r}_i$ and $\mathbf{v}_i$ for $i = 1, 2, \cdots, N$ in the order

$$x_1, y_1, z_1, v_{1x}, v_{1y}, v_{1z}; x_2, y_2, z_2, v_{2x}, v_{2y}, v_{2z}; \cdots, x_N, y_N, z_N, v_{Nx}, v_{Ny}, v_{Nz}.$$

The other vector variables are also treated similarly. Since

$$J = \begin{vmatrix} J_{11} & J_{12} & \cdots & J_{16N} \\ J_{21} & J_{22} & \cdots & J_{26N} \\ \cdots & \cdots & \cdots & \cdots \\ \cdots & \cdots & \cdots & \cdots \\ J_{6N_1} & J_{6N_2} & \cdots & J_{6N6N} \end{vmatrix}, \tag{3.13.4}$$

the contribution to the elements common to the first six rows and columns come from the equations of (3.13.1) and (3.13.2) for $i = 1$. The non-zero elements of this portion of the $(6N)$th order Jacobian determinant J of (3.13.4) are

$$J_{11} = \frac{\partial R_{1x}}{\partial x_1} = 1, \quad J_{22} = \frac{\partial R_{1y}}{\partial y_1} = 1, \quad J_{33} = \frac{\partial R_{1z}}{\partial z_1} = 1,$$

$$J_{41} = \frac{\partial p_{1x}}{\partial x_1} = 0 - \frac{e}{c}\frac{\partial A_{1x}}{\partial x_1}, \quad J_{42} = -\frac{e}{c}\frac{\partial A_{1y}}{\partial y_1}, \quad J_{43} = -\frac{e}{c}\frac{\partial A_{1x}}{\partial z_1},$$

$$J_{44} = \frac{\partial p_{1x}}{\partial v_{1x}} = m, \quad J_{52} = \frac{\partial p_{1y}}{\partial v_{1y}} = m, \quad J_{63} = \frac{\partial p_{1z}}{\partial v_{1z}} = m. \tag{3.13.5}$$

Similarly the contributions to the elements common to the next six rows and columns come from (3.13.1) and (3.13.2) for $i = 2$.

We can write (3.13.4) as

$$J = \begin{vmatrix} J_1 & 0 & 0 & \cdots & \cdots \\ 0 & J_2 & 0 & \cdots & \cdots \\ \cdots & \cdots & \cdots & \cdots & \cdots \\ \cdots & \cdots & \cdots & \cdots & \cdots \\ 0 & 0 & 0 & \cdots & J_N \end{vmatrix}, \tag{3.13.6}$$

where J_k has six rows and six columns for $k = 1, 2, \cdots, N$, and

$$\det J_k = \begin{vmatrix} J_{6k-5\,6k-5} & J_{6k-5\,6k-4} & \cdots & J_{6k-5\,6k} \\ J_{6k-4\,6k-5} & J_{6k-4\,6k-4} & \cdots & J_{6k-4\,6k} \\ \cdots & \cdots & \cdots & \cdots \\ \cdots & \cdots & \cdots & \cdots \\ J_{6k\,6k-5} & J_{6k\,6k-4} & \cdots & J_{6k\,6k} \end{vmatrix}. \tag{3.13.7}$$

Evaluating the elements of this determinant from (3.13.1) and (3.13.2) for $i = k$ we find that

$$\det J_k = \begin{vmatrix} 1 & 0 & 0 & 0 & 0 & 0 \\ 0 & 1 & 0 & 0 & 0 & 0 \\ 0 & 0 & 0 & 0 & 0 & 0 \\ -\frac{e}{c}\frac{\partial A_{kx}}{\partial x_k} & -\frac{e}{c}\frac{\partial A_{kx}}{\partial y_k} & -\frac{e}{c}\frac{\partial A_{kx}}{\partial z_k} & m & 0 & 0 \\ -\frac{e}{c}\frac{\partial A_{ky}}{\partial x_k} & -\frac{e}{c}\frac{\partial A_{ky}}{\partial y_k} & -\frac{e}{c}\frac{\partial A_{ky}}{\partial z_k} & 0 & m & 0 \\ -\frac{e}{c}\frac{\partial A_{kz}}{\partial x_k} & -\frac{e}{c}\frac{\partial A_{kz}}{\partial y_k} & -\frac{e}{c}\frac{\partial A_{kz}}{\partial z_k} & 0 & 0 & m \end{vmatrix}. \tag{3.13.8}$$

$$J = m^{3N}, \tag{3.13.9}$$

$$f_N(r_i, v_i, t) = m^{3N} F_N(r_i, p_i, t), \tag{3.13.10}$$

where p_i is the canonical momentum of (3.13.1). Then (3.14.14) becomes

$$\frac{\partial F_N}{\partial t} + \sum_i \left(v_i \cdot \frac{\partial F_N}{\partial r_i} \right) + \sum_i \left(F_i + \sum_j F_{ij} - \frac{e}{c}\frac{dA_i}{dt} \right) \frac{\partial F_N}{\partial p_i} = 0 \tag{3.13.11}$$

with

$$\frac{\partial H.}{\partial p_i} = \frac{dr_i}{dt}, \quad \frac{\partial H}{\partial r_i} = e(E_i + E_i') + \frac{e}{c}(v_i \times B_i) + \frac{e}{c}\frac{dA_i}{dt},$$

where

$$F_i + \sum_{j, i \neq j} F_{ij}$$

is the total force exerted on the ith particle. So

$$F_i + \sum_{j, i \neq j} F_{ij} = -e(e_i + E_i') - \frac{e}{c}(v_i \times B_i). \tag{3.13.12}$$

But, according to (3.13.10),

$$m^{3N}\frac{\partial F_N}{\partial t} = \frac{\partial}{\partial t}\left[f_N\left(r_i, \frac{1}{m}\left(p_i + \frac{e}{c}A_i\right), t\right)\right] = \frac{\partial F_N}{\partial t} + \frac{e}{c}\left(\frac{\partial F_N}{\partial v_i}\cdot\frac{\partial A_i}{\partial t}\right).$$

$$m^{3N}\frac{\partial F_N}{\partial x_i} = \frac{\partial f_N}{\partial x_i} + \frac{e}{c}\left(\frac{\partial f_N}{\partial v_i}\cdot\frac{\partial A_i}{\partial x_i}\right). \tag{3.13.13}$$

$$m^{3N}\frac{\partial F_N}{\partial p_{ix}} = \frac{1}{m}\frac{\partial f_N}{\partial v_{ix}}. \tag{3.13.14}$$

Multiplying all the terms of (3.13.11) by m^{3N}, and using (3.13.13)–(3.13.15) we obtain for f_N the equation

$$\frac{\partial f_N}{\partial t} + \sum_i\left(v_i\frac{\partial f_N}{\partial r_i}\right) + \sum_i\frac{1}{m}\left(\left(F_i + \left(\sum_{j,i\neq j}F_{ij}\right)\right)\cdot\frac{\partial f_N}{\partial v_i}\right)$$

$$-\frac{e}{mc}\left(\frac{\partial f_N}{\partial v_i}\cdot\left[-\frac{\partial A_i}{\partial t} - \left(\sum_{x,y,z}v_{ix}\frac{\partial A_i}{\partial x_i}\right) + \frac{\partial A_i}{\partial t}\right]\right) = 0$$

Since the last term is identically zero this equation reduces to

$$\frac{\partial f_N}{\partial t} + \sum_i\left(v_i\cdot\frac{\partial f_N}{\partial r_i}\right) + \sum\frac{1}{m}\left(\left(F_i + \sum_{j,j\neq i}F_{ij}\right)\cdot\frac{\partial f_N}{\partial v_i}\right) = 0. \tag{3.13.15}$$

More general study of the Liouville theorem for plasma constituents is available in Simon and Harris (1960), Kalman (1962), Balescu (1963) and Montgomery and Tidman (1964, Chapter 4).

3.14 The Liouville Theorem

A set of conjugate scalar variables, for the state of a system of N particles, are $q_1, q_2, \cdots, q_{3N}$, $p_1, p_2, \cdots, p_{3N}$. The volume element in the $6N$ dimensional phase space is

$$dz = dq_1 dq_2 dq_{3N} dp_1 dp_2 dp_{3N} \tag{3.14.1}$$

N being large, and for quantum mechanical indeterminacy, the phase cannot be determined exactly. The concept of probability density in the phase space gives

$$F_N(q_i, p_i)dz \quad (i = 1, 2, \cdots, 3N) \tag{3.14.2}$$

for the probability of the representative point of the system in the volume dz of the phase space (q_i, p_i). $F_N(q_i, p_i)$ is the probability density at the phase point co-ordinates q_1, q_2, q_{3N}, p_1, p_2, $\cdots$, p_{3N}.

At the instant t the phase point is inside the volume element dz. At the instant $t + \delta t$, every point following the trajectory has a new position. All these positions, together, fill a volume element dz' of the phase space. The relation between the volume elements dz and dz', containing the Jacobian of transformation J, is

$$dz' = Jdz. \tag{3.14.3}$$

We now find a systems of differential relations from co-ordinate transformation. So, at the instant t, a pair of scalar conjugate variables is denoted by q_i and p_i, and at the instant $t'(= t + \delta t)$ by q_i' and p_i' respectively. Then

$$q_i' = q_i + \frac{dq_i}{dt}\delta t, \quad p_i' = p_i + \frac{d}{dt}p_i \delta t, \tag{3.14.4}$$

where for $H(q_i, p_i)$, the Hamiltonian of the system, we have

$$\frac{dq_i}{dt} = \frac{\partial H}{\partial p_i}, \quad \frac{dp_i}{dt} = \frac{\partial H}{\partial q_i},$$

$$q_i' = q_i + \frac{\partial H}{\partial p_i}\delta t, \quad p_i' = p_i - \frac{\partial H}{\partial q_i}\delta t, \tag{3.14.5}$$

$$dq_i' = dq_i + \sum_j \left(\frac{\partial^2 H}{\partial p_i \partial q_j} dq_j \delta t\right) + \sum_j \left(\frac{\partial^2 H}{\partial p_i \partial p_j} dp_j \delta t\right), \tag{3.14.6}$$

$$dp_i' = dp_i - \sum_j \left(\frac{\partial^2 H}{\partial q_i \partial q_j} dq_j \delta t\right) - \sum_j \left(\frac{\partial^2 H}{\partial q_i \partial q_j} dp_j \delta t\right). \tag{3.14.7}$$

The Jacobian J of transformation in the $6N$ dimensional volume element of (3.14.3), from the instant t to the primed system of time t', therefore, is

$$J = \begin{vmatrix}
\frac{dq_1'}{dq_1} & \frac{dq_1'}{dq_2} & \cdots & \frac{dq_1'}{dq_{3N}} & \frac{dq_1'}{dp_1} & \frac{dq_1'}{dp_2} & \cdots & \frac{dq_1'}{dp_{3N}} \\
\frac{dq_2'}{dq_1} & \frac{dq_2'}{dq_2} & \cdots & \frac{dq_2'}{dq_{3N}} & \frac{dq_2'}{dp_1} & \frac{dq_2'}{dp_2} & \cdots & \frac{dq_2'}{dp_{3N}} \\
\cdots & \cdots & \cdots & \cdots & \cdots & \cdots & \cdots & \cdots \\
\cdots & \cdots & \cdots & \cdots & \cdots & \cdots & \cdots & \cdots \\
\frac{dq_{3N}'}{dq_1} & \frac{dq_{3N}'}{dq_3} & \cdots & \frac{dq_{3N}'}{dq_{3N}} & \frac{dq_{3N}'}{dp_1} & \frac{dq_{3N}'}{dp_2} & \cdots & \frac{dq_{3N}'}{dp_{3N}} \\
\frac{dp_1'}{dq_1} & \frac{dp_1'}{dq_2} & \cdots & \frac{dp_1'}{dq_{3N}} & \frac{dp_1'}{dp_1} & \frac{dp_1'}{dp_2} & \cdots & \frac{dp_1'}{dp_{3N}} \\
\frac{dp_2'}{dq_1} & \frac{dp_2'}{dq_2} & \cdots & \frac{dp_2'}{dq_{3N}} & \frac{dp_2'}{dp_1} & \frac{dp_2'}{dp_2} & \cdots & \frac{dp_2'}{dp_{3N}} \\
\cdots & \cdots & \cdots & \cdots & \cdots & \cdots & \cdots & \cdots \\
\cdots & \cdots & \cdots & \cdots & \cdots & \cdots & \cdots & \cdots \\
\frac{dp_{3N}'}{dq_1} & \frac{dp_{3N}'}{dq_2} & \cdots & \frac{dp_{3N}'}{dq_{3N}} & \frac{dp_{3N}'}{dp_1} & \frac{dp_{3N}'}{dp_2} & \cdots & \frac{dp_{3N}'}{dp_{3N}}
\end{vmatrix}$$

$$= \begin{vmatrix} 1+\frac{\partial^2 H\delta t}{\partial p_1 \partial q_1} & \frac{\partial^2 H\delta t}{\partial p_1 \partial q_2} & \cdots\cdots & \frac{\partial^2 H\delta t}{\partial p_1 \partial p_{3N}} \\ \frac{\partial^2 H\delta t}{\partial p_2 \partial q_1} & 1+\frac{\partial^2 H\delta t}{\partial p_2 \partial q_2} & \cdots\cdots & \frac{\partial^2 H\delta t}{\partial p_2 \partial p_{3N}} \\ \cdots\cdots & \cdots\cdots & \cdots\cdots & \cdots\cdots \\ \cdots\cdots & \cdots\cdots & \cdots\cdots & \cdots\cdots \\ \frac{\partial^2 H\delta t}{\partial p_{3N} \partial q_1} & \frac{\partial^2 H\delta t}{\partial p_{3N} \partial q_2} & \cdots\cdots & \frac{\partial^2 H\delta t}{\partial p_{3N} \partial p_{3N}} \\ -\frac{\partial^2 H\delta t}{\partial q_1 \partial q_1} & -\frac{\partial^2 H\delta t}{\partial q_1 \partial q_2} & \cdots\cdots & -\frac{\partial^2 H\delta t}{\partial q_1 \partial p_{3N}} \\ \cdots\cdots & \cdots\cdots & \cdots\cdots & \cdots\cdots \\ \cdots\cdots & \cdots\cdots & \cdots\cdots & \cdots\cdots \\ -\frac{\partial^2 H\delta t}{\partial q_{3N} \partial q_1} & -\frac{\partial^2 H\delta t}{\partial q_{3N} \partial q_2} & \cdots\cdots & 1-\frac{\partial^2 H\delta t}{\partial q_{3N} \partial p_{3N}} \end{vmatrix}. \tag{3.14.8}$$

Keeping terms up to first power of δt, after expansion of (3.14.8), we obtain

$$J = 1 + \sum_j \left(\frac{\partial^2 H}{\partial p_i \partial q_i} - \frac{\partial^2 H}{\partial q_i \partial p_i} \right) \delta t, \tag{3.14.9}$$

$$dz' = dz \tag{3.14.10}$$

because only the main diagonal terms are not of second order in δt. The probability of finding the system in dz' at the instant $t + \delta t$ equals the probability of finding it in dz at time t. So

$$F_N(q'_i, p'_i, t')dz' = F_N(q_i, p_i, t)dz \tag{3.14.11}$$

and F_N is conserved during motion, because

$$F_N(q'_i, p'_i, t + \delta t) = F_N(q_i, p_i, t). \tag{3.14.12}$$

Hence

$$\frac{d}{dt} F_N = 0 = \frac{\partial F_N}{\partial t} + \sum_i \left(\frac{dq_i}{dt} \frac{\partial F_N}{\partial q_i} \right) + \sum_i \left(\frac{dp_i}{dt} \frac{\partial F_N}{\partial p_i} \right). \tag{3.14.13}$$

With the help of the Hamiltonian equations of motion, this equation becomes

$$\frac{\partial F_N}{\partial t} + \sum_i \left(\frac{\partial H}{\partial q_i} \cdot \frac{\partial F_N}{\partial q_i} \right) - \sum_i \left(\frac{\partial H}{\partial q_i} \cdot \frac{\partial F_N}{\partial p_i} \right) = 0. \tag{3.14.14}$$

The state (or phase) of the system at the instant t is defined by the sets of vectors

$$r_1, r_2, \cdots, r_N, v_1, v_2, \cdots, v_N \tag{3.14.15}$$

of a point in the $6N$ dimensional phase space; moreover,

$$p_i = mv_i. \tag{3.14.16}$$

The state of the ith particle of the system of N identical particles, free from rotation and vibration, is specified by its position vector $\mathbf{r}_i$, velocity $\mathbf{v}_i$ and momentum $\mathbf{p}_i$, and probability of a phase of the system, at the time t, in the volume element $dr_1 dr_2 \cdots dr_N dv_1 dv_2 \cdots dv_N$, is

$$f_N dr_1 dr_2 \cdots dr_N dv_1 dv_2 \cdots dv_N. \tag{3.14.17}$$

The ith particle is in the volume element

$$dr_i = dx_i dy_i dz_i \tag{3.14.18}$$

in the position space, and possesses a velocity whose tip is in the volume element

$$dv_i = dv_{ix} dv_{iy} dv_{iz} \tag{3.14.19}$$

of the velocity space. Using (3.14.15) and (3.14.1) we write (3.14.2) as

$$F_N dr_1 dr_2 \cdots dr_N d(mv_1) d(mv_2) \cdots d(mv_N). \tag{3.14.20}$$

Since (3.14.17) and (3.14.20) are identical quantities, the Liouville equation for f_N is

$$f_N(r_i, v_i, t) = m^{3N} F_N(r_i, mv_i, t), \tag{3.14.21}$$

where f_N is the probability density at the phase point.

Here the relativistic effects are neglected, the force acting on the particles are independent of velocity and these forces are derivable from potential functions. The Hamiltonian of the system is

$$H = \sum_i \left(\frac{1}{2m} p_i^2\right) + \sum_i (\phi_i) + \sum_{i<j}\sum (\phi_{ij}), \tag{3.14.22}$$

where ϕ_i is the potential energy of the ith particle due to the field of external forces, and ϕ_{ij} is the potential energy of the interaction between the ith and jth particles. Differentiating H with respect to the components of $\mathbf{p}_i$ and $\mathbf{r}_i$, we obtain

$$\frac{\partial H}{\partial p_{ix}} = v_{ix} = \frac{dx_i}{dt}, \tag{3.14.23}$$

$$\frac{\partial H}{\partial x_i} = \frac{\partial \phi}{\partial x_i} + \sum_j \frac{\partial \phi_{ij}}{\partial x_i} = -\frac{dp_{ix}}{dt}. \tag{3.14.24}$$

If the total force acting on the ith particle is

$$F_i + \sum_j F_{ij},$$

then

$$\frac{\partial \phi}{\partial x_i} + \sum_j \frac{\partial \phi_{ij}}{\partial x_i} = F_{ix} + \sum_j F_{ijx} \tag{3.14.25}$$

$$\nabla H = -\frac{dp_i}{dt} = -F_i - \sum F_{ij}. \tag{3.14.26}$$

Hence, r_i and mv_i form a pair of conjugate vector variables. The vector form of the Liouville equation for f_N is

$$\frac{\partial f_N}{\partial t} + \sum_i \left[\left(v_i \cdot \frac{\partial f_N}{\partial r_i}\right) + \sum_i \left(\frac{1}{m}\left(F_i + \sum_i \frac{\partial}{\partial u_i}\right)\frac{\partial f_N}{\partial v_i}\right)\right] = 0, \tag{3.14.27}$$

where the summation is taken over the N particles of the system, and (3.14.14), (3.14.21) and (3.14.26) have been used.

3.15 Concepts and the Basic Equations of the Lorentz Theory

The microscopic theory of Lorentz (1915) for electrodynamics is a sequel to the phenomenological theory of Maxwell. It is based on the atomic concepts of matter, so that the classical theory of electromagnetics is assumed to be valid through out all space, including matters interior of an atom, molecule or electron. The equations of Lorentz predict the behaviour of particles and their fields, seek the origin of the macroscopically observable effects, and the phenomena-based constitutive relations, and express the vector and scalar potentials, in terms of charges and currents. Multipole expansions in the internal co-ordinates of molecules give the average current for the averaged fields of the macroscopic theory.

Statistical mechanics assumes the existence of atoms as mechanical systems. But the microscopic theory goes deep into the structure of atoms, and into the nature of their charged constituents, assuming that all events occur in vacuum due to motion of a large number of *highly irregularly* distributed discrete charges. Non-uniformly distributed matter exists in the neighbourhood of each particle. At every point either inside or outside the molecules, fields vary very rapidly in space and time, and the field equations remain valid. The macroscopic averages E and B of the microscopic electric field and magnetic induction follow from the laws of statistical mechanics for averaging. The vectors D and H are macroscopic quantities; they have no microscopic meaning. They are derived from the average quantities E and B with the help of the *electrical* and *magnetic* multipole distributions in matter. In the inhomogeneous Maxwell equations the averages $\langle \rho^c \rangle$ and $\langle \mathbf{j} \rangle$ (denoted by ρ^c and $\mathbf{j}$ in these equations) obtain from the charge and current density of the electrons and ions of the material.

The concepts of classical fundamental particles do not go beyond the concepts for electrons; so the scope of the dynamics of charged particles do not go beyond the study of electron motion. This theory explains the interaction of charged particles with one another and with radiation. The macroscopic Maxwell theory depends on distribution of charges, and on the electric current of their motion. The concept of charge as an aggregate of elementary charged particles is foreign to this theory.

Microscopically, the charge is a sphere carrying a spherically symmetric distribution of charge. Thus, an electron is a microscopic charged body, much smaller in size than macroscopic structures. So, there is nothing elementary about it. The repulsive Coulomb force acts between its parts. This force causes symmetric expansion of the spherically charged electron configuration like a balloon to arbitrarily large radius. The microscopic charged structures are not purely electromagnetic in nature; the minute particle is only a macroscopic particle of small size. Stable macroscopic charged spheres do not "explode" due to interatomic and inter molecular forces which hold such a sphere rigidly together and bind the charged distribution into a rigid sphere. The internal forces, providing the solid-like rigidity, are included in the microscopic theory, without the associated explanation of the origin of these forces. They have only the adhoc function of furnishing the "glue" which prevents the microscopic charge elements from expansion.

A satisfactory description of elementary charged particles, on the classical level, is not possible without reference to the structure of the particles involved. Classical physics is an asymptotic limit of more fundamental levels of physics like quantum physics, quantum field theory and still more fundamental levels of physics which may be discovered in future, for explanation of unsolved difficulties.

3.15.1 The problem of the Lorentz Theory

The problem[2], is to establish the equations of the phenomenological macroscopic theory of Maxwell for electrodynamics, obtain and interpret the average E, H, D, B, in terms of the macroscopic quantities like the polarization density, electric current etc., of the macroscopic theory, from the Lorentz equations of electrodynamics, due to a given assembly of atomic, molecular and free charge system. The charge energy is partly spent on radiation. So the system of charges is not a conservative one.

The Lorentz theory depends on the ideas of statistical description of matter, the Newtonian methods of analytical mechanics for motion of electrons, both free and within each atom or molecules, in the presence of the Lorentz force and other forces, and the fundamental laws of electrodynamics. Even the relativistically correct equation of motion of the charges in empty space are valid.

3.15.2 Concepts of Lorentz Theory

In the Lorentz theory material bodies contain enormous number of electrons (smallest particles of negative charge) and ions, which are also extremely small in size. All these particles occupy minute portions of space; ions contain practically the whole atomic mass, and electrons are very light particles. In material bodies the space between particles is filled up with electromagnetic fields excited by charged particles. *Currents* are either of *displacement* or *convection*, and magnetic fields are from motion of electric charges. There exists no *conduction current* and magnetism. Space acts as a carrier of electric fields, charges move in space, motion of charges takes place in vacuum (so that $\varepsilon = 1$ and $\mu = 1$), the vectors B and D cannot be distinguished from H and E respectively, the displacement current is $\partial E/c\partial t$, and the convection current is $\rho^c v$, where ρ^c is the net charge density and $\mathbf{v}$ is the average velocity. The force per unit charge for a charge moving with velocity v is $\mathbf{E} + (\mathbf{v} \times \mathbf{H})/c$.

3.15.3 Basic equations of the Lorentz theory

The basic equations of the Lorentz theory for a system of charged particles, in vacuum, are

$$\nabla \times E = -\frac{1}{c}\frac{\partial H}{\partial t}, \tag{3.15.1}$$

$$\nabla \times E = -\frac{1}{c}\frac{\partial H}{\partial t} + \frac{4\pi}{c} j, \tag{3.15.2}$$

$$(\nabla \cdot E) = 4\pi\rho^c \tag{3.15.3}$$

$$(\nabla \cdot E) = 0. \tag{3.15.4}$$

Since j is the convection current, $j = \rho^c\bar{v}$ and the charge conservation condition is

$$\frac{\partial \rho^c}{\partial t} + \nabla \cdot (\rho\bar{v}) = 0 \quad \text{or} \quad \frac{d\rho^c}{dt} = 0, \tag{3.15.5}$$

where $\bar{v}$ is the microscopically averaged charge velocities, $\partial/\partial t$ is the partial time derivative operator and d/dt is the substantial or total time derivative operator. So, the charge is

[2]See Fock (1964), p. 77, section 26.

conserved, and charge is a primary concept. In the macroscopic theory the total charge in the universe is conserved (section 1.4).

For the discrete charged particles in motion in matter-free space, delta functions (Appendix 1) express the charge density ρ^c and current density $\mathbf{j}$:

$$\rho^c = \frac{1}{V}\sum_i e_i\delta(r - r_i(t)) \tag{3.15.6}$$

$$\mathbf{j} = \frac{1}{V}\sum_i e_i v_i\delta(\mathbf{r} - \mathbf{r}_i(t)) \tag{3.15.7}$$

in a given volume V, N is the total number of charges in V, $\mathbf{r}_i$ is the position vector of the ith particle at time t, e_i is the charge of the ith particle (with $i = 1, 2, \cdots, N$), $\mathbf{v}_i$ is the velocity of the ith particle at the point $\mathbf{r} = \mathbf{r}_i$, and summation is carried out over V. Both convection current and the polarization current are represented by j. Phenomenologically, $\varepsilon \neq 1$ and $\mu \neq 1$; so the quantities j and ρ^c are not identical to these j and ρ^c.

The relativistic equation of motion of the ith particles is

$$\frac{dp_i}{dt} = e_i E + \frac{\rho_i}{c_i}(v_i \times H) + F_i, \tag{3.15.8}$$

where $p_i = m_i v_i / \left(1 - v_i^2/c^2\right)$, and F_i represents the gravitational force and other forces. Since (3.15.6) and (3.15.7) satisfy a continuity condition, (3.15.2) implies (3.15.3). Hence the system of equations (3.15.1), (3.15.3), (3.15.4) and (3.15.8) give a total of $3N + 6$ linearly independent equations. The variables being the components of $\mathbf{E}$, $\mathbf{H}$, $\mathbf{r}_i$ for $i = 1, 2, \cdots, N$, their total (all $= 3N + 6$), is sufficient, in principle, to solve problems of classical electrodynamics. So, it is a closed system of equations.

Equations (3.15.1)–(3.15.8), for a system of N charges in motion in vacuum do not contain material constants, these contain the charges e_i and masses m_i with $i = 1, 2, \cdots, N$, which are universal constants. The material constants ε, μ, σ and the field vectors D and B do not appear in these equations. The Lorentz equations are expanded for obtaining dipole and higher order (multipole) moments of the charges, the distributions of which depend on the material density and state of thermal motion of the constituents of the material. Also, some material constants appear when the delta functions are smoothed out to obtain field equations which are much reduced for practical suitability. It is necessary to know, a priori, the pressure change law. For the adiabatic law of thermal change, grad p_i is replaced by C_{is}^2 grad ρ_i^c, where C_{is} is the sound velocity in the ith fluid.

In continuum mechanics the mass density ρ^m is generally used instead of the particle number density N. For relativistic systems, or for systems with chemical reactions, or for multispecies systems, whose species have different velocities, the mass per particle is not a simple constant, the use of mass density ρ^m is not necessarily more convenient than the number density $N(r, t)$, which, for chemical reactions, is the product of the Avogadro number and the number of moles. Thus, the concept of density of matter is distinguished from the concept of mass, and $N(r, t)$ is a bonafide continuum variable.

3.15.4 The analytical and phenomenological approaches for electrodynamics

Physical systems, commented Ter Haar (1958), are described in macroscopic, phenomenological terms. But, he writes, "if we try to start from first principles—granting for a moment that we really know what first principles to use—and use a microscopic description we are led to introduce simplifying assumptions at practically every stage. In view of all these simplifications, it is really remarkable how often one is able to obtain results which bear some resemblance to experimentally verifiable facts. The reason is that the treatment of so many systems of non-interacting (or, quasi interacting) particles and a system of non-interacting particles can be dealt with using the ordinary methods of statistical mechanics. However, it is well-known that, as soon as interactions come into play the discussion of the statistical behaviour of a system becomes extremely complicated, even though, in using statistical mechanics one has already simplified the problem by considering the average behaviour in the system instead of the detailed behaviour of all the particles".

The macroscopic theory gains from the notions about the microscopic structure of matter, regarding the electromagnetic properties like polarization and conduction currents. The electrical origin, before the advent of quantum mechanics., implied that by probing down to subatomic distances, it should be possible to formulate the equations of classical electrodynamics in vacuum without introduction if ponderable dielectric and magnetic media. The charges and their currents are replaced in the phenomenological theory by the ponderable current **j** due to the introducition of ponderable dielectric and magnetic media. Applied fields, of high frequency, remove the bonds holding the particles of atoms and molecules together, and consequently let the predominance of long range interactions in their behaviour. The higher harmonics generated by masers and lasers, due to non-linear polarizations, are considered with the help of models of the microscopic theory of Lorentz, Armstrong et al (1962), Bloembergen and Pershan (1962), Bloembergen and Shen (1964), Lamb (1964).

The microscopic theory is impractical, because equations of motion are set up for individual charges in the external field and the field of these charges, a large number of particles are involved in gases and liquids, and because experimental observations are not concerned with the variations in field from molecule to molecule but with aggregate effects involving large number of molecules. The irregularities, smoothed out by averaging, are important in the macroscopic theory. The microscopic equations are similar in structure to the macroscopic equations in vacuum and are supposed to hold at every point either inside or outside the molecules. So (Van Vleck (1932)), these equations are more fundamental than the macroscopic equations which are statistical in nature and so are not exact.

According to Maxwell, events occur in media which differ electrically and magnetically; but according to Lorentz all events take place in vacuum. Up to the times of Faraday and Maxwell attention was away from the discrete charges and was toward the lines of force. Lorentz established the basis for the new electrodynamics based on electrons and ions. However, the basic questions regarding the nature of electron remains unclarified in the Lorentz theory, Caldirola (1979).

In magnetism and optical dispersion quantum theory plays an essential role. The classical theory of electron is valid for quantum descriptions of these features. But, for metals, electrons follow the exclusion principle of Pauli. So for metallic properties classical lines are not suitable.

If the point at which the potentials are evaluated is close to some of the molecules, then the multipole expansion becomes invalid. So the analysis is correct for points outside a material. Since this method starts with microscopic charges and currents, deduces the potentials and finally obtains average macroscopic current and charge density, it is an indirect method.

The microscopic theory deals with many beautiful results of the classical electron theory of matter, but all the modern developments cannot be covered by this theory, Stratton ((1941), Preface). Although many laws of classical electrodynamics apply directly to submicroscopic domains, there is no basis of selection, Stratton (1941). So, the transition from quantum electrodynamics towards the classical limit, is necessary rather than the reverse transition since the statistical average of the equations of quantum electrodynamic must lead to the Maxwell equations.

3.16 Microscopic Coulomb's Law

An electric charge at a point O, fig. 3.16.1, exerts on a unit charge at P, the field $E = ie/OP^2$, of Coulomb's law (Two charges of opposite sign attract, and of same sign repel each other, according to the law of inverse square of the distance), where $\mathbf{i}(= \mathbf{OP}/|OP|)$ is the unit vector along **OP**. On the surface element $d\Sigma$, then :

$$(E \cdot d\Sigma) = \frac{c}{r^2} d\Sigma \cos\theta,$$

where the vector element $d\mathbf{\Sigma} = \mathbf{n} d\Sigma$, n is the unit normal to $d\mathbf{\Sigma}$ at P, drawn outward to a volume V, containing the charge P, $\mathbf{r} = \mathbf{OP}$ and θ is the angle between $\mathbf{i}$ and $\mathbf{n}$. If $d\Omega$ is the solid angle subtended at O by $d\Sigma$, then $(E \cdot d\Sigma) = ed\Omega$, $r^2 d\Omega = d\Sigma \cos\theta$,

$$\iint_\Sigma (E \cdot d\Sigma) = e \iint_\Sigma d\Omega = 4\pi e \text{ if } O \text{ lies inside of } \Sigma$$

$$= 2\pi e \text{ if } O \text{ lies on } \Sigma$$

$$= 0 \text{ if } O \text{ lies outside of } \Sigma$$

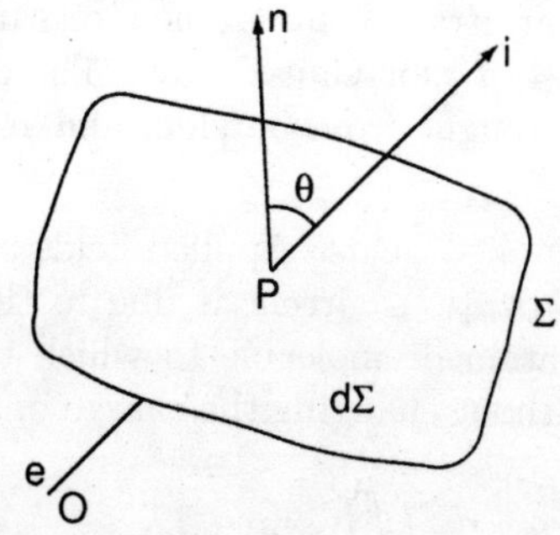

Fig. 3.16.1 Flux of field across the surface element dΣ at P due to the charge e at O.

When several charges are held fixed, instantaneously, the sum force

$$E = \sum_i (e_i r_i / r_i^3)$$

acts on unit charge at P, and

$$\iint_\Sigma (E \cdot d\Sigma) = \iint_\Sigma \sum_i e_i \frac{(r_i \cdot d\Sigma)}{r_i^3},$$

where O_i is the position of the ith charge e_i and $O_iP = r_i$. Then

$$\iint_\Sigma (E\cdot d\Sigma) = \sum_i \iint_\Sigma \frac{e_i}{r_i^2}\cos\theta_i d\Sigma = \sum_i \iint_\Sigma e_i d\Omega_i$$

$$= \sum_p e_p \iint_\Sigma d\Omega_p + \sum_q e_q \iint_\Sigma d\Omega_q,$$

where the indices p and q stand for charges inside and outside Σ, θ_i is the angle between r_i and $d\Sigma$, and $d\Omega_i$ is the solid angle element subtended by $d\Sigma$ at O_i. Since the points O_q are outside Σ, and the points O_p are inside Σ,

$$\iint_\Sigma d\Omega_q = 0, \quad \iint_\Sigma d\Omega_p = 4\pi, \quad \iint_\Sigma (E\cdot d\Sigma) = 4\pi \sum_p e_p.$$

If ρ^c is the density of a charge distribution, and dV is a volume element at O, in a volume V, then

$$\iint_\Sigma (E\cdot d\Sigma) = 4\pi \iiint_V \rho^c dV.$$

This is the integral form of Gauss's law for normal flux of the electric field. This macroscopic law shows that the total charge of a body is nothing but the surface integral of E over the entire surface of it. Green's theorem (Appendix 2) gives

$$\iint_\Sigma (E\cdot d\Sigma) - 4\pi \iiint_V \rho^c dV = 0 = \iiint_V ((\nabla\cdot E) - 4\pi\rho^c)dV.$$

Since this relation is valid in any arbitrary volume, the integrand vanishes at each point, giving the differential equation $\nabla\cdot \mathbf{E} = 4\pi\rho^c$, for Gauss's law. The charge density ρ^c consists of the sum $\rho_e^c + \rho_i^c$ (ρ_e^c is the density brought from outside, and ρ_i^c is the field induced density of redistribution).

Dielectrics are polarized by applied fields. Applied fields also reorient molecules that have permanent dipole moment. Polarization current is due to the field induced displacement of charged particles bound to the atom or molecule to which they belong. When the average charge density is redistributed without changing the charge in dielectrics, by an applied electric field, we can write

$$\iiint_V \rho_i^c dV = 0,$$

where V is the dielectric volume. Writing $\rho_i^c = -\nabla\cdot P$, and applying Green's theorem for reduction of a triple integral into a double integral (Appendix 2), we obtain

$$\iiint_V \rho_i^c dV = 0 = -\iiint_V (\nabla\cdot P)dV = -\iint_\Sigma (P\cdot d\Sigma).$$

P is the dielectric polarization vector. Hence the normal component of the polarization at all points of the dielectric boundary must vanish, giving

$$\iiint_V r\rho_i^c dV = -\iiint_V P dV,$$

where $\mathbf{r}\rho_i^c$ is the dipole density. The integral in the left hand side is the electric dipole moment ν of a radiating body. In general, the total dipole moment of volume distribution of charges with density ρ^c, is

$$\nu(t) = \iiint\limits_V r\rho^c(r,t)dV$$

and the polarization P is the dipole moment per unit volume. If, in addition, charges from other sources of density ρ_e^c exist then the divergence equation reads $\rho_i^c = \text{div } P$, and

$$\text{div } \mathbf{E} = 4\pi(\rho_e^c - \text{div } \mathbf{P}).$$

The well-known form of this law, when the electric polarization properties are taken into account, is

$$\text{div } \mathbf{D} = 4\pi\rho^c, \tag{3.16.1}$$

where $D = E + 4\pi P$. D is the electric displacement vector, and the subscript e is dropped from ρ_e^c. See the derivation of (3.16.1) from multipole expansion and spatial averaging done in section 3.18.

For a continuous distribution of charges of density $\rho^c(\mathbf{r})$ at the field point $P(\mathbf{r})$ in a volume V, Coloumb's law gives the electric field

$$E(\mathbf{r}) = \iiint\limits_V \frac{(\mathbf{r}-\mathbf{r}')}{|r-r'|^3}\rho^c(\mathbf{r}_i)dV, \tag{3.16.2}$$

where r' is the position vector of the integration point Q inside V and $\rho^c(\mathbf{r}')d\mathbf{r}'$ is the number of charges inside the volume $d\mathbf{v}_r$, at $\mathbf{r}'$. The quantity $(r-r')/|r-r'|^3$ is the only part inside integration which depends on co-ordinates of the observation point $\mathbf{r}$, the other part depends only on co-ordinates of Q. We can, therefore, write

$$\frac{(r-r')}{|r-r'|^3} = -\nabla_r\left(\frac{1}{|r-r'|}\right) = \nabla_{r'}\left(\frac{1}{|r-r'|}\right), \tag{3.16.3}$$

$$E = \nabla_r\phi(r), \tag{3.16.4}$$

$$\phi(r) = -\iiint\limits_V \frac{\rho(r')}{|r-r'|}dV_{r'}, \tag{3.16.5}$$

where $\nabla_r = (\partial/\partial x, \partial/\partial y, \partial/\partial z)$. These relations give

$$\nabla\times E = \nabla\times(\nabla\phi) = 0, \tag{3.16.6}$$

due to the central nature of the force of Coulomb between charges. Since

$$\nabla^2(1/|r-r'|) = -4\pi\delta(r-r'),$$

divergence of (3.16.4) gives

$$\nabla\cdot E = \nabla_r^2\phi = -\nabla_r^2\iiint\limits_V \frac{\rho(r')dV_{r'}}{|r-r'|} = -\iiint\limits_V \rho(r')\nabla^2\left(\frac{1}{|r-r'|}\right)dV_{r'}$$

$$= 4\pi\iiint\limits_V \rho(r')\delta(r-r')dV_{r'} = 4\pi\rho(r). \tag{3.16.7}$$

3.17 Multipole Expansion of Electric Field

Each molecule, in a volume V, containing a large number of molecules, is a cloud of charges in a specific molecular volume, having a given center of mass. The jth charge spreads over the volume δV_j. In fig. 3.17.1, the center of mass of δV_j is Q_j, and R is the integration point in δV_j; the origin O is away from the molecule, and the position vectors P, Q and R are $\mathbf{r}$, $\mathbf{r}_j$ and $\mathbf{r}_j + \mathbf{r}'$, respectively. Only space averaging or ensemble averaging is sufficient and time averaging is not necessary. So, the time factor t is dropped here from the list of independent variables, because expansion and other operations of fields and variables involved are not affected by t.

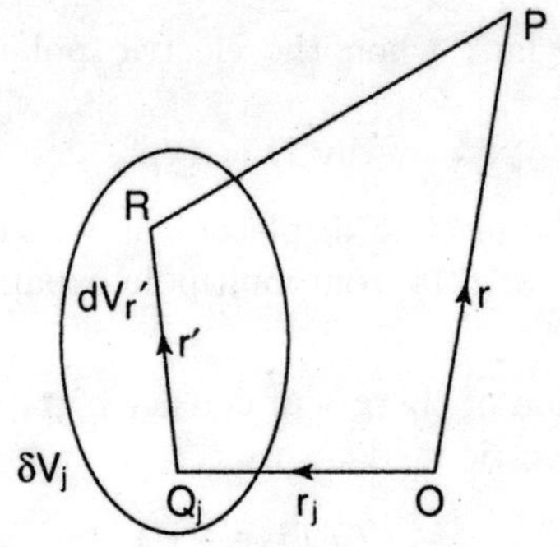

Fig. 3.17.1 It shows the integration point P inside the volume δVj of the jth molecule.

The charge density of the jth molecule, $\rho_j^c(\mathbf{r}_j + \mathbf{r}')$, and its electric field $\mathbf{E}_j(r)$ at P, are given by

$$\mathbf{E}_j(r) = -\nabla_r \iiint\limits_{\text{mol}} \frac{\rho_j^c(r_j + r')dV_{r'}}{|\mathbf{r} - \mathbf{r}_j - \mathbf{r}'|}, \tag{3.17.1}$$

$$\phi_j(\mathbf{r}) = -\iiint\limits_{\text{mol}} \frac{\rho_j^c(\mathbf{r}_j + \mathbf{r}')dV_{r'}}{|\mathbf{r} - \mathbf{r}_j - \mathbf{r}'|}, \tag{3.17.2}$$

with $|\mathbf{r} - \mathbf{r}_j| \neq 0$, when the integration point P is outside V. Since $|\mathbf{r} - \mathbf{r}_j| \gg |\mathbf{r}'|$, expansion of $|\mathbf{r} - \mathbf{r}_j - \mathbf{r}'|^{-1}$ in power of $\mathbf{r}'$ gives

$$\begin{aligned}\frac{1}{|\mathbf{r} - \mathbf{r}_j - \mathbf{r}'|} &= \frac{1}{|\mathbf{r} - \mathbf{r}_j|} + \frac{(\mathbf{r}' \cdot (\mathbf{r} - \mathbf{r}_j))}{|\mathbf{r} - \mathbf{r}_j|^3} + \cdots \\ &= \frac{1}{|\mathbf{r} - \mathbf{r}_j|} + (\mathbf{r}' \cdot \nabla_{\mathbf{r}}) \left(\frac{1}{|\mathbf{r} - \mathbf{r}_j|} \right) + \cdots \\ &= \frac{1}{|\mathbf{r} - \mathbf{r}_j|} + (\mathbf{r}' \cdot \nabla_{r_j}) \left(\frac{1}{|\mathbf{r} - \mathbf{r}_j|} \right) + \cdots \end{aligned} \tag{3.17.3}$$

$$\begin{aligned}\phi_j(\mathbf{r}) = &-\iiint\limits_{\text{mol}} \frac{\rho_j^c(\mathbf{r}_j + \mathbf{r}')dV_{r'}}{|\mathbf{r} - \mathbf{r}_j|} - \iiint\limits_{\text{mol}} (\mathbf{r}' \cdot \nabla_{r_j}) \left(\frac{1}{|\mathbf{r} - \mathbf{r}_j|} \right) \rho_j^c(\mathbf{r}_j + \mathbf{r}')dV_{r'} \\ &- \iiint\limits_{\text{mol}} (\mathbf{r}' \cdot \nabla_{r_j}) \left(\frac{1}{|\mathbf{r} - \mathbf{r}_j|} \right) \rho^c(\mathbf{r}_j + \mathbf{r}')dV_{r'}. \end{aligned} \tag{3.17.4}$$

Factors other than $\rho^c(\mathbf{r}_j+\mathbf{r})$ are independent of $\mathbf{r}'$. So, these are taken out of the integration. In the first term of the right hand side we use (3.10.3), in the second term we use (3.10.8) and in the third term (3.12.3) and like that, in the expansion, and obtain

$$\phi_j(r) = \left[\frac{e_j(r_j')}{|r-r_j|} + (p_j\cdot\nabla_j)\left(\frac{1}{|r-r_j|}\right) + Q^0_{lm}\frac{\partial^2}{\partial r_{jl}\partial r_{jm}}\left(\frac{1}{|r-r_j|}\right) + \cdots\right] \tag{3.17.5}$$

$$\mathbf{E}_j = \nabla\phi_j, \tag{3.17.6}$$

where r_{jl} is the component of r_j parallel to the lth co-ordinate direction. Summing over all molecules in unit volume at $\mathbf{r}$, we obtain

$$\phi(\mathbf{r}) = \sum_{j=1}^{N(\mathbf{r})}\phi_j, \quad \mathbf{E}(\mathbf{r}) = \nabla\phi, \tag{3.17.7}$$

where $N(\mathbf{r})$ is the average number of molecules per unit volume at r at time t. The delta functions of (3.10.4), (3.10.7) and (3.12.4) are introduced in (3.17.7) and the assumption is made that the distribution of molecules within the volume V is continuous; then we obtain

$$\begin{aligned}\phi(r) = &-\iiint_V \frac{\rho(r')dV_{r'}}{|r-r''|} - \iiint_V (P(r'')\cdot V_{r''})\left(\frac{1}{|r-r''|}\right)dV_{r''} \\ &-\iiint_V Q_{lm}\frac{\partial^2}{\partial r_l''\partial r_m''}\left(\frac{1}{|r-r''|}\right)dV_{r''} - \cdots .\end{aligned} \tag{3.17.8}$$

Averaging, as in section 3.10, and dropping the averaging over time, give

$$\begin{aligned}\langle\phi(\mathbf{r})\rangle &= -\frac{1}{\delta V}\iiint_V \phi(\mathbf{r}+\boldsymbol{\xi})dV_\xi \\ &= -\frac{1}{\delta V}\iiint_{\delta V}\left[\iiint_V \frac{\rho(\mathbf{r}'')dV_{r''}}{|\mathbf{r}+\boldsymbol{\xi}-\mathbf{r}''|} + \iiint_V (P(\mathbf{r}'')\cdot\nabla_{r''})\left(\frac{1}{|\mathbf{r}+\boldsymbol{\xi}-\mathbf{r}''|}\right)dV_{r''}\right. \\ &\quad\left. + \iiint_V Q_{lm}\frac{\partial^2}{\partial r_l''\partial r_m''}\left(\frac{1}{|\mathbf{r}+\boldsymbol{\xi}-\mathbf{r}''|}\right)dV_{r''} + \cdots\right]dV_\xi.\end{aligned} \tag{3.17.9}$$

Replacing the variable of integration $\mathbf{r}''$ by $\mathbf{r}'(=\mathbf{r}''-\boldsymbol{\xi})$, we obtain

$$\begin{aligned}\langle\phi(\mathbf{r})\rangle = &-\frac{1}{\delta V}\iiint_{\delta V}\left[\iiint_V \frac{\rho^c(\mathbf{r}'+\boldsymbol{\xi})}{|\mathbf{r}-\mathbf{r}'|}dV_{r'} + \iiint_V (P(\mathbf{r}'+\boldsymbol{\xi})\cdot\nabla_{r'}\left(\frac{1}{|\mathbf{r}-\mathbf{r}'|}\right)dV_{r'}\right. \\ &\left. + \iiint_V Q_{lm}\frac{\partial^2}{\partial r_l'\partial r_m'}\left(\frac{1}{|\mathbf{r}-\mathbf{r}'|}\right)dV_{r'} + \cdots\right]dV_\xi.\end{aligned} \tag{3.17.10}$$

Using (3.10.7) in the first term, (3.10.6) in the second term and (3.12.5) in the third term we obtain

$$\langle\phi(r)\rangle = -\iiint\limits_V \frac{N(r')\langle e(r')\rangle}{|r-r'|}dV_{r'} - \iiint\limits_V N(r')(\langle p(r')\rangle\cdot\nabla_{r'})\left(\frac{1}{|r-r'|}\right)dV_{r'}$$

$$-\iiint\limits_V N(r')\langle q_{lm}\rangle\frac{\partial^2}{\partial r'_l\partial r'_m}\left(\frac{1}{|r-r'|}\right)dV_{r'} - \cdots \tag{3.17.11}$$

$$\langle E(r)\rangle = \nabla\langle\phi(r)\rangle, \tag{3.17.12}$$

$$\nabla\cdot\langle E(r)\rangle = \nabla^2\langle\phi(r)\rangle \tag{3.17.13}$$

for the averaged scalar potential of the electric field.

3.18 Macroscopic Coulomb's Law

Using (3.10.4) and (3.10.5), we obtain

$$\nabla\cdot\langle\mathbf{E}(r)\rangle = 4\pi\left[\iiint\limits_V N(\mathbf{r}')\langle e(\mathbf{r}')\rangle\delta(\mathbf{r}-\mathbf{r}')dV_{r'} + \iiint\limits_V N(\mathbf{r}')(\langle p(\mathbf{r}')\rangle\cdot\nabla_{r'})\delta(\mathbf{r}-\mathbf{r}')dV_{r'}\right.$$

$$\left.+\iiint\limits_V N(\mathbf{r}')\langle q_{lm}\rangle\frac{\partial^2}{\partial r'_l\partial r'_m}\delta(\mathbf{r}-\mathbf{r}')dV_{r'} + \cdots\right]. \tag{3.18.1}$$

Delta function properties 9(b)–(d) of Appendix 1, when applied to the first, second and third terms in the right hand side, give

$$\nabla\cdot\langle\mathbf{E}(\mathbf{r})\rangle = 4\pi N(\mathbf{r})\langle e(\mathbf{r})\rangle - 4\pi\nabla\cdot(N(\mathbf{r})\langle p(\mathbf{r})\rangle) + 4\pi\frac{\partial^2}{\partial r_1\partial r_m}(N(\mathbf{r})\langle q_{lm}\rangle). \tag{3.18.2}$$

The first term here is the average charge per unit volume of the medium. For non-uniform polarization charge neutrality will be violated. Uniform polarization cannot sustain any macroscopically effective accumulation of charges. The second term gives the charge density. Third term contains the second order derivatives of the quadrupole moments. Divergence of the dipole moment per unit volume at $\mathbf{r}$ is non-zero when some charges are lost due to change of positions of some molecules. The substitutions

$$\rho^c(r) = N(r)\langle e(r)\rangle,\quad E(r) = \langle E(r)\rangle,\quad D = E + 4\pi P.$$

$$P(r) = N(r)\langle p(r)\rangle,\quad Q_{lm} = \langle Q^0_{lm}\rangle \tag{3.18.3}$$

give the macroscopic divergence law for the electric displacement vector :

$$\nabla\cdot\mathbf{D} = 4\pi\rho^c + 4\pi\frac{\partial^2 Q_{lm}}{\partial r_l\partial r_m} + \cdots. \tag{3.18.4}$$

3.19 Law of Biot-Savart, Ampere and Maxwell

Essentially the law of Biot-Savart and Ampere is that the volume current I flowing in a conducting wire gives rise to a magnetic induction B. The observation point is P, with position vector $\mathbf{r}$ and Q is a point in the wire with position vector $\mathbf{r}'$, both relative to the fixed origin O fig. 3.19.1. For the increment $d\mathbf{B}$ of the magnetic induction, due to the current element $Id\mathbf{l}$ quantitatively this law gives

$$d\mathbf{B} = \frac{I}{c}\frac{[d\mathbf{l} \times (\mathbf{r} - \mathbf{r}')]}{|\mathbf{r} - \mathbf{r}'|^3} = \frac{1}{c}\frac{[\mathbf{J}(r') \times (\mathbf{r} - \mathbf{r}')]}{|\mathbf{r} - \mathbf{r}'|^3} dV_{r'} \tag{3.19.1}$$

$$= -\frac{1}{c}(\mathbf{J} \times \boldsymbol{\nabla}_r)\left(\frac{1}{|\mathbf{r} - \mathbf{r}'|}\right)\left(\frac{1}{|\mathbf{r} - \mathbf{r}'|}\right) dV_{r'} = \frac{1}{c}\left[\nabla_r \times \frac{J(r')}{|\mathbf{r} - \mathbf{r}'|}\right] dV_{r'}. \tag{3.19.2}$$

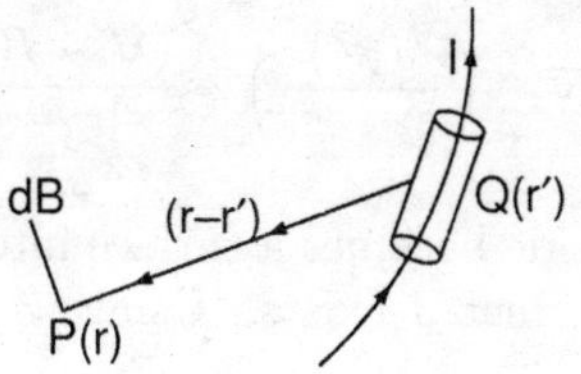

Fig. 3.19.1 It shows the induction field element dB at the point P due to the volume current Idr at Q.

Here the volume current is I in stat amperes (in electromagnetic units), B is in gausse, C in cm/sec. and dl in cm, and since $\mathbf{J}(\mathbf{r})$ is the current density at $\mathbf{r}'$, it can be taken within the curl operators which operates at $\mathbf{r}$. The continuity condition for $\mathbf{J}$ and charge density ρ^c is

$$(\boldsymbol{\nabla}\cdot \mathbf{J}) + \frac{\partial \rho^c}{\partial t} = 0. \tag{3.19.3}$$

Integrating (3.19.2) we obtain

$$\mathbf{B} = \frac{1}{c}\boldsymbol{\nabla}\times \iiint_V \frac{\mathbf{J}(\mathbf{r}')dV_{r'}}{|r - r'|}. \tag{3.19.4}$$

Taking divergence the solenoidal law for magnetic induction is established

$$\boldsymbol{\nabla}\cdot \mathbf{B} = 0. \tag{3.19.5}$$

Putting $\mathbf{B} = $ curl $\mathbf{A}$ in (3.19.4) gives the vector potential

$$\mathbf{A} = \frac{1}{c}\iiint_V \frac{J(\mathbf{r}')dV_{r'}}{|\mathbf{r} - \mathbf{r}'|}. \tag{3.19.6}$$

Note that this A is not unique because the substitution $\mathbf{B} = \boldsymbol{\nabla} \times (\mathbf{A} + \boldsymbol{\nabla}\phi)$ also satisfies (3.19.4), where ϕ is any smooth scalar function of position co-ordinates. Taking curl of both sides of (3.19.4) we find that

$$\boldsymbol{\nabla} \times \mathbf{B} = \frac{1}{c}\nabla \times \left(\nabla\times \iiint_V \frac{J(r')dV_{r'}}{|r - r'|}\right)$$

$$= \frac{1}{c}\nabla_r \left[\nabla_r\cdot \iiint_V \frac{J(r')dV_{r'}}{|r - r'|}\right] - \frac{1}{c^2}\nabla_r^2 \cdot \iiint_V \frac{J(r')dV_{r'}}{|r - r'|}$$

$$= \frac{1}{c}\nabla_r \iiint\limits_V \left(\nabla_{r'}\cdot \frac{J(r')dV_{r'}}{|r-r'|}\right) + \frac{4\pi}{c}\iiint\limits_V J(r')\delta(r-r')dV_{r'}$$

$$= \frac{1}{c}\nabla_r \iiint\limits_V (J(r')\cdot \nabla_r)\left(\frac{1}{|r-r'|}\right) dV_{r'} + \frac{4\pi}{c}J(r)$$

$$= -\frac{1}{c}\nabla_r \iiint\limits_V (J(r')\cdot \nabla_{r'})\left(\frac{1}{|r-r'|}\right) dV_{r'} + \frac{4\pi}{c}J(r)$$

$$= -\frac{1}{c}\nabla_r \iiint\limits_V \left[\nabla_{r'}\cdot\left(\frac{J(r')}{|r-r'|}\right) - \frac{(\nabla_{r'}\cdot J(r'))}{|r-r'|}\right] dV_{r'} + \frac{4\pi}{c}J(r). \tag{3.19.7}$$

The first term in the right hand side becomes a surface integral, which vanishes for integration outside the region, where the current **J** exists. Using (3.19.3) in the second term and **E** of (3.16.3), we obtain

$$\nabla\times\mathbf{B} = -\frac{1}{c}\nabla_r \iiint\limits_V \frac{\partial\rho}{\partial t}\frac{dV_{r'}}{|r-r'|} + \frac{4\pi}{c}\mathbf{J}(r) = \frac{1}{c}\frac{\partial\mathbf{E}}{\partial t} + \frac{4\pi}{c}\mathbf{J}(r) \tag{3.19.8}$$

for the Ampere Maxwell law for magnetoelectric induction.

3.20 Macroscopic Ampere-Maxwell Law in the Far Field Approximation

Inside each molecular volume the field of the dipoles formed by the ions and electrons and their variation are appreciably large. Expanding the right hand side of (3.19.6) in powers of $\mathbf{r}'$, and putting $\mathbf{R} = \mathbf{r} - \mathbf{r}_j$ in that, we find the vector potential

$$\mathbf{A}_j = \frac{1}{cR}\iiint\limits_{\text{mol}} \mathbf{J}_j(\mathbf{r}_j+\mathbf{r}')dV_{r'} + \frac{1}{cR^3}\left(\mathbf{R}\cdot\iiint\limits_{\text{mol}}\mathbf{r}'\right)\mathbf{J}_j(\mathbf{r}_j+\mathbf{r}')dV_{r'} + \cdots \tag{3.20.1}$$

for the current in the jth molecule, $\mathbf{r}'$ of (3.19.6) is replaced by $\mathbf{r}_j + \mathbf{r}'$. The current vector $\mathbf{j}_j$ is the vector sum $\hat{x}_k J_{jk}$, where $\hat{x}_k$ is the unit vector parallel to the kth space direction. The coefficient of $1/CR$ in the first term in the right hand side of (3.20.1) is

$$= \hat{X}_k \iiint\limits_{\text{mol}} J_{jk}(r_j+r')dV_{r'} = \hat{X}_k\left[\iiint\limits_{\text{mol}} \{\nabla\cdot(x_k' J_j) - x_k'(\nabla\cdot J)\}\, dV_{r'}\right]$$

$$= 0 + \iiint\limits_{\text{mol}} r'\frac{\partial\rho_j^c}{\partial t}dV_{r'}. \tag{3.20.2}$$

The first term vanishes because the integrand is an exact differential. And since in the volume of integration the independent variables are the components of $\mathbf{r}_j$ and not $\mathbf{r}'$, we can take the time differential outside the integration, in this term. Then it is

$$= \frac{\partial}{\partial t}\iiint\limits_{\text{mol}} \mathbf{r}'\rho_j^c(\mathbf{r}_j+\mathbf{r}',t')dV_{r'}, \tag{3.20.3}$$

where t' is the retarded time $t - |\mathbf{r} - \mathbf{r}'|/c$. In the second term of the right hand side of (3.20.1) the coefficient of $1/CR^3$

$$= \hat{X}_k R_i \iiint\limits_{\text{mol}} x'_i J_{jk}(r_j + r', t') dV_{r'}$$

$$= \hat{X}_k R_i \iiint\limits_{\text{mol}} x'_i [(\nabla \cdot (x'_k J_j) - x'_k (\nabla \cdot j_j)] dV_{r'}$$

$$= \hat{X}_k R_i \iiint\limits_{\text{mol}} [\nabla \cdot (x'_i x'_k J_j) - x'_k J_{ji} - x'_i x'_k (\nabla \cdot j_j)] dV_{r'}$$

$$= 0 - \hat{X}_k R_i \iiint\limits_{\text{mol}} \left[x'_k J_{ji} - x'_i x'_k \frac{\partial \rho^c_j}{\partial t} \right] dV_{r'}$$

$$= -\left(R \cdot \iiint\limits_{\text{mol}} J_j \right) r' dV_{r'} + \left(R \cdot \iiint\limits_{\text{mol}} r' \right) \frac{\partial \rho^c_j}{\partial t} r' dV_r. \tag{3.20.4}$$

Here R_i is the ith component of R. The first term

$$= \left(\mathbf{R} \cdot \iiint\limits_{\text{mol}} \mathbf{r}' \right) J_j dV_{r'}$$

$$= -\frac{1}{2} \left[\iiint\limits_{\text{mol}} (\mathbf{r}' \times \mathbf{J}_j) \right] dV_{r'} + \frac{1}{2} \left(\mathbf{R} \cdot \frac{\partial}{\partial t} \iiint\limits_{\text{mol}} \mathbf{r}' \right) r' \rho^c_j dV_{r'}$$

$$= -C(\mathbf{R} \times \mathbf{m}_j) + \frac{1}{2} \left(\mathbf{R} \cdot \iiint\limits_{\text{mol}} \mathbf{r}' \right) \frac{\partial \rho^c_j}{\partial t} dV_{r'}, \tag{3.20.5}$$

where the vector result

$$(\mathbf{R} \times (\mathbf{r}' \times \mathbf{J}_j)) = (\mathbf{R} \cdot \mathbf{j}_j) \mathbf{r}' - (\mathbf{R} \cdot \mathbf{r}') \mathbf{J}_i$$

has been applied. Hence equations (3.20.1)–(3.20.5) give

$$A_j = \frac{1}{|r - r_j|} \frac{\partial}{\partial t} \iiint\limits_{\text{mol}} r' \rho^c_j dV_{r'} + \frac{(\mu_j \times R)}{|r - r'|^3} + \frac{1}{2} \left(R \cdot \frac{\partial}{\partial t} \iiint\limits_{\text{mol}} r' \right) r' \rho^c_j dV' + \cdots, \tag{3.20.6}$$

where $\boldsymbol{\mu}_j$ is the magnetic moment for the current $\mathbf{J}_j$. The third and the subsequent terms are the negligible quadrupole and higher order moments of charge distributions. The first term

$$= \frac{1}{c} \frac{1}{|r - r_j|} \frac{\partial}{\partial t} \iiint\limits_{\text{mol}} \rho^c_j (\mathbf{r}_j + \mathbf{r}') r' dV_{r'} = \frac{1}{c|\mathbf{r} - \mathbf{r}'|} \frac{\partial p_j}{\partial t}. \tag{3.20.7}$$

Hence

$$\mathbf{A}_j(\mathbf{r}, t) = \frac{1}{c|\mathbf{r} - \mathbf{r}'|} \frac{\partial \mathbf{p}_j}{\partial t} + \frac{[\boldsymbol{\mu}_j \times (\mathbf{r} - \mathbf{r}_j)]}{|\mathbf{r} - \mathbf{r}'|^3}. \tag{3.20.8}$$

The total current density is the sum of the conduction current $\mathbf{J}$ of transport of charges, and $\sum_j \mathbf{J}_j$ is the sum of currents inside all the molecules in the volume. Hence (3.20.4) gives

$$\mathbf{A}(r,t) = \frac{1}{c}\iiint_V \frac{\mathbf{J}(\mathbf{r}',t')}{|\mathbf{r}-\mathbf{r}'|}dV_{r'} + \frac{1}{c}\sum_j \iiint_{\text{mol}} \frac{\mathbf{J}(\mathbf{r}_j+\mathbf{r}'_j,t')}{|\mathbf{r}-\mathbf{r}_j-\mathbf{r}'|}dV_{r'} \tag{3.20.9}$$

$$= \frac{1}{c}\iiint_V \frac{\mathbf{J}(\mathbf{r}',t')}{|\mathbf{r}-\mathbf{r}'|}dV_{r'} + \frac{1}{c}\iiint_V \frac{\partial}{\partial t}p\frac{dV_{r''}}{|\mathbf{r}-\mathbf{r}''|}$$

$$+ \iiint_V \frac{[\mathbf{M}(\mathbf{r}'')\times(\mathbf{r}-\mathbf{r}'')}{|\mathbf{r}-\mathbf{r}''|}dV_{r''}. \tag{3.20.10}$$

The last term of this equation

$$= \iiint_V \left[\mathbf{M}(\mathbf{r}'')\times\boldsymbol{\nabla}_{r''}\left(\frac{1}{|\mathbf{r}-\mathbf{r}''|}\right)\right]dV_{r''}$$

$$= \iiint_V \left[\frac{(\nabla_{r''}\times\mathbf{M})}{|\mathbf{r}-\mathbf{r}''|} - \boldsymbol{\nabla}\times\left(\frac{M(\mathbf{r}'')}{|\mathbf{r}-\mathbf{r}''|}\right)\right]dV_{r''}$$

$$= \iiint_V \frac{1}{|\mathbf{r}-\mathbf{r}''|}[\nabla_{r''}\times\mathbf{M}]dV_{r''} + 0, \tag{3.20.11}$$

where the vector formula curl $(\phi\mathbf{A}) = \phi$ curl $\mathbf{A} + (\boldsymbol{\nabla}\phi\times\mathbf{A})$ has been used in the second step in the first term, and the volume integral

$$\iiint_V \left[\boldsymbol{\nabla}_{r''}\times\left\{\frac{\mathbf{M}(r'')}{|\mathbf{r}-\mathbf{r}''|}\right\}\right]dV_{r''} \tag{3.20.12}$$

on reduction to a surface integral vanishes outside the region, where $\mathbf{M} \neq 0$. Hence,

$$\mathbf{A}(r,t) = \frac{1}{c}\iiint_V \frac{1}{|\mathbf{r}-\mathbf{r}''|}\left\{\mathbf{J}(r',t') + \frac{\partial}{\partial t}(\mathbf{N},\mathbf{p}) + \boldsymbol{\nabla}\times\mathbf{M}\right\}dV_{r''} \tag{3.20.13}$$

$$\langle\mathbf{A}(\mathbf{r},t)\rangle = \frac{1}{c}\iiint_V \frac{\langle\mathbf{J}(\mathbf{r}',t')\rangle}{|\mathbf{r}-\mathbf{r}'|}dV_{r'}, \tag{3.20.14}$$

where

$$\langle\mathbf{J}\rangle = \left\langle\mathbf{J} + \frac{\partial}{\partial t}(N\mathbf{p}) + c\boldsymbol{\nabla}\times\mathbf{M}\right\rangle. \tag{3.20.15}$$

Replacing $\mathbf{J}$ of (3.19.8) by

$$\mathbf{J}(\mathbf{r}) + \frac{\partial}{\partial t}(Np) + c\boldsymbol{\nabla}\times\mathbf{M},$$

and then averaging that equation we find that

$$\boldsymbol{\nabla}\times\langle\mathbf{B}\rangle = \frac{1}{c}\frac{\partial}{\partial t}\langle\mathbf{E}\rangle + \frac{4\pi}{c}\left\{\langle\mathbf{J}\rangle + \frac{\partial}{\partial t}\langle N\mathbf{p}\rangle + c\langle\boldsymbol{\nabla}\times\mathbf{M}\rangle\right\} + \cdots. \tag{3.20.16}$$

If we write $\langle \mathbf{B} \rangle - 4\pi \langle \mathbf{M} \rangle = \mathbf{H}$, $\langle \mathbf{E} \rangle + 4\pi \langle N\mathbf{p} \rangle = \mathbf{D}$ and $\mathbf{j} = \langle \mathbf{J} \rangle$ we obtain the Maxwell equation in the standard form :

$$\nabla \times \mathbf{H} = \frac{1}{c}\frac{\partial D}{\partial t} + \frac{4\pi}{c}\mathbf{j}. \tag{3.20.17}$$

3.21 Relativistic Nature of Fluctuating Magnetic Fields

Omitting the terms containing $1/c^2$, the equations for the vector potential $\mathbf{A}$ and the scalar potential ϕ are

$$\nabla^2 \phi - \frac{1}{c}\frac{\partial}{\partial t}(\nabla \cdot \mathbf{A}) = \frac{4\pi}{V}\sum_i e_i \delta(\mathbf{r}_i - \mathbf{r}),$$

$$\nabla^2 \mathbf{A} - \nabla\left(\nabla \cdot \mathbf{A} - \frac{1}{c}\frac{\partial \phi}{\partial t}\right) = -\frac{4\pi}{cV}\sum_i \mathbf{e}_i \mathbf{r}_i \delta(\mathbf{r}_i - \mathbf{r}).$$

For the Lorentz gauge, these equations reduce to

$$\nabla^2 \phi = \frac{4\pi}{V}\sum_i \mathbf{e}_i \delta(\mathbf{r}_i - \mathbf{r}), \quad \nabla^2 A = \frac{4\pi}{cV}\sum_i \mathbf{e}_i \dot{\mathbf{r}}_i \delta(\mathbf{r}_i - \mathbf{r}),$$

where also terms proportional to $1/c^2$ have been rejected. The solutions, using the delta function property 12 of Appendix 1, are

$$\mathbf{E} = \frac{1}{V}\sum_i \left\{\nabla\left(\frac{e_i}{|\mathbf{r}_i - \mathbf{r}|}\right)\right\}, \quad \mathbf{H} = \frac{1}{cV}\sum_i \left\{\nabla\left(\frac{e_i r_i'}{|r_i - r|}\right)\right\},$$

because $\mathbf{E} = \nabla\phi$ and $\mathbf{H} = \nabla \times \mathbf{A}$. Therefore, the magnetic field is of the order of $1/c$ and $\partial\beta/c\partial t$ is of the order of $1/c^2$ when E is of the order of unity. So, Faraday law of EM induction, basically is a relativistic equation. Non-relativistically, it should be replaced by the conservative field of $\nabla \times \mathbf{E} = 0$.

3.22 Important Works on the Lorentz Theory

Lorentz (1915) initiated work on the formulation of the microscopic equations of electrodynamics for charges and their currents in empty space, and then on the derivation of the phenomenological macroscopic EM equation of the Maxwell theory from these by averaging. Formalisms of statistical mechanics are different for classical and quantum mechanical treatments of the atomic systems. The atomistic interpretation of the electric, magnetic and optical properties of matter was developed by Rosenfeld (1951) and extended to explain the high electrical conductivity and strong magnetic properties in some substances. The electromagnetic field averages and averages of atomic charges and current densities, both over space and time over physically infinitesimal elements, have been identified.

Mazur and Nijboer (1953), Mazur (1958), De Groot (1965) and De Groot and Suttorp (1972) have generalized and extended the Lorentz theory by a systematic use of averaging of statistical mechanics. This method does not give as detailed a physical picture of the underlying processes as the method laid out by Rosenfeld (1951).

De Groot (1965) used the multipole expansion to derive the atomic field equations from the microscopic field equations and the Maxwell equations by means of statistical averaging of the equations, which are assumed to be valid in atomic levels of matter. The relativistic theory is developed in a covariant manner, by defining the multipole moments in atomic proper frames and by using a covariant averaging for taking into account the retardation.

De Groot and Suttorp (1972) started with composite particles like atoms, molecules or ions which consist of point charges without structure. The microscopic Maxwell-Lorentz equations and the Newtonian equations of motion (or the relativistically correct equations of motion) containing the Lorentz force, together with the other forces, are used for these point particles. Multipole series expansion of the field variables of the Lorentz equation of electrodynamics and the equations for the momentum, energy and angular momentum of the composite particles gives the phenomenological equations of Maxwell. Their Taylor series expansions in the internal molecular co-ordinates of delta functions, prior to the spatial averaging, are without justification. These authors also discussed the relativistic corrections and provided a statistical mechanical treatment of the averaging procedure.

Robinson (1973) has distinguished between the spatial averaging and the statistical mechanical averaging over various sorts of ensemble. Moreover, for each macroscopic problem he has assigned an appropriate lower limit of relevant lengths which sets criterion for the statistical averaging procedure.

Russakoff (1970) adopted a modified version of spatial averaging and worked on unification of several earlier approaches. The spatial averaging is done with the help of a test function, which is real, non-zero in some neighbourhood of a reference point and normalized to unity over all space. Jackson (1975), second edition, section 6.7) has followed Russakoff; but in his first edition (1962), Jackson followed Rosenfeld (1951). We have followed Jackson (1962) for its simplicity and suitability in introductory treatments.

Chapter 4

Electrodynamic Fields in media and from Particle Motion

4.1 Introduction

Electrodynamics of media and particles include the force aspects, the energy aspects, the stress-strain aspects, the acceleration consequences, and their dependence on charges, currents, and intensity of electric field and magnetic field. Different types of electric current have been detailed and discussed. The Maxwell stress elements, generally studied earlier in section 1.7, are considered further for application. The Debye shielding potential has been derived. The mean free path method of Townsend, being useful in electronics and ionospheric physics of weak wave interaction processes, is explained. The other topics covered are : the Lagrangian formalism, dynamics of relativistic electrons in presence of fields, EM radiation from fields and media, etc.

The treatment at places is some what non-conventional. The intention is to establish a new emphasis in the exposure of electrodynamics to learners. The new inclusions are as important as the usual topics of the standard texts.

4.1.1 Effects of fields on materials

Atoms break up into ions and electrons when the average kinetic energy per particle exceeds their ionizing potential. There is no sharp melting point-like, or boiling point like[1] temperature for this change, because, simply, by heating, the thermal energy of some particles exceeds the energy binding the molecules, thus dissociating molecular gas into atomic gas. If the thermal energy is strong enough to overcome the atomic binding energy, then some electrons become free and ionization starts. The temperature for complete ionization depends on the material constitution. Ultimately, all the atoms get stripped of their electrons until the gas becomes a mixture of freely moving electrons, and nuclei, if the heat applied is sufficient for that change. Electrons and ions, thus generated, are forced to move together to prevent growth of huge

[1] The amount of heat energy introduced into a system to bring about a phase transition at a fixed temperature and pressure, is the latent heat. This transition occurs for change from a solid state to a liquid state and from liquid to vapour (gas).

electrostatic force of even small charge separation. This transition from neutral gas to ionized gas is gradual, and is not a phase transition in the thermodynamic sense.

Coupling between classical and quantum phenomena is generally very weak, because the characteristic lengths or times and quantum mechanical behaviour, such as the deBroglie wavelengths, or times to emit a photon, are usually very short compared to the characteristic lengths and times of development of field induced material properties.

Ionized media can be given a many fluid continuum description, or a single fluid continuum description. Eulerian concepts of motion in position space, or the Lagrangian concepts of motion are useful. Moreover, there is the particle-kinetic approach, or the statistical distribution approach, providing an Eulerian description of motion of matter in phase space. So, for understanding electrodynamics of media and particles, a basic knowledge of particle dynamics, fluid mechanics, kinetic theory, statistical mechanics and thermodynamics, besides electromagnetics, is necessary. The suitability of a description depends on the distinct time scales and length scales specifying a phenomenon. Fluid dynamical methods depend on properties of packets of moving fluids, and on the concepts of electrical conductivity. However, orbits of the packets are not to be confused with the orbits of individual particles, of the particle orbit approach. Moreover, this type of compartmentalization can give rise to misbeliefs like (i) pressure pressure gradient can exist without any inter particle collision, and (ii) The individual particles may be subject to pressure gradient forces. So, the usefulness of such divisions of study may be questioned.

Charged particle trajectory is studied in presence of fields (electrical, magnetic, gravitational, etc.) when inelastic collisional encounters are rare, and the mean free path (average distance between two collisions) is large, and also when collisional encounters are too-frequent at an instant. Then the collective classical long range interactions dominate over short range, individual, large angle, inelastic collisions of quantum mechanical nature. Study of trajectories of many particles on the basis of guiding center drift equations require vast computation. A simpler approximate method is to use the adiabatic invariant formalism.

Dynamics of charged particles is supplementary to the macroscopic fluid theory for fluid-like collective behaviour, and the kinetic theory for many particle interactions. The fluid elements follow the trajectory of charged particles since all such particles move in concert in a fluid.

Delta functions of Dirac appear in the expressions for the discrete charges and their currents. These functions are avoided for collisional processes which are quite slow compared with the collective ones. Then the individuality of the particles is of little importance, and each particle can be subdivided into smaller and smaller subparticles, such that the delta functions reduce to some smooth and continuous functions of space and time. The discrete distributions then reduce to a continuum of fluid mixtures or a distribution in multidimensional phase space. This smearing out process keeps unchanged the characteristics frequency of the medium, the Debye length, gyro frequencies and gyro radii; and the collision mean free path becomes infinite.

Fields are either the applied macroscopic (slowly varying) fields or the microscopic self fields of the particles, and are of a short-range (fluctuating) collisional type. When the gravitational force is negligible, neutral particles are force-free between collisions with other particles. So, the force acting on charged and neutral particles are of different origin. The macroscopic force affects all the particles in a given fluid element similarly. The collisional force has a non-random component that tends to drive particles down the pressure gradient, and a random component that scatters the particles. In a neutral gas, the pressure gradient force and scattering forces are transmitted impulsively by hard collisions. The Coulomb collision force is a continuously acting grazing collision; so, the particles experience a continuous force, and are gradually scattered through a series of small angles.

Energy is dissipated by electrostatic Coulomb interactions of the charges and also by interaction with electromagnetic field. The static and low frequency parts of wave fields excite magnetostatic interactions between the particles, and the high frequency components constitute the radiation part. The electrostatic and magnetostatic interactions are of long range nature; the Debye shielding phenomenon provides a natural cut-off distance for the short range Coulomb force in a collision.

4.1.2 Forces on charges and electric currents

Electrodynamical laws of magnetoelectric induction, in the names of Biot, Savart and of Ampere and Oersted state that current of charges through a closed circuit Γ generates an axial magnetic field. The final form of this law, before the advent of the displacement current of Maxwell, states that the magneto motive force

$$\oint_{\Gamma} (B \cdot dr) \tag{4.1.1}$$

is proportional to the flux of electric current through the surface Σ enclosed by the circuit Γ :

$$\oint_{\Gamma} B \cdot dr = \alpha \iint_{\Sigma} (j \cdot d\Sigma), \tag{4.1.2}$$

where α is a constant. This law in the form of a differential equation in CGS Gaussian units is

$$\nabla \times B = \frac{4\pi}{c} j. \tag{4.1.3}$$

It implies that a current element $I_1 dr_1$ in a non-rectilinear path produces an axial increment of magnetic field; the differential dB of the field is given by

$$dB = \frac{I_1}{cr^3}(dr_1 \times r), \tag{4.1.4}$$

where r is the position vector from the position of the length element dr_1 to that of an observation point Q.

Ampere found the force experienced by one current in presence of another current. The force is that of attraction under the law of inverse square of the distance when the currents in the two circuits Γ_1 and Γ_2 flow in opposite directions. When the two currents are in the same direction the force is that of repulsion also under the inverse square distance law.

The elemental force $d\mathbf{F}$ experienced by the current element $I_1 dr_1$ in presence of the magnetic induction field $\mathbf{B}$ is

$$d\mathbf{F} = \frac{1}{c} I_1 (d\mathbf{r}_1 \times \mathbf{B}). \tag{4.1.5}$$

When the field B is due to a current I_2 in a closed circuit Γ_2, then the total force experienced by the current I_1 in the closed circuit Γ_1 is experimentally found to be

$$F_{12} = \oint_{\Gamma_1} \oint_{\Gamma_2} \frac{I_1 dr_1 \times (I_2 dr_2 \times \hat{r}_{12})}{c^2 r_{12}^2}, \tag{4.1.6}$$

where the vector. So, $\mathbf{r}_2 - \mathbf{r}_1 = \mathbf{r}_{12} = |\mathbf{r}_{12}| \hat{\mathbf{r}}_{12}$ and $\hat{\mathbf{r}}_{12}$ is the unit vector along $\mathbf{r}_{12}$.

In terms of differentials of curve lengths, the force differential is given by

$$d\mathbf{F}_{12} = \frac{I_1 I_2}{c^2 r_{12}^2}\left[(d\mathbf{r}_1 \cdot \hat{\mathbf{r}}_{12})\, d\mathbf{r}_2 - (d\mathbf{r}_1 \cdot d\mathbf{r}_2)\hat{\mathbf{r}}_{12}\right]. \tag{4.1.7}$$

This relation is not symmetric in $I_1 d\mathbf{r}_1$ and $I_2 d\mathbf{r}_2$ because the vector triple products $A\times(B\times C)$ and $B \times (A \times C)$ are not identical in magnitude and direction. So $d\mathbf{F}_{12} \neq d\mathbf{F}_{21}$. This seems to contradict Newton's third law of motion which states that the forces of action and reaction are equal in magnitude and opposite in direction. For clarification we note that the steady currents cannot be isolated. But all steady currents must flow around continuous loops. Hence, if the entire loops Γ_1 and Γ_2 are considered, as in (4.1.6), Newton's third law is satisfied, because the first term of (4.1.7) being

$$d\mathbf{r}_2 \cdot \frac{I_1 I_2}{c^2}\left(\nabla\left(\frac{1}{r_{12}}\right) \cdot d\mathbf{r}_1\right) = \frac{I_1 I_2}{c^2}\left(d\mathbf{r}_1 \cdot \nabla\left(\frac{1}{r_{12}}\right) \cdot d\mathbf{r}_2\right)$$

is an exact differential. So, this part in (4.1.6) will cancel out, making the integral in (4.1.6) from only the second term of (4.1.7) which is symmetric in $I_1 d\mathbf{r}_1$ and $I_2 d\mathbf{r}_2$.

Equation (4.1.7) can be written as

$$\mathbf{F}_{21} = \oint_{\Gamma_2} I_2 (d\mathbf{r}_2 \times \mathbf{B}_1), \tag{4.1.8}$$

where

$$\mathbf{B}_1 = \oint_{\Gamma_1} I_1 \frac{(d\mathbf{r}_1 \times \hat{\mathbf{r}})}{r^2}. \tag{4.1.9}$$

Thus a charge q moving with a velocity $\mathbf{v}$ is equivalent to an element of current $I d\mathbf{r} = q\mathbf{v}$. Hence (4.1.8) takes the standard form for the Lorentz force

$$\mathbf{F} = \frac{q}{c}(\mathbf{v} \times \mathbf{B}). \tag{4.1.10}$$

Similar conclusions are valid for motion of a free charged particle, instead of a wire carrying current. So, an electric current is associated with a magnetic field which pervades the surrounding space. Some other currents or any moving charged particle, under the influence of this field, experiences a force proportional to the strength of the magnetic field. This force is always perpendicular to the velocity of a charged particle, and acts on the test charge in motion. So, the force on a particle carrying a charge q is given by

$$\mathbf{F} = q\mathbf{E} + \frac{q}{c}(\mathbf{v} \times \mathbf{B}). \tag{4.1.11}$$

Thus the magnetic induction field B, like E, is a device for describing interaction of charged particles with one another. One can, therefore, predict uniquely the velocity dependent part of the force of any charged particle. This magnetic action of electric current is a corollary of Coulomb's law. Considering validity of Coulomb's law and utilising the postulates of relativity theory and charge invariance, one can deduce Ampere's law. Thus, magnetic effects emerge from electric interactions between a moving charge and other moving charges, and lead to Ampere's hypothesis, which states that Magnetic substances contain permanently circulating electric currents.

4.1.3 Conductivity and dielectric—like properties

Matter becomes electrically conducting when it contains free charged particles. The conductivity depends on the concentration of electrons and ions, and on their mobility. For evaluation of the electrical conductivity, and for finding its stability in a partially ionized medium, the ionization process must be known in *advance*. Dense, electrically conducting, fluids behave like a set of closely packed dipoles. The plasma dielectric properties are determined by free charges, and not by dipoles. A metal, or a semiconductor can have these properties, although its mechanical properties of compressibility and rheology resembles those of solids. Dense interior of stars has compressibility of solids or liquids and plasma properties.

4.1.4 Collision and ionization

Ionization occurs when the kinetic energy $(mv^2/2) > I$, where I is the ionization energy. Thus, a neutral particle absorbs so much energy that it becomes positively charged by loosing one or more electrons. Electrons are also liberated from a solid. A collision between an excited molecule and a neutral particle creates two particles in a ground state with enhanced kinetic energy. All collisions do not result in ionization or excitation. The energy exchange depends on certain probabilities that are very small positive fractions.

Collisions between an electron and an excited atom from neutral atoms and electrons of energy high enough to cause ionization upon collision with other neutral atoms. Potential energy exchanged in collisions between two excited atoms, ionizes one of them. Exchange of kinetic energy (collisions of first order) is possible by collisional ionization, and when a particle delivers part of its potential energy for this cause (second order collisions). A heavy particle ionizes a particle in the ground state, its kinetic energy before impact is at least twice the ionization or excitation energy of the ground state particle. Also, the kinetic energy of an electron equals the ionization or excitation energy of a neutral particle to ionize it by loosing all its kinetic energy.

4.1.5 Plasmas

A physical plasma is a state of matter of high electrical conductivity and has mostly gaseous mechanical property. It has also dielectric properties which are determined by free charges and not dipoles. Plasmas have a tendency to become electrically neutral. Long range Coulomb interaction dominates over short range individual encounters in plasmas.

The ratio of average ion energy and ionization energy is much greater than unity for hot plasmas, of the order of unity for temperate plasmas, and much less than unity for cold plasmas. Cold plasmas (low electron-energy plasmas) are found in interstellar and interplanetary spaces, in the ionosphere and in some alkali vapours. Hot plasmas, at temperatures of many million degrees centigrade, are found in cosmology and in thermonuclear fusion power generation programmes.

Laboratory plasmas are produced by techniques of silent discharges, direct current and low frequency glow discharges. These are sources of UV and other electromagnetic fields. Glow discharges are radiations in the visible regions. Experiments determine the current density, electron density, including the number of free electrons present, temperature, specific heat, viscosity, thermal and electrical conductivity, of a plasma.

For temperatures less than a few KeV, relativistic non-linear effects are not important. The charge density, temperature, magnetic field, ionization percentage, etc. specifying the equilibrium state of a plasma are its average, non-fluctuating properties. A small disturbance is a harmonic function of the time and space co-ordinates, so it propagates as a wave.

High pressure plasmas around a probe have layers like those surounding a test charge. A non-neutral space charge sheath exists near the probe. The transition regions, beyond the space charge region, of a probe have field dependent mobilities. Beyond this is the diffusion region, where mobilities and diffusion coefficients are constant and the influence of the probe is small compared to that of the density gradient. Lastly, there is the undisturbed region of neutral plasma.

4.2 Electric Current

Electric current is due to motion of electric charges. The flow is along the direction of motion of positive charges even though current is largely due to the electron motion. Mobile charges have random motion on the average. Motion depends on the thermal state of the body. But, it is convenient to think of moving charges as a fluid in motion, or as a mixture of several species of charged and uncharged fluids in motion.

The current I of electricity, defined in amperes or Coulombs per second, is the rate of flow of electric charges, through a wire or a conductor. Hence, if q is the charge in units of Coulomb, its time rate of variation is the current I; so

$$I = \frac{dq}{dt}.$$

If u is the velocity of a charged particle, σ is the area of cross section of the conducting wire and N is the number of free electrons per unit volume, then

$$I = \sigma N q u.$$

The flow of one ampere current is the flow of one Coulomb of charge per second. The charge of an electron being 1.6×10^{-19} Coulomb, the one ampere current means $1/(1.6\times10^{-16})$ electrons per second, which is about 6×10^{16} electrons/sec. So, a current of one ampere means a flow of vast number of electrons per second.

The charge current density is defined as the rate of flow of charges per unit area, $J = I/\sigma$. In terms of motion of several species of charges, we can define the surface current vector $\mathbf{j}$ by

$$\mathbf{j} = \Sigma N_\alpha q_\alpha \mathbf{u}_\alpha,$$

where N_α is the number of charges per unit volume of the αth species, q_α is the charge per particle and $\mathbf{u}_\alpha$ is the average velocity of the αth species. So, $\mathbf{j}$ is the current density vector at a point on the plane perpendicular to the direction of the flow, and

$$j = \lim_{\delta\Sigma\to0}\left(\frac{\delta I}{\delta\Sigma}\right),$$

where δI is the current passing through an element of area $\delta\Sigma$. $\delta I(= j\cos(j,n)\delta\Sigma)$ is maximum when this area is oriented perpendicular to the lines of current flow, where $(\mathbf{j},\mathbf{n})$ is the angle between $\mathbf{j}$ and the unit vector $\mathbf{n}$ along the normal to the surface element $\delta\Sigma$. We can write

$$\delta I = (j \cdot \delta\Sigma), \quad \frac{\partial I}{\partial \Sigma_x} = j_x$$

because $\boldsymbol{\delta\Sigma} = \mathbf{n}\delta\Sigma$. The current passing through any given area is

$$I = \iint_{\Sigma} (\mathbf{j} \cdot d\Sigma) = \iint_{\Sigma} j \cos(j, n)\, d\Sigma = \iint_{\Sigma} j_n d\Sigma,$$

where j_n is the component of $\mathbf{j}$ parallel to $\mathbf{n}$. If dl is the length element along a conducting wire at a point P and dV is the volume element at P, then $Id\mathbf{l} = \mathbf{j}dv$.

4.2.1 Different types of electric current

Several types of currents due to electric charges and fields have been specified in the literature. Mainly these are : the conduction current, the convection current, the polarization current, the displacement current, the magnetization current, the advection current, the Hall current, the Pedersen current.

Conduction current

In electrical conductors the conduction current is due to the motion of free or quasi-free charges through distances large compared to atomic dimensions, in presence of applied electric fields. Thus, bound electrons can, on the atomic scale, move through considerable distance between two collisions within the crystal lattice. Electric field produces dynamical effects, including bulk motion of the medium. This mass motion affects the fields also. The conduction current is not due to any mass motion. Ohm's law $\mathbf{j} = \sigma\mathbf{E}$ provides the balance between the applied force field and the frictional drag, or stands for dynamical equilibrium between these forces. And, collisional forces are proportional to the relative velocity, over a wide range of velocities. The collisional resistive force

$$m_i n_i \sum_j \gamma_{ij}(\mathbf{v}_i - \mathbf{v}_j),$$

opposes the motion of charges in the direction of the applied field, where N_i, m_i and $\mathbf{v}_i$ are the number density, mass per particle and average velocity of the ith species of the charged constituents, γ_{ij} is the frequency of collisions of the ith species elements with the elements of the jth species. If the applied force frequency exceeds the collision frequency, the charges have time to accelerate and decelerate in the time between two successive changes from collision. Then the electrical conductivity becomes complex. So, the conduction current, defined by Ohm's law, takes account of the heat dissipation without any transport of matter, through the drift of free electrons and ions in presence of applied fields. The net charge density for conduction currents may or may not be equal to zero.

Convection current

Convection[2] current results from convective effects due to heat if a material medium in motion contains charges of various types. For this current the density and velocity of charges should

[2]Convection (or convective motion) is the tendency of a travelling fluid element to have constant characteristics (like constancy of temperature) in the absence of diffusion.

be explicitly stated, as in a vacuum tube in which it originates from a space charge, or as the electron beam in a cathode-ray tube or from the motion of charged conductor. These currents are derived from the motion of both true and polarization charges contained in the medium. If ρ^c is the density of the moving charge at a given point, in a stationary system of co-ordinates, and v is its velocity, the convection current is

$$\mathbf{j} = \rho^c \mathbf{v}.$$

Polarization current arises from the change due to an applied field of force, of polarization with time. And charges of polarization remain bound to the atom or molecule to which they belong. The charge continuity condition follows from the inclusion of the displacement current (which is not a true current) in the Ampere-Maxell law. While the current through a resister (when a voltage is applied to a resister and a capacitor in parallel, for instance) is a conduction current, the current through the capacitor may be called a *displacement current*. Although current does not flow through the capacitor, the external effect is as though it did, since, as much current flows out of one plate as flows into the opposite plate. The vacuum displacement current does not have the significance of a current in the sense of that from the motion of charge. This current ensures electromagnetic radiation, shows that light is an electromagnetic wave effect, and that electromagnetic waves of all frequencies exist.

Dielectrics are polarized by electric field, and conductors become carriers of electric current by electric field. When the conduction current is greater than the displacement current, the material is a conductor of electricity. When the displacement current is greater than the conduction current, the medium behaves like a dielectric. When the two currents are of the same order of magnitude, as in plasma, in presence of high frequency waves, the material exhibits properties of both dielectric polarization and electrical conductivity.

In MHD processes, currents from transport of momentum and energy of electrically conducting fluids due to motion, are called *advective currents,* Hide and Roberts (1962). In presence of crossed electrostatic and magnetostatic fields E and H, a current is generated in electrical conductors, which is perpendicular to the E field in the plane of E and H. This is the Hall current. Then the current parallel to E is called the *Pederson current,* Pederson (1927).

4.3 Poynting Theorem and Energy Transport Aspects

The rate of work done by the Lorentz force density $\mathbf{F}$ on the charges and currents in a medium is

$$(\mathbf{F}\cdot \mathbf{v}) = \rho^c(\mathbf{E}\cdot \mathbf{v}) + \mathbf{v}\cdot \frac{(\mathbf{j}\times\mathbf{B})}{c}.$$

Putting $\mathbf{j} = \rho^c\mathbf{v}$, this relation reduces to

$$(\mathbf{F}\cdot \mathbf{v}) = (\mathbf{E}\cdot \mathbf{j}). \tag{4.3.1}$$

We can then write

$$(\mathbf{F}\cdot dr) = (\mathbf{E}\cdot \mathbf{j})\, dt = dW. \tag{4.3.2}$$

It is the work done for an infinitesimal change of charge and current configuration in the interval dt, and for the displacement dr by the Lorentz force $\mathbf{F}$; dw is the infinitesimal change in the mechanical energy density w. With the help of the dynamical Maxwell equations, we express the right side of (4.3.1) as a function of **E**, **D**, **B** and **H** only. So, using the values of **j** from (1.4.4) we get

$$(\mathbf{E}\cdot\mathbf{j}) = \frac{c(\mathbf{E}\cdot \operatorname{curl}\mathbf{H})}{4\pi} - \frac{(\mathbf{E}\cdot \partial\mathbf{D}/\partial t)}{4\pi}.$$

Eliminating the first term, with the help of the vector identity (A3.8) for **E** and **H**, and using equation (1.4.3), we obtain

$$\begin{aligned}(\mathbf{E}\cdot\mathbf{j}) &= \frac{c}{4\pi}(\mathbf{H}\cdot\nabla\times\mathbf{E}) - \frac{c}{4\pi}\nabla\cdot(\mathbf{E}\times\mathbf{H}) - \frac{1}{4\pi}\left(\mathbf{E}\cdot\frac{\partial\mathbf{D}}{\partial t}\right)\\ &= -(\nabla\cdot\mathbf{S}^t) - \frac{1}{4\pi}\left(\mathbf{H}\cdot\frac{\partial\mathbf{B}}{\partial t}\right) - \frac{1}{4\pi}\left(\mathbf{E}\cdot\frac{\partial\mathbf{D}}{\partial t}\right),\end{aligned} \tag{4.3.3}$$

where

$$\mathbf{S}^t = \frac{c}{4\pi}(\mathbf{E}\times\mathbf{H}) \tag{4.3.4}$$

$\mathbf{S}^t$ (the Poynting flux) is the flux of energy of the electromagnetic field. In linear isotropic media, using (1.6.2) and (1.6.3) this relation simplifies to

$$(\mathbf{E}\cdot\mathbf{j}) = -\nabla\cdot\mathbf{S}^t - \frac{\partial}{\partial t}\left(\frac{\varepsilon E^2}{8\pi} + \frac{\mu H^2}{8\pi}\right). \tag{4.3.5}$$

The quantity $(\mathbf{E}\cdot\mathbf{j})$ is an energy source or sink. Electrically conducting matter gains this amount of energy per unit volume per unit time at the cost of energy of the field. Since, the time rate of variation of $(\varepsilon E^2 + \mu H^2)/8\pi$ is a power density, this quantity is the sum of energy densities in electric and magnetic fields. So, writing

$$\mathcal{W}_E = \frac{\varepsilon E^2}{8\pi}, \quad \mathcal{W}_H = \frac{\mu H^2}{8\pi}, \quad (\mathbf{E}\cdot\mathbf{j}) = \frac{\partial \mathcal{W}}{\partial t}, \tag{4.3.6}$$

where w is the energy density lost through dissipation, $\mathcal{W}_E$ and $\mathcal{W}_H$ are the energy densities of electric and magnetic forms, we can express (4.3.5) as

$$\nabla\cdot\mathbf{S}^t + \frac{\partial}{\partial t}(\mathcal{W} + \mathcal{W}_E + \mathcal{W}_H) = 0, \tag{4.3.7}$$

which resembles the charge conservation relation (1.4.5). Now, integrating (4.3.5) over a volume V (Appendix 2), we obtain the integral relation

$$-\frac{\partial}{\partial t}\iiint_V (\mathcal{W}_E + \mathcal{W}_H)\, dv = \iint_\Sigma S_n^t\, d\Sigma + \iiint_V (\mathbf{E}\cdot\mathbf{j})\, dv \tag{4.3.8}$$

for the work-energy theorem of electrodynamics, known as the *Poyntings theorem* (The time rate of loss of energy in electric and magnetic forms, inside a body, equals the flow of electromagnetic energy through its surface, plus the rate at which the field does work on the matter of the body), where S_n^t is the normal component of the electromagnetic energy flux density, S^t, at a point on the surface Σ of V. In this form the theorem is valid in linear isotropic dielectric media.

The integrals

$$\frac{1}{4\pi}\iiint_V (\mathbf{E}\times\mathbf{H})\, dv, \quad \frac{1}{4\pi\rho}\iiint_V [\mathbf{r}\times(\mathbf{E}\times\mathbf{H})] dV \tag{4.3.9}$$

represent the linear momentum and the angular momentum, respectively, of the electromagnetic force field. The Poynthing flux S^t is non-zero when an electromagnetic force field is crossed by a uniform magnetic field. But then $(\nabla \cdot \mathbf{S}^t) = 0$, so such fields do not have any influence on the electromagnetic energy balance condition. For conversion of EM energy into mechanical energy in the thermal form (the joule heat), we put $j = \sigma E$, and get the useful relation

$$(E \cdot j) = \sigma E^2 = \frac{1}{\sigma} j^2,$$

where σ is the electrical conductivity of the medium, Lorentz force and other forces influence the dynamics of electrically responsive materials. To include these influence in the theorem of conservation of electromagnetic energies and their fluxes, involvement of conditions of conservation of mass momentum, etc. in necessary, see Chakraborty (1997, Chapter III). Energy and the energy flux aspects have been further discussed in section 5.15.

4.4 Maxwell Stresses

Electromagnetic force, like other forces, generate stress and pressure on matter, for finding which consider the momentum transfer condition of continuum mechanics in the general form

$$F = \frac{dG}{dt} + \hat{X}_j \frac{\partial T_{ij}}{\partial x_i}, \tag{4.4.1}$$

where G must be the momentum of an element in a volume, T_{ij} must be the ijth component of the pressure tensors, $\hat{X}_j$ is the unit vector and X_j is the co-ordinate both in the jth direction, and $\mathbf{F}$ is the applied force per unit volume, including the Lorentz force. We would find G, T_{ij} and $\mathbf{F}$, in terms of the field values, with the help of the Maxwell equations, and replace the expression for the force $\mathbf{F}$ by time rate of change of momentum, and space rate of variation of pressure tensor. Thus, the action-at-distance electrodynamics of a distribution of currents and charges will be shown to lead to the Faraday concept of a continuous distribution of stresses and a momentum density.

To find the Lorentz force density $\rho\mathbf{E} + (\mathbf{j} \times \mathbf{B})/c$ in terms of the continuous field variables $\mathbf{E}, \mathbf{H}, \mathbf{D}, \mathbf{B}$, we replace ρ by $(\nabla \cdot \mathbf{D})/4\pi$ and j by $[(c\nabla \times H) - (\partial \mathbf{D}/\partial t)]$, and obtain

$$\mathbf{F}_0 = \rho\mathbf{E} + \frac{1}{c}(\mathbf{j} \times \mathbf{B}) = \frac{1}{4\pi}\mathbf{E}(\nabla \cdot \mathbf{D}) + \frac{1}{4\pi}((\nabla \times \mathbf{H}) \times \mathbf{B}) - \frac{1}{4\pi}\left(\frac{\partial \mathbf{D}}{\partial t} \times \mathbf{B}\right). \tag{4.4.2}$$

Using vector results (A3.10) we find that

$$((\nabla \times \mathbf{E}) \times \mathbf{E})_x = \frac{\partial}{\partial x}\left(E_x^2 - \frac{1}{2}E^2\right) + \frac{\partial}{\partial y}(E_x E_y) + \frac{\partial}{\partial z}(E_z E_x) - E_x(\nabla \cdot \mathbf{E}),$$

$$((\nabla \times \mathbf{H}) \times \mathbf{H})_x = \frac{\partial}{\partial x}\left(H_x^2 - \frac{1}{2}H^2\right) + \frac{\partial}{\partial y}(H_x H_y) + \frac{\partial}{\partial z}(H_z H_x), \tag{4.4.3}$$

where $(\nabla \cdot \mathbf{H}) = 0$ in the second identity. The simplest state relations $\mathbf{D} = \varepsilon\mathbf{E}$, $\mathbf{B} = \mu\mathbf{H}$ make

$$F_{0x} = \frac{\partial}{\partial x}\left(\frac{1}{4\pi}(\varepsilon E_x^2 + \mu H_x^2) - \frac{1}{8\pi}(\varepsilon E^2 + \mu H^2)\right) + \frac{\partial}{\partial y}\left(\frac{1}{4\pi}(\varepsilon E_x E_y + \mu H_x H_y)\right)$$
$$+ \frac{\partial}{\partial z}\left(\frac{1}{4\pi}(\varepsilon E_z E_x + \mu H_z H_x)\right) - \frac{c}{4\pi}((\nabla \times E) \times E)_x$$
$$- \frac{\varepsilon\mu}{4\pi c}\left(\frac{\partial E}{\partial t} \times H\right)_x . \tag{4.4.4}$$

Replacing $c\boldsymbol{\nabla} \times \mathbf{E}$ by $-\mu\partial\mathbf{H}/\partial t$, we can write

$$F_{0x} = \frac{\partial}{\partial x}T_{xx} + \frac{\partial}{\partial y}T_{xy} + \frac{\partial}{\partial z}T_{xz} + \frac{\varepsilon\mu}{4\pi c}\left(\frac{\partial H}{\partial t} \times E\right)_x - \frac{\varepsilon\mu}{4\pi c}\left(\frac{\partial E}{\partial t} \times H\right)_x , \tag{4.4.5}$$

where

$$T_{xx} = \frac{1}{4\pi}(\varepsilon E_x^2 + \mu H_x^2) - \frac{1}{8\pi}(\varepsilon E^2 + \mu H^2),$$
$$T_{xy} = \frac{1}{4\pi}(\varepsilon E_x E_y + \mu H_x H_y),$$
$$T_{xz} = \frac{1}{4\pi}(\varepsilon E_x E_z + \mu H_x H_z).$$

Hence writing

$$\mathbf{g} = \frac{1}{4\pi}(\mathbf{E} \times \mathbf{B}). \tag{4.4.6}$$

$$F_{0i} = \frac{\partial}{\partial t}g_i + \frac{\partial}{\partial x_j}T_{ij}. \tag{4.4.7}$$

We find that

$$T_{ij} = \begin{bmatrix} \frac{1}{4\pi}(\varepsilon E_x^2 + \mu H_x^2) - \frac{1}{8\pi}(\varepsilon E^2 + \mu H^2) & \frac{1}{4\pi}(\varepsilon E_x E_y + \mu H_x H_y) & \frac{1}{4\pi}(\varepsilon E_x E_z + \mu H_x H_z) \\ \frac{1}{4\pi}(\varepsilon E_x E_y + \mu H_x H_y) & \frac{1}{4\pi}(\varepsilon E_y^2 + \mu H_y^2) - \frac{1}{8\pi}(\varepsilon E^2 + \mu H^2) & \frac{1}{4\pi}(\varepsilon E_y E_z + \mu H_y H_z) \\ \frac{1}{4\pi}(\varepsilon E_x E_z + \mu H_x H_z) & \frac{1}{4\pi}(\varepsilon E_y E_z + \mu H_y H_z) & \frac{1}{4\pi}(\varepsilon E_z^2 + \mu H_z^2) - \frac{1}{8\pi}(\varepsilon E^2 + \mu H^2) \end{bmatrix}, \tag{4.4.8}$$

where $\mathbf{g}$ is the electromagnetic momentum per unit volume, and T_{ij} is the ijth component of the Maxwell stress tensor. A compact version of the relation (4.4.8) is

$$T_{ij} = \frac{1}{4\pi}(\varepsilon E_i E_j + \mu H_i H_j) - \frac{1}{8\pi}(\varepsilon E^2 + \mu H^2)\delta_{ij}, \tag{4.4.9}$$

where the Kronecker delta

$$\delta_{ij} = \begin{cases} 1 \text{ if } i = j \\ 0 \text{ if } i \neq j \end{cases} \tag{4.4.10}$$

This stress is symmetric. Integration of

$$F^0 = \hat{x}_i \frac{\partial T_{ij}}{\partial x_j}, \tag{4.4.11}$$

over the volume V of an electrically responding body, with the help of Gauss's theorem (Appendix 2), gives

$$\hat{x}_i \iiint\limits_V \frac{\partial}{\partial x_j} T_{ij}\, dv = \hat{x}_i \iint\limits_\Sigma T_{ij} d\Sigma_j, \tag{4.4.12}$$

where $d\Sigma_j$ is the projection of the surface element vector $d\Sigma$ on the plane whose normal is parallel to the jth co-ordinate direction. Hence, charges and currents inside a material create electromagnetic stress on its boundary surface, and T_{ij} is the stress along the ith space direction, on a surface whose normal, at a point of the surface Σ is along the jth space direction.

In the 3 + 1 space-time continuum notation of Minkowski (section 2.13), equation (4.4.1) becomes

$$F_\mu = \frac{\partial}{\partial x_4} G_\mu + \frac{\partial}{\partial x_i} T_{i\mu}, \quad \mu = 1, 2, 3, 4; i = 1, 2, 3,$$

where $x_4 = ict, T_{\mu 4} = T_{4\mu} = 0$, and T_{ij} is given in (4.4.8). Here G_μ is the 4-momentum with $G_4 = 0$ and $T_{\mu\gamma}$ represents the stress elements.

Further joint discussion of the Maxwell stress elements of electric and magnetic fields is not possible because forces of purely electric origin and electromagnetic forces are not similar. So, we would now treat the magnetic stress elements in some detail because of their importance in physics, and then briefly consider the electrostatic stress elements, because their further analysis requires the dynamics of time varying fields, which means the involvement of the whole of electrodynamics. This is not necessary.

The diagonal elements ($i = j = 1, 2, 3$) and the non-diagonal elements ($i \neq j$) of the stress tensor T_{ij} of (4.4.9) are

$$\begin{aligned} T_{\|} &= \frac{1}{4\pi}(\varepsilon E_1^2 + \mu H_1^2) - \frac{1}{8\pi}(\varepsilon E^2 + \mu H^2) \\ &= \frac{1}{8\pi}(\varepsilon E_1^2 + \mu H_1^2) - \frac{1}{8\pi}\left\{(\varepsilon E_2^2 + \mu H_2^2) + (\varepsilon E_3^2 + \mu H_3^2)\right\}. \end{aligned}$$

$$T_{22} = \cdots, \quad T_{33} = \cdots. \tag{4.4.13}$$

$$T_{ij} = \frac{1}{4\pi}(\varepsilon E_i E_j + \mu H_i H_j), \quad i \neq j \tag{4.4.14}$$

T_{ii} is the normal stress acting at a point P, on the surface Σ of a body, parallel to the ith co-ordinates axis, on the co-ordinate plane whose normal is the ith co-ordinate axis. T_{ij} $(i \neq j)$ is the tangential (grazing) stress, acting parallel to the ith co-ordinate direction, on the plane whose normal is along the jth co-ordinate direction. If T_{ii} is negative, then it is a pressure force, if positive, it is a force of tension. Positive, or negative, T_{ij}, for $i \neq j$, is a tension only; only the direction of action is reversed by the change of sign. The outward drawn normal to the surface Σ at P, with respect to the body of volume V, is positive. So, the normal stress is a tension if it is directed outward at P, with respect to the closed surface Σ. If this stress is inward, it is a pressure force.

The matrix of the stress relation (4.4.8), for convenience of interpretation, is resolved into two matrices,

$$T_{ij} = \begin{bmatrix} T^0_{xx} & T^0_{xy} & T^0_{xz} \\ T^0_{yx} & T^0_{yy} & T^0_{yz} \\ T^0_{zx} & T^0_{zy} & T^0_{zz} \end{bmatrix} - p^0 \begin{bmatrix} 1 & 0 & 0 \\ 0 & 1 & 0 \\ 0 & 0 & 1 \end{bmatrix} \tag{4.4.15}$$

with

$$T^0_{ij} = \frac{1}{4\pi}(\varepsilon E_{ij} + \mu H_i H_j), \quad p^0 = \frac{1}{8\pi}(\varepsilon E^2 + \mu H^2). \tag{4.4.16}$$

Therefore, the fluid scalar pressure p^0 is supplemented by the stress elements T_{ij}, so that, the force expression of (4.4.11) reads

$$F^0 = \hat{x}_i \frac{\partial T^0_{ij}}{\partial x_j} - \nabla p^0. \tag{4.4.17}$$

The sign of T^0_{ii} indicates whether the normal stress components are like pressure, or tension, where $T_{ii} = -p^0 + T^0_{ii}$. When $T^0_{\|} = T^0_{22} = T^0_{33} = T$, the field pressure is scalar, otherwise it is anisotropic.

The pressure of electromagnetic radiation is a small quantity; its measurement is difficult experimentally. For sun, it is 4.5×10^{-5} dynes/cm^2. For the whole earth, the force of this pressure is 5.7×10^7 k gm wt. Compared to the gravitational attraction, which is 3.6×10^{21} k gm wt. per unit volume, the earth pressure force of EM radiation is very small.

The radiation pressure is a surface force, and gravity is a volume force; their ratio becomes greater if the volume density of the body is smaller. For this reason, a comet is deflected by solar radiation, when it comes near the sun.

4.4.1 Magnetostatic stress elements

The magnetostatic field induced force F^0_M of the Maxwell stress components, is the relation (4) of Exercise 4.4.2,

$$F^0_M = \frac{\mathbf{H}}{H^2}(\mathbf{H}\cdot\nabla)T^0 - \nabla p^0 = \frac{\mathbf{H}}{H^2}(\mathbf{H}\cdot\nabla)\left(\frac{\mu H^2}{4\pi}\right) - \nabla\left(\frac{\mu H^2}{8\pi}\right). \tag{4.4.18}$$

Thus, a magnetostatic field $\mathbf{H}$ generates a scalar pressure $p^0 (= \mu H^2/8\pi$ per cm^2) and a tension $T (= \mu H^2/4\pi)$ parallel to the direction of H increasing. A closed magnetic line of force is analogous to an endless elastic string (or spring), having tension $\mu H^2/4\pi$ and lateral pressure $\mu H^2/8\pi$. Owing to the lateral pressure, magnetic lines do not collapse like elastic strings on release from a stretched state. For $H = 5000$ gauss, $H^2/8\pi = 10^6$ dynes/cm^2 = one atmosphere. So, the stiffness (loss of mobility) of a body and the body pressure on it increase, and transparency is lost when it is immersed in a magnetostatic field. This is the "frozen-in effect", or the "glue effect", or the "fossilization effect". H. Alfven first noted around 1940 that the pressure of the sunspot magnetic field (of the order of 2000 to 3000 gauss or even more), considerably decreases the transparency of the sunspot region.

We find that

$$H_{\parallel} = H, \quad H_{\perp} = 0, \quad p_{\parallel} = p^0 - T_{\parallel} = -\frac{\mu H^2}{8\pi}, \quad p_{\perp_1} = p_{\perp_2} = p^0 - 0 = -\frac{\mu H^2}{8\pi}, \tag{4.4.19}$$

where $p_{\parallel}$ is parallel to H, $p_{\perp_1}$, and $p_{\perp_2}$ are along two orthogonal directions relative to H (for instance, $\perp_1$ may be along the principal normal and $\perp_2$ along the binormal at a point in a curved line of force). So, this induced pressure is anisotropic, the parallel component is a tension, the perpendicular components are pressure-like stress elements. The anisotropic compressibility of magnetized plasma is studied by the double-adiabatic theory[3], also known as the CGL (*Chew-Goldberger-Low*) approximation.

Shape of the boundary of a body is affected by a magnetostatic field. Concavity is reduced by the normal pressure and the tangential tension along the field lines. Convexity is increased by the normal field pressure, but decreased by the tangential field tension, see figures 4.4.1(a) and (b). A large quasi-static magnetic field is accompanied by waves and radiations, which are fast processes; so these effects are treated by other techniques.

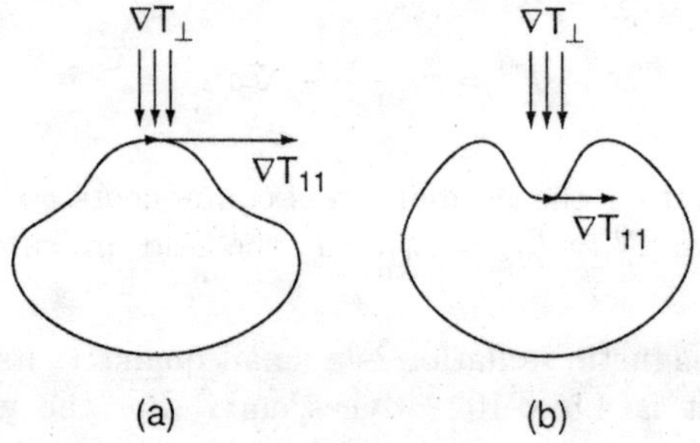

Fig. 4.4.1 Magnetic pressure on the surface of a body: (a) On a concave portion of a surface; (b) On a convex portion.

4.4.2 Electrostatic stress elements

The Maxwell stress elements of the electrostatic field generate electrostatic oscillation due to macroscopic space charge separation which is guided by the relation $\mathbf{\nabla} \cdot \mathbf{D} = 4\pi\rho$ in gaseous media and due to vibrations in dielectrics. The analogous situation is absent in magnetism, because the magnetic induction vector is solenoidal, $\mathbf{\nabla} \cdot \mathbf{B} = 0$. Anisotropic crystals are studied in optics, with the help of the anisotropic linear constitutive relations $D_i = \varepsilon_{ij} E_j$ for $(i, j) = (1, 2, 3)$, and the solenoidal conditions $\mathbf{\nabla} \cdot \mathbf{D} = 0, \mathbf{\nabla} \cdot \mathbf{E} = 0$.

► **Example 4.4.1** *Explain the Maxwell stress distribution on the sides of a rectangular box.*

[3]In adiabatic processes (1) The regions of compression and rarefaction of a gas do not have time to exchange energy during propagation of the acoustic wave; (2) Compression is so slow that the reactions are reversible; (3) As wave length of the wave is increased, the adiabaticity condition is better satisfied; (4) The particles must hit the boundary walls several times; (5) But the compression also must not be so slow that the particles find time to scatter in different directions.

This double adiabatic theory (also called the *intermediate adiabatic theory*) has two different equations of state for the transverse and longitudinal (parallel) pressure components, because collisions are not effective enough to keep pressure isotropic; but collisions are sufficient for neglecting the heat flow. So, the term adiabaticity has different implications in the adiabatic and double-adiabatic theories.

When the collision frequency is very large, the gas remains isotropic, and the pressure is a scalar. In the collisionless limit the ion gyration radius and the Debye shielding length are smaller than the characteristic lengths of the problem, and the magnetic moment of particle motion is an adiabatic invariant.

Solution : The rectangular box PQDEFLMN of figure 4.4.2 has sides PD, PE, PF parallel to the co-ordinate axes OX, OY, OZ, at the point $P(x, y, z)$. The stress component T_{xx} is a normal stress, acting on the plane PELF, parallel to OX. The component T_{xy} is a tangential stress, parallel to OX, in the plane PDNE, T_{xz} is the tangential stress, acting parallel to OX, in the plane PFMD. The other six components of T are either tangential or normal stresses, acting on the planes through P and Q. The diagonal elements T_{xx}, T_{yy}, T_{zz} are normal stresses, and the non-diagonal elements are the tangential stresses.

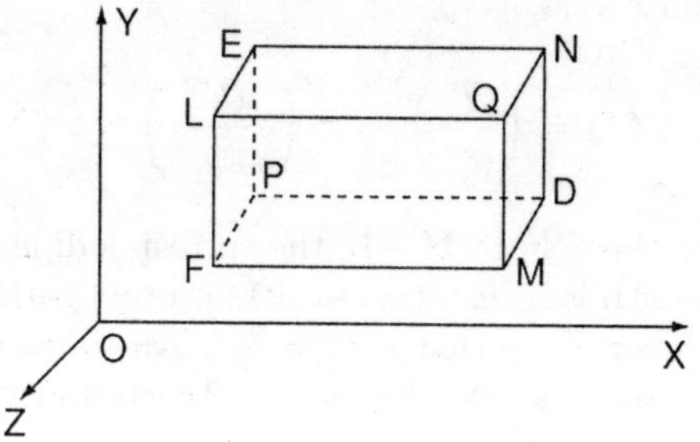

Fig. 4.4.2 The Maxwell stresses on the surface of a box in a medium.

▶ **Example 4.4.2** *It determines the magnetic Maxwell stresses for an unidirectional magnetic field.*

Solution : The Maxwell stress matrix for a static, strong, unidirectional (along the x-direction) applied magnetic field, is

$$T = \begin{bmatrix} \frac{\mu H^2}{4\pi} - \frac{\mu H^2}{8\pi} & 0 & 0 \\ 0 & -\frac{\mu H^2}{8\pi} & 0 \\ 0 & 0 & -\frac{\mu H^2}{8\pi} \end{bmatrix} = \begin{bmatrix} T^0 & 0 & 0 \\ 0 & 0 & 0 \\ 0 & 0 & 0 \end{bmatrix} - p^0 \begin{bmatrix} 1 & 0 & 0 \\ 0 & 1 & 0 \\ 0 & 0 & 1 \end{bmatrix}, \tag{1}$$

where $T^0 = \mu H^2/4\pi$ and $p^0 = \mu H^2/8\pi$. Hence the magnetic stress force is

$$\begin{aligned} F_M^0 &= \left[\frac{\partial}{\partial x}\left(\frac{\mu H^2}{4\pi}\right) - \frac{\partial}{\partial x}\left(\frac{\mu H^2}{8\pi}\right), \quad -\frac{\partial}{\partial y}\left(\frac{\mu H^2}{8\pi}\right), \quad -\frac{\partial}{\partial z}\left(\frac{\mu H^2}{8\pi}\right)\right] \\ &= \left[\frac{\partial}{\partial x}\left(\frac{\mu H^2}{4\pi}\right), 0, 0\right] - \left(\frac{\partial}{\partial x}, \frac{\partial}{\partial y}, \frac{\partial}{\partial z}\right)\left(\frac{\mu H^2}{8\pi}\right). \end{aligned} \tag{2}$$

In the first term we replace the parallel differential operator $\partial/\partial x$ by $\partial/\partial r_\parallel$ and obtain

$$\hat{x}_0 \frac{\partial}{\partial x} = \frac{\partial}{\partial r_\parallel} = \frac{1}{H^2}\mathbf{H}(\mathbf{H}\cdot\nabla), \quad \hat{x}_\parallel = \frac{\mathbf{H}}{H}. \tag{3}$$

Then the general vector relation, in place of (2), is

$$\mathbf{F}_H^0 = \frac{\mathbf{H}}{H^2}(\mathbf{H}\cdot\nabla)\left(\frac{\mu H^2}{4\pi}\right) - \nabla\left(\frac{\mu H^2}{8\pi}\right). \tag{4}$$

► **Example 4.4.3** *Prove the existence of Alfven waves in MHD media from analogy with transverse waves of disturbance in an elastic string, and hence find its phase velocity.*

Solution : In elastic strings, dispersive transverse oscillations propagate at the phase velocity $(T/\rho^m)^{1/2}$, where T is the tension per unit length and ρ^m is the mass per unit volume of the medium. The tension $T(=\mu H^2/4\pi)$, of the magnetic lines of force, generates, similarly, in MHD media, Alfven waves, which are dispersive transverse oscillations, propagating at the Alfven velocity C_A (wave phase velocity),

$$C_A = \left(\frac{T}{\rho^m}\right)^{1/2} = \left(\frac{\mu H^2}{4\pi\rho^m}\right)^{1/2} . \tag{1}$$

The disturbance perturbs the field to $\mathbf{H}+\mathbf{h}$, the corresponding current increment is C curl $\mathbf{h}/4\pi$, and the wave solution for $\mathbf{h}$ is $\mathbf{h}_0$ exp $(ikz-i\omega t)$ for propagation parallel to the z direction at frequency ω and wave number K, so that $w/K = C_A$. Alfven waves are a consequence of the "frozen-in" effect, where the magnetic field provides the elasticity, and the MHD continuum provides the inertia.

► **Example 4.4.4** *Find the radiation pressure of a plane polarized EM wave which is incident normally from vacuum on a surface when* (a) *it absorbs, completely, the wave;* (b) *reflects, completely, the wave.*

Solution : The electric field of the light wave, $E = (E_0 \cos\theta, 0, 0)$ with $\theta = KZ - \omega t$, gives $H = (0, E_0 \cos\theta, 0)$, because $KC = w$ in vacuum,

$$T_{xx} = \frac{1}{4\pi}E_x^2 - \frac{1}{8\pi}(E_x^2 + H_y^2) = \frac{1}{8\pi}(E_x^2 - H_y^2) = 0,$$

$$T_{yy} = \frac{1}{4\pi}H_y^2 - \frac{1}{8\pi}(E_x^2 + H_y^2) = 0,$$

$$T_{zz} = 0 - \frac{1}{8\pi}(E_x^2 + H_y^2) = -\frac{1}{4\pi}E_0^2\cos^2\theta, \quad T_{ij} = 0 \text{ when } i \neq j.$$

The negative sign in T_{zz} indicates that the surface force is a pressure; the average of T_{zz} over the time period $T(=2\pi/\omega)$,

$$\langle T_{zz}\rangle = \frac{1}{T}\int_0^T T_{zz}\,dt = -\frac{E_0^2}{8\pi}$$

is the electric energy density. Per unit area of the beam, the energy

$$\xi = \frac{\langle E^2 + H^2\rangle}{8\pi} = \frac{\langle E_x^2 + H_y^2\rangle}{8\pi} = \frac{E_0^2}{4\pi}.$$

Hence, the intensity of light [the average (over the time period T) amount of energy transported across unit area in unit time] $= \xi c = cE_0^2/8\pi$.

(a) The radiation pressure, on a perfectly absorbing surface, equals the intensity of light divided by c (its velocity of propagation in vacuum).

(b) The opposite of a perfectly absorbing surface is a perfectly reflecting surface. For a perfectly reflecting surface the intensity of the radiation pressure is double, because the intensity of light is twice as great as in a perfectly absorbing surface, for normal incidence.

▶ **Example 4.4.5** *Find the radiation pressure for oblique incidence of light from vacuum on a perfectly absorbing surface.*

Solution : For oblique incidence the force on the surface element $d\Sigma$ is $pd\Sigma\cos\psi$, where p is the pressure on $d\Sigma$ and ψ is the angle between the direction of incidence and the direction normal to $d\Sigma$. Along the normal to $d\Sigma$ the component of this force $= pd\Sigma\cos^2\psi$. Hence, on $d\Sigma$ the pressure per unit area $= p\cos^2\psi$. For isotropic radiation[4], the pressure on $d\Sigma$ is the average value of $p\cos^2\psi$ for-all directions. On the curved surface of a sphere of unit radius, this average

$$= \frac{1}{2\pi}\int_{\phi=0}^{2\pi}\int_{\psi=0}^{\pi} p\cos^2\psi\sin\theta d\psi d\phi = \frac{1}{3}\frac{E_0^2}{4\pi} = \frac{2}{3}\xi$$

because $p = E_0^2/8\pi$ and also $\xi = \langle E^2 + H^2\rangle/8\pi = E_0^2/8\pi$, ξ being the average energy density.

▶ **Example 4.4.6** *Establish the double-adiabatic invariance relations* $(p_{\|}B^2)/\rho^m =$ *constant,* $p_{\perp}/(\rho^m B) =$ *constant for the parallel and the perpendicular components of the pressure tensor* p, *where* ρ^m *is the mass density.*

Solution : We can write, in a local cartesian co-ordinate system

$$p_{ij} = p_{\|}l_i l_j + p_{\perp}(\delta_{ij} - l_i l_j), \tag{1}$$

$$\varepsilon = \varepsilon_{\|} + \varepsilon_{\perp}, \quad \varepsilon_{\|} = p_{\|}/2\rho^m, \quad \varepsilon_{\perp} = p_{\perp}/\rho^m, \tag{2}$$

where l_i, l_j, l_k are the direction cosines of B with respect to these axes, ρ^m is the mass density and ε is the internal energy per unit mass. When collisions are not frequent, $\varepsilon_{\|}$ and $\varepsilon_{\perp}$ are not interdependent. By the first law of thermodynamics,[5]

$$d\varepsilon = -pd\lambda, \tag{3}$$

where $\lambda(= 1/\rho^m)$ is the specific volume of the fluid (volume per unit mass). This law is separately valid for the longitudinal and transverse components of p. So,

$$d\varepsilon_{\|} = -p_{\|}\, d\lambda_{\|}, \quad d\varepsilon_{\perp} = -p_{\perp}d\lambda_{\perp}, \tag{4}$$

---

[4]The radiation is isotropic when it is propagated in all directions such that the probability of the direction of propagation of any ray, selected at random, existing within a given solid angle, is proportional to the solid angle.

[5]It relates the energy of work with the energy of heat, and, is essentially, the law of conservation of energy. One statement of this law is : "The amount of heat produced by work is proportional to the work done"; or, conversely, "If heat is transformed into work, then heat lost by a system is proportional to the work done by it". The constancy (universal) of the ratio of these two quantities is the essence of the first law of thermodynamics. Helmholtz's general definition of this law is "The different forms of energy system are interchangeable. However, when a quantity of one kind of energy disappears, an equivalent amount, of another kind, makes its appearance. Hence, energy cannot be annihilated, nor can energy be created from nothing." According to Clasius, varied changes and transformations might occur, but the total energy of the universe is constant.

Conventionally, the work done by a system is positive, and the work done on a system by the surrounding, is negative; also, the increase in the internal energy of a system is positive, and the decrease is negative. The heat gained by a system is positive, and lost is negative.

The change (physical, chemical or mechanical) of a state A of a system, having internal energy ε_A, to a state B is specified by the internal energy ε_0. It absorbs a small quantity of heat, q, say, for doing some work, W say. The energy before the change is $\varepsilon_n + q$, and that after the change to the state B is $\varepsilon_B + W$. The principle of conservation of energy, $\varepsilon_A + q = \varepsilon_B + W$, can be written as $q = \varepsilon + W$ with $\varepsilon = \varepsilon_B - \varepsilon A$. For infinitesimal such changes, quantitatively, the first law of thermodynamics is $\delta q = \delta\varepsilon + \delta W$.

and the second and third relations of (2), give

$$d\varepsilon_{\|} = -p_{\|}\, d\lambda_{\|} = \frac{1}{2} d(p_{\|}\lambda), \quad d\varepsilon_{\perp} = d(p_{\perp}\lambda), \tag{5}$$

$$\lambda dp_{\|} = -p_{\|}\, d\lambda - 2p_{\|}\, d\lambda_{\|}. \tag{6}$$

The length element l parallel to $\mathbf{B}$, and the element of area Σ perpendicular to B give

$$\lambda = \frac{1}{\rho} = \Sigma l \tag{7}$$

and, due to a small displacement ξ, the change is

$$dl = [\hat{x}\cdot \xi(r + R)] - [\hat{x}\cdot \xi(r)] \approx (x\cdot \nabla)(\xi\cdot R), \tag{8}$$

where

$$\hat{x} = B/B, \quad R = l\hat{x}. \tag{9}$$

for total change in volume, the condition of continuity $(d\rho^m/dt) + \rho^m(\nabla\cdot v) = 0$ gives

$$d\lambda = -\frac{dl}{\rho^2} = \lambda(\nabla\cdot v)\, dt = \lambda(\nabla\cdot \xi), \tag{10}$$

where $\xi = vdt$. Then the changes in volume, due to changes $d\lambda_{\|}$ and $d\lambda_{\perp}$ are

$$d\lambda_{\|} = \Sigma dl = \Sigma(\hat{x}\cdot \nabla)(v\cdot \hat{x})ldt = \lambda(\hat{x}\cdot \nabla)(v\cdot \hat{x})\, dl \tag{11}$$

$$d\lambda_{\perp} = d\lambda - d\lambda_{\|} = \lambda[(\nabla\cdot v) - (x\cdot \nabla)(v\cdot \hat{x})]dt. \tag{12}$$

Using (10) and (11), equation (6) can be written as

$$\lambda dp_{\|} = -p_{\|}\lambda(\nabla\cdot v)\, dt - 2p_{\|}\lambda(\hat{x}\cdot \nabla)(v\cdot \hat{x})\, dt. \tag{13}$$

So, the differential equation for $p_{\|}$ is

$$\frac{dp_{\|}}{dt} + p_{\|}\, \mathrm{div}\, v = -2p_{\|}(\hat{x}\cdot \nabla)(\hat{x}\cdot v) = -2p_{\|} l_i l_j \frac{\partial v_i}{\partial x_j}. \tag{14}$$

Similarly proceeding with the second relations of (4) and (5), for finding $p_{\perp}$, we obtain

$$\lambda dp_{\perp} = -p_{\perp}\lambda(\nabla\cdot v)\, dt - p_{\perp}[\lambda(\nabla\cdot v) - \lambda(\hat{x}\cdot \nabla)(v\cdot \hat{x})]dt \tag{15}$$

and the differential equation for $p_{\perp}$ reads

$$\frac{dp_{\perp}}{dt} + 2p_{\perp}(\nabla\cdot v) = p_{\perp}(\hat{x}\cdot \nabla)(\hat{x}\cdot v) = p_{\perp} l_i l_j \frac{\partial v_i}{\partial x_j}. \tag{16}$$

In non-viscous MHD media, the evolution equation for $\mathbf{B}$,

$$\frac{\partial B}{\partial t} = \boldsymbol{\nabla} \times (\mathbf{v} \times \mathbf{B}), \tag{17}$$

can be written as

$$\frac{d\mathbf{B}}{dt} + \mathbf{B}(\boldsymbol{\nabla}\cdot \mathbf{v}) = (\mathbf{B}\cdot \boldsymbol{\nabla})\mathbf{v}. \tag{18}$$

Scalar multiplication of terms of (18) with B/B^2 gives

$$\frac{1}{B} = \frac{dB}{dt} + (\nabla \cdot v) = -l_i l_j \frac{\partial v_i}{\partial x_j} \tag{19}$$

for the convective rate of change of magnitude of $\mathbf{B}$. The adiabatic law

$$\frac{d}{dt}\left(\frac{B^2}{\rho^m} p_\parallel\right) = 0 \tag{20}$$

for the pressure component $p_\parallel$ obtains by eliminating $l_i l_j (\partial v_i / \partial x_j)$ between (14) and (19), and using the mass continuity condition. This elimination between (16) and (19) gives the adiabatic law

$$\frac{d}{dt}\left(\frac{p_\perp}{\rho^m B}\right) = 0 \tag{21}$$

for the perpendicular component $p_\perp$.

► **Example 4.4.7** *Establish the double-adiabatic state relations from the macroscopic averages of the Boltzmann equation in neutral, unmagnetized media.*

Solution : The equation for the second order velocity moment p_{ii} is

$$\frac{d}{dt}\left(\frac{1}{N} p_{ii}\right) + \frac{2}{N} p_{li}\left(\frac{\partial U_l}{\partial x_l}\right) + \frac{2}{N}\frac{\partial}{\partial x_l}(Q_{iil}) = \frac{2eB_k}{Nmc}\varepsilon_{ink} p_{ni} + \frac{1}{N} L_{ii}, \tag{1}$$

where $(i, j, k) = (x, y, z)$ or (1,2,3) in cyclic order, ε_{ijk} is the alternate unit tensor (Appendix 4) and L_{ii} is the contribution from the collision term. It obtains from the macroscopic equation

$$\iiint_V \left[\frac{\partial f}{\partial t} + (\mathbf{v} \cdot \nabla_r) f + (\mathbf{a} \cdot \nabla_v) f\right] \phi(\mathbf{v})\, dv = \iiint_V \left(\frac{\partial f}{\partial t}\right)_{\text{coll}} (\mathbf{v})\, dv\phi$$

of kinetic theory, for the average of the function $\phi(v)$ in the velocity space, by putting in it

$$\phi(v) = (v_i - u_i)(v_j - u_j),$$

$$u = \frac{1}{N}\iiint_V v f dv, \quad N = \iiint_V f dv,$$

$$p_{ij} = m \iiint_V (v_i - u_i)(v_j - u_j) f dv,$$

$$Q_{ijk} = \frac{1}{2} m \iiint_V (v_i - u_i)(v_j - u_j) f dv.$$

It follows from the Boltzmann equation when all its terms are multiplied by the velocity function $\phi(v)$ and integrated in the velocity space. The Boltzmann distribution function in the phase space of position co-ordinates and velocity co-ordinates, is $f(r, v, t)$, and a is the Lorentz force per unit mass. Assuming that this distribution function is symmetric about the z-axis, we neglect the influence of the heat flow tensor Q_{ijk}, the Lorentz force, and collisional dissipation of energy, and write

$$p_{xx} = p_{yy} = p_{\perp}, \quad p_{zz} = p_{\parallel}, \quad p_{ij} = 0 \text{ if } i \neq j, \quad \frac{\partial}{\partial x} = 0, \quad \frac{\partial}{\partial y} = 0. \tag{2}$$

Then, from (1) we obtain the two equations

$$\frac{d}{dt}\left(\frac{1}{N}p_{\parallel}\right) + \frac{2}{N}p_{\parallel}\frac{du_{\parallel}}{dz} = 0, \tag{3}$$

$$\frac{d}{dt}\left(\frac{p_{\perp}}{N}\right) = 0. \tag{4}$$

Eliminating $du_{\parallel}/dz$ in (3) by the continuity condition, we integrate these relations, and obtain the adiabatic laws

$$(p_{\parallel}/N^3) = \text{ constant.} \tag{5}$$

$$(p_{\perp}/N) = \text{ constant.} \tag{6}$$

Another form of this double adiabatic law is

$$\frac{T_{\parallel}}{N^2} = \text{constant}, \quad T_{\perp} = \text{constant}, \tag{7}$$

where the kinetic temperatures $T_{\parallel}$ and $T_{\perp}$ are defined by $p_{\parallel} = NK_{13}T_{\parallel}$, $p_{\perp} = NK_BT_{\perp}$, K_B being the Boltzmann constant.

► **Example 4.4.8** *Show that for rotating magnetized structures, the material along a magnetic field line rotates at a constant rate, and that, surfaces of constant magnetic field values are also surfaces of constant mass density, and vice-versa (Ferraro's law of isorotation).*

Solution : For rotation about the z-axis, all the layers at different axial distances will have different angular velocity Ω, so this is a non-rigid rotation (also called *differential rotation,* or *non-uniform rotation*), because Ω is a function of r and Z. In the steady state, the velocity components are $(0, r\Omega, 0)$ at (r, θ, z), and the MHD equation

$$\begin{aligned}\nabla \times (\mathbf{v} \times \mathbf{B}) &= \left[0, \frac{\partial}{\partial z}(r\Omega B_z) + \frac{\partial}{\partial r}(r\Omega B_r), 0\right] \\ &= \left[0, r\Omega\left(B_r + \frac{\partial B_z}{\partial z}\right) + r\left(B_r\frac{\partial \Omega}{\partial r} + B_z\frac{\partial \Omega}{\partial z}\right), 0\right] \\ &= [0, r\Omega(\nabla \cdot B) + r(B \cdot \nabla)\Omega, 0] = 0 \end{aligned} \tag{1}$$

is valid in infinitely conducting media, since $v \times B = (r\Omega B_z, 0, -r\Omega B_r)$. Hence

$$(B \cdot \nabla)\Omega = 0 \tag{2}$$

and Ω is a constant along a line of force of $\mathbf{B}$; so, Ω is a constant on a surface of constant B. The period of rotation of a magnetized revolving star is different at equator from that of other latitudes. This theorem of isorotation is applicable only for electrically highly conducting materials, like dense cosmic bodies, in which the magnetic lines are frozen properly.

4.5 Debye Potential Field

Due to random thermal motion, the region surrounding an ion is sometimes excess in ions and sometimes in electrons. In an average time the electron number, per unit volume at a point, slightly exceeds the ion number because of the Coulomb action and smallestness of the electronic mass. So, every charge is surrounded by an atmosphere of opposite charges, and the difference between electron and ion charge densities decreases rapidly with increasing distance from an ion charge. The thickness of this screening atmosphere is the Debye shielding distance or the Debye-Huckel radius. The charge neutrality is maintained over distances large compared to this screening distance. Weak random thermal motion cannot overcome the electrostatic attraction between opposite charges within the Debye sphere. This allows only the dipole polarization of dielectrics to occur. At high temperatures ($> 10^6\,^0K$) the Debye shielding distance is modified by quantum effects.

The continuous potential solution, for the screening effect, is determined with the help of the distribution function

$$f(r,v) = N\left(\frac{m}{2\pi K_B T}\right)^{3/2} \exp\left[-\frac{e\phi}{K_B T} - \frac{mv^2}{2K_B T}\right],$$

which is the steady state solution of the Boltzmann equation, where m is the electronic mass, T is the kinetic temperature, K_B is the Boltzmann constant, ϕ is the potential of the Coulomb electric field and $-e$ is the electronic charge. The velocity space integration of $f(\mathbf{r},\mathbf{v})$ gives, for electrons, the number density

$$N_e(r) = Ne \times P\left(-\frac{e\phi}{K_B T_e}\right)$$

and for ions the number density

$$N_i(r) = N\exp\left(\frac{e\phi}{K_B T_i}\right).$$

The total charge density surrounding the test charge e at the origin is

$$e\delta(r) + eN\exp\left(\frac{e\phi}{K_B T_i}\right) - eN\exp\left(-\frac{e\phi}{K_B T_e}\right),$$

where $\delta(r)$ is the delta function (Appendix 1).

When the potential energy of the electromagnetic field is much less than the kinetic energy per particle of thermal motion, we write $|e\phi| \ll K_B T$. Then keeping terms up to the first power of ϕ in the Poisson equation

$$\nabla^2\phi = 4\pi e\delta(r) + 4\pi eN\left[\exp\left(\frac{e\phi}{K_B T_i}\right) - \exp\left(-\frac{e\phi}{K_B T_e}\right)\right]$$

we obtain

$$\nabla^2\phi - \frac{\phi}{\lambda^2 D} = 4\pi e\delta(r),$$

where

$$\lambda_D^2 = \frac{K_B T_e T_i}{4\pi N e^2}(T_i + T_e).$$

For radial symmetry ϕ satisfies the equation

$$\frac{1}{r^2}\frac{d}{dr}\left(r^2\frac{d\phi}{dr}\right)=\frac{4\pi Ne^2}{K_B}-\left(\frac{1}{T_e}+\frac{1}{T_i}\right)\phi.$$

The boundary condition $\phi\to 0$ as $r\to\infty$ gives the solution

$$\phi(r)=-\frac{e}{r}\exp\left(-\frac{r}{\lambda_D}\right).$$

Hence the potential due to the charge e is reduced by the factor $\exp(-r/\lambda_D)$ from the value it has in vacuum. Neglecting ion thermal motion we write $N_i(r)=N$ ($=$ constant) and obtain

$$\lambda_D^2=\frac{K_BT}{4\pi Ne^2}.$$

The distance from the charge e, beyond which its electrical potential falls off rapidly, is λ_D; so, the numerical factor is 4 for $T_e > T_i$, and 8 for $T_e\approx T_i$. The number of electrons or ions in the Debye sphere of radius λ_D, is,

$$N_D=\frac{4\pi\lambda_D^3}{3}N.$$

For $N=10^5/\text{cm}^3$ and $T=10$ keV, we get $\lambda_D=7.5\times10^{-3}$ cm. In a fusion plasma, at 10 keV and electron density $10^{16}/\text{cm}^3$, it is found that $\lambda_D=10^{-3}$ cm, and $N_D=1.8\times10^9$ particles of each sort. So, within the Debye sphere, a particle interacts with a large number of particles, of both sign. For classical study, a region must be greater in size than λ_D, and λ_D should be large compared to the average interparticle distance. In equilibrium, the point charge attracts enough charges of opposite sign (and repels charges of the same sign), so that the potential is reduced by the factor $\exp(-r/\lambda_D)$ from its vacuum value at a point.

4.5.1 Negative potential near walls of vacuum discharge chambers

Electrons are lost much faster than ions at the walls of vacuum chambers of finite size and shape in laboratory experiments. To equalize the rate at which electrons and ions strike it, a negative potential develops there. Charges are not reflected at the walls, and the wall surface does not collect any net current. A layer of negative charges is set up there, which drives back most of the electrons into the discharge region, and accelerates the ions towards the surface. The potential of these charges is screened, and attracts the ions, establishing an equilibrium such that the electron flux equals the ion flux and the wall negative potential has a fixed value. The thickness of the resulting space-charge sheath is of the order of the Debye shielding distance and the potential across it is a few multiples of the mean electron energy.

Such sheaths are formed in front of missiles, in trails of jets from rockets and on surfaces of space vehicles at the time of their reentry into the atmosphere. Low frequency signals cannot penetrate these sheaths. The bow shock ahead of a nose cone becomes steep, and is displaced by turning magnets towards it. In diodes a sheath forms close to each electrode. At low gas pressures and very low current densities (as in cold cathode) the electron sheath adjacent to the anode is pronounced and most of the voltage (in the kilo volt range) appears as an anode drop.

► **Example 4.5.1** *Find the expression for the Debye screening distance from energy considerations.*

Solution : Consider two parallel conducting planes, separated along the x-axis, by a distance $2x_0$, and held at the same potential ϕ. The electric field E is everywhere parallel to the x-axis. The uniform number density of electrons, each of charge $-e$ is N. If ions, each of charge e, are removed from the region between the planes, then

$$E = \nabla\phi = \left(\frac{\partial\phi}{\partial x}, 0, 0\right).$$

Integration of Gauss's law

$$\frac{d^2\phi}{dx^2} = -4\pi N\rho$$

twice, and the assumption that $\phi = \phi_0$ and $d\phi/dx = 0$ when $x = 0$, gives

$$\phi - \phi_0 = -2\pi Nex^2.$$

The energy required to transfer a particle from the surface to the mid plane is the change δw of the potential energy w of an electron, across the thickness x_0 of the slab, so

$$\delta w = -e(\phi - \phi_0) = 2\pi Ne^2x_0^2.$$

The kinetic energy of thermal motion, per particle, is $K_BT/2$, in unidirectional motion. Equating δw with $K_BT/2$ we get

$$x_0 = \lambda_D = \left(\frac{K_BT}{4\pi Ne^2}\right)^{1/2} = 6.9\left(\frac{T}{N}\right)^{1/2} \text{ cm},$$

where T is in absolute degree Kelvin, and the distance in which the collective space charge effect is of the order of the random behaviour due to thermal motion is λ_D. So, equating the potential energy gained by a charge in moving through a distance, with the kinetic energy lost per particle in the direction of its motion, gives the expression for the Debye shielding distance.

► **Example 4.5.2** *For a high negative voltage V_0, placed at the edge of a plasma, show that a transient sheath of ions, called the ion wave sheath, is formed of thickness*

$$\lambda_i = \frac{(2eV_0)^{1/2}}{(4\pi N_i^2 e)^{1/2}}.$$

Alexeff and Jones, Phys. Fluids **12**, 345–346 (1969).

Solution : Plane geometry is used because the transmitting electrode is a flat plate. Also, the duration of the transmitter pulse is so short ($\approx$ 0.1 μsec) that the ions can be considered fixed, so that the ion density N_i is a constant. The rise time is so long compared to $1/\omega_{pe}$($\approx$ 0.003 μsec) that the electrons can be considered at rest relative to the leading edge of the sheath. The applied voltage is so much greater than the electron thermal energy that the electrons can be considered as cold. For a narrow negative pulse having a duration shorter than $1/\omega_{pi}$ and moderate voltage, the ions neither have time to move during the pulse, nor do they obtain enough momentum to flow appreciably after the pulse is over.

Then the solution of the Poisson equation is assumed to be

$$V = a + bx + 2\pi N_i e^2 x^2,$$

where x is the position variable; at $x = 0$, $V = V_0$; $V = 0$ at $x = x_0$ (say); and the electrostatic field $E(= -\partial V/\partial x) = 0$ when $x = x_0$. These conditions give

$$x_0 = \lambda_i = \frac{(2eV_0)^{1/2}}{(4\pi N_i e^2)^{1/2}}.$$

Thus the sheath has the same formal structure as the Debye shidding distance λ_D except that the externally induced potential energy V_0 is replaced by the electron kinetic energy $K_BT/2$. For $T_e = 1ev$, $N_e = 10^9/\text{cm}^3$, $\lambda_D = 0.03$ cm, for $V_0 = 250$ V (negative), $\lambda_i = 1$ cm.

► **Example 4.5.3** *What is the role of the Debye shidding distance λ_D in the theory of waves in plasmas?*

Answer : Collective modes exist in plasmas, at the cut-off frequency $\omega_p[= (4\pi N_0 e^2/m)^{1/2}$, where N_0 is the electron concentration per unit volume]. The time of one complete oscillation of an external field at frequency ω is about $1/\omega$. The average time required by the particles having thermal motion to move through the distance λ_D, is of the order of the time elapsed for one complete oscillation at frequency ω_p, because $\omega_p \lambda_D = (K_BT/m)^{1/2}$. If q/ω is greater than $1/\omega_p$, the particles prevent the wave from passing through the plasma. So, a wave at $\omega < \omega_p$ cannot penetrate into a plasma, but at $\omega > \omega_p$ the coherent response of the plasma does not affect its dispersion completely.

4.6 Ionization processes

4.6.1 Mean Free Path Method of Townsend for Electronics

For gas discharge investigations, in the beginning of 20th Century, Townsend developed the mean free path method empirically, which is applicable to steady electron motion in gases at a given pressure p, in presence of an electric field and a transverse magnetic field.

All electrons are assumed to have the same mean free path l in presence of an electric field in the time between two collisions. Diffusion velocity of electrons in the gas depends on the mean distance l travelled by an ensemble of electrons per unit time for their binary collisions with neutral particles. Motion of particles consists of series of short and long periods of motion. Short periods are for times of changes due to inelastic collisions. Long periods are the times of domination of the field influence; then the short range inelastic encounters become negligible.

The relevant spatial diffusion equations are

$$\mathbf{F} = N\mathbf{v} - D\boldsymbol{\nabla}N, \quad (\boldsymbol{\nabla}\cdot\mathbf{F}) = 0, \tag{4.6.1}$$

where $\mathbf{F}$ is the mass flow velocity, $\mathbf{v}$ is the mean free drift velocity in the direction of the electric field $\mathbf{E}$, and $\mathbf{D}$ is the diffusion coefficient. When $\mathbf{E}$ is parallel to the X-axis and constant, these relations give the diffusion equation

$$\nabla^2 N = \frac{v}{D}\frac{\partial N}{\partial x}. \tag{4.6.2}$$

The Townsend relation

$$\frac{v}{D} = \frac{1}{40.3} \frac{E}{K_0} \tag{4.6.3}$$

is useful, where ξ and ξ_0 are the mean kinetic energies of an electron and a molecule, respectively, and $K_0\xi_0 = \xi$. K_0 is the Townsend energy factor. The diffusion equation now is

$$\nabla^2 N = \frac{E}{40.3 K_0} \frac{\partial N}{\partial x}. \tag{4.6.4}$$

Hence the spatial distribution of electron density depends only on E and K_0, and the boundary conditions. Thus K_0 is estimated by measuring the spatial variation of N in presence of the applied field. Electrons also possess some free kinetic energy of thermal origin; the related random thermal velocity is u. Putting $\xi = mu^2/2$, we obtain $u = (2K_0\xi_0/m)^{1/2}$. Moreover, equating the mean rate of gain and loss of momentum due to collisions, gives the relation $v = eE/m\nu$, where ν is the elastic collision frequency. So, the mean free path $l = u/\nu$, and at any given pressure, ν increases with E because

$$\frac{\nu}{D} = \frac{1}{40.3} \frac{E\xi_0}{(mu^2/2)}. \tag{4.6.5}$$

The mean gain in energy per particle of charge e is $e(\mathbf{E}\cdot\mathbf{v})$ and the mean collisional loss of kinetic energy is $\nu(mu^2/2)$. If α is the mean proportion of electron energy lost by collision with a molecule, then $\alpha\nu mu^2/2 = eEv$, so

$$\alpha = \frac{2v^2}{u^2}. \tag{4.6.6}$$

Thus, from observed values of k_0 and v, the values of u, ν, l, α are determined.

A constant and uniform magnetic field H, applied perpendicular to E, deflects the electron stream through angle θ, which is measured to find v. Taking the direction of the resultant force parallel to $\mathbf{E}$, and $(\mathbf{v} \times \mathbf{H})/C$ along the direction of electron beam, by adjusting $\mathbf{H}$, the angle θ of deflection is given by

$$\theta = \frac{vH}{EC} \tag{4.6.7}$$

The gas pressure p is maintained at the order of a few millimeter of mercury and a constant electric field is applied to measure k_0 and v. k_0 is found to be greater than unity and v to depend on the ratio E/p. The collision frequency ν increases by an alternating electric field E. This generalization for alternating electric field explained the Luxembourg effect of radio waves which was later known as the *ionospheric cross modulation effect.* It belongs to the weak wave interaction process in plasma. So, Townsend's mean free path method for alternating electric fields is also applicable to ionospheric cross modulation.

The speciality of this elegant phenomenological method is that, without complicating the mathematics, it skillfully deals with three velocities to study the physics of gas discharge process, namely the bulk diffusion velocity F, the thermal velocity u and the applied electric field driven velocity v. This empirical Townsend relation has the number 40.3, as a coefficient, in (4.6.5).

4.6.2 Non-equilibrium ionization

Non-equilibrium ionization is a *two temperature process* in a *three-element device,* the electron temperature T_e and the gas temperature T_g being the two temperatures the three elements are the electrons, positively charged particles and neutral particles. Though the gaseous medium, as a whole, is relatively cool, the electron temperature T_e is much elevated above the gas temperature T_g due to acceleration of electrons by the applied field. Generally, $T_e - T_g \approx 300^0$ K or more. Even a uniform increase of electron temperature by 20 per cent is useful.

The amount of ionization depends on T_e. Non-equilibrium ionization preferentially heats the electrons of the system without appreciably raising the temperature of the working fluid. Even for a large current flow, only a small fraction of the energy acquired by the electrons, in each collision, is exchanged with the heavy particles, because the electronic mass is much smaller than that of the scatterer. The electrons do not individually acquire from the applied electric field a large amount of energy to become high energy particles for escape and for energy loss of the plasma. In fact, due to high rates of recombination, the electrons are distributed over almost the entire volume of the generator.

Neutral particles and charged particles collide among themselves and with one another. For inelastic collisions the internal energy of the colliding particles is charged in quantas. The internal energy does not change for elastic collisions. If P_i is the sum of all the lost energies in inelastic collisions, we can write

$$P_i = P_{\text{rot}} + P_v + P_r, \tag{4.6.8}$$

where P_{rot} is the power loss due to rotational degrees of freedom of the heavy particles; P_v is the vibrational loss and P_r is the radiation loss. The energy balance relation can be expressed as

$$(E \cdot j) = \frac{j^2}{\sigma} = \sigma E^2 = P_e + P_L, \tag{4.6.9}$$

where P_e is the energy lost by elastic collinsions, P_i is the sum of the inelastic losses, E is the field measured in the co-ordinate system in which the gas is at rest and has a maximum value vB/C, where v is the flow velocity, B is the magnetic field in gauss and c is the vacuum speed of light in cm per sec. Internal dissipation due to the current is j^2/σ per unit volume, where σ is the electrical conductivity of the gas. The power density $(E \cdot j)$, which is produced by the induced emf, is dissipated in the gas and increases the electron temperature, for achieving non-equilibrium ionization. Electrons do not couple to the rotational degrees of freedom, because their angular momentums, at a given collision impact parameter, are very small. Sometimes, σE^2 in (4.6.9) is replaced by $\sigma E^2(1 - k^2)$ because the effective field is less than the induced field by the factor $(1 - K)^{-1}$, where K is the load factor of the generator.

Non-equilibrium ionization is not uniform. The non-uniformity creates internal short circuits due to the inherent non-linearity existing near the electrodes and in generators employing non-atomic gases. In deriving (4.6.9) the losses through diffusion and heat conduction to walls, and by radiation from the gas, have been ignored. Actually, radiation loss is negligible for electron densities at $10^{13}/\text{cm}^3$ and more; so its exclusion is a valid approximation in some cases.

The electron number density N_e varies between 10^{12} and 10^{17} per cm^3. At relatively low energies (less than 0.5 ev) and these high values of the electron density, encounters among electrons themselves are very frequency. So, the electron energy distribution is approximated to the Maxwellian at the temperature T_e, and the electron temperature differs in value with the gas temperature T_g. Encounters between electrons and ions are then negligible.

The energy loss P_e, due to elastic collisions, is given by

$$P_e = \frac{3}{2} K_B (T_e - T_g) N_e \sum_i \left(\frac{2m}{M_i} \right) N_i Q_{ie} C_e, \tag{4.6.10}$$

where K_B is the Boltzmann constant, M_i is the mass per particle and N_i is the number density of the ith species, Q_i is the cross section of elastic energy transfer for the electrons and C_e is the mean thermal velocity of the electrons at the temperature T_e. The collision frequency γ_i of an electron with the particles of the ith species, is

$$\gamma_i = N_i Q_i C_e. \tag{4.6.11}$$

Since the mass of heavy particles $M_i \gg m$, the average energy lost per collision is $2m/M_i$ times the energy difference $3K_B(T_e - T_g)/2$. The energy loss factor $G(\approx \delta(2m/M_i))$ can be introduced to take account of the inelastic losses over the elastic losses. Then $\delta = 1$ for ideal elastic collisions.

The maximum electric field measured in the co-ordinate system in which the gas is at rest being $E + uB/c$, equation (4.6.9) gives

$$\frac{1}{\sigma} j^2 = \sigma \left[E + \frac{1}{c} uB \right]^2 = P_e. \tag{4.6.12}$$

Since the electrical conductivity is

$$\sigma = \frac{Ne^2}{m\nu} \tag{4.6.13}$$

in a one-species plasma (electron plasma), equation (4.6.10) becomes

$$\frac{3}{2} K_B(T_e - T_g) = \frac{1}{2} M_i \left[\frac{e}{m\nu} \left(E + \frac{u}{c} B \right) \right]^2. \tag{4.6.14}$$

If v_e is the ordered velocity acquired by the electron from the applied wave field, then

$$v_e = eE + \left(\frac{ev}{c} \right) B. \tag{4.6.15}$$

$$\frac{3}{2} K_B(T_e - T_g) = \frac{1}{2} M_i v_e^2. \tag{4.6.16}$$

Hence the excess electron energy equals the kinetic energy of the heavy particles in the co-ordinate system in which the electrons are at rest. It can also be written as

$$\frac{3}{2} K_B(T_e - T_g) = \frac{1}{2} M_i \left(\frac{e}{m} \right)^2 \frac{(K_B T_0)^2}{(Q_{ie} v_e)^2} \left(\frac{E'}{p_0} \right)^2, \tag{4.6.17}$$

where

$$E' = E + \frac{\dot{v}B}{c}; \quad T_0 = 273^0\ K \tag{4.6.18}$$

and the reduced pressure

$$p_0 = 273 \frac{p}{T_g}, \tag{4.6.19}$$

where p is the gas pressure and T_g (in degrees Kelvin) is the gas temperature. Equation (4.6.17) expresses kinetically, the electron temperature as a function of E/p_0; which was empirically conceived in the Townsend theory for gas discharge.

4.6.3 Saha ionization theory

Gasses are generally partially ionized and not fully ionized by fields. This is because by balancing many effects thermal equilibrium is difficult to achieve. Actually many factors are responsible for ionization and recombination, Cambel (1963, Ch. 5), the effects of which cannot be balanced by thermal equilibrium.

If I is the ionization potential of an atom and p is the free electron momentum, then the energy ξ needed to remove an electron from the atom, and for giving it to the kinetic energy $p^2/2m$, is given by $\xi = I + (p^2/2m)$.

The ratio of the densities of different states of ionization is given by the Saha ionization equation

$$\frac{N_{i+1}Ne}{N_i} = \frac{g_{i+1}\, g_e}{g_i} \left(\frac{mK_BT}{2\pi\hbar^2}\right)^{3/2} \exp\left(-\frac{I}{K_BT}\right),$$

where $\hbar$ is the Planck constant, N_i is the number density of atoms in the next higher state, N_e is the electron number density; g_e, g_i, g_{i+1} are statistical weights which are assigned to the electrons and to the different states of ionization, I is the ionization energy required to ionize the atom from the ith state to the $(i+1)$th state, m is the electron mass, T is its kinetic temperature and K_B is the Boltzmann constant. The Saha ionization equation is valid only for dilute gases.

The weight factors g_e, g_i, g_{i+1} define the number of different ways that the various configurations can be filled. Since electrons have two states of polarization, we put $g_e = 2$. the factors g_i and g_{i+1} are not so easy to estimate.

4.7 Lagrangian for Motion in presence of Electromagnetic Field

The Lagrangian description is possible of electromagnetic field, interacting with specified charges and currents in a continuous medium. The Lagrangian formalism for continuous fields follows that for a discrete system of n particles. For this, the finite number of co-ordinates, $q_k(t)$ and $\dot{q}_k(t)$ for $k = 1, 2, \cdots, n$, are replaced by an infinite number of degrees of freedom. In the continuous 4-space treatment, generalized co-ordinate q_k is replaced by a continuous field $\phi_i(\mathbf{x})$ for $i = 1, 2, \cdots, n$, where $x \equiv (x_1, x_2, x_3, x_4)$; the generalized velocity $\dot{q}_k$ is replaced by the gradient $\partial\phi_i/\partial x_\mu$ for $\mu = 1, 2, 3, 4$. The position co-ordinates x_μ are independent of the velocity (or momentum) co-ordinates $\dot{x}_\mu$. So, the spatial gradient operator ∇_q does not operate on $\dot{q}$. The particle velocity $u = \dot{q}$, because the position vector of the particle $r = q$. Since x_μ is independent of $\dot{x}_\nu$, the relations

$$(A\cdot\nabla)u = 0 \quad \text{and} \quad \operatorname{curl} u = 0 \tag{4.7.1}$$

are valid, where u is the velocity and A is the vector potential. The Lorentz force density is

$$F = \rho^c\nabla\left(\phi + \frac{1}{c}(u.A)\right) - \frac{\rho^c}{c}\frac{dA}{dt}, \tag{4.7.2}$$

where

$$\frac{d}{dt} = \left(\frac{\partial}{\partial t}\right) + (u\cdot\nabla). \tag{4.7.3}$$

Partial derivatives of A and ϕ, with respect to the components of u vanish because these are independent of u. So, for a single particle of charge q,

$$F = -q\frac{d}{dt}\nabla_u\left\{\frac{1}{c}(u.A)+\phi\right\}+q\left\{\phi+\frac{1}{c}(u.A)\right\}. \tag{4.7.4}$$

The Lagrangian $\mathcal{L}$ of the particle is the difference between the kinetic energy and the potential energy, due to the Lorentz force. The Lagrangian $\mathcal{L}_F$ of a free, non-relativistic, particle is

$$\mathcal{L}_f = -m_0c^2\left(1-\frac{u^2}{2c^2}\right). \tag{4.7.5}$$

For relativistic motion,

$$\mathcal{L}_F = -m_0c^2\left(1-\frac{u^2}{c^2}\right)^{1/2}, \tag{4.7.6}$$

where m_0 is the rest mass. The Euler-Lagrange equations of motion are

$$\frac{d}{dt}\left(\frac{\partial\mathcal{L}_F}{\partial u_i}\right)-\frac{\partial\mathcal{L}_F}{\partial x_i}=F_i \quad \text{for } i=1,2,\cdots, \tag{4.7.7}$$

where F_i is the ith component of the force acting on the particle. In presence of the Lorentz force, a new Lagrangian $\mathcal{L}$ is defined as

$$\mathcal{L} = \mathcal{L}_F + q\phi + \frac{q}{c}(u.A). \tag{4.7.8}$$

The additional terms are $(u.A)/c$ for the electromagnetic field, and ϕ for the electrostatic potential. The Euler-Lagrange equation for $\mathcal{L}$ is

$$\frac{d}{dt}\left(\frac{\partial\mathcal{L}}{\partial u_i}\right)-\frac{\partial\mathcal{L}}{\partial x_i}=0. \tag{4.7.9}$$

The relativistic kinetic energy and momentum for a particle, are

$$\xi_F = -m_0c^2\left(1-\frac{u^2}{c^2}\right)^{1/2}. \tag{4.7.10}$$

$$p=\frac{\partial\mathcal{L}}{\partial u} = -\frac{m_Fu}{\sqrt{1-\dfrac{u^2}{c^2}}}+\frac{q}{c}A = p_F+\frac{q}{c}A, \tag{4.7.11}$$

where

$$p_F = -\frac{m_0u}{\sqrt{1-\dfrac{u^2}{c^2}}}. \tag{4.7.12}$$

Differentiating p totally with respect to time, we get

$$\frac{dp}{dt}=\frac{dp_F}{dt}+\frac{q}{c}\left(\frac{\partial A}{\partial t}+(u.A)\right). \tag{4.7.13}$$

So, p_F is the relativistic ordinary momentum for the free particle, p is the generalized (canonical) momentum;

$$u = \frac{(cp - qA)}{\sqrt{m_0^2c^2 + \left(q - \frac{qA}{c}\right)^2}} = \frac{cp_F}{\sqrt{m_0^2c^2 + p_F^2}}, \tag{4.7.14}$$

$$1 - \frac{u^2}{c^2} = \frac{m_0^2c^2}{m_0^2c^2 + \left(p - \frac{qA}{c}\right)^2}, \tag{4.7.15}$$

$$\frac{m_0c}{\sqrt{1 - \frac{u^2}{c^2}}} = \sqrt{m_0^2c^2 + \left(p - \frac{qA}{c}\right)^2}. \tag{4.7.16}$$

The Hamiltonian $\mathcal{H}$ is a function of the position vector $\mathbf{r}$ and the canonical momentum p, and is a constant of motion if $\mathcal{L}$ is not an explicit function of time. So

$$\mathcal{H} = (p.u) - \mathcal{L} = c\sqrt{m_0^2c^2 + p_F^2} - q\phi = c\sqrt{m_0^2c^2 + (p - qA)^2} - q\nabla\phi, \tag{4.7.17}$$

$$\nabla\mathcal{H} = \frac{e}{\sqrt{m_0^2c^2 + p_F^2}}\left[(p_F\cdot\nabla)\left(-\frac{q}{c}A\right) + \left(p \times \nabla \times \left(-\frac{q}{c}A\right)\right)\right]$$

$$= -\frac{q}{c}(u\cdot\nabla)A - \frac{q}{c}(u \times (\nabla \times A)) - q\nabla\phi \tag{4.7.18}$$

$\mathbf{r}$ and $\mathbf{p}$ are, therefore, the two independent variables of the system. The Hamilton equation

$$\frac{dp_F}{dt} + \frac{q}{c}\left(\frac{\partial A}{\partial t} + (u\cdot\nabla)A\right) = \frac{q}{c}(u.A) + \frac{q}{c}(u \times B) + q\nabla\phi \tag{4.7.19}$$

recovers easily the equation of motion

$$\frac{dp_F}{dt} = q\mathbf{E} + \frac{q}{c}(\mathbf{u} \times \mathbf{A}). \tag{4.7.20}$$

The total Lagrangian, for a system of particles and fields produced by them, is the integral in the relation

$$L = \iiint_V \left[\frac{1}{8\pi}(E^2 - H^2) + \frac{1}{c}(A\cdot j) + \rho\phi\right] dv - \sum_i \left(m_ic^2\sqrt{1 - \frac{u_i^2}{c^2}}\right), \tag{4.7.21}$$

where V is the finite volume of the universe. The Lagrangian density is

$$\mathcal{L}(r,t) = \frac{1}{8\pi}(E^2 - H^2) + \frac{1}{c}(A\cdot j) + \rho\phi \sum_i \left(N_im_ic^2\sqrt{1 - \frac{u_i^2}{c^2}}\right), \tag{4.7.22}$$

where N_i is the uniform number density of the ith species. L contains the kinetic energy, the energy of interaction of the EM field with matter and the energy of the free space EM field.

4.8 Dynamics of Relativistic Electrons

The relativistic equation of motion of an electron of charge $-e$ and rest mass m, in a given electromagnetic field, is

$$\frac{dp}{dt} = -eE - \frac{e}{c}(v \times H) \tag{4.8.1}$$

with $p = mv/(1 - v^2/c^2)^{1/2}$. The Hamiltonian $\mathcal{H}$ then assumes the form

$$\mathcal{H} = \sqrt{m^2c^4 + (pc + eA)^2}. \tag{4.8.2}$$

The Lagrangian of fields and motion, and the energy equation, respectively, are

$$\mathcal{L} = -mc^2\sqrt{\left(1 - \frac{u^2}{c^2}\right)} - \frac{e}{c}(v\cdot A) + e\phi, \tag{4.8.3}$$

$$\frac{d}{dt}\left(\frac{mc^2}{\sqrt{1 - \frac{v^2}{c^2}}}\right) = \left(v\cdot \frac{dp}{dt}\right) = -e(E\cdot v), \tag{4.8.4}$$

$$\frac{mc}{\sqrt{1 - \frac{v^2}{c^2}}} + e\phi = \text{constant}. \tag{4.8.5}$$

When $\phi = 0$, $|v|$ is constant. Then v produces the centrifugal force v^2/ρ, where ρ is the principal radius of curvature, a time varying magnetic field cannot exist without an associated electric field, v must be supported by a magnetostatic force and

$$p = ip, \quad \frac{dp}{dt} = 0 + p\frac{di}{dt} = pj\frac{d\psi}{dt} = \frac{pv}{\rho}j,$$

where i and j are unit vectors along the tangent and principal normal, respectively, to the trajectory, ψ is the angle made by the geometrical tangent to the trajectory at the point P at time t with a fixed line, and $\psi + d\psi$ is the same at the time $t + dt$. So, the equation of motion becomes

$$\frac{pv}{p} = \frac{|e|}{c}vH\sin\theta, \tag{4.8.6}$$

giving

$$\rho = \frac{v}{\Omega_0 \sin\theta\sqrt{1 - \frac{v^2}{c^2}}}, \tag{4.8.7}$$

where $\Omega_0 = |e|H/mc$, and θ is the angle between v and H. Hence, in a constant magnetic field H, the motion consists of an arbitrary constant velocity parallel and a gyration perpendicular to H at the Larmor frequency $\Omega_0\sqrt{1 - v^2/c^2}$, which is inversely proportional to the relativistic energy $mc^2/\sqrt{1 - v^2/c^2}$ and the gyration radius increases by the factor $(1 - v^2/c^2)^{1/2}$.

Writing (4.8.1) as

$$\frac{dp}{dt} = ma, \quad a = -\frac{eE}{m} - \frac{e}{mc}(v \times H), \tag{4.8.8}$$

and

$$\mathbf{u} = \frac{v}{\sqrt{c^2 - v^2}}, \tag{4.8.9}$$

we find that $\mathbf{u}$ is a dimensionless velocity and

$$\mathbf{a} = -\frac{e}{m}\mathbf{E} - \frac{e}{m}\frac{(\mathbf{u}\times H)}{\sqrt{1+u^2}}, \tag{4.8.10}$$

$$\sum_i \left(\frac{\partial a_i}{\partial u_i}\right) = \text{div}_u\, a = -\frac{2e}{m}((\mathbf{u}\times\mathbf{H}).u)\frac{1}{(1+u^2)^{3/2}}. \tag{4.8.11}$$

Neglecting the induced electric field, the relativistic electron motion, in presence of a strong, constant magnetic field H is

$$\frac{d}{dt}\left[\frac{m\mathbf{v}}{\sqrt{1-\dfrac{v^2}{c^2}}}\right] = -\frac{e}{c}(\mathbf{v}\times\mathbf{H}). \tag{4.8.12}$$

For $\phi = 0$, (4.8.5) gives

$$\frac{m}{\sqrt{1-\dfrac{v^2}{c^2}}} = \text{constant}. \tag{4.8.13}$$

Hence $m\gamma[\gamma = (1 - v/c^2)^{1/2}]$ can be taken outside the differentiation, and the equation of motion becomes

$$\frac{dv}{dt} = -(\mathbf{v}\times\boldsymbol{\Omega}), \quad \boldsymbol{\Omega} = \frac{e\mathbf{H}}{m\gamma c}. \tag{4.8.14}$$

The relativistically correct magnetic moment and drifts in presence of inhomogeneous magnetic field, are

$$\boldsymbol{\mu} = -\frac{\gamma W_\perp}{H^2}\mathbf{H}, \quad U_G = \frac{c\gamma W_\perp}{eH^3}(\mathbf{H}\times\boldsymbol{\nabla}_\perp\mathbf{H}), \tag{4.8.15}$$

$$U_R = \frac{2c\gamma}{eH^4}W_\parallel[H\times(H\cdot\nabla)H], \tag{4.8.16}$$

where

$$W_\perp = \frac{mu_\perp^2}{2}, \quad W_\parallel = \frac{mu_\parallel^2}{2},$$

U_g is the transverse gradient drift and U_R is the curvature drift, Chakraborty (1996, Section 2.10 and 2.11). The parallel force $F_\parallel$ of the longitudinal gradient of H is

$$F_\parallel = m\gamma\frac{dv_\parallel}{dt} = -\gamma\frac{W_\perp}{B}\frac{\partial H}{\partial x_3}. \tag{4.8.17}$$

The one-particle equation (4.8.1) of motion should be supplemented by the Maxwell equations containing the average current $-eN\mathbf{v}$ per unit volume and the charge density $e(N_0 - N)$, where N and N_0 are the number densities of electrons and ions (static) respectively. The influence of the self induced field is excluded here.

4.9 Electron Motion in the Field of an Electromagnetic Wave

Consider the elliptically polarized electric and magnetic wave fields

$$E \equiv (E_x, E_y, 0), \quad H \equiv (H_x, H_y, 0), \tag{4.9.1}$$

propagating parallel to the z-axis, where

$$E_x = A\cos\theta, \quad E_y = B\sin\theta, \quad \theta = \omega t - kz. \tag{4.9.2}$$

A and B are the amplitudes, θ is the wave phase, ω is the wave frequency and K is the wave number.

The Lorentz transformation is

$$t = \gamma\left(t' + \frac{V}{c^2}z'\right), \quad z = \gamma(z' + Vt'). \tag{4.9.3}$$

The primed system moves relative to the unprimed system with velocity V, parallel to the z-axis. We put

$$\beta = \frac{V}{c}, \quad \beta = \tanh\psi, \quad \gamma = (1-\beta^2)^{-1/2} \tag{4.9.4}$$

and find that

$$t = t'\cosh\psi + \frac{z'}{c}\sinh\psi, \quad z = z'\cosh\psi + ct'\sinh\psi, \tag{4.9.5}$$

$$\omega t - kz = (\omega\cosh\psi - kc\sinh\psi)t' - \left(k\cosh\psi - \frac{\omega}{c}\sinh\psi\right)z'. \tag{4.9.6}$$

The coefficient of z' vanishes when

$$kc = \omega\tanh\psi, \tag{4.9.7}$$

$$\omega t - kz = \omega\sqrt{1-\beta^2}\, t', \tag{4.9.8}$$

$$\omega' T = \omega t - kz, \quad V_z = \dot{z} = V - \frac{V}{\gamma}\frac{\partial T}{\partial t}, \tag{4.9.9}$$

where the symbol t' is replaced by T, and

$$\omega' = \omega\sqrt{1-\beta^2} = \frac{\omega}{\gamma}, \quad V = \frac{\omega}{k}. \tag{4.9.10}$$

The relativistically correct equation of motion of an electron is

$$\frac{dp}{dt} = -eE - \frac{e}{c}(V \times B), \quad \mathbf{p} = m\gamma\mathbf{v}. \tag{4.9.11}$$

It can be written as

$$\frac{d}{dt}(\gamma V) = \frac{e\gamma_0}{mc}\left(\hat{x}\frac{\partial A_x}{\partial T} + \hat{y}\frac{\partial A_y}{\partial T} + \hat{z}\frac{\partial A_z}{\partial T}\right) - \frac{e\gamma_0}{mcV}V \times \left(\hat{x}\frac{\partial A_x}{\partial T} - \hat{y}\frac{\partial A_y}{\partial T}\right)$$

$$= \frac{e\gamma_0(V-c)}{mcV}\left(\hat{x}\frac{\partial A_x}{\partial T} + \hat{y}\frac{\partial A_x}{\partial T}\right) + \frac{e\gamma_0\hat{z}}{mcV}\left(V_x\frac{\partial A_x}{\partial T} + V_y\frac{\partial A_y}{\partial T} + V_z\frac{\partial A_z}{\partial T}\right), \tag{4.9.12}$$

where $\gamma_0 = \sqrt{1 - c^2/V^2}$.

And since

$$\frac{d}{dt}(\gamma v) = \hat{x}\frac{d}{dt}(\gamma v_x) + \hat{y}\frac{d}{dt}(\gamma v_y) + \hat{z}\frac{d}{dt}(\gamma v_z),$$

we obtain

$$\left.\begin{aligned} \frac{d}{dt}(\gamma v_x) &= \frac{e\gamma_0(V - v_z)}{mcv}\frac{\partial A_x}{\partial T} \\ \frac{d}{dt}(\gamma v_y) &= \frac{e\gamma_0(v - v_z)}{mcV}\frac{\partial A_y}{\partial T} \end{aligned}\right\}. \tag{4.9.13}$$

$$\frac{d}{dt}(\gamma v_z) = \frac{e\gamma_0}{mcV}\left(v_x\frac{\partial A_x}{\partial T} + v_y\frac{\partial A_y}{\partial T} + v_z\frac{\partial A_z}{\partial T}\right). \tag{4.9.14}$$

Scalar product of (4.9.11) with V gives

$$\frac{d}{dt}(\gamma c) = -\frac{e}{mc}(\mathbf{E}\cdot \mathbf{v}) = \frac{c\gamma_0}{mc^2}\left(v\cdot \frac{\partial A}{\partial T}\right). \tag{4.9.15}$$

Putting the value of $V - v_z$ from (4.9.9) in (4.9.13) we obtain

$$\left.\begin{aligned} \frac{d}{dt}(\gamma v_x) &= \frac{e}{mc}\frac{\partial T}{\partial t}\frac{\partial A_x}{\partial T} \\ \frac{d}{dt}(\gamma v_y) &= \frac{e}{mc}\frac{\partial T}{\partial t}\frac{\partial A_y}{\partial T} \end{aligned}\right\}. \tag{4.9.16}$$

$$\frac{d}{dt}\left[\gamma\left(V - \frac{V}{\gamma_0}\frac{\partial T}{\partial t}\right)\right] = \frac{e}{mcv}\left(v_x\frac{\partial A_x}{\partial T} + v_y\frac{\partial A_y}{\partial T} + V_z\frac{\partial A_z}{\partial T}\right). \tag{4.9.17}$$

When $A_z = 0$, elimination of the common factor

$$v_x\left(\frac{\partial A_x}{\partial T}\right) + v_y\left(\frac{\partial A_y}{\partial T}\right)$$

gives

$$\frac{d}{dt}\left[\gamma c^2 - \gamma\gamma_0\left(V - \frac{V}{\gamma_0}\frac{\partial T}{\partial t}\right)\right] = 0, \tag{4.9.18}$$

$$\text{or,}\quad \gamma c^2 - \gamma V^2\left(\gamma_0 - \left(\frac{\partial T}{\partial t}\right)\right) = \text{constant}. \tag{4.9.19}$$

Solving for $\partial T/\partial t$ from this equation we get

$$\frac{\partial T}{\partial t} = \frac{a}{\gamma} + b, \tag{4.9.20}$$

where a is arbitrary, and

$$b = \frac{1}{V^2}(V^2\gamma_\theta - c^2). \tag{4.9.21}$$

If the electron is initially at rest in the co-ordinate system used here, then (4.9.9) initially gives

$$t = 0, \quad V = 0, \quad \frac{\partial T}{\partial t} = \gamma, \tag{4.9.22}$$

$$a = \gamma_0 - b = \frac{c^2}{\gamma^2}, \tag{4.9.23}$$

$$\frac{\partial T}{\partial t} = \frac{c^2}{V^2\gamma} + \gamma_0 - \frac{c^2}{V^2} = \gamma_0\left(1 - \frac{V^2}{V^2}\right). \tag{4.9.24}$$

Solving for v_z from this equation, we find that

$$v_z = \frac{c^2}{V\gamma_0}\left(1 - \frac{1}{\gamma}\right). \tag{4.9.25}$$

The relations of (4.9.13) can be written as

$$\frac{d}{dt}(\gamma v_x) = \frac{e}{m}\frac{\partial A_x}{\partial t}, \quad \frac{d}{dt}(\gamma v_y) = \frac{e}{mc}\frac{\partial A_y}{\partial t}. \tag{4.9.26}$$

Replacing, approximately, the total time derivatives in the left hand side of these two relations by the corresponding partial time derivatives, we obtain

$$\frac{\partial}{\partial t}\left(\gamma v_x - \frac{eA_x}{mc}\right) = 0, \quad \frac{\partial}{\partial t}\left(\gamma v_y - \frac{eA_y}{mc}\right) = 0. \tag{4.9.27}$$

Integration gives

$$\gamma v_x = \frac{eA_x}{mc}, \quad \gamma v_y = \frac{eA_y}{mc}. \tag{4.9.28}$$

The relations of (4.9.25) can be written as

$$\frac{V\gamma_0}{c}\frac{v_z}{c} = 1 - \sqrt{1-\beta^2} = \frac{1}{2c^2}(v_x^2 + v_y^2) = \frac{1}{2}\left(\frac{e}{mc^2}\right)^2(A_x^2 + A_y^2), \tag{4.9.29}$$

where

$$v_x = \frac{eA_x}{mc}, \quad v_y = \frac{eA_y}{mc} \tag{4.9.30}$$

and terms up to the first power of v^2/c^2, in the expansion of γ_0, only, have been retained. This wave being elliptically polarized, we write

$$A_x = a(T)\cos\omega T, \quad A_y = b(T)\sin\omega T. \tag{4.9.31}$$

Then,

$$\frac{\partial a}{\partial T} = 0, \quad \frac{\partial b}{\partial T} = 0, \tag{4.9.32}$$

$$\frac{v_z}{c} = \frac{c}{2V\gamma_0}\left(\frac{e}{mc^2}\right)^2(a^2\cos^2\omega T + b^2\sin^2\omega T). \tag{4.9.33}$$

Denoting the average of v_z, in one time period $2\pi/\omega$ by $\langle v_z\rangle$, we have

$$\langle v_z\rangle = \frac{\int_0^{2\pi/\omega} v_z dt}{\int_0^{2\pi/\omega} v_z dt} = \frac{c}{4V\gamma_0}\frac{\left(\frac{e}{mc^2}\right)^2(a^2+b^2)\frac{2\pi}{\omega}}{\int_0^{2\pi}\left(\frac{dT}{\gamma} + \frac{1}{V}\frac{\partial z}{\partial T}dT\right)}. \tag{4.9.34}$$

Since the denominator of this relation is

$$= \frac{2\pi}{\gamma_0\omega} + \frac{1}{V}\int_0^{2\pi} v_z \frac{dt}{dT}\, dT = \frac{2\pi}{\gamma_0\omega} + \frac{1}{V}\int_0^{2\pi/\omega} \frac{v_z dT}{\gamma_0(1 - v_z/c)}$$

$$= \frac{2\pi}{\gamma_0\omega} + \frac{1}{\gamma_0 V}\int_0^{2\pi/\omega} v_z dT + \cdots$$

$$= \frac{2\pi}{\gamma_0\omega} + \frac{c^2}{2V^2\gamma_0^2}\left(\frac{e}{mc^2}\right)^2 \left(\frac{a^2 + b^2}{2}\right)\frac{2\pi}{\omega},$$

equation (4.9.34) reduces to

$$\left\langle \frac{v_z}{c} \right\rangle = \frac{c}{4V\gamma_0} \frac{\left(\frac{e}{mc^2}\right)^2 (a^2 + b^2)}{\left\{\frac{1}{\gamma_0} + \frac{c^2}{4V^2\gamma_0^2}\left(\frac{e}{mc^2}\right)^2 (a^2 + b^2)\right\}}$$

$$= \frac{\left(\frac{e}{mc^2}\right)^2 \frac{(a^2 + b^2)c}{4V}}{\left\{1 + \left(\frac{e}{mc^2}\right)^2 \frac{c^2(a^2 + b^2)}{4V^2}\right\}}. \tag{4.9.35}$$

Averaging the Poynting flux $\mathbf{S}(= c(\mathbf{E} \times \mathbf{H})/4\pi)$ over the time period $2\pi/\omega$, we obtain

$$|\langle S \rangle| = \frac{c}{4\pi}\langle E_x H_y - E_y H_x \rangle$$

$$= \frac{c}{4\pi}\left\langle \frac{\gamma_0^2}{cV}\left\{\left(\frac{\partial A_x}{\partial T}\right)^2 + \left(\frac{\partial A_y}{\partial T}\right)^2\right\}\right\rangle$$

$$= \frac{\gamma_0^2\omega^2}{2cV}(a^2 + b^2). \tag{4.9.36}$$

Elimination of $a^2 + b^2$ gives

$$\left\langle \frac{v_z}{c} \right\rangle = \frac{\frac{c^2}{2\omega^2\gamma_0^2}\left(\frac{e}{mc^2}\right)^2 |\langle \mathbf{E} \times \mathbf{H} \rangle|}{1 + \frac{c^3}{2\omega^2 V\gamma_0^2}\left(\frac{e}{mc^2}\right)^2 |\langle(\mathbf{E} \times \mathbf{H})\rangle|}. \tag{4.9.37}$$

4.10 Radiation Solution for the Field Potentials

The solutions of the wave equations (1.9.6) and (1.9.7) are

$$\left[\nabla^2 - \frac{\partial^2}{c^2}\partial t^2\right][A, \phi] = 4\pi\left[-\mu\frac{j}{c}, \frac{\rho^c}{\varepsilon}\right]$$

are

$$A(r, t) = \frac{4\pi\mu}{c}\iiint_V \int_t G(r', t'; r, t) j(r', t')\, dv' dt' = \frac{\mu}{c}\iiint_V \frac{1}{R} j(r', t')\, dt', \tag{4.10.1}$$

$$\phi(\mathbf{r},t) = -\frac{4\pi}{\varepsilon}\iiint_V \int_t G(\mathbf{r}',t';\mathbf{r},t)\,dv'dt' = -\frac{1}{\varepsilon}\iiint_V \frac{1}{R}\rho^c(\mathbf{r}',t')\,dv', \tag{4.10.2}$$

where G is Green's function, $\mathbf{r}'$ is the position vector of the integration point Q, r is the position vector of the observation point P, $\mathbf{R} = \mathbf{r} - \mathbf{r}'$, dV' is the volume element at Q, and $t'(= t - R(t)/c)$ is the retarded time at which the effect occurring at Q is observed at P at time t, Fig. 4.10.1.

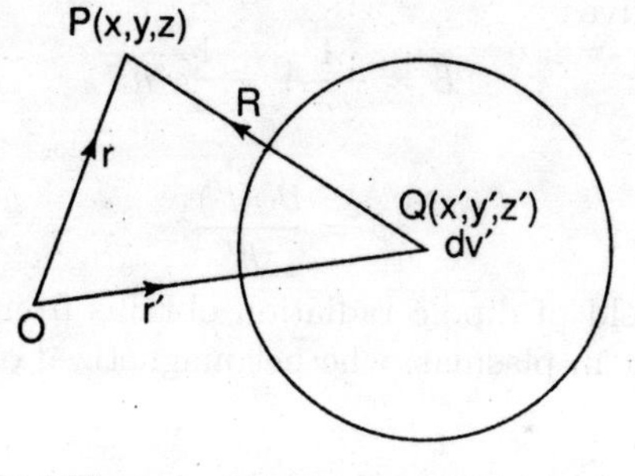

Fig. 4.10.1 The source volume V, for charges and currents, the origin O, observation point P and integration point Q are shown.

Integration has therefore to be performed at the retarded time,

$$t' = t - \frac{1}{c}|\mathbf{r} - \mathbf{r}'| = t - \frac{1}{c}R(t'), \tag{4.10.3}$$

and over those regions, where j and ρ^c are non-zero in the retarded time solutions. The advanced solution, at the time $t + PQ/C$, is non-physical. Since A and ϕ decrease inversely as distance from the source, there is no source at infinity. The components of $\mathbf{r}$ are denoted either as (x, y, z) or as (x_1, x_2, x_3); the $\mathbf{r}'$ components are also either (x', y', z') or (x_1', y_1', z_1').

The expanded form of the gradient operator ∇ is

$$\begin{aligned}\nabla &= \hat{x}_i\frac{\partial}{\partial x_i} = \hat{x}_i\frac{\partial t'}{\partial x_i}\frac{\partial}{\partial t'}\\ &= \hat{x}_i\frac{\partial}{\partial x_i}\left[t' - \frac{1}{c}\{(x_1 - x_1')^2 + (x_2 - x_2')^2 + (z - z')^2\}\frac{\partial}{\partial t'}\right]\\ &= -\hat{x}_i\frac{l_i}{c}\frac{\partial}{\partial t'},\end{aligned} \tag{4.10.4}$$

where repeated index means summation over that (Einstein convention),

$$l_i = \frac{x_i - x_i'}{R}, \qquad \boldsymbol{\nabla} = -\frac{1}{cR}\mathbf{R}\frac{\partial}{\partial t'} \tag{4.10.5}$$

(x_1, x_2, x_3) are the co-ordinates of P, (x_1', x_2', x_3') are the co-ordinates of Q and $(\hat{x}_1, \hat{x}_2, \hat{x}_3)$ are the unit vectors along the three mutually perpendicular co-ordinate axis. Also, since $\partial t'/\partial t = 1$, we have

$$\frac{\partial}{\partial t} = \frac{\partial t'}{\partial t}\frac{\partial}{\partial t'} = \frac{\partial}{\partial t'}. \tag{4.10.6}$$

Using (1.9.1), we obtain

$$H = \frac{1}{\mu}(\nabla \times A) = -\frac{1}{\mu}\left[\left(\hat{x}_i \frac{l_i}{c}\frac{\partial}{\partial t'}\right) \times A\right] = -\frac{1}{\mu c R}\left(R \times \frac{\partial A}{\partial t'}\right) - \frac{1}{c}(A' \times n), \tag{4.10.7}$$

where

$$A' = \frac{\partial}{\partial t} A(t') - \frac{1}{c}(A' \times M). \tag{4.10.8}$$

Hence (1.9.3) and (4.10.4) give

$$E = -\frac{1}{c}A' - \frac{1}{c}\phi' n, \tag{4.10.9}$$

where

$$\phi' = \frac{\partial \phi(t')}{\partial t'}. \tag{4.10.10}$$

Neglecting ϕ and ρ^e, the field of dipole radiation obtains from the vector potential solution. But, ρ^c and ϕ are important in plasmas, whether magnetized or not.

4.11 The Far Field Approximation

For the far field radiation solution, the largest length of the source volume V is small compared to its distance from the observation point P, in the radiation zone, Fig. 4.10.1. If the origin O is taken within V, then $OP \approx OQ$,

$$\mathbf{n} \approx \frac{\mathbf{r}}{|r|}, \quad \frac{1}{R} = \frac{1}{r}, \quad R \approx |r - r'| = r - (\mathbf{n} \cdot \mathbf{r}'), \tag{4.11.1}$$

$$\mathbf{A}(\mathbf{r}, t) = \frac{\mu}{cr} \iiint_V \mathbf{j}\left(t - \frac{r}{c} + \frac{(n \cdot r')}{c}\right) dV' \tag{4.11.2}$$

$$= \frac{\mu}{cr} \iiint_V \mathbf{j}\left(t' + \frac{1}{c}(n \cdot r')\right) dV', \tag{4.11.3}$$

$$\phi(\mathbf{r}, t) = -\frac{1}{\varepsilon r} \iiint_V \zeta\left(t - \frac{r}{c} + \frac{(\mathbf{n} \cdot \mathbf{r}')}{c}\right) dV \tag{4.11.4}$$

$$\approx -\frac{1}{\varepsilon r} \iiint_V \zeta\left(t' + \frac{1}{c}(n \cdot r')\right) dV', \tag{4.11.5}$$

where $t' = t - r/c$; t' is a retarded time. Retaining the first and second terms of the expansion of the integrands in powers of $(\mathbf{n} \cdot \mathbf{r}')/c$, we obtain

$$\mathbf{A}(\mathbf{r}, t) = \frac{\mu}{cr} \iiint_V \mathbf{j}(t')\, dV' + \frac{\mu}{c^2 r}\frac{\partial}{\partial t} \iiint_V \mathbf{j}(t')(\mathbf{n} \cdot \mathbf{r}')\, dV', \tag{4.11.6}$$

$$\phi(\mathbf{r}, t) = -\frac{1}{\varepsilon r} \iiint_V \zeta(t')\, dV' - \frac{1}{\varepsilon c r} \iiint_V \zeta(t')(\mathbf{A} \cdot r')\, dV. \tag{4.11.7}$$

The dipole moment $\mathbf{p}$ of the radiating volume V is defined as

$$\mathbf{p}(t') = \iiint_V \mathbf{r}'\zeta(\mathbf{r}', t')\, dV'. \tag{4.11.8}$$

Then,

$$\dot{p}(t') = \frac{\partial p}{\partial t'} = \iiint_V j dV', \tag{4.11.9}$$

$$\mathbf{A}(\mathbf{r}, t) = \frac{1}{cr}\dot{\mathbf{p}} + \frac{1}{c^2 r}\frac{\partial}{\partial t'}\iiint_V \mathbf{j}(t')(\mathbf{n}\cdot\mathbf{r}')\, dV', \tag{4.11.10}$$

where a dot denotes differentiation with respect to the retarded time t', at which radiation occurs in V, and not with respect to the time t at the distant observation point P. Resolution of the integrand in the second term into the sum of a symmetric and antisymmetric parts, gives

$$\frac{1}{c}(\mathbf{n}\cdot\mathbf{r}')\mathbf{j}(t') = \frac{1}{2c}\left[(\mathbf{n}\cdot\mathbf{r}')\mathbf{j} + (\mathbf{n}\cdot\mathbf{j})\mathbf{r}'\right] + \frac{1}{2c}(\mathbf{r}' \times \mathbf{j}) \times \mathbf{n}. \tag{4.11.11}$$

The integral of the symmetric part is the quadrupole moment, which is negligible. The same of the antisymmetric part is the magnetic moment $\mathbf{m}$ of the body;

$$\mathbf{m} = \frac{1}{2c}\iiint_V (\mathbf{r}' \times \mathbf{j})\, dV'. \tag{4.11.12}$$

Neglecting the unimportant quantities, we find that

$$\mathbf{A} = \frac{1}{cr}\dot{\mathbf{p}} + \frac{1}{cr}(\dot{\mathbf{m}} \times \mathbf{n}) = \frac{1}{cr}\dot{p} + \frac{1}{c}\nabla \times \mathbf{M}, \tag{4.11.13}$$

$$\mathbf{H} = \frac{1}{c^2 r}(\ddot{\mathbf{p}} \times \mathbf{n}) + \frac{1}{cr}((\ddot{\mathbf{m}} \times \mathbf{n}) \times \mathbf{n}). \tag{4.11.14}$$

The intensity dP of radiation in the elementary solid angle $d\Omega$ determines the energy flowing in unit time through the elementary surface $d\Sigma(= r^2 d\Omega)$ of the sphere having center at the origin O of the co-ordinate system in V, and the radial distance r. Hence, if $\mathbf{S}$ is the Poynting energy flux, $dP = |S|r^2 d\Omega$, equations (4.10.7) and (4.10.8) show that

$$(n\cdot H) = 0, \quad \mathbf{S} = \frac{CH^2}{4\pi}\mathbf{n}.$$

Moreover, in the radiation zone, since $\mathbf{E} = (\mathbf{H} \times \mathbf{n})$, putting $m = 0$ we find that

$$dP = \frac{1}{4\pi c^3}|\ddot{\mathbf{p}} \times \mathbf{n}|^2 d\Omega \tag{4.11.15}$$

$$= \frac{1}{4\pi c^3}|\ddot{p}|^2 \sin^2\theta \cdot 2\pi \sin\theta d\theta, \tag{4.11.16}$$

where θ is the angle between $\mathbf{p}$ and $\mathbf{n}$, and $d\Omega = 2\pi \sin\theta d\theta$. Integration with respect to θ from 0 to π, gives

$$P = \frac{|\ddot{p}|^2}{4\pi c^3}\int_0^\pi 2\pi \sin^3\theta d\theta = \frac{2}{3c^3}|\ddot{p}|^2 \tag{4.11.17}$$

for the flow of energy of radiation. Its value depends on the acceleration of the charges inside the source V.

4.12 Lienard-Weichert Potentials

The retarded radiative process of the field of a single electron of charge e at the point $\mathbf{r}'(t')$, moving in vacuum with velocity $\mathbf{v}(= dr'/dt')$, along the trajectory $\mathbf{r} = \mathbf{r}'(t')$, is described by two-body Coulomb interaction in the absence of anisotropy in the dielectric behaviour. At any time, only a negligible fraction of the particles is interacting mutually, so there is no correlation in their motion, and long range interactions, in dilute gases, dominate.

The delta function representations of charge and current, for a single charge, are

$$q(\mathbf{r}, t) = e\delta(\mathbf{r} - \mathbf{r}'(t)), \tag{4.12.1}$$

$$\mathbf{j}(\mathbf{r}, t) = e\mathbf{r}'(t')\delta(\mathbf{r} - \mathbf{r}'(t')). \tag{4.12.2}$$

Then

$$\mathbf{A}(\mathbf{r}, t) = \frac{e}{c}\iiint\limits_V \int_t \frac{1}{R}\dot{\mathbf{r}}(t')\delta\left(T - \frac{R}{c}\right)\delta(\mathbf{r} - \mathbf{r}'(t'))\, dV' dt', \tag{4.12.3}$$

where $T = t - t'$, $\mathbf{R}(t') = \mathbf{r}(t) - \mathbf{r}'(t') = R\mathbf{n}(t')$, $\mathbf{n} = \mathbf{R}/R$. Integrating over r' we obtain

$$\mathbf{A}(\mathbf{r}, t) = \frac{1}{c}\int \frac{1}{R(t')}\dot{r}'(t')\delta\left(t' - t + \frac{1}{c}R(t')\right) dt'. \tag{4.12.4}$$

The delta function property $(A1 - 11)$,

$$\int_{-\infty}^{\infty} \delta\{\psi(x)\} f(x)\, dx = \sum_j \frac{f(x_j)}{|d\psi/dx|_{x=x_j}}, \tag{4.12.5}$$

where the x_js are the roots of $\psi(x) = 0$, then gives

$$\mathbf{A}(r, t) = \frac{e}{c}\left[\frac{\dot{\mathbf{r}}'(t')\cdot}{R(t')\left[1 + \left(\dfrac{dR}{dt'}\right)\right]}\right]_{t'=t-R(t)/c} \tag{4.12.6}$$

Since

$$R^2 = c^2(t - t')^2 = (x - x')^2 + (y - y')^2 + (z - z')^2; \tag{4.12.7}$$

we find that

$$\begin{aligned}\frac{dR}{dt'} &= \frac{\partial R}{\partial x'}\frac{dx'}{dt'} + \frac{\partial R}{\partial y'}\frac{dy'}{dt'} + \frac{\partial R}{\partial z'}\frac{dz'}{dt'} \\ &= -\left[(x - x')\frac{dx'}{dt'} + (y - y')\frac{dy'}{dt'} + (z - z')\frac{dz'}{dt'}\right]\Big/ R \\ &= -\frac{1}{R}(R\cdot v) = -(n\cdot v) = -v_R,\end{aligned} \tag{4.12.8}$$

$$\mathbf{A}(r, t) = \left[\frac{e\mathbf{v}}{(R - (v\cdot R))}\right]_{t'} = \left[\frac{e\beta}{S}\right]_{t'}, \tag{4.12.9}$$

$$\phi(\mathbf{r}, t) = -\left[\frac{e}{R - (v\cdot R)/c}\right]_{t'}, \tag{4.12.10}$$

where

$$\mathbf{S} = gR = R - \frac{Rv_R}{c} = R - \frac{(\mathbf{R}\cdot \mathbf{v})}{c} = R[1 - (\mathbf{n}\cdot \mathbf{B})], \tag{4.12.11}$$

$$g = 1 - \frac{v_R}{c} = 1 - (\mathbf{n}\cdot \boldsymbol{\beta}), \quad \boldsymbol{\beta} = \frac{\mathbf{v}}{c}. \tag{4.12.12}$$

The Lienard-Weichert potential solutions for radiation of an accelerated charge are (4.12.9) and (4.12.10). The term $(\mathbf{n}\cdot \boldsymbol{\beta})$ in S is relativistic. So non-relativistic results are

$$\mathbf{A}(\mathbf{r}, t) = \left[\frac{e}{R}\boldsymbol{\beta}\right]_{t'}, \quad \phi(r, t) = -\left[\frac{e}{R}\right]_{t'}. \tag{4.12.13}$$

To include the retarded effects of radiation from all points in a small volume V, in which the electronic charge is continuously distributed, at times differing slightly from each other, and reaching the observation point P at time t, Fig. 4.10.1, take an elementary volume dv in V, containing the charge element de, at the point Q. Actually, the charges in V are those found in the respective volume elements at the corresponding retarded times $t - PQ/C$. So, due consideration should be given to these retarded times in the expressions for $\mathbf{A}$ and ϕ at P at time t.

The radiation from the surface element $d\Sigma$ of a sphere of radius R with center at Q, at time t' reaches P at time t. At time $t' + dt'$ this radiation is on the surface of sphere of radius $R - Cdt'$. If the electron is at rest, the amount of charge that this sphere will cross during the time dt', as it shrinks by dR, is $qd\Sigma dR$. But, if $\mathbf{V}$ is the electron velocity, a charge which is less than $qd\Sigma dR$ by the amount $qd\Sigma v_R dt'$ will be crossed by the sphere. Since $cdt' = dR$,

$$dq = qd\Sigma dR - qd\Sigma U_R dt' \tag{4.12.14}$$

$$= qdv - qdv\frac{U_R}{c} = qdv\left[1 - \frac{1}{c}(\mathbf{R}\cdot \mathbf{v})\right], \tag{4.12.15}$$

$$qdv = \frac{de}{1 - \beta\cos\theta}, \tag{4.12.16}$$

$$\mathbf{A} = \frac{1}{c}\iiint\limits_V \left[\frac{\mathbf{v}de}{R(1 - \beta\cos\theta)}\right]_{t'}, \tag{4.12.17}$$

$$\phi = -\iiint\limits_V \left[\frac{de}{R(1 - \beta\cos\theta)}\right]_{t'}, \tag{4.12.18}$$

because de is the observed charge. These far-field solutions easily recover (4.12.9) and (4.12.11) for the point-like charge.

A three dimensional consideration is possible of the distortion, at a particular instant, in the charge-carrying volume element, due to the retarded effects at different points in it. For this retarded solution, following Ray (1957), in addition to Q at the retarded time t' with co-ordinates x', y', z', consider another point Q_1 in the neighbourhood, with co-ordinates $(x' + x_1', \; y' + y_1', \; z' + z_1')$, also at the same retarded time t'. Since $PQ \neq PQ_1$, radiation from Q_1 at the time t' cannot reach P at the time t. But, the motion of the electron being known, we find the position Q_1^e, of Q_1 at another time, from where the radiation reached P at time t. Let the co-ordinates of Q_1^e be $(x' + x_e', \; y' + y_e', \; z' + z_e')$. Since the electron volume is small, the difference in time $\delta t (= t' - t')$ is small, and both QQ_1 and QQ_1^e are small; so

$$Q_1^e P = c(t - t_e') = R - c\delta t, \tag{4.12.19}$$

$$x'_e = x'_1 + v_x\delta t, \quad y'_e = y'_1 + v_y\delta t, \quad z'_e = z'_1 + v_z\delta t, \tag{4.12.20}$$

$$Q_1^e P^2 = (x - x' - x'_1 - v_x\delta t)^2 + (y - y' - y'_1 - v_y\delta t)^2 + (z - z' - z'_1 - v_z\delta t)^2. \tag{4.12.21}$$

Since QQ_1 and QQ_1^e are small, we write

$$\begin{aligned} Q_1^e P^2 &= \Sigma(x - x')^2 - 2\Sigma(x - x')(x'_1 + v_z\delta t) \\ &\approx R^2 - 2(R.(\delta R' + v\delta t)), \end{aligned} \tag{4.12.22}$$

where $\delta R' = (x'_1, y'_1, z'_1)$. Also, squaring (4.12.20) gives

$$Q_1^e P^2 = R^2 - 2Rc\delta t. \tag{4.12.23}$$

Equating these two values of $Q_1^e P^2$, we obtain

$$\delta t = \frac{(\mathbf{R}\cdot \delta\mathbf{R}')}{cR[1 - (\beta\cdot n)]} = \frac{(\mathbf{R}\cdot \delta\mathbf{R})}{cs}. \tag{4.12.24}$$

Differentiating this relation with respect to x'_1, we have

$$\frac{\partial}{\partial x'_1}(\delta t) = \frac{x - x'}{cs}. \tag{4.12.25}$$

Let the volume elements at Q_1 and Q_1^e be dv and dv_e, respectively. Then $dv_e = Jdv$, where the Jacobian of the volume transformation of the relations of (4.12.20) is

$$J = \begin{vmatrix} 1 + v_x\dfrac{\partial\delta t}{\partial x'_1} & v_x\dfrac{\partial\delta t}{\partial y'_1} & v_x\dfrac{\partial\delta t}{\partial z'_1} \\ v_y\dfrac{\partial\delta t}{\partial x'_1} & 1 + v_y\dfrac{\partial\delta t}{\partial y'_1} & v_y\dfrac{\partial\delta t}{\partial z'_1} \\ v_z\dfrac{\partial\delta t}{\partial x'_1} & v_z\dfrac{\partial\delta t}{\partial z'_1} & 1 + v_z\dfrac{\partial\delta t}{\partial z'_1} \end{vmatrix} \tag{4.12.26}$$

$$\begin{aligned} &= 1 + v_x\frac{\partial\delta t}{\partial x'_1} + v_y\frac{\partial\delta t}{\partial y'_1} + v_z\frac{\partial\delta t}{\partial z'_1} + \cdots \\ &= 1 + \frac{1}{cs}\{v_x(x - x') + v_y(y - y') + v_z(z - z')\} \\ &= 1 + \frac{1}{cs}(R\cdot v). \end{aligned} \tag{4.12.27}$$

Hence

$$dv_e = \left[1 + \frac{1}{cs}(R\cdot v)\right] dv = \frac{R}{s}\, dv. \tag{4.12.28}$$

The volume element dv' in (4.12.3) is the effective volume dv_e of (4.12.28). So

$$A = \iiint_V \left[\frac{1}{R}\rho\beta dv_e\right]_{t'}, \quad \phi = -\iiint_V \left[\frac{1}{R}\rho dv_e\right]_{t'}. \tag{4.12.29}$$

▶ **Example 4.12.1** *Develop a 4-vector method for the L-W solution.*

Solution : In the co-ordinate system moving with the electron, since $v = 0$, we find that

$$J = 0, \quad A = 0, \quad \phi = -\frac{\rho}{R(t')}.$$

Hence, in the laboratory system, the 4-potential is

$$A_\mu \equiv (\mathbf{A}_1 - i\phi) = \frac{eU_\mu}{R_\lambda U_\lambda},$$

in terms of the 4-velocity $U_\mu \equiv (v, ic)$ and the 4-co-ordinate $R_\mu \equiv (R, ic(t-t'))$. Since $t' = t - R(t')/c$, $R_\mu\, R_\mu = 0$,

$$R_\mu U_\mu = \gamma\{(v\cdot R) - CR\},$$

and so

$$A_\mu = \left[\frac{e\mathbf{v}}{cs}, \frac{ie}{s}\right].$$

4.13 Radiation from an Accelerated Electron

Using (4.12.9) and (4.12.11), for $\mathbf{A}$ and ϕ in (1.9.3) and (1.9.1) for $\mathbf{E}$ and $\mathbf{B}$, we find that

$$\mathbf{E} = -\left[\frac{e}{c^2}\frac{\partial}{\partial t}\left(\frac{\mathbf{v}}{s}\right) + e\nabla\left(\frac{1}{s}\right)\right]_{t'} = -\left[\frac{e}{c^2 s}\frac{\partial \mathbf{v}}{\partial t} + \frac{e\mathbf{v}}{c^2 S^2}\frac{\partial s}{\partial t} + \frac{e}{S^2}\nabla S\right]_{t'}, \tag{4.13.1}$$

$$\mathbf{B} = \nabla\times A = \frac{e}{c}\left[\nabla\times\left(\frac{\mathbf{v}}{s}\right)\right]_{t'} = \left[\frac{e}{cs}(\nabla\times\mathbf{v}) + \frac{e}{c}\left(\nabla\frac{1}{s}\right)\times\mathbf{v}\right]_{t'}$$

$$= \left[\frac{e}{cs}(\nabla\times\mathbf{v}) + \frac{e}{cS^2}(\mathbf{v}\times\nabla S)\right]_{t'}, \tag{4.13.2}$$

where, because of the involvement of two-times t and t', we ignore the variation of t' while differentiating with respect to t, and then replace t by t'. To consider the influence of t' in these results, we write

$$\frac{\partial}{\partial t}\mathbf{v}(t') = \dot{\mathbf{v}}(t')\frac{\partial t'}{\partial t}, \tag{4.13.3}$$

where dot denotes derivative with respect to t'. Thus

$$\frac{ds}{dt} = \dot{s}\frac{\partial t'}{\partial t} = \frac{dR}{dt} - \frac{1}{c}(\mathbf{R}\cdot\dot{\mathbf{v}}) + \frac{v^2}{c}, \tag{4.13.4}$$

where we have used (4.12.10), and the fact that

$$\mathbf{v} = -\frac{d}{dt'}\mathbf{R}. \tag{4.13.5}$$

Since

$$[\nabla\times\mathbf{v}]_x = \frac{\partial v_z}{\partial y} - \frac{\partial v_y}{\partial z} = \dot{v}_z\frac{\partial t'}{\partial y} - \dot{v}_y\frac{\partial t'}{\partial z},$$

we find that

$$(\nabla\times\mathbf{v}) = (\nabla t'\times\mathbf{v}) \tag{4.13.6}$$

$$(\mathbf{R}\cdot\mathbf{v}) = -\left(\mathbf{R}\cdot\frac{d}{dt'}\mathbf{R}\right) = -\frac{1}{2}\frac{dR^2}{dt'} = -R\frac{dR}{dt'}, \tag{4.13.7}$$

$$\frac{dR}{dt'} = \frac{1}{R}(\mathbf{R}\cdot\mathbf{v}) = -v\cos\theta = -v_R = -\frac{c}{R}(R - \dot{S}), \tag{4.13.8}$$

$$\mathbf{v} = -\frac{dR}{dt'}\mathbf{n} = -\mathbf{n}v\cos\theta \tag{4.13.9}$$

$$\nabla S(t') = \nabla S(R,t) + \frac{dS}{dt'}\nabla t' \tag{4.13.10}$$

$$= \mathbf{n} - \boldsymbol{\beta} + \frac{dS}{dt'}\nabla t', \tag{4.13.11}$$

where θ is the angle between $\mathbf{R}$ and $\mathbf{v}$, and in (4.13.8) and (4.13.11) the relation (4.12.10) has been used. Using (4.13.8) in the first term of (4.13.4), and taking gradient of t' we obtain

$$\frac{dS}{dt'} = -v_R = \frac{1}{c}(\mathbf{R}\cdot\mathbf{v}') + \frac{v^2}{c}, \tag{4.13.12}$$

$$\nabla t' = \nabla\left(t - \frac{1}{c}R(t')\right) = 0 - \frac{1}{c}\nabla R(t')$$

$$= -\frac{1}{c}\nabla R - \frac{1}{c}(\nabla t')\frac{dR}{dt'} = -\frac{\mathbf{R}}{c} + \frac{v_R}{c}\nabla t'. \tag{4.13.13}$$

Solving for grad t' from this relation we obtain

$$\nabla t' = -\frac{1}{cS}\mathbf{R}. \tag{4.13.14}$$

Differentiating of (4.12.31) with respect to t, gives

$$\frac{dR}{dt} = c - c\frac{dt'}{dt} = \frac{dR}{dt'}\frac{dt'}{dt}, \tag{4.13.15}$$

$$\frac{dt'}{dt} = \frac{R}{S}, \tag{4.13.16}$$

$$\frac{dS}{dt'} = \frac{R}{S}\left\{-v_R - \frac{1}{c}(\mathbf{R}\cdot\mathbf{v}) + \frac{v^2}{c^2}\right\}. \tag{4.13.17}$$

Then (4.13.3) and (4.13.11) transform to

$$\frac{\partial\mathbf{v}}{\partial t} = \frac{\mathbf{R}}{S}\dot{v}, \tag{4.13.18}$$

and

$$\nabla S = \mathbf{n} - \boldsymbol{\beta} - \frac{\mathbf{R}}{cS}\left\{-v_R - \frac{1}{c}(\mathbf{R}\cdot\mathbf{v}) + \frac{v^2}{c}\right\}. \tag{4.13.19}$$

The changes in the time scale of the three terms of (4.13.1) for **E** and the two terms of (4.13.2) for **B** are now complete. We find that

$$\mathbf{E} = -\left[\frac{e\mathbf{R}}{c^2S^2}\dot{\mathbf{v}}\right]_{t'} + \left[\frac{eR\mathbf{v}}{c^2S^3}\left\{-v_R - \frac{1}{c}(\mathbf{R}\cdot\dot{\mathbf{v}}) + \frac{v^2}{c}\right\}\right]_{t'}$$

$$+\frac{e}{S^2}\left[-\boldsymbol{\beta} + \mathbf{n} - \frac{\mathbf{R}}{cS}\left\{-v_R - \frac{1}{c}(\mathbf{R}\cdot\dot{\mathbf{v}}) + \frac{v^2}{c}\right\}\right]_{t'} \quad (4.13.20)$$

$$= \left[\frac{eR}{cS^2}\dot{\boldsymbol{\beta}}\right]_{t'} + \left[\frac{eR}{S^3}\boldsymbol{\beta}\left\{-(\mathbf{n}\cdot\boldsymbol{\beta}) - \frac{R}{c}(\mathbf{n}\cdot\dot{\boldsymbol{\beta}}) + \beta^2\right\}\right]_{t'}$$

$$+\frac{e}{S^2}\left[-\boldsymbol{\beta} + \mathbf{n} - \frac{\mathbf{R}}{\mathbf{s}}\mathbf{n}\left\{-(\mathbf{n}\cdot\boldsymbol{\beta}) - \frac{\mathbf{R}}{c}(\mathbf{n}\cdot\dot{\boldsymbol{\beta}}) + \beta^2\right\}\right]_{t'}, \quad (4.13.21)$$

$$\mathbf{B} = \left[\frac{e}{cS^2}(\dot{\mathbf{v}} \times \mathbf{R}) + \frac{e}{cS^2}(\mathbf{v} \times \nabla S)\right]_{t'}. \quad (4.13.22)$$

Collection of the terms containing $\boldsymbol{\beta}, \dot{\boldsymbol{\beta}}$ and $\mathbf{n}$ together, in groups, gives

$$\mathbf{E} = \frac{eR^2}{cS^2}\left[(\mathbf{n} - \boldsymbol{\beta})(\mathbf{n}\cdot\dot{\boldsymbol{\beta}}) - \frac{s}{R}\dot{\boldsymbol{\beta}}\right] + \frac{eR}{S^3}\boldsymbol{\beta}\left[\beta^2 - (n\cdot\beta) - \frac{s}{R}\right]$$
$$+ \frac{eR}{S^3}\mathbf{n}[1 - (\mathbf{n}\cdot\boldsymbol{\beta}) + (\mathbf{n}\cdot\dot{\boldsymbol{\beta}}) - \beta^2].$$

Since $1 - (\mathbf{n}\cdot\boldsymbol{\beta}) = (\mathbf{n}\cdot(\mathbf{n} - \boldsymbol{\beta}))$, relativistically correct expression for the radiation electric field becomes

$$\mathbf{E}(\mathbf{r}, t) = \left[\frac{eR^2}{cS^2}\mathbf{n} \times ((\mathbf{n} - \boldsymbol{\beta}) \times \dot{\boldsymbol{\beta}}) + \frac{eR}{S^3}(\mathbf{n} - \boldsymbol{\beta})(1 - \beta^2)\right]_t. \quad (4.13.23)$$

A standard form of this formula is

$$\mathbf{E}(\mathbf{r}, t) = \left[\frac{eR^2}{cS^2}\mathbf{n} \times ((\mathbf{n} - \boldsymbol{\beta}) \times \dot{\boldsymbol{\beta}}) + \frac{eR}{S^3}(\mathbf{n} - \boldsymbol{\beta})(1 - \beta^2)\right]_{t'}. \quad (4.13.24)$$

Cross product of **E** of (4.13.20) with **n** gives

$$(\mathbf{n} \times \mathbf{E}) = \left[-\frac{eR}{cS^2}(\mathbf{n} \times \dot{\boldsymbol{\beta}}) + \frac{eR}{S^3}(\mathbf{n} \times \boldsymbol{\beta})\left\{-(\mathbf{n}\cdot\boldsymbol{\beta}) - \frac{R}{c}(\mathbf{n}\cdot\dot{\boldsymbol{\beta}}) + \beta^2\right\}\right.$$
$$\left. - \frac{e}{S^2}(\mathbf{n} \times \boldsymbol{\beta})\left\{1 - \beta^2 + \frac{R}{c}(\mathbf{n}\cdot\dot{\boldsymbol{\beta}})\right\}\right]_{t'}$$

$$= \left[-\frac{eR}{cS^2}(\mathbf{n} \times \dot{\boldsymbol{\beta}}) - \frac{eR}{S^3}(\mathbf{n} \times \boldsymbol{\beta})\left\{1 - \beta^2 + \frac{R}{c}(\mathbf{n}\cdot\dot{\boldsymbol{\beta}})\right\}\right]_{t'}. \quad (4.13.25)$$

And since

$$(\boldsymbol{\beta} \times \nabla S) = -(\mathbf{n} \times \boldsymbol{\beta}) + \frac{R}{cS}(\mathbf{n} \times \boldsymbol{\beta})\{-(\mathbf{n}\cdot\mathbf{v}) - (\mathbf{R}\cdot\dot{\boldsymbol{\beta}}) + c\beta^2\}$$

$$= -\frac{R}{S}(\mathbf{n} \times \boldsymbol{\beta})\left\{1 - \beta^2 + \frac{R}{c}(\mathbf{n}\cdot\dot{\boldsymbol{\beta}})\right\}, \quad (4.13.26)$$

we can write (4.13.25) and (4.13.22) as

$$(\mathbf{n}\times\mathbf{E}) = \left[-\frac{eR}{cS^2}(\mathbf{n}\times\dot{\boldsymbol{\beta}}) + \frac{e}{S^2}(\boldsymbol{\beta}\times\boldsymbol{\nabla}S)\right]_{t'}, \tag{4.13.27}$$

$$\mathbf{B} = \left[-\frac{eR}{c^2S^2}(\mathbf{n}\times\dot{\boldsymbol{\beta}}) + \frac{e}{S^2}(\boldsymbol{\beta}\times\boldsymbol{\nabla}S)\right]_{t'}, \tag{4.13.28}$$

$$(\mathbf{n}\times\mathbf{E}) = \mathbf{B}. \tag{4.13.29}$$

Hence the electric field, the magnetic field and the direction of radiation at a point are three mutually perpendicular vectors.

With the help of the relation (4.13.19) for $\boldsymbol{\nabla}S$, the formula (4.13.28) becomes

$$\begin{aligned}\mathbf{B} &= \left[\frac{eR}{cS^2}(\dot{\boldsymbol{\beta}}\times\mathbf{n})\right]_{t'} + \frac{e}{S^2}\left[1 - \frac{R}{s}\left\{(\mathbf{n}\cdot\boldsymbol{\beta}) + \frac{R}{c}(\mathbf{n}\cdot\dot{\boldsymbol{\beta}}) - \beta^2\right\}(\boldsymbol{\beta}\times\mathbf{n})\right]_{t'} \\ &= \left[\frac{eR}{cS^2}(\dot{\boldsymbol{\beta}}\times\mathbf{n}) + \frac{eR}{S^3}\rho_0(\boldsymbol{\beta}\times\mathbf{n})\right]_{t'},\end{aligned} \tag{4.13.30}$$

where

$$\rho_0 = 1 + \frac{R}{c}(\mathbf{n}\cdot\dot{\boldsymbol{\beta}}) - \beta^2. \tag{4.13.31}$$

▶ **Example 4.13.1** *Find an expression for* $\mathbf{E}(\mathbf{r},t)$ *in term of the time* t *of the observation point* P.

Solution : Using the relations (4.10.9), (4.11.3) and (4.11.4) we find that

$$\begin{aligned}\mathbf{E}(\mathbf{r},t) &= -\frac{e}{c^2}\frac{\partial}{\partial t}\int\frac{\mathbf{v}(t')}{R(t')}\delta(t'-t+R(t')/c)\,dt' \\ &\quad -e\mathbf{n}\frac{\partial}{\partial R}\int\frac{1}{R(t')}\delta(t'-t+R(t')/c)\,dt' \\ &= -\frac{e}{c^2}\frac{\partial}{\partial t}\int\frac{1}{R(t')}\mathbf{v}(t')\delta(t'-t+R(t')/c)\,dt' \\ &\quad +e\mathbf{n}\left[\int\frac{1}{R^2}\delta(t'-t+R/c)\,dt' - \int\frac{1}{R}\delta(t'-t+R/c)\,dt'\right],\end{aligned} \tag{1}$$

where the second term has been differentiated with respect to R, and the primed delta function stands for its derivative with respect to the argument T, defined by

$$T = t' - t + R(t')/C. \tag{2}$$

The first term of of (1) is differentiated with respect to t, and not t', so $v(t')$ and $R(t')$ are not affected and

$$\mathbf{E}(\mathbf{r},t) = e\int\left[\frac{\mathbf{n}}{R^2}\delta T + \frac{1}{cR}(\boldsymbol{\beta}-\mathbf{n})\delta' T\right]\frac{dT}{g}, \tag{3}$$

where (4.12.12) has been used for g. The second integral of $\mathbf{E}$

$$= \int \frac{1}{gcR}(\boldsymbol{\beta}-\mathbf{n})\delta' T dT = \int \left[\frac{d}{dT}\left\{(\boldsymbol{\beta}-\mathbf{n})\frac{\delta T}{cRg}\right\} - \delta T \frac{d}{dT}\left(\frac{\boldsymbol{\beta}-\mathbf{n}}{cRg}\right)\right] dT$$

$$= 0 - \int \left\{\frac{d}{dT}\left(\frac{\boldsymbol{\beta}-\mathbf{n}}{cRg}\right)\right\} \delta T dT = -\int \delta\left(t' - t + \frac{R(t')}{c}\right)\left[\frac{dn}{dt'}\left(\frac{\boldsymbol{\beta}-\mathbf{n}}{cRg}\right)\right] dt'$$

$$= -\left[\frac{1}{cg}\frac{d}{dt'}\left(\frac{\boldsymbol{\beta}-\mathbf{n}}{Rg}\right)\right]_{t'}, \tag{4}$$

where the delta function relation (4.12.5) has been used. Hence

$$\mathbf{E}(\mathbf{r}, t) = e\left[\frac{\mathbf{A}}{R^2 g} + \frac{1}{cg}\frac{d}{dt'}\left(\frac{\mathbf{n}-\boldsymbol{\beta}}{Rg}\right)\right]_{t'}. \tag{5}$$

We now differentiate $R\mathbf{n}$ with respect to t', and, using (4.12.8) for dR/dt', obtain

$$\frac{d}{dt'}(R\mathbf{n}) = R\frac{d\mathbf{n}}{dt'} + \frac{dR}{dt'}\mathbf{n} = R\frac{d\mathbf{n}}{dt'} - \mathbf{n}(\mathbf{n}\cdot\mathbf{v}) = \frac{d}{dt'}[\mathbf{r}(t) - \mathbf{r}(t')], \tag{6}$$

$$\frac{1}{c}\frac{d\mathbf{n}}{dt'} = \frac{1}{Rc}\left[\frac{d}{dt'}\{\mathbf{r}(t) - \mathbf{r}(t')\} + \mathbf{n}(\mathbf{n}\cdot\mathbf{v})\right] = \frac{1}{Rc}[-\mathbf{v} + \mathbf{n}(\mathbf{n}\cdot\mathbf{v})] - \frac{1}{R}[-\boldsymbol{\beta} + \mathbf{n}(\mathbf{n}\cdot\boldsymbol{\beta})]$$

$$= \frac{1}{R}(\mathbf{n}\times(\mathbf{n}\times\boldsymbol{\beta})), \tag{7}$$

$$\frac{1}{R}\boldsymbol{\beta} = \frac{1}{R}\mathbf{n}(\mathbf{n}\cdot\boldsymbol{\beta}) - \frac{1}{c}\frac{d\mathbf{n}}{dt'}, \tag{8}$$

$$\frac{1}{c}\frac{d}{dt}\left(\frac{\mathbf{n}}{R}\right) = \frac{\mathbf{R}}{c}\left[\frac{d}{dt}\left(\frac{1}{R^2}\boldsymbol{\beta}\right) - \frac{1}{R}\mathbf{n}\frac{d}{dt}\left(\frac{1}{R}\right)\right] = \frac{R}{c}\frac{d}{dt}\left(\frac{1}{R^2}\mathbf{n}\right) - \frac{\mathbf{n}}{c}\frac{d}{dt'}\left(\frac{1}{R}\right)\frac{dt'}{dt}$$

$$= \frac{R}{c}\frac{d}{dt}\left(\frac{1}{R^2}\mathbf{n}\right) - \frac{(\mathbf{n}\cdot\boldsymbol{\beta})\mathbf{n}}{R^2 g}, \tag{9}$$

$$E(\mathbf{r}, t) = \frac{e\mathbf{n}}{R^2 g} + \frac{e}{c}\frac{d}{dt}\left\{\frac{1}{Rg}\mathbf{n} - \frac{(\mathbf{n}\cdot\boldsymbol{\beta})}{Rg}n + \frac{1}{cg}\frac{d\mathbf{n}}{dt'}\right\}$$

$$= \frac{e\mathbf{n}}{R^2 g} + \frac{e}{c}\frac{d}{dt}\left(\frac{1}{R}\mathbf{n} + \frac{1}{c}\frac{d}{dt}\mathbf{n}\right)$$

$$= \frac{e\mathbf{n}}{R^2 g} + \frac{e\mathbf{R}}{c}\frac{d}{dt}\left(\frac{1}{R^2}\mathbf{n}\right) - \frac{e(\mathbf{n}\cdot\boldsymbol{\beta})}{R^2 g}n + \frac{e}{c^2}\frac{d^2\mathbf{A}}{dt^2}, \tag{10}$$

where (4.13.8), (4.13.17) have been used for finding (9) and equation (5) has been transformed in terms of time t of the observation point P. Using $\boldsymbol{\beta}$ from (8) to obtain (10), g is given by (4.12.12),

$$\frac{d}{dt} = \frac{1}{g}\frac{d}{dt'}. \tag{11}$$

► **Example 4.13.2** *Recover the formula for* $\mathbf{E}(\mathbf{r}, t)$ *of* (4.13.24), *which is in terms of* t', *from the formula* (5) *of Exercise* 4.13.1.

Solution : To get the results in terms of the retarded time t', we carry out the differentiation in the second term of (5) of Example 4.13.1 and denoting differentiation with respect to t' by dot, with the help of (4.12.8) we obtain

$$\mathbf{E}(r,t) = \left[\frac{e\mathbf{n}}{R^2 g}\right]_{t'} + \frac{e}{c}\left[\frac{1}{g}(\mathbf{n}-\boldsymbol{\beta})\left\{\frac{1}{Rg^2}((\dot{\mathbf{n}}\cdot\boldsymbol{\beta}+(\mathbf{n}\cdot\dot{\boldsymbol{\beta}}))+\frac{c}{R^2 g}(\mathbf{n}\cdot\boldsymbol{\beta})\right\}+\frac{1}{Rg}(\dot{\mathbf{n}}-\dot{\boldsymbol{\beta}})\right]_{t'} . \quad (1)$$

Collecting all the terms containing $\dot{\beta}$ in one side, we find that

$$\begin{aligned}
\mathbf{E}(r,t) &= \left[\frac{e}{cg}\left\{\frac{1}{Rg^2}(\mathbf{n}-\boldsymbol{\beta})(\mathbf{n}\cdot\dot{\boldsymbol{\beta}})-\frac{1}{Rg}\dot{\boldsymbol{\beta}}\right\}+\frac{e\dot{\mathbf{n}}}{R^2 g}+\frac{e}{Rcg^2}(\dot{\mathbf{n}}\cdot\boldsymbol{\beta})(\mathbf{n}-\boldsymbol{\beta})\right.\\
&\qquad \left.+\frac{e}{R^2 g^2}(\mathbf{n}\cdot\boldsymbol{\beta})(\mathbf{n}-\boldsymbol{\beta})+\frac{e\dot{\mathbf{n}}}{Rcg^2}\right]_{t'}\\
&= \left[\frac{e\mathbf{n}\times((\mathbf{n}-\boldsymbol{\beta})\times\dot{\boldsymbol{\beta}})}{cRg^3}\right]_{t'}+\left[\frac{e\mathbf{n}}{R^2 g}+\frac{e}{R^2 g^3}(\mathbf{n}-\boldsymbol{\beta})\{(\mathbf{n}\times(\mathbf{n}\times\boldsymbol{\beta}))\cdot\mathbf{n}\}\right.\\
&\qquad \left.+\frac{e}{R^2 g^2}(\mathbf{n}-\boldsymbol{\beta})(\mathbf{n}\cdot\boldsymbol{\beta})+\frac{e}{R^2 g^2}(\mathbf{n}\times(\mathbf{n}\times\boldsymbol{\beta}))\right]_{t'} , \qquad (2)
\end{aligned}$$

where we have used (4.12.8) in the first term and (9) of Exercise 4.13.1, for finding $\mathbf{n}$. Hence, the second term of (2) becomes

$$\begin{aligned}
&= \left[\frac{e\mathbf{n}}{R^2 g}-\frac{e}{R^2 g^3}(\mathbf{n}\times\boldsymbol{\beta})^2(\mathbf{n}-\boldsymbol{\beta})+\frac{e}{R^2 g^2}\{\mathbf{n}(\mathbf{n}\cdot\boldsymbol{\beta})-\boldsymbol{\beta}(\mathbf{n}\cdot\boldsymbol{\beta})+(\mathbf{n}\cdot\boldsymbol{\beta})\mathbf{n}-\boldsymbol{\beta}\}\right]_{t'}\\
&= \left[\frac{e}{R^2}\left\{\frac{1}{g}\mathbf{n}-\frac{(\mathbf{n}\times\boldsymbol{\beta})^2}{g^3}(\mathbf{n}-\boldsymbol{\beta})+\frac{2}{g^2}(\mathbf{n}\cdot\boldsymbol{\beta})\mathbf{n}-\frac{1}{g^2}\boldsymbol{\beta}\{1+(\mathbf{n}\cdot\boldsymbol{\beta})\}\right\}\right]_{t'}\\
&= \left[\frac{e}{R^2}\left\{\frac{\mathbf{n}}{g}(1+(\mathbf{n}\cdot\boldsymbol{\beta}))-\frac{(\mathbf{n}\times\boldsymbol{\beta})^2}{g^3}(\mathbf{n}-\boldsymbol{\beta})-\frac{1}{g^2}\boldsymbol{\beta}\{1+(\mathbf{n}\cdot\boldsymbol{\beta})\}\right\}\right]_{t'}\\
&= \left[\frac{e}{R^2}\left\{\frac{1}{g^2}(\mathbf{n}-\boldsymbol{\beta})(1+(\mathbf{n}\cdot\boldsymbol{\beta}))-\frac{(\mathbf{n}\times\boldsymbol{\beta})^2}{g^3}(\mathbf{n}-\boldsymbol{\beta})\right\}\right]_{t'}\\
&= \left[\frac{e}{R^2 g^3}(\mathbf{n}-\boldsymbol{\beta})\{1-|n|^2|\beta|^2\}\right]_{t'} = \left[\frac{e}{R^2 g^2}(\mathbf{n}-\boldsymbol{\beta})(1-|n|^2|\beta|^2\right]_{t'}\\
&= \left[\frac{e(\mathbf{n}-\boldsymbol{\beta})}{R^2 g^3}(1-\beta^2)\right]_{t'} \qquad (3)
\end{aligned}$$

and the desired expression for $\mathbf{E}$ in terms of t' is

$$\mathbf{E}(r,t) = \left[\frac{e}{R^2 g^3}(\mathbf{n}-\boldsymbol{\beta})(1-\beta^2)\right]_{t'} + \left[\frac{e}{cRg^2}\mathbf{n}\times\{(\mathbf{n}-\boldsymbol{\beta})\times\dot{\boldsymbol{\beta}}\}\right]_{t'} . \qquad (4)$$

4.14 Radiation Power

The first term of (4.13.24), inversely proportional to R^2, is independent of the acceleration $\dot{v}$, the second term, inversely proportional to R, depends on $\dot{\mathbf{v}}$. So the first term gives the static part of the field of the moving charge, and at large distances, vanishes faster than the second term. Hence

$$\mathbf{E}(\mathbf{r}, t) = \left[\frac{e}{cRg^3}\mathbf{n} \times \{(\mathbf{n} - \boldsymbol{\beta}) \times \dot{\boldsymbol{\beta}}\}\right]_{t'} . \tag{4.14.1}$$

The Poynting flux of energy, using (4.13.29), is

$$\mathbf{S} = \frac{c}{4\pi}(\mathbf{E} \times \mathbf{H}) = \frac{c}{4\pi}(\mathbf{E} \times (\mathbf{n} \times \mathbf{E})) = \frac{CE^2}{4\pi}\mathbf{n}. \tag{4.14.2}$$

This energy flux is radial in direction and the field spreads out as spherical wave radially from the accelerated electron. The total radiation passing through the surface Σ of a sphere of radius R is

$$\iint_{\Sigma} (\mathbf{s} \cdot d\boldsymbol{\Sigma}) = \frac{c}{4\pi} \iint_{\Sigma} R^2 E^2 d\Omega \tag{4.14.3}$$

$$= \frac{c}{4\pi} \iint R^2 E^2 \sin\theta d\theta d\Omega, \tag{4.14.4}$$

where the integral of $\mathbf{S}$ is taken over Σ, and $d\Omega (= (d\Sigma/R^2) = \sin\theta d\theta d\phi)$ is the solid angle element subtended at the seat Q of the charge, Fig. 4.10.1.

If $dP(t)/d\Omega$ is the energy flowing out, per unit time, measured by the observer at time t, due to emission by the charge at time t', then

$$\frac{dP(t)}{d\Omega} = R^2(S \cdot n). \tag{4.14.5}$$

Further, if $dP(t')/d\Omega$ is the energy flowing out per unit time at time t' at the charge, then

$$\frac{dP(t')}{d\Omega} = (S \cdot n)R^2\frac{dt}{dt'} = R^2 g(\mathbf{S} \cdot \mathbf{n}), \tag{4.14.6}$$

where g is the given by (4.12.12), and (4.13.16) has been used. Hence,

$$\frac{dP(t')}{d\Omega} = \frac{e^2}{4\pi c g^2}|\mathbf{n} \times ((\mathbf{n} - \boldsymbol{\beta}) \times \dot{\boldsymbol{\beta}})|^2. \tag{4.14.7}$$

The conditions $\beta = 0$, $\dot{\beta} \neq 0$ are valid approximately when $v \ll c$, giving

$$\frac{dP}{dt} = 1, \quad \mathbf{E} = \frac{e}{cR}\{(\mathbf{n} \cdot \dot{\boldsymbol{\beta}})\mathbf{n} - \dot{\boldsymbol{\beta}}\},$$

$$E^2 = \frac{e^2}{c^2R^2}\{\dot{\beta}^2 - (n \cdot \dot{\beta})^2\} = \frac{e^2}{c^2R^2}(\mathbf{n} \times \dot{\boldsymbol{\beta}})^2, \quad \frac{dP}{d\Omega} = \frac{e^2\dot{v}^2}{4\pi c^2}\sin^2\theta, \tag{4.14.8}$$

where θ is the angle made by $\mathbf{n}$ about $\boldsymbol{\beta}$ (polar axis) and $dP/d\Omega$ is the instantaneous power radiated per unit solid angle. The power radiated in all directions (the Larmor formula) is found by integrating $dP/d\Omega$ over all solid angles;

$$P = \frac{2e^2}{3c^3}\dot{v}^2 \tag{4.14.9}$$

$$= \frac{2e^2}{3m^2c^3}\left(\frac{dp}{dt}\cdot\frac{dp}{dt}\right), \tag{4.14.10}$$

where p is the particle momentum mv.

The relativistically correct formula for P, in 4-space, Boyd and Sanderson (1969, p. 256) and Jackson (1975, p. 660), is

$$P = \frac{2e^2}{3m^2c^3}\left(\frac{dp_\mu}{dT'}\frac{dp_\mu}{dT'}\right), \tag{4.14.11}$$

where dT' is the proper time element at the retarded time in the source, and p_μ is energy momentum 4-vector. Since $\mathbf{p} = m\gamma\mathbf{v}$, m is the rest mass, and $mc\gamma = (m^2c^2 + p^2)^{1/2}$, we find that

$$\frac{dp_\mu}{dT'}\frac{dp_\mu}{dT'} = \left(\frac{dp}{dT'}\right)^2 - \frac{1}{c^2}\left(\frac{d}{dT'}(\gamma mc^2)\right)^2 \tag{4.14.12}$$

$$= \left(\frac{dp}{dT'}\right)^2 - \left(\beta\cdot\frac{dp}{dT'}\right)^2 \tag{4.14.13}$$

$$\left\{\frac{d}{dT'}(mc\gamma)\right\}^2 = \left\{\frac{1}{mc\gamma}\left(p\cdot\frac{dp}{dT'}\right)^2\right\} = \left(\beta\cdot\frac{dp}{dT'}\right)^2, \tag{4.14.14}$$

$$\frac{dp}{dT'} = \gamma\frac{dp}{dt'} = mc\gamma^2\{\beta + \gamma^2(\beta\cdot\dot{\beta})\beta\}, \tag{4.14.15}$$

where again the derivative with respect to the retarded time t' is denoted by dot. Hence

$$\begin{aligned}\left(\frac{dp}{dT'}\right)^2 - \left(\beta\cdot\frac{dp}{dT'}\right)^2 &= m^2c^2\gamma^4\{\dot{\beta}^2 + 2r^2(\beta\cdot\dot{\beta})^2 + \gamma^4(\beta\cdot\dot{\beta})^2\beta^2\} \\ &\quad -m^2c^2\gamma^4\{(\beta\cdot\dot{\beta}) + \gamma^2\beta^2(\beta\cdot\dot{\beta})\}^2 \\ &= m^2c^2\gamma^4\{\dot{\beta}^2 + (\beta\cdot\dot{\beta})^2(2\gamma^2 + \gamma^4\beta^2 - \gamma^4)\} \\ &= m^2c^2\gamma^4\{\dot{\beta}^2 + \gamma^2(\beta\cdot\dot{\beta})^2\} \\ &= m^2c^2\gamma^6(\dot{\beta}^2 - \beta^2\dot{\beta}^2\sin^2\psi) = m^2c^2\gamma^6\{\dot{\beta}^2 - (\beta\times\dot{\beta})^2\},\end{aligned} \tag{4.14.16}$$

where ψ is the angle between β and $\dot{\beta}$. Hence (4.14.11) becomes

$$P = \frac{2e^2\gamma^6}{3c}\{\dot{\beta}^2 - (\beta\times\dot{\beta})^2\} = \frac{2e^2}{3c}\gamma^4\{\dot{\beta}^2 + \gamma^2(\beta\cdot\dot{\beta})^2\}. \tag{4.14.17}$$

Also, the variation of (4.14.7), with respect to the solid angle Ω, gives this formula, Clemmow and Dougherty (1969, section 2.2.6). A more complete 4-vector deduction of (4.14.17) is available in Landau and Lifshitz (1975, section 76).

Here we consider only the classical theory of radiation from free charges. Since electrons repel each other, their accelerations are equal in magnitude and opposite in direction. So, to a first approximation, the resultant radiation from two closely situated electrons, though both are accelerated, is zero. The radiation from electron gyration in presence of static magnetic field, is the cyclotron radiation or synchroton radiation. In the radiation through the Cerenkov process (section 2.35) the particle velocity is faster than the velocity of light in a medium.

Changes in vibration, rotation, and electronic states, of di-or polyatomic gaseous matter, on heating, emit spectral bands. Further heating dissociates the molecules into atoms. At MHD power generation temperatures, the radiation is mainly from changes in vibration and rotational energy of the molecules, and is of quantum mechanical origin.

The conditions in the radio sources emitting a continuous spectrum of waves, cannot be determined uniquely from the nature of the radiation received. Synchrotron radiation is more important than bremsstrahlung and inverse Compton effect for study of astrophysical bodies. Bremsstrahlung is the radiation from a hot ionized medium at high frequencies and when it is free from any strong applied static magnetic field. In other words, this is the radiation from collisionally accelerated electrons.

4.14.1 Synchrotron radiation

Charged particles have circle-like, or spiral-like, ordered velocity around static magnetic fields, which is an accelerated motion. Here we consider only the emission from gyrating electrons.

In the presence of an applied constant magnetic field H, the relativistic equation of motion is

$$\frac{d}{dt}(\gamma \mathbf{v}) = -(\mathbf{v} \times \mathbf{\Omega}),$$

where $\Omega_0 = eH/mc$ and $\gamma^2 = 1/(1 - v^2/c^2)$. When this motion is in the plane perpendicular to the direction of the magnetic field $\mathbf{H}$, $(\boldsymbol{\beta} \cdot \dot{\boldsymbol{\beta}}) = 0$ and $\dot{\boldsymbol{\beta}} = -(\boldsymbol{\beta} \times \mathbf{\Omega}_0)/\gamma$, $\boldsymbol{\beta} = \mathbf{v}/c$. Putting this value of $\dot{\beta}$ in (4.14.17) we obtain the synchrotron radiation formula

$$P = \frac{2e^2}{3c}\gamma^4 \frac{(\beta \times \Omega_0)^2}{\gamma^2} = \frac{2e^2 \Omega_0^2 \beta^2}{3c(1 - \beta^2)}.$$

4.14.2 Radiation from plasma

A radiation can be absorbed by matter when the dipole oscillations are excited. The excited dipoles can re-radiate the absorbed energy at frequencies convenient to its environment, which means scattering of other radiations. The loss of energy of the excited dipoles, through re-radiation, causes damping and exponential decay of their oscillation intensity.

Radiation from a plasma determines its thermal conditions and the degree of ionization. It is the only source of information on the processes that take place in interstellar and solar plasma.

We can write

$$\sum_i (e_i \dot{\mathbf{v}}_i) = \frac{d^2}{dt^2} \sum_i \left\{ \left(\frac{e_i}{m_i}\right) m_i \mathbf{r}_i(t) \right\} = \left(\frac{e}{m}\right)\left(\frac{d^2}{dt^2}\right) \sum_i \{m_i \mathbf{r}_i(t)\} = \frac{e}{m}\frac{d^2}{dt^2}(\mathbf{RM})$$

if the ratio e/m of charge and mass has the same value, for a dynamical system, where

$$\sum_i (m_i r_i) = R\left(\sum_i m_i\right) = RM;$$

so the total mass is M, and $\mathbf{R}$ is the position vector of the center of gravity (c.g.) of the system. When the c.g. is at rest or in uniform motion,

$$\sum_i (e_i \dot{\mathbf{v}}_i) = 0,$$

and equation (4.14.9), for the radiation power, gives $P = 0$. Then, moreover, the dipole radiation power also vanishes, Bekefi (1966, p. 67).

Including the radiation damping term $(2e^2/3c^3)\ddot{v}$ of equation (4.15.4), in a Lorentz force field, the equation for electron motion is

$$m\dot{\mathbf{v}} = -e\mathbf{E} - \frac{e}{c}(\mathbf{v} \times \mathbf{H}) + \frac{2e^2}{3c^2}\ddot{\mathbf{v}}.$$

In the classical model of a bound electron, executing a simple harmonic oscillation of angular frequency ω_0, and under the influence of an applied electric field $\mathbf{E}$, the equation of motion

$$m\ddot{\mathbf{r}} = -m\omega_0^2\mathbf{r} + \left(\frac{2e^2}{3c^3}\right)\dddot{\mathbf{r}} - e\mathbf{E}$$

gives the Rayleigh scattering cross section.

4.14.3 The bremsstrahlung

Bremsstrahlung is the emission by decelerated charged particles. The deceleration may be from collisions of Coulomb interaction with particles of the neighbourhood. The motion is slowed down by the emission of a gamma photon.

When electrons are distributed uniformly, and isotopically, they approach the heavy ions from all directions with all possible speeds. This radiation is isotropic, and the polarization is random. Bremsstrahlung, in plasma, is the radiation from electrons, due to the electric field of the nearby positive charges, and due to encounters between electrons and atoms. The former is dominant when ionization is very high, and the latter when ionization is weak. We only consider, classically, the former type of radiation here, for non-relativistic energies in low density plasmas in the absence of strong static magnetic fields. The minimum distance of impact is the de Broglie wave length of the particles.

Putting $\dot{v} = f = e^2/mR^2$ in the equation (4.14.9) for P, for non-relativistic energies, we obtain

$$P = \frac{2e^6}{3m^2c^3R^4},$$

where R is the distance of the electron from an ion.

Let N_e and N_i be the number densities of electrons and ions respectively. Multiplying this P by $4\pi R^2 dRN_e$, and integrating with respect to R, gives the total radiation from the electrons,

$$P_1 = \int_{R_{\min}}^{\infty} P.4\pi R^2 N_e dR = \frac{2\pi}{3}\frac{N_e e^6}{c^3m^2R_{\min}}.$$

Here $R_{\min}(= h/p = \lambda_1)$ is the distance up to which the classical approximation holds good, where λ_1 is the de Broglie wave length, h is the Planck constant and $p(= \sqrt{mk_BT}$, k_B is the

Boltzmann constant and T is electron kinetic temperature) is the particle momentum. For high electron energies λ_1 is larger than the minimum collision impact parameter. The total radiation, per unit volume of plasma, P_T, is N_i times P_1; so

$$P_T = \frac{8\pi}{3}\frac{e^6 N_e N_i}{m^2 c^3 R_{\min}}.$$

The principal source of radiation from a hot and fully ionized unmagnetized plasma is this bremsstrahlung. In thermonuclear reactors, the energy loss by bremsstrahlung is considerable, compared to the rate of energy production by nuclear fuels.

4.15 Damping Force of Radiation

Radiation from a source means a loss of its energy, momentum and angular momentum. So, the electron motion is damped by radiation. This radiation reaction generates a self-force of damping, F_s, which is evaluated non-relativistically, here, from the condition of conservation of energy.

Assume that the motion is the same at two prescribed times t_1 and t_2; then the work done by the force F_s, on the electron, equals the negative of the energy radiated in the interval (t_1, t_2); so

$$\int_{t_1}^{t_2} (F_s \cdot v)\, dt = -\frac{2e^2}{3c^3}\int_{t_1}^{t_2} \dot{v}^2 dt. \tag{4.15.1}$$

If $\dot{v}$ is the same at the two times t_1 and t_2, we obtain

$$\int_{t_1}^{t_2} (\dot{\mathbf{v}} \cdot \dot{\mathbf{v}})\, dt = \int_{t_1}^{t_2} \left\{ \frac{d}{dt}(\dot{\mathbf{v}} \cdot \mathbf{v}) - (\ddot{\mathbf{v}} \cdot \mathbf{v}) \right\} dt = 0 - \int_{t_1}^{t_2} (\ddot{\mathbf{v}} \cdot \mathbf{v})\, dt, \tag{4.15.2}$$

$$\int_{t_1}^{t_2} (\mathbf{F}_s \cdot \mathbf{v})\, dt = \frac{2e^2}{3c^3}\int_{t_1}^{t_2} (\ddot{\mathbf{v}} \cdot \mathbf{v})\, dt. \tag{4.15.3}$$

Then the self-force of radiation, at any instance, is

$$F_s = \frac{2e^2}{3c^3}\ddot{v}. \tag{4.15.4}$$

This formula can also be derived by other methods, Dutta (1977, section 10.2).

In the classical model of a bound electron, executing a simple harmonic oscillation of angular frequency ω_0, and under the influence of an applied electric field $\mathbf{E}$, including this force, the equation of motion of a particle of charge $-e$ and mass m, in presence of the electric field E, for the classical model of a bound electron, executing a simple harmonic oscillation at the angular frequency ω_0 is

$$m\ddot{\mathbf{r}} = -m\omega_0^2\mathbf{r} + \frac{2e^2}{3c^3}\dddot{\mathbf{r}} - e\mathbf{E} - \frac{e}{c}(\dot{\mathbf{r}} \times \mathbf{H}), \tag{4.15.5}$$

where r is its field-induced displacement from the respective nucleus. The Rayleigh scattering cross section in plasmas and other gaseous media and modification in Rutherford scattering are studied with the help of this equation.

The radiation loss is very slow, and the radiation reaction force is a small perturbation, because the radiation damping time factor is about seven orders of magnitude larger than the oscillation time period. The condition of smoothing over many, oscillation, namely

$$\int_0^T \cos^2 \omega t dt = \int_0^T \sin^2 \omega t dt$$

is then valid, if $T \gg T_0$, where $T_0 (= 2\pi/\omega)$ is the time period of the radiation field, because the difference between these two integrals, from integration by parts of the left hand side, is the very small term

$$\frac{1}{\omega}\Big[\sin \omega t \cos \omega t\Big]_0^T$$

compare to the term retained on the right hand side which grows linearly with T and approaches $T/2$. If the radiation damping force per charged particle is denoted by f_r then $f_r = -m\gamma v$, where $\gamma = \mu_0 e^2\omega^2/(6\pi m^3 c) = 2e^2\omega^3/3mc^3$, where γ is the radiation damping constant. At optical frequencies γ is about 10^8/sec.

Relativistically correct equation of motion, Sokolov and Termov (1968, equation (1.2)), and radiation damping, Caldirola (1979) are

$$m\ddot{x}_\mu = -\frac{e}{c}\dot{x}_\nu F_{\mu\nu} + \frac{2e^2}{3c^3}(\dddot{x}_\mu - \frac{1}{c^2}\dot{x}_\mu \dot{x}_\nu \dddot{x}_\nu), \tag{4.15.6}$$

$$F_s = \frac{2e^2}{3c^3}\left[\ddot{\beta} + \gamma^2(\beta\cdot\ddot{\beta})\beta + 3\gamma^2(\beta\cdot\dot{\beta})\dot{\beta} + 3\gamma^4(\beta\cdot\dot{\beta})^2\beta\right], \tag{4.15.7}$$

where m is the rest mass, $x_\mu (\equiv r, ict)$ is the 4-vector of co-ordinates, the electromagnetic field tensor

$$F_{\mu\nu} = \begin{matrix} \to \nu \\ \downarrow \mu \end{matrix} \begin{vmatrix} 0 & H_3 & -H_2 & -iE_1 \\ -H_3 & 0 & H_1 & -iE_2 \\ H_2 & -H_1 & 0 & -iE_3 \\ iE_1 & iE_2 & iE_3 & 0 \end{vmatrix} \tag{4.15.8}$$

$\gamma^2 = 1/(1-\beta^2)$, and differentiation is done with respect to the proper time of equation (4.14.11). Some conceptual difficulties exist in the theory of radiation damping of an accelerated electron, Jackson (1975, Chapter 17) and Caldirola (1979).

Chapter 5

Fields and Motion in Electrodynamics

5.1 Introduction

Comparison conclusions, in electrodynamics involved processes, reveal the relative importance of effects in a problem. Scaling laws, and series expansion of fields and independent variables for specific purposes are some examples of the comparison conclusions. The number of field variables can be reduced for convenience of solution. Wave polarization, wave coherence, ray theory and wave theory of transmission of fields, geometrical optics, the eikonal equations, the group path, phase path of waves, Larmor precession theory, electron spin, scaling laws and expansion of variables for quasilinear coherent wave process, are discussed. Also considered are the co-ordinate expansion techniques and the transformation of variables, as well as their field equations. The Maxwell equations have been symmetrized by involving in these the fields of magnetic monopoles and their currents and the Bateman vectors.

5.2 Ray Theory

In ray theory the Maxwell equations, together with the constitutive relations are solved initially for study of wave properties in a material medium. For example, behaviour of the refractive index n and the wave amplitudes, the phase and group paths of propagation of disturbances from one point to another, time of flight, absorption, wave polarization etc. are discussed. The underlying assumption is that the properties of the medium vary sufficiently slowly, so that these behaviours do not change appreciably as the wave travels through the medium. The path of a wave packet (limited in space and time) is then specified from a knowledge of the refractive index properties of the medium.

There are certain conditions under which the ray approximation is not valid. Such conditions hold when the refractive index n changes appreciably within a wave length. To obtain the electric and magnetic fields under these conditions it is necessary to use the full-wave theory, which provides the solution of the Maxwell equations, together with the properties of the medium, and any boundary conditions, for every given point. This process is much more elaborate than that involved in ray theory, but it is necessary for sharply bounded media and near the level of reflection even in slowly varying media.

Energy flows between the transmitter and the receiver along the ray path. When the medium is anisotropic the direction of energy flow departs from that of phase propagation, so that distorted ray paths are obtained. Geometrical theory of propagation actually operates with bundles of rays, all of which are normal to some surface. Actually the wave nature of transmission of a disturbance is based on the phenomena of interference, the consequence of which is the location, under certain conditions, of regions of maximum intensity and zero intensity in space and time, of the perturbation. Geometrical theory would predict either only the prevalence of the maximum intensity or of zero intensity of such disturbance.

Rays are curves described in space such that the tangents to them passing through a space point are along the direction of propagation of the disturbance through the point. Hence a ray path is perpendicular to the transverse wave field and parallel to the longitudinal wave field. Bundles of rays obey the geometrical laws of propagation which should be expressed in terms of behaviour of individual rays. These bundles obey the laws of geometrical optics for electromagnetic and other types of transverse wave field; and the laws of geometrical acoustics for acoustic type of dispersive vibrations.

5.2.1 Basic elements of the ray theory

Consider the solution in the form

$$\phi = a(r,t)\exp(i\psi(r,t)) \tag{5.2.1}$$

of the Klein Gordon type of two-parameter wave equation

$$\nabla^2\phi - \frac{1}{c^2}\frac{\partial^2\phi}{\partial t^2} - \frac{\omega_0^2}{c^2}\phi = 0,$$

where ω_0 and c are the two parameters. It is more general than the one-parameter wave equation

$$\nabla^2\phi - \frac{1}{c^2}\frac{\partial^2\phi}{\partial t^2} = 0.$$

The intensity of the field is a measure of the energy flux and is proportional to the magnitude of the Poynting energy flux S^t for the EM field and of the mechanical energy flux S^l for the electro acoustic field. This flux being quadratic in the field variables should be proportional to $|\phi|^2$.

The amplitude 'a' is a more slowly varying function of space and time than ψ. The family of surfaces $\psi =$ constant are called *wave surfaces* and ψ is called the *phase* or the *eikoal* (or the eiconal). The direction of propagation of the wave at each point (or at each instant of time) is the direction of the normal to the wave surface through the given point (or at that instant). The phase function ψ has a large value because it changes by the dimensionless number 2π in the space of one wave length or in the time of one period. If several ray paths pass through a point simultaneously then instead of (5.2.1) we take

$$\phi = \sum_n [a_n \exp(i\psi_n)], \tag{5.2.2}$$

where the sum includes one term for each ray through the point $\mathbf{r}$.

If ψ is linearly dependent on t we can write

$$\psi(r,t) = \psi_0(r) - \omega t, \tag{5.2.3}$$

where ω is called the *frequency of the oscillation.* Then $\psi_0(r)$, like $\psi(r,t)$, is also called the *eikonal function.* The points, where $\psi = 0$, the wave is at its equilibrium position. At each instant of time these points form a set of continuous surfaces in space. The amplitude function 'a' is the envelope of the local maxima. The general definition of the wave frequency ω is

$$\frac{\partial \psi}{\partial t} = -\omega \tag{5.2.4}$$

ω can therefore be a function of $\mathbf{r}$ and t. The wave vector $\mathbf{K}$, in general, a function of r and t, is given by

$$\text{grad}\,\psi = K, \quad \text{or,} \quad \frac{\partial \psi}{\partial x_i} = K_i. \tag{5.2.5}$$

K must be irrotational because (5.2.5) gives curl $K = 0$. For the plane-wave front approximation the surfaces $\psi =$ constant are a system of parallel planes having normal parallel to $\mathbf{K}$ and moving with the phase velocity v_p. The direction of propagation of the ray through the point $\mathbf{r}$ is along $\mathbf{K}$. The dispersion relation for the fields satisfying the Klein Gordon equation being

$$K^2c^2 = \omega^2 - \omega_0^2,$$

we get

$$\sum_i \left(\frac{\partial \psi}{\partial x_i}\right)^2 = \frac{1}{c^2}\left(\frac{\partial \psi}{\partial t}\right)^2 - \frac{\omega_0^2}{c^2}\psi^2. \tag{5.2.6}$$

This is the equation for the eikonal when ψ is linearly dependent on ω and on the components of K. Occasionally the substitution

$$\psi_0(r) = K_0 S(r) \tag{5.2.7}$$

is used, where K_0 is the wave number in a standard medium. Also one uses

$$c_0 = \frac{1}{K_0}(\omega^2 - \omega_0^2)^{1/2}, \quad n = \frac{K}{K_0} = \frac{\omega}{K_0 c}\left(1 - \frac{\omega_0^2}{\omega^2}\right)^2 = \frac{c_0}{c}, \tag{5.2.8}$$

where n is the dimensionless local refractive index at the point $\mathbf{r}$. Then equation (5.2.6) becomes

$$\sum_i \left(\frac{\partial S}{\partial x_i}\right)^2 = n^2. \tag{5.2.9}$$

If the values of n and an initial surface $S(r) =$ constant are given, solutions of this equation can be obtained. For example, in uniform media n is constant. Then initially if the wave surface is a plane having direction cosines l_0, m_0, n_0 for its normals, a solution of (5.2.9) is

$$S(r) = n(l_0 x + m_0 y + n_0 z); \tag{5.2.10}$$

so

$$\text{grad}\, S = in, \tag{5.2.11}$$

where $\mathbf{i}$ is the unit vector along the direction of wave propagation. In general, also we can write

$$\text{grad}\, S = \mathbf{ni}(\mathbf{r}), \tag{5.2.12}$$

where $\mathbf{i}$ is the unit vector along the ray direction at the point $\mathbf{r}$; so, $\mathbf{i}$ is normal to S at $\mathbf{r}$. For surfaces more complicated than the plane given by (5.2.10) the relation (5.2.11) may be used because continuous curves of rays can be drawn in space such that at the point $\mathbf{r}$ the tangent is along the direction of $\mathbf{i}(\mathbf{r})$.

5.2.2 The eikonal equations

To find the full set of equations determining the behaviour of $\psi(\mathbf{r}, t)$ and $\mathbf{a}(\mathbf{r}, t)$ the value of ϕ given in (5.2.1) is substituted into the Klein Gordon equation. The result is the relation

$$\left[\frac{1}{a}\nabla^2 a + \frac{2i}{a}\sum_i\left(\frac{\partial a}{\partial x_i}\frac{\partial \psi}{\partial x_i}\right) + i\nabla^2\psi - \sum_i\left(\frac{\partial \psi}{\partial x_i}\right)^2 - \frac{\ddot{a}}{c^2 a} - \frac{i}{c^2}\ddot{\psi}\right.$$

$$\left. + \frac{1}{c^2}\dot{\psi}^2 - \frac{\omega_0^2}{c^2} - \frac{2}{c^2}i\dot{a}\dot{\psi}\right]\psi = 0. \tag{5.2.13}$$

Assuming a and ψ to be real, using the fact that ϕ is not zero for all the values of $\mathbf{r}, t$ and separating the real and the imaginary parts we obtain

$$a(\nabla\psi)^2 + \frac{1}{c^2}\ddot{a} - \frac{1}{c^2}a\dot{\psi}^2 + \frac{1}{c^2}a\omega_0^2 - \nabla^2 a = 0, \tag{5.2.14}$$

$$a\nabla^2\psi + 2(\nabla a \cdot \nabla\psi) - \frac{a}{c^2}\ddot{\psi} - \frac{2}{c^2}\dot{a}\dot{\psi} = 0. \tag{5.2.15}$$

Multiplication of (5.2.15) by 'a' gives

$$a^2\nabla^2\psi + (\nabla a^2 \cdot \nabla\psi) - \frac{1}{c^2}(a^2\ddot{\psi} + 2\dot{a}\dot{\psi}) = 0.$$

Using (5.2.4) and (5.2.5) this equation can be written as

$$a^2(\boldsymbol{\nabla}\cdot\mathbf{K}) + (\mathbf{K}\cdot\boldsymbol{\nabla}a^2) + \frac{\partial}{\partial t}\left(\frac{a^2}{c^2}\omega\right) = 0$$

because a^2 is independent of t. In other words,

$$\boldsymbol{\nabla}\cdot(\mathbf{K}a^2) + \frac{\partial}{\partial t}\left(\frac{a^2}{c^2}\omega\right) = 0. \tag{5.2.16}$$

This equation looks like the equation of continuity if Ka^2 is proportional to the flow of mass or charge per unit volume per unit time and $a^2\omega/c^2$ is proportional to the mass density. If Kc^2/ω is proportional to the velocity v of flow and $a^2\omega/c^2$ to the material density the continuity relation is the conservation of $a^2\omega/c^2$ in time. Instead, if $a^2\omega/c^2$ is a measure of the energy density and Kc^2/ω proportional to the velocity of the energy flux at the point $\mathbf{r}$, then the rays appear as paths of propagation of energy, a thin pencil of rays being analogous to a tube through which the energy flows.

The condition for which the relation (5.2.9) is approximately true around the point $\mathbf{r}$ in an inhomogeneous medium would now be determined. Equation (5.2.14) shows that this approximation tantamounts to a smallness of $(\nabla^2 a)/a$ and $\ddot{a}/(c^2 a)$ compared to the remaining three terms. Since the variation of 'a' in time is small in comparison with that of ψ,

$$(\ddot{a}/a) \pm \dot{\psi}^2 \approx \pm\dot{\psi}^2;$$

so $\ddot{a}/c^2$ is neglected in (5.2.14). To show the smallness of $(\nabla^2 a)/a$ the amplitude 'a' is expanded in Taylor's series. Using the notation of repeated index to denote summation over that index, one can write

$$a(\mathbf{r} + \delta\mathbf{r}) - a(\mathbf{r}) = \frac{\partial a}{\partial x_i}\delta x_i + \frac{1}{2}\frac{\partial^2 a}{\partial x_i \partial x_j}\delta x_i \delta x_j + \cdots.$$

Let the distance between the points $\mathbf{r}+\delta\mathbf{r}$, $\mathbf{r}$ be of the order of the wave length $\lambda_0(=2\pi/k_0)$. Then

$$\lambda_0^2 = \delta x_i \delta x_i = (\delta r)^2.$$

Taking average over a sphere of radius λ_0 with center at the point $\mathbf{r}$ and denoting such an average of x by $\langle x \rangle$ we have

$$\langle \delta x_i \rangle = 0, \quad \langle \delta x_i \delta x_j \rangle = 0, \quad i \neq j,$$

$$\langle (\delta x_1)^2 \rangle = \langle (\delta x_2^2) \rangle = \langle (\delta x_3)^2 \rangle = \frac{1}{3}\langle \delta x_i \delta x_i \rangle = \frac{1}{3}\lambda_0^2.$$

Hence

$$\langle a(\mathbf{r}+\delta\mathbf{r}) - a(\mathbf{r}) \rangle = \frac{1}{2}\left[\frac{\partial^2 a}{\partial x_i^2}\langle (\delta x_i)^2 \rangle\right] = \frac{1}{6}\lambda_0^2 \sum_i \left(\frac{\partial^2 a}{\partial x_i^2}\right)$$

because the quantities $\partial^2 a/\partial x_i^2$ $(i = 1, 2, 3)$ are evaluated at the center of the sphere and so are not affected by the averaging process. Therefore, when the inequality $\lambda_0^2 \nabla^2 a \ll a$ is true we would have

$$\langle a(\mathbf{r}+\delta\mathbf{r}) - a(\mathbf{r}) \rangle = (\lambda_0^2/6)\nabla^2 a \ll a/6.$$

In this situation the average of the fluctuations of 'a' around the point $\mathbf{r}$, within the distance λ_0, is much smaller than 'a' itself. This is the condition of validity of geometrical optics (or geometrical acoustics) because it is the condition of validity of (5.2.9) which is equivalent to the Hamilton-Jacobi partial differential equation for an imaginary particle yielding the Hamiltonian H of (5.2.4) and the particle momentum p of (5.2.5). Hence the equation (5.2.9) can be regarded as the approximation of geometrical optics (or of geometrical acoustics) of the more general equation (5.2.14).

Ignoring the time dependence of 'a' and regarding ω as constant, equation (5.2.15) can be written as

$$-\frac{1}{2}\operatorname{div}\operatorname{grad}\psi = \frac{1}{a}(\operatorname{grad} a \cdot \operatorname{grad}\psi).$$

Now replacing ψ by $S(\mathbf{r})$ with the help of (5.2.3), (5.2.7)–(5.2.9) and (5.2.12) we get

$$\frac{n}{a}\frac{da}{ds} = -\frac{1}{2}\boldsymbol{\nabla}\cdot(n\mathbf{i}),$$

where ds is an element of arc along $\operatorname{grad} S$. Ignoring $\operatorname{div}\mathbf{i}$ this relation becomes

$$\frac{1}{a}\frac{da}{ds} = -\frac{1}{2n}(\mathbf{i}\cdot\boldsymbol{\nabla}n). \tag{5.2.17}$$

So the change of amplitude 'a' can be determined only along a ray and not in the plane perpendicular to the tangent to the ray.

To eliminate S from the ray trajectory equation it becomes necessary to find the rate of change of $n\mathbf{i}$ along a ray. For this using (5.2.11) we get

$$\frac{d}{ds}(n\mathbf{i}) = \frac{d}{ds}(\boldsymbol{\nabla}S) = (\mathbf{i}\cdot\boldsymbol{\nabla})\boldsymbol{\nabla}S. \tag{5.2.18}$$

Again, using (5.2.11) to eliminate $\mathbf{i}$ from the right hand side we find that

$$\frac{d}{ds}(n\mathbf{i}) = \frac{1}{n}(\boldsymbol{\nabla}S\cdot\boldsymbol{\nabla})(\boldsymbol{\nabla}S) = \frac{1}{2n}\boldsymbol{\nabla}(\boldsymbol{\nabla}S^2) = \frac{1}{2n}\boldsymbol{\nabla}n^2. \tag{5.2.19}$$

So the trajectory equation is

$$\frac{d}{ds}(n\mathbf{i}) = (\nabla n). \tag{5.2.20}$$

This ray curve equation can be solved when the local value of the refractive index n and the initial direction $\mathbf{i}_0$ are given.

If $\mathbf{j}$ is the unit vector along the principal normal and ρ is the radius of curvature, then by differential geometry

$$\frac{d\mathbf{i}}{ds} = \frac{\mathbf{j}}{\rho}. \tag{5.2.21}$$

Therefore,

$$\frac{1}{\rho} = \left(\mathbf{j}\cdot\frac{d\mathbf{i}}{ds}\right) = \frac{1}{n}(\mathbf{j}\cdot\nabla n) + 0 = (\mathbf{j}\cdot\nabla\log n);$$

so the rays bend towards the region of increasing n.

The parametric equations to the ray curve are

$$x = x(T), \quad y = y(T), \quad z = z(T), \tag{5.2.22}$$

where T is the parameter. Direction ratios of the normal to the wave surface $S(\mathbf{r}) =$ constant at the point $\mathbf{r}$ are the components of grad S. Also, if $\mathbf{r}$, $\mathbf{r} + d\mathbf{r}$ are the position vectors of two points on the ray curve, then $d\mathbf{r}$ is proportional to grad S. Hence

$$\frac{dx}{dT} = \lambda\frac{\partial S}{\partial x}, \quad \frac{dy}{dT} = \lambda\frac{\partial S}{\partial y}, \quad \frac{dz}{dT} = \lambda\frac{\partial S}{\partial z}, \tag{5.2.23}$$

where λ is the constant of proportionality. Eliminating components of grad S between these equations and using (5.2.9) we get

$$\left(\frac{dx}{dT}\right)^2 + \left(\frac{dy}{dT}\right)^2 + \left(\frac{dz}{dT}\right)^2 = \lambda^2 n^2 \tag{5.2.24}$$

so

$$|ds| = |\lambda n dT|. \tag{5.2.25}$$

For getting the parametric equations of the ray path in terms of the eikonal function and not in terms of $S(r)$, the first of the equations (5.2.23) is divided by λ and differentiated with respect to T. This gives

$$\frac{d}{dT}\left(\frac{1}{\lambda}\frac{dx}{dT}\right) = \frac{dx}{dT}\frac{\partial^2 S}{\partial x^2} + \frac{dy}{dT}\frac{\partial^2 S}{\partial x \partial y} + \frac{dz}{dT}\frac{\partial^2 S}{\partial x \partial z}. \tag{5.2.26}$$

Now using again (5.2.23) this equation reduces to

$$\frac{d}{dT}\left(\frac{1}{\lambda}\frac{dx}{dT}\right) = \lambda\left(\frac{\partial S}{\partial x}\frac{\partial^2 S}{\partial x^2} + \frac{\partial S}{\partial y}\frac{\partial^2 S}{\partial x \partial y} + \frac{\partial S}{\partial z}\frac{\partial^2 S}{\partial x \partial z}\right)$$

$$= \frac{\lambda}{2}\frac{\partial}{\partial x}\left\{\left(\frac{\partial S}{\partial x}\right)^2 + \left(\frac{\partial S}{\partial y}\right)^2 + \left(\frac{\partial S}{\partial z}\right)^2\right\}.$$

Eliminating S with the help of (5.2.9) gives

$$\frac{2}{\lambda}\frac{d}{dT}\left(\frac{1}{\lambda}\frac{dx}{dT}\right) = \frac{\partial}{\partial x}(n^2).$$

Therefore, the set of three second order differential equations for the ray curve, free from $S(r)$, is

$$\frac{2}{\lambda} - \frac{d}{dT}\left(\frac{1}{\lambda}\frac{dx_i}{dT}\right) = \frac{\partial}{\partial x_i}(n^2). \tag{5.2.27}$$

On integration they yield six constants all of which cannot take up independent values because at least one restriction is given by condition (5.2.24).

The quantity T can be chosen in such a way that $\lambda = 1$. Then equations (5.2.27) and (5.2.24) can be expressed as

$$\ddot{x}(T) = \frac{\partial}{\partial x}\left(\frac{n^2}{2}\right), \quad \ddot{y}(T) = \frac{\partial}{\partial y}\left(\frac{n^2}{2}\right), \quad \ddot{z}(T) = \frac{\partial}{\partial z}\left(\frac{n^2}{2}\right), \tag{5.2.28}$$

where $\dot{x}(T)$ means derivative of x with respect to T. Therefore, in a region characterized by an index of refraction $n(\mathbf{r})$ the rays are curves which satisfy the vector equation

$$\ddot{\mathbf{r}} = \text{grad}\left(\frac{n^2}{2}\right), \tag{5.2.29}$$

where the parameters T is normalized by the condition

$$\dot{r}^2 = n^2. \tag{5.2.30}$$

If now in (5.2.25) the elementary curve length dS along the ray trajectory is equated to cdt, then the parameter T is linearly related to the time variable t according to the relation

$$ct = nT. \tag{5.2.31}$$

At each point of the ray

$$\phi = a\exp\{ik_0 S(r) - i\omega t\}, \tag{5.2.32}$$

where the phase function $S(r)$ is given in terms of the ray length by the relation

$$S(r) = S_0 + \int_0^S n[r(s)]ds. \tag{5.2.33}$$

Here the phase S_0 is constant and is the value of S at the point P_0 from which the arc length S is measured along the ray.

The system of equations (5.2.29) and (5.2.31) can have various types of solution depending on the nature of the problem. When resolved in spherical polar co-ordinates r, θ, ϕ these vector equations give the four equations

$$\ddot{r} - r\dot{\theta}^2 - r\sin^2\theta\dot{\phi}^2 = n\frac{\partial n}{\partial r}, \tag{5.2.34}$$

$$\frac{1}{r}\frac{d}{dT}(r^2\dot{\theta}) + r\sin\theta\cos\theta\dot{\phi}^2 = \frac{n}{r}\frac{\partial n}{\partial \theta}, \tag{5.2.35}$$

$$\dot{r}\dot{\phi}\sin\theta + \frac{d}{dT}(r\sin\theta\dot{\phi}) + r\dot{\theta}\dot{\phi}\cos\theta = \frac{n}{r\sin\theta}\frac{\partial n}{\partial \phi}, \tag{5.2.36}$$

$$\dot{r}^2 + r^2\dot{\theta}^2 + r^2\sin^2\theta\dot{\phi}^2 = n^2, \tag{5.2.37}$$

where dot on a quantity means the derivative of the corresponding undotted quantity with respect to T.

Keller and Levy (1963) have studied the case in which the ray trajectory lies in the plane $\phi = 0$ and the medium is spherically stratified. See Landau and Lifshitz (1960, Chapter X; 1959, p. 66); Kline and Kay (1965, Chapter II, p. 4, 6); Keller and Levy (1963, pages 1–24), Brandstatter (1963, pages 7–13).

5.2.3 Rays as paths of corpuscles

From the knowledge of particle dynamics it is evident that the motion of a particle of unit mass, moving in the conservative field of force of potential

$$U = -\frac{n^2}{2} \tag{5.2.38}$$

in the time scale

$$T = \frac{ct}{n}$$

is given by the set of equations (5.2.29) and (5.2.30). The total energy of the particle, being equal to the sum of the kinetic and potential energies, is

$$\frac{\dot{r}^2}{2} + U = 0. \tag{5.2.39}$$

So the rays are paths of imaginary particles moving with zero total energy. In (5.2.14), ignoring $\ddot{a}$ and $\nabla^2 a$ and putting

$$\psi = \frac{W}{\alpha}, \tag{5.2.40}$$

where α is constant we get

$$(\nabla W)^2 = \frac{1}{c^2}\left(\frac{\partial W}{\partial t}\right)^2 - \frac{W_0^2}{c^2}\alpha^2.$$

The value of α would be adjusted in such a way that W becomes the work function. Regarding W as 'action' of particle mechanics, we can define the Hamilton energy function

$$H = -\frac{\partial W}{\partial t} \tag{5.2.41}$$

and the particle momentum vector $\mathbf{p}$ as

$$\mathbf{p} = \text{grad}\, W. \tag{5.2.42}$$

Then the Hamiltonian

$$H = (c^2 p^2 + \alpha^2 W_0^2)^{1/2}.$$

The Hamiltonian equations of particle mechanics now are

$$\dot{p} = \nabla\left(\frac{\partial W}{\partial t}\right) = -\nabla H = -p\nabla\left(c^2 + \frac{1}{p^2}\alpha^2 W_0^2\right)^{1/2} \tag{5.2.43}$$

$$v = \dot{r} = \frac{\partial H}{\partial p} = \frac{cp}{\left(p^2 + \frac{1}{c^2}\alpha^2 W_0^2\right)^{1/2}}. \tag{5.2.44}$$

Comparison of (5.2.41) and (5.2.42) with the ray theory results gives

$$\mathcal{W} = -\frac{\partial \psi}{\partial t}, \quad k = \nabla \psi. \tag{5.2.45}$$

Hence H becomes analogous to the frequency $\mathcal{W}$ and p to k when the action W is related to ψ by (5.2.40). Then the equations to the ray theory, corresponding to the Hamiltonian equations of mechanics, are

$$\dot{K} = -\nabla w, \quad \dot{r} = \frac{\partial \mathcal{W}}{\partial K}. \tag{5.2.46}$$

The rate of variation of $\mathcal{W}$ along a ray path, being the total time derivative of $\mathcal{W}$, is

$$\frac{d\mathcal{W}}{dt} = \frac{\partial \mathcal{W}}{\partial t} + (\dot{r}\cdot \nabla)\mathcal{W} + \left(\dot{K}\cdot \frac{\partial \mathcal{W}}{\partial K}\right). \tag{5.2.47}$$

Using (5.2.46) this equation reduces to

$$\frac{d\mathcal{W}}{dt} = \frac{\partial \mathcal{W}}{\partial t}. \tag{5.2.48}$$

Since, under steady conditions, $\partial \mathcal{W}/\partial t = 0$ we get

$$\frac{d\mathcal{W}}{dt} = 0. \tag{5.2.49}$$

So $\mathcal{W}$ remains constant along a ray when properties of the medium at each point of space do not vary with time. This happens when perturbation propagates under steady conditions. Writing now (5.2.49) as

$$\frac{d\mathcal{W}^2}{dt} = 0 \tag{5.2.50}$$

we get

$$\mathcal{W}^2 - \mathcal{W}_0^2 = K^2 c^2(r), \tag{5.2.51}$$

where $C(r)$ and $\mathcal{W}_0(r)$ are functions of space only and not of time. Hence, in inhomogeneous media, at rest, the phase velocity of propagation of waves can only be functions of space and not of time..
The relations (5.2.45) and (5.2.46) therefore give

$$\dot{r} = \frac{\partial \mathcal{W}}{\partial k} = i\frac{KC^2}{\mathcal{W}},$$

$$\dot{k} = -\nabla \mathcal{W} = -\frac{KC}{\mathcal{W}}\left\{K\nabla C + \frac{\mathcal{W}}{KC}\nabla \mathcal{W}_0\right\}, \tag{5.2.52}$$

where $\mathbf{i} = \mathbf{K}/|K|$. The value of r shows that the velocity of the particle corresponding to the limits of geometrical optics of a wave motion is the group velocity $\partial \mathcal{W}/\partial K$ and not c or the wave phase velocity $\mathcal{W}/K$.

The wave number K varies along a ray simply according to (5.2.51), such that $\mathcal{W}$ is constant. Putting

$$K = \mathbf{i}\cdot \frac{1}{c}(\mathcal{W}^2 - \mathcal{W}_0^2)^{1/2}$$

in the second relation in (5.2.52) for studying the change in the direction of the ray we get

$$K\frac{\partial i}{\partial t} + (\dot{\mathbf{r}}\cdot\nabla)(iK) = -\frac{\mathbf{K}c}{\omega}\left[k\nabla c + \frac{\omega}{kc}\nabla\omega_0\right]. \tag{5.2.53}$$

Using the first of the equations (5.2.52) in the last term of the left hand side and (5.2.51) in the right hand side of this equation, after some simplifications, we get

$$\begin{aligned}\frac{di}{dt} &= -\frac{kc}{\omega}\nabla c - \frac{i}{\omega}c^2\frac{dk}{ds} - \frac{\omega_0}{k\omega}\nabla\omega_0 \\ &= -\frac{kc}{\omega}\nabla c + i\frac{kc}{\omega}\frac{dc}{ds} + \frac{i\omega_0}{k\omega}\frac{d\omega_0}{ds} - \frac{\omega_0}{k\omega}\nabla\omega_0.\end{aligned} \tag{5.2.54}$$

Putting $dt = (\omega/kc^2)\,ds$ in this equation, and using (5.2.8) we obtain

$$\begin{aligned}\frac{kc^2}{\omega}\frac{di}{ds} &= -\frac{kc}{\omega}\nabla c + i\frac{kc}{\omega}\frac{dc}{ds} + \frac{i\omega_0}{k\omega}\frac{d\omega_0}{ds} - \frac{\omega_0}{k\omega}\nabla\omega_0, \\ \text{or,}\quad \frac{di}{ds} &= c\nabla\left(\frac{1}{c}\right) - ic\frac{d}{ds}\left(\frac{1}{c}\right) + \frac{i\omega_0}{\omega^2-\omega_0^2}\frac{d\omega_0}{ds} - \frac{\omega_0}{\omega^2-\omega_0^2}\nabla\omega_0 \\ &= c\nabla\left(\frac{1}{c}\right) - ic\frac{d}{ds}\left(\frac{1}{c}\right) - \frac{i}{2}\frac{d}{ds}\log(\omega^2-\omega_0^2) + \frac{1}{2}\nabla\log(\omega^2-\omega_0^2) \\ &= c\nabla\left(\frac{1}{c}\right) - ic\frac{d}{dt}\left(\frac{1}{c}\right) + \frac{kc + c\nabla k}{kc} - \frac{i}{kc}\left(k + c\frac{dk}{ds}\right) \\ &\doteq \frac{1}{k}\nabla k - \frac{i}{k}\frac{dk}{ds} = \frac{1}{n}\nabla n - \frac{i}{n}\frac{dn}{ds},\end{aligned} \tag{5.2.55}$$

$$n\frac{di}{ds} + i\frac{dn}{ds} = \nabla n$$

which is the equation (5.2.54).

The wave field amplitude 'a' can be determined from the principle of conservation of energy through any cross sectional area of a tube of rays. For this consider the point P, whose current co-ordinate is r. P_0 is a standard point having co-ordinate r_0, dA is the elementary vectorial cross sectional area at P, and dA_0 at P_0. Then considering ω, a^2, c^2 as independent of time and integrating (5.2.53) within the volume of the bundle between P and P_0, we get

$$a^2(r)(K\cdot dA) = a_0^2(K\cdot dA_0),$$

where K is the wave vector at P and K_0 is that at P_0. Therefore,

$$a = a_0\left[\frac{(K_0\cdot dA_0)}{(K\cdot dA)}\right]^{1/2}.$$

Since K is proportional to the refractive index n, by (5.2.8) we can write

$$a^2(r)n(r)\,dA = a_0^2 n_0 dA_0, \tag{5.2.56}$$

$$\text{or,}\quad a = a_0\left(\frac{n}{n_0}\frac{dA_0}{dA}\right)^{1/2}. \tag{5.2.57}$$

▶ **Example 5.2.1** *Find the eikonal solution of the non-linear equations :*

(1) $\nabla^2\phi - \dfrac{1}{c^2}\dfrac{\partial^2\phi}{\partial t^2} - \dfrac{\omega_0^2}{c^2}\phi + \phi^3 = 0,$

(2) $\nabla^2 A - \dfrac{1}{c^2}\dfrac{\partial^2 A}{\partial t^2} - \dfrac{\omega_0^2}{c^2}A + A^2(A \times B) = 0,$

(3) $\nabla^2 A - \dfrac{1}{c^2}\dfrac{\partial^2 A}{\partial t^2} - \dfrac{\omega_0^2}{c^2}(A \times B) + \alpha A^2 A = 0,$

where $(A \cdot K) = 0$ *and* B *and* α *are constants in time and space.*

5.3 Geometrical Optics

5.3.1 A standard treatment of geometrical optics

Light rays deviate from their rectilinear path when the medium is heterogeneous because the index of refraction n depends on the co-ordinates of the space point $P(x, y, z)$. The path of the photon of light (of definite frequency n) from the point P to the point P' is given by Fermat's principle

$$\int_P^{P'} n(x, y, z)\, ds = \text{extremum.}$$

We can derive the equation of motion of a photon from Fermat's principle

$$\int_P^{P'} nds = \int_t^{t'} nsdt = \text{extremum.}$$

This follows Hamilton' principle of least action of mechanics,

$$\int_t^{t'} L(x, y, z, \dot{x}, \dot{y}, \dot{z})\, dt = \text{extremum},$$

where $L = n(x, y, z)\dot{s}$, and the equation of motion is

$$\frac{d}{dt}\left(\frac{\partial L}{\partial \dot{x}}\right) = \frac{\partial L}{\partial x}.$$

It can be written as

$$\frac{d}{dt}\left(\frac{\partial}{\partial \dot{x}}(n\dot{S})\right) = \frac{\partial}{\partial x}(n\dot{S}).$$

So, the equation of motion becomes

$$\frac{d}{dt}\left(n\frac{dx}{ds}\right) = \dot{S}\frac{\partial N}{\partial x}$$

because[1]

$$\frac{\partial \dot{S}}{\partial \dot{x}} = \frac{\partial S/\partial t}{\partial x/\partial t} = \frac{dx}{ds}.$$

[1]For proof, consider the example $r^2 = x^2 + y^2$ which shows that $\partial r/\partial x = x/r$; hence

$$\frac{\partial \dot{S}}{\partial \dot{x}} = \frac{\dot{x}}{\dot{S}} = \frac{dx}{ds}.$$

To bring out the analogy with mechanics, we compare these calculations with Hamilton's principle in the form

$$\int_t^{t'} (T - V)\, dt = \text{extremum},$$

where T is the kinetic energy and V is the potential energy. Since $\xi = T + V$ = total energy = constant, we get

$$\int_t^{t'} (2T - \xi)\, dt = \text{extremum}$$

which means

$$\int_t^{t'} 2T dt = \text{extremum}.$$

Therefore

$$2T = m\dot{S}^2 = \dot{S}(2Tm)^{1/2} = \text{maximum}.$$

Hence

$$\int_P^{P'} [2m(\xi - V)]^{1/2} ds = \text{extremum}.$$

If C is the constant vacuum velocity and the energy ξ = constant, then Fermat's principle gives

$$\int_P^{P'} \frac{n}{c}\, ds = \text{extremum},$$

$$\text{or,} \quad \int_P^{P'} (1/\xi)[2m(\xi - V)]^{1/2} ds = \text{extremum}.$$

The two quantities n/c and $(2m(\xi - V))^{1/2}/\xi$ have the dimension of time/length.

A simplified wave theory of light obtains if we conceive of a wave to spread along rays of geometrical optics with constant frequency $\nu(= c/n\lambda$, where the refractive index n is a function of x, y, z). The wave function may be written as $\psi = \exp[2\pi i\gamma(t - s)]$, along with

$$\frac{1}{\nu}\int_0^P \frac{1}{\lambda}\, ds = \frac{1}{c}\int_0^P nds, \quad \text{optical path} = \int_0^P cds.$$

We consider a continuous current surface and the system of rays orthogonal to this surface. Let O be any point on the curved surface corresponding to rays with phase $2\pi\nu t$. Then another point P of the ray, at an optical distance CS from the point O has the phase $2\pi\nu(t - s)$. On this ray let the wave function be

$$\psi(x, y, z, t) = \exp(2\pi i\nu(t - s)).$$

The surfaces $S = 0$ and S = constant (non-zero) are orthogonal to the rays with equal phase, which are the eikonal surfaces. This analogy allows us to introduce a wave function for a moving particle. Along each such path we introduce a wave frequency $\nu = h/mv = h/\sqrt{2m(\xi - V)}$. The phase velocity for a particle is

$$v_p = \frac{\xi}{\sqrt{2m(\xi - V)}} = \frac{h\nu}{\sqrt{2m(h\nu - V)}}.$$

So, v_p depends on the frequency ν. This velocity is different from the particle velocity $\dot{r} = v = p/m$. Hence the maximum part of the wave moves with the group velocity v_g, which is the particle velocity

$$v_g = 1 \Big/ \frac{\partial}{\partial \nu}\left(\frac{\nu}{n}\right) = 1 \Big/ \frac{\partial}{\partial \nu}\left(\frac{1}{n}\sqrt{2m(h\nu - V)}\right) = \sqrt{\frac{2}{m}(h\nu - V)} = \frac{p}{m} = \dot{s} = v.$$

Schroedinger then concluded that what we call a particle is nothing but the maximum point of a group of waves. Consider a wave function for a material particle, satisfying

$$\nabla^2\psi - \frac{1}{v_g^2}\frac{\partial^2\psi}{\partial t^2} = 0.$$

Putting $\psi = \psi_0 \exp(2\pi i\nu t)$ in this equation we obtain

$$\nabla^2\psi + \frac{4\pi^2\nu^2 m}{c^2}\psi = 0.$$

From optical analogy we write

$$\frac{n}{c} = \frac{1}{\xi}\sqrt{2m(\xi - V)},$$

so that the wave equation of a particle is

$$\nabla^2\psi + \frac{8\pi^2 m\nu^2}{\xi^2}\psi = 0.$$

And since $\xi = h\nu$, it reads

$$\nabla^2\psi + \frac{8\pi^2 m}{h^2}(\xi - V)\psi = 0.$$

This is the Schroedinger equation.

These waves are not to be identified with the EM waves. To understand De Broglie's conception of waves we consider a moving particle of mass m and velocity v. Then we can write $m = m_0/(1 - \nu^2/c^2)^{1/2}$, m_0 is the rest mass in a rest system $K_0(x_0, y_0, z_0, t_0)$. The rest frequency of the wave (standing), in the K_0 system is given by $\nu_0 = (m_0/\hbar)c^2$ and $m_0c^2 = \xi_0$ (rest energy). So, the wave system in K_0 can be expressed in the form

$$\psi = f(x_0, y_0, z_0, t_0)\sin 2\pi\nu_0 t.$$

In the K system moving with the velocity v with respect to K_0, the Lorentz transformation relations

$$x_0 = \frac{x - vt}{\sqrt{1 - \nu^2/c^2}}, \quad y_0 = y, \quad z_0 = z, \quad t_0 = \frac{t - (vx/c^2)}{\sqrt{1 - \nu^2/c^2}},$$

show that

$$\psi = f\left(\frac{x - vt}{\sqrt{1 - v^2/c^2}}, y, z, \frac{(t - vx/c^2)}{\sqrt{1 - v^2/c^2}}\right)\sin\left(2\pi v_0\left(\frac{(t - vx/c^2}{\sqrt{1 - v^2/c^2}}\right)\right).$$

Hence in K, there is a progressive wave of frequency $\nu = \dfrac{v_0}{\sqrt{1 - \nu^2/c^2}}$, and the phase velocity $v_p = c^2/v > c$ because $v < c$. Since the amplitude f of the wave varies with time t, ψ does not represent a single wave, but a group of waves and, consequently, the phase velocity $v_0(> c)$ is not of much significance. The velocity of the wave group is given by its group velocity v_g from the relation $x - vt = 0$. So $v_g = v$, $v_p v_g = c^2$.

The frequency ν in K is different from ν_0, and this must satisfy the De Broglie relation $\xi = h\nu$. Since

$$\xi = mc^2 = \frac{m_0 c^2}{\sqrt{1 - \nu^2/c^2}}, \quad \text{or,} \quad \xi = \frac{h\nu_0}{\sqrt{1 - \nu^2/c^2}},$$

because $h\nu_0 = m_0 c^2$, we obtain $\nu = \dfrac{\nu_0}{\sqrt{1 - \nu^2/c^2}}$. The relation $v_g = \dfrac{c^2}{v_p}$ gives

$$v_p = \frac{mc^2}{mv} = \frac{h\nu}{mv}.$$

Since $\nu = v/\lambda$, we find that

$$mv = \frac{mv_0}{\sqrt{1 - \nu^2/c^2}} = \frac{h}{\lambda},$$

which is the De Broglie relation between the momentum and the wave length. Finally, for the law of dispersion,

$$n = \frac{c}{v_p} = \frac{v}{c} = \sqrt{1 - \frac{\nu^2}{\nu_0^2}},$$

where n is the refractive index.

5.3.2 Geometrical optics for light sources in rotating rings

The eikonal ϕ is the phase of the wave front of a light beam. After one circulation in a closed loop Γ of arbitrary shape it is given by

$$\phi = \frac{1}{2\pi} \oint_\Gamma (K \cdot dr) - \frac{1}{2\pi} \int_0^T \omega dt, \tag{5.3.1}$$

where T is the circulation time for light of the stationary loop. The first term is the spatial part of the line integral and so is the number of wave-lengths in the closed path. The second term is the time part and is the number of radians over which the monochromatic signal advances during the time needed for one circulation and is not closed.

In the 4-space of wave vector and frequency or of space co-ordinate and time, the space time integral of (5.3.1) can be written in the form

$$\int K_\mu dx^\mu. \tag{5.3.2}$$

Quantum mechanically the 4-vector of momentum energy is $(p, -\mathcal{H})$, where

$$p = \hbar K, \quad \mathcal{H} = \hbar\omega \tag{5.3.3}$$

$\hbar$ is the Planck constant. The phase and action are therefore related invariants for the general space-time transformation in the manner

$$\oint (p \cdot dr) - \int \mathcal{H}\, dt = \int \mathcal{L}\, dt, \tag{5.3.4}$$

where $\mathcal{L}$ is the Lagrangian. The quantities K_μ form a 4-vector because a scalar (the phase ϕ of the wave) is obtained when they are multiplied by dx^μ and integrated. So $K^\mu K_\mu$ and $K_\mu dx^\mu$ are both scalar invariants.

5.3.3 The Sommerfeld-Runge law in 4-dimensional space

The rays of light are curves with tangents at each point coincident with the direction of wave propagation. The conditions for the approximation of geometrical optics are that this direction and amplitude remain constant over distances of the order of the wave-length λ and that λ is small. For this approximation the phase ϕ does not have a simple form, but ϕ is large because it changes by the amount 2π in one wavelength. Hence, over small regions and time intervals,

$$\phi = \frac{1}{2\pi}\oint(\nabla\psi\cdot dr) - \frac{1}{2\pi}\int\frac{\partial\psi}{\partial t}\,dr. \tag{5.3.5}$$

Comparing this expression with (5.3.1) we can write

$$K = \nabla\psi, \quad \omega = \frac{\partial\psi}{\partial t} \tag{5.3.6}$$

and obtain the Sommerfeld-Runge law

$$\operatorname{curl} K = 0. \tag{5.3.7}$$

Introducing the 4-vector

$$\mathbf{r} \equiv (\mathbf{r}, ict), \quad \mathbf{K} \equiv \left(\mathbf{K}, \frac{i\omega}{c}\right) \tag{5.3.8}$$

the 4-dimensional analogue of (5.3.1) is obtained

$$\phi = \frac{1}{2\pi}\oint(\mathbf{K}\cdot d\mathbf{r}). \tag{5.3.9}$$

The 4-dimensional generalization of the Sommerfeld-Runge law therefore is

$$\Box \times K = 0, \tag{5.3.10}$$

where

$$\Box = \left(\frac{\partial}{\partial x}, \frac{\partial}{\partial y}, \frac{\partial}{\partial z}, -\frac{i}{c}\frac{\partial}{\partial t}\right). \tag{5.3.11}$$

Equation (5.3.10) consists of six component equations, three of which represent the three-dimensional Sommerfeld-Runge law of (5.3.7), the other three, in three-dimensional vector notation, read

$$\nabla\omega + \frac{\partial K}{\partial t} = 0. \tag{5.3.12}$$

The variation of frequency may be due to propagation in time-varying media. Poverlain (1962) has discussed the Sommerfeld-Runge law in some detail.

The Hamiltonian equations for a particle are

$$\dot{p} = -\operatorname{grad}\mathcal{H}, \quad v = \dot{r} = \frac{\partial\mathcal{H}}{\partial p}. \tag{5.3.13}$$

Putting the values of p and $\mathcal{H}$ from (5.3.3) in these relations we obtain

$$K = -\operatorname{grad}\omega, \quad \dot{r} = \frac{\partial\omega}{\partial K}. \tag{5.3.14}$$

The Hamiltonian equation (5.3.13) plus the Sommerfeld-Runge law (5.3.7) are mathematically the necessary and sufficient conditions for the validity of (5.3.10). These equations are equivalent to the statement that the 4-vector of wave number and frequency is free in a space-time sense. This property makes the integrand of (5.3.1) a total differential for the permissible trajectories, the conditions for which, in other words, are those of (5.3.6). For light rays the relations of (5.3.13) become operational only if the medium is dispersive. The Sommerfeld-Runge law is the operationally dynamical condition for existence of light rays. In ring lasers, for simplicity, the trajectories of light beams are assumed to be straight lines between the mirrors, so ω and K are not functions of position and time.

5.3.4 Variation of a line integral in a moving system

To find the time rate of variation of the line integral $\oint (F\cdot dr)$ over a closed loop in a moving system of co-ordinates we assume that the radius vector $\mathbf{r}(t)$ is a function of one parameter, time say. The velocity v of motion of the field is then given by

$$\delta\psi = ((v\cdot\Sigma)\phi\delta l)v = \frac{dr}{dt}. \tag{5.3.15}$$

A scalar field ψ on its own does not depend on time if evaluated for a fixed point of the moving system. So its differential $\delta\psi$ is given by

$$\delta\psi = (v\cdot\nabla)\phi\delta t. \tag{5.3.16}$$

The variation of the line integral of a moving closed loop must take into account the motion of the circuit with a velocity v and so

$$\begin{aligned}\frac{d}{dt}\oint(\mathbf{F}\cdot d\mathbf{r}) &= \oint\left[\left(\frac{d\mathbf{F}}{dt}\cdot d\mathbf{r}\right)+\left(\mathbf{F}\cdot\frac{d}{dt}(d\mathbf{r})\right)\right]\\ &= \oint\left[\left(\frac{d\mathbf{F}}{dt}\cdot d\mathbf{r}\right)+\mathbf{F}(d\mathbf{r}\cdot\boldsymbol{\nabla})v\right]\\ &= \oint\left[\left(\frac{\partial\mathbf{F}}{\partial t}+(\mathbf{v}\cdot\boldsymbol{\nabla})\mathbf{F}\right)\cdot dr + F\cdot(d\mathbf{r}\cdot\boldsymbol{\nabla})v\right].\end{aligned} \tag{5.3.17}$$

Treating $\mathbf{v}$ as a fixed vector we can write

$$(\mathbf{v}\cdot\boldsymbol{\nabla})\mathbf{F} + (\mathbf{v}\times\operatorname{curl}\mathbf{F}) = \boldsymbol{\nabla}_{\mathbf{F}}(\mathbf{v}\cdot\mathbf{F}), \tag{5.3.18}$$

where in $\boldsymbol{\nabla}_{\mathbf{F}}(\mathbf{v}\cdot\mathbf{F})$ only the components of $\mathbf{F}$ have to be differentiated. Putting (5.3.18) in (5.3.17) we obtain

$$\begin{aligned}\frac{d}{dt}\oint(\mathbf{F}\cdot d\mathbf{r}) &= \oint\left[dr\cdot\left(\frac{\partial\mathbf{F}}{\partial t}+\boldsymbol{\nabla}_{\mathbf{F}}(\mathbf{v}\cdot\mathbf{F})-(\mathbf{v}\times\operatorname{curl}\mathbf{F})\right)+(\mathbf{F}\cdot(d\mathbf{r}\cdot\boldsymbol{\nabla})\mathbf{v}\right]\\ &= \oint d\mathbf{r}\cdot\left(\frac{\partial F}{\partial t}+\boldsymbol{\nabla}(\mathbf{v}\cdot\mathbf{F})-(\mathbf{v}\times\operatorname{curl}\mathbf{F})\right).\end{aligned} \tag{5.3.19}$$

Here we have used the identity

$$\nabla_{\mathbf{F}}(\mathbf{v}\cdot\mathbf{F})\cdot d\mathbf{r} + \mathbf{F}\cdot(d\mathbf{r}\cdot\nabla)\mathbf{v} = \nabla_F(\mathbf{v}\cdot\mathbf{F})\cdot d\mathbf{r} + \left(F_1\frac{\partial v_1}{\partial x} + F_2\frac{\partial v_2}{\partial x} + F_3\frac{\partial v_3}{\partial x}\right) dx + \cdots$$

$$= \nabla(\mathbf{v}\cdot\mathbf{F})\cdot d\mathbf{r} + \cdots. \quad (5.3.20)$$

Hence we find that

$$\frac{d}{dt}\int_1^2 F_i dr_i = \int_1^2 dr_i \left[\frac{\partial \mathbf{F}}{\partial t} + \nabla(\mathbf{v}\cdot\mathbf{F}) - (\mathbf{v}\times\nabla)\times\mathbf{F}\right]_i \quad (5.3.21)$$

Here the limits 1 and 2 denote the values of the integral at the end and at the beginning of one circulation around the loop.

5.4 Waves and Rays through Moving Media

Introduce two co-ordinate frames, in one of which, the L frame, the medium is moving with velocity u and in the other, the L' frame, the medium is seen at rest. Let $\mathbf{r}(x, y, z)$ and $\mathbf{r}'(x', y', z')$ be the co-ordinates of a point in the L and L' systems respectively. In the L' system the plane wave solution of the wave equation for the field variable ϕ is

$$\phi = a\exp\{i(K'\cdot r') - i\omega' t\}, \quad (5.4.1)$$

where

$$\mathbf{r}' = \mathbf{r} - \mathbf{u}t. \quad (5.4.2)$$

The oscillation being dispersive in space with finite speed, ω' and k' are connected by a dispersion relation. So,

$$v_p' = \frac{\omega}{k'}, \quad v_g' = \frac{\partial\omega'}{\partial k'}, \quad (5.4.3)$$

where v_p' is the phase velocity and v_g' is the group velocity. In the L system

$$\phi = a\exp[i(k'\cdot r) - i\{\omega' + (k'\cdot u)t\}], \quad (5.4.4)$$

ω, the Doppler shifted frequency of the perturbation, is given by

$$\omega = \omega' + (k'\cdot u). \quad (5.4.5)$$

Hence

$$v_g = \frac{\partial\omega}{\partial k'} = v_g' + u. \quad (5.4.6)$$

When the source of perturbation and the medium are at rest and the observer's system is moving with velocity $\mathbf{u}$,

$$\mathbf{r}' = \mathbf{r} + \mathbf{u}t, \quad (5.4.7)$$

$$\omega = \omega' - (\mathbf{K}'\cdot\mathbf{u}) = \omega'\left(1 - \frac{u\cos\theta}{v_p'}\right), \quad (5.4.8)$$

where θ is the angle between $\mathbf{K}'$ and $\mathbf{u}$.

Let now ω be the frequency of perturbation in the system fixed with the source when the medium and the observer are at rest and the source of perturbation is moving relative to them with velocity $\mathbf{u}$. Then

$$\omega = Kv_p. \tag{5.4.9}$$

In the system in which the medium is at rest

$$\phi = a\exp(i(K\cdot r) - i\omega t), \tag{5.4.10}$$

where

$$\mathbf{r} = \mathbf{r}' + \mathbf{u}t, \tag{5.4.11}$$

provided $\mathbf{r}'$ is the position vector of the current point in the system moving with the source. If ω' is the frequency in the system in which the medium is at rest

$$\omega' = \omega - (K\cdot u). \tag{5.4.12}$$

Hence, using (5.4.9) we get

$$\omega = \omega' / \left(1 - \frac{u\cos\theta}{v_p}\right). \tag{5.4.13}$$

For $U\cos\theta > v_p$ the disturbance received by the observer at any given instant of time is received earlier than that emitted at a previous instant because then $\omega < 0$.

In case of slow streaming motion we include, in the expression for the linearized Lorentz force per unit charge, the term $1/c - [U \times H]$, where U is the streaming velocity vector and H is the perturbation magnetic field. When the drift velocity U varies in space slowly, simple transformation of results from the L' system to the L system is not sufficient. Then using the value of ω from (5.4.5) in (5.2.46) we get

$$K = -\nabla\omega' - \nabla(K'\cdot u). \tag{5.4.14}$$

Simplifying the second term in the right hand side with the help of the vector result for $\boldsymbol{\nabla}(\mathbf{A}\cdot\mathbf{B})$, and regarding ω' as constant the above equation becomes

$$\dot{\mathbf{K}} = -(\mathbf{K}'\boldsymbol{\nabla})\mathbf{u} - (\mathbf{K}' \times (\boldsymbol{\nabla} \times \mathbf{u})). \tag{5.4.15}$$

Also the second of the equations (5.2.46) yields

$$\dot{\mathbf{r}} = \mathbf{v} = \frac{\partial\omega'}{\partial\mathbf{K}} + \frac{\partial}{\partial\mathbf{K}}(\mathbf{K}'\cdot\mathbf{u}),$$

where $\mathbf{v}$ is the velocity of the ray at the point under study. The quantity $\dot{\mathbf{r}}$ reduces to the group velocity for $\mathbf{u} = 0$. Since the wave vector is same in the L and L' systems, we write

$$\dot{\mathbf{r}} = \mathbf{v}'_g + \mathbf{U}. \tag{5.4.16}$$

Moreover,

$$\frac{d\mathbf{u}}{dt} = (\mathbf{v}\cdot\boldsymbol{\nabla})\mathbf{u} = (\mathbf{v}'_g\cdot\boldsymbol{\nabla})\mathbf{u} + (\mathbf{u}\cdot\boldsymbol{\nabla})\mathbf{u}.$$

because $\mathbf{u}$ being steady $\partial\mathbf{u}/\partial t = 0$. Also

$$\frac{d\mathbf{u}}{dt} = (\mathbf{v}'_g\cdot\boldsymbol{\nabla})\mathbf{u} \tag{5.4.17}$$

because $|\mathbf{u}| \ll |\mathbf{v}'_g|$. Differential of Kv with respect to time gives

$$\frac{d}{dt}(\mathbf{K}v) = \frac{\partial}{\partial t}(\mathbf{K}v) + (\mathbf{u}\cdot\boldsymbol{\nabla})(\mathbf{K}v). \tag{5.4.18}$$

Since the ray propagation speed is uniform in time the first term on the right hand side can be ignored. Simplifying then the second term of the right hand side with the help of the vector formula (A.3.10) we get

$$\frac{d}{dt}(\mathbf{Kv}) \approx -\mathbf{Kv}[\mathbf{i}\times(\boldsymbol{\nabla}\times\mathbf{u})], \tag{5.4.19}$$

where $\mathbf{i}$ is the unit vector along $\mathbf{v}$ and other terms of (A.3.10) are neglected because $|v| \gg |u|$. Also

$$\frac{d}{dt}(\mathbf{Kv}) = \frac{d}{dt}(iKv) = i\frac{d}{dt}(Kv) + Kv\frac{di}{dt}. \tag{5.4.20}$$

The equation (5.4.9) and (5.4.20), therefore, yield

$$i\frac{d}{dt}(Kv) + Kv\frac{di}{dt} = -Kv[i\times(\nabla\times u)]. \tag{5.4.21}$$

Since $d\mathbf{i}/dt$ is perpendicular to $\mathbf{i}$ (because $i^2 = 1$) and coefficient of $\mathbf{i}$ on the left side of (5.4.21) must vanish, we obtain

$$\frac{d\mathbf{i}}{dt} = \frac{j}{\zeta} = -[\mathbf{i}\times(\boldsymbol{\nabla}\times\mathbf{u})]. \tag{5.4.22}$$

It dt is replaced by dS/v, then

$$\frac{d\mathbf{i}}{dS} = \frac{j}{\rho} = -\frac{[\mathbf{i}\times \mathrm{curl}\,\mathbf{u}]}{v}, \tag{5.4.23}$$

where $\mathbf{j}$ is the unit vector along the principal normal and ρ is the radius of curvature of the ray path. This equation for determining the ray trajectory, can be approximately written as

$$\frac{d\mathbf{i}}{ds} = -\frac{1}{v'_g}[\mathbf{i}\times(\boldsymbol{\nabla}\times\mathbf{u})] \tag{5.4.24}$$

because, in (5.3.6) $|u| \ll |v'_j|$. The unit vector $\mathbf{i}$ is along the tangent to the ray. For $u \neq 0$, $\mathbf{i}$ is not parallel to $\mathbf{K}$. ω need not be large if $u > v'_g$. In that case ignoring ω in (5.4.5) and using the relation $\omega' = Kv_p$ for steady small perturbations we have

$$kv'_p = -(K\cdot U). \tag{5.4.25}$$

Let the perturbation move along OX and let U be uniform in space and time. Also, let the direction of propagation be in the XY plane. Then equation (5.4.25) gives

$$k_x^2(u^2 - v_p'^2) = v_p'^2 k_y^2. \tag{5.4.26}$$

Using now from (5.2.46) the relations

$$\dot{x} = \frac{\partial\omega}{\partial k_x}, \quad \dot{y} = \frac{\partial\omega}{\partial k_y}, \tag{5.4.27}$$

the trajectory equation can be written as

$$\frac{dy}{dx} = \left(\frac{\partial \omega}{\partial k_x}\right) \bigg/ \left(\frac{\partial \omega}{\partial k_y}\right). \tag{5.4.28}$$

Expanding ω as a function of K in the form

$$\omega(k + dk) = \omega(k) + \left(\frac{\partial \omega}{\partial k_x}\right) dk_x + \left(\frac{\partial \omega}{\partial k_y}\right) dk_y + \cdots, \tag{5.4.29}$$

and regarding ω as constant this expansion approximately gives

$$\left(\frac{\partial \omega}{\partial k_y}\right) \bigg/ \left(\frac{\partial \omega}{\partial k_x}\right) = -\frac{dk_x}{dk_y}. \tag{5.4.30}$$

The trajectory equation therefore becomes

$$\frac{dy}{dx} = \pm \frac{v'_p}{(u^2 - v'^2_p)^{1/2}}. \tag{5.4.31}$$

Regarding u as a constant and writing $v'_p = u \sin \alpha$ this equation simplifies to

$$\frac{dy}{dx} = \pm \tan \alpha. \tag{5.4.32}$$

So, the ray paths are approximately two straight lines intersecting the direction of $\mathbf{u}$ at angles $\pm\alpha$.

5.5 Interference of CW and CCW Rays in Rotating Rings

Consider a light source inside a rotating close ring—like cavity emitting two radiations of extreme monochromacity, one propagating in the clockwise (CW) sense, and the other, in the counter-clockwise (CCW) sense at frequency ω. Mode degeneracy is removed by rotation of the cavity containing the light source. The resulting split in frequency can be measured to high accuracy by using frequency standards.

Any closed geometrical path can be used as the ring, square, rectangular and triangular configurations with perimeters several meter are used. The dielectric part and the source part can be kept in two different segments of the ring. Then these are separated by one or more reflecting mirrors. The beats are observed by photo conductors. The cavity is rotated by means of an external drive about its input axis or by submitting the apparatus to a torsional vibration. Either mechanical or optical techniques, or a combination of the two is used to produce beats between the CW and CCW rays of the light source inside the closed ring.

Sometime, a dielectric moving along the path of the rays is inserted inside the ring. The moving dielectric can be a solid (say a quartz crystal), a liquid (say CCl_4), or a gaseous medium (say dry air or even plasma). The ring is rotated about an axis normal to the flat surface in such a manner that the motion has a component along the path of ring; and the path of refracted light and the axis of rotation are not coplanar. The ring is fixed such that any ray in either direction retraces the same path during successive round trips and at such angles that a maximum of the circulating light flux is retraced within the circuit. The CW and CCW modes are extracted through one of the corner mirrors which is made collinear by external combining optics and mixed on a photo cathode.

Rotation and other accelerated motions affect radiation from the light source inside the closed optically resonant cavity. When there is no source inside, rotation still affects EM resonance of closed optical cavities, and coherent CW and CCW beams are derived from a rotating beam splitter. When these are combined the phase differences due to path differences between the beams are manifested as a series of fringes of interference. Since the external light source moving with an interferometer unambiguously determines the light frequency occurring in the interferometer, the mode degeneracy prevails. As a result the frequency difference between the CCW and CW travelling rays vanishes, and

$$\delta\omega = 0. \tag{5.5.1}$$

A Doppler shift can equally affect these beams if the light source is stationary. Then the fringe shift is only affected in higher order by rotation. Equal fringes, due to (5.5.1), are still necessary for interference.

The optical path length inside the cavity of the ring must equal an integral number times the wave lengths of the CW and CCW oscillating modes. So the circumferential optical path of the circuit of total length p determines the frequency ν of the preferred mode resonance of the optical field. For the stationary system the paths p of the CW and CCW rays are equal; so the wave lengths are equal and the eigen modes are degenerate in frequency. For the rotating system, since the wave fronts of the CW and CCW modes travel the lengths $p+\Delta p$ and $p-\Delta p$, respectively, of fig. 5.5.1 to close on themselves, the paths are unequal, and consequently a frequency shift $\Delta\nu$ is obtained between the two modes.

Consider a circular ring of radius R, rotating in the clockwise sense (fig. 5.5.1), Post (1967). The wave fronts are constrained to follow the circular path of the ring, and the two beams leave the beam splitter when it is in position C. The CCW circulation is opposite to the sense of rotation of the ring, and meets the beam splitter again in the new position C' which is shifted from C by the amount Δp. As seen by a stationary observer, the CW beam travelling in the sense of the rotation of the ring, meets the beam splitter when it is shifted through a distance $\Delta p''$ to c''.

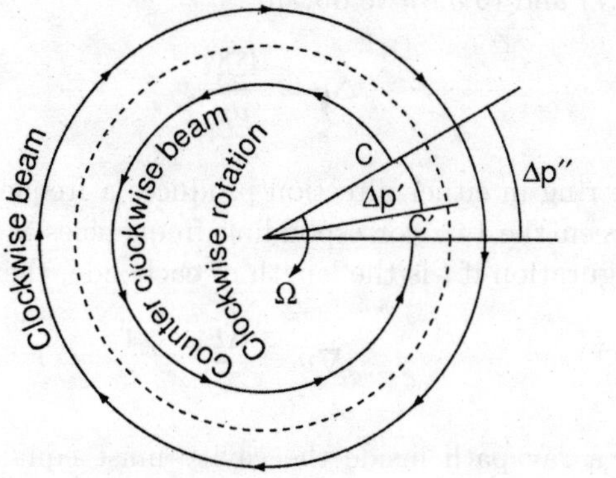

Fig. 5.5.1 It shows the path difference of clockwise and counter-clockwise beams from a source of light in a circular ring rotating in its own plane.

If Ω is the constant rate of rotation of the ring, T' is the time of the CCW motion, T'' is the time of the CW motion and C is the free space velocity of light, then

$$T' = \frac{1}{2}(2\pi R - \Delta p') = \frac{\Delta p'}{R\Omega}, \quad T'' = \frac{1}{2}(2\pi R + \Delta p'') = \frac{\Delta p''}{R\Omega}. \tag{5.5.2}$$

In the stationary frame of the observer the time interval between the consecutive positions C' and C'' of the beam-splitter. denoted by $(\Delta T)_s$, is given by

$$(\Delta T)_s = T'' - T' = \frac{4\Sigma R}{c^2 - R^2\Omega^2}, \tag{5.5.3}$$

where Σ is the area enclosed by the optical path of the ring, so $\Sigma = \pi R^2$, and for all practical purposes,

$$(\Delta T)_s = \frac{1}{c^2}4\Sigma\Omega. \tag{5.5.4}$$

Hence if Δp is the fringe shift, measured in fringe numbers, then

$$\Delta p = \frac{c}{\lambda}(\Delta T)_s = \frac{4\Sigma\Omega}{c\lambda}, \tag{5.5.5}$$

where λ is a nominal wave-length in vacuum for zero rotation. More generally, when the axis of rotation is not perpendicular to the plane of the ring, we obtain

$$(\Delta T)_s = \frac{4(\Sigma\cdot\Omega)}{c^2 - R^2\Omega^2}, \tag{5.5.6}$$

$$\Delta p = \frac{4(\Sigma\cdot\Omega)}{c\lambda}. \tag{5.5.7}$$

The phase agreement criterion for light source in a ring gives

$$\frac{\Delta p}{p} = \frac{\Delta\nu}{\nu}, \tag{5.5.8}$$

where $\Delta\nu$ is the frequency difference and ν is the oscillation frequency for zero rotation. The beat frequency $\Delta\nu$ is therefore obtained by combining the highly coherent, oppositely circulating beams. Combining (5.5.7) and (5.5.8) we obtain

$$\Delta\nu = \frac{4\Sigma\Omega}{\nu p}. \tag{5.5.9}$$

Thus rotation of the ring in either direction produces a frequency split in each axial mode and the difference between the two corresponding frequencies is proportional to the angular rate. For a square configuration if l is the length of each side, then

$$\nabla\nu = \frac{l\Omega}{\lambda}. \tag{5.5.10}$$

Since the length of a ray path inside the cavity must equal an integral multiple of the wave-length, so that the returning wave is in phase with the initiated energy, the round trip phase must be a multiple of 2π radians. Since the length of a ray path inside the cavity (approximately equals the ring perimeter) is millions of times greater than the optical wave length, this requirement is met by many closely spaced wavelengths. The number of axial modes, which can be sustained concurrently, depends upon (1) and the spectral distribution curve of energy emitted by the discharge tube, (2) the threshold energy level required for laser oscillation, and (3) frequency separation between the adjacent axial modes.

The usuable energy lies between the cut-off points established by intersection of the threshold level and gain curve. Oscillations occur between these limits at discrete axial modes, with a frequency separation $\Delta\nu'$ (or spectral free range) nearly given by

$$\Delta\nu' = \frac{c}{p}, \tag{5.5.11}$$

where p is the ring perimeter. If $p = 100$ cm; then $\Delta\nu' = 300$ MC.

By the time the wave emitted from a point Q and propagating CW, completes one round trip, Q will move ahead, because of rotation of the ring. So, with respect to the phase, the optical path is lengthened, and since it is a feed back loop driving a sustained oscillation, the wavelength must increase for the returning wave to be in phase with the initiating oscillation. The converse is true for the wave travelling in a CCW sense, that is, the optical path is shortened. The two effects will be interchanged if the ring is rotated in the CCW sense.

5.5.1 Modifications in presence of a moving dielectric in the ring path

The moving dielectric can be a solid (say a quartz piece), a liquid (say CCl_4) or a gaseous medium (say dry air or plasma). The CW oscillation is obtained by directing the travelling waves around the complete optical circuit and not by retroreflecting them. Owing to the Fresnel drag of the moving dielectric a part of the splitting $\Delta\nu$ of the frequencies of the CW and CCW modes is proportional in frequency to its axial velocity component and the ratio of the path length through the media and the total cavity perimeter.

If v' is the velocity of light in the frame of the dielectric, n is its refractive index and c is the free space speed of light, then,

$$v' = c/n. \tag{5.5.12}$$

Moreover, if u is the speed of the moving dielectric then the special-relativistic addition rule of velocities gives

$$v = \frac{(v' \pm u)}{1 \pm \dfrac{v'u}{c^2}}. \tag{5.5.13}$$

Expanding to lowest order and using (5.5.12), we get

$$v = \frac{c}{n} \pm u\left(1 - \frac{1}{n^2}\right). \tag{5.5.14}$$

Since n is a function of the wave length λ, additional terms due to the Doppler shift $\delta\lambda$ and the dispersive effect of the dielectric appear in the expression for the wavelength in the moving dielectric, because n is assumed to depend on λ. Hence

$$\delta\lambda \approx \pm\frac{\lambda n v}{c}, \tag{5.5.15}$$

$$\frac{c}{n(\lambda + \delta\lambda)} \approx \frac{c}{n(\lambda)} - \frac{c}{n^2}\left(\frac{dn}{d\lambda}\right)\delta\lambda, \tag{5.5.16}$$

$$v \approx \frac{c}{n} \pm u\left\{1 - \frac{1}{n^2} - \left(\frac{\lambda}{n}\right)\left(\frac{dn}{d\lambda}\right)\right\}. \tag{5.5.17}$$

The drag coefficient α is the alternation of the light path. It is given by

$$\alpha = \left(v - \frac{c}{n}\right)\frac{1}{u} = 1 - \frac{1}{n^2} - \frac{d\ln n}{d\ln\lambda}. \tag{5.5.18}$$

The condition of cavity resonance is that the wave must change its phase by an integral multiple of 2π in traversing the closed loop. So the cavity path length must equal an integral multiple of wavelengths. If N is this integral number in the cavity, p is the ring perimeter, n_0 is the effective refractive index, l is the distance travelled by the light beam through the moving dielectric and n' is its refractive index, we have

$$N\lambda = (p - l)n_0 + ln'. \tag{5.5.19}$$

Assuming viscosity of the medium to be small we can write for the CW and CCW beams

$$n' = n \pm \frac{u}{c}(n^2 - 1), \tag{5.5.20}$$

$$N\lambda = pn_0 + l(n - n_0) \pm \frac{lu}{c}(n^2 - 1). \tag{5.5.21}$$

Because of the Fresnal drag the difference in velocities of the two waves, propagating parallel and anti-parallel to the motion of the dielectric, produces an effective fringe shift Δp given by

$$\Delta p = p_- - p_+ = \frac{2lu(n^2 - 1)}{cn_0}. \tag{5.5.22}$$

With the help of (5.5.18) the corresponding shift in frequency, $\Delta\nu_1$, is evaluated :

$$\Delta\nu_1 = \frac{2lu(n^2 - 1)\nu}{cn_0 p}. \tag{5.5.23}$$

Since $n_0 = kc/\nu = c/\lambda\nu$ and $(n^2 - 1)/n_0^2 = 1 - 1/n^2$, equation (5.5.23) can be written as

$$\Delta\nu_1 \approx \frac{2lu}{\lambda p}\left(1 - \frac{1}{n^2}\right). \tag{5.5.24}$$

Refractive index of most gases at usual mass densities is slightly greater than unity. So putting $n = 1 + \varepsilon$ and regarding $|\varepsilon| \ll 1$ in equation (5.5.24) we obtain

$$\Delta\nu_1 \approx \frac{4lu}{\lambda p}(n - 1). \tag{5.5.25}$$

Since n is proportional to the number of gas molecules per unit volume (and so to mass density), $\Delta\nu_1$ should be proportional to the mass flow rate, and one cycle of pulse from the ring light corresponds to the transfer of a particular mass of gas. The flow can be followed because the fluid is contacted optically, without imposition of mechanical parts. From measurement of $\Delta\nu_1$ the density of even rarefied gases can be determined if the flow velocity is known or the flow rate is determined if mass density is known. Using a static bias to prevent mode locking, air flow sensitives of less than 3 mm sec^{-1} have been measured.

(1) Putting $l = 15$ cm. $\lambda = 1.1528 \times 10^{-4}$ cm and $\Omega = 1^0 \text{ sec}^{-1}$ in the formula (15.5.9) we find that

$$\Delta\nu = 2278.1 \text{ cps.} \tag{5.5.26}$$

(2) For water $n = 1.33$. Hence if $u = 0.1$ cm. sec^{-1} and if equation (5.5.25) is used we find that

$$\Delta\nu = \frac{l}{p} \times 11.42 \times 10^2 \text{ cps.} \tag{5.5.27}$$

So a beat frequency of about 100 cps is generated if $p \approx 10l$.

(3) For quartz if $l = 1.7$ cm. and $u = 12.5$ to 250 cm. sec^{-1}, then

$$\Delta\nu_1 = 10^3 \text{ to } 2 \times 10^4 \text{ cps.} \tag{5.5.28}$$

(4) For CCl_4 if $l = 0.58$ cm., and $u = 25$ to 28 cm. sec^{-1}, then

$$\Delta\nu_1 = 650 \text{ to } 1350 \text{ cps.} \tag{5.5.29}$$

(5) For dry air if $l = 62$ cm. and $u = 520$ to 950 cm. sec^{-1}, then

$$\Delta\nu_1 = 850 \text{ to } 1530 \text{ cps.} \tag{5.5.30}$$

5.5.2 Integration of a closed-loop line integral of a velocity field

The closed-loop line integral of velocity field is studied in an approximate calculation of the path-length changes due to motion of an interferometer as a whole, Post (1972). The velocity field v that appears in the integrand of the line integral describes rigid motion of the instrument. The expression for path length change or frequency change in self-oscillating devices has something like a distributed Doppler effect. The index of refraction n varies along the path of the beam due to material properties of the latter. The optical loop is initially at rest in an inertial frame after which it is set in motion. For self-oscillating systems the dimensionless optical path length z (measured in terms of the wave length λ) is also called the *mode number*. If Δz denotes the change of the optical path length due to motion of the closed loop, then it is also the number of lost or gained phase reversals along the loop. Hence,

$$\Delta z = \frac{\omega}{2\pi c^2} \oint (v \cdot dr), \tag{5.5.31}$$

where ω is its spectral frequency in circular measure. The material and wave guide properties are completely absent in (5.5.31) because Δz depends solely on the velocity $\mathbf{v}$ and on the geometry of the loop and not on its optical properties.

The optical path length change is proportional to the circulation time of light around the loop. An increase of this time due to the index n of refraction, when $n > 1$, exposes the circuit proportionately longer to the field of v, and the actual length change Δl is also increased by the presence of n. The absence of n in (5.5.31) is due to the definition of Δz as a number of phase reversals gained or lost. From the relation

$$\oint (\mathbf{v} \cdot d\mathbf{r}) = \iint_{\Sigma} (\nabla \times \mathbf{v} \cdot d\Sigma)$$

it is evident that the path length change Δl vanishes if v is a uniform velocity of translation. So, from the view point of special relativity, $\Delta z = 0$. Hence when $\Delta z \neq 0$ equation (5.5.31) violates the principle of special relativity.

The unperturbed light path in the original inertial frame is used in (5.5.31). So all parameters refer to the local wavelength which equals $2\pi/k$, where k is the local wave number. The free space wavelength λ_0 is given by $\lambda_0 = \lambda/n$. Hence

$$\Delta z = \frac{1}{\lambda_0 c} \oint (\mathbf{v} \cdot d\mathbf{r}) = \frac{1}{c} \oint \frac{n}{\lambda} (\mathbf{v} \cdot d\mathbf{r}). \tag{5.5.32}$$

This is the distributed Doppler effect because the element of optical length dr/λ, expressed in terms of local λ, is affected by the Doppler ratio of local velocities, namely nv/c, the integrated results of which give Δz.

Since Δz is the number of phase reversals due to motion and T is the time needed for light to go around the loop once, it is obvious that $\lambda_0 \Delta z$ is the free space change of path length and $(\lambda_0/c)\Delta z$ is the change δT of the circulation time T. These results, for the equivalent free space change of path length, remain valid both for $n = 1$ and $n \neq 1$. Hence (5.5.32) gives

$$\delta T = \frac{1}{c^2} \oint (\mathbf{v} \cdot d\mathbf{r}) \tag{5.5.33}$$

δT is independent of the optical properties of the path, though T depends on that, because

$$T = \oint \frac{n}{c}\, dl, \tag{5.5.34}$$

where $dl = d|r|$. Therefore,

$$\frac{\delta T}{T} = \frac{\oint (\mathbf{v} \cdot d\mathbf{r})}{c \oint n dl}. \tag{5.5.35}$$

For uniform translation

$$\oint (\mathbf{v} \cdot d\mathbf{r}) = \iint_{\Sigma} (\text{curl}\, \mathbf{v} \cdot d\mathbf{\Sigma}) = 0$$

because $\text{curl}\, \mathbf{v} = 0$, so $\delta T = 0$.

When $\delta\omega \neq 0$ but the number m of nodal points along the closed path remains an adiabatic invariant, as in ring lasers,

$$\delta m = 0, \tag{5.5.36}$$

where $m = T\nu = T\omega/2\pi$. Putting (5.5.36) in (5.5.35) we obtain

$$\frac{\delta T}{T} = -\frac{\delta\omega}{\omega} = \frac{\oint (v \cdot dr)}{c \oint n dl}. \tag{5.5.37}$$

Equation (5.5.34) gives the unperturbed circulation time T for light in a stationary loop in terms of the path length and propagation velocity through the Fermat integral. In that relation

dl depends on the optical properties of the beam path. If L is the ring perimeter and l is the length of the dielectric inside it, then

$$\oint ndl = L - l + nl \quad \text{and} \quad T = c(L - l + nl). \tag{5.5.38}$$

Since n and the dielectric slug length l occur as a product, an increase of either has the same effect. Also it is obvious that the value of the integral in (5.5.38) is independent of the position of the slug in the beam path. When another section of length a is filled up with a laser active medium of refractive index n_a, then dispersions of n and n_a are also included and the total optical length of the resonator, Khromykh (1966), is given by

$$\oint ndl = L - l - a + wl\left(\frac{dn}{d\omega}\right) + wa\left(\frac{dn_a}{d\omega}\right). \tag{5.5.39}$$

Since l/L is small, dispersion in dielectrics yields a contribution of the order of 10^{-3} to 10^{-4} to the main shift. The refractive index of the active gaseous medium is of the order of unity. Even then the dispersion term for that should be retained because of the anomalous dispersion in the region of laser action.

5.5.3 Phase of light in a moving closed loop

In the integration in (5.3.1) the upper and lower limits are the values of the independent variable at the end and beginning of one circulation around the loop. Only these limits are affected by the displacement q generated by the velocity v of the field and ϕ is not affected by that. So the variation of the limits of (5.3.21) should be incorporated in the integration of (5.3.1).

Since the phase ϕ is a general invariant in 4-space we have

$$\delta\phi = \delta z - \delta T = 0, \tag{5.5.40}$$

where

$$2\pi\delta z = \delta \oint(\mathbf{k}\cdot d\mathbf{r}) = \oint\{\delta\mathbf{k} - (\mathbf{q}\times \text{curl}\,\mathbf{k}) + \nabla(\mathbf{k}\cdot\mathbf{q})\}\cdot d\mathbf{r} \tag{5.5.41}$$

and

$$2\pi\delta T = \delta\int_0^T \omega dt = \int_0^T \delta\omega dt + \omega\delta T. \tag{5.5.42}$$

The relation (5.3.21) has been used to obtain (5.5.41). The fringe shift associated with the interference of CW and CCW beams is δz, and δT is the change of circulation time with respect to the stationary case.

The variation of a line integral due to the deformation field is expressed by (5.5.41) and (5.5.42) in terms of parameters of an inertial frame of reference in which the stationary system of the phase ϕ is described. Hence, before introducing material properties the first order change of z of (5.5.42), with the help of the result (5.3.7), is given by

$$2\pi\delta z = \oint(\delta\mathbf{k}\cdot d\mathbf{r}) + (\mathbf{k}\cdot\mathbf{q})_2 - (\mathbf{k}\cdot\mathbf{q})_1. \tag{5.5.43}$$

The subscripts 1 and 2 in the last two terms of the right hand side of (5.5.43) denote the value of $(\mathbf{k}\cdot\mathbf{q})$ at the end and beginning of one circulation around the loop. As q is not single valued for this circulation, the deformation increases with time. The increment δz is the fringe shift due partly to the intrinsic change of the wave vector by the motion $\mathbf{v}$ and partly to the contribution from the change of the boundaries of the integral.

In a linear medium $k = \omega/u = \omega n/c$, where u is the phase velocity. Hence if $\delta\omega = 0$ we have

$$\delta k = -\left(\frac{k}{u}\right)\delta u = -\frac{\omega n^2}{c^2}\delta u, \tag{5.5.44}$$

where δu is the change in effective velocity of propagation of the phase of light in the moving medium due to Fresnel drag as seen by the stationary observer. So if α is the drag coefficient we can put

$$\delta u = \alpha v, \tag{5.5.45}$$

where v is the velocity field describing motion of the optical loop. Therefore, using (5.5.45) in (5.5.44) we obtain the relation

$$\delta k = -\frac{\omega}{c^2}\left\{n^2 - 1 + kn\left(\frac{dn}{dk}\right)\right\}v, \tag{5.5.46}$$

which will be useful in the evaluation of the first term in the right hand side of (5.5.43). Since $d\mathbf{r}/dl$ is the unit vector along the tangent to the beam direction, in place of (5.5.45) we can write

$$\delta u = \alpha\left(\mathbf{v}\cdot\frac{d\mathbf{r}}{dl}\right). \tag{5.5.47}$$

So, now (5.5.44) is replaced by

$$\delta k = -\frac{k\alpha}{u}\left(v\cdot\frac{dr}{dt}\right) = -\frac{k\alpha}{u}v_t, \tag{5.5.48}$$

where v_t is the component of v along the beam direction. Hence

$$(\delta\mathbf{k}\cdot d\mathbf{r}) = \delta k dl = -\frac{k\alpha}{u}v_t dl = -\frac{k\alpha}{u}(\mathbf{v}\cdot d\mathbf{r}). \tag{5.5.49}$$

So the change of k and $d\mathbf{r}$ in the direction of $\delta\mathbf{k}$ are obtained. The last term in (5.5.43) can be expressed in terms of the velocity field v if the fact that $(\mathbf{k}\cdot\mathbf{q})$ changes over the interval of time dt by the amount $(\mathbf{k}\cdot\mathbf{v})\,dt$ is used. Hence if the complete loop is covered we find that

$$(\mathbf{k}\cdot\mathbf{q})_2 - (\mathbf{k}\cdot\mathbf{q})_1 = \int_0^T (\mathbf{k}\cdot\mathbf{v})\,dl = \oint(\mathbf{k}\cdot\mathbf{v})\frac{dl}{u} = \oint\frac{k}{u}(\mathbf{v}\cdot d\mathbf{r}), \tag{5.5.50}$$

the increment dl of the light path is udt, and $\mathbf{k}$ and $d\mathbf{r}$ are assumed to have the same direction. So putting (5.5.50) in (5.5.44) we obtain

$$2\pi\delta z = \oint\left\{\left(\delta\mathbf{k} + \frac{k}{u}\mathbf{v}\right)\cdot d\mathbf{r}\right\}. \tag{5.5.51}$$

We can write (5.5.51) with the help of (5.5.49) as

$$2\pi\delta z = \oint\frac{k}{u}(1-\alpha)(\mathbf{v}\cdot d\mathbf{r}). \tag{5.5.52}$$

Since the sign of $\mathbf{v}$, for the two oppositely circulating beams, differs we can write (5.5.52) as

$$2\pi\delta z_{\pm} = \pm \oint \frac{k}{u}(1-\alpha)(\mathbf{v}\cdot d\mathbf{r}). \tag{5.5.53}$$

In moving type interferometers the observed fringe shift is twice the amount δz of that given by (5.5.51) in comparison with the fringe position for stationary interferometers. If the fringe shift due to interference of the CW and CCW beams is Δz, then

$$\Delta z = \delta z_{+} - \delta z_{-} = \frac{1}{\pi}\oint \frac{k}{u}(1-\alpha)(\mathbf{v}\cdot d\mathbf{r}). \tag{5.5.54}$$

This relation can also be written as

$$\Delta z = \frac{2}{c\lambda_0}\oint n^2(1-\alpha)(\mathbf{v}\cdot d\mathbf{r}), \tag{5.5.55}$$

where λ_0 is the free space wavelength of light and we have put $\omega/2\pi = k_0 c = c/\lambda_0$. If δT is the change in the time of circulation in the loop for the CW beam, compared with the stationary case, then putting $\delta\omega = 0$ in (5.5.42) and using (5.5.40) we obtain

$$\omega\delta T = 2\pi\delta z. \tag{5.5.56}$$

Denoting now by ΔT the difference in circulation time between the CW and CCW beams we have

$$\omega\Delta T = 2\pi\Delta z, \tag{5.5.57}$$

where Δz is given by (5.5.54) and (5.5.55). Hence in the interferometer co-moving with the optical medium

$$\Delta T = \frac{2}{c^2}\oint (1-\alpha)n^2(\mathbf{v}\cdot d\mathbf{r}). \tag{5.5.58}$$

Since $\mathbf{v} = (\mathbf{\Omega}\times\mathbf{r})$ and $\operatorname{curl}\mathbf{v} = 2\mathbf{\Omega}$ we have

$$\oint(\mathbf{v}\cdot d\mathbf{r}) = \iint_{\Sigma} (\operatorname{curl}\mathbf{v}\cdot d\mathbf{\Sigma}) = 2(\mathbf{\Omega}\cdot\mathbf{\Sigma}). \tag{5.5.59}$$

The constancy of Ω therefore shows that the interference effects of closed beams and the ring laser effects are independent of the center around which the equipment is being rotated. When the dielectric slug is stationary and the interferometer or ring laser moves, the velocity of propagation is measured in a frame of reference in which the medium lies. So,

$$\delta k = 0, \quad \delta u = 0. \tag{5.5.60}$$

The Fresnel drag is then zero. Putting, therefore, $\alpha = 0$ in (5.5.55) we find that

$$\Delta z = \frac{2}{c\lambda_0}\oint n^2(\mathbf{v}\cdot d\mathbf{r}). \tag{5.5.61}$$

When the ring interferometer and ring laser are stationary and the dielectric slug is moving, the integrand of (5.5.60) vanishes; hence,

$$\oint(\operatorname{grad}(\mathbf{k}\cdot\mathbf{q})\cdot d\mathbf{r}) = (\mathbf{k}\cdot\mathbf{q})_2 - (\mathbf{k}\cdot\mathbf{q})_1 \tag{5.5.62}$$

because the loop is then stationary. Equation (5.5.55), therefore, becomes

$$\Delta z = -\frac{2}{c\lambda_0}\oint n^2\alpha(\mathbf{v}\cdot d\mathbf{r}). \tag{5.5.63}$$

This is the Fresnel drag effect for translational motion. It is used to determine the extent to which this drag coefficient can be extrapolated to cases of non-uniform motion.

The initial displacement q_1 of the circulation considered in (5.5.43) can be assumed zero; so $(k\cdot q)_2$ comes about gradually by the velocity. Hence if v_g is the group velocity and T is the circulation time we obtain

$$(\mathbf{k}\cdot\mathbf{q})_2 = \int_0^T (\mathbf{k}\cdot\mathbf{v})\,dt = \oint \frac{k}{v_g}(\mathbf{v}\cdot d\mathbf{r}), \tag{5.5.64}$$

where $\mathbf{k}$ and $d\mathbf{r}$ have the same direction. Since

$$v_g = u\left\{1 - \frac{k}{u}\left(\frac{du}{dk}\right)\right\} \quad \text{and} \quad \frac{k}{n}\frac{du}{dk} \ll 1$$

we have

$$\frac{1}{v_g} = \frac{1}{u}\left\{1 + \frac{k}{n}\frac{dn}{dk}\right\}. \tag{5.5.65}$$

Hence

$$(\mathbf{k}\cdot\mathbf{q})_2 = \frac{w}{c^2}\oint\left\{n^2 + kn\frac{dn}{dk}\right\}(\mathbf{v}\cdot d\mathbf{r}). \tag{5.5.66}$$

Putting the value of $\delta\mathbf{k}$ from (5.5.46) and $(\mathbf{k}\cdot\mathbf{q})_2$ from (5.5.66) in (5.5.43) we find that

$$2\pi\delta z = \frac{w}{c^2}\oint(\mathbf{v}\cdot d\mathbf{r}) \tag{5.5.67}$$

because the material contribution of refraction and dispersion cancel each other. Hence δz does not depend on the physical properties of the refractive material co-moving with the optical loop.

5.5.4 Frequency change due to rotation of rings

The phase going around the closed loop must be univalued for sustained oscillation. So the number of wavelengths in the loop should remain constant, or in other words, the condition (5.5.36) is imposed. Consequently

$$\delta z = 0. \tag{5.5.68}$$

Equation (5.5.40) then gives

$$\delta T = 0 \tag{5.5.69}$$

and equation (5.5.42) reduces to

$$\int_0^T \delta w dt + w\delta T = 0. \tag{5.5.70}$$

Therefore, for a uniform and stationary medium we have

$$\frac{\delta\omega}{\omega} = -\frac{\delta T}{T}. \tag{5.5.71}$$

Following the doubling procedure for frequency shift and time difference between the CW and CCW modes, we have

$$\left|\frac{\Delta\omega}{\omega}\right| = \frac{\Delta T}{T} \tag{5.5.72}$$

Since ω varies, relations $k = \omega/u = \omega n/c$ which are valid in a linear medium give

$$\frac{\delta k}{k} = \frac{\delta\omega}{\omega} - \frac{\delta u}{u}. \tag{5.5.73}$$

Then using (5.5.68) in (5.5.56) and following the arguments used for obtaining the relations (5.5.46)–(5.5.49) we obtain

$$-\int \frac{k}{u}(\mathbf{v}\cdot d\mathbf{r}) = \oint (\delta\mathbf{k}\cdot d\mathbf{r}) = \oint \delta k dr = \oint \left(\frac{\delta\omega}{\omega} - \frac{k\alpha}{u}v_t\right) dr. \tag{5.5.74}$$

Putting equation (5.5.34) in (5.5.73) we obtain

$$-\oint \frac{k}{u}(\mathbf{v}\cdot d\mathbf{r}) = \frac{\delta\omega}{\omega}\int ndl - \oint \frac{k\alpha}{u}v_t dr, \tag{5.5.75}$$

$$\frac{\delta\omega}{\omega} = -\frac{\oint n^2(1-\alpha)(\mathbf{v}\cdot d\mathbf{r})}{c\oint ndl}. \tag{5.5.76}$$

Hence equation (5.5.72) gives

$$\left|\frac{\Delta\omega}{\omega}\right| = \frac{\Delta T}{T} = \frac{2\oint n^2(1-\alpha)(\mathbf{v}\cdot d\mathbf{r})}{c\oint ndl}. \tag{5.5.77}$$

Since the difference in the circulation times between the CW and CCW beams does not depend on whether the optical circuit is being used as an interferometer or as a ring laser, this relation can be obtained by putting (5.5.38) and (5.5.58) in (5.5.72). For the two special cases corresponding to the conditions (5.5.60) and (5.5.62) the ring laser effect can be easily evaluated.

5.6 Polarization and Coherence of Waves

The photons are mutually uncorrelated since the atomic processes by which they are created are stochastically independent of each other. So, the photons superpose in an incoherent manner. A classical wave may be considered as an ensemble of coherent quantities. Particle polarization is identified with incoherent superposition of mutually uncorrelated photons. The averaging over a statistical ensemble of photons represents incoherent superposition of the photon states, when the averaging is done after forming the bilinear products of the wave fields, so that the phase factor $\exp(-i\omega t)$ of the different photons disappear. If the averaging is done before forming these products, then that is valid for a coherent superposition of the states.

A 100% polarized EM radiation obtains from coherent superposition of EM waves. The polarization may be elliptic with arbitrary shape and orientation of the polarization ellipse.

Quantum mechanically this means that each photon is always 100% polarized. The scattered radiation can also be 100% elliptically polarized when the incident radiation is 100% polarized, or partially polarized, or unpolarized even when the incident radiation is 100% linearly polarized. Moreover, both the incoming and scattered photons are always 100% polarized for each individual scattering event. An incoherent superposition of stochastically uncorrelated wave trains is possible. An incoherent superposition of an ensemble of uncorrelated photon states can have an arbitrary degree of partial polarization.

From polarization due to anisotropic radiative excitation and some other causes, an atomic state acquires magnetic dipole moment through an orientation, and an alignment from electric quadrupole moment.

The radiation becomes partially ionized due to incoherent superposition of the photon states. Consider two electric fields E_1 and E_2 of waves. From coherent superposition, the resultant field intensity is

$$|E|^2 = |E_1 + E_2|^2,$$

so that the effect of interference from superposition of two waves is possible for coherent superposition. For incoherent superposition the resultant intensity $|E^2| = |E_1^2| + |E_2^2|$. So that the wave interference effect is not possible for incoherent superposition.

Bilinear products of complex combination of electric vectors are introduced in the classical treatment of a statistical ensemble of uncorrelated wave trains for the study of radiative transfer processes. Bilinear products between all the different combinations of coefficients, in the expansion of the wave function, fix the relative phases within each individual quantum system of a statistical ensemble, before the different uncorrelated individual quantum systems are statistically superposed. These products describe the statistical ensemble of uncorrelated photons in the study of diagnostic of electromagnetic radiations.

The coherences refer to phase relations between the two different polarization components of an electromagnetic wave mode travelling along a given direction at a given frequency. Contribution from coherence between wave fields of different frequencies or directions generally are not considered.

5.7 Magnetization Current

The magnetization current $\mathbf{j}_M$ is the curl of the magnetic moment per unit volume, of charges gyrating round a quasistatic magnetic field H_0. If the electron density N, or the ambient field H_0, or the perpendicular kinetic energy $W_\perp$ is non-uniform, then also the magnetization is produced by gyratory motion without any drift of the guiding center. In a volume of the order of the dimensions of the Larmor radius ρ, there will be many particles with random phases. The fluctuating magnetic field of this current from these particles is small compared to the average field and is determined from the average moment produced by the distribution of the orbits, or from their magnetic moments. This field is small compared to the average field.

An element of area Σ, bounded by a curve Γ of fig. 5.7.1 is crossed by the orbits of the gyrating charges and also by the orbits of their guiding centers. Orbits, which are completely within Γ, such as b and d, penetrate twice the surface Σ. So they contribute nothing to the current of charges in the guiding center approximation. Orbits, like those of a and c, contribute to the current, because these penetrate Σ only once. Consider the length element dl at a point P along Γ. If their are N current loops, per unit volume, each carrying a current I around an area A, then the number of current loops penetrated by dl, in the periphery of Σ is $N(A\cdot dl)$ because $A\cdot dl$ is a volume element at P and the current from the gyrating particles,

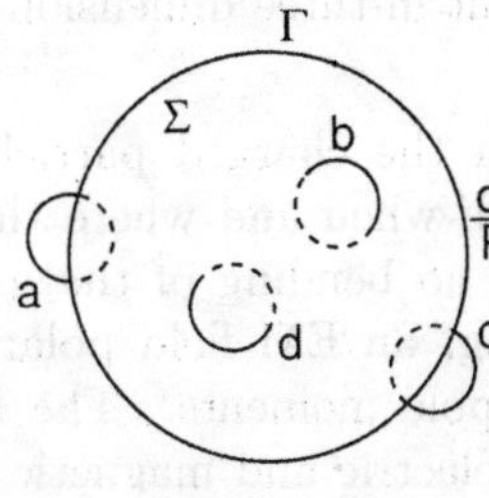

Fig. 5.7.1 It shows the particle orbits b and d which contribute nothing to the magnetization current j_M and particle orbits a and c which contribute to j_M.

$$I_m = \int_\Gamma NI(A\cdot dl) = c\int_\Gamma (M\cdot dl) = c\iint_\Sigma ((\nabla\times M)\cdot d\Sigma) = \iint_\Sigma (j_M\cdot d\Sigma),$$

where $M = NIA/c = Nm = -N(W/H_0)n_{11}$, $n_{11} = H_0/H_0$, and $j_M(= C(\nabla\times M))$ is the surface current vector of magnetization. Since this relation is valid for every element of area, even when the guiding centers of the particles are at rest, it acts on the particles in the eyes of the observer, moving with that center. The total electric current j is the sum of $\mathbf{j}_M$ and the material current of conduction and convection, j_0. Then the Ampere-Maxwell equation for magnetoelectric induction is

$$\nabla\times H = \frac{1}{c}\frac{\delta E}{\delta t} + \frac{4\pi}{c} j_0,$$

where $H = B - 4\pi M$. Since $\nabla\cdot j_M = 0$, the corresponding charge density vanishes. The net current

$$j_M = \nabla\times M = -\nabla\times\left(\frac{1}{B}W_\perp N\hat{n}_{11}\right)$$

$$= \frac{1}{B^2}W_\perp N(\hat{n}_{11}\times\nabla B) - \frac{1}{B}(\hat{n}_{11}\times\nabla(NW_\perp)) - \frac{1}{B}NW_\perp(\nabla\times\hat{n}_{11}).$$

5.8 Dielectric Polarization and Magnetic Moment

The primary source of EM fields are the charges and currents and the secondary sources are the polarizable and magnetizable matter. Matter is polarized by the Coulomb field and magnetized by electric current. The dielectric polarization from electrification of matter depends on the translational displacement and the magnetic moment depends on the rotational part of the displacement. So, polarization is related to linear momentum, and magnetization to angular momentum. The linear momentum and angular momentum of particle mechanics are the mechanical equivalents of the dielectric polarization and magnetic moment of charged particles of electrodynamics of media. Polarization, linear velocity, linear momentum, translational displacement are polar vectors; and angular momentum, rotational velocity, magnetic moment are axial vectors[2].

The field induced translational displacement is the polarization variable of the charges of a system. It is the time integral of the linear momentum per unit mass. The magnetic moment

[2]See footnotes, section 1.10.

is an angular momentum like quantity. Any material displacement, induced by a force field, consists of the sum of a translational displacement and a rotational displacement. They form the set of elements of the displacement in three dimensional position space, and are mutually complementary.

Force of an electric field acts on the charged particles at all points of their motional paths. The magnetic moments vanish when and where the material displacement is parallel to the electric current, which means no bending of the motional paths of the charges, then or three. Macroscopically considering, an EM field polarizes and magnetizes a medium by exciting electric and magnetic multipole moments. The magnetic monopole vanishes in all situations. The dominant molecular electric and magnetic multipoles are the electric and the magnetic dipoles. The averages of these molecular dipoles are taken over many molecules for use in the macroscopic theory. In the absence of applied fields, the multipole moments vanish. The influence of these multipole moments on the EM fields is represented in the constitutive relations which are conceived phenomenologically. These supplementary relations depend on the parameters of the medium and the force field applied. These are necessary to form a close system of equations of electrodynamics for sc ution. These constitutive relations are formulated completely independent of the derivation of the macroscopic Maxwell equations from the microscopic analytical formulation by Lorentz. The magnetic moments may be induced by fields exerted on the atoms, or these may be permanent magnetic moments (as in magnetic materials, example; ferromagnets).

For a particle of mass m and charge q, the linear momentum $p = m\dot{\xi}$, the angular momentum $L = m(\mathbf{r} \times \dot{\boldsymbol{\xi}})$, dielectric polarization $P = q\xi$ and the magnetic moment $\mathbf{m} = q(\mathbf{r} \times \dot{\boldsymbol{\xi}})/2c$. Here $\boldsymbol{\xi}$ is the field induced displacement and $\dot{\boldsymbol{\xi}}$ is the field induced velocity.

5.9 Magnetic Moment Physics

The three sources of the magnetic moment of an atom are : (1) the electron spin; (2) the electron orbital angular momentum about its nucleus; (3) applied electrostatic fields, magnetostatic fields, electromagnetic wave fields, and other forces. Classical study is tolerable so long as the field applied is not so large that the electron gyroradius (section 4.8) is of the order of atomic dimensions (10^{-8} cm).

A strictly classical system, in thermal equilibrium, displays no magnetic moment even in presence of an ambient magnetic field. The magnetic moments of atomic nuclei are about 10^{-3} times smaller than those of orbital electrons, so their neglect is justified for high frequency waves. But, in calculations on outputs from low frequency phenomena, the approximation that ion dynamics is dominant in presence of only inertial motion of electrons, is valid. So, at low frequencies the effect of magnetic moment of the ions is not negligible.

The magnetic moment field follows from the basic laws of Ampere and Biot-Savart, of electrodynamics, which states that a current, through a closed loop, produces an axial magnetic field. Macroscopically considering, the closed loops of the current are formed by the bending of the direction of motion of the mobile charged constituents of electrically responsive matêrials. These bendings instantaneously generate the magnetic moment.

Any change in the direction of a motion is a consequence of the disturbance which is not symmetric in all directions. The asymmetry means a gain or loss of the mechanical moment (the angular momentum, or the moment of momentum), from the interaction of applied fields with local noise fields. The fluctuations in the charged particle trajectories generate the asymmetry responsible for the net gain (or loss) of the magnetic moment, which is proportional to its angular momentum. The features of this excited field, which is a volume effect of density M, depend on the thermal state prevailing during the interaction, and on the local modes of energy transfer in the medium.

The analytical formula for the magnetic moment M, per unit volume, section 4.11, is

$$M = \frac{1}{2c}(\boldsymbol{\xi} \times \mathbf{j}), \tag{5.9.1}$$

where $\boldsymbol{\xi}$ is the displacement and $\mathbf{j}$ is the current of electricity, both being field induced. It vanishes when (or where) ξ is parallel to j, or, in other words when (or where) there is no bending of the path of the charges. $\boldsymbol{\xi}$ and $\mathbf{j}$ are determined from the equations of motion, or from the Maxwell equations. The direction of M is reversed on reversal of rotation of the electric field vector $\mathbf{E}$. This formula for $\mathbf{M}$ is a function of space co-ordinates and time. So, it is the local expression for M.

For a particle of charge q, the magnetic moment

$$\mathbf{m} = \frac{q(\boldsymbol{\xi} \times \dot{\xi})}{2c}. \tag{5.9.2}$$

Currents of charges in a volume give rise to a magnetic field, which is determined by the magnetic moment of a system of particles,

$$M = \sum_\alpha m_\alpha = \frac{1}{2c}\sum_\alpha \left(\frac{1}{m_\alpha} q_\alpha L_\alpha\right) = \frac{1}{2c}\sum_\alpha [q_\alpha(\xi_\alpha \times v_\alpha)]. \tag{5.9.3}$$

We can write $\mathbf{r}_\alpha = \mathbf{r}_0 + \boldsymbol{\xi}_\alpha$, where $\mathbf{r}_0$ is the prefield position P of a point whose position vector is $\mathbf{r}_\alpha$. For study of volume effects we can put $r_\alpha = 0$ and obtain

$$M = \frac{1}{2c}\sum_\alpha (\boldsymbol{\xi}_\alpha \times \mathbf{j}_\alpha), \tag{5.9.4}$$

where $\mathbf{U}_\alpha = \dot{\boldsymbol{\xi}}_\alpha$, dot denotes the time derivative, N_α is the number density of the αth species of charges, ξ_α is its wave field induced average displacement at the prefield point P and q_α is the charge per particle of this species.

These relations are valid in presence of a single force field acting in a medium, consisting of several species of mobile charge distributions, say a multispecies plasma. In presence of several force fields the non-relativistic magnetic moment density is given by

$$M = \sum_\alpha M_\alpha = \frac{1}{2c}\sum_\alpha \left[\left(\sum_\beta \boldsymbol{\xi}_\alpha^\beta\right) \times \left(\sum_\beta \mathbf{j}_\alpha^\beta\right)\right], \tag{5.9.5}$$

where

$$M_\alpha = \frac{1}{2c}\left[\left(\sum_\beta \boldsymbol{\xi}_\alpha^\beta\right) \times \left(\sum_\beta \mathbf{j}_\alpha^\beta\right)\right]. \tag{5.9.6}$$

Here the superscript β denotes the βth force field and the subscript α stands for the αth species of charge (or mass); $\boldsymbol{\xi}_\alpha^\beta$ is, therefore, the displacement and $\mathbf{j}_\alpha^\beta$ is the current induced by the βth force field on the αth species of charge.

The domain formula for the magnetic moment is

$$M = \frac{(IS\hat{\mathbf{n}})}{c}, \tag{5.9.7}$$

where S is the area of the loop carrying the current I and $\hat{\mathbf{n}}$ is the unit vector normal to the plane of the loop. Its direction follows the right handed screw rule. Unlike the local analytical formula (5.9.1), it is not a function of the position co-ordinates of the field point and time. The domain of integration is either the area of the loop Γ or the time of describing the loop once completely in the neighbourhood of a point. If q is the charge per particle and T is the time of a complete gyration, then per particle, the magnetic moment

$$M = \frac{qS\hat{\mathbf{n}}}{cT}. \tag{5.9.8}$$

If N is the electron density, then $I = -Nef$, where $f(=\omega/2\pi)$ is the frequency of the oscillation of the electrons in presence of the applied field. More generally,

$$\left.\begin{aligned} I &= \sum_\alpha (N_\alpha q_\alpha f_\alpha), \\ M &= \sum_\alpha \frac{(N_\alpha q_\alpha f_\alpha S_\alpha \hat{\mathbf{n}}_\alpha)}{c} = \sum_\alpha \frac{(\boldsymbol{\xi}_\alpha \times \mathbf{j}_\alpha)}{2c}, \end{aligned}\right\}, \tag{5.9.9}$$

where $\boldsymbol{\xi}_\alpha$ is the field induced displacement of the αth species. So, the two formulas are equivalent. For the same body we can write

$$M_1 = \frac{1}{2c}\iiint_V (\xi \times j)\, dV, \quad M_2 = \frac{1}{c}\iint_\Sigma I d\Sigma, \tag{5.9.10}$$

where $d\Sigma = (\xi \times d\xi)/2$; $d\Sigma$ is the length element at a point in the periphery of the open surface Σ of a closed circuit Γ carrying the current I. Since $\mathbf{j}dv = Id\boldsymbol{\xi}$, we find that

$$dM_1 = \frac{1}{2c}(\xi \times j)\, dV = \frac{I}{2c}(\xi \times d\xi) = \frac{1}{c}Id\Sigma = dM_2. \tag{5.9.11}$$

For N electrons, equation (5.9.7) is replaced by

$$\mathbf{M} = \frac{NIS\hat{\mathbf{n}}}{c}. \tag{5.9.12}$$

The delta function representation of some quantities is :

(a) polarization density

$$P = \sum_n P_n \delta(r - r_n);$$

(b) charge density

$$\rho^c = \sum_n q_n \delta(r - r_n);$$

(c) magnetization density

$$M = \sum_n m_n \delta(r - r_n);$$

(d) current density

$$\mathbf{j} = \sum_n q v_n \delta(\mathbf{r} - \mathbf{r}_n).$$

The equation of motion and the mass conservation relation for the αth species of charges are

$$N_\alpha \frac{d}{dt}(\mathbf{p}_\alpha) = N_\alpha \left(\frac{\partial}{\partial t} + (\mathbf{v}_\alpha \cdot \boldsymbol{\nabla})\right) \mathbf{p}_\alpha$$

$$= N_\alpha q_\alpha \left(\mathbf{E} + \frac{1}{c}(\mathbf{v} \times \mathbf{B})\right) - \boldsymbol{\nabla} p_\alpha - m_\alpha v_\alpha \sum_\beta (\boldsymbol{\nu}_\alpha - \boldsymbol{\nu\beta}), \tag{5.9.13}$$

$$\frac{\partial}{\partial t}\rho_\alpha^m + (\mathbf{v}_\alpha \cdot \boldsymbol{\nabla})\rho_\alpha^m = 0. \tag{5.9.14}$$

For $\mathbf{H}$ and $\mathbf{E}$ the Maxwell equations give the wave equations

$$\nabla^2 \mathbf{H} - \frac{1}{c^2}\frac{\partial^2 \mathbf{H}}{\partial t^2} = \frac{4\pi}{2c^2}\frac{\partial^2}{\partial t^2}\left(\sum_\alpha (N_\alpha q_\alpha (\boldsymbol{\xi}_\alpha \times \dot{\boldsymbol{\xi}}_\alpha)\right) - \frac{4\pi}{c}\boldsymbol{\nabla}$$

$$\times \frac{\partial}{\partial t}\left(\sum_\alpha (N_\alpha q_\alpha \boldsymbol{\xi}_\alpha)\right) + \frac{4\pi}{c}\boldsymbol{\nabla} \times \mathbf{j}_c. \tag{5.9.15}$$

$$\nabla(\nabla \cdot E) + \nabla^2 E - \frac{1}{c^2}\frac{\partial^2 E}{\partial t^2} = \frac{4\pi}{c^2}\frac{\partial^2}{\partial t^2}\left(\sum_\alpha (N_\alpha q_\alpha \xi_\alpha)\right) + \frac{4\pi}{2c^2}\nabla$$

$$\times \frac{\partial}{\partial t}\left(\sum_\alpha (N_\alpha q_\alpha \xi_\alpha) \times \xi_\alpha\right) + \frac{4\pi}{c^2}\frac{\partial}{\partial t} j_c. \tag{5.9.16}$$

The field of a magnetic material reorients the axes of its magnetic dipoles, like the polarization of electric dipoles by electric field. The total magnetic moment, due to all the charged particles, moving in a body, is

$$\frac{1}{2c}\iiint_V (\mathbf{r} \times \rho\bar{\mathbf{v}})\, dV = \frac{1}{2}\iiint_V [\mathbf{r} \times (\nabla \times M)]\, dV, \tag{5.9.17}$$

where $\rho\bar{v}$ is the average current, M is the magnetization, and

$$\rho\bar{\mathbf{v}} = (\nabla \times M). \tag{5.9.18}$$

Since $\rho\bar{\mathbf{v}} = 0$ outside V, integration is done over any volume which includes the body. The vector formula for $\boldsymbol{\nabla}(\mathbf{r} \cdot M)$, Appendix 3, formula 10, shows that

$$\iiint_V (\mathbf{r} \times (\nabla \times M))\, dV = \iint_\Sigma [(\mathbf{r} \cdot M)\, d\Sigma - (\mathbf{r} \cdot d\Sigma)M] + 2\iiint_V M dv, \tag{5.9.19}$$

where $d\Sigma$ is a surface element of any cross section Σ. Since $M = 0$ outside V, the surface integrals vanish and for the magnetization M (the magnetic moment per unit volume), we obtain

$$\frac{1}{2c}\iiint_V (\mathbf{r} \times \rho\bar{v})\, dV = \iiint_V M dV. \tag{5.9.20}$$

M is not uniquely defined thus, because gradient of any scalar field could be added to it there. The total outward flux of currents in the body is

$$\iint_{\Sigma} (\rho\bar{\mathbf{v}}\cdot d\mathbf{\Sigma}) = c\iint_{\Sigma} ((\nabla\times M)\cdot d\Sigma) = c\oint_{\Gamma} (M\cdot d\mathbf{r}), \tag{5.9.21}$$

where we have integrated both sides over a surface bounded by a curve which encloses V and nowhere enters it. This relation is valid for every cross section of V.

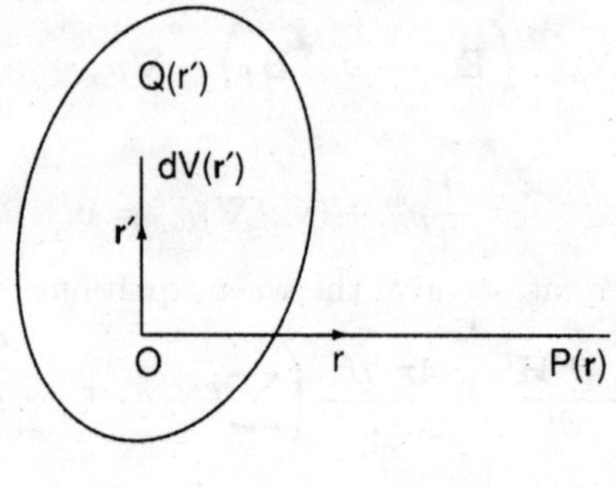

Fig. 5.9.1 It shows the magnetic moment μ at P(r), due to the current j(r′) at the volume point Q(r′).

The magnetic moment $\boldsymbol{\mu}$, at the point P of fig. 5.9.1, with position vector $\mathbf{r}$, due to currents in the volume and the magnetization density M, respectively, therefore, are

$$\boldsymbol{\mu}(\mathbf{r}) = \frac{1}{2c}\iiint_{V} (\mathbf{r}'\times\mathbf{j}(\mathbf{r}'))\,dV', \tag{5.9.22}$$

$$M = \frac{1}{2c}(\mathbf{r}\times\mathbf{j}), \tag{5.9.23}$$

when $\mathbf{j}$ is due to a discrete distribution of charges q_i, moving with velocity $\mathbf{v}_i$,

$$\mathbf{j} = \sum_i q_i\mathbf{v}_i\delta(\mathbf{r}-\mathbf{r}_i) \tag{5.9.24}$$

and, the orbital angular momentum of the ith particle is

$$L_i = m_i(\mathbf{r}_i\times\mathbf{v}_i), \tag{5.9.25}$$

where m_i is the particle mass. Hence (5.9.1) becomes

$$\boldsymbol{\mu} = \sum_i\left(\frac{(q_iL_i}{2m_ic}\right). \tag{5.9.26}$$

For particles having the same ratio q/m of charges and mass,

$$\boldsymbol{\mu} = \frac{q}{2mc}\sum_i L_i = \frac{qL}{2mc}, \tag{5.9.27}$$

where L is the total angular momentum. This relation is valid even for charges inside atoms and molecules for non-relativistic motion in non-quantized fields. For this reason for electrons the right hand side of (5.9.27) is multiplied by a factor g, called the *Lande factor.* Approximately $g = 2$ when L is replaced by the spin S.

Magnetic materials tend to adjust a dipole magnetic field to a new direction. The induction $\mathbf{B}$ from a current loop Γ acts on a small probing current loop Γ_1, forming a magnetic moment $\boldsymbol{\mu}$. The dipole is given a rotational motion in various direction, such that the flux through it remains constant. The fall in energy, for a given angular displacement $\delta\theta$ is greatest when the dipole is turned towards the direction of B. The torque acts so as to turn $\boldsymbol{\mu}$ (here the dipole moment of the current loop Γ_1) in that direction. The corresponding angular displacement $N\delta\theta = \mu(B\cdot n) = \mu B \sin\theta\delta\theta$, where N is the torque on the dipole, θ is the angle between $\boldsymbol{\mu}$ and $\mathbf{B}$, and n is the unit vector normal to $\boldsymbol{\mu}$. Therefore,

$$N = (\boldsymbol{\mu} \times B). \tag{5.9.28}$$

The force and its torque on currents, in an external magnetic field $\mathbf{H}(\mathbf{r})$, are:

$$F = \frac{1}{c} \iiint\limits_V (\mathbf{j} \times B)\, dV \tag{5.9.29}$$

$$N = \frac{1}{c} \iiint\limits_V (\mathbf{r} \times (\mathbf{j} \times B))\, dV, \tag{5.9.30}$$

where V is the volume of the system and dV is the volume element at a point in V. On a localized current distribution the total torque is

$$N = \frac{1}{c} \iiint\limits_V (\mathbf{r}' \times (\mathbf{j} \times B))\, dV',$$

where $\mathbf{r} = \mathbf{r}_0 + \mathbf{r}'$, and $\mathbf{r}_0$ is a fixed point on the body. Writing out the vector product we obtain

$$N = \frac{1}{c} \iiint\limits_V [(\mathbf{r}'\cdot B)j - (\mathbf{r}', \mathbf{j}')\mathbf{B}]dV'.$$

For a volume current I in a single contour Γ,

$$F = \frac{I}{c} \oint (d\mathbf{r} \times B), \quad N = \frac{I}{c} \oint_\Gamma (\mathbf{r} \times (d\mathbf{r} \times \mathbf{B})). \tag{5.9.31}$$

because, for a steady solenoidal current distribution, the local current obtains from the relation

$$\mathbf{j}(\mathbf{r})\, dv = \sum_i I_i d\mathbf{r}_i. \tag{5.9.32}$$

Then, by Stoke's theorem (Appendix 2),

$$F = \frac{1}{c} I \iint\limits_\Sigma (d\Sigma \times \nabla) \times B. \tag{5.9.33}$$

When $H(r)$ changes slowly in Σ it is taken out of integration, giving

$$F = (\mu \times \nabla) \times B = \nabla(\mu \cdot B) - \mu(\nabla \cdot B), \tag{5.9.34}$$

where

$$\mu = \frac{I}{c} \iint_{\Sigma} d\Sigma = \frac{1}{c} I\Sigma. \tag{5.9.35}$$

For a volume distribution of current,

$$F = \sum_i \frac{1}{c} I_i \iint_{\Sigma} (d\Sigma_i \times \nabla) \times B, \tag{5.9.36}$$

$$\boldsymbol{\mu} = \sum_i \boldsymbol{\mu}_i = \frac{1}{c} \sum_i (I_i \Sigma_i). \tag{5.9.37}$$

Since H is solenoidal, the lowest order force of (5.9.34) on a localized current distribution in the presence of the magnetic field, is

$$F = \nabla(\boldsymbol{\mu} \cdot B). \tag{5.9.38}$$

This result holds even for time varying applied magnetic fields. This field gradient dependent force tends to drive the system to regions of strong fields. It vanishes in a homogeneous magnetic field. In the absence of magnetic field the angular momentum is constant. For steady state fields, since $\nabla \times H = 0$, equation (5.9.38) gives

$$F = (\boldsymbol{\mu} \cdot \nabla) B. \tag{5.9.39}$$

Equation (5.9.38) shows that the potential energy of a permanent magnet of dipole moment $\boldsymbol{\mu}$, in an applied magnetic field, is

$$W_m = -(\boldsymbol{\mu} \cdot \mathbf{B}), \tag{5.9.40}$$

because the conservative force is the gradient of a potential energy.

The equation for N, (5.9.31), is written as

$$N = \frac{1}{c} I \left(B \cdot \oint_{\Gamma} r \right) d\mathbf{r} - \frac{1}{c} IB \oint_{\Gamma} (r \cdot d\mathbf{r}). \tag{5.9.41}$$

The second integral being exact, vanishes over Γ. Therefore,

$$\begin{aligned} N &= \frac{1}{c} I \left[\left(B_x \oint_{\Gamma} x \right) d\mathbf{r} + \left(B_y \oint_{\Gamma} y \right) d\mathbf{r} + \left(B_z \oint_{\Gamma} z \right) \right] d\mathbf{r} \\ &= \frac{1}{c} \left[B_x \left\{ \oint_{\Gamma} d(x\mathbf{r}) - \oint_{\Gamma} \mathbf{r} dx \right\} + \cdots \right]. \end{aligned} \tag{5.9.42}$$

Dropping the exact integrals over Γ, we have

$$N = -\frac{1}{c} I \left[B_x \oint_{\Gamma} \mathbf{r} dx + B_y \oint_{\Gamma} \mathbf{r} dy + B_z \oint_{\Gamma} \mathbf{r} dz \right]. \tag{5.9.43}$$

Adding (5.9.41) and (5.9.43) we obtain, for the torque,

$$N = -\frac{1}{2c} \left(B \times \oint_{\Gamma} (\mathbf{r} \times dr) \right) = (\boldsymbol{\mu} \times B). \tag{5.9.44}$$

Putting $\rho^c\bar{v} = c\nabla \times M$, integration of both sides over a surface Σ_1, of volume V_1, bounded by a curve Γ, which encloses the body of the dielectric, and nowhere enters it, gives

$$0 = \iint_{\Sigma_1} ((\rho^c\bar{\mathbf{v}})\cdot d\Sigma) = c\iint_{\Sigma} ((\nabla \times M)\cdot d\Sigma) = c\oint_{\Gamma} ((\nabla \times M)\cdot d\mathbf{r}), \tag{5.9.45}$$

where $d\mathbf{r}$ is a length element at the integration point in Γ. Since Σ_1 is outside V, this relation is an identity, $\rho^c\bar{\mathbf{v}}$ is not zero in dielectrics and conductors, and

$$\iiint_{V_1} \frac{1}{2c}(\mathbf{r} \times \rho^c\bar{\mathbf{v}})\,dV = -\frac{1}{2c}\iiint_{V_1} (\mathbf{r} \times (\nabla \times M))\,dV, \tag{5.9.46}$$

$$\frac{1}{2c}\iiint_{V_1} [\mathbf{r} \times (\nabla \times M)]dV = \iint [(\mathbf{r}\cdot M)\,d\Sigma - (r\cdot d\Sigma)M] + 2\iiint_{V} M dV \tag{5.9.47}$$

$$\frac{1}{2c}\iiint_{V_1} [(\mathbf{r} \times (\rho^c\bar{\mathbf{v}})]dv = \frac{1}{2c}\iiint_{V} [(\mathbf{r} \times (\rho^c\mathbf{v})]dv = \iiint_{V_1} M dV = \iiint_{V} M dV. \tag{5.9.48}$$

because M vanishes at every point of Σ_1 and at every point in the volume enclosed between Σ_1 and Σ. Hence, the magnetization vector per unit volume, in v, at $\mathbf{r}$, is

$$M = \frac{1}{2c}(\mathbf{r} \times \rho^c\mathbf{v}). \tag{5.9.49}$$

Thus M is not uniquely defined, because its sum with the gradient of all smooth scalar functions of position co-ordinates also satisfies the relation

$$\rho^c\bar{\mathbf{v}} = C(\nabla \times M). \tag{5.9.50}$$

The similar form of the magnetic moment and angular momentum indicates that these should be related through a dyadic. However, for a system having the same ratio q/m of charge and mass per particle, M is directly proportional to L, where the gyromagnetic ratio $\Gamma = (q/2)mc$, and $\Gamma_0(=\Gamma^{-1})$ is the magnetogyric ratio. Other forces of quantum mechanical nature act in the atomic scale, so the gyromagnetic ratio Γ differs from $q/2mc$.

The Lorentz force, for a distribution of currents in a volume V, is

$$\iiint_V F dV = \frac{1}{c}\iiint_V (\mathbf{j} \times B)\,dV. \tag{5.9.51}$$

Since $jdV = Id\mathbf{r}$, where $d\mathbf{r}$ is a line element along the loop Γ carrying the current I. Using Stoke's theorem, we can write

$$\iiint_V F dV = \frac{1}{c}I\oint_{\Gamma} (d\mathbf{r} \times B) = \frac{1}{c}I\iint_{\Sigma} (d\Sigma \times \nabla) \times B. \tag{5.9.52}$$

For small loops B is taken out of integration, so

$$F = (M \times \nabla) \times B = M \times (\nabla \times B) - M(\nabla\cdot B) + (M\cdot\nabla)B \tag{5.9.53}$$

because, vectorially $(A \times \nabla) \times B - A \times (\nabla \times B) = (A \cdot \nabla)B - A(\nabla \cdot B)$. In quasistatic processes, in the absence of applied currents, working as the source of generation of B, we write

$$F = (M \cdot \nabla)B. \tag{5.9.54}$$

For permanent magnets and for systems of atomic or smaller dimensions, the magnitude of the magnetic moment is not affected by the motion it undergoes. In presence of a uniform magnetic field B, the field of the magnetic dipole, then, is the potential $U = -(M \cdot B)$; U depending only on the orientation of M relative to B when the moment is force-free.

The torque on a current distribution within a volume V is

$$N = \iiint_V \frac{1}{c}[\mathbf{r} \times (j \times B)]dV = \frac{1}{c}I \oint_\Gamma (r \cdot B)\, d\mathbf{r} - \frac{1}{c}\oint_\Gamma (\mathbf{r} \cdot d\mathbf{r})\mathbf{B} \tag{5.9.55}$$

B can be taken out of integration over the small volume of the current loop Γ. Then the second term becomes an exact differential for integration over the complete loop, and vanishes, giving

$$N = \frac{I}{c}\oint_\Gamma (\mathbf{r} \cdot B)\, d\mathbf{r}, \tag{5.9.56}$$

$$\text{or,} \quad N = \frac{1}{c}I\left[\oint_\Gamma d(\mathbf{r}(\mathbf{r} \cdot \mathbf{B})) - (\mathbf{B} \cdot d\mathbf{r})\mathbf{r}\right]. \tag{5.9.57}$$

Since the first integral has an exact differential, it vanishes, so

$$N = -\frac{1}{c}I \oint_\Gamma \mathbf{r}(B \cdot d\mathbf{r}). \tag{5.9.58}$$

Adding the two expressions for N the torque exerted by a magnetic field on the magnetic moment is obtained :

$$\begin{aligned} N &= \frac{1}{2c}I \oint_\Gamma [(\mathbf{r} \cdot B)\, d\mathbf{r} - (B \cdot d\mathbf{r})\mathbf{r}] = \frac{1}{2c}B \times \oint_\Gamma (d\mathbf{r} \times \mathbf{r}) \\ &= \frac{1}{2c}IB \times \iiint_V (\mathbf{j} \times \mathbf{r})\, dV = M \times B. \end{aligned} \tag{5.9.59}$$

The size of the region over which the average is taken for finding the force F and torque N depends on the problem. Over a magnetic dipole, or for an extended current distribution

$$F = \iiint_V (M \times \nabla) \times B dV, \quad B = \iiint_V (M \times B)\, dV. \tag{5.9.60}$$

Equating dL/dt with this N, we obtain $dL/dt = M \times B = \Gamma L \times B$, where $\Gamma = q/2mc$, Γ is the gyromagnetic ratio Γ. The equation for the evolution of M is

$$\frac{dM}{dt} = \Gamma(M \times B). \tag{5.9.61}$$

If E_d is the field of an electric dipole of moment μ_e, then

$$N = \mu_e \times E_d. \tag{5.9.62}$$

5.9.1 Angular momentum

The angular momentum of a system of particles is

$$L = \sum_i (r_i \times p_i), \tag{5.9.63}$$

where r_i is the position vector of the ith particle and p_i is its momentum. The law of motion

$$F_i = \frac{d}{dt}\mathbf{p}_i \tag{5.9.64}$$

gives, for the net torque N, the equation

$$N_1 = \sum_i (r_i \times F_i) = \sum_i \left(r_i \times \frac{d}{dt} p_i \right) = \frac{d}{dt} \sum_i (r_i \times p_i) = \frac{d}{dt} L. \tag{5.9.65}$$

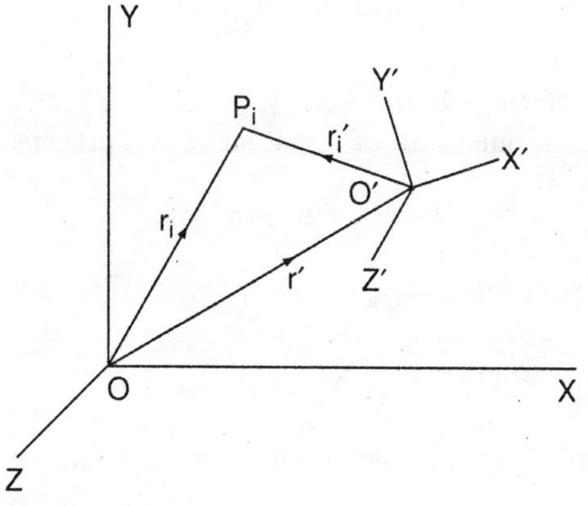

Fig. 5.9.2 It shows the change from the coordinate system OXYZ to the system O′X′Y′Z′. The position vector of the ith particle at the point P_i, with respect to the two systems, are r_i and r_i', where $OO' = r'$, so that $r_i = r_i' + r'$.

So, the torque applied on the system equals the rate of changes of the angular momentum. Since $N = 0$ for the central forces ($F_i = \hat{r}_i F(r_i)$, where $\hat{r}_i$ is the unit vector along r_i) the angular momentum is a constant of such motion. L depends on the choice of origin of the co-ordinate system, and not on the change in the directions of the co-ordinate axes due to rotation, fig. 5.9.2. A shift of the origin from O to O', replacing r_i by $r_i - r^1$, where $r^1 = OO^1$, affects the angular momentum L_i, but not the linear momentum $\mathbf{p}_i$. The angular momentum becomes independent of the origin when

$$\mathbf{r}^1 \times \sum_i \mathbf{p}_i = \mathbf{r}^1 \times \mathbf{p} = 0, \tag{5.9.66}$$

where $\mathbf{p}\left(= \sum_i p_i\right)$ is the momentum of the system. Only in the frame moving with the center of gravity of the system (the rest frame of the system) the mechanical moment has a unique value. So, it is defined at a time only in one particular system of co-ordinate axes, namely the rest system of co-ordinate axes of it, in which the linear momentum vanishes. the angular momentum does not transform.

If the angular momentum per unit mass is G_i, per particle of the ith species then L and M are two linear combinations of G_i, where

$$\left.\begin{aligned} L &= \sum_i m_i(r_i \times \dot{r}_i) = \sum_i (m_i G_i) \\ M &= \frac{1}{2c}\sum_i q_i(r_i \times \dot{r}_i) = \frac{1}{2c}(q_i \cdot G_i) \end{aligned}\right\}. \tag{5.9.67}$$

The gyromagnetic ratio tensor Γ and its inverse Γ_0 are defined by the relations $M = \Gamma : L$ and $L = \Gamma_0 : M$.

The relativistically correct angular momentum is defined by

$$L = \sum_i m_i^0 \gamma_i (\mathbf{r}_i \times \mathbf{v}_i), \tag{5.9.68}$$

where m_i^0 is the rest mass of the ith particle, $\gamma_i = (1 - \beta_i^2)^{-1/2}$, $\boldsymbol{\beta}_i = \mathbf{v}_i/c$. The 4-vector representation of the angular momentum of a system of N particles is

$$L^{\mu\gamma} = (x^\mu p^\gamma - x^\gamma p^\mu), \tag{5.9.69}$$

where $x^\mu = (\mathbf{r}, ct)$, $p^\mu = (\mathbf{p}, \varepsilon_v/c)$, $\varepsilon_v = m_i c^2 = m_i^0 \gamma_i c^2$. The 4-vector magnetic moment is similarly defined by

$$M^{\mu\gamma} = \frac{1}{2c}\sum (x^\mu j^\gamma - x^\gamma j^\mu), \tag{5.9.70}$$

where $j^\mu = (j, c\rho^c)$. In terms of the complex quantities $x_\alpha \pm iy_\alpha = r_{\alpha\pm}$, $v_{\alpha x} \pm iv_{\alpha y} = r_{\alpha\pm}$, we can write

$$L_z = \sum_\alpha \frac{m_\alpha(r_{\alpha+} r_{\alpha-} - r_{\alpha+} r_{\alpha-})}{2i}$$

for the z component of L, where $i = \sqrt{-1}$. Similarly proceeding, we can also develop the other components of L. The complex notations $r_\pm = x \pm iy$, $u_\pm = x + iy$ for the angular momentum give

$$L_z = \frac{u_+ r_- - u_- r_+}{2i} \tag{5.9.71}$$

in terms of the complex field variables.

5.10 Spin Kinematics

The spin s of a system is its angular momentum. It is a pseudo vector. Its rate of change with time equals the applied torque N. So

$$N = \frac{dS}{dt}. \tag{5.10.1}$$

Relativistically, the spin S of a system is its angular momentum, in the inertial frame (the rest frame of the system, as a whole), in which its total momentum is zero. In this frame, many particles have intrinsic angular momentum that is combined with the orbital $\mathbf{r} \times \mathbf{p}$ term in (3.11.3), so the law (3.11.1) defines L. In the classical non-relativistic limit $L = S$.

When the applied field is switched off, a classical electron of charge q, orbiting round a nucleus of positive charge, can be considered as a current loop. If this current, in electromagnetic units, is I and T is the time period to describe the complete orbit once, then

$$I = \frac{q}{cT}. \tag{5.10.2}$$

By Keepler's law of planetary motion, for central orbits,

$$\frac{dA}{dt} = \frac{L}{2m}, \tag{5.10.3}$$

where dA is the area of the orbit swept way by the electron in time dt, and m is its mass. Hence

$$A = \frac{LT}{2m}, \tag{5.10.4}$$

and since the magnetic moment is the product of the area of the current loop and the current,

$$I = \frac{qL}{2mcA} = \frac{q}{cA}\frac{dA}{dt} \tag{5.10.5}$$

$$\mu = IA \tag{5.10.6}$$

$$\mu = \frac{qL}{2mc} \tag{5.10.7}$$

$$N = \frac{dL}{dt} = \frac{q}{2mc}(L \times B) \tag{5.10.8}$$

The kinematic equation for the precession of angular momentum at the angular rates

$$\Omega = -\frac{qB}{2mc}, \tag{5.10.9}$$

called the *Larmor precession frequency*, is given by

$$\frac{dL}{dt} = (\Omega \times L). \tag{5.10.10}$$

For spinning elementary particles, (5.10.7) is replaced by

$$\mu = \frac{q}{2mc} gL \tag{5.10.11}$$

$(qg/2mc)$ is the gyromagnetic ratio. For electrons, when both orbital motion and spin are included, $g - 2 \approx 1/400$. So, practically $g = 2$. For an orbital electron, $2mc/q = 1.13 \times 10^{-7}$, using (5.10.1), (5.10.8) and (5.10.11) we find that

$$N = \frac{dL}{dt} = \frac{dS}{dt} = \frac{q}{2mc} L \times B = (\mu \times B) \tag{5.10.12}$$

for the intrinsic spin angular momentum S of an electron, in a magnetic induction field B. For purely orbital motion $g = 1$.

The electron rest frame K rotates with angular velocity Ω of the spin. In this frame, the unit vectors, $\hat{\mathbf{x}}$ and $\hat{\mathbf{y}}$, satisfy the kinematical relations

$$\frac{d\hat{\mathbf{x}}}{dt} = (\Omega \times \hat{\mathbf{x}}) = \Omega \hat{\mathbf{y}}, \quad \frac{d\hat{\mathbf{y}}}{dt} = (\Omega \times \hat{\mathbf{y}}) = \Omega \hat{\mathbf{x}}. \tag{5.10.13}$$

According to Minkowski's special theory of relativity, (section 1.14), for moving media, in the non-relativistic limit, the magnetic induction is

$$B = B^0 - (\beta \times E^0), \tag{5.10.14}$$

where B^0 and E^0 are the magnetic induction and electric field in the laboratory inertial frame K^0, and $\beta = v/c$. This relation follows from the relativistic relations of transformation, section 12.3,

$$E = \gamma(E^0 + (\beta \times B^0)) - \frac{\gamma^2}{\gamma+1}(\beta \cdot E^0)\beta, \tag{5.10.15}$$

$$B = \gamma(B^0 - (\beta \times E^0)) - \frac{\gamma^2}{\gamma+1}(\beta \cdot B^0)\beta. \tag{5.10.16}$$

The torque, $\boldsymbol{\mu} \times B$, of the dipole moment $\boldsymbol{\mu}$ in the field of B, is the time rate of change of angular momentum $\mathbf{L}$. So the equation for the change of angular momentum of the electron, in its rest frame, is

$$\begin{aligned}\frac{d\mathbf{S}}{dt} &= \frac{gq}{2mc}(\mathbf{S} \times \beta) + (\Omega \times \mathbf{S}) = \mathbf{S} \times \left(\frac{gq}{2mc}B - \Omega\right)\\ &= \frac{gq}{2mc}(\mathbf{S} \times (B^0 - (\beta \times E^0)) + (\Omega \times \mathbf{S}).\end{aligned} \tag{5.10.17}$$

The energy $\mathcal{W}$ of interaction of the electron spin with the magnetic induction field, is then given by

$$\begin{aligned}\mathcal{W} &= -\boldsymbol{\mu}(B^0 + (\beta \times E^0)) + (\mathbf{S} \cdot \Omega)\\ &\doteq -(\mathbf{S} \cdot B^0)\frac{gq}{2mc} - \frac{qg}{2mc}(\mathbf{S} \cdot (\beta \times E^0)) + (\mathbf{S} \cdot \Omega).\end{aligned} \tag{5.10.18}$$

The spin of a space vehicle is its angular momentum in its center of mass frame. The spin of an atom is its angular momentum (which is spin plus orbit) in its center of mass frame. The spin of one of its electrons or of the nucleus is its angular momentum in its rest frame.

The linear momentum has definite components in all reference frames, but angular momentum is defined only in one particular reference frame which is the rest frame at every instant. So, spin of an electron is the spin in the rest frame of the electron making the angle ϕ with the line of velocity of the laboratory frame. The concept of spin satisfies the law of conservation of angular momentum of an isolated system.

5.11 Relativistic Spin Kinematics

The relativistic spin S of a system is its angular momentum in the rest frame which is an inertial frame. In this frame its total linear momentum is zero. The angular momentum sum in (3.11.3) includes both intrinsic angular momentum of particles and the orbital term $(\mathbf{r} \times \mathbf{p})$. The law of motion (3.11.1) defines it. Only in the classical relativistic limit, $L = S$.

If N_0 is an applied torque, then (3.11.1) is replaced by

$$N = (\mathbf{r} \times F) + N_0. \tag{5.11.1}$$

For a single particle,

$$L = (\mathbf{r} \times \mathbf{p}) + \mathbf{S}. \tag{5.11.2}$$

Since $N = dL/dt$,

$$\frac{dS}{dt} + (v \times p) = N_0. \tag{5.11.3}$$

In presence of a strong applied static magnetic field, we ignore E in the Lorentz force and find that

$$\left(\mathbf{p} \cdot \frac{d\mathbf{p}}{dt}\right) = \frac{e}{c}(\mathbf{p} \cdot (\mathbf{v} \times \mathbf{H})) = \frac{e}{c}(\mathbf{H} \cdot (\mathbf{p} \times \mathbf{v})). \tag{5.11.4}$$

In the absence of an applied torque N_0, this equation becomes

$$\left(\mathbf{p} \cdot \frac{dp}{dt}\right) = \frac{d}{dt}\left(\frac{e}{c} H \cdot S\right). \tag{5.11.5}$$

Integration gives

$$\frac{p^2}{2m} - (\boldsymbol{\mu} \cdot \mathbf{H}) = \text{constant}, \quad \boldsymbol{\mu} = \frac{eS}{mc}. \tag{5.11.6}$$

Non-relativistically, $\mathbf{p} = m\mathbf{v}$, so (5.11.4) reduces to $dS/dt = N_0$, and the magnetic moment $\boldsymbol{\mu}$, induced on particles by the magnetic field H, is

$$\boldsymbol{\mu} = gS. \tag{5.11.7}$$

Then, following (5.10.12), we write

$$N_0 = (\boldsymbol{\mu} \times H) = g(S \times H) \tag{5.11.8}$$

and writing $dS/dt = N_0$, we find that

$$\frac{dS}{dt} = g(S \times H) \tag{5.11.9}$$

for the non-relativistic precession of the spin of a particle in a magnetic field, due to its magnetic moment $\boldsymbol{\mu}$. The factor g makes the result valid for spinning elementary particles.

5.11.1 Some Remarks

The angular momentum is defined only in the rest frame at every instant. The spin satisfies the law of conservation of angular momentum of an isolated system. The electron spin, at angle ϕ with its velocity in the laboratory frame, is the spin in the rest frame of the electron, at angle ϕ with the line of motion of the laboratory. A rotating charge distribution has a magnetic moment. A finite moment of angular momentum is associated with a point particle. Anomalous magnetic moment arises from the assumption that the internal composition of the spinning object remains constant during the motion, and from the distribution of matter inside the particle (which replaces the equivalent parameters like the moments of inertia).

5.12 Magnetic Moment of Orbital Electrons

Atoms and molecules have their own internal motion like rotation about their axis or vibration about their equilibrium structures. The equilibrium distribution of the negative and positive charges of molecules and atoms is disturbed by electric field. These are then said to be polarized. The magnitude of the dipole moment is determined by the polarizability of the molecules, and direction by the relative orientation of it and the incident field. A rotating charged particle is known as a *magneton.* The equilibrium charge distribution in the molecule is modulated by external influences, and the oscillation of the induced dipole moment is modified by the internal frequencies of rotation and vibration.

Orbital electrons are like electric currents of moving charges. These have magnetic moments, and behave as small magnets or, magnetic elements. Of the two sources of magnetic moment and angular momentum of electrons (the orbital motion and spin), the orbital motion is described well classically but the spin effect requires the quantum description. Sometimes, the classical results for orbital motion are applied to the electron spin.

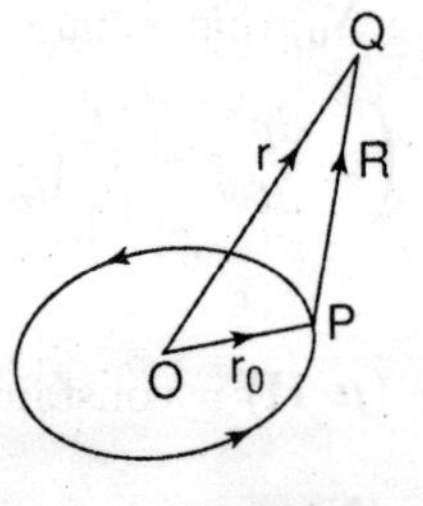

Fig. 5.12.1 It shows the orbital electron at P, nucleus at O and the observation point at Q.

The focus of an electron orbit, about its nucleus, is O of fig. 5.12.1. The electron is at P at time t; Q is a point outside the orbit. Then

$$\frac{1}{R} \approx \frac{1}{r} + \frac{(r_0 \cdot r)}{r^3} \tag{5.12.1}$$

with $OP = r_0$, $OQ = r$, $PQ = R$, because $\mathbf{R} = \mathbf{r} - \mathbf{r_0}$ and $R^2 = r^2 + r_0^2 - 2(r \cdot r_0)$.

In the Lienard-Wiechart formula for the vector potential, (section 4.12),

$$A = \frac{q\mathbf{v}}{C(R - (v \cdot R)/c)} \tag{5.12.2}$$

$$= \frac{v}{cr}\left\{1 + \frac{(\mathbf{r} \cdot \mathbf{r_0})}{r^2}\right\} = \frac{q}{cr}\left\{\dot{\mathbf{r}}_0 + \frac{(\mathbf{r} \cdot \mathbf{r_0})}{r^2}\dot{\mathbf{r}}_0\right\}, \tag{5.12.3}$$

where $v = \dot{r}_0$, we put $\dot{r} = 0$ and obtain

$$A = \frac{q}{cr}\frac{\partial r_0}{\partial t} + \frac{q}{2cr^3}\left\{\frac{\partial}{\partial t}\mathbf{r}_0(\mathbf{r}_0 \cdot \mathbf{r}) + \mathbf{r} \times (\dot{\mathbf{r}}_0 \times \mathbf{r}_0)\right\} \tag{5.12.4}$$

for the orbital motion of the electron because O and Q are now fixed, where the vector results

$$(r \times (\dot{p} \times p)) = (r \cdot p)p - (r \cdot \dot{p})p,$$

$$\frac{\partial}{\partial t}\{p(r \cdot p)\} = \dot{p}(r \cdot p) + p(\dot{r} \cdot p) + p(r \cdot \dot{p}).$$

have been used. In the mean value $\langle A \rangle$ of A, obtained by integration over the time period T, of orbital motion, some terms vanish, giving

$$\langle A \rangle = \frac{1}{T}\int_0^T A dt = \frac{q}{2cr^3} \quad (\mathbf{r} \times (\dot{\mathbf{r}}_0 \times \mathbf{r}_0)). \tag{5.12.5}$$

The moment of an element of magnetic pole intensity m_H is

$$\mu = m_H l, \tag{5.12.6}$$

where l is the distance of separation of the magneticpole m_H from the other pole of opposite polarity. The magnetic moment $\boldsymbol{\mu}$, due to currents from motion of charges, is given by (5.12.10). So, replacing $\langle A \rangle$ by $\boldsymbol{A}$ we find that

$$A = -\boldsymbol{\mu} \times \nabla \left(\frac{1}{r}\right) = \frac{(\boldsymbol{\mu} \times \mathbf{r})}{r^3}, \tag{5.12.7}$$

$$H = \nabla \times A = -\nabla \times \left(\boldsymbol{\mu} \times \nabla\left(\frac{1}{r}\right)\right) = \nabla\left\{\boldsymbol{\mu}\cdot \nabla\left(\frac{1}{r}\right)\right\} = \nabla\phi_H, \tag{5.12.8}$$

where the magnetic potential is

$$\phi_H = (\mu\cdot \nabla)\left(\frac{1}{r}\right) \tag{5.12.9}$$

and the magnetic moment is

$$\boldsymbol{\mu} = \frac{q}{2c}(\mathbf{r}_0 \times \dot{\mathbf{r}}_0) \tag{5.12.10}$$

because an electron moving in an orbit is found to be equivalent to this value. The magnitude of the vector $(r_0 \times \dot{r}_0)/2$ being dA/dt of (5.10.3), we write

$$\frac{1}{2}(r_0 \times \dot{r}_0) = \frac{1}{T}An; \tag{5.12.11}$$

where A is the area of the orbit swept away by the electron in the time of the time period T, and $\mathbf{n}$ is the unit vector normal to the orbital plane. By Keepler's law for central orbits, A/T is constant, and

$$\boldsymbol{\mu} = \frac{qA}{cT}\mathbf{n}. \tag{5.12.12}$$

This relation was derived from the equivalents of magnetic shells and current circuits in Section 5.10. By Biot-Savart's law, Landau and Liftshitz (1975, p. 93),

$$H = \frac{q}{CR^3}(\mathbf{r}_0 \times R) \tag{5.12.13}$$

$$= \frac{q}{cr^3}\dot{\mathbf{r}}_0 \times (\mathbf{r} - \mathbf{r}_0)\left\{1 + \frac{2}{r^2}(\mathbf{r}\cdot \mathbf{r}_0)\right\}, \tag{5.12.14}$$

$$H_x = \frac{q}{cr}\left[\dot{y}_0 \left\{(z - z_0) + \frac{3z}{r^2}(\mathbf{r}\cdot \mathbf{r}_0)\right\} - \dot{z}_0\left\{(y - y_0) + \frac{3y}{r^2}(\mathbf{r}\cdot \mathbf{r}_0)\right\}\right], \text{ etc.} \tag{5.12.15}$$

for the field of the moving charges, where (5.12.1) has been used, when the co-ordinates of P and Q are (x_0, y_0, z_0) and (x, y, z) respectively. The time averaging gives

$$\langle H_x \rangle = \frac{q}{cr^3}\left\{-\dot{y}_0 z_0 + y_0\dot{z}_0 + \frac{3}{r^2}(z\dot{y}_0 - y\dot{z}_0)(r\cdot r_0)\right\} \tag{5.12.16}$$

because

$$\int_0^T \left\{\dot{x}_0, x_0\dot{x}_0, x_0\dot{y}_0, \frac{d}{dt}(x_0\dot{y}_0)\right\} dt = 0. \tag{5.12.17}$$

Further, denoting $\langle H \rangle$ by H, we obtain

$$H_x = \frac{q}{cr^3}\left[y_0\dot{z}_0 - \dot{y}_0 z_0 + \frac{3}{r^2}\{(y^2+z^2)\dot{y}_0 z_0 + y_0 zx\dot{x}_0 - x\dot{y}x_0\dot{z}_0\}\right] \tag{5.12.18}$$

$$= \frac{q}{cr^3}\left\{\dot{y}_0 z_0 + y_0\dot{z}_0 + \frac{3x}{r^2}(xy_0\dot{z}_0 + x_0\dot{x}_0 y + \dot{y}_0 z_0 z)\right\}. \tag{5.12.19}$$

since $r^2 = x^2 + y^2 + z^2$. Considering mean values, we obtain

$$\mu_x = \frac{q}{2c}(y_0\dot{z}_0 - \dot{y}_0 z_0) \tag{5.12.20}$$

$$H_x = -\frac{\mu_x}{r^3} + \frac{3x}{r^5}(\boldsymbol{\mu}\cdot\mathbf{r}), \text{ etc.}$$

$$\mathbf{H} = -\frac{\mu}{r^3} + \frac{3}{r^5}(\boldsymbol{\mu}\cdot\mathbf{r})r, \tag{5.12.21}$$

where $\boldsymbol{\mu}$ is given by (5.12.10). The formula (5.12.21) also follows from the curl of (5.12.7). When r is perpendicular to the plane of the orbit passing through the center of force,

$$H = \frac{q}{cr^5}(\mathbf{r}_0 \times \dot{\mathbf{r}}_0). \tag{5.12.22}$$

This H can be compared with the field $2\mu/r^3$ for a magnet in the end on position.

5.13 Dielectric Polarization Theory

Charges in a body are displaced from their positions in the field free state by applied fields and fields of other charges, and by collisions with uncharged particles. If q_α is the charge of a particle, N_α is the number density and ξ_α is the field induced displacement of the αth species, then the polarization density and the current density are

$$P = \sum_\alpha (N_\alpha q_\alpha \boldsymbol{\xi}_\alpha). \tag{5.13.1}$$

$$j_p = \sum_\alpha (N_\alpha q_\alpha \dot{\xi}_\alpha). \tag{5.13.2}$$

For a continuous distribution of charges of both sign, if the electrons in a unit volume, at the position $\mathbf{r}_0$, are displaced to the position $r_0 + \xi$, due to their relative velocity V, and ρ_e^c is the charge density of these electrons, the dipole moment in a volume V is the integral

$$\iiint_V \rho_e^c r dV. \tag{5.13.3}$$

For internal redistribution of bound charges of density ρ^c_{in} by an applied field in a body, having macroscopically no net distribution of charges of any kind, then

$$\iiint_V \rho^c_{\text{in}} dV = 0. \tag{5.13.4}$$

Using Gauss's theorem for transforming a volume integral into a surface integral, we obtain

$$\iiint_V \rho^c_{\text{in}} dV = -\iiint_V (\nabla\cdot P)\, dV = -\iint_\Sigma (P\cdot d\Sigma) = -\iint_\Sigma P_N d\Sigma = \iint_\Sigma \sigma d\Sigma, \tag{5.13.5}$$

where P_N is the normal component of P at a point on the surface Σ, and σ is the surface density of charges on Σ of the body. So

$$P_N = \sigma \text{ or } \Sigma. \tag{5.13.6}$$

Integration of the dipole density $\rho^c_{\text{in}} r$ of the internal distribution of charges, over the volume V, assuming that $P_N = 0$ at every point of Σ, gives

$$\iiint_V \rho^c_{\text{in}} r dV = -\iiint_V r(\nabla\cdot \rho)\, dV = \iiint_V P dV. \tag{5.13.7}$$

So, P is the dipole moment per unit volume of the internal redistribution of charges by an applied electric field. If, moreover, there is a charge separation density ρ^c_{ex} from causes other than that of polarization, say the space charge separation from application of an electrostatic field, or from compressibility of the medium, the total electric charge density

$$\rho^c = \rho^c_{\text{in}} + \rho^c_{\rho x}. \tag{5.13.8}$$

Then Gauss's law for electric field, using (5.13.4), reads

$$\nabla\cdot E = 4\pi\rho^c = 4\pi(\rho^c_{\text{in}} + \rho^c_{ex}). \tag{5.13.9}$$

Writing

$$\nabla\cdot D = 4\pi\rho^c_{ex} \tag{5.13.10}$$

this equation yields

$$D = E + 4\pi P. \tag{5.13.11}$$

It can be written as

$$D = \varepsilon : E, \tag{5.13.12}$$

or as

$$P = \eta : E \tag{5.13.13}$$

or as

$$(\varepsilon - I) : E = 4\pi P, \tag{5.13.14}$$

where ε is the dielectric permittivity tensor, η is the electric susceptibility and I is the 3×3 unit diagonal matrix, η and ε are dimensionless material parameters. These and $\varepsilon - I$ are non-singular matrices. These parameters depend on temperature, material density, chemical constitution, frequency and strength of the applied field.

The continuity condition

$$(\nabla \cdot j_{\text{in}}) + \frac{\partial \rho_{\text{in}}^c}{\partial t} \doteq 0$$

for the polarization charge ρ_{in} and current j_{in} can be written as

$$j_{\text{in}} = \frac{\partial P}{\partial t} + cA, \tag{5.13.15}$$

where c is the vacuum velocity of light and A is a divergence free vector. When the polarized body is moving with velocity v, the polarization current

$$j_{\text{in}} = \rho_{\text{in}}^c v = -v(\nabla \cdot P). \tag{5.13.16}$$

For a combined system of particles and fields, the linear momentum is conserved, Jackson (1975, section 6.8). So, assuming that

$$\frac{dp}{dt} = 0 = \frac{\partial P}{\partial t} + (v \cdot \nabla)P, \tag{5.13.17}$$

$$\frac{dP}{dt} = 0 = \frac{\partial P}{\partial t} + (v \cdot \nabla)P \tag{5.13.18}$$

we find that

$$CA = j_{\text{in}} - \frac{\partial P}{\partial t} = -v(\nabla . P) + (v \cdot \nabla)P = -\nabla \times (v \times P). \tag{5.13.19}$$

Since A is a divergence free vector, we can write

$$A = \nabla \times M; \tag{5.13.20}$$

and using (5.13.19) obtain

$$j_{\text{in}} = \frac{\partial P}{\partial t} + c\nabla \times M, \tag{5.13.21}$$

$$M = \frac{1}{c}(P \times \mathbf{v}). \tag{5.13.22}$$

Hence a moving dielectric has a magnetization when its velocity is not parallel to the displacement by the applied field.

The mean value of the microscopic current density $\rho^c \mathbf{v}$ (where $\mathbf{v}$ is the microscopically averaged velocity) is obtained by averaging the current term in the Ampere-Maxwell equation (3.15.2) of the Lorentz theory. It is not zero in dielectrics and conductors. In dielectrics $\rho^c v$ vanishes at every point of the surface Σ. So

$$\iint_{\Sigma} \rho^c(\mathbf{v} \cdot d\Sigma) = 0. \tag{5.13.23}$$

In conductors this surface integral need not be zero.

The non-linear material properties are described by expanding P in powers of E. For unidirectional electric dipoles

$$P = \alpha E(1 + a_1 E + a_2 E^2 + a_3 E + 0) \tag{5.13.24}$$

in one dimensional examples, where $\alpha, a_1, a_2, \cdots$ are constants and the a's represent non-linearities of increasingly higher orders. so, $P^L = \alpha E$ and $P^N \doteq P - P^L$. When E is not unidirectional, and only electric dipoles are formed, for small optical non-linearities, the expansion is

$$P^L = \chi^{(1)} : E, \quad P = \chi^{(1)} : E + \chi^{(2)} : EE + \chi^{(3)} EEE + \cdots \tag{5.13.25}$$

$\chi^{(1)}$ is the linear electrical susceptibility, $\chi^{(2)}$ is the second order non-linear susceptibility, and so on. This field of polarization, excited by an applied EM field, is an additional source of the EM field.

The atomic mechanism of this type of weakly non-linear susceptibility is better understood by it quantum mechanical derivation, than by the semiclassical treatment. The second harmonic longitudinal field, in the photon theory, for example, follows from a three-quantum scattering process. If the number of quanta passing per cm^2 per sec in the direction of wave propagation at 2ω is increased by one, the number of quanta in the power flux density at the frequency of the fundamental (incident transverse wave) is decreased by two. The second-harmonic field intensity builds up by depletion of power at the fundamental. Experimentally complete conversion is possible for acoustic waves but not for light waves.

In a dense medium, the field, acting on an atom or ion, is the local field E^{Lo}. For cubic symmetry of the material, the state relations are

$$E^{Lo} = E + 4\pi P_{\text{tot}} = E + 4\pi P^L + 4\pi P^N. \tag{5.13.26}$$

If α is the polarizability factor, then

$$E^{Lo} = D = E + 4\pi P^L + 4\pi P^N, \quad P^L = N\alpha E^L, \tag{5.13.27}$$

where N is the number of molecules per unit volume. Then, by a Lorentz formula, Van Vleek (1932) section 1.5,

$$P^{Lo} = \frac{N\alpha E}{1 - (4\pi N\alpha/3)} + \frac{4\pi}{3} N\alpha \left(\frac{P^N}{1 - (4\pi N\alpha/3)} \right) = \frac{\varepsilon - 1}{4\pi} E + \frac{1}{3}(\varepsilon - 1) P^N. \tag{5.13.28}$$

$$D = E + \frac{1}{3}(\varepsilon + 2) 4\pi P^N = E + 4\pi P^{Ns}, \tag{5.13.29}$$

where

$$\varepsilon E = E + 4\pi P^L, \quad P^{Ns} = \frac{1}{3}(\varepsilon + 2) P^N. \tag{5.13.30}$$

Here P^{NS} is the effective non-linear polarization source term. The presence of a non-linear dipole moment at one site changes the local field and, therefore, the linear dipole moment at other sites, changing, thereby the dielectric displacement. The dipolar energy of the non-linear dipolar moment at one site is different from that of the linear dipole moment at another site. So, P^{NS} is distinct from the true non-linear polarization P^N. However, this difference is negligible in the semiclassical non-linear theory.

The electric field is not strictly monochromatic at the frequency ω. It contains some near frequencies which may be denoted by $\omega + \delta\omega$. Then, for a linear dispersive medium,

$$E(\omega, t) = \int_{-\delta\omega_0}^{\delta\omega_0} E(t) \exp(-i(\omega + \delta\omega)t)\, d(\delta\omega). \tag{5.13.31}$$

$$\frac{\partial}{\partial t}E(\omega,t) = -i\omega E(t) - i\int E(\delta\omega)\exp(-i(\omega+\delta\omega)t)\,d(\delta\omega)). \tag{5.13.32}$$

$$P(\omega,t) = \int \chi(\omega+\delta\omega)E(\delta\omega)\exp(-i(\omega+\delta\omega)t)\,d(\delta\omega). \tag{5.13.33}$$

Now expanding the susceptibility tensor χ about the frequency ω we retain terms only of lowest power in $\delta\omega$, and

$$\begin{aligned} P(\omega,t) &= \chi(\omega)E(t) + \left(\frac{\partial\chi}{\partial\omega}\right)\int \delta\omega E(\delta\omega t)\exp(-i(\omega+\delta\omega)\,d(\delta\omega) \\ &= \chi(\omega,t)E(t) - \omega\left(\frac{\partial\chi}{\partial\omega}\right)E(t) + i\left(\frac{\partial E}{\partial t}\right)\left(\frac{\partial\chi}{\partial\omega}\right). \end{aligned} \tag{5.13.34}$$

5.14 Magnetomechanical Effects

The electron spin (section 5.10) generates magnetic moment and angular momentum. A spinning top precesses in a gravitational field. The orbital motion or the spin of the electron, analogously, gives rise to a similar precessional effect in presence of an applied magnetic field, or rotation. Rotation gives rise to a magnetic field; conversely, magnetization of the electron is accompanied by rotational impulses.

Freely suspended ferromagnetic materials (iron bars) have angular momentum of rotation, on magnetization. This is the Einstein-de Hass effect. The rotation is observed by magnetization of a small ferromagnetic rod suspended on a torsion fiber. Conversely, magnetic field develops when ferromagnets are rotated. This is the Barnett effect. Hence, a body can be magnetized without an external magnetic field. As in gyroscopes, rotation tends to align on the axes of the carriers along the axis of rotation; this effect is equivalent to a magnetic field. Piezo electric effects are from electric polarization by deformation of crystals by forces of non-electromagnetic origin.

The ratio of the magnetic moment,

$$\mu = \frac{1}{2c}\iiint_V (r\times j)\,dv, \tag{5.14.1}$$

and the mechanical angular momentum

$$L = \iiint_V \rho(r\times v)\,dv \tag{5.14.2}$$

denoted by Γ, is given by

$$\Gamma\mu = L, \quad \Gamma = \frac{q}{2mc}. \tag{5.14.3}$$

In experiments with ferromagnets, Γ is twice this value. The reason for this discrepancy is that the nature of the electrons inside atoms is more complicated than the picture of orbital motion around the respective nucleus. Besides the charge q, the electron has an inner mechanical angular momentum, due to an intrinsic spin. This spin generates a magnetic moment, which is twice the large classical value, from orbital motion.

When an iron bar is rotated about an axis the kinematical relation between the resulting torque N and the angular momentum L of the bar is (5.4.3). The angular momentum will change by the amount

$$\int_0^{\delta t} N dt = \delta L \tag{5.14.4}$$

by a rotational impulse, given to the bar for a time δt.

5.14.1 Rotation of magnetized rods

Magnetization of an initially unmagnetized rod, at rest, will partially, or completely align the elementary carriers. The magnetic moment M_0 of the rod, and the magnetic moment $\boldsymbol{\mu}$ and angular momentum L of the carriers, are given by

$$M_0 = \sum(\boldsymbol{\mu}\cdot \hat{\mathbf{z}}), \tag{5.14.5}$$

where $\hat{z}$ is the unit vector in the axial direction of the rod. The change in the resultant angular momentum of these carriers is met by an equal and opposite change in the angular momentum of the rod as a whole. So the total angular momentum L, generated by the field, is zero; hence

$$L = -\Sigma(l\cdot z). \tag{5.14.6}$$

Since the ratio q/m of charge and mass is the same for all carriers, using (5.10.7) we obtain

$$\frac{L}{M_0} = -\frac{l}{\mu} = \frac{2mc}{q}, \tag{5.14.7}$$

$$L = -\frac{l}{\mu}M_0, \quad \delta L = -\frac{l}{\mu}\delta M_0 \tag{5.14.8}$$

because

$$N = -\frac{l}{\mu}\frac{d}{dt}M_0. \tag{5.14.9}$$

Measurement of the change δM_0 in the magnetization of the rod by the rotational impulse, δL, gives an estimate of the gyromagnetic ratio l/μ of the carriers.

5.14.2 Magnetization of rotating bodies

The co-ordinates system K, fixed in the rod, rotates at the angular velocity Ω, so (5.10.10) gives

$$\frac{dL}{dt} = -(\Omega \times L). \tag{5.14.10}$$

Rotation is therefore, equivalent to a field H such that

$$\frac{dM}{dt} = (\mu \times H) = L \times \Omega, \tag{5.14.11}$$

where the effective torque on the carriers is $\mu \times H$. Hence, the magnetization H, of rotation, is given by

$$H = \frac{l}{\mu}\Omega = \frac{2mc}{q}\Omega, \tag{5.14.12}$$

because the precessional velocity imposed on the carriers by the rotation of the rod relative to the K system is $-\Omega$. The magnetic field is, therefore, linearly dependent on the uniform angular velocity, which is the Barnett effect.

When the axes of M and L are not identical, μ and L are not parallel, and the gyromagnetic ratio g is a square matrix[3]. Since q/m is the same for all the elementary carriers of magnetism, we sum (5.10.11) over the elements and write, for the whole system, the tensor relation

$$M_i = \frac{q}{2mc} g_{ij} L_j \tag{5.14.13}$$

g_{ij} is the ijth element of the gyromagnetic ratio tensor.

M (magnetization) and Ω (angular velocity) should be linearly related since both change sign when the sign of time is reversed. The linear relation holds even in an isotropic body for which M and Ω are simply proportional, since both are axial vectors.

If ξ is the free energy of the body in a co-ordinate system rotating with the body (the K system), then

$$L = -\frac{\partial \xi}{\partial \Omega}. \tag{5.14.14}$$

The gyroenergy, ξ_g, depending on the magnetic moment, is given by

$$\xi_g = -\iiint\limits_V \lambda_{ij} \Omega_i M_j dv = -\nu_{ij} \Omega_i M_j, \tag{5.14.15}$$

where λ_{ij} and ν_{ij} are non-symmetric tensors; the tensor g_{ik} is the inverse of ν_{ik} because

$$g_{ik} \nu_{kl} = \frac{2mc}{q} \delta_{il}, \tag{5.14.16}$$

where δ_{il} is the Kronecker delta (Appendix 4). So, the magnetic effect of rotation of the body is given by

$$H_i = \nu_{ji} \Omega_j = \frac{2mc}{q} g_{ij}^{-1} \Omega_j, \tag{5.14.17}$$

while (5.14.13) represents the Einstein-de Hass effect, the Barnett effect is (5.14.17). Both these effects are determined by the same gyromagnetic tensorial coefficients g_{ik}. When the magnetic susceptibility χ_{ij} of the body is small, the magnetic moment M acquired by it is independent of its shape; so we can write

$$M_i = \chi_{ij} H_j = \frac{2mc}{q} \chi_{ij} \nu_{jl} \Omega_l. \tag{5.14.18}$$

5.15 Energies and Energy Fluxes

In the Maxwell equations

$$\nabla \times E = -\frac{\partial B}{c \partial t}, \quad \nabla \times H = \frac{\partial D}{c \partial t} + \frac{4\pi j}{c},$$

[3] Both the symbols g and Γ have been used in the text for the gyromagnetic ratio.

the volume averaged current density $\mathbf{j}$ is expanded into a sum of a multipole series and a part $\mathbf{j}_c$ due to convection current,

$$\mathbf{j} = \mathbf{j}_c + \frac{\partial P}{\partial t} + c\nabla \times M - \frac{\partial}{\partial t}(\nabla : Q),$$

where M is the magnetization and Q is the quadrupole moment tensor. The polarization vector P is uniquely defined if the unit volume element at the current point $\mathbf{r}$ has overall charge neutrality. The magnetization M is unique only if $\partial P/\partial t = 0$. Q is not uniquely defined unless $P = 0$. This expansion, however, is useful if the volume can be divided into unit cells for which these moments follow as successive approximations to $\mathbf{J}$. They all may contain non-linear parts besides the linear terms.

The constitutive relations being now

$$H = B - 4\pi M, \quad D = E + 4\pi P,$$

the Maxwell equations can be written as

$$\nabla \times H = \frac{1}{c}\frac{\partial B}{\partial t} + \frac{4\pi}{c}J_c - \frac{4\pi}{c}\nabla : \dot{Q}, \quad \nabla \times E = -\frac{1}{c}\frac{\partial B}{\partial t}.$$

And the Poynting theorem (section 4.3) reads

$$\nabla\cdot\left(\frac{c}{4\pi}(E \times H)\right) = -\frac{1}{4\pi}\left\{\left(H\cdot\frac{\partial B}{\partial t}\right) + \left(E\cdot\frac{\partial D}{\partial t}\right)\right\} - (E\cdot j_c) + E : \nabla\dot{Q}.$$

The Joule heat energy loss per unit volume is at the rate $(E\cdot j)$. It is from the iteration of the Lorentz force with the electric current density due to material motion. Since,

$$E : \nabla : \dot{Q} = \nabla(E : \dot{Q}) - \dot{Q} : \nabla E,$$

this equation becomes

$$\nabla\cdot\left[\frac{c}{4\pi}(E \times H) - (E : \dot{Q})\right] = -\frac{1}{4\pi}\left\{(E\cdot\dot{D}) + (H\cdot\dot{B})\right\} - (E\cdot j_c) + \nabla E : \dot{Q}.$$

Hence, in presence of a time-varying quadrupole moment field, the Poynting flux S^t is the sum in

$$S^t = \frac{c}{4\pi}(E \times H) - (\nabla : \dot{Q}).$$

And the rate of energy flow out of a volume element, plus the rate of work done by the fields on a unit volume of the material, equals the decrease in energy density ξ_0 of the field.

The net flux of energy into a finite volume V, bounded by a surface Σ, must equal the sum of the change in the stored energy of the linear medium and the work done by the non-linear polarization P^n on the linear wave E at the frequency ω. So, for only real field expressions we have

$$-\frac{c}{4\pi}\iint_{\Sigma}((E \times H)\cdot d\Sigma) = \frac{1}{8\pi}\iiint_{V}\{(E : \varepsilon : E) + (H : \mu : H)\}\,dV.$$

The left hand side represents the total power flow through the surface Σ. If ξ_0 is the stored energy density of the field, then

$$\xi_0 = -\frac{1}{8\pi}(E^2 + H^2),$$

$$\frac{\partial \xi_0}{\partial t} = -\frac{1}{4\pi}\{(E\cdot \dot{E}) + (H\cdot \dot{H})\} = -\frac{1}{4\pi}\{(E\cdot \dot{D}) + (H\cdot \dot{B})\} + (E\cdot \dot{P}) + (H\cdot \dot{M})$$
$$= (\nabla.S^t) + (E\cdot j_c) - \nabla E\dot{Q} + (E\cdot \dot{P}) + (H\cdot \dot{M}).$$

Hence, for a loss free medium, the rate of work done per unit volume equal the rate at which the stored field energy density ξ_0 increases, that is

$$\frac{\partial \xi_0}{\partial t} = \left(E\cdot \frac{\partial P}{\partial t}\right) + \left(H\cdot \frac{\partial M}{\partial t}\right) - (\nabla E) : \dot{Q}.$$

Hence,

$$d\xi_0 = (E\cdot dP) + (H\cdot dM) - (\nabla E : \dot{Q})$$

and we can write

$$P = \frac{\partial \xi_0}{\partial E}, \quad M = \frac{\partial \xi_0}{\partial H}, \quad \dot{Q} = -\frac{\partial \xi_0}{\partial \nabla E}.$$

Analytically, for a distribution of only pure electric dipoles of all the charged species in a medium,

$$P = \sum_i (N_i q_i \xi_i),$$

where ξ_i is the displacement of the ith species, induced by the electric field. So, P is the sum of the product of the charge per particle, q_i, and the displacement $N_i\xi_i$ per unit volume of the ith species, and

$$(E\cdot P) = \sum_i (N_i q_i (\xi_i.E)) = \sum_i (F_i\cdot \xi_i),$$

where F_i is the electric force per unit volume acting on the ith species of charges at a point. Hence

$$(E\cdot \dot{P}) = \sum_i (f_i\cdot v_i) = \sum_i (F_i\cdot u_i) + \sum_i \{N_i(f_i\cdot \xi_i)\},$$

where $f_i (= q_i E)$ is the force per charge and

$$v_i = \frac{\partial}{\partial t}(N_i \xi_i), \quad u_i = \frac{\partial \xi_i}{\partial t}.$$

Hence, if

$$W = \int_0^T \frac{1}{T}(E\cdot P)\, dt,$$

W is the average of work done per unit volume over the time T.

5.15.1 Thermal energy involvement

The theory of non-linear polarization permits the definition of the non-linear part of the time averaged free energy density of dielectrics, though the instantaneous value of the free energy has no meaning, because in a dispersive medium the total polarization is not uniquely determined by the total field at each instant. Thermodynamic considerations are important because heat deposition by strong EM wave incident on matter is more important than optical effects.

If S is the entropy and T is the absolute temperature (in degrees Kelvin), the stored energy density ξ_0 and the free energy density ξ are given by

$$\xi = \xi_0 - TS - (E \cdot P)$$

$$d\xi_0 = (E \cdot dP) + TdS$$

so that

$$d\xi = -SdT - (P \cdot dE).$$

The thermodynamic potential function ξ is identified with the free enthalpy, or the capability of the medium to do work at constant entropy on the generators of the field. If the medium is loss free, no heat is developed, and the entropy of the medium is constant. The enthalpy of molecular vibration in an electric field is

$$\xi = -\frac{1}{2}\alpha E^2 - \frac{1}{2}\frac{\partial \alpha}{\partial R} RE^2,$$

where R is the nuclear vibrational co-ordinates and α is the polarizability of the field. Here the free enthalpy ξ is the energy associated with the work done by the generators of the field in the medium, so

$$\xi = \xi - (E \cdot P) - (H \cdot M) + \nabla E : \dot{Q}.$$

It has the exact differential

$$d\xi = -P \cdot dE - M \cdot dH + \dot{Q} : \nabla E.$$

So, the polarization may be derived from differentiation of ξ or ξ_0. In general, ξ is a joint power series in E, H and ∇E in a medium with inversion symmetry, the lowest order non-linear terms in ξ take the form

$$\langle \xi^n_{\text{Sym}} \rangle - \chi_1 EEH^* - \chi_2 EE^* H - \chi_3 EE\nabla E^* - \chi_4 EE^* \nabla E$$
$$- \chi_5 EEEE^* - \chi_6 EEE^* E^* + CC = 0.$$

Here the sum of the negative frequencies of the starred quantities and the positive frequencies of the instarred Fourier components equals zero.

The non-linear part of the free energy density ξ^n is proportional to the product $E_1E_2E_3$ of the amplitudes. It is given by the line integral

$$\xi^n = -\frac{1}{2}\int_0^{E_1E_2E_3} [dE_3 : \chi(\omega_3 = \omega_1 + \omega_2)E_1E_2 + dE_2 : \chi(\omega_2 = \omega_3 - \omega_1)E_3 : E_1$$
$$+ dE_1 : \chi(\omega_1 = \omega_3 - \omega_2)E_2E_3].$$

Permutation symmetry relations exist for anisotropic crystals. These show that the result of the integration is independent of the path of interaction. Also the result does not depend on the order in which the three waves are switched on. Partial differentiation of ξ^n with respect to the respective fields E_1, E_2, E_3 gives P_n at the three frequencies. The work done by non-dissipative dielectrics on the wave at the frequency ω_i, is

$$W_i = \frac{1}{T_i}\int^{T_i} \left(E_i \cdot \frac{dP_i}{dt}\right) dt = \frac{\omega_i}{2\pi}\int_0^{2\pi/\omega_i} (E_i \cdot dP_i),$$

where the wave time period $T_i = 2\pi/\omega_i$. Since

$$\frac{1}{T_i}\int_0^{T_i} d(E_i.P_i) = 0 = \frac{\omega_i}{2\pi}\oint(E\cdot dP_i) + \frac{\omega_i}{2\pi}\oint(P_i\cdot dE_i),$$

we obtain

$$\omega_i \oint(E_i\cdot dP_i) = -\omega_i\oint(P_i\cdot dE_i).$$

The net work done by a non-dissiparative dielectric on the waves is zero. So, is the total work done by the waves on the medium, giving

$$\sum_i \omega_i \oint(E_i\cdot dP_i) = -\sum_i \omega_i \oint\left(P_i\cdot\frac{dE_i}{dt}\right) dt.$$

5.15.2 Thermodynamic energy considerations

In laser produced plasmas and in other applications of very strong wave fields of electrodynamics on matter thermodynamics becomes important because much heat is deposited on matter. The internal energy of the particles, which equals the amount of heat added to the medium plus the energy radiated into the medium and the work done on the medium, is increased by an applied field. Here, the adiabatic state relation[4] for isentropic changes of state of a fluid element, due to negligibly small dissipation, is assumed.

The state of a compressible fluid is defined by the internal energy ε, the entropy s, the absolute temperature T (in degree Kelvin), the mass density ρ^m and the specific volume λ of the fluid (volume per unit mass, $1/\rho$). The law of conservation of energy is equivalent to the first law of thermodynamics (See footnote of Example 4.4.6), which for the two-parameter media, considering the thermodynamical variables as function of ρ^m or λ) and s, gives

$$d\varepsilon = Tds - pd\lambda$$

$$dQ = Tds + \frac{1}{\rho^m}dp$$

$$\nabla p = \rho^m\nabla Q - \rho^m T\nabla s,$$

where Q is the heat function per unit mass, and

$$Q = \varepsilon + \frac{1}{\rho^m}p.$$

Equating coefficients of $d\lambda$ and ds from both sides of

$$Tds = \frac{\partial\varepsilon}{\partial s}ds + \left(\frac{\partial\varepsilon}{\partial\lambda} + p\right)d\lambda$$

we obtain

$$T = \frac{\partial\varepsilon}{\partial s} \quad \text{and} \quad p = -\frac{\partial\varepsilon}{\partial\lambda}.$$

[4]In thermally excited adiabatic process (i) the regions of compression and rarefaction do not have time to exchange energy during propagation of acoustic waves of long wave length; (ii) compressions are generally slow, but not so slow that the particles find time to scatter in different direction; and (iii) the reactions are reversible. See also footnote of section 5.20.

In adiabatic processes, conservation of the entropy of the system $ds/dt = 0$, means

$$\frac{\partial s}{\partial t} + (v \cdot \nabla)s = 0.$$

For negligible viscosity and thermal conductivity (ideal gas conditions) we have

$$p = R\rho^m T,$$

where R is the universal gas constant. For polymorphic gases, therefore,

$$\varepsilon = C_v T \quad \text{and} \quad C_p = C_v + p\frac{d\lambda}{dT},$$

$$p\frac{d\lambda}{dT} = K, \quad R = C_p - C_v,$$

$$Tds = C_v dT + pd\lambda,$$

$$ds = C_v\left(\frac{dT}{T}\right) + \left(\frac{P}{T}\right) d\lambda = C_v\left(\frac{dP}{p}\right) + C_p\left(\frac{d\lambda}{\lambda}\right),$$

where C_p and C_v are the specific heats of constant pressure and constant volume respectively. Solution of this equation is

$$P = \left(\frac{\rho^{m^\gamma} p_0}{\rho_c^m \lambda}\right) \exp\left(\frac{s - s_0}{C_v}\right), \quad \gamma = \frac{C_p}{C_v},$$

P_0, ρ_0^m and S_0 denote the initial values of p, ρ^m and s, respectively. The adiabatic law gives the differential equation

$$\frac{d}{dt}\left(P\rho^{m^{-v}}\right) = 0.$$

Its solution is

$$\frac{p}{p_c} = \left(\frac{\rho^m}{\rho_0^m}\right)^\gamma = \left(\frac{T}{T_0}\right)^{\gamma/\gamma - 1}.$$

5.16 Larmor Precession of Bound Electrons

Electrons orbiting round their nuclei, due to central binding forces, holding them, experience a precessional rotation when acted on by a homogeneous magnetic field. This is the Larmor precession frequency of electrons. When the central binding force is $r\phi(r)/r$, the equation of motion of the charge q is

$$m\ddot{r} = -\phi(r)\frac{r}{r} + \frac{q}{c}(\dot{r} \times H), \tag{5.16.1}$$

where r is the electron position vector from the associated nucleus. For the force of gravitation, the potential is

$$\phi(r) = \frac{MmG}{r^2},$$

where M is the ionic mas. For an elastic binding force, proportional to the distance from the ion,

$$\phi(r)\frac{r}{r} = kr,$$

where k is an elastic constant, and the equation of motion is

$$m\ddot{r} = -kr + \frac{q}{c}(\dot{r} \times H). \tag{5.16.2}$$

We have considered the Larmor precession effect here in a simple case only. The gravitational force case is briefly treated in Exercise 5.16.3. Before introduction of the field H, the elastic force balances the centrifugal force of rotation. Electrons of the outer most orbits of heavy atoms are loosely bound (classically) in this manner to their respective nucleus, which being very heavy, is some what fixed.

If ω is the angular velocity of oscillation (harmonic) proportional to $e^{i\omega t}$, then

$$(-m\omega^2 + k)r = \frac{q}{c}(v \times H). \tag{5.16.3}$$

When the magnetic field is applied, the force $q(\mathbf{v} \times H)/c$ acts perpendicular to both $\mathbf{v}$ and H. Before introduction of the applied field, if $\omega = \omega_0$ we obtain

$$(-m\omega_0^2 + K) = 0.$$

So the frequency of the natural orbit is

$$\omega_0 = \left(\frac{k}{m}\right)^{1/2}. \tag{5.16.4}$$

Taking Oz parallel to H, equation (5.16.2) gives

$$(-\omega^2 + \omega_0^2)(x, y) = \Omega_0(y, -x), \quad \Omega_0 = \frac{qH}{mc}, \tag{5.16.5}$$

$$\omega_0^2 - \omega^2 = \pm w\Omega_0. \tag{5.16.6}$$

Its admissible solution is

$$\omega = \frac{1}{2}\Omega_0 + \left\{\frac{1}{4}\Omega_0^2 + \omega_0^2\right\}^{1/2}. \tag{5.16.7}$$

The other solution is rejected because the condition that $\omega = \Omega_0$ when $\omega_0 = 0$ must be satisfied. For small fields

$$\omega = \left(\frac{\Omega_0}{2}\right) + \omega_0, \tag{5.16.8}$$

and the motion around a central nucleus is the superposition of motion at the natural frequency ω_0 and the Larmor precessional rotation at the frequency $\Omega_0/2$. This frequency is half of the electron gyration frequency ($= qH/mc$) which is also referred to as the Larmor frequency, but definitely not as the Larmor precession frequency, which is the gyration frequency of a free electron.

For a system of charges, this precession has a common value only when the system has the same value of q/m in equilibrium under the influence of a single central force. The Larmor frequency of N such electrons is equivalent to the current

$$I = -\frac{Nq}{2\pi c}\frac{qH}{2mc}. \tag{5.16.9}$$

The magnetic moment of a current loop of area $\pi\rho^2$ being $I\pi\rho^2/c$, for N electrons per unit volume, we obtain

$$\mu = -\frac{Nq^2H}{4mc^2}\langle\rho^2\rangle, \tag{5.16.10}$$

where $\langle \rho^2 \rangle$ is the mean of the square of the perpendicular distance of the electrons from the field axis through the nucleus. If N_1 is the number of atoms per unit volume, then the magnetic susceptibility is

$$\chi = \frac{N_1 \mu}{H} = -\frac{N N_1 q^2 H}{4mc^2} \langle c^2 \rangle. \tag{5.16.11}$$

The Larmor precessional rotation gives a diamagnetic field, and is known as the *de Hass Van Alphen effect,* in solid state physics, Kittel (1979, Chapters 9 and 14).

5.16.1 Classical notions of bound electrons

The assumption of electrons, classically bound to the respective nucleus of positive charge, is not new in classical physics. This concept is applied to explain some physical effects, including the Larmor precession effect. The existence of the magnetic moment and other results from the binding forces cannot be guessed from material models having only free electrons.

Orbits of electrons in atoms are distorted non-linearly by a strong monochromatic radiation. These distortions are studied as the sum of the higher harmonics of the linear solution with the help of this bound electron model, Bloembergen (1965). Electrons are bound loosely thus, to positively charged nuclei, in a macroscopically neutral system of charges by a force of the simple harmonic motion, which is proportional to the wave field induced displacement. The impact of strong radiation at a frequency which is higher than the characteristics frequency of the system, on the atom, is an electron displacement for which the quantized field concepts may be replaced by such classical models. Actually, this simple harmonic force follows from the centrifugal force of the electrons in their orbital planes, in plasmas. In crystals this is the one dimensional simple harmonic force about the lattice points. So, only the transverse waves (both elliptically and circularly polarized) are affected thus in plasmas, whereas longitudinal waves, propagating along the directions of the unidirectional lattice vibrations, are thus affected in crystals.

► **Example 5.16.1** *Show that the electric field of a slowly changing magnetic field generates the velocity of Larmor precession.*

Answer. As the slowly changing magnetic field grows from zero to the value H in time t, the generated electric field establishes a new equilibrium state. Since x and y are small, in an inertial co-ordinate system, expand E_x and E_y about their values $(E_x)_0$ and $(E_y)_0$ at the origin, and obtain

$$\left.\begin{aligned} E_x &= (E_x)_0 + x\left(\frac{\partial E_x}{\partial x}\right)_0 + y\left(\frac{\partial E_y}{\partial y}\right)_0 + \cdots \\ E_y &= (E_y)_0 + x\left(\frac{\partial E_y}{\partial x}\right)_0 + y\left(\frac{\partial E_y}{\partial y}\right)_0 + \cdots \end{aligned}\right\}. \tag{1}$$

The momentum transfer condition is

$$(\ddot{x}, \ddot{y}) = \frac{q}{m}(E_x, E_y). \tag{2}$$

Hence if Ω is the angular velocity about the z axis and H_z is the magnetic field in that direction, we have

$$\frac{d\Omega}{dt} = \frac{d}{dt}\left(\frac{x\dot{y} - \dot{x}y}{r^2}\right) = \frac{x\ddot{y} - \ddot{x}y}{r^2}, \tag{3}$$

$$-\frac{1}{c}\frac{\partial H_z}{\partial t} = \frac{\partial E_x}{\partial y} - \frac{\partial E_y}{\partial x}. \tag{4}$$

Putting $x = r\cos\theta$, $y = r\sin\theta$ (r is constant) and $d\theta/dt = \omega$, and using (3) we obtain

$$\frac{d\Omega}{dt} = -\frac{q}{mr^2}\left[(yE_x - xE_y) + \left(y^2\frac{\partial E_x}{\partial y} - x^2\frac{\partial E_y}{\partial x}\right)_0 + xy\left(\frac{\partial E_x}{\partial x} - \frac{\partial E_y}{\partial y}\right)_0 + \cdots\right]. \tag{5}$$

The mean value is

$$\left\langle\frac{d\Omega}{dt}\right\rangle = \frac{q}{2mc}\frac{\partial H_z}{\partial t},$$

since

$$\langle x^2\rangle = \langle y^2\rangle = \frac{r^2}{2}, \quad \langle x, y, xy\rangle = 0. \tag{6}$$

Hence the electric field of a slowly changing magnetic field generates the velocity of Larmor precession, and the motion in the rotating reference system is the same as the original motion in the rest system before its application.

► **Example 5.16.2** *Show that for rotation about the direction of a constant H, in presence of a conservative electric field, at the rate of the Larmor precession frequency, the results for non-rotating motion in the presence of H are recovered.*

Answer. Consider the Lagrangian L of a system of N particles,

$$L = \sum\left[\frac{1}{2}mv^2 - q\left\{\phi - \frac{1}{c}(v.A)\right\}\right]. \tag{1}$$

For a constant magnetic field H, we can write

$$H = \frac{1}{2}\nabla \times (H \times r). \tag{2}$$

Then the vector potential is

$$A = \frac{1}{2}(H \times r). \tag{3}$$

For a system of charges in equilibrium, under a central force (say, for classical electrons bound to their respective atomic nuclei), we have

$$L = \sum\left[\frac{1}{2}mv^2 - q\left\{\phi - \frac{1}{2c}(r.(v \times H))\right\}\right], \tag{4}$$

$$H = -\frac{2mc}{q}\Omega = \sum\left[\frac{1}{2}m(v - (\Omega \times r))^2 - \frac{1}{2}m(\Omega \times r)^2 - qQ\right], \tag{5}$$

where

$$H = -\frac{2mc}{q}\Omega. \tag{6}$$

When Ω and r are so small that $|v| \gg |\Omega \times r|$ we neglect the second of the three terms, within the summation, and obtain

$$L = \sum\left[\frac{1}{2}(\mathbf{v} - (\Omega \times r))^2 - qQ\right]. \tag{7}$$

Hence, if v' is the velocity of the particle as seen in the laboratory system in which only the electrostatic potential force $q\nabla\phi$ acts on the particle, evidently

$$\mathbf{v}' = \mathbf{v} - (\Omega \times r), \tag{8}$$

$$L = \sum \left[\frac{1}{2}mv'^2 - qQ\right]. \tag{9}$$

The particle velocity is v in the system of co-ordinates in which the particle is uniformly rotating about an axis at the rate Ω. In the laboratory inertial frame k_0, the equation of motion is

$$m\ddot{r}_0 = qE + \frac{q}{c}(v \times H), \tag{10}$$

where r_0 is the radius vector from the center of force to the electron. Let now $r, \dot{r}, \ddot{r}$ be the values of $r_0, \dot{r}_0, \ddot{r}_0$ in a system K which is rotating about the center of force with uniform angular velocity Ω. Then

$$\dot{r}_0 = \dot{r} + (\Omega \times r) = \left(\frac{\partial}{\partial t} + \Omega\right) \times r,$$
$$\ddot{r}_0 = \ddot{r} + 2(\Omega \times \dot{r}) + (\Omega \times (\Omega \times r)). \tag{11}$$

Neglecting squares of Ω and H, the equation of motion becomes

$$\ddot{r} + 2\left(\Omega + \frac{qH}{2mc}\right) \times \dot{r} = \frac{q}{m}E, \tag{12}$$

because for central forces E transforms in the same way as r_0. So the equation of motion in the rotating system is the same as in the original system in the absence of H if Ω is given by (6).

► **Example 5.16.3** *Find the Larmor precession effect when the binding force is the gravitational field of the ion.*

Answer. In this case the equation of motion after application of the field H, is

$$m\ddot{r} = -\frac{mMG}{(r_0 + r)^3}r + \frac{q}{c}(\dot{r} \times H) = -\frac{MmG}{r_0^3}r + \frac{q}{c}(\dot{r} \times H)$$
$$= -\omega_0^2 r + \frac{q}{c}(\dot{r} \times H),$$

because the field induced displacement r is small in magnitude compared to r_0, where $r = r_0$ when $H = 0$, so $\omega_0^2 = MmG/r_0^3$.

5.17 Comparison conclusions

5.17.1 On relative strength of some forces

For two point charges q_1 and q_2, of masses m_1 and m_2, the Coulomb force F_c and the gravitational force F_G are

$$F_c = \frac{q_1 q_2}{r^2}, \quad F_G = G\frac{m_1 m_2}{r^2}, \tag{5.17.1}$$

where $G(= 6.67 \times 10^{-8}$ cm^3/gm sec^2) is the universal gravitational constant. Also

$$F_E = qE, \quad F_H = \frac{q}{c}(v \times H). \tag{5.17.2}$$

So that F_E is the force of E, and F_H is the force of motion of the charge q in presence of the magnetic field. The Coulomb force of repulsion between two-equal charges and Biot-Savart's law are

$$F_E = \frac{q^2}{r^2}, \tag{5.17.3}$$

$$H = \frac{j}{cr^2}, \quad j = qv. \tag{5.17.4}$$

These give the force relations

$$F_H = \frac{q}{c}vH = \frac{q^2v^2}{c^2r^2} = \frac{v^2}{c^2}F_E. \tag{5.17.5}$$

So, the force of E is smaller than the force of **v** and H by a factor of the order of v^2/c^2.

Consider the circular motion of a particle (the hydrogen atom) at earth's solar distance $R(= 2 \times 10^{13}$ cm) at the rate of earth's orbital velocity $v(= 3 \times 10^6$ cm/sec). If M is the solar mass ($= 10^{33}$ gms) and m is the hydrogen atom mass ($= 167 \times 10^{-24}$ gm), then

$$F_G = \frac{(6.67 \times 10^{-8} \times 10^{33} \times 1.67 \times 10^{24})}{(2 \times 10^{13})^2}.$$

The electron of the hydrogen atom experiences the force

$$F_H = \frac{(4.8 \times 10^{-10} \times 3 \times 10^6 \times 1.2 \times 10^{-6})}{(3 \times 10^{10})},$$

where $H = 1.2 \times 10^{-6}$ gauss, because the solar magnetic field is due to a dipole moment 0.42×10^{34} gauss cm^2 situated at the solar center. We here neglect the effect of solar magnetic field upon the atomic magnetic moment. Then $F_H = 2 \times 10^5 \times F_G$. So, the force of the solar magnetic field is enormously large, compared to the force of gravitation, even at earth's distance from the sun. The motions of ions and electrons are separated in the magnetic field, but the resulting polarization generates a strong electric field which works against this space charge separation.

The pressure of compression by the gravitational force in the center of Sun (mass 10^{33} gms) is produced by about 10^{-3} gm of charged matter, Meierovich (1982). The potential energy per charge of such strong compression is of order of magnitude of the electron rest mass, and is higher than the electron binding energy in atoms; so the compressed matter forms a fully ionized (non-neutral) material. It is impossible to compress such a plasma by electrostatic forces only, because this force is attractive between charges of opposite sign, and repulsive for charges of same sign.

The situation is different for motion of identical charges with respect to the charges of the opposite sign. When the average electron velocity (drift), with respect to the ions is non-zero, a current arises in a medium creating a magnetic field. The force between carriers of parallel electric current produced by a magnetic field, is attractive. When, like that, the pinch effect of the magnetic compression is stronger than the collective electrostatic force of the EM field, it increases until equilibrium is established.

5.17.2 On Coriolis force field

A pipe carrying a flowing fluid generates a Coriolis force that encircles it just as a wire carrying current generates a magnetic field around it. Also, a gyro rotor has a Coriolis dipole moment, which is the angular momentum, just as a current carrying loop of wire has a magnetic dipole moment. The equation for the torque of a current carrying loop, in a magnetic field, holds good for the torque developed by a gyro rotor in a Coriolis field. A molten metal, flowing through two parallel pipes, with spacing $d = 0.1$ meter, generates a gravitational interaction which is due to mass currents J_1 and J_2. The equivalent magnetic case of two parallel wires with currents I_1 and I_2, gives rise to a pinch effect force per unit length of amount, Forward (1961),

$$\frac{F}{L} = \frac{\mu I_1 I_2}{2\pi d}.$$

The equivalent gravitational relation is

$$\frac{F}{L} = \frac{J_1 J_2}{2\pi d}.$$

The Coriolis field, induced by a gyro rotor, having, for instance, an angular momentum 10^7 gm cm^2/sec, at a distance of 10 cm, is equivalent to a physical rotation of about 10^{-19} angular degree per hour. Each meter of pipe containing molten iridium weighs about 200 kg. For motion at the rate of a kilometer per second, the mass current in a pipe is

$$J_1 = J_2 = MV = (2 \times 10) \text{ Km/sec}.$$

The force between the two pipes, due to the pinch effect of gravitation, is about 2×10^{-15} Newton per meter of pipes. The Newtonian gravitational force of attraction between the same two pipes is about 3×10^{-1} Newton per meter of pipe. So, the forces due to the gravitational pinch effect are very much hidden by the Newtonian gravitational attraction effect which is itself not usually observable.

The power radiated from a l-meter dumbell, weighing 1 metric ton, and spinning at about 10,000 rpm, is only 4.5×10^{-33} watts, Forward (1961). For a high rotational velocity, with a low mass, the effect is insignificant. For a low rotational speed and a great mass, as in the case of a rotating double star system, the effect is significant. The rotational speed could be on the order of 1 month per see, so $\Omega = 10^{-4}$ rad/sec, and the power radiated from a binary star, due to this effect, is about 10^7 watts. This power would take 10^{10} billion years for the system to damp out due to the radiation loss. These Coriolis force fields, therefore, are too small for laboratory detection. This force is possibly significant in neutron stars and their evolution, and in the relative motion of the galaxies.

5.18 Scaling Laws

Typical speeds (speeds of matter and in matter and speed-like parameters) in media are compared with the light speed c (a universal constant), to assess the importance of an effect with the help of scaling laws, in electrodynamics related problems. Scaling laws cannot be formulated when c is infinite. The scaling laws determine the dominating terms of field equations for a physical process, compared to the non-dominating terms.

When the linear call is changed by a factor ξ, the other quantities should be changed to keep unchanged the characteristic features of a process. Scales can be changed from one experiment to the other in several ways. For change of a linear dimension by a factor ξ, the electric field E must be changed by the factor $1/\xi$, and D, H, B must also be changed by $1/\xi$, because (1) the potential $\phi \sim lE$, (2) these fields have the dimensions of E and (3) they appear in the Maxwell equations. The displacement current

$$\frac{\partial D}{\partial t} = \frac{D}{t} = \frac{D}{l}v,$$

where the speed v is of the order of the particle velocity; so v is unchanged because the kinetic energy of the particle is unchanged. Hence, $\partial D/\partial t$ must be changed by the factor $1/\xi^2$, the total current $I(= jl^2)$ remains unchanged, $\sigma(= j/E)$ changes by the factor $1/\xi$, the inductance $L_0(= \phi/(DT/dt) = 1\phi/Iv)$ and the capacity $c(= L)$ changes by the factor ξ, the mean free path changes as ξ, the density ρ (inversely proportional to the mean free path) changes by the factor $1/\xi$.

When $B_0(t)$ is increased by a factor ξ, at all times, keeping $v_0(= L/T)$ and L unchanged, to keep B' unchanged, the magnetic field $B(r,t)$ must be multiplied by ξ. And since $E' = cE/v_0 B_0$, the electric field $E(r,t)$ must be multiplied by ξ at all positions and times to increase it to keep E' unchanged. Since g' does not contain B_0, the gravitational field is unchanged, as are the initial position $\mathbf{r}_0$ and the unit vector $\hat{\mathbf{v}}_0$. Since the dimensionless time $t'(= v_0 t/L = (\xi v_0)(t/\xi)L)$ contains v_0, it is not easy to decrease v_0 by a factor $\xi(< 1)$ while keeping B_0 and L unchanged from one experiment to the next. When $E'(r', t')$ remain unchanged, E' must be reduced by ξ, and time in which it varies is increased by $1/\xi$, that is, $E'(r, t') = \xi E(r,t)$. The g field is reduced by ξ^2, and the time scale is increased without changing $\mathbf{r}_0$ and $\mathbf{v}_0$, so that

$$g'(\mathbf{r}', t') = \xi^2 g(\mathbf{r}, \xi, t).$$

To keep $B'(r', t')$ unchanged, for a given t', the time in which $B(r,t)$ varies is increased by the factor $1/\xi$, so that $B'(r't') = B(r, \xi t)$. The scaling decreases both $v_0/L\Omega_0(= \rho_0/L$, where ρ_0 is the charge gyration radius) and the ratio T/T_0 (where $T = 2\pi/\Omega_0$ and $T_0 = v_0/2\pi l$, so that T is the time period of gyration and T_0 is the time rate in which the field changes effectively) just as decreasing m/e does. So, any one of the four parameters m/e, $1/B_0$, v_0, $1/L$ can be made smaller for making the guiding center approximation closer to the actual particle motion. For appearance of the parameter m/e in the field equations it is not necessary to write these in dimensionless forms. But the other three parameters appear only in the dimensionless field equations. So, the former has some advantage when the question of choice arises.

5.19 Comparison Conclusions in Flow Problems

The important speeds of MHD processes are v, C_A and C_S, where v is a typical material flow speed, $C_A\left(= \sqrt{H^2/4\pi\zeta}\right)$ is the Alfven speed and $C_S\left(= \sqrt{dp/d\zeta} = \sqrt{kT/m}\right)$ is the sound speed for adiabatic thermalization in the medium. The ratios v^2/c^2, c_A^2/c^2, c_S^2/c^2 measure the macroscopic kinetic energy, magnetic energy and thermal energy in terms of the rest energy of a fluid particle. In dense, electrically conducting media, the mean free path and the Debye shielding distance are of the order of molecular distances, and the macroscopic motion does not differ from the microscopic thermal motion. High frequency electromagnetic waves (a consequence of the non-trivial existence of the displacement current of the Maxwell equations) then propagate at infinite speed; so their contribution is filtered out of the field equations. The conduction current is most effective in MHD processes.

In gas dynamics the flow is supersonic if the Mach number $M(= v/c_s) > 1$, subsonic if $M < 1$ and transonic when $M \approx 1$. The Alfven number $A = (v/c_A)$ determines the nature of flow of electrically conducting fluids in channels in presence of a parallel magnetic field. An MHD flow is super Alfvenic if $A > 1$, sub Alfvenic when $A < 1$, and trans Alfvenic for $A \approx 1$.

5.19.1 Order of magnitude determination in electromagnetic processes

The order of magnitude of a physical quantity relative to another quantity determines its importance in a physical process. A quantity P is of the order of magnitude of Q (or symbolically $P \approx Q$) when P and Q have the same dimensions and $P = \lambda Q$, where λ is of the order of unity, say lying between 10 and 1/10. The time derivative and space derivative (along the Z direction, say) of a field variable ψ, for practical purposes can be expressed as

$$\frac{\partial \psi}{\partial t} = \frac{\psi}{T}, \quad \frac{\partial \psi}{\partial z} = \frac{\psi}{L_z},$$

where L_z and T are called the *characteristic length* (along the Z direction) and time, respectively, of its variation in that process. So, ψ varies appreciably in times and lengths of the order of T and L_z and not less. Then $v(= L_z/T)$ is a characteristic speed of the process.

The time scale of variation of electromagnetic wave fields in stationary media is L/c, and L/v for bulk motion with a typical speed v. The time L/v is much greater than the time L/c because maximum v is atmost the sound speed C_S (of the order of 330 cm/sec). So, if $Q \approx (v^2/c^2)$ then $P + Q \approx P$. The Maxwell equation $c\nabla \times E = -\dfrac{\partial B}{\partial t}$ gives

$$\frac{E}{L} = \frac{B}{cT}.$$

Comparing the displacement current $\partial D/\partial t$ with $c\nabla \times H$, we obtain

$$\frac{1}{c}\frac{\partial D}{\partial t} = \frac{E}{cT} = \frac{BL}{(cT)^2} = \frac{B}{L}\left(\frac{L}{cT}\right)^2 = |\nabla \times H|\frac{v^2}{c^2}.$$

So, the displacement current is negligible compared to $\nabla \times H$, because then

$$\nabla \times H = \frac{4\pi j}{c} \quad \text{and} \quad (\nabla \cdot j) = 0.$$

The divergence equation $(\nabla \cdot D) = 4\pi q$ gives $q = E/L$, and comparison of the two terms of the Lorentz force gives

$$qE = \frac{E^2}{L} = \left(\frac{B^2}{L}\right)\left(\frac{v^2}{c^2}\right) = \left|\frac{1}{c}(j \times B)\right|\left(\frac{v^2}{c^2}\right).$$

Hence the body force qE is negligible compared to $(j \times B)/c$. The viscous force term $\nu\nabla^2 v$, on comparison with $(\mathbf{j} \times B)/c$, gives

$$|\nu\nabla^2 v| = \left|\frac{1}{c}(j \times B)\right|\left(\frac{\nu v}{H^2 L}\right).$$

Assuming $H = 100$ gauss, $L = 100$ cm, $v = 1$ cm/sec. and $\nu = 10^{-2}$ gm/cm sec (for mercury), we find that $\nu v/H^2L = 10^{-7}$. In astrophysical problems L is much larger, so this coefficient is still smaller, and the viscous force is negligible. When the inequality $|\nabla p| > |\nu\Delta^2 v|$ gives $|\nu v/L| < p$, the viscous force is negligible, relative to the pressure gradient. Since $qv = E(v/L) = Hv/cT = (v^2/c^2)j$, the convective term qv, causing EM waves and plasma oscillations, in Ohm's law $j = qv + \sigma E + \sigma(v \times B)/c$, is negligible. The viscous effects can not be neglected in some high speed Laboratory MHD experiments. In the Lorentz theory, the dispersive magnetic field is relativistic, section 3.2.1. Then, non-relativistically, the Faraday law of induction is $\nabla \times E = 0$, E and H are determined from the gradient of the scalar potential and curl of the vector potential. So, for charges inside a volume V,

$$E = \frac{1}{V}\sum_i \nabla\left(\frac{q_i}{|r_i - r|}\right), \quad H = \frac{1}{cV}\sum_i \nabla \times \left(\frac{q_i \dot{r}_i}{|r_i - r|}\right).$$

Then

$$\frac{H^2}{8\pi} = \frac{E^2}{8\pi}\frac{v^2}{c^2},$$

so the magnetic energy density is negligible compared with the electrical energy density, in this electrodynamics.

5.20 Expansion of Variables for Quasilinear Processes

Expansion of field variables in powers of a parameter ε also (enables removal of) the unimportant terms in the field equations depending on the size of the body and other characteristic features. This removal is necessary in study of complicated interactions between atoms, electrons and quanta in media, depending upon the electromagnetic fields and electrostatic potentials. For ε, the numerically small, non-dimensional ratio m/e of the rest mass and charge of an electron is chosen. This ε measures the slowness of the field variables for motion in a given field.

The gyro period $2\pi/\Omega$ and the gyroradius v/Ω, for electron motion around a constant magnetic field H, at a point P at time t, (where $\Omega = eH/mc$, and v is the gyrovelocity, which depends on the initial velocity) vary directly with the ratio m/e; $v/\Omega t$ varies over the characteristic distance L of variation of the field variables in a process. ε can be reduced without change of the fields, the initial velocity, and position of the particle. Though the ratio m/e appears in the natural dimensional form of the field equations for motion of a charged particle, the quantities H, v and L do not appear thus.

5.21 Co-ordinate Expansion Techniques in Quasilinear Processes

Perturbation techniques are suitable for finding closed form solution of weakly non-linear systems, by successive approximation, through asymptotic expansion of variables, including an independent variable. The criterion of weak non-linearity is that the maximum of the dimensionless electric field (applied) intensity parameter is very small, that is, $|eE_{\max}/m\omega c| \ll 1$. This condition ensures rapid convergence of the infinite series expansion of the field variable $f(r,t)$ of the type

$$f(r,t) = f_0 + \sum_{n=1}^{\infty}\left[f_n(r)e^{-in\omega t} + \bar{f}_n(r)e^{in\omega t}\right],$$

where $\bar{f}_n$ is the complex conjugate (cc) of f_n, the incident wave is at the frequency ω, f_0 is a constant, and, for convergence,

$$|f_0| \gg |f_1| \gg |f_2| \gg |f_3| \cdots \gg |f_n|.$$

The infinite series expansion of $f(r,t)$ is of the type

$$U = U_0 + U_1 + U_2 + \cdots + U_n + \cdots,$$

where U_0 is the field-free state value of U, U_1 contains only the first harmonic terms, which are of the linearized approximation, U_2 stands for the sum harmonics, difference (beat) harmonics which includes the zero frequency or the zero harmonic terms, etc. U_n (for $n = 2, 3, \cdots$) is a linear combination of the nth harmonic of all the waves and other sum and difference harmonics, each of which is of total order n. The condition of convergence is $|U_n|/|U_{n-1}| < 1$. The product $U_r\ U_{n-r}$ is of the order of U_n for $n = 1, 2, 3$, where $r \leq n$. An equivalent alternative formal expansion scheme is

$$u = u_0 + \varepsilon u_1 + \varepsilon^2 u_2 + \cdots + \varepsilon^n u_n + \cdots.$$

The expansion parameter ε equals unity in the solution. Then, the artifice ε is used merely for book keeping purposes in a systematic fashion, for grouping together terms of comparable degree of approximation. For these problems ε should necessarily be dimensionless, and need not be specified. Saturated contribution to the solution of non-linear systems of equations is achieved through processes of accelerated convergence of the expansion of both field variables and independent variables. In temporal evolution problems the independent variable of expansion is time, and in space evolution problems this variable is a space co-ordinate. In temporal evolution due to non-linearity, the expansion of the independent variable t is

$$t = T + \varepsilon a_1(T) + \varepsilon^2 a_2(T) + \cdots.$$

The parameter ε is the ratio of a field increment of the nth orders of approximation to that of the $(n-1)$th order. The second and other subsequent terms are auxiliary functions of the fundamental time variable T. The relations between the derivative operators

$$\frac{\partial}{\partial t}, \frac{\partial^2}{\partial t^2} \quad \text{and} \quad \frac{\partial}{\partial T}, \frac{\partial^2}{\partial T^2}, \cdots, \quad \text{etc.}$$

are

$$\frac{\partial}{\partial t} = \frac{1}{1 + \varepsilon a_1' + \varepsilon^2 a_2' + \cdots} \frac{\partial}{\partial T},$$

$$\frac{\partial^2}{\partial t^2} = \frac{1}{(a + \varepsilon a_1' + \varepsilon^2 a_2' + \cdots)^2} \frac{\partial^2}{\partial T^2} - \frac{(\varepsilon a_1'' + \varepsilon^2 a_2'' + \cdots)}{(1 + \varepsilon a_1' + \varepsilon^2 a_2' + \cdots)^2} \frac{\partial}{\partial T},$$

where $a_1' = da_1/dT$, etc. For weak cubic non-linearity it is enough to write

$$t = T + \varepsilon b_1 T + \varepsilon^2 b_2 T,$$

where b_1 and b_2 are small constants. For quadratic non-linearities, the field induced shifts do not occur in the first harmonic of the wave fields; so, for these $b_1 = 0$, $b_2 = 0$. For cubic non-linearities $b_1 = 0$, $b_2 \neq 0$ and $t = T + \varepsilon^2 b_2 T$, $(\partial/\partial t) \approx (1 - \varepsilon^2 b_2)\partial/\partial T$, $(\partial^2/\partial t^2) \approx (1 - 2\varepsilon^2 b_2)\partial^2/\partial T^2$.

A weakly non-linear closed system of differential equations admits stationary wave solution. The words secular-free solution, solution having an accelerated convergence, most saturated solution, etc., have the same meaning and less rigorous solutions have secularly growing terms. The secular growth is removed by the introduction of a frequency shift in time evolution problems, and a wave number shift in space evolutions problems. Both shifts are due to cubic and higher order non-linearities. For cubic non-linearities the shifts depend on the squares and other double products of the field amplitudes.

For secular-free wave train solutions of quasi-linear process, the important methods are : (a) Method of co-ordinate stretching of Lindstedt and others (Bellman (1964)); (b) Method of expansion of the wave parameter of Bogoliubov, Krilov and Mitropolsky (Minorsky (1962)); (c) A direct method of successive approximation, (d) Normalization technique (Arons and Max (1974)).

5.22 Systems having Different Degrees of Freedom

Macroscopic field variables are distinct from the corresponding quantity obtained from an appropriate velocity moments of distribution functions. Asymptotic methods of expansion of the equations of a fluid continuum, and of the Boltzmann system of equations, for their solution give qualitatively different results. Whitham (1965) remarked "... the detailed microscopic structure of (a field variable) ϕ is not necessarily governed by hyperbolic equations, but the equations for the 'microscopic' variables, such as amplitude, wave number, etc. are hyperbolic ... the Boltzmann equation for a gas is not hyperbolic ..., and sound waves appear from appropriate averaged equations .... The Maxwell distribution describes a uniform state in term of density ρ, temperature T, mean velocity u, ... This is a solution for constant values of ρ, T, u, but it is used to derive continuous equations for ρ, T, u when they are slowly varying functions relative to the mean free path and collision time".

The expansion of the distribution function $f(r, v, t)$ and the microscopic field variables, which depend on its velocity moments, are different. To show this difference we write

$$f(r,v,t) = N_0 f_0(v^2) + \varepsilon f_1(r,v,t) + \varepsilon^2 f_2(r,v,t) + \cdots,$$

$$\langle v \rangle = \frac{\iiint\limits_V v f dv}{\iiint\limits_V f dv}.$$

Since

$$\langle v \rangle \iiint\limits_V f dv = \iiint\limits_V v f dv,$$

equating coefficients of same powers of ε from both sides, we obtain

$$u_0 = \frac{\iiint\limits_V v f_0 dv}{\iiint\limits_V f_0 dv}, \quad u_1 = \frac{\iiint\limits_V v f_1 dv}{N_0 \iiint\limits_V f_0 dv} - \frac{\left(\iiint\limits_V v f_0 dv\right)\left(\iiint\limits_V f_1 dv\right)}{N_0 \left(\iiint\limits_V f_0 dv\right)^2},$$

$$u_2 = \frac{\iiint\limits_V v f_2 dv}{N_0 \iiint\limits_V f_0 dv} - \frac{\left(\iiint\limits_V v f_0 dv\right)\left(\iiint\limits_V f_2 dv\right)}{N_0 \left(\iiint\limits_V f_0 dv\right)^2} - \frac{\left(\iiint\limits_V f_0 dv\right)\left(\iiint\limits_V v f_1 dv\right)}{N_0^2 \left(\iiint\limits_V f_0 dv\right)^2}$$

$$+ \frac{\left(\iiint\limits_V v f_0 dv\right)\left(\iiint\limits_V f_1 dv\right)^2}{N_0^2 \left(\iiint\limits_V f_0 dv\right)^3}, \text{ etc.}$$

The pair of velocity integrals of the distribution function of the zeroth order, first order and second order, for relating with u_0, u_1, u_2 (the average velocity of the zeroth, first and second order) are

$$\iiint\limits_V f_i dv, \quad \iiint\limits_V v f_i dv \quad \text{for } i = 0, 1, 2.$$

The relations between the integrals of the distribution functions of different orders of approximation and the average velocities are not reversible, and solution is not possible for the velocity integrals of different orders in terms of the average velocities of same order from these relations without the help of connecting relations between the integrals

$$\iiint\limits_V f_i dv \quad \text{and} \quad \iiint\limits_V v f_i dv$$

for all orders of magnitude, without some assumptions. Expansion of the number density N,

$$N = N_0 + \varepsilon N_1 + \varepsilon^2 N_2 + \cdots = \iiint\limits_V f dv = \iiint\limits_V (N_0 f_0 + \varepsilon f_1 + \varepsilon^2 f_2 + \cdots) \, dv,$$

gives the relations

$$\iiint\limits_V f_0 dv = 1, \quad N_1 = \iiint\limits_V f_1 dv, \quad N_2 = \iiint\limits_V f_2 dv, \text{ etc.}$$

$$u_0 = \iiint\limits_V v f_0 dv, \quad N_0 u_1 + N_1 u_0 = \iiint\limits_V v f_1 dv,$$

$$N_0 u_2 + N_1 u_1 + N_2 u_0 = \iiint\limits_V v f_0 dv, \text{ etc.}$$

Expansion of E and $\iiint\limits_V v f dv$ shows the same absence of one-to-one correspondence between E_i and $\iiint\limits_V v f_i dv$ for $i = 1, 2, 3, \cdots$. Consider the Ampere-Maxwell equation in the two forms

$$\nabla \times H = \frac{1}{c}\frac{\partial E}{\partial t} - \frac{4\pi\rho N}{c}v,$$

$$\nabla \times H = \frac{1}{c}\frac{\partial E}{\partial t} + \frac{4\pi\rho}{c}\iiint\limits_V v(N_0 f_0 - f)\, dv$$

for electron current, and static ions in the two-models, namely the hydrodynamical model, and the kinetic theory model. Equating coefficients of ε^2 (for instance) from both sides for the two models we obtain

$$\nabla \times H_3 = \frac{1}{c}\frac{\partial}{\partial t}E_3 - \frac{4\pi\zeta}{c}(N_0 u_3 + N_1 u_2 + N_2 u_1 + N_3 u_0),$$

$$\nabla \times H_3 = \frac{1}{c}\frac{\partial E_3}{\partial t} - \frac{4\pi\zeta}{c}\iiint\limits_V v f_3 dv.$$

The non-linear terms in the right hand side of the fluid model are linear combinations of lower order terms, but there is no such non-linearity in the second equation. Hence, it will be improper to regard H_3 of the two equations as completely identical quantities, though both are of the same order of magnitude. Thus, some qualitatively non-identical results of higher order non-linear corrections are obtained; their quantitative distinction may be insignificant, because the series expansions converge very rapidly.

5.23 Transformation of Variables and Equations to Dimensionless Form

The variables $\mathbf{r}$ (displacement vector), t (time), ϕ_g (gravitational potential), ϕ (scalar field potential) and A (vector potential) have different dimensions of mass, length and time. But the corresponding primed quantities r', t', ϕ' and A' are dimensionless if they are defined as

$$r = Lr', \;\; t = Tt', \;\; \phi_{go} = \phi_{g'}\phi_g, \;\; \phi = \phi_0\phi', \;\; A = A_0 A',$$

where L, T, ϕ_{go}, ϕ_0, A_0 are constant characteristic values of $\mathbf{r}$, t, ϕ_g, ϕ, A respectively; the particular choice of these constant values is not important.

We write

$$v_0 = L/T, \;\; g = g_0 g' = g'(v^2/L), \;\; E(r,t) = E_0 E' = (v_0 B/c)E', \;\; B(r,t) = B_0 B',$$

where g is the acceleration due to gravity. The initial conditions are :

$$t' = 0, \;\; r' = r_0/L, \;\; dr'/dt' = \hat{v}_0,$$

where $\mathbf{r}_0$ is the initial position of the particle and $\mathbf{v}_0 = v_0\hat{v}_0, \hat{v}_0$ being the unit vector of the initial velocity. Then the equation of motion of a particle of mass m and charge e,

$$\frac{mc}{e}[\ddot{r} - g(r,t)] = CE(r,t) + r \times B(r,t),$$

in the non-dimensional form, is

$$\left(\frac{mcv_0}{eB_0L}\right)\left[\frac{d^2r'}{dt'^2} - g'(r',t')\right] = E'(r',t') + \left(\frac{dr'}{dt'}\right) \times B'(r',t')$$

with

$$\frac{v_0}{\Omega_0 L} \to \frac{mc}{e}, \ r' \to r, \ t' \to t, \ g' \to g, \ B' \to B, \ E' \to cE$$

when $\Omega_0 = eB_0/mc$. Thus, any solution of the equation in r', when expressed as a function of t', gives a solution of $\mathbf{r}$ as a function of t. So, in the first equation (which is not non-dimensional) we use m/e as the smallness parameter. The solution of the second equation depends on $v_0/\Omega L$ for given dimensionless initial conditions r_0/L and $\hat{v}_0$, and fields $E'(r', t')$, $B'(r', t')$, $g'(r', t')$. Therefore, $v_0/L\Omega_0$ is decreased by decreasing $v_0/B_0 L$, instead of m/e, and at the same time by keeping $E', B', g', r_0/L$ and v_0 unchanged.

For the relativistic equation of motion

$$\frac{dp}{dt} = eE + \frac{e}{c}\left(\frac{d\mathbf{r}}{dt} \times B\right) + mg(r, t)$$

we write

$$E' = \frac{mc}{p_0 B_0} E(\mathbf{r}, t), \quad g' = \frac{m^2 L}{p_0^2} g(\mathbf{r}, t), \quad t = \frac{p_0 t}{mL},$$

where p is the momentum of the particle, $\mathbf{r}$ is its position vector, dr/dt is the velocity, E and B are the electric field and magnetic field respectively, g is the vector sum of the other forces, m is the rest mass of the electron, e its charge and p_0 is the initial momentum. So, $p_0 = mv_0\gamma_0$, $\gamma_0^2 = 1/(1 - \beta_0^2)$, $\beta_0 = v_0/c$. Then this equation becomes

$$\left(\frac{p_0 c}{eB_0 L}\right) \frac{d}{dt'} \left\{ \left(\frac{dr'}{dt'}\right) \left[1 - \left(\frac{p_0}{mc}\right)^2 \left(\frac{dr'}{dt'}\right)^2\right]^{-1/2} \right\} - g'(r'; t')$$
$$= \left(\frac{dr'}{dt'}\right) \times B'(r', t') + E'(r', t').$$

The initial conditions are $t' = 0$, $r' = r_0/L$, $dr'/dt' = v_0/\sqrt{1 + p^2/m^2c^2}$. For a given r_0/L and p_0, the equation of motion and the initial conditions contain the dimensionless parameters $p_0 c/e\beta_0 L$ and $p_0/mc\left[= \beta_0/(1 - \beta_0^2)\right]^{1/2}$. In the non-relativistic limit, $(p_0/mc)^2 \ll 1$, and the equation of motion in the dimensionless form and initial conditions are recovered easily.

5.24 Wave Polarization

A dispersive wave, the electric field intensity vector of which lies in a plane through the direction of its propagation, is a plane polarized wave, and that plane is the plane of its polarization. A group of waves is plane polarized if their electric vectors for different times or space points lie in the same plane. A wave whose electric vector always lies parallel to a given direction, is polarized in that direction. When two waves have the same frequency, but their electric vectors lie in planes at right angles to one another, the resultant is an elliptically polarized wave, because the tip of the resultant electric field vector (or the magnetic field vector) describes an ellipse in one time period or in one wave length as the wave proceeds along the direction of its propagation. In special cases, the ellipse degenerates into a circle or a straight line. In other words, consider a transverse wave propagating parallel to the x-axis. When there is displacement in the XY-plane, in addition to the displacement in the XZ-plane, and both such displacements always remain perpendicular to the wave normal, the wave with a plane front is then circularly or elliptically polarized. An elliptically polarized wave can be resolved into the sum of one right and one left circularly polarized waves of different amplitudes, but same phase velocity. A plane polarized wave is the sum of one left and one right circularly polarized waves of same amplitude and phase velocity.

The displacements of the wave undergo rapid changes in directions as well as in form, but always remain perpendicular to the direction of propagation. So, a wave is said to be polarized when the transverse nature of the corresponding vibration does not change in direction or form. In order to show the different polarized waves the wave displacements perpendicular to a wave-normal are split into two components along the y- and the z-directions, both normal to the wave normal. Then, considering the simplest form of wave equation of an electric field E, different sets of polarized wave with $\theta = kx - wt$ and plane fronts differing in phase value of the component waves can be written as

$$E_y = E_0 \sin\theta, \quad E_z = E_0 \sin\left(\theta \pm \frac{\pi}{2}\right), \tag{5.24.1}$$

for a circularly polarized wave when phase difference of the components is $\pm\frac{\pi}{2}$;

$$E_y = E_1 \sin\theta, \quad E_z = E_2 \sin(\theta + \delta), \tag{5.24.2}$$

for an elliptically polarized wave when the phase difference of the component is δ, not equal to $\pm\pi$ or any integral multiple of $\pm\pi$;

$$E_y = E_0 \sin\theta, \quad E_z = 0, \tag{5.24.3}$$

for a plane polarized wave, polarized in the XY-plane;

$$E_y = 0, \quad E_z = E_1 \sin\theta, \tag{5.24.4}$$

for a plane polarized wave polarized in the ZX-plane;

$$E_y = E_1 \cos\theta, \quad E_z = E_2 \cos\theta, \tag{5.24.5}$$

for a plane polarized wave, polarized in a plane parallel to the x-axis and making the angle $\tan^{-1}(E_2/E_1)$ with the XY-plane,

$$E_y = E_1 \sin\theta, \quad E_z = E_2 \sin(\theta + \pi), \tag{5.24.6}$$

for a plane polarized wave, when the field components are in opposite phases.

By putting $E_1 = E_2 = E_0$ and $\delta = \pm\pi/2$, a circularly polarized wave is obtained as a particular case of the elliptically polarized wave. Similarly, a plane polarized wave is obtained when either of the amplitudes of the components is put equal to zero or $\delta = 0$ or $\delta = \pi$, in (5.24.2).

The characteristic ratio E_y/E_z determines the polarization of the wave; for instance, in a magneto ionic non-isotropic medium the ratios $E_y/E_z, E_x/E_z.H_y/H_z$ give the wave polarization, corresponding to the electric and magnetic fields E and H which change in direction and magnitude as the wave travels in a given direction. Studies of anisotropic media require a representative wave, known as *the characteristic wave* whose polarization is not changed; it should satisfy the condition $P_y/E_y = P_z/E_z$, where P_y, P_z are the components of volume polarization. The polarization of the wave field is given by

$$R = \frac{E_y}{E_z} = -\frac{H_z}{H_y}. \tag{5.24.7}$$

It shows that both the waves E and H are elliptically polarized if the ratios are complex. When R is a real quantity, both magnetic and electric wave fields are linearly polarized.

5.25 Group Path and Phase Path

The group travel path measures the variations in the signal amplitude propagated at the group velocity, which is given by $v_g = \partial w/\partial k$. In a highly dispersive medium the phase velocity changes rapidly with the frequency, but the group velocity is some times quite different for the different Fourier components of the composite wave, so that the wave becomes distorted after travelling a large distance. In such cases the concept of group velocity becomes quite useless.

If dt_g be the time of travel at the group velocity v_g, of a dispersive wave, then the distance element ds travelled is given by

$$ds = v_g dtg.$$

The time t_g required for the pulse to travel over a given finite path is the path integral in

$$t_g = \int_{\text{path}} \frac{ds}{v_g}.$$

This time t_g, called the *group travel time*, can be measured. The group travel path P' is defined by

$$P' = ctg = c\int_{\text{path}} \frac{ds}{v_g} = \int_{\text{path}} n_g ds,$$

where the group refractive index

$$n_g = \frac{c}{v_g} = c\frac{dk}{dw}.$$

Since

$$\frac{1}{v_g} = \frac{\partial k}{\partial w} = \frac{\partial}{\partial w}\left(\frac{w}{v_p}\right) = \frac{1}{v_p} + w\frac{\partial}{\partial w}\left(\frac{1}{v_g}\right),$$

where the phase velocity $v_p = w/k$, we find that

$$n_g = n_p + w\frac{\partial v_p}{\partial w},$$

where the phase refractive index $n_p = kc/w$. Then

$$P' = P + w\frac{\partial P}{\partial w}.$$

where the phase travel path P is defined by

$$P = c\int_{\text{path}} \frac{ds}{v_p} = \int_{\text{path}} n_p ds.$$

Similarly it follows that

$$t_g = t_p + w\frac{\partial t_p}{\partial w},$$

where the phase travel time

$$t_p = \int_{\text{path}} \frac{ds}{v_p}.$$

5.26 Magnetic Field, Rotation and Vorticity

Rotation (of particles and continuous bodies), and its consequence, vorticity, are somewhat analogus to magnetism. Their study would clarify questions on the "how" and "why" of different aspects of rotation and magnetization and their combined action, which is evident from Ferraro's law of isorotation and from the dependence of magnetic moment on the wave induced rotational displacement. Curl of the MHD Navier-Stoke's equation

$$\frac{\delta \mathbf{v}}{\delta t} + (\mathbf{v}\cdot\nabla)\mathbf{v} = -\nabla\left(\frac{P}{\rho}\right) + \nu\nabla^2\mathbf{v} \tag{5.26.1}$$

gives, for evolution of the vorticity $\boldsymbol{\omega}$, the equation

$$\frac{\delta\boldsymbol{\omega}}{\delta t} = \nabla\times(\mathbf{v}\times\boldsymbol{\omega}) + \nu\nabla^2\boldsymbol{\omega}, \tag{5.26.2}$$

where ν is the fluid viscosity. The MHD equation for evolution of magnetic field in a fluid of electrical conductivity σ is

$$\frac{\delta H}{\delta t} = \nabla\times(\mathbf{v}\times H) + \eta\nabla^2 H, \tag{5.26.3}$$

where $\eta = c^2/4\pi\sigma$. Comparing we can write $\boldsymbol{\omega} = \alpha\mathbf{H}$, where α is constant. The vorticity also satisfies the relations $\nabla\times\mathbf{v} = 2\boldsymbol{\omega}$ and $\mathbf{v} = (\Omega\times\mathbf{r})$, where Ω is the angular velocity.

Analogy exists between (5.26.2) and (5.26.3) though there is the difference that vorticity $\boldsymbol{\omega}$ obtains from the velocity $\mathbf{v}$ of the Navier-Stoke's equation (5.26.1), and for momentum transfer H in (5.26.3) does not follow them. However, $\mathbf{H}$ is related to the electric current $\mathbf{j}$, as $\mathbf{v}$ to vorticity. Relations (5.26.2) and (5.26.3) show that the local rate of change of B and $\boldsymbol{\omega}$ are due to the effects of conduction (the term $\nabla\times(v\times H)$ in (5.26.2) and $\nabla\times(\mathbf{v}\times\boldsymbol{\omega})$ in (5.26.2) and diffusion (the terms $\nabla^2 H$ and $\nu\nabla^2\boldsymbol{\omega}$). When diffusion of the magnetic field dominates over convection, (5.26.3) is replaced by

$$\frac{\delta H}{\delta t} = \eta\nabla^2 H. \tag{5.26.4}$$

For $\boldsymbol{\omega}$ the corresponding relation is

$$\frac{\delta\boldsymbol{\omega}}{\delta t} = \nu\nabla^2\boldsymbol{\omega}. \tag{5.26.5}$$

It is valid in a configuration in which no quantity varies along a stream line, Shercliff (1965, p. 33). So, this equation obtains not by putting $v = 0$, but by making convection irrelevant. Then the equations for H and $\boldsymbol{\omega}$ are like the heat conduction equation

$$\frac{\delta T}{\delta t} = \zeta\nabla^2 T, \tag{5.26.6}$$

where T is the temperature and ζ is the thermal diffusivity, except the fact that T is a scalar, whereas H and ω are vectors. The equations (5.26.4)–(5.26.6) show that B, T, ω cannot change their distribution in a medium instantaneously but must diffuse at a specific rate.

The steady flow of electrically conducting fluids, involving the magnetic field and electric current, is similar to the steady vortex motion in incompressible, non-viscous fluids, Shafranov (1958), Yoler (1961), Shercliff (1965, section 3.2), Moffat (1978, section 3.2). In this analogy the magnetostatic variables are related by the Biot-Savart law like the fluid variables of Newton's laws of motion. So, we can write

Equations for the kinetics of vortex motion	Magnetohydrostatic equations
$(\nabla\cdot v) = 0$	$(\nabla\cdot B) = 0$
$(\nabla \times v) = 2\omega$	$(\nabla \times B) = \frac{4\pi}{c} j$
$(\omega \times v) = \nabla\left(\frac{P}{\rho} + \frac{1}{2}v^2\right)$	$\frac{1}{c}(j \times B) = \nabla p$

Thus, a vector sheet is analogus to a current sheet, and a current ring, in a perfect conductor, to a vortex ring, and the fluid vorticity corresponds to the magnetic field. But, conclusions on stability of one system cannot be transferred to its analogue in the other system, because, for stability study, one must consider flows in presence of perturbations which make the flow unsteady. Moreover, since $\omega = \nabla \times \mathbf{v}$, the equation for the evolution of ω is non-linear, whereas the equation for H is linear when the velocity $\mathbf{v}$ is given. So, ω is restricted through the relation $\omega = \nabla \times \mathbf{v}$ and, therefore, the results of the equation for B are more general than those for ω. Evidently, therefore, magnetic moment, mechanical moment (angular momentum), rotation and vorticity have many interesting common features.

5.27 Bateman Vectors

The four Maxwell equations, described with the help of the field vectors E, D, B and H and the sources ρ^c (charge density) and j (electric current vector), are not symmetrical in any sense. But these can be made symmetric by introducing a fictitious magnetic monopole density and the current due to these pole elements, Jackson (1975, section 6.12). To show this it is necessary to use a suitable complex combination electric and magnetic field vectors which is known as the *Bateman vector,* Silverstein (1907), Bateman (1915), Bremmer (1956), Stratton (1941), section 8.4). The Bateman vector can be introduced in the macroscopic theory only in linear isotropic dielectrics without any difficulty. In the Lorentz theory it is seems to be possible to introduce it without restrictions.

Symmetry principles are useful and important in electrodynamics, but not in classical mechanics, because the mathematical frame work of this mechanics was not properly defined for a long time. The quantitative dynamics, as developed by Lagrange, Hamilton, Jacobi and others, during the nineteenth century, was not sufficient to solve many problems of celestial mechanics. Poincare pointed out the necessity of introducing new methods and concepts, giving rise to a qualitative dynamics which makes use of differential forms (Chapter 9) on manifolds. Theoretical physicists make use of developments in mathematics which include new search for symmetry principles to obtain useful results of classical mechanics.

It will now be assumed that each particle has a charge q and a magnetic pole of intensity q_m; the densities of average charge and the magnetic polarity are ρ and ρ_M respectively, the average velocity of such a particle is u.

5.27.1 Bateman vectors in macroscopic theory

In the phenomenological theory of Maxwell for electrodynamics we introduce the fictitious magnetic pole density ρ_M and the associated current density j_M and obtain

$$\nabla \times H = \frac{1}{c}\frac{\partial D}{\partial t} + \frac{4\pi}{c} j, \tag{5.27.1}$$

$$\nabla \times E = -\frac{1}{c}\frac{\partial H}{\partial t} + \frac{4\pi}{c} j_M, \tag{5.27.2}$$

$$\nabla \cdot D = 4\pi\rho^c, \quad \text{or,} \quad \frac{\partial \rho^c}{\partial t} + (\nabla \cdot j) = 0, \tag{5.27.3}$$

$$\nabla \cdot B = 4\pi\rho_M, \quad \text{or,} \quad \frac{\partial \rho_M}{\partial t} + (\nabla \cdot j_M) = 0. \tag{5.27.4}$$

We next write $D = \varepsilon E$ and $B = \mu H$ and introduce the complex combinations

$$M = \frac{\sqrt{\mu}H + I\sqrt{\varepsilon}E}{(\varepsilon\mu)^{1/4}}, \quad J = \frac{\sqrt{\mu}j + i\sqrt{\varepsilon}j_M}{(\varepsilon\mu)^{1/4}}, \quad \sigma = \frac{\sqrt{\mu}\rho^c + i\sqrt{\varepsilon}\rho_M}{(\varepsilon\mu)^{1/4}}, \tag{5.27.5}$$

M is called the *Bateman vector;* J and σ are the associated current and charge density, respectively. Since ε and μ are constants, M is linearly related to E and H, J to j and j_M and σ to ρ^c and ρ_M. The Maxwell equations (5.27.1)–(5.27.4) then reduce to the two equations :

$$\nabla \times M = -\frac{i\sqrt{\varepsilon\mu}}{c}\frac{\partial M}{\partial t} + \frac{4\pi}{c}\bar{J}, \tag{5.27.6}$$

$$\nabla \times \bar{M} = \frac{i\sqrt{\varepsilon\mu}}{c}\frac{\partial \bar{M}}{\partial t} + \frac{4J}{c}J \tag{5.27.7}$$

and the divergence equations, consistent with these, are

$$(\nabla \cdot M) = \frac{4\pi i\sigma}{(\varepsilon\mu)^{1/2}}, \quad \frac{\partial \sigma}{\partial t} + \nabla \cdot J = 0, \tag{5.27.8}$$

where $\bar{M}$ and $\bar{J}$ are the cc of M and J respectively. The Lorentz force density, in terms of M, etc. is

$$F = (\varepsilon\mu)^{1/2}\left[-\frac{M}{4\pi}(\nabla \cdot M) + \frac{1}{c}(J \times M)\right]. \tag{5.27.9}$$

Its real part is $(F + \bar{F})/2$, where $\bar{F}$ is the cc.

The vector and scalar potentials, for the Lorentz gauge, are given by

$$B = \nabla \times A + \frac{\mu}{c}\frac{\partial A_M}{\partial t} + \mu\nabla\phi_M, \tag{5.27.10}$$

$$E = \frac{1}{c}\nabla \times A_M - \frac{1}{c}\frac{\partial A}{\partial t} + \nabla\phi, \tag{5.27.11}$$

$$\nabla \cdot A - \frac{\varepsilon\mu}{c}\frac{\partial\phi}{\partial t} = 0, \quad \nabla \cdot A_M + \frac{\varepsilon\mu}{c}\frac{\partial\phi_M}{\partial t} = 0, \tag{5.27.12}$$

and the wave equations for A_M and ϕ_M are

$$\nabla^2 A_M - \frac{1}{c_1^2}\frac{\partial^2 A_M}{\partial t^2} = \frac{4\pi}{c}\varepsilon J_M, \tag{5.27.13}$$

$$\nabla^2 \phi_M - \frac{1}{c_1^2}\frac{\partial^2 \phi_M}{\partial t^2} = \frac{4\pi}{\mu}\rho_M. \tag{5.27.14}$$

The complex Bateman vector and scalar potentials **G** and ψ of the Lorentz gauge, are defined by

$$G = \frac{1}{(\varepsilon\mu)^{1/4}}\left[\frac{1}{\sqrt{\mu}}A + \frac{i}{\sqrt{\mu}}A_M\right], \tag{5.27.15}$$

$$\psi = \frac{1}{(\varepsilon\mu)^{1/4}}\left[\sqrt{\mu}\phi_M + i\sqrt{\varepsilon}\phi\right], \tag{5.27.16}$$

$$M = \nabla \times G - \frac{i}{c}\sqrt{\varepsilon\mu}\frac{\partial G}{\partial t} + \nabla\psi, \tag{5.27.17}$$

$$\nabla \cdot G + \frac{1}{c}(\varepsilon\mu)^{1/2}\frac{\partial\psi}{\partial t} = 0, \tag{5.27.18}$$

$$\nabla^2 G - \frac{1}{c_1^2}\frac{\partial^2 G}{\partial t^2} = -\frac{4\pi}{c}\bar{J}, \tag{5.27.19}$$

$$\nabla^2 \psi - \frac{1}{c_1^2}\frac{\partial^2 \psi}{\partial t^2} = \frac{4\pi i\bar{\sigma}}{(\varepsilon\mu)^{1/2}}. \tag{5.27.20}$$

The transformations, Jackson (1975, section 6.12)

$$E = E'\cos\xi + H'\sin\xi, \quad D = D'\cos\xi + B'\sin\xi, \tag{5.27.21}$$

$$H = -E'\sin\xi + H'\cos\xi, \quad B = -D'\sin\xi + B'\cos\xi, \tag{5.27.22}$$

$$\rho = \rho'\cos\xi + \rho'_M\sin\xi, \quad J = J'\cos\xi + J'_M\sin\xi, \tag{5.27.23}$$

$$\rho_M = -\rho'\sin\xi + \rho'_M\cos\xi, \quad J_M = -J'\sin\xi + J'_M\cos\xi, \tag{5.27.24}$$

show that, for a real single value of ξ, the equations for the primed field variables are the same as the equations for the unprimed field variables. Thus equations (5.27.1)–(5.27.4), the quadratic forms $(E \times H)$, $(E.D) + (B.H)$ and the Maxwell stress elements T_{ij} of section 4.4, remain invariant under the transformations (5.27.21)–(5.27.24). Equations (5.27.2) and (5.27.4) show that ρ_M is a pseudoscalar density, odd under time reversal and j_M is a pseudovector density, even under time reversal; the symmetries of ρ_M under both spatial inversion and time reversal are opposite to those of ρ^c. Therefore, for a particle with both electric charges and magnetic monopoles, space inversion and time reversal are no longer valid symmetries of the laws of physics.

The Poynting theorem for energy flux reads

$$\nabla \cdot \left\{\frac{c}{4\pi}(M \times \bar{M})\right\} = -\frac{ic}{4\pi c_1}\frac{\partial}{\partial t}(M, \bar{M}) + (\bar{M} \cdot j) - (M \cdot j). \tag{5.27.25}$$

For finding the Maxwell stress tensor elements T_{ij} and the EM momentum density g per unit volume, we write

$$F = \hat{X}_i \left(\frac{\partial T_{ij}}{\partial x_j} - \frac{\partial g_i}{\partial t} \right), \tag{5.27.26}$$

where summation is implied for a repeated index and $\hat{\mathbf{X}}_i$ is the unit vector along the ith space direction. Then the computation of the terms of F, obtained from (5.27.9), shows that

$$\left[(\nabla \times \bar{M}) \times M + (\nabla \times M) \times \bar{M}\right]_x = \frac{\partial}{\partial x}\left[2M_x\bar{M}_x - (\bar{M}\cdot M)\right] + \frac{\partial}{\partial y}(M_x\bar{M}_y + \bar{M}_x M_y)$$
$$+\frac{\partial}{\partial z}(M_x\bar{M}_z + \bar{M}_x M_z) - M_x(\nabla\cdot\bar{M}) - \bar{M}_x(\nabla\cdot M). \tag{5.27.27}$$

Hence we find that

$$T_{ij} = \frac{(\varepsilon\mu)^{1/2}}{8\pi}\left[M_i\bar{M}_j + \bar{M}_i\bar{M}_j - (M\cdot\bar{M})\delta_{ij}\right], \tag{5.27.28}$$

$$g = -\frac{i(\varepsilon\mu)^{3/4}}{8\pi c}(M \times \bar{M}). \tag{5.27.29}$$

Then using (5.27.5) F simplifies to

$$F = \rho^c F + \frac{1}{c}(j \times B) - \rho_M H - \frac{1}{c}(j_M \times D). \tag{5.27.30}$$

Putting now

$$j = \rho^c v \quad \text{and} \quad j_M = \rho_M u. \tag{5.27.31}$$

We obtain

$$F = \rho^c E + \frac{\rho^c}{c}(u \times B) - \rho_M H - \frac{1}{c}\rho_M(u \times D). \tag{5.27.32}$$

5.27.2 Bateman vectors in Lorentz theory

Putting $\varepsilon = 1$, $\mu = 1$, $B = H$, $D = E$, the symmetrical Maxwell-Lorentz equations obtain in the form

$$\nabla \times H = \frac{1}{c}\frac{\partial E}{\partial t} + \frac{4\pi}{c}j, \tag{5.27.33}$$

$$\nabla \times E = -\frac{1}{c}\frac{\partial H}{\partial t} - \frac{4\pi}{c}j_M, \tag{5.27.34}$$

$$\nabla\cdot E = 4\pi\rho^c, \quad \text{or,} \quad \frac{\partial \rho^c}{\partial t} + \nabla\cdot j = 0, \tag{5.27.35}$$

$$\nabla\cdot H = 4\pi\rho_M, \quad \text{or,} \quad \frac{\partial \rho_M}{\partial t} + \nabla\cdot j_M = 0. \tag{5.27.36}$$

So, M, J and σ are now defined as

$$M = H + iE, \quad J = j + ij_M, \quad \sigma = \rho + i\rho_M \tag{5.27.37}$$

and (5.27.25)–(5.27.28) reduces to

$$\nabla \times M = -\frac{i}{c}\frac{\partial M}{\partial t} + \frac{4\pi}{c}J, \quad \nabla \times \bar{M} = \frac{i}{c}\frac{\partial \bar{M}}{\partial t} + \frac{4\pi}{c}J, \tag{5.27.38}$$

$$\nabla \cdot M = -4\pi i \bar{\sigma}, \quad \nabla \cdot \bar{M} = -4\pi i \sigma. \tag{5.27.39}$$

▶ **Example 5.27.1** *Find the field tensor $F_{\mu\nu}$ and the Poynting theorem in terms of the Bateman vector and the related potentials of section* 5.27.1.

▶ **Example 5.27.2** *Extend the symmetric Lorentz theory of electrodynamics further, and obtain the momentum transfer relations with the help of the force term of* (5.27.9). *For the Lorentz model find quantities like the radiation damping force and powerloss due to acceleration of the particles. Obtain the related system of the Boltzmann-Vlasov equations as far as possible in the six dimensional phase space of position co-ordinates and velocity.*

Chapter 6

Variational Aspects of Electrodynamics

6.1 Introduction

Variational aspects of classical electrodynamics are helpful for a proper understanding of electrodynamics in non-inertial frames. Variational techniques involving Hamilton's principle of least action, Lagrangian, Hamiltonian and other canonical, formalisms, are suitable for formulation of the basic equations of electrodynamics and mechanics. The role of forces of the Newtonian formulation of mechanics is replaced by potentials, or by the expressions of a Lagrangian or a Hamiltonian. There is actually no rule for finding the Lagrangian density L which generates a given set of field equations and only certain structural features can be assumed in advance. The invariance of L under the same invariance group, as that for the field equations, is useful to find L itself, which is determined only by guess work. "A theory" writes Rohrlick (1965, Section 4.9), "necessarily involves setting down a set of basic equations. Hamilton's principle and the associated Lagrangian are an after-thought, a poste facto construct, by which these equations can be reduced to variational principle; physically, it adds nothing that is not contained in the basic equations. But the Lagrangian is of great convenience in the study of symmetry properties and conservation laws and in suggesting possible generalizations of the theory at hand".

These formalisms do not require the use of concepts of force and differential dynamical properties of the dynamics in the small, or the local point dynamics. These formalisms rather depend on the integral principle for the whole of motion, which is the dynamics in the large or the domain dynamics or the dynamics when after integration the boundary conditions are imposed on the field solutions.

The Lagrangians which differ by only a gauge transformation of the potentials are equivalent. This property follows from invariance, which states that the Lagrangians should have the same form when a 4-divergence or a total time derivative is added to it. Thus the field potentials of the Lorentz gauge, A and ϕ, are introduced which couple the densities of charge current and net charge, respectively, but which are not gauge invariant. The corresponding Lagrangian depends on the potentials. A unified field theory may be obtained through a gauge theory, Kobe (1978). Only those fundamental interactions are meaningful that arise by making the

Lagrangian form invariant under local gauge transformation. Fundamental interactions are studied by making the Lagrangian form invariant by the introduction of a compensating term of the 4-divergence type, as a local gauge transformation. The Maxwell equations follow from the principle of the maintenance of symmetry of the Lagrangian under such a transformation, though these equations have originally developed as a synthesis of experimental laws. This type of transformation of the EM Lagrangian depicts the connection between mechanics and EM theory. Derivative of the potential field generate the physical fields through elimination of the arbitrary gauge function.

Solution of problems of motion is often conveniently obtained with the help of the Lagrangian and Hamiltonian of motion. Also, these alternative approaches sometimes lend new insight into the physical problems. But these add nothing new to electrodynamics. The Lagrangian is useful in studying symmetry properties and conservation laws, and in suggesting possible generalizations of the existing theory. These topics are not discussed in text books on electrodynamics, though these exist in other kinds of literature of physical sciences in a scattered manner.

The Lagrangian function and the canonical formalisms were first developed for a system of discrete particles, and then modified for a continuous distribution of fields in a medium for study of various types of fields of modern physics. The classical theory of field is essentially a generalization of classical mechanics of point particles, having, at any time, a well-defined mass and a velocity. Microscopic (discrete) particle view points can lead to the limit of a continuum for finding a Lagrangian for long wave length phenomena. The continuum limit introduces the material co-ordinate systems for transformation from the position co-ordinates of the charged particles to the center-of-mass co-ordinate system in problems. The field functions and their time derivatives, thus introduced, at each 4-point, play the roles of the familiar, canonically conjugate, generalized co-ordinates and momentums of the Hamiltonian mechanics. The entire Hamiltonian-Lagrangian theory of point mechanics for a system of very numerous discrete particles, can be thus taken over in the field theory only if the primary quantities with which we operate are now termed densities, e.g., momentum density, energy density, the Lagrangian density, the Hamiltonian density. Since the physical quantities of interest are obtained from the corresponding densities by integration over position space of the continuum, the ordinary derivative is replaced by its natural generalization which is the substantial, or the variational derivative.

A closed system, in interaction with EM fields, obeys the conditions of conservation of energy, momentum and angular momentum. For these types of interactions sets of internal co-ordinates are useful in description of various internal motions of the systems, and in the study of the constitutive relations, Nelson (1979). There are analogues, such as the kinetic energy is analogous to the magnetic energy, and the potential energy to the electrical energy, which establish proper relationship between the discrete and the continuous systems. In the EM theory it is necessary to discard the discrete picture.

The auxiliary fields of the potentials A and ϕ introduce compensating fields to preserve the form of the Lagrangian of a classical particle under the addition of a total time derivative. The gauge principle or the principle of form invariance under gauge transformation is that the Lagrangian retains the same form when a total time derivative is added to it.

The symmetry principles guide the discovery of new dynamical equations. The Maxwell equations can be deduced on the basis of the maintenance of the symmetry of the Lagrangian, Kobe (1980). A compensating potential field ensures the form invariance of the Lagrangian under a transformation. This potential field generates the physical field by differentiation in such a way that the arbitrary gauge function is eliminated. The equations of motion for the field are obtained by constructing the simplest non-trivial Lagrangian, and using the principle of least action. For conservative systems of mass points with holonomic constraints, the Lagrangian is the difference between kinetic energy and potential energy.

The action-at-a-distance basis, or the point particle approach, is marked by a well defined mass and velocity at any time. The two concepts, fields and particles, have become the instrument of description of various aspects of physical theory, and particularly of electrodynamics related physics. Variational methods allow functional differentials of integrals of Lagrangian density which generate differential equations for study of dynamics of a process.

Gauge transformation and form invariance of the EM Lagrangian, in the variational methods, determine the field equations of electrodynamics. These methods and conservation methods and the concepts of gauge invariance and form invariance of some entities, and the consequences of the non-uniqueness of the Lagrangian density, are useful for study of classical electrodynamics. In the formalism for EM Lagrangian these concepts are complementary, i.e., are applied simultaneously.

EM field configurations, and their changes, are represented by the field potentials and the displacement vector (the polarization vector), for displacement of the charge assembly due to fields. The Lagrangian functions for fields, particles and their interaction, are developed in terms of these field-like dynamical variables. The principle of least action gives Newton's second law, including the expression for the Lorentz force. The electric and magnetic fields are obtained in terms of the potentials with the help of the Lorentz force. The kinematical Maxwell equations (section 1.10) then obtain from the electric and magnetic fields in terms of the field potentials. The dynamical Maxwell equation (section 1.10) obtain from application of the principle of least action on the simplest gauge invariant scalar Lagrangian for the field.

Vector methods, and pseudo-cartesian tensor methods are used to find the EM Lagrangian. Hamilton's principle obtains the Euler-Lagrange (EL) equations, which determine the basic equations of electrodynamics in continuous media, and the relativistically covariant form of the energy-momentum tensor of the field particle assembly, in pseudo-Euclidean 4-space. The property, that the Lagrangians which differ by a gauge transformation of the field potentials A and ϕ are equivalent, leads to the energy conservation theorem and to a derivation of the field equations. The expressions for the EM Lagrangian and Hamiltonian density functions obtain in terms of the EM field energy, the kinematic and potential energies of material motion and the effects of interaction of fields and matter.

6.2 Lagrangian and Hamiltonian Formalisms for a Systems of Particles

6.2.1 Lagrangian formalism

The Lagrangian L, for small velocities, is the difference between the kinetic energy T, and the potential energy V,

$$L = T - V, \tag{6.2.1}$$

both for a single particle and for a system of N discrete particles. For conservative systems of mass points, in presence of integrable displacements for constraints[1] (also called *holonomic*

[1] Equations of constraints are holonomic when by integration these are made free from differentials. These equations are non-holonomic (or anholonomic) when integration is not possible to make them completely free from differentials.

constraints), and for non-relativistic motion, the Lagrangian has this form. In presence of EM fields the distinction between these two categories of energy is lost even for holonomic constraints. In the co-ordinate system with axes OX, OY, OZ, the position vector of the ith particle is denoted by $r_i (= x_i, y_i, z_i)$ with $i = 1, 2, 3, \cdots, N$. The constraints are represented by l independent equations of the form

$$f_j(r_1, r_2, \cdots, r_N, t) = 0, \quad j = 1, 2, \cdots, l. \tag{6.2.2}$$

There are then the $3N - L(= K \text{ say})$ remaining degrees of freedom. Hence, for the complete description of the configuration of the system, we introduce K independent variable, which are denoted by $q_1, q_2, \cdots, q_k$. These are called the *generalized position co-ordinates.* The corresponding K generalized velocities are defined by

$$\dot{q}_i = \frac{dq_i}{dt}. \tag{6.2.3}$$

In presence of arbitrary motions these are independent of the position co-ordinates. So, for all admissible values of i and j, including these for $i = j$, we have

$$\frac{\partial \dot{q}_i}{\partial q_j} = 0, \quad \frac{\partial q_i}{\partial \dot{q}_j} = 0. \tag{6.2.4}$$

The position co-ordinates $\mathbf{r}_j$ for the system are then written as

$$r_j = r_j(q_1, q_2, \cdots, q_k), \quad j = 1, 2, \cdots, N. \tag{6.2.5}$$

Let now

$$r_j^i = \phi_{ij}(q_1, q_2, \cdots, q_k), \tag{6.2.6}$$

where r_j^i is the ith components of the position vector $\mathbf{r}_j$ of the jth particles. The time derivative of r_j^i is

$$U_j^i = \dot{r}_j^i = \left(\frac{\partial}{\partial q_l}\phi_{ij}\right)\dot{q}_l, \tag{6.2.7}$$

where the Einstein summation convention is implied for repeated index.

Since U_j is the velocity of the particles of the jth species,

$$U_j^2 = \dot{x}_j^2 + \dot{y}_j^2 + \dot{z}_j^2, \tag{6.2.8}$$

and the kinetic energy of the system is

$$T_0 = \frac{1}{2}(m_j U_j^2). \tag{6.2.9}$$

For a conservative field with holonomic constraints equation (6.2.1) becomes

$$L = \frac{1}{2}\{p_{ij}(q)\dot{q}_i\dot{q}_j\} - V(q), \tag{6.2.10}$$

where p_{ij} are some functions of the K generalized co-ordinates q_i and the kinetic energy is quadratic in the generalized velocities.

For the ith species particle, if $F_i^{(a)}$ is the applied force and $F_i^{(c)}$ is the force of the constraints, the equation of motion, given by the laws of motion of Newton, is

$$\frac{dp_i}{dt} = F_i^{(a)} + F_i^{(c)}. \tag{6.2.11}$$

Then, D'Alembert's principle, Goldstein ((1980), Section 1.4) or Moiseiwitsch (1966, section 1.3), gives the relation

$$\left[(\dot{p}_i - F_i^{(a)})\cdot \delta r_i\right] = 0. \tag{6.2.12}$$

Employing generalized co-ordinates, $q_i s$, and dropping the superscript from $F_i^{(a)}$, we have write

$$(F_i\cdot \delta r_i) = (q_j \delta q_j) \tag{6.2.13}$$

with

$$Q_j = \left(F_i\cdot \frac{\partial r_i}{\partial q_j}\right). \tag{6.2.14}$$

Here Q_j is the jth component of the generalized forces. Moreover, (6.2.12) can be transformed, in terms of the q_is and $\dot{q}_i$s as

$$(\dot{p}_i\cdot \delta r_i) = \left[\frac{d}{dt}\left(\frac{\partial T_0}{\partial q_i}\right) - \frac{\partial T_0}{\partial q_i}\right]\delta q_j, \tag{6.2.15}$$

where T_0 is given by (6.2.9). The Lagrangian equation of motion

$$\frac{d}{dt}\left(\frac{\partial T_0}{\partial \dot{q}_j}\right) - \frac{\partial T_0}{\partial q_j} = Q_j, \tag{6.2.16}$$

follows from (6.2.12)–(6.2.15).

When the forces represented by F_i are derived from a scalar potential V, which depends on the $3N$ position co-ordinates only, as in the case of quasi-static problems, we write

$$F_i = -\nabla_i V. \tag{6.2.17}$$

Then equation (6.2.14) becomes

$$Q_j = -\left[\left(\nabla_i V\cdot \frac{\partial r_i}{\partial q_j}\right)\right] = -\frac{\partial V}{\partial q_j} \tag{6.2.18}$$

and equation (6.2.16) reads

$$\frac{d}{dt}\left(\frac{\partial L}{\partial \dot{q}_j}\right) - \frac{\partial L}{\partial q_j} = 0, \tag{6.2.19}$$

where L is given by (6.2.1). Here the partial derivative of L with respect to q_i means a derivative taken regarding all other qs and all $\dot{q}s$ as constants. Similarly, in the partial derivative of L with respect to $\dot{q}_i$, all the other $\dot{q}$'s and the qs are regarded as constants. These assumptions ensure that the qs and $\dot{q}s$ are independent variables.

In a closed system, consisting of N weakly interacting particles, the N mass points interact only among themselves and not with those of other systems. So, evidently,

$$L = \left(\frac{1}{2}m_i U_i^2\right) - V(r_1, r_2, \cdots, r_N, t), \tag{6.2.20}$$

together with

$$\frac{\partial V}{\partial \dot{q}_j} = 0. \tag{6.2.21}$$

However, if F_i^0 represents the non-conservative part, due to friction, or EM effects involving circulation (or rotation) or both, acting simultaneously, or due to other causes, we write

$$F_i = -\nabla_i V + F_i^0, \tag{6.2.22}$$

along with

$$Q_j^0 = \left(F_i^0 \cdot \frac{\partial r_i}{\partial q_j}\right), \tag{6.2.23}$$

and obtain the Lagrangian equation in the form

$$\frac{d}{dt}\left(\frac{\partial L}{\partial \dot{q}_j}\right) - \frac{\partial L}{\partial q_j} = Q_j^0. \tag{6.2.24}$$

In some cases of velocity dependent non-conservative forces, the generalized force q_j^0 is expressed as

$$Q_j^0 = \frac{d}{dt}\left(\frac{\partial \phi}{\partial \dot{q}_j}\right) - \frac{\partial \phi}{\partial q_j}, \tag{6.2.25}$$

where ϕ is a function of both q_is and q_is. Equation (6.2.25) can be put in the form (6.2.19). Then (6.2.24) reads

$$\frac{d}{dt}\left(\frac{\partial L_T}{\partial \dot{q}_i}\right) - \frac{\partial L}{\partial q_i} = 0 \tag{6.2.26}$$

with

$$L_T = L - \phi. \tag{6.2.27}$$

The Lorentz force, being non-conservative, because it is velocity dependent, the force potential of (6.2.25) can be considered for its Lagrangian formalism. When L is a function of the q_is, $\dot{q}_i$s and time, we have

$$\delta L = \left(\frac{\partial L}{\partial q_i}\delta q_i\right) + \left(\frac{\partial L}{\partial \dot{q}_i}\delta \dot{q}_i\right) + \frac{\partial L}{\partial t}\delta t. \tag{6.2.28}$$

The last term in this relation exists so long as L depends explicitly on time. In the non-relativistic Lagrangian formulation for a system with n degrees of freedom, the EL equation of motion, numbering n, are (6.2.19) or (6.2.26). All these n differential equations being of the second order, the motion of the system is determined for all time only when $2n$ initial values are specified. For example, at a particular time t_0, these are the q_is, numbering n, and the $\dot{q}_i$s, also numbering n. This is the representation by a point in an n-dimensional configuration space, of the state of the system. The co-ordinates in this space are the n generalized co-ordinates, q_i. These co-ordinates follow the motion of the system point in time as it traverses its path in this space. Physically, the problem of a system with n degrees of freedom, in the Lagrangian view point, is a problem in n independent variables $q_i(t)$, and $\dot{q}_i$s appear only merely for the time derivative of q_i.

6.2.2 The Hamiltonian and Hamilton's equations of motion

The Hamiltonian $\mathcal{H}$ of motion is defined as

$$\mathcal{H} = (p_i\dot{q}_i) - L = \left(\dot{q}_i\frac{\partial L}{\partial \dot{q}_i}\right) - L, \tag{6.2.29}$$

where the canonical (or the generalized) momentum, conjugate to the generalized co-ordinate q_i is

$$p_i = \frac{\partial L}{\partial \dot{q}_i}. \tag{6.2.30}$$

The Einstein convention of summation for repeated subscripts is adopted here. Essentially, the Hamiltonian is the energy of a system expressed in terms of momentum. The equation (6.2.19) reads

$$\dot{p}_i = \frac{\partial L}{\partial p_i}. \tag{6.2.31}$$

For a single particle,

$$\mathcal{H} = (p \cdot U) - L, \tag{6.2.32}$$

where the Lagrangian L is that for a single particle. The relation (6.2.29) shows that equation (6.2.19) can be replaced by the pairs of equations

$$\dot{q}_i = \frac{\partial \mathcal{H}}{\partial q_i}, \tag{6.2.33}$$

$$\dot{p}_i = -\frac{\partial \mathcal{H}}{\partial q_i}. \tag{6.2.34}$$

These are called *Hamilton's equations* (or the *canonical equations*). They form a set of $2n$ first order partial differential equations, replacing the Lagrangian equation (6.2.19) and are expressed in terms of the q_is and the p_is, whereas the Lagrangian equation (6.2.19), expressed in terms of the q_is and $\dot{q}_i$s, whereas the Lagrangian equation (6.2.19), expressed in terms of the q_is and $\dot{q}_i$s, are a set of n second order partial differential equations.

When L does not depend on the time t explicitly, equation (6.2.28) with the help of (6.2.30) and (6.2.31), simplifies to

$$\delta L = (\dot{p}_i \delta q_i) + (p_i \delta \dot{q}_i). \tag{6.2.35}$$

Then, Moiseiwitseh (1966, section 1.10), together with equations (6.2.28) and (6.2.29), this relation yields

$$\delta\left[(p_i \dot{q}_i) - L\right] = (\dot{q}_i \delta p_i - \dot{p}_i \delta q_i). \tag{6.2.36}$$

This relation can be written as

$$\delta \mathcal{H} = (\dot{q}_i \delta p_i - \dot{p}_i \delta q_i). \tag{6.2.37}$$

When both the Hamiltonian and Lagrangian are explicit functions of the other parameters, which relate to the mechanical properties of the mechanical system itself, or to the external forces on it, following Landau and Lifshitz (1975, section 40), let λ be one such parameter; λ can be the time t as well. The differential of $\mathcal{H}$, now regarded as a function of q_i, p_i, λ, is

$$\delta \mathcal{H} = \left(\frac{\partial \mathcal{H}}{\partial q_i} \delta q_i + \frac{\partial \mathcal{H}}{\partial p_i} \delta p_i\right) + \frac{\partial \mathcal{H}}{\partial \lambda} \delta \lambda. \tag{6.2.38}$$

Relations (6.2.36) and (6.2.37), and the fact that $\mathcal{H}$ and L are functions of λ, as well, show that

$$\delta \mathcal{H} = \left(\dot{q}_i \delta p_i - \frac{\partial L}{\partial q_i} \delta q_i\right) - \frac{\partial L}{\partial \lambda} \delta \lambda. \tag{6.2.39}$$

Equating coefficients of $\delta q_i, \delta p_i$, and $\delta\lambda_i$, from both sides, in addition to (6.2.33) we obtain the relations

$$-\frac{\partial \mathcal{H}}{\partial q_i} = \frac{\partial L}{\partial q_i}, \tag{6.2.40}$$

$$\frac{\partial \mathcal{H}}{\partial \lambda} = -\frac{\partial L}{\partial \lambda}. \tag{6.2.41}$$

These, with the help of (6.2.24), give

$$-\frac{\partial \mathcal{H}}{\partial q_i} = \frac{d}{dt}\left(\frac{\partial L}{\partial \dot{q}_i}\right) - Q_i^0. \tag{6.2.42}$$

This relation can be written as

$$-\frac{\partial \mathcal{H}}{\partial q_i} = \dot{q}_i - Q_i^0. \tag{6.2.43}$$

Differentiating now $\mathcal{H}(p_i, q_i, t)$ totally with respect to t, replacing λ by t in (6.2.38), and using that relation we find that

$$\frac{d\mathcal{H}}{dt} = \left(\frac{\partial \mathcal{H}}{\partial q_i}\dot{q}_i + \frac{\partial \mathcal{H}}{\partial p_i}\dot{p}_i\right) + \frac{\partial \mathcal{H}}{\partial t} \tag{6.2.44}$$

$$\text{or} \quad \frac{d\mathcal{H}}{dt} = (\dot{q}_i Q_i^0) + \frac{\partial \mathcal{H}}{\partial t}. \tag{6.2.45}$$

Hence, energy changes at a rate determined by the non-conservative force Q_i^0, in addition to the effect of change due to the explicit dependence of $\mathcal{H}$ on time. If $Q_i^0 = 0$ and both $\mathcal{H}$ and L do not explicitly depend on time, then the Hamiltonian is a constant of motion. Equation (6.2.41) shows that, for $\lambda = t$,

$$\frac{\partial \mathcal{H}}{\partial t} = -\frac{\partial L}{\partial t}. \tag{6.2.46}$$

When the Lagrangian is of the form

$$L = L_0 + L', \tag{6.2.47}$$

where L' is a small correction on L_0, the corresponding additional term is $\mathcal{H}'$ in the Hamiltonian $\mathcal{H}(= \mathcal{H}_0 + \mathcal{H}')$. $\mathcal{H}'$ is given by

$$\frac{\partial \mathcal{H}'}{\partial \lambda} = -\frac{\partial L'}{\partial \lambda}, \tag{6.2.48}$$

as may be seen from (6.2.41). So, the small additions to $\mathcal{H}$ and L are equal in magnitude and opposite in sign.

In the Hamiltonian formulation of mechanics the motion of a system, with n degrees of freedom, is described in terms of $2n$ independent first order differential equations of motion. These are expressed in terms of $2n$ independent variables, and the number of initial conditions are again $2n$. The $2n$ independent variables of the system may be regarded as a point having $2n$ co-ordinates in a $2n$ dimensional phase space. Half of this set of variables is the n generalized co-ordinates q_i. The remaining half of these is the set of generalized or conjugate moments defined by (6.2.30). The set of $2n$ variables (q, p) is called the *canonical variables,* where q is a n dimensional vector, the component of which along the ith direction is q_i, and similarly, the component of p along the ith direction is p_i. The Hamiltonian formulation, following from the Lagrangian formulation, means changing the variables from the set $(q, \dot{q}, t)$ to the set (q, p, t) by a Legendre transformation which is discussed in section 6.5 following Goldstein (1980, p. 339).

6.3 The Principle of Least Action

Consider the time integral

$$S = \int_{t_1}^{t_2} L\{q_i(t), \dot{q}(t)\}dt, \tag{6.3.1}$$

where L is the Lagrangian of motion of a system of N particles (section 6.2.1). L is a function of the co-ordinates of the elements of this system and of the first order derivatives of these with respect to time t. This relation is also called the *action integral.* The action S is a scalar quantity formed out of the common quadrupole : mass m, charge q, length l and time t. The unit of action, dimensionally, is $[S] = [ml^2/t]$, which is for work done for motion through a distance l against a force. Both S and q are invariant because both occur in the quantized manner, and so are countable entities.

Hamilton's principle of least action is the variational principle of mechanics. It states that the motion of a mechanical system is such that in going from a configuration 'a' at time t_1 to a configuration 'b' at time t_2 the action integral has an extreme value. In other words, the action integral takes on an extreme value on the world line along which the ith particle of the system actually moves under the action of the forces. Therefore, small variations of the co-ordinates and velocities, away from the actual path, should be restricted by the condition

$$\delta S = \delta \int_{t_1}^{t_2} L\left\{\dot{q}_1(t), \quad q_i(t)\right\} dt = 0. \tag{6.3.2}$$

Since t_1 and t_2 are two given times,

$$\delta S = \int_{t_1}^{t_2} \left[L\left(q_i + \delta q_i, \dot{q}_i + \delta \dot{q}_i\right) - L\left(q_i, \dot{q}_i\right)\right] dt = 0. \tag{6.3.3}$$

For a free particle the principle of least action gives

$$\delta S = -m_0 c \delta \int_a^b \sqrt{-ds^2} = 0, \tag{6.3.4}$$

where dS is the 4-space subinterval. Then, using the Einstein summation convention we obtain

$$\delta S = -m_0 c \int_a^b \frac{dx_\mu \delta(dx^\mu)}{\sqrt{-dx^\mu dx_\mu}} = -m_0 \int_0^b U_\nu d(\delta x^\mu). \tag{6.3.5}$$

Integration by parts gives

$$\delta S = \Big[-m_0 U_\mu \delta x^\mu\Big]_a^b + m_0 \delta x^\mu \frac{dU_\mu}{dT} dT. \tag{6.3.6}$$

At the world point limits a and b

$$(\delta x^\mu)_a = 0, \quad (\delta x^\mu)_b = 0. \tag{6.3.7}$$

And the actual trajectory is determined by comparing different trajectories between these limits and from the condition (6.3.2). Hence, for the free particle, the equation of motion,

$$\frac{dU_\mu}{dT} = 0, \tag{6.3.8}$$

shows that the 4-velocity is constant. To determine the variation of S as a function of the co-ordinates, the point 'a' should be considered fixed and b as variable, so that

$$(\delta x^{\mu})_a = 0, \quad (\delta x^{\mu})_b \neq 0. \tag{6.3.9}$$

Using (6.3.8) and (6.3.9), equation (6.3.6) becomes

$$\delta S = m_0 \delta x^{\mu} U_{\mu}, \tag{6.3.10}$$

where δx^{μ} is written for $(\delta x^{\mu})_b$. Another form of it is

$$\delta S = -p^{\mu} \delta x_{\mu}, \tag{6.3.11}$$

where using (11.2.30) the contravarient 4-momentum reads

$$p^{\mu} = -\frac{\delta S}{\delta x_{\mu}}. \tag{6.3.12}$$

For a system of N free particles

$$\delta S = -\sum_N (p^{\mu} \delta x_{\mu}). \tag{6.3.13}$$

The assembly of particles is described by Lagrangian variables, and the fields, on the other hand, by Eulerian variables, respectively. Field equations for trajectories of particles moving in specified fields of force follow from the variational principles. Interaction of charged particles with EM fields also can be studied by the variational principles.

The action function S, for an assembly of charged particles interacting with an EM field, is written as the sum of three terms :

$$S = S_f + S_p + S_i, \tag{6.3.14}$$

where S_f is the action for the field, S_p is that for particles and S_i for the interaction. The variational principle gives

$$\delta S = \delta \iiint\limits_R \int_t L dx = 0, \tag{6.3.15}$$

where L, the Lagrangian density, is a function of space co-ordinates and time, dx is the 4-volume element $dx^1 dx^2 dx^3 dx^4$. The action integral S can also be written as

$$S = \int_{t_1}^{t_2} L_0 dt \tag{6.3.16}$$

with

$$L_0 = \iiint\limits_V L dV, \tag{6.3.17}$$

where L is the Lagrangian density. Hence

$$S = \iiint\limits_V \int_{t_1}^{t_2} L dV dt. \tag{6.3.18}$$

We note that L contains the generalized variables $\mathbf{r}(\mathbf{r}_0, t)$, $\mathbf{A}(\mathbf{r}, t)$ and $\phi(\mathbf{r}, t)$ for electrodynamical problems, where A and ϕ are the field potentials.

6.4 Density Concepts

The Lagrangian density L, per unit volume per unit time, at a point of position co-ordinates $\mathbf{r}_0$ at time t, is regarded as a function of the field variables $\boldsymbol{\xi}, \phi, A$. So we write

$$L = L(\mathbf{A}, \phi; \boldsymbol{\xi}; \mathbf{r}_0, t), \tag{6.4.1}$$

where r_0 and t are independent variables along a prescribed reference trajectory and ξ is the polarization variable. Then the action integral S can be expressed as

$$S = \iiint_R \int L(\xi, \phi, A; \dot{\xi}, \phi', \dot{A}; \ddot{\xi}, \phi'', \ddot{A}, A'')dV dt, \tag{6.4.2}$$

where dV is the three dimensional volume element in position space, dt is the time differential, R is the four dimensional volume in the space time continuum, the prime denotes derivate with respect to Z when the wave propagates parallel to the Z-axis, and dot denotes derivative with respect to time. The limits of integration here are taken from one given configuration R_1 to another configuration R_2 in space-time. The Euler-Lagrange equations follow from Hamilton's principle of least action. For the variation of ξ, ϕ, A, the Euler-Lagrange equations are

$$\frac{\partial L}{\partial \xi} = \frac{\partial}{\partial t}\left(\frac{\partial L}{\partial \dot{\xi}}\right) - \frac{\partial^2}{\partial t^2}\left(\frac{\partial L}{\partial \ddot{\xi}}\right), \tag{6.4.3}$$

$$\frac{\partial L}{\partial \phi} = \frac{\partial}{\partial z}\left(\frac{\partial L}{\partial \phi'}\right) + \frac{\partial}{\partial t}\left(\frac{\partial L}{\partial \dot{\phi}}\right) - \frac{\partial^2}{\partial z^2}\left(\frac{\partial L}{\partial \phi''}\right) - \frac{\partial^2}{\partial t^2}\left(\frac{\partial L}{\partial \ddot{\phi}}\right), \tag{6.4.4}$$

$$\frac{\partial L}{\partial A} = \frac{\partial}{\partial t}\left(\frac{\partial L}{\partial \dot{A}}\right) + \frac{\partial}{\partial z}\left(\frac{\partial L}{\partial \dot{A}}\right) - \frac{\partial^2}{\partial t^2}\left(\frac{\partial L}{\partial \ddot{A}}\right) - \frac{\partial^2}{\partial z^2}\left(\frac{\partial L}{\partial A''}\right). \tag{6.4.5}$$

For variations δA and $\delta\phi$ which vanish on the boundary of the volume V in position space, and at times t_1 and t_2, standard forms of the EL equations are

$$\frac{\partial}{\partial t}\left[\frac{\partial L}{\partial(\partial A/\partial t)}\right] - \frac{\partial L}{\partial A} = 0. \tag{6.4.6}$$

$$\frac{\partial}{\partial t}\left[\frac{\partial L}{\partial(\partial \phi/\partial t)}\right] - \frac{\partial L}{\partial \phi} = 0, \tag{6.4.7}$$

where the functional derivatives of L with respect to A and ϕ, in Eulerian co-ordinates, are defined by

$$\frac{\partial L}{\partial A} = \frac{\partial L}{\partial A} - \frac{\partial}{\partial r_i}\left[\frac{\partial L}{\partial(\partial A/\partial r_i)}\right], \tag{6.4.8}$$

$$\frac{\partial L}{\partial(\partial A/\partial t)} = \frac{\partial L}{\partial(\partial A/\partial t)} - \frac{\partial}{\partial r_i}\left(\frac{\partial L}{\partial(\partial^2 A/\partial r_i \partial t)}\right) \tag{6.4.9}$$

$$\frac{\partial L}{\partial \phi} = \frac{\partial L}{\partial \phi} - \frac{\partial}{\partial r_i}\left[\frac{\partial L}{\partial(\partial \phi/\partial r_i)}\right] \tag{6.4.10}$$

and the Einstein summation convention is implied for repeated subscripts.

Since $\mathbf{r}$ is the integration variable, these variations were carried out with respect to Eulerian Co-ordinates. However, transformation is possible of the integration variable from $\mathbf{r}$ to $\mathbf{r}_0$ for that part of the Lagrangian L which describes the particles so that the integration is effectively performed at $t = 0$. If the time derivative $\dot{\mathbf{r}}(= d\mathbf{r}/dt)$ is taken with fixed $\mathbf{r}_0$, the fields A and ϕ are considered to be dependent on r_0 through their explicit dependence on $\mathbf{r}(\mathbf{r}_0, t)$.

With this choice of integration variables, the EL equations, resulting from variation with respect to A and ϕ, need not be rederived to terms of the Lagrangian co-ordinates (r_0, t). These should agree with those in the Eulerian co-ordinates since Hamilton's principle of least action is invariant under transformation of integration variables, Courant and Hilbert (1953) (p. 231–242).

The EL equation for the variation of $\mathbf{r}$ is

$$\frac{d}{dt}\left(\frac{\mathcal{D}L}{\mathcal{D}\dot{\mathbf{r}}}\right) - \frac{\partial L}{\partial \mathbf{r}} = 0. \tag{6.4.11}$$

In Lagrangian co-ordinates the functional derivative of L, with respect to $\mathbf{r}$ and $\dot{\mathbf{r}}$, have the forms

$$\frac{\mathcal{D}L}{\mathcal{D}r_i} = \frac{\partial L}{\partial r_i} - \frac{\partial}{\partial r_{0j}}\left(\frac{\partial L}{\partial r_{ij}}\right), \tag{6.4.12}$$

$$\frac{\mathcal{D}L}{\mathcal{D}\dot{r}_i} = \frac{\partial L}{\partial \dot{r}_i} - \frac{\partial}{\partial r_{0j}}\left(\frac{\partial L}{\partial \dot{r}_i}\right), \tag{6.4.13}$$

with

$$r_{ij} = \frac{\partial r_i}{\partial r_{0j}}. \tag{6.4.14}$$

These may be compared to those of (6.4.8) and (6.4.10) for Eularian co-ordinates, for $\mathcal{D}L/\mathcal{D}A$ and $\mathcal{D}L/\mathcal{D}\phi$. In particular, equation (6.4.13) should be replaced by

$$\frac{\mathcal{D}L_I}{\mathcal{D}\dot{r}_i} = \frac{\partial L_p}{\partial \dot{r}_i} - \frac{\partial}{\partial r_{0j}}\left(\frac{\partial L_p}{\partial \dot{r}_{ij}}\right) \tag{6.4.15}$$

and (6.4.9) by

$$\frac{\mathcal{D}L_i}{\mathcal{D}(\partial A/\partial t)} = \frac{\partial L_L}{\partial(\partial A/\partial t)} - \frac{\partial}{\partial r_i}\left(\frac{\partial L_L}{\partial(\partial^2 A/\partial r_i \partial t)}\right). \tag{6.4.16}$$

Consider the action integral

$$S = \iiiint_R L[(x, \psi_i(\mathbf{x}), \psi_{i,\mu}(\mathbf{x})]\mathbf{dx}, \tag{6.4.17}$$

where R is the 4-space domain (volume V and the time t taken together) for integration, a comma before a subscript means a partial derivative with respect to the independent variable bearing that subscript; so

$$\psi_{i,\mu}(x) = \frac{\partial \psi_i}{\partial x_\mu} \tag{6.4.18}$$

L is evidently a function of x_μ, the state functions $\psi(x)$ and the derivative function $\psi_{i,\mu}(x)$.

The domain R being well-defined, the limits of integration are fixed, so the variations of $\psi_i(x)$ on the surface of V, or at the beginning and end of the time integral, vanish. Actually, the variation of the integral in (6.4.17) is induced from the following two sources : (1) The variation of the state functions $\psi_i(x)$; and (2) The variation of the domain of integration R, which also induces variation on $\psi_i(x)$, because these are dependent on the co-ordinates x_μ.

Consider the three dimensional integral

$$\mathcal{H}_0 = \iiint\limits_V \mathcal{H} dv \tag{6.4.19}$$

in the position space, where $\mathcal{H}$ is the Hamiltonian density

$$\mathcal{H} = \sum_i (P_i \dot{\psi}_i) - L, \tag{6.4.20}$$

where P_i is the canonical momentum density corresponding to the field variable ψ_i. So,

$$P_i = \frac{\partial L}{\partial \dot{\psi}_i}. \tag{6.4.21}$$

The variation of $\mathcal{H}_0$ is given by the relation

$$\delta \mathcal{H}_0 = \iiint\limits_V \left[\sum_i (P_i \delta \dot{\psi}_i + \dot{\psi}_i \delta P_i) - \left\{ \frac{\partial L}{\partial \psi_i} - \left(\frac{\partial L}{\partial \psi_{i,j}} \right) \right\} \delta \psi_i - \frac{\partial L}{\partial \dot{\psi}_i} \delta \dot{\psi}_i \right\} dV. \tag{6.4.22}$$

Using (6.4.21) the first term in the right hand side cancels out with the last term. Moreover, using the Euler-Lagrangian equation (6.7.8) on the two terms within the brackets { } we find that

$$\delta \mathcal{H}_0 = \iiint\limits_V \left[\dot{\psi}_i \delta P_i - \frac{\partial}{\partial t} \left(\frac{\partial L}{\partial \dot{\psi}_i} \right) \delta \psi_i \right] dV. \tag{6.4.23}$$

Applying (6.4.21) on the (6.4.23) we obtain

$$\delta \mathcal{H}_0 = \iiint\limits_V \{ \dot{\psi}_i \delta P_i - \dot{P}_i \delta \psi_i \} dV \tag{6.4.24}$$

for the variation of the $\mathcal{H}_0$. Hence

$$\frac{\partial \mathcal{H}_0}{\partial P_i} = \dot{\psi}_i, \quad \frac{\partial \mathcal{H}_0}{\partial \psi_i} = -\dot{P}_i. \tag{6.4.25}$$

Hence if $\psi = x_i$, where x_i is the co-ordinate of a point in space along the ith co-ordinate direction,

$$\frac{\partial \mathcal{H}_0}{\partial P_i} = \dot{x}_i, \quad \frac{\partial \mathcal{H}_0}{\partial x_i} = -\dot{P}_i. \tag{6.4.26}$$

These can also be written vectorially as

$$\nabla \mathcal{H}_0 = -\dot{P}, \quad \frac{\partial \mathcal{H}_0}{\partial P} = \dot{r}. \tag{6.4.27}$$

6.5 Legendre Transformations

The Lagrangian $L(q, \dot{q}, t)$ is a function of $q, \dot{q}, t$; the corresponding Hamiltonian $\mathcal{H}(q, p, t)$ is given by

$$\mathcal{H}(q, p, t) = q_i p_i - L(q, \dot{q}, t), \tag{6.5.1}$$

where the components $q_1, q_2, \cdots, q_n$ of q are the generalized position co-ordinates, the components $\dot{q}_1, \dot{q}_2, \cdots, \dot{q}_n$ of $\dot{q}$ are the time derivatives of the q_is, and $p_1, p_2, \cdots, p_n$ are the components of the generalized momentum p, called the *momentum co-ordinates.* The Hamiltonian is obtained by the transformation from the set of variables $q, \dot{q}, t$ to the set p, q, t. The rule of this transformation follows the simple theory of the Legendre transformation, Goldstein (1980, p. 33). To explain this technique consider the differential df of a function of f of x and y :

$$df = udx + vdy \tag{6.5.2}$$

so that

$$u = \frac{\partial f}{\partial x}, \quad v = \frac{\partial f}{\partial y}. \tag{6.5.3}$$

Changing next from x and y to the set of variables u and y, we have

$$df = adu + bdy, \tag{6.5.4}$$

where

$$a = \frac{\partial f}{\partial x}, \quad b = \frac{\partial f}{\partial y}. \tag{6.5.5}$$

We now define another function g as

$$g = f - ux \tag{6.5.6}$$

and obtain

$$dg = df - vdx - xdu. \tag{6.5.7}$$

With the help of (6.5.2) this relation becomes

$$dg = vdy - xdu. \tag{6.5.8}$$

Hence we obtain the relations of the Legendre transformation

$$v = \frac{\partial g}{\partial y}, \quad x = -\frac{\partial g}{\partial u}, \tag{6.5.9}$$

which are converse to those of (6.5.3). The expression for the Hamiltonian follows (6.5.6) except for a minus sign.

6.6 Lagrangian and Hamiltonian Densities

Both for a single particle, and for a system of N particles and for a continuous distribution of matter in a volume V, for small translational velocities, the Lagrangian L, in presence of conservative fields, is defined as the difference between the kinetic energy T_0 and the potential energy V,

$$L = T_0 - V. \tag{6.6.1}$$

For a continuous distribution, the total Lagrangian $\mathcal{L}$ may be replaced by the density L, so that

$$\mathcal{L} = \iiint_V Ldv. \tag{6.6.2}$$

Then we can write

$$L = \xi - \phi, \tag{6.6.3}$$

where ξ and ϕ are the densities of kinetic energy and potential energy. For relativistic motion of a particle in conservative fields,

$$L = -\frac{m_0}{\gamma}c^2 - \phi. \tag{6.6.4}$$

To show the justification of the first term we denote it by ξ_r and assume that

$$\frac{\partial \xi_r}{\partial \dot{x}_i} = p_i, \tag{6.6.5}$$

where p is the relativistic momentum $m_0\gamma u$. Then, since

$$\frac{d}{dt}\xi_r = \sum\left(\frac{\partial \xi_r}{\partial \dot{x}_i}\ddot{x}_i\right) = \sum_i (p_i \ddot{x}_i) = m_0\gamma u \frac{du}{dt}, \tag{6.6.6}$$

integration gives $\xi_r = -m_0c^2/\gamma$. Then kinetic energy of the particles is $m_0u^2/2$ and not this relativistic kinetic energy-like quantity.

The Hamiltonian density is

$$\mathcal{H} = \left(U \cdot \frac{\partial}{\partial U} L\right) - L. \tag{6.6.7}$$

6.7 The Euler Lagrange equations

The principle of least action, when applied on the action integral (6.4.17), gives

$$\delta S = \delta \iiiint_R L[x, \psi_i(x), \psi_{i,\mu}(x)]d\mathbf{x} = 0$$

$$= \iiiint_R \left[\sum_i \left(\frac{\partial L}{\partial \psi_i}\right) + \sum_{i,\mu}\left\{\frac{\partial L}{\partial \psi_{i,\mu}}\delta\psi_{i,\mu}\right\}\right] d\mathbf{x}. \tag{6.7.1}$$

To replace the variation $\delta\psi_{i,\mu}$ by the variation $\delta\psi_i$ in the second term, the operations δ and $\partial/\partial x_\mu$ are interchanged, and a partial integration is performed. Thus, since (1) $\delta\psi_i(x)$ vanishes on the periphery of R, and since (2) $\partial/\partial x_\mu$ operates on both explicit and implicit dependences on x_μ, we find that

$$\iiiint_R \frac{\partial L}{\partial \psi_{i,\mu}}\delta\left(\frac{\partial \psi_i}{\partial x_\mu}\right) d\mathbf{x} = -\iiiint_R \frac{\partial}{\partial x_\mu}\left\{\frac{\partial}{\partial \psi_{i,\mu}}\right\}\delta\psi_i d\mathbf{x}. \tag{6.7.2}$$

Hence (6.7.1) becomes

$$\iiiint_R \sum_i \left\{\delta\psi_i\left[\frac{\partial L}{\partial \psi_i} - \sum_\mu \frac{\partial}{\partial x_\mu}\left(\frac{\partial L}{\partial \psi_{i,\mu}}\right)\right]\right\} dx = 0. \tag{6.7.3}$$

Each variation $\delta\psi_i$ being independent and arbitrary, and since the time interval and the system volume are also arbitrary, the integral in (6.7.3) vanishes if the coefficients of $\delta\psi_i$ vanish, giving the EL equation for the ψ_i field :

$$\frac{\partial L}{\partial \psi_i} - \frac{\partial}{\partial x_\mu}\left(\frac{\partial L}{\partial \psi_{i,\mu}}\right) = 0. \tag{6.7.4}$$

If ψ_i is replaced by the generalized co-ordinate q_i, then it is the EL equation involving these co-ordinates. Here, the Lagrangian density L is used, and not the total Lagrangian L itself for the ψ_i field because these are functions of the co-ordinates x_μ, and so represent a continuous system. Moreover, for each state function ψ_i, thee is only one such equation of motion. It assumes that L depends on the 4-vector $x, \psi_x(\mathbf{x})$ and $\psi_{i,\mu}(x)$ only. But, if the dependence of L on fields is extended to include $\psi_{i,\mu\nu}$, where

$$\psi_{i,\mu\nu} = \frac{\partial^2 \psi_i}{\partial x_\mu \partial x_\nu}, \tag{6.7.5}$$

then this derivation yields the more general EL equation

$$\frac{\partial L}{\partial \psi_i} - \frac{\partial}{\partial x_\mu}\left(\frac{\partial L}{\partial \psi_{i,\mu}}\right) + \frac{\partial^2}{\partial x_\mu \partial x_\nu}\left(\frac{\partial L}{\partial \psi_{i,\mu\nu}}\right) = 0. \tag{6.7.6}$$

Since both the Greek indices μ and ν run from 1 to 4, we replace these by j which runs from 1 to 3 and by $x_4(= -ct)$. Thus (6.7.4) and (6.7.6) become

$$\frac{\partial L}{\partial \psi_i} - \frac{\partial}{\partial x_j}\left(\frac{\partial L}{\partial \psi_{i,j}}\right) = \frac{\partial}{\partial t}\left(\frac{\partial L}{\partial \dot{\psi}_i}\right), \tag{6.7.7}$$

$$\frac{\partial L}{\partial \psi_i} - \frac{\partial}{\partial x_j}\left(\frac{\partial L}{\partial \psi_{i,j}}\right) + \frac{\partial^2}{\partial x_j \partial x_\nu}\left(\frac{\partial L}{\partial \psi_{i,j\nu}}\right) = \frac{\partial}{\partial t}\left(\frac{\partial L}{\partial \dot{\psi}_i}\right) - \frac{\partial^2}{\partial t \partial x_\nu}\left(\frac{\partial L}{\partial \dot{\psi}_i}\right). \tag{6.7.8}$$

Since $x_4 = -ct$ and $x^4 = ct$. We can write

$$\frac{\partial}{\partial x_4}\left(\frac{\partial L}{\partial \psi_{i,4}}\right) = \frac{\partial}{\partial t}\left(\frac{\partial L}{\partial \dot{\psi}_i}\right) \tag{6.7.9}$$

For this reason $\partial/\partial t$ and $\dot{\psi}_i$ appear in the right hand side of (6.7.7) and (6.7.8) in place of $\partial/\partial x_4$ and $\psi_{i,4}$.

For motion of a single particle in the presence of continuous fields, the EL equation is

$$\frac{d}{dt}\left(\frac{\partial L}{\partial u_i}\right) - \frac{\partial L}{\partial x_i} = 0, \tag{6.7.10}$$

where u_i and x_i are the components of the 3-velocity and 3-position vector, respectively, of it. If the accelerations $a_i = \dot{u}_i$ also enter into the Lagrangian, the equations of motion take the form

$$-\frac{d^2}{dt^2}\left(\frac{\partial L}{\partial a_i}\right) + \frac{d}{dt}\left(\frac{\partial L}{\partial u_i}\right) - \frac{\partial L}{\partial x_i} = 0. \tag{6.7.11}$$

When all the forces acting on the system are not derivable from a potential or have not been fully taken into account in $\mathcal{L}$, then (6.7.10) is replaced by

$$\frac{d}{dt}\left(\frac{\partial L}{\partial u_i}\right) - \frac{\partial L}{\partial x_i} = F_i, \tag{6.7.12}$$

where F_i is the ith component of the sum of these forces.

6.8 Variation of the Total Hamiltonian

The integral for the total Hamiltonian $\mathcal{H}_0$ has been developed in equation (6.4.22). For analysis following from variation of the integral for $\mathcal{H}_0$, regard the Hamiltonian density $\mathcal{H}$ as a function of the field components ψ_i, the canonical momentum density component P_i, the spatial derivatives $\psi_{i,j}$ and time t. Then

$$\delta\mathcal{H}_0 = \iiint_V \sum \left[\frac{\partial\mathcal{H}}{\partial\psi_i}\delta\psi_i + \frac{\partial\mathcal{H}}{\partial P_i}\delta P_i + \frac{\partial\mathcal{H}}{\partial\psi_{i,j}}\delta\psi_{i,j}\right] dV. \tag{6.8.1}$$

Integration by parts and use of the terminal conditions that $\delta\psi_i$ and δP_i vanish at the boundary Σ of V, transform (6.8.1) into

$$\delta\mathcal{H}_0 = \iiint_V \left[\left\{\frac{\partial\mathcal{H}}{\partial\psi_i} - \left(\frac{\partial\mathcal{H}}{\partial\psi_{i,j}}\right)\right\}\delta\psi_i + \frac{\partial\mathcal{H}}{\partial P_i}\delta P_i\right] dV. \tag{6.8.2}$$

Comparison of this $\delta\mathcal{H}_0$ with (6.4.24) yields the Hamilton equations or the canonical equations

$$\dot{\psi}_i = \frac{\partial\mathcal{H}}{\partial P_i}, \tag{6.8.3}$$

$$\dot{P}_i = -\frac{\partial\mathcal{H}}{\partial\psi_i} + \sum_i\left(\frac{\partial\mathcal{H}}{\partial\psi_{i,j}}\right). \tag{6.8.4}$$

These are more general than those of (6.2.42) and (6.2.43). The total time derivative of $\mathcal{H}$ gives

$$\frac{d\mathcal{H}_0}{dt} = \iiint_V \sum_i \left[\left\{\frac{\partial\mathcal{H}}{\partial\psi_i} - \sum_j\left(\frac{\partial\mathcal{H}}{\partial\psi_{i,j}}\right)\right\}\dot{\psi}_i + \frac{\partial\mathcal{H}}{\partial P_i}\dot{P}_i\right] dV + \iiint_V \frac{\partial\mathcal{H}}{\partial t} dV. \tag{6.8.5}$$

Elimination of $\dot{\psi}_i$ and $\dot{P}_i$ with the help of (6.8.3) shows that the term under the square brackets vanishes. Hence, if $\mathcal{H}$ does not depend explicitly on t,

$$\frac{d\mathcal{H}_0}{dt} = \iiint_V \frac{\partial\mathcal{H}}{\partial t} dV = 0, \tag{6.8.6}$$

so that $\mathcal{H}_0$ is a constant of motion.

6.9 Non-uniqueness of the Lagrangian Density

The Lagrangian density L is not uniquely defined for the action integral of a problem. For instance, equation (6.7.6), obtained from (6.4.17) with the help of Hamilton's principle, is form invariant up to the inclusion of a 4-divergence

$$\frac{\partial}{\partial x_\mu}(a^\mu\wedge) = (a^\mu\wedge),\mu. \tag{6.9.1}$$

Here a^μ is a 4-vector, and both a^μ and $\wedge$ are well behaved unspecified functions of x_μ. Consider two Lagrangian densities L_m and L_n such that

$$L_n = L_m + (a^\mu\wedge),\mu. \tag{6.9.2}$$

If the action integrals for L_n and L_m are denoted by S_n and S_m, then

$$S_n = \iiiint_R L_n d\mathbf{x} = \iiiint_R L_m d\mathbf{x} + \iiiint_R (a^\mu \wedge)_\mu d\mathbf{x} = S_m + \iiint_\Sigma \wedge a^\mu d\Sigma_\mu, \quad (6.9.3)$$

where Σ is the hyper surface enclosing the 4-volume R, $d\Sigma$ is surface element of this Σ. The variation of a^μ on Σ vanishes since the end points of the domain which lie on Σ are fixed. Hence

$$S_m = S_n \quad (6.9.4)$$

and both S_m and S_n give the same EL equations. So, the action integrals S_m and S_n are equivalent if the corresponding integrands L_m and L_n differ by the 4-divergence $(a^\mu \wedge), \mu$.

6.10 The EM Lagrangian

When the wave phase velocity v_p is infinite, the Lagrangian of a system of particles depends only on the co-ordinates and velocities of these particles at one and the same time. However, in the EM theory this velocity is finite; so the field must be considered as an independent system with its own degree of freedom. For a system of interacting charged particles, both system and the field of interaction must be considered simultaneously.

When the particle velocity $u \ll c$, a system of charges can be described by a certain approximate Lagrangian (see section 6.11). However, when all powers of u/c are neglected, we obtain the classical Lagrangian

$$L^{(0)} = \sum_i \left(\frac{1}{2} m_i u_i^2\right) - \sum_{i=1, j>i}^{N} \left(\frac{q_i q_j}{|r_j - r_i|}\right), \quad (6.10.1)$$

which is that of the zeroth order. The second term is the potential energy of interaction for the system of charges at rest.

The total EM Lagrangian density L can be written as

$$L = L_m + L_f + L_I, \quad (6.10.2)$$

where L_m is the matter Lagrangian, L_f is the EM field Lagrangian and L_I is the interaction Lagrangian. Then the action integral is

$$S = S_m + S_f + S_I = \iiiint_R (L_m + L_f + L_I) d\mathbf{x}. \quad (6.10.3)$$

The variation of S_m and S_f yield the equations of motion of a free particle and of a free field respectively. The variation of S_I with respect to particle co-ordinates gives the equation of motion in the field, and the variation with respect to the field variables gives the inhomogeneous terms in the field equations. Thus the transition from classical mechanics to quantum mechanics can be studied, and conservation Laws are defined unambiguously.

Since mass and energy are related and all fields are carriers of energy, the separation of L into the parts L_m, L_f, L_1 is not very rigid. In fact, L_M, L_f, L_I contain common terms in the sum Lagrangian L; these common terms are to be taken once only. The same rule applies to the division of L.

For fields in vacuum

$$L_f = \iiint_V \frac{1}{8\pi}(E^2 - H^2)dV, \tag{6.10.4}$$

where E and H are the long range electric and magnetic field vectors. For dielectrics E^2 should be replaced by $(E \cdot D)$ and H^2 by $(B \cdot H)$. The field potentials A and ϕ of the Lorentz gauge depend on the set of position co-ordinates of all the particles. So, the electrostatic potential is that of all the short range forces only for charges residing in vacuum. The quantum mechanical effects of short range interactions are not considered here. The electric and magnetic forces on the particles due to long range electric and magnetic fields (having wave lengths greater than the mean free path) existing in the vacuum around the particles arise from the interaction Lagrangian

$$L_I = \sum_i [q_i(\dot{r}_i(t) \cdot A(r,t) + q_i\phi(r,t)] . \tag{6.10.5}$$

It produces the EL equations for discrete particles, where the potentials A and ϕ are evaluated at the position of each particle. For continuous distributions of charges, the corresponding density,

$$L_I = (A \cdot j) + \rho^c\phi \tag{6.10.6}$$

should be used in

$$L_I = \iiint_V L_I dv, \tag{6.10.7}$$

where

$$\rho^c(r,t) = \frac{1}{v}\sum_i [q_i\delta\{r - r_i(t)\}]$$

$$j(r,t) = \frac{1}{v}\sum_i [q_i\dot{r}_i(t)\delta\{r - r_i(t)\}] \tag{6.10.8}$$

$\rho^c(\mathbf{r},t)$ is the charge density and $\mathbf{j}(\mathbf{r},t)$ is the current density at the 4-point $(\mathbf{r},t)$; $\delta(\mathbf{r}-\mathbf{r}_i)$ is the delta function of Dirac. L_I is usually expanded in series of multipole moments. In dielectrics the monopole or free charge density vanishes, but not in plasma. Also, in many cases, the electric quadrupole, the magnetic dipole and higher moments are negligible.

The Lorentz force for a charged particle, $F = qE + q(u \times B)/c$, is velocity dependent and non-conservative in nature. However, it can be expressed as Q_j^0 of (6.2.34) in terms of A and ϕ. For a single particle of charge q,

$$F = -\frac{q}{c}\frac{\partial A}{\partial t} + q\nabla\phi + \frac{q}{c}u \times (\nabla \times A) = q\nabla\left[\frac{1}{c}(u \cdot A) + \phi\right] - \frac{q}{c}\frac{dA}{dt}, \tag{6.10.9}$$

where the velocity co-ordinates are supposed to be independent of the position co-ordinates, and

$$\frac{dA}{dt} = \frac{\partial A}{\partial t} + (u \cdot \nabla)A. \tag{6.10.10}$$

Hence

$$F = q\left[\nabla \cdot \left\{\frac{1}{c}(u \cdot A) + \phi\right\} - \frac{d}{dt}\nabla_u\left\{\frac{1}{c}(u \cdot A) + \phi\right\}\right], \tag{6.10.11}$$

where

$$\nabla_u = \left(\frac{\partial}{\partial U_x}, \frac{\partial}{\partial U_y}, \frac{\partial}{\partial U_z}\right)$$

and because A and ϕ are assumed to be independent of the velocity co-ordinates u. Another form of F is

$$F_i = -q\frac{d}{dt}\frac{\partial}{\partial U_i}\left\{\frac{1}{c}(U\cdot A)+\phi\right\} - q\frac{\partial}{\partial x_i}\left\{\frac{1}{c}(U\cdot A)+\phi\right\} = -\frac{d}{dt}\left(\frac{\partial L_I}{\partial \dot{x}_i}\right)+\frac{\partial L_I}{\partial x_i}, \tag{6.10.12}$$

where

$$L_I = \frac{q}{c}(U\cdot A)+q\phi. \tag{6.10.13}$$

Here L_I is the total Lagrangian of a single particle in a continuous distribution of fields. Hence the Lagrangian of a charge q in an external field is

$$L_1 = L_m + L_I = -\frac{m_0}{\gamma}c^2 + \frac{q}{c}(u\cdot A)+q\phi, \tag{6.10.14}$$

where (6.6.4) with the potential $\phi = 0$ has been used. Differentiation of L_1 with respect to u_i and x_i gives

$$\frac{\partial L_1}{\partial u_i} = m_0\gamma u_i + \frac{q}{c}A_i, \tag{6.10.15}$$

$$\frac{\partial L_1}{\partial x_i} = q\frac{\partial \phi}{\partial x_i} + q\frac{1}{c}\frac{\partial A_j}{\partial x_j}u_j, \tag{6.10.16}$$

where the summation convention of Einstein is implied over the repeated index j. Then the EL equation yields the equation of motion

$$\frac{d}{dt}(m_0\gamma u) = qE + \frac{q}{c}(U\times B). \tag{6.10.17}$$

Taking scalar product of both sides with u we obtain the relation between the rate of work done by the field on the particle and the rate of increase of its kinetic energy,

$$\frac{d}{dt}(m_0\gamma c^2) = q(E\cdot U). \tag{6.10.18}$$

The generalized moment,

$$P = \frac{\partial L_1}{\partial U}, \tag{6.10.19}$$

for the EM field, evidently is

$$P = m_0\gamma u + \frac{q}{c}A. \tag{6.10.20}$$

The Lagrangian density L_{pf} for the particle motion (non-relativistic) in presence of prescribed EM fields is

$$L_{pf} = q\phi + \frac{q}{c}(u\cdot A) + \frac{1}{2}mu^2. \tag{6.10.21}$$

The corresponding Hamiltonian

$$\mathcal{H}_{pf} = \left(u\cdot\frac{\partial}{\partial u}L_{pf}\right) - L_{pf} = q\phi + \frac{1}{2}mu^2 = q\phi + \frac{1}{2m}\left(p-\frac{q}{c}A\right)^2. \tag{6.10.22}$$

In free space, the Lagrangian density, Landan and Lifshitz (1975, p. 209), is

$$L_f = \frac{1}{8\pi}(E^2-H^2) = \frac{1}{8\pi}\left(-\frac{1}{c}\frac{\partial A}{\partial t}+\nabla\phi\right)^2 - \frac{1}{8\pi}(\nabla\times A)^2. \tag{6.10.23}$$

For a single particle the combined field particle non-relativistic Lagrangian and the Hamiltonian are

$$\mathcal{L} = \iiint_V \frac{1}{8\pi}\{(\varepsilon E^2 - \mu H^2)\}dV + q\phi + \frac{q}{c}(U\cdot A) + \frac{1}{2}mu^2. \tag{6.10.24}$$

$$H = \iiint_V \frac{1}{8\pi}(\varepsilon E^2 + \mu H^2)dV + \frac{1}{2}mu^2. \tag{6.10.25}$$

For an electron plasma, in the relativistic cold plasma limit,

$$L = \frac{1}{8\pi}(\varepsilon E^2 - \mu H^2) - eN_0\phi(r,t) - \frac{e}{c}N_0(A\cdot U) - mN_0c^2\sqrt{1 - \frac{u^2}{c^2}}, \tag{6.10.26}$$

where m is the mass and $-e$ is the charge per electron and N_0 is the electron number density.

6.11 Approximate Evaluation of L_1 for Two Charges

Consider the retarded solution

$$\phi(r,t) = -\iiint_V \frac{\rho^c(r',t')dV}{|r - r'|} \tag{6.11.1}$$

of the wave equation (4.10.2) for the scalar potential ϕ, where $t' = t - |r - r'|/c$. The expansion of $\rho^c(r',t')$ in positive integral powers of $|r - r'|$ gives

$$\rho^c(r',t') = \rho^c(r',t) - \frac{1}{c}|r - r'|\dot{\rho}^c(r',t) + \frac{1}{2c^2}|r - r'|^2\ddot{\rho}^c(r',t) + \cdots, \tag{6.11.2}$$

where a dot stands for a derivative with respect to t. For this Taylor expansion $v^2 \ll c^2$, the motion of the particles should be sufficiently smooth, so that quantities like the acceleration and higher order time derivatives of the displacements must exist and be continuous. The expression for the charge q, independent of time, is given by the volume integral

$$q = \iiint_V \rho^c(r,t)dV, \tag{6.11.3}$$

where V is the volume of a particle in a system, we find that

$$\phi(r,t) = -\iiint_V \frac{\rho(r',t')}{|r - r'|}dV' - \frac{1}{2c^2}\frac{\partial^2}{\partial t^2}\iiint_V \rho(r',t')|r - r'|dV', \tag{6.11.4}$$

which follows when (6.11.3) is used. If the point charge q_i is concentrated in the neighbourhood of r_i, then $r' = r_i(t)$ and

$$\phi(r,t) = -\frac{q_i}{|r - r_i|} - \frac{q_i}{2c^2}\frac{\partial}{\partial t}\left\{\frac{u_i(r - r_i)}{(r - r_i)}\right\}\dot{r}_i(t) = u_i. \tag{6.11.5}$$

The first term is the Coulomb potential for q_i at rest at r_i; the second term is a retardation correction, and it involves both velocity U_i and the acceleration $\dot{U}_i$ of the particle which produces the field.

For generalization, replace q by q_i, q_i by $q_j, r - r_i$ by $r_i - r_j$ put

$$r_i - r_j = R_{ij}, \quad n_{ij} = \frac{R_{ij}}{|R_{ij}|} \tag{6.11.6}$$

and sum over all particles of both types i and j. Then the Lagrangian of the whole system, symmetric with respect to the particle of charge q_i on which the field acts, is obtained

$$L_1 = -\sum_i \left[m_0 c^2 \left(1 - \frac{u_i^2}{c^2}\right)^{1/2} \right] - \sum_{i,j} \frac{1}{2} \left(\frac{q_i q_j}{R_{ij}} \right)$$

$$+ \frac{1}{4c^2} \sum_{\substack{i,j \\ i \neq j}} \left\{ \frac{q_i q_j}{R_{ij}} \right\} [(u_i - u_j) + (u_i \cdot n_{ij})(u_j - n_{ij})] \tag{6.11.7}$$

Actually, for a charge q, under a given motion of all the others, this L_1 goes over into the L_1 of (6.10.14). Neglecting terms higher then of order u^2/c^2 everywhere in (6.11.7) we find that

$$L_1 = c^2 \sum_i (m_i) + \sum_i \left(\frac{1}{2} m_i u_i^2 \right) + \sum_i \frac{1}{8} \left(m_i \frac{u_i^4}{c^2} \right) - \frac{1}{2} \sum_{i,j} \left(\frac{q_i q_j}{R_{ij}} \right)$$

$$+ \frac{1}{4c^2} \sum_{i,j} \left[\frac{1}{R} q_i q_j \{ (u_i u_j) + (u_i \cdot n_{ij})(u_j \cdot n_{ij}) \} \right], \quad j \neq i. \tag{6.11.8}$$

Here the third and the fifth term are small corrections to the zeroth order Lagrangian L^0 of (6.10.1). The generalized momentum P on the particle i in presence of the particle of type j, is given by

$$P = \frac{\partial L_1}{\partial u_i} = m_i u_i + \frac{1}{2c^2} m_i u_i^2 u_i + \frac{q_1}{4c^2} \sum_j \left(\frac{1}{R_{ij}} q_i u_j \right)^2 + \frac{q_1}{4c^2} \sum_j \left\{ \frac{1}{R_{ij}} q_i (u_j \cdot u_{if}) n_{ij} \right\}, \quad j \neq i. \tag{6.11.9}$$

Hence, for two charges, approximately

$$L_1 = c^2(m_1 + m_2) + \frac{1}{2}\mu u^2 + \frac{1}{8c^2}\mu u^4 \cdot \frac{(m_1^3 + m_2^3)}{(m_1 + m_2)^3} - \frac{\mu^2 q_1 q_2}{4c^2 r m_1 m_2} \{ u^2 + (u \cdot n)^2 \}, \tag{6.11.10}$$

where

$$\mu = \frac{m_1 m_2}{m_1 + m_2}, \quad -m_1 u_1 = m_2 u_2 = \mu u = p \tag{6.11.11}$$

μ is the reduced mass. Then the canonical momentum

$$P = \frac{\partial L_1}{\partial u} = \mu u + \frac{\mu u^2}{2c^2} \cdot \frac{(m_1^3 + m_2^3)}{(m_1 + m_2)^3} u - \frac{\mu^2 q_1 q_2}{2c^2 m_1 m_2 r} \{ u + (u \cdot n) n \}. \tag{6.11.12}$$

When expressed in terms of the current density j, the potential A contains the factor $1/c$. In the Lagrangian it is once more multiplied by $1/c$. So, for a Lagrangian correct up to terms of the second order, the second order time derivatives are neglected. Moreover, if j is replaced by $\rho^c u$, for a concentrated charge,

$$A = \frac{q u_i}{c|r - r_i|}. \tag{6.11.13}$$

The gauge transformation relations for the potentials A and ϕ,

$$\phi' = \phi - \frac{1}{c}\frac{\partial \chi}{\partial t}, \tag{6.11.14}$$

$$A' = A + \nabla\chi \tag{6.11.15}$$

are considered to see whether the fields are changed thereby, where χ is an arbitrary function of x_μ. Putting

$$\chi = \frac{1}{2cR}q'(u_i\cdot R), \quad R = r - r_i, \tag{6.11.16}$$

we find that

$$\phi'(r,t) = -\frac{q}{R}, \tag{6.11.17}$$

$$A' = \frac{1}{cR}q_i u_i + \frac{1}{2c}q_i\nabla\left\{\frac{(u_i\cdot R)}{R}\right\}. \tag{6.11.18}$$

Since

$$\nabla\left\{\frac{(u_i\cdot R)}{R}\right\} = \frac{u_i}{R} - \frac{(u_i\cdot R)}{R^2}R. \tag{6.11.19}$$

we obtain

$$A' = \frac{q}{2c}\frac{u_i}{R} - \frac{q_i}{2cR}(u_i\cdot R)R, \tag{6.11.20}$$

$$(\nabla\cdot A') = 0. \tag{6.11.21}$$

Putting the value of ϕ' and A' from (6.10.12) and (6.11.20) into the Lagrangian L_1 we find that

$$L_1 = -\frac{m_0}{\gamma}c^2 - \frac{1}{R}qq_i + \frac{qq_i}{2c^2R}[(u\cdot u_i) + (u\cdot n)(u_i\cdot n)], \tag{6.11.22}$$

where

$$n = \frac{R}{|R|}, \quad R = r - r_i. \tag{6.11.23}$$

This L_1 represents, approximately, the effect of interaction between two particles.

6.12 EM Hamiltonian

The Hamiltonian $\mathcal{H}$ was defined as a function of the Canonical momentum P and position vector $\mathbf{r}$, in (6.4.19) and (6.4.20). The total Hamiltonian $\mathcal{H}_0$ is given by

$$\mathcal{H}_0 = (P\cdot U) - L_1 \tag{6.12.1}$$

and

$$\nabla\mathcal{H}_0 = -\frac{dP}{dt} = -\frac{d}{dt}\left(\frac{\partial L_1}{\partial u}\right), \tag{6.12.2}$$

where L_1 is defined in (6.10.4). Thus we obtain

$$\mathcal{H}_0 = c(m_0c^2 + p^2)^{1/2} - q\phi = [(cP - qA)^2 + m_0^2c^2]^{1/2} - q\phi. \tag{6.12.3}$$

Solving equation (6.10.20) of the canonical moment P, for u we find that

$$u = (cP - qA)\left[\left(P - \frac{q}{c}A\right)^2 + m_0^2c^2\right]^{-1/2}. \tag{6.12.4}$$

Then (6.12.2) shows that

$$\nabla\mathcal{H}_0 = \frac{1}{(m_0^2c^2+p^2)^{1/2}}\left[(p\cdot\nabla)\left(\frac{1}{c}qA\right)+(p\cdot\nabla)\times\left(-\frac{q}{c}A\right)\right]+q\nabla\phi$$

$$= -\frac{q}{c}(u\cdot\nabla)A - \frac{q}{c}(u\times(\nabla\times A)) + q\nabla\phi. \tag{6.12.5}$$

The total energy ξ of the particle is the sum of the free particle energy

$$\left(m_0^2c^2\gamma^2u^2 + m_0^2c^4\right)^{1/2} \tag{6.12.6}$$

and the potential energy $-q\phi$. So $\xi = -H$,

$$(\xi + q\phi)^2 - (cP - qA)^2 = m_0^2c^4 \tag{6.12.7}$$

and Hamilton's equations of motion, (6.4.26), are satisfied.

► **Example 6.12.1** *Find an approximate Hamiltonian for two charges.*

Answer. With the help of (6.2.29) and (6.10.1) the approximate expression for the zeroth order Hamiltonian $\mathcal{H}_0^{(0)}$ is determined :

$$\mathcal{H}_0^{(0)} = \sum_i\left(\frac{1}{2}m_iu_i^2\right) + \sum_{i\neq j}\left(\frac{q_iq_j}{R_{ij}}\right) - \sum_i\left(\frac{1}{2m_i}p_i^2\right), \tag{1}$$

where the velocity u_i is replaced by the momentum $p_i(= m_iu_i)$. For finding $\mathcal{H}_1$ we note that the third and fifth terms of (6.11.8) are small corrections over $L^{(0)}$ of (6.10.1), the first term of (6.11.8) being a constant, is a trivial presence. Since these small terms can be represented as the Lagrangian L' of (6.2.56), we write

$$\mathcal{H}_1 = \sum_i\left(\frac{p_i^2}{2m_i}\right) + \sum_{\substack{i,j\\ i\neq j}}\left(\frac{q_iq_j}{R_{ij}}\right) - \sum_i\left(\frac{p_i^4}{8m_i^3c^2}\right) - c^2\sum_i(m_i)$$

$$\frac{1}{4c^2}\sum_{\substack{i,j\\ i\neq j}}\left[\frac{q_iq_j}{m_im_jR_{ij}}\{(p_i\cdot p_j)(p_i\cdot n_{ij})(p_j\cdot n_{ij})\}\right], \tag{2}$$

where the fact noted after (6.2.57) that small additions to L and $\mathcal{H}$ are equal in magnitude and opposite in sign has been used here.

For a system of two particles of charges q_1 and q_2 and masses m_1 and m_2, we can choose the co-ordinate system in which the total momentum of the two particles vanishes. Then the motion of the center of mass is omitted, we can write $-p_1 = p_2 = p$ and $r = r_2 - r_1$, where p is the momentum of the relative motion of the particles, and the formula (2) simplifies to

$$\mathcal{H}_1 = \frac{1}{2}\left(\frac{p^2}{m_1}+\frac{p^2}{m_2}\right) - \frac{p^4}{8c^2}\left(\frac{1}{m_i^3}+\frac{1}{m_2^3}\right) + \frac{q_1q_2}{r} + \frac{q_1q_2}{2c^2m_1m_2r}[p^2+(p\cdot n)^2], \tag{3}$$

where $n = r/|r|$.

6.13 Tensor Formulation of the EM Lagrangian

We start with the equation of motion of a particle of charge q and mass m in the form

$$\frac{d}{dT}P^{\mu} = \frac{q}{c}\eta_{\nu\xi}F^{\mu\xi}U^{\nu} = \frac{q}{c}F^{\mu\xi}U\xi, \tag{6.13.1}$$

where $\eta_{\mu\gamma}$ is given in (11.2.16), $F^{\mu\gamma}$ in (12.2.10) and P^{μ} is the canonical momentum. Since $F^{\mu\gamma} = \partial^{\mu}A^{\nu} - \partial^{\nu}A^{\mu}$, which follows from the conditions of the Lorentz gauge, we have

$$\frac{dP^{\mu}}{dT} = \frac{q}{c}\left(\frac{\partial A^{\nu}}{\partial x_{\mu}} - \frac{\partial A^{\mu}}{\partial x_{\nu}}\right)u_{\nu}. \tag{6.13.2}$$

We find that

$$\frac{dA^{\mu}}{dT} = \frac{dA^{\mu}}{dx_{\nu}}\cdot\frac{dx_{\nu}}{dT} = \frac{\partial A^{\mu}}{\partial x_{\nu}}U_{\nu}. \tag{6.13.3}$$

A^{μ} is a function of x^{μ},

$$\frac{dA^{\mu}}{dT} = \frac{dA^{\mu}}{dx_{\nu}}\frac{dx_{\nu}}{dT} = \frac{\partial A^{\mu}}{\partial x_{\nu}}U_{\nu}. \tag{6.13.4}$$

Hence (6.13.1) becomes

$$\frac{dP^{\mu}}{dT} = \frac{q}{c}\frac{\partial A^{\nu}}{\partial x_{\mu}}U_{\mu}, \tag{6.13.5}$$

where following (6.10.20) we write the tensor form of the canonical 4-vector momentum,

$$P^{\mu} = \left(p^{\mu} + \frac{q}{c}A^{\mu}\right). \tag{6.13.6}$$

Hence

$$L_1 = \left(p^{\mu} + \frac{q}{c}A^{\mu}\right)U_{\mu} = \frac{m_0 c U^{\mu}U_{\mu}}{\sqrt{-U^{\mu}U_{\mu}}} + \frac{1}{c}A^{\mu}j_{\mu}, \quad j_{\mu} = qU_{\mu}. \tag{6.13.7}$$

Differentiation of L_1 partially with respect to U_{μ} gives

$$\frac{\partial L_1}{\partial u_{\mu}} = \frac{m_0 c\eta^{\mu\nu}(u_{\nu} + u_{\xi}\delta^{\xi}_{\gamma})}{2\sqrt{-u_{\alpha}u_{\beta}\eta^{\alpha\beta}}} + \frac{q}{c}A^{\mu} \tag{6.13.8}$$

$$= \frac{m_0 c\eta^{\mu\nu}u_{\nu}}{\sqrt{-U_{\alpha}u^{\alpha}}} + \frac{q}{c}A^{\mu} = m_0\gamma u^{\mu} + \frac{q}{c}A^{\mu} \tag{6.13.9}$$

$$= P^{\mu}. \tag{6.13.10}$$

Hence,

$$L_1 = P^{\mu}u_{\mu}, \tag{6.13.11}$$

$$\frac{\partial L_1}{\partial x_{\mu}} = \frac{1}{c}\frac{\partial A^{\mu}}{\partial x_{\mu}}u_{\mu}. \tag{6.13.12}$$

Since x^{μ} and u^{μ} (or p^{μ}) are independent of each other,

$$\frac{\partial x^{\mu}}{\partial p^{\mu}} = 0, \quad \frac{\partial p^{\mu}}{\partial x^{\mu}} = 0 \tag{6.13.13}$$

and the EL equation

$$\frac{dP^{\mu}}{dt} = \frac{d}{dt}\left(\frac{\partial L_1}{\partial u_{\mu}}\right) = \frac{\partial L_1}{\partial x_{\mu}} \tag{6.13.14}$$

recovers (6.13.5).

6.14 EM Energy Momentum Tensors by Variational Principle

The action integral

$$S = \iiiint\limits_R L\left(A_\mu, \frac{\partial A_\mu}{\partial x^\mu}\right) d\mathbf{x} \tag{6.14.1}$$

does not depend explicitly on x^μ (so the system is closed), and is assumed to be a function of A_μ and $\partial A_\mu/\partial x^\mu$ only. Hamilton's principle gives the EL equation

$$\frac{\partial}{\partial x^\mu}\left[\frac{\partial L}{\partial(\partial A_\nu/\partial x^\mu)}\right] - \frac{\partial L}{\partial A_\nu} = 0. \tag{6.14.2}$$

Since

$$\frac{\partial L}{\partial x^\mu} = \frac{\partial L}{\partial A_\nu}\frac{\partial A_\nu}{\partial x^\mu} + \frac{\partial L}{\partial(\partial A_\nu/\partial x^\lambda)}\frac{\partial}{\partial x^\mu}\left(\frac{\partial A_\nu}{\partial x^\lambda}\right), \tag{6.14.3}$$

interchanging the indices λ and μ in the second factor of the last term, and in the first term using (6.14.2) for $\partial L/\partial A_\nu$ we obtain

$$\begin{aligned}\frac{\partial L}{\partial x^\mu} &= \frac{\partial}{\partial x^\lambda}\left[\frac{\partial L}{\partial(\partial A_\nu/\partial x^\lambda)}\right]\frac{\partial A_\nu}{\partial x^\mu} + \frac{\partial L}{\partial(\partial A_\nu/\partial x^\lambda)}\frac{\partial}{\partial x^\lambda}\left(\frac{\partial A_\nu}{\partial x^\mu}\right)\\ &= \frac{\partial}{\partial x^\lambda}\left[\frac{\partial L}{\partial(\partial A_\nu/\partial x^\lambda)}\frac{\partial A_\nu}{\partial x^\mu}\right].\end{aligned} \tag{6.14.4}$$

Hence

$$\frac{\partial}{\partial x^\lambda}\left[\frac{\partial L}{\partial(\partial A_\nu/\partial x^\lambda)}\frac{\partial A_\nu}{\partial x^\mu} - L\delta_\mu^\lambda\right] = 0, \tag{6.14.5}$$

where δ_μ^λ (Kronecker delta) $= 1$ when $\lambda = \mu$ and $= 0$ when $\lambda \neq \nu$. Putting now

$$T_\mu^\lambda = \frac{\partial L}{\partial(\partial A_\nu/\partial x^\lambda)}\frac{\partial A_\nu}{\partial x^\mu}\frac{\partial A_\nu}{\partial x^\mu} - L\delta_\mu^\lambda, \tag{6.14.6}$$

we obtain the conservation relation

$$\frac{\partial T_\mu^\lambda}{\partial x^\lambda} = 0. \tag{6.14.7}$$

Integration over the 4-volume R and application of the Gauss theorem to reduce that to a triple integral, gives

$$\iiint\limits_S T_\mu^\nu dS_\nu = \text{constant}. \tag{6.14.8}$$

The component

$$T_4^4 = \frac{\partial L}{\partial A_\nu}A_\nu - L \tag{6.14.9}$$

is the energy density of the system occupying the volume V of the position space. Hence the total energy of the system is

$$\iiint\limits_V T_4^4 dV. \tag{6.14.10}$$

Consider the variation of

$$L = -\frac{1}{16\pi} F_{\mu\gamma} F^{\mu\gamma}, \tag{6.14.11}$$

which is the first term of (6.16.2). For this purpose, using the Lorentz gauge relations we have

$$\delta L = -\frac{1}{8\pi} F^{\mu\nu} \delta F_{\mu\nu} = -\frac{1}{8\pi} F^{\mu\nu} \left[\delta \left(\frac{\partial A_\nu}{\partial x^\mu} - \frac{\partial A_\mu}{\partial x^\nu} \right) \right]. \tag{6.14.12}$$

Since $F_{\mu\gamma}$ is antisymmetric, this relation can be written as

$$\delta L = -\frac{1}{4\pi} F^{\mu\nu} \delta \left(\frac{\partial A_\nu}{\partial x^\mu} \right). \tag{6.14.13}$$

Hence

$$\frac{\partial L}{\partial(\partial A_\nu / \partial x^\mu)} = -\frac{1}{4\pi} F^{\mu\nu} \tag{6.14.14}$$

and

$$T_\mu^\nu = -\frac{1}{4\pi} \frac{\partial A_\nu}{\partial x^\mu} F^{\nu\lambda} + \frac{1}{16\pi} \delta_\mu^\nu F_{\lambda\xi} F^{\lambda\xi}. \tag{6.14.15}$$

6.15 Tensor Form Equation of Motion of a Charge

The Lagrangian L_1 of (6.13.7) can be written as

$$L_1 = \left[-m_0 c \sqrt{-ds^2} + \frac{q}{c} A_\mu dx^\mu \right] \Big/ dt. \tag{6.15.1}$$

So, the action integral for this L_1 is

$$S = \int_a^b \left[-m_0 c \sqrt{-ds^2} + \frac{q}{c} A_\mu dx^\mu \right]. \tag{6.15.2}$$

Hamilton's principle of least action gives

$$\begin{aligned} \delta S &= \int_a^b \left[m_0 c \frac{dx_\mu d(\delta x^\mu)}{\sqrt{-ds^2}} + \frac{q}{c} \{ (A_\mu d(\delta x^\mu) + \delta A_\mu dx^\mu \} \right] = 0 \\ &= \int_a^b \left[-m_0 c dW_\mu \delta x^\mu - \frac{q}{c} \delta x^\mu dA_\mu + \frac{q}{c} \delta A_\mu dx^\mu \right] + \left[\left(m_0 W_\mu + \frac{q}{c} A_\mu \right) \delta x^\mu \right]_a^b \\ &= 0, \end{aligned} \tag{6.15.3}$$

where

$$W^\mu = \frac{dx^\mu}{\sqrt{-ds^2}}.$$

Since the second term vanishes due to the conditions of variation, it becomes

$$\int_a^b \left[m_0 c dW_\mu \delta x^\mu + \frac{q}{c} \delta x^\mu \frac{\partial A_\mu}{\partial x^\nu} dx^\nu - \frac{q}{c} \frac{\partial A_\mu}{\partial x^\nu} dx^\mu \partial x^\nu \right] = 0, \tag{6.15.4}$$

where

$$\delta A_\mu = \frac{\partial A_\mu}{\partial x^\nu} \delta x^\nu. \tag{6.15.5}$$

Since

$$dW_\mu = \left(\frac{dW_\mu}{\sqrt{-ds^2}}\right)\sqrt{-dS^2} \quad \text{and} \quad dx^\mu = W^\mu\sqrt{-s^2}.$$

Equation (6.15.4) becomes

$$\delta\int_a^b \left[m_0 c\frac{dW_\mu}{\sqrt{-dS^2}} - \frac{q}{c}\left(\frac{\partial A_\nu}{\partial x^\mu} - \frac{\partial A_\mu}{\partial x^\nu}\right)W^\mu\right]\delta x^\mu\sqrt{-dS^2} = 0 \tag{6.15.6}$$

δx^μ being arbitrary, this equation gives the equation of motion in the form

$$m_0 c\frac{dW_\mu}{\sqrt{-dS^2}} = \frac{q}{c}F_{\mu\nu}W^\nu, \quad \text{or,} \quad m_0 c\frac{dW^\mu}{\sqrt{-dS^2}} = \frac{q}{c}F^{\mu\nu}W_\nu. \tag{6.15.7}$$

Now equation (6.15.2) can be written as

$$S = -\int_a^b \left[m_0 c\sqrt{\left\{-g_{\mu\nu}\frac{dx^\mu}{\sqrt{-dS^2}}\frac{dx^\nu}{\sqrt{-dS^2}}\right\}} - \frac{q}{c}A_\mu\frac{dx^\mu}{\sqrt{-dS^2}}\right]\sqrt{-dS^2}. \tag{6.15.8}$$

So the Lagrangian is

$$L = -m_0\sqrt{\left\{-\frac{dx^\mu}{\sqrt{-dS^2}}\frac{dx^\nu}{\sqrt{-dS^2}}g_{\mu\nu}\right\}} - \frac{q}{c}A_\mu\frac{dx^\mu}{\sqrt{-dS^2}}. \tag{6.15.9}$$

The EL equation and the conjugate 4-momentum $\overset{=}{P}^\mu$, are therefore given by

$$\frac{d}{\sqrt{-dS^2}}\left[\frac{\partial L}{\partial(dx^\mu/\sqrt{-dS^2}}\right] - \frac{\partial L}{\partial x^\mu} = 0. \tag{6.15.10}$$

$$\overset{=}{P}^\mu = -\frac{\partial L}{\partial(dx^\mu/\sqrt{-dS^2})} = -\frac{\partial L}{\partial W^\mu} = m_0 W_\mu + \frac{q}{c}A_\mu. \tag{6.15.11}$$

6.16 Total Lagrangian and Total Hamiltonian

Since A^μ satisfies a second order wave equation, L_1 should contain only the first derivatives $\partial A^\mu/\partial x^\nu$. So, one possible choice is

$$L_1 = -\frac{1}{8\pi}(\partial^\mu A^\nu)(\partial_\mu A_\nu) + \frac{1}{c}A^\mu j_\mu, \tag{6.16.1}$$

where ∂^μ and ∂_μ have been defined in (11.2.30) and (11.2.31). An alternative choice is

$$L_1 = -\frac{1}{16\pi}F_{\mu\nu}F^{\mu\nu} + \frac{1}{c}A^\mu j_\mu. \tag{6.16.2}$$

This L_1 and L_1 of (6.13.7) have the common term $A^\mu j_\mu$. So, taking it once only in the sum and summing over all the species of charges per unit volume, we write

$$\begin{aligned} L &= \sum(p^\mu U_\mu) + \frac{1}{c}A^\mu j_\mu - \frac{1}{8\pi}(\partial^\mu A^\nu)(\partial_\mu A_\nu) \\ &= \sum(p_\mu U^\mu) + \frac{1}{c}A_\mu j^\mu - \frac{1}{16\pi}F_{\mu\nu}F^{\mu\nu} \\ &= -\sum\left(m_0 c^2\sqrt{1-u^2/c^2}\right) + \rho\phi + \frac{(A\cdot j)}{c} + \frac{(E^2-B^2)}{8\pi}. \end{aligned} \tag{6.16.3}$$

The last term depends only on the properties of the EM Lagrangian itself, and should exist even in the absence of charges. Hence, in a volume V,

$$\mathcal{L}_T = -\sum_i \left\{ m_i c^2 \sqrt{1 - \frac{u^2}{c^2}} \right\} + \iiint_V \left[\frac{1}{8\pi}(E^2 - B^2) + \rho\phi + \frac{1}{c}(A \cdot j) \right] dV. \tag{6.16.4}$$

Differentiation of terms of equation (6.16.1) with respect to A^μ and $\partial^\mu A^\nu$ gives

$$\frac{\partial L_1}{\partial A^\mu} = \frac{1}{c} j_\mu, \tag{6.16.5}$$

$$\frac{\partial L_1}{\partial(\partial^\mu A^\nu)} = \frac{1}{16\pi} \frac{\partial(F_{\alpha\beta}F^{\alpha\beta})}{\partial(\partial^\mu A^\nu)} = \frac{1}{4\pi} F_{\mu\nu}. \tag{6.16.6}$$

The variation of the action integral for L_1, therefore, gives the EL equation

$$\frac{\partial L_1}{\partial A^\mu} = \frac{d}{dx_\mu} \left\{ \frac{\partial L_1}{\partial(\partial^\mu A^\nu)} \right\} = 0. \tag{6.16.7}$$

The Hamiltonian density $\mathcal{H}$, when expressed in term of A_μ, is

$$\mathcal{H} = P^\mu A_\mu - L_1, \tag{6.16.8}$$

where the Canonical momentum

$$P_\mu = \frac{\partial L_1}{\partial A^\mu}, \quad \text{or,} \quad P^\mu = \frac{\partial L_1}{\partial A_\mu}. \tag{6.16.9}$$

Then

$$P_\mu = -\frac{\partial H}{\partial A^\mu} + \frac{\partial}{\partial x^\nu} \left[\frac{\partial H}{\partial(\partial A^\mu / \partial x^\nu)} \right], \tag{6.16.10}$$

$$\frac{\partial A_\mu}{\partial t} = \frac{\partial H}{\partial P^\mu}. \tag{6.16.11}$$

$$\mathcal{H} = \frac{1}{8\pi}(E^2 - B^2) - \frac{1}{4\pi}(A \cdot E) + \frac{1}{8\pi}\left(\nabla \cdot A - \frac{\dot{\phi}}{c} \right)^2 + \frac{\phi}{4\pi e}\left(\nabla \cdot A - \frac{1}{c}\dot{\phi} \right)$$
$$- \rho\phi - \frac{1}{c}(A \cdot j). \tag{6.16.12}$$

The Lorentz gauge conditions reduce it to

$$\mathcal{H} = \frac{1}{8\pi}(E^2 + B^2) - \frac{1}{4\pi}(E \cdot \nabla\phi) + \frac{1}{8\pi}\left\{ (\nabla \cdot A)^2 - \frac{1}{c}\dot{\phi}^2 \right\} - \zeta\phi - \frac{1}{c}(A \cdot j). \tag{6.16.13}$$

Further reduction gives

$$\mathcal{H} = \frac{1}{8\pi}(E^2 + B^2) - \frac{1}{4\pi}(E \cdot \nabla\phi) - \rho\phi - \frac{1}{c}(A \cdot j). \tag{6.16.14}$$

Integration gives the total Hamiltonian

$$\mathcal{H}_0 = \iiint_V \mathcal{H} dV = \iiint_V \left[\frac{1}{8\pi}(E^2 + B^2) - \frac{1}{c}(A \cdot j) \right] dV, \tag{6.16.15}$$

where it has been assumed that the surface integral

$$\iint_{\Sigma} \phi(E\cdot d\Sigma) \tag{6.16.16}$$

vanishes at infinity with respect to the system and Gauss's law $\nabla\cdot D = 4\pi\rho^c$ has been used.

For dielectrics the first term of (6.16.2), denoted by L_{em}, is replaced by $L_{em} = -(F^{\mu\nu}\mathcal{H}_{\mu\nu})/16\pi$, where

$$F^{\mu\nu} = \begin{bmatrix} 0 & B_z & -B_y & E_x \\ -B_z & 0 & B_x & E_y \\ B_y & -B_x & 0 & E_z \\ -E_x & -E_y & -E_z & 0 \end{bmatrix}, \quad \mathcal{H}^{\mu\nu} = \begin{bmatrix} 0 & H_z & -H_y & -D_x \\ -H_z & 0 & H_z & -D_y \\ H_y & -H_x & 0 & -D_z \\ D_x & D_y & D_z & 0 \end{bmatrix}. \tag{6.16.17}$$

Then

$$L = \sum(p_\mu U^\mu) + \frac{(A_\mu j^\mu)}{c} - \frac{(F_{\mu\nu}\mathcal{H}^{\mu\nu})}{16\pi} \tag{6.16.18}$$

and the vector form of L is

$$L = -\sum m_0 c^2\sqrt{1-\frac{u^2}{c^2}} + \rho\phi + \frac{(A\cdot j)}{c} + \frac{[(E\cdot D)-(B\cdot H)]}{8\pi}. \tag{6.16.19}$$

Since

$$\frac{\partial L}{\partial(\partial_\mu A_\nu)} = \frac{1}{4\pi}\mathcal{H}^{\mu\nu}, \quad \frac{\partial L}{\partial A_\mu} = \frac{1}{c}j^\mu, \tag{6.16.20}$$

the Ampere Maxwell law, together with Gauss's law for D is tensorially represented as

$$\frac{\partial \mathcal{H}^{\mu\nu}}{\partial x^\nu} = \frac{4\pi}{c}j^\mu. \tag{6.16.21}$$

It gives the EL equation

$$\frac{\partial L}{\partial A_\mu} = \frac{d}{dx^\nu}\left[\frac{\partial L}{\partial(\partial_\mu A_\nu)}\right] = 0 \tag{6.16.22}$$

which is an alternate form of (6.16.7). The total Lagrangian then is

$$\mathcal{L} = -\sum_i\left(m_0c^2\sqrt{1-\frac{u_i^2}{c^2}}\right) - \frac{1}{8\pi c}\frac{d}{dt}\iiint_V (D\cdot A)dV + \frac{1}{2}\iiint_V\left[\rho\phi + \frac{1}{c}(A\cdot j)\right]dV$$

$$+\frac{1}{8\pi}\iint_\Sigma (\{\phi D + (H\times A)\}\cdot d\Sigma). \tag{6.16.23}$$

For anisotropic dielectrics

$$\mathcal{H}^{\lambda\mu} = \frac{1}{2}\chi^{\lambda\mu\nu\eta}, \tag{6.16.24}$$

where $\chi^{\lambda\mu\nu\eta}$ is a fourth order constitutive tensor. Then

$$L_{em} = \frac{1}{32\pi}\chi^{\mu\nu\xi\eta}F_{\mu\nu}F_{\xi\eta} = -\frac{1}{32\pi}\chi^{\mu\nu\xi\eta}(\partial_\mu A_\nu - \partial_\nu A_\mu)(\partial_\xi A_\eta - \partial_\eta A_\xi). \tag{6.16.25}$$

6.17 Field Induced Displacement

6.17.1 Two types of variables and their synthesis

Motion of an assembly of particles is described by Lagrangian variables and the fields by Eulerian variables. However, in the fluid continuum approximation, the "field-like" Eulerian variables of hydrodynamics lead to a variational formulation of the equations of motion of non-interacting fluids. In the fluid approximation (also called the *dispersed carrier approximation*) the constituents of all species of charged and uncharged particles are regarded as finely sub-divided (of infinitesimally small dimensions), the individuality of the particles becomes unimportant, and a limiting process involving the density concepts becomes effective, so that the medium becomes mixture of some species of fluids.

The Lagrangian concept of fluid motion is that a particular fluid element is chosen to describe the location of a fluid element and its state in terms of the time co-ordinate alone. The field variables are then regarded as functions of the initial location r_0 of a fluid element and of time t.

The Eulerian variables alone are not sufficient for use in variational problems of electrodynamics, where for representation of the charges and the EM field and their interaction, the "field-like" Lagrangian functions are also necessary. The fact is that the proper choice of variables is a problem in the development of the Lagrangian and Hamiltonian formalisms for electrodynamics, because the generalized (dependent) variables have to be distinguished from the integration (independent) variables. The charged particle trajectories are conventionally expressed in Lagrangian co-ordinates following the particles; whereas the EM field is conventionally expressed in Eulerian co-ordinates which are defined in relation to fixed axes.

The position vector $\mathbf{r}$ plays a dual role in the formulation of the Lagrangian formalism. To find the particle equation of motion from the single particle Lagrangian density description, variation with respect to r is necessary, which requires that r is a dependent variable, and time t is the independent variable. However, in the variations of the potentials A and ϕ, which lead to the field equations of Maxwell, both r and t are to be regarded as independent variables.

The position vector $\mathbf{r}$ is the Integration variable for both the EM field and the particle displacement $\xi(r, t)$ of polarization which measures the distance between a specified particle trajectory $(\mathbf{r}, t)$ and the actual trajectory $(\mathbf{r}+\xi, t)$, Sturrock (1958). Low (1958) introduced two types of co-ordinates into the Lagrangian for the distribution of charges in the six-dimensional phase space of velocity co-ordinates and position co-ordinates. One of these was composed for study of motion of particles in a specified EM field and the other for analysis of the Maxwell equations for specified particle trajectories. The field is treated as a given function of $\mathbf{r}$ in the variation for the particle motion results. However, $\mathbf{r}$ is an integration variable, in the variation leading to the Maxwell equations. So, $\mathbf{r}$ plays the roles of both generalized and integration variables, while the field play the dual role of specified functions and generalised variables, in this "dual role approach", Pang and Grawford (1983). In the Legendre transformation theory (section 6.5) variable u has a dual role; it is a dependent variable in the relations (6.5.2) and (6.5.3), but it plays the role of an independent variable in (6.5.9).

The macroscopic Lagrangian of Peng and Crawford (1983), including a pressure tensor, and effects of heat conduction and elastic collision, is free from this distinction between dependent and independent variables. An expanded view of the macroscopic potential clarifies the physical basis of the approach of Peng and Crawford (1983). The Maxwell equations and the macroscopic force law follow from Hamilton's canonical equations when the microscopic and macroscopic field variables of electrodynamics are properly distinguished. The concept of the macroscopic Lagrangian and Hamiltonian apply equally to both dissipative and non-dissipative media.

6.17.2 Field induced displacement of points in a continuum

In the Lagrangian concept of fluid motion the behaviour of a particular fluid element, specifically its location and its state, are studied as a function of time alone. The field variables are then regarded as functions of the initial location $\mathbf{r}_0$ of a fluid element and of time t. So, the position vector $\mathbf{r}$ is a dependent variable, and time t is the independent variable in this interpretation of fluid motion.

The concept of a field induced displacement vector, for each species of charged fluids, provides a field-like description of its motion, which is a departure from a given (normally steady) state. The polarization vector ξ is given by

$$\mathbf{r} = \mathbf{r}_0 + \boldsymbol{\xi}(\mathbf{r}_0, t). \tag{6.17.1}$$

It measures the displacement of an element of a species, in an arbitrary motion, from its position in a specified (possibly physically not realizable) motion. So, the relations for the particle velocity in terms of the polarization variable ξ, are

$$\dot{r} = u = \dot{r}_0 + \dot{\xi} = u_0 + \dot{\xi}. \tag{6.17.2}$$

Here $\mathbf{r}_0$ is the position vector of a point along a prescribed reference trajectory and $\mathbf{r}$ is the corresponding point in the actual trajectory. So now $\mathbf{r}_0$ and t are independent variables along a prescribed reference trajectory, and the field variables are functions of the initial location $\mathbf{r}_0$ of a fluid element and of time t.

In many-particle models charges are regarded as fully pulverized, so that a full Eulerian treatment, involving the polarization variable ξ, is possible. The Lagrangian L is then expanded by Taylor's series in powers of both ξ and a suitable field component ψ_i say, regarding ξ as small, though there is no mathematical requirement that ξ be small.

Lagrangian co-ordinates are required for setting up a variational principle for the trajectories of an assembly of particles. But, for study of electromagnetism through variational techniques, the charged particles should be described by "field-like" co-ordinates. Both these requirements can be satisfied by the introduction of the displacement vector.

However, this synthesis of field-like and particle-like behaviour is generally more formal than real, because the Lagrangian function is in an unusual and inconvenient form. The approximation of small departure from a given steady state leads to a Lagrangian function which is of normal form.

Since ξ is the displacement vector of the fluid element about the equilibrium position r_0, ξ vanishes initially and at the reference point $\mathbf{r}_0$; so

$$\xi(r, 0) = 0, \quad \xi(0, t) = 0. \tag{6.17.3}$$

Then the macroscopic fluid velocity $u(r, t)$ is given in terms of ξ by the relations

$$u(r, t) = \frac{dr}{dt} = U(r_0 + \xi, t). \tag{6.17.4}$$

Since the displacement amplitude is small, we neglect higher order terms, and obtain

$$u(r, t) = U(r_0, t). \tag{6.17.5}$$

To the lowest order, ξ is regarded as given at a point, but to higher orders it is left undefined through the relations

$$\dot{r} = U(r_0, t) = \frac{\partial}{\partial t}\xi(r, t). \tag{6.17.6}$$

The usual chain rule of differentiation yields the operator relation

$$\nabla = \nabla_0 - (\nabla_0 \cdot \xi)\nabla_0, \tag{6.17.7}$$

where ∇_0 is the gradient operator with respect to $\mathbf{r}_0$.

The equilibrium relations for systems which are passing through a configuration of static equilibrium at time zero, are

$$\nabla_0 p_0 = \frac{1}{c}(j_0 \times B_0) + \rho_0 \nabla_0 \phi_0 = 0. \tag{6.17.8}$$

$$\nabla_0 \times B = \frac{4\pi}{c} j_0 \tag{6.17.9}$$

$$\nabla_0 \cdot B_0 = 0. \tag{6.17.10}$$

The equations for the various perturbed field quantities at $\mathbf{r}$, to first order in ξ, are determined by linearizing the MHD equations

$$\zeta \frac{dU}{dt} = -\nabla p + \frac{1}{c}(j \times B) - \rho \nabla \phi. \tag{6.17.11}$$

$$\frac{\partial \rho}{\partial t} + \nabla \cdot (\rho v) = 0. \tag{6.17.12}$$

$$E + \frac{1}{c}(v \times B) = 0. \tag{6.17.13}$$

$$\left[\frac{\partial}{\partial t} + (v \cdot \nabla)\right](p\rho^{-\gamma}) = 0. \tag{6.17.14}$$

$$\nabla \times E = -\frac{1}{c}\frac{\partial B}{\partial t}, \tag{6.17.15}$$

$$\nabla \times B = \frac{4\pi}{c} j. \tag{6.17.16}$$

Combining (6.17.13) and (6.17.15) and integrating in time, we obtain

$$B = B_0 + Q + (\xi \cdot \nabla)B_0, \tag{6.17.17}$$

where

$$Q = \nabla_0 \times (\xi \times B_0). \tag{6.17.18}$$

Taylor expansion of the external potential field yields the relations

$$\frac{4\pi}{c} j = \frac{4\pi}{c} j_0 - [(\nabla_0 \cdot \xi)\nabla_0] \times B_0 + \nabla_0 \times Q + \nabla_0 \times (\xi \cdot \nabla)B_0, \tag{6.17.19}$$

$$\rho = \rho_0 - \rho_0(\nabla_0 \cdot \xi). \tag{6.17.20}$$

$$p = p_0 - \gamma p_0(\nabla_0 \cdot \xi). \tag{6.17.21}$$

$$\phi = \phi_0 + (\xi \cdot \nabla_0)\phi_0, \tag{6.17.22}$$

which are the first order parts of (6.17.12)–(6.17.16). These involve ξ, but not its time derivative $\dot{\xi}$.

6.17.3 The displacement 4-vector

Though the relativistic formulation in covariant form is possible, the 4-dimensional displacement vector is, to some extent, arbitrary. It is specialized to be a spatial vector (which is the "polarization vector" in continuum mechanics) or to have a spatial part transverse to a given direction. This formulation yields the transition from the canonical energy-momentum tensor to the "physical" energy-momentum tensor.

For defining the displacement 4-vector we define the scalar and vector functions $N(x)$ and $v^\mu(x)$ by

$$N(x) = \frac{n}{\gamma} \tag{6.17.23}$$

$$v^\mu = \frac{dx^\mu}{dT} = \gamma u^\mu, \tag{6.17.24}$$

where

$$\gamma = \left(1 - \frac{u^2}{c^2}\right)^{-1/2}, \quad u^\mu = \frac{dx^\mu}{dt}, \tag{6.17.25}$$

$$ds^2 = -c^2 dT^2 = \eta_{\mu\nu} dx^\mu dx^\nu, \tag{6.17.26}$$

$$\eta_{\mu\nu} = \begin{bmatrix} 1 & 0 & 0 & 0 \\ 0 & 1 & 0 & 0 \\ 0 & 0 & 1 & 0 \\ 0 & 0 & 0 & -1 \end{bmatrix} \tag{6.17.27}$$

$N(\mathbf{x})$ is the proper number density of charges of a species and n is their number density, measured with respect to the given co-ordinate system.

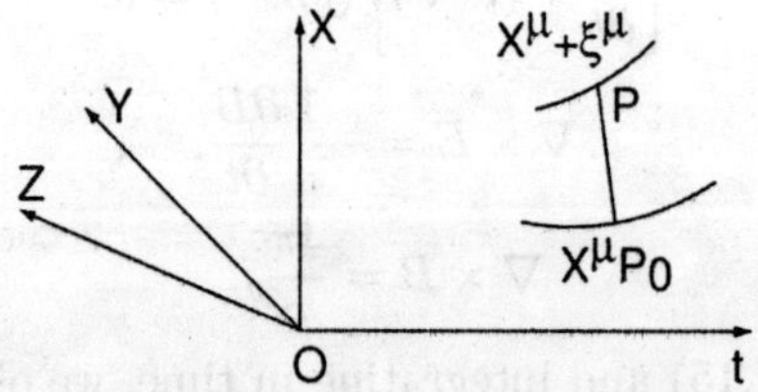

Fig. 6.17.1 Shows the displacement vector in the space-time continuum.

These functions characterize, as the base of the system of variables, an assembly of world lines, which is subsequently identified with a possible set of particle trajectories. The world lines of perturbation (perturbations not necessarily small) are traced out by the same set of particles as the unperturbed assembly of world lies. The perturbed set of world lines is characterized by the function $\xi^\mu(\mathbf{x})$ (where $\mathbf{x}(=\mathbf{x}^\mu)$ is a 4-vector). As shown in fig. 6.17.1, the particle which in the unperturbed system, passed through the event with co-ordinates x^μ is, in the perturbed system, found to pass trough the event with co-ordinates $x^\mu + \xi^\mu(x)$. The description of the perturbed system by the function $\xi^\mu(\mathbf{x})$ is not unique. Only three, and not four functions, are needed for describing the perturbed system. The rule of transformation from ξ^μs to η^μs is

$$\eta^\mu(\mathbf{x}) = \xi^\mu(\mathbf{x}) + \left(v^\mu + v^\nu \cdot \frac{\partial \xi^\mu}{\partial x^\nu}\right) d\sigma = \xi^\mu(\mathbf{x}) + (v^\mu + v^\nu \partial_\nu \xi^\mu)\, d\sigma, \tag{6.17.28}$$

where the differential operator $\partial\nu$ is explained in (11.2.31) and may be regarded as an arbitrary, infinitesimal function of $\mathbf{x}$, as in fig. 6.17.2. This transformation generates a set of representations of the same physical system.

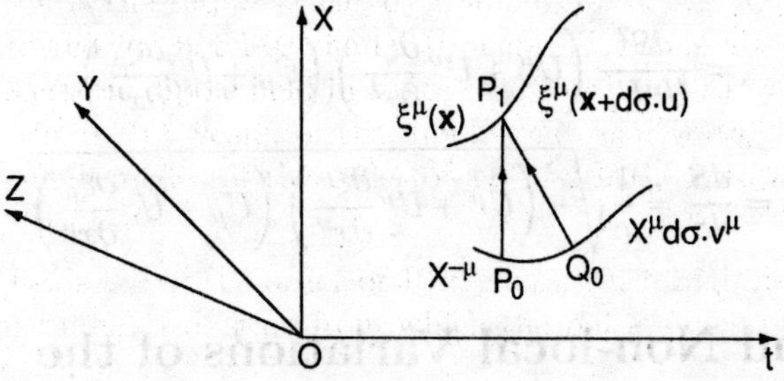

Fig. 6.17.2 Shows the displacement vector in the perturbed system.

The action integrals of section 6.3 are modified when the integrals refer to the perturbed system rather than the unperturbed system. In the unperturbed system the hypervolume of fig. 6.17.3 is broken down into disks of three dimensional area dV and proper time thickness cdT. Under perturbation this hypervolume assumes area dV and thickness cdT. Also, the proper number density N changes to $\bar{N}$. Then the differential contribution changes under perturbation from

$$dxN(= dv\cdot cdt\cdot N) \text{ to } dvcdT\cdot N(= dxN). \tag{6.17.29}$$

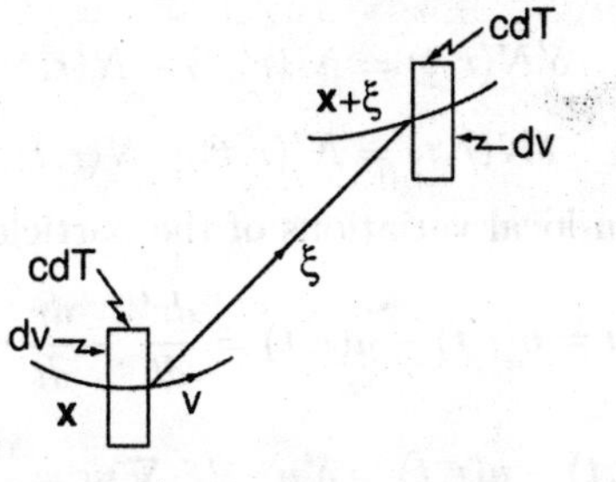

Fig. 6.17.3 Shows the distortion of volume elements due to perturbation.

However, since the number of particles intercepting the hyper area dV cannot change under the perturbation, we see that

$$NdV = \bar{N}d\bar{V}. \tag{6.17.30}$$

Hence the transition (6.17.29) reduces to the transition from

$$d\mathbf{x}\cdot N \text{ to } d\mathbf{x}\cdot N\frac{dT}{d\bar{T}}\left(= dx\cdot N\cdot\frac{d\bar{S}}{dS}\right). \tag{6.17.31}$$

And since

$$d\bar{x}^{\mu} = \bar{v}^{\mu}dT = dT\left(v^{\mu} + \bar{v}^{\mu}\frac{\partial\xi^{\mu}}{\partial x^{\nu}}\right), \tag{6.17.32}$$

it follows that

$$\begin{aligned} d\bar{S}^2 &= d\bar{x}^2 + d\bar{y}^2 + d\bar{z}^2 - c^2d\bar{t}^2 = -c^2d\bar{T}^2 = d\bar{x}^1d\bar{x}_1 + d\bar{x}^2d\bar{x}_2 + d\bar{x}^3d\bar{x}_3 + d\bar{x}^4d\bar{x}_4 \\ &= dT^2\left(v^{\mu} + v^{\nu}\frac{\partial\xi^{\mu}}{\partial x^{\nu}}\right)\left(v_{\mu} + v_{\nu}\frac{\partial\xi_{\mu}}{\partial x^{\nu}}\right). \end{aligned} \tag{6.17.33}$$

Using (6.4.2) this relation gives

$$d\bar{S}^2 = -\frac{1}{c^2} d\bar{S}^2 r^2 \left(U^\mu + U^\nu \frac{\partial \xi^\mu}{\partial x^\nu}\right)\left(U_\mu + U_\nu \frac{\partial \xi_\mu}{\partial x^\nu}\right)$$

$$= \frac{dS^2}{U^\mu U_\mu}\left(U^\mu + U^\nu \frac{\partial \xi^\mu}{\partial x^\nu}\right)\left(U_\mu + U_\nu \frac{\partial \xi_\mu}{\partial x^\nu}\right), \tag{6.17.34}$$

$$\gamma = \frac{d\bar{S}}{dS} = \frac{1}{c}\sqrt{-\left(U^\mu + U^\nu \frac{\partial \xi^\mu}{\partial x^\nu}\right)\left(U_\mu + U_\nu \frac{\partial \xi_\mu}{\partial x^\nu}\right)}. \tag{6.17.35}$$

6.18 Local and Non-local Variations of the Position Vector r

The variation with respect to **r** involves definition of local and non-local variations due to the virtual displacement, Bobroff (1959) and Peng and Crawford (1983),

$$\xi = r' - r. \tag{6.18.1}$$

Actually, the non-local variation $\delta' N$ in the particle number density N is defined by comparing N' (the non-local number density) with N (the local number density) for the same cell, while the local variation δN is defined for the same r, before and after the displacement :

$$\delta' N(r,t) = N'(r',t) - N(r,t). \tag{6.18.2}$$

$$\delta N(r,t) = N'(r,t) - N(r,t). \tag{6.18.3}$$

Similarly, for the local and non-local variations of the particle velocity we have

$$\delta' u = u(r't) - u(r,t) = \frac{dr'}{dt} - \frac{dr}{dt} = \frac{d\xi}{dt}. \tag{6.18.4}$$

$$\delta u = u'(r,t) - u(r,t) = \delta' u - (\xi\cdot\nabla)u = \frac{d\xi}{dt} - (\xi\cdot\nabla)u, \tag{6.18.5}$$

where, for the final expression, the Taylor expansion up to the first order in ξ has been used.

A similar expansion of N is possible with the constraint that the same particles are contained by the cell before and after displacement along ξ. Then, following Peng and Crawford (1983), we can write

$$\frac{\delta' N}{\Delta t} + N\nabla\cdot\left(\frac{1}{\Delta t}\xi\right) = 0. \tag{6.18.6}$$

Cancelling Δt from both sides we obtain

$$\delta' N = -N(\nabla\cdot\xi). \tag{6.18.7}$$

Taylor expansion to first order, in ξ, gives

$$\delta N = N'(r,t) - N(r,t) = \delta' N - (\xi\cdot\nabla)N = -\nabla\cdot(\xi N). \tag{6.18.8}$$

These concepts of local and non-local variations are developed for application in the variations of the electromagnetic total Lagrangian L with respect to the particle co-ordinates, field potentials and the canonical momentum conjugate to these variables, Peng and Crawford (1983).

6.19 Symmetry Principles, Variational Principles, Gauge Theories

In modern physics much emphasis is given to the symmetry principles as a guide, because these lead to new dynamical relations and their uses. Only those fundamental interactions are meaningful which arise by making the Lagrangian form invariant under local gauge transformations. The deep connection that exists between mechanics and EM theory in STR, is illustrative of the approach of the modern gauge theories, which are believed to be followed by all fields of nature.

Transformation theory of physical fields provides the a priori knowledge essential to the use of Lagrangian methods. The uniqueness of the Lagrangian procedure hinges on the behaviour of the transformation of the EM potentials. This behaviour leads to the field equations and the stress energy relations for unspecified media.

The Lagrangian L of a system of charged particles consists of the sum of three Lagrangians L_m, L_I and L_f, where L_m is the Lagrangian for motion of the particles, L_I is that for interaction between the charges and fields and L_f is that for the fields which exist in the matter-free space.

The principle of gauge invariance (or form invariance under gauge transformation) states that the scalar Lagrangian should have the same form when a total time derivative, or a 4-divergence, is added to it, or when the Lagrangians differing by a gauge transformation of the A^μs, are equivalent. This principle gives rise to the gauge transformation properties of the field potentials A^μ. It is applied to the Lagrangian of the EM field, including the field of interaction with the motion of the charged particles, to find the Maxwell equations of the Lorentz theory. The 4-potentials A^μs, which transform according to the usual gauge transformation rule of the EM theory, are, moreover, used as compensating fields in order to preserve the form of the Lagrangian of a classical particle. The fields E and H, and other observable fields do not depend on the arbitrary gauge function, and are defined in the usual way in terms of the A^μs.

The Maxwell equations will be derived in the following sections, on the basis of the symmetry of L under a basic transformation process. This symmetry is emphasized in the study of classical mechanics from the gauge theory point of view. The compensating field is obtained in such a manner that the arbitrary gauge function is eliminated.

Though the field variables of the Maxwell equations are gauge invariant, the potentials A^μ are not. However, the Lagrangians of the EM field, differing only by a gauge transformation on A^μ, are equivalent. The two dynamical Maxwell equations (1.4.1) and (1.4.4) are derived from energy considerations, and also with the help of the principle of least action from the simplest gauge invariant Lagrangian for the EM field. The two kinematical Maxwell equations (1.4.2) and (1.4.3) follow from the definition of the EM field strength tensor in terms of the potentials. Applications of the principle of least action to the Lagrangian of the charged particles leads to the law of motion under the Lorentz force.

The EM field being an observable field, an energy density is associated with it. Since the Lorentz theory involves only two field vectors E and H, it is simpler to adopt, compared to the phenomenological Maxwell theory, for various studies. The reason is that the Maxwell theory involves the four field vectors E, H, D, B and so some additional relations in the form of state equations for the study of quantities like the densities of energy, energy flux and rate of work done for study of variational processes connected to the EM theory.

6.20 Maxwell Equations by a Variational Method

The macroscopic Maxwell equations (1.4.1)–(1.4.4) follow from variational techniques, with the help of the constitutive relations (1.6.2) and (1.6.3), and the Lagrangian, which is regarded as an unspecified function of A, ϕ, E, B. Equations thus derived are valid for all fields with potentials A and ϕ. And, only after the Lagrangian has been specified further, and completely, for our purpose, do they acquire a meaning. Otherwise, they are nothing but empty forms.

Since the field vectors B and E are certain linear combinations of the first order partial derivatives of the potentials A and ϕ, with respect to the position co-ordinates r and time t, a Lagrangian function, depending on A and ϕ, B and E, is the integrand of the action S, in Hamilton's principle of least action,

$$\delta S = \delta \iiiint_R L(\phi, A, E, B) dx = 0. \tag{6.20.1}$$

Applying variation with respect to ϕ, A, E, B, we obtain

$$\iiiint_R \left[\left(\frac{\partial L}{\partial \phi}\delta\phi\right) + \left(\frac{\partial L}{\partial A}\cdot\delta A\right) + \left(\frac{\partial L}{\partial E}\cdot\delta E\right) + \left(\frac{\partial L}{\partial B}\cdot\delta B\right)\right] dx = 0. \tag{6.20.2}$$

We can put

$$\frac{\partial L}{\partial \phi} = -\rho^c, \quad \frac{\partial L}{\partial A} = \frac{1}{c} j \tag{6.20.3}$$

and obtain

$$\iiiint_R \left[-\rho^c \delta\phi + \frac{1}{c}(j\cdot\delta A) + \left(\frac{\partial L}{\partial E}\cdot\delta E\right) + \left(\frac{\partial L}{\partial B}\cdot\delta B\right)\right] dx = 0. \tag{6.20.4}$$

Using the relation (1.9.3) between E and the potentials A and ϕ, in the third term of (6.20.2), it becomes

$$\iiiint_R \left(\frac{\partial L}{\partial E}\cdot\delta E\right) = -\iiiint_R \left(\frac{\partial L}{\partial E}\cdot\delta\left(\frac{1}{c}\frac{\partial A}{\partial t} - \nabla\phi\right)\right) \mathbf{dx}$$

$$= \iiiint_R \sum_i \left[\frac{\partial L}{\partial E_i}\left\{\frac{\partial}{\partial x_i}(\delta\phi) - \frac{1}{c}\frac{\partial}{\partial t}(\delta B_i)\right\}\right] \mathbf{dx}, \tag{6.20.5}$$

where the order of the operators δ and $\partial/\partial x_i$ have been interchanged. Similarly, using $B = \nabla \times A$ in the fourth term of (6.20.4), it reads

$$\iiiint_R \left[\frac{\partial L}{\partial B_x}\delta\left(\frac{\partial A_z}{\partial y} - \frac{\partial A_y}{\partial z}\right) + \cdots\right] \mathbf{dx}$$

$$= \iiiint_R \left[\left\{\frac{\partial L}{\partial B_y}\delta\left(\frac{\partial A_x}{\partial z}\right) - \frac{\partial L}{\partial B_z}\delta\left(\frac{\partial A_x}{\partial y}\right)\right\} + \cdots\right] \mathbf{dx}, \tag{6.20.6}$$

where, in the last step we have brought together the terms containing A_x, A_y, A_z. We now write

$$D = 4\pi\frac{\partial L}{\partial E}, \quad H = -4\pi\frac{\partial L}{\partial B}. \tag{6.20.7}$$

The signs and the constants are chosen so that the Maxwell equations in the CGS units are obtained with the help of these relations. Then (6.20.5) gives

$$\frac{1}{4\pi}\iiiint\limits_R \left[\sum_i D_i\left(\frac{\partial}{\partial x_i}\delta\phi - \frac{1}{c}\frac{\partial}{\partial t}\delta A_i\right)\right] d\mathbf{x}$$

$$= -\frac{1}{4\pi}\iiiint\limits_R \left[\delta\phi\sum_i\left(\frac{\partial D_i}{\partial x_i}\right)\right] d\mathbf{x} + \frac{1}{4\pi c}\iiiint\limits_R \left[\sum_i \frac{\partial D_i}{\partial t}\delta A_i\right] d\mathbf{x}, \tag{6.20.8}$$

where the property that the field variations vanish at the limits of integration at infinity has been used and partial integration by parts with respect to x_i and t have been carried out. Similarly proceeding, (6.20.6) gives

$$-\frac{1}{4\pi}\iiiint\limits_R \left[\left\{H_y\frac{\partial}{\partial z}(\delta A_x) - H_z\frac{\partial}{\partial y}(\delta A_x)\right\} + \cdots + \right] d\mathbf{x}$$

$$= \frac{1}{4\pi}\iiiint\limits_R \left[\delta A_x\left(\frac{\partial H_y}{\partial z} - \frac{\partial H_z}{\partial y}\right) + \cdots\right] d\mathbf{x}$$

$$= \frac{1}{4\pi}\iiiint\limits_R \left[\delta A_x\left(\frac{\partial H_y}{\partial z} - \frac{\partial H_z}{\partial y}\right) + \cdots\right] d\mathbf{x}$$

$$= \frac{1}{4\pi}\iiiint\limits_R (\delta A\cdot\nabla\times H) d\mathbf{x}. \tag{6.20.9}$$

Hence (6.20.4) is equivalent to

$$\iiiint\limits_R \left[-\rho^c\delta\phi + \frac{1}{c}(j\cdot\delta A) - \frac{1}{4\pi}\delta\phi(\nabla\cdot D) + \frac{1}{4\pi}\left(\delta A\cdot\frac{\partial D}{\partial t}\right) - \frac{1}{4\pi}(\delta A\cdot\nabla\times A)\right] d\mathbf{x} = 0. \tag{6.20.10}$$

It can be written as

$$\iiiint\limits_R \left[\left\{-\rho^c + \frac{1}{4\pi}(\nabla\cdot D)\right\}\delta\phi + \left\{\frac{1}{c}j + \frac{1}{c}\frac{\partial D}{\partial t} - \frac{1}{4\pi}(\nabla\times H)\right\}\delta A\right] d\mathbf{x} = 0. \tag{6.20.11}$$

Equating the coefficients of $\delta\phi$ and δA from both sides we recover the dynamical pair of Maxwell equations (1.4.1) and (1.4.4).

6.21 Gauge Transformation and Form Invariance of the Lagrangian of EM Fields

The gauge principle states that the Lagrangian should have the same form when a total time derivative is added to it, and is thus the same as the principle of form invariance under a gauge transformation. Here the symmetry of the Lagrangian is emphasized in the study of classical mechanics from the gauge theory point of view. The field potentials are used as compensating fields to preserve the form of the Lagrangian of a classical particle under the addition of a total derivative.

The Lagrangian of a single classical particle of mass m in presence of a conservative field of potential $\psi(r)$ is

$$L^0 = -mc^2\sqrt{1-\frac{u^2}{c^2}} + \psi(r), \tag{6.21.1}$$

where $u = \dot{r}$. The action S^0 associated with this L^0 is,

$$S^0 = \int_{t_1}^{t_2} L^0 dt. \tag{6.21.2}$$

By varying S^0 with respect to the displacement $\mathbf{r}(t)$, keeping the end points fixed, we obtain

$$\frac{d}{dt}(m\gamma u) = \nabla\psi(r). \tag{6.21.3}$$

We now add the function $-((e/c)(d\wedge/dt))$ to L^0 and obtain a new Lagrangian,

$$L'^0 = L^0 - \frac{e}{c}\frac{d\wedge}{dt}, \tag{6.21.4}$$

where e and c are constants. Then following (6.9.4) we write

$$S'^0 = S^0, \tag{6.21.5}$$

where S'^0 is the corresponding action. Since the displacement $\mathbf{r}$ is a function of time, the total time derivative in (6.21.4) is

$$\frac{d\wedge}{dt} = \frac{d\wedge}{dt} + (\dot{r}\cdot\nabla)\wedge. \tag{6.21.6}$$

For the point particle the density of charge is given by the delta function in

$$P(R,t) = -\rho\dot{r}\delta(R - r(t)), \tag{6.21.7}$$

where R is an arbitrary point in space and $-e$ is the charge of the particle of mass m. The particle vanishes everywhere except at the point $r(t)$. The corresponding charge current is

$$j(R,t) = -e\dot{r}\delta(R - r(t)). \tag{6.21.8}$$

Multiplying both sides of (6.21.4) by $\delta(R - r(t))$ and using (6.21.7) and (6.21.8), and then integrating with respect to r, we find that the new Lagrangian can be written in the form

$$L'^0 = L^0 + \frac{1}{c}\iiint\limits_V \left[j(R,t)\cdot\nabla\wedge(R,t) + CP(R,t)\frac{1}{c}\frac{\partial\wedge}{\partial t}\right] dV, \tag{6.21.9}$$

where V is the material volume, dV is a volume element in V, $\nabla = \partial/\partial R$ and integration is over V. Integration by part gives

$$L'^0 = L^0 - \frac{1}{c}\iiint\limits_V \left((\nabla\cdot j) + \frac{\partial P}{\partial t}\right)\wedge dV + \frac{1}{c}\frac{d}{dt}\iiint\limits_V P\wedge dV, \tag{6.21.10}$$

where it is assumed that the current at infinity is zero. The first integral in the right side vanishes due to the condition of continuity (1.4.5). So (6.21.10) is reduced to (6.21.4), and the Lagrangians L^0 and L^{10} are equivalent.

We now write

$$L_i^0 = L^0 + L_i, \tag{6.21.11}$$

where

$$L_i^0 = L^0 + \frac{1}{c}\iiint\limits_V [(A\cdot j) + \rho^c cA_4]dV. \tag{6.21.12}$$

$$L_i = \frac{1}{c}\iiint\limits_V [(A\cdot j) + \rho^c cA_4]dV. \tag{6.21.13}$$

Then L_i^0 represents the Lagrangian of charged particles interacting with a new field, called the *EM field*, which couples the charges and currents in it. The Lagrangian equivalent to L_i^0 is

$$L_i'^0 = L_i^0 - \frac{e}{c}\frac{d\wedge}{dt}. \tag{6.21.14}$$

Since the total time derivative in (6.21.14) is equal to the integral in (6.21.10), we find that

$$L_i'^0 = L^0 + \frac{1}{c}\iiint\limits_V \left[(j\cdot(A+\nabla\wedge)) - \rho^c c\left(A_4 + \frac{1}{c}\frac{\partial\wedge}{\partial t}\right)\right] dV. \tag{6.21.15}$$

This relation can be written as

$$L_i'^0 = L^0 + \frac{1}{c}\iiint\limits_V [(j\cdot A) + \rho^c cA_4']\, dV \tag{6.21.16}$$

with

$$A' = A + \nabla\wedge, \quad A_4' = A_4 + \frac{1}{c}\frac{\partial\wedge}{\partial t}, \quad \text{or,} \quad A_\mu' = A_\mu + \frac{\partial\wedge}{\partial x^\mu}. \tag{6.21.17}$$

The fields A and A_4 are compensating fields. These are introduced to maintain the form invariance of the Lagrangian under the addition of a total time derivative. Thus, an arbitrary constant vector can be added to the potentials A and ϕ (or A_4). The definitions of E and H contain only derivatives of A and ϕ. So, the addition of constants to these quantities does not affect the field quantities. Only those quantities have physical meaning which are invariants with respect to the transformation relations (6.21.17) for these potentials, which are called the *gauge transformation.* All field equations having significance are gauge invariants under this transformation. The non-uniqueness of the potentials is used to find an auxiliary condition, due to which the potentials become unique.

To find the equation of motion for a charged particle with the help of the principle of least action, we write the action integral

$$S_i^0 = \int_{t_1}^{t_2} L_i^0 dt. \tag{6.21.18}$$

Then the relativistically correct equation of motion, obtained by setting the variation of S_i^0 to zero, is

$$m_0\frac{d}{dt}\left[\frac{\dot{r}}{\sqrt{1-\dot{r}^2/c^2}}\right] = \nabla\psi - e\left(\nabla A_4 - \frac{1}{c}\frac{\partial A}{\partial t}\right) - \frac{e}{c}(r\times(\nabla\times A)). \tag{6.21.19}$$

When L_i^0 is replaced by $L_i^{\prime 0}$, the corresponding action integral $S_i^{\prime 0}$ is defined like S_i^0, as in (6.21.18). Using $L_i^{\prime 0}$ the similar equation of motion, obtained by replacing A and A_4 by A' and A_4', is

$$m\frac{d}{dt}\left[\frac{\dot{r}}{\sqrt{1-\dot{r}^2/c^2}}\right] = \nabla\psi - e\left(\nabla A_4' - \frac{1}{c}\frac{\partial A'}{\partial t}\right) - \frac{e}{c}(\dot{r}\times(\nabla\times A')). \tag{6.21.20}$$

Since A and A_4 are not unique, they are not measurable fields. So, now the magnetic field is defined as

$$H = \nabla\times A. \tag{6.21.21}$$

Since curl of $\nabla\wedge$ is zero, H remains unchanged when A is replaced by A'. Likewise, the electric field E is defined as

$$E = \nabla A_4 - \frac{1}{c}\frac{\partial A}{\partial t}. \tag{6.21.22}$$

It remains unchanged when A and A_4 are replaced by A' and A_4' because space and time derivative of $\wedge$ commute. Hence E and H are invariant under the gauge transformation (6.21.17), and the relativistic equation of motion in presence of the Lorentz force reads

$$m\frac{d}{dt}\left[\frac{\dot{r}}{\sqrt{1-\dot{r}^2/c^2}}\right] = \nabla\psi - eE - \frac{e}{c}(\dot{r}\times H). \tag{6.21.23}$$

This second law of motion can be used as an operational definition to determine whether an electric or magnetic field is present by measuring the acceleration of a test body of mass m and charge $-e$.

Taking curl of (6.21.22) and using (6.21.21) we obtain the Faraday law of EM induction, and the divergence of (6.21.21) gives Gauss's law for magnetic induction. These two vector relations are not sufficient to find E and H completely. So, the principle of least action is used for the fields E and H for finding a proper Lagrangian L_{em} for the EM field, which is essentially a scalar quantity. The simplest way to construct a quadratic scalar quantity from the vectors E and H, is to have a linear combination of their products taking two at a time. So we write

$$\mathcal{L}_{em} = \iiint_V (\alpha E^2 + \beta H^2 + \gamma(E\cdot H))dV, \tag{6.21.24}$$

where the constants are to be evaluated from the results of the variation of the action integral calculated from the total Lagrangian L, where

$$\mathcal{L} = \mathcal{L}^0 + \mathcal{L}_i + \mathcal{L}_{em}, \tag{6.21.25}$$

where $\mathcal{L}^0$ is defined in (6.21.1) for a single particle, $\mathcal{L}_i$ in (6.21.13) and $\mathcal{L}_{em}$ in (6.21.24). The corresponding action is

$$S = \int_{t_1}^{t_2} \mathcal{L}dt. \tag{6.21.26}$$

We now use (6.21.21) and (6.21.22) for H and E in (6.21.24). Then for variation with respect to A_4 we obtain

$$2\alpha(\nabla\cdot E) = \rho^c - \gamma(\nabla\cdot H). \tag{6.21.27}$$

The Gaussian law (1.4.1) then follows from (1.4.2) if we put $\alpha = 1/8\pi$. Next varying (6.21.24) with respect to A we obtain

$$\frac{1}{4\pi c}\frac{\partial E}{\partial t} + \frac{1}{c}j + 2\beta(\nabla \times H) = -\gamma\nabla \times E - \frac{\gamma}{c}\frac{\partial H}{\partial t}. \tag{6.21.28}$$

Then using the Faraday law (1.4.3) the right side vanishes, so we are left with

$$-8\pi\beta\nabla \times H = \frac{1}{c}\frac{\partial E}{\partial t} + \frac{4\pi}{c}j. \tag{6.21.29}$$

For $j = 0$ this equation and (1.4.3) give the EM wave solution in vacuum propagating with velocity C, only when $\beta = -1/8\pi$. Hence (6.21.29) gives the Ampere-Maxwell law, and (6.21.24) now reads

$$\mathcal{L}_{em} = \iiint\limits_V \left[\frac{1}{8\pi}(E^2 - H^2) + \gamma(E\cdot H)\right] dV. \tag{6.21.30}$$

The quantity $(E\cdot H)$ is a pseudo-scalar because H, being equal to $\nabla \times A$, is a pseudo-vector and E which is related to the momentum $\mathbf{p}$ and the potential A, is a polar vector. But $\mathcal{L}_{em}$ should be a true scalar. So, $\gamma = 0$ and

$$\mathcal{L}_{em} = \iiint\limits_V \frac{1}{8\pi}(E^2 - H^2)dV. \tag{6.21.31}$$

In the new Lorentz gauge, the total Lagrangian would be $\mathcal{L}'$ with

$$\mathcal{L}' = \mathcal{L}'^0 + \mathcal{L}_{em}. \tag{6.21.32}$$

Since the fields E and H remains unaffected by the gauge variance, L_{em} remains unchanged. So $\mathcal{L}'$ has the same form as $\mathcal{L}$ in (6.21.25). The corresponding new action S' would be varied with respect to the generalized co-ordinates A' and A'_4. Since the equations, obtained by varying S with respect to A and A_4, involve only E and H, the same results would be obtained by varying S' with respect to A' and A'_4. Hence, the same dynamical equations, as earlier, would be obtained. The formulation is therefore completely gauge invariant, because of the form invariance of the Lagrangian.

Associated with the EM field is a local energy density $\mathcal{W}_F(r,t)$ which is a scalar quantity. The only way a scalar quantity can be constructed from the vectors E and H, as for $\mathcal{L}_{em}$ of (6.21.24) is to write

$$\mathcal{W}_F = \alpha E^2 + \beta H^2 + \gamma(E\cdot H). \tag{6.21.33}$$

Since energy conservation must hold locally, there must be an energy flux S_F associated with $\mathcal{W}_F$. The only way to construct a bilinear vector from E and H is to take their cross product

$$S_F = \zeta(E \times H), \tag{6.21.34}$$

where ζ is a constant.

There should also be sinks or sources for energy due to the interaction of the fields E and H with matter. The only way to couple these vectors with the motion of particles to give a scalar quantity is to take their scalar product with j. Thus the simplest form of the sink or source is Q_F, where

$$Q_F = \xi(E\cdot j) + \eta(H\cdot j) \tag{6.21.35}$$

for the energy per unit volume per unit time, which is removed from (or added to) the field. The equation of continuity for local conservation of energy is

$$\frac{\partial W_F}{\partial t} + \nabla\cdot S_F = Q_F. \tag{6.21.36}$$

Using (6.21.33) to (6.21.35) in (6.21.36) we obtain the equation

$$\left(E\cdot\left[2\alpha\frac{\partial E}{\partial t} + \gamma\frac{\partial H}{\partial t} - \zeta\nabla\times H - \xi j\right]\right)$$

$$+\left(H\cdot\left[2\beta\frac{\partial H}{\partial t} + \gamma\frac{\partial E}{\partial t} + \zeta(\nabla\times E) - \eta j\right]\right) = 0, \tag{6.21.37}$$

where the vector identity (A.3.8) has been used. If we now assume that the EM field equations are linear, we can set the vectors dotted into E and H in equation (6.21.37) equal to zero, because both these vectors are, in general, non-zero. Thus for the dynamical behaviour of fields we obtain

$$2\alpha\frac{\partial E}{\partial t} + \gamma\frac{\partial H}{\partial t} - \zeta(\nabla\times H) - \xi^j = 0. \tag{6.21.38}$$

$$2\beta\frac{\partial H}{\partial t} + \gamma\frac{\partial E}{\partial t} + \zeta(\nabla\times E) - \eta^j = 0. \tag{6.21.39}$$

The potentials A and ϕ, given by (1.9.1) and (1.9.3), satisfying the Lorentz gauge condition (1.9.5), yields the homogeneous pair of Maxwell equations (1.4.2) and (1.4.3). In one sense (1.4.2) and (1.4.3) are inherent in the definitions of the observable fields E and H in terms of A and ϕ. The inhomogeneous pair, (1.4.1) and (1.4.4), follows from (6.21.38) and (6.21.39). To show that we substitute (1.4.3) into (6.21.39) to eliminate $\nabla\times E$, and obtain

$$\left(2\beta - \frac{1}{c}\zeta\right)\frac{\partial H}{\partial t} + \gamma\frac{\partial E}{\partial t} - \eta j = 0. \tag{6.21.40}$$

Since this relation should be valid even for static fields, $\eta j = 0$. This gives $\eta = 0$ because, in general $j \neq 0$. Hence

$$Q_F = \xi(E\cdot j). \tag{6.21.41}$$

So, the magnetic field H does not transfer energy to the particles. Therefore, taking divergence of (6.21.40) we obtain

$$\gamma\frac{\partial}{\partial t}(\nabla\cdot E) = 0 \tag{6.21.42}$$

because (1.9.1) gives $\nabla\cdot B = 0$. Also divergence of (6.21.38) gives

$$2\alpha\frac{\partial}{\partial t}(\nabla\cdot E) - \xi(\nabla\cdot j) = 0. \tag{6.21.43}$$

Hence $\gamma = 0$, because otherwise either $(\nabla\cdot j) = 0$ or $\xi = 0$. But the charge continuity condition shows that $(\nabla\cdot j) = -\partial\rho^c/\partial t \neq 0$ and (6.21.41) shows that $\xi \neq 0$. Hence the energy density W_F of (6.21.33) does not have the cross term $(E\cdot H)$. Another argument for rejection of $(E\cdot H)$ is given after (6.21.30).

Putting now $\eta = 0$ and $\gamma = 0$ in (6.21.40) we obtain

$$2\beta = \frac{\zeta}{c} = 0 \tag{6.21.44}$$

because, in general, $\partial H/\partial t \neq 0$. Now, taking divergence of (6.21.38) we obtain

$$2\alpha\frac{\partial}{\partial t}(\nabla\cdot E) + \xi\frac{\partial\rho}{\partial t} = 0, \tag{6.21.45}$$

where the continuity condition $\nabla\cdot j = -\partial\rho/\partial l$ has been used. Since, on integration, this relation should give $\nabla\cdot E = 4\pi\rho^c$ we write $\xi = -8\pi\alpha_0$. Since the rate of absorption of energy by the particles per unit volume equals the rate of work done by the Lorentz force, this rate must be $(E\cdot j)$ in (6.21.41). We must put $\xi = -1$, and obtain $\alpha = 1/8\pi$. Then (6.21.38) becomes

$$\nabla\times H = \frac{1}{4\pi\zeta}\frac{\partial E}{\partial t} - \frac{1}{\zeta}j = 0. \tag{6.21.46}$$

For $j = 0$, this equation, together with (1.4.3), must give the vacuum wave equation for waves propagating at phase velocity c. So, we must have $4\pi\zeta = c$ and $\beta = 1/8\pi$. Hence, the energy density $\mathcal{W}_F$, the Poynting flux S_F and the energy sink Q_F are

$$\mathcal{W}_F = \frac{1}{8\pi}(E^2 + H^2), \quad S_F = \frac{1}{8\pi}(E\times H), \quad Q_F = -(E\cdot j). \tag{6.21.47}$$

Chapter 7

Lorentz Transformation

7.1 Introduction

The Lorentz transformation (LT) was originally derived for two inertial frames when initially their co-ordinate axes coincide, and the relative velocity is parallel to one of the co-ordinate directions. Such an LT is called a *pure LT*, or a *special LT*, or a *boost transformation.* An LT for which the origins coincide initially is called a *homogeneous LT;* when this condition is violated the LT is inhomogeneous. An inhomogeneous LT is made homogeneous by shifting the space-time origin properly. The whole class of inhomogeneous LTs form a group, which is called an *inhomogeneous Lorentz group* or the *Poincare group.* The homogeneous LT is a subgroup of the Poincare group.

If it is assumed that the two co-ordinate system do not coincide initially, then for uniform relative velocity of the two systems, starting with the LT, results of transformation to systems rotating at uniform angular rate are obtained. These results are more general than those for Newtonian (or Galilean) rotation with absolute time; these have the local time dilation and length contraction, which vary with distance perpendicular to the axis of rotation. For this reason, we derive the classical of theory Thomas precession, and study the general features of the LT, including those of successive LT, and for infinitesimal changes of velocity along an arbitrary direction.

This chapter is a supplementary to Chapter 2 which is on special theory of relativity, for convenience in study of electrodynamics in rotating and other accelerating frames of reference.

7.2 Special Lorentz Transformation

7.2.1 Geometrical relations

All LTs are linear transformations from one co-ordinate system K^0 to another system K, for the same point P, when the frame k moves relative to K^0 with the uniform speed v along a fixed direction. The co-ordinates of P are (x^0, y^0, z^0, t^0) in the system K^0 and (x, y, z, t) in K. The speed of light in vacuum, $c(= 3 \times 10^{10}$ cm/sec), is preserved in the two systems. This gives the invariant relation

$$x^2 + y^2 + z^2 - (ct)^2 = x^{0^2} + y^{0^2} + z^{0^2} - (ct^0)^2 \tag{7.2.1}$$

The notations $x_1(= x), x_2(= y), x_3(= z), x_4(= ct)$, etc. for the co-ordinates of P in the two systems gives

$$x_1^2 + x_2^2 + x_3^2 - x_4^2 = x_1^{0^2} + x_2^{0^2} + x_3^{0^2} - x_4^{0^2}. \tag{7.2.2}$$

A point source of light, located at the origin of the K system, emits a light wave at time $t = 0$. The equation of the wave front is the sphere

$$x_1^2 + x_2^2 + x_3^2 = x_4^2. \tag{7.2.3}$$

For two physical events P and Q, denoted by the four co-ordinates (x_1, y_1, z_1, t_1) for P and by (x_2, y_2, z_2, t_2) for Q in the K system, and by $(x_1^0, y_1^0, z_1^0, t_1^0)$ for P and by $(x_2^0, y_2^0, z_2^0, t_2^0)$ for Q in the K^0 system, this invariant relation reads

$$\begin{aligned}(x_2 - x_1)^2 + (y_2 - y_1)^2 + (z_2 - z_1)^2 - c^2(t_2 - t_1)^2 \\ = (x_2^0 - x_1^0)^2 + (y_2^0 - y_1^0)^2 + (z_2^0 - z_1^0)^2 - c^2(t_2^0 - t_1^0)^2.\end{aligned} \tag{7.2.4}$$

If Q is only infinitesimally separated from P, then assuming that in K the co-ordinates of Q are

$$(x_1 + dx, y_1 + dy, z_1 + dz, t_1 + dt) \tag{7.2.5}$$

and in K^0 these are

$$(x_1^0 + dx^0, y_1^0 + dy^0, z_1^0 + dz^0, t_1^0 + dt^0), \tag{7.2.6}$$

this invariant condition becomes

$$ds^2 = dx^2 + dy^2 + dz^2 - (cdt)^2 = dx^{0^2} + dy^{0^2} + dz^{0^2} - c^2 dt^{0^2}, \tag{7.2.7}$$

where ds is a 4-space sub-interval. The tensorial form of this relation in pseudo cartesian systems is

$$ds^2 = \eta_{\mu\nu}\, dx^\mu dx^\nu. \tag{7.2.8}$$

Evidently, ds^2 thus defined is not a positive definite. The co-ordinates satisfying the condition of variance of (7.2.1) are the preudocartesian Lorentz co-ordinates. Their space is called the *Minkowski space.* Any co-ordinate transformation which preserves the invariance of ds of (7.2.5) is linear in x, y, z and t.

If (i) both the co-ordinate systems K^0 and K are inertial, (ii) the relative velocity v is parallel to one of the axes of the position co-ordinates (x, y, z) (say the z-axis), and is a constant quantity, (iii) all the co-ordinate axes of K^0 and K are parallel, and (iv) the origins of K and K^0 coincide initially (at $t = t^0 = 0$), then the LT satisfied by (7.2.5) consists of the following four relations

$$dx = dx^0, \quad dy = dy^0, \quad dz = \gamma(dz^0 - vdt^0), \quad cdt = \gamma(cdt^0 - \beta dz^0), \tag{7.2.9}$$

where

$$\beta = \frac{v}{c}, \quad \gamma^2 = \frac{1}{(1 - \beta^2)}. \tag{7.2.10}$$

The invariant condition (7.2.1) is satisfied by the relations

$$x = x^0, \quad y = y^0, \quad z = \gamma(z^0 - \beta ct^0), \quad ct = \gamma(ct^0 - \beta z^0). \tag{7.2.11}$$

The reverse relations of transformation are

$$x^0 = x, \quad y^0 = y, \quad z^0 = \gamma(z + \beta ct), \quad ct^0 = \gamma(ct + \beta z). \tag{7.2.12}$$

The vector form (7.2.11),

$$r = r^0 - vt^0 + (\gamma - 1)\frac{v}{v^2}\{(v \cdot r^0) - v^2 t^0\}, \quad ct = \gamma\{ct^0 - (\beta \cdot r^0)\}, \tag{7.2.13}$$

can be written also tensorially as

$$x^\mu = \Lambda^\mu_{0\nu} x^{0\nu}, \tag{7.2.14}$$

where

$$\Lambda^\mu_{0\nu} = \begin{bmatrix} 1 & 0 & 0 & 0 \\ 0 & 1 & 0 & 0 \\ 0 & 0 & \gamma & -\beta\gamma \\ 0 & 0 & \beta\gamma & \gamma \end{bmatrix}. \tag{7.2.15}$$

Its inverse obtains by changing the sign of β. The Jacobian determinant of this linear transformation of the co-ordinates of a point, on using (7.2.10), is unity :

$$J = \frac{\partial(x, y, z, ct)}{\partial(x^0, y^0, z^0, ct^0)} = \begin{vmatrix} 1 & 0 & 0 & 0 \\ 0 & 1 & 0 & 0 \\ 0 & 0 & \gamma & -\beta\gamma \\ 0 & 0 & -\beta\gamma & \gamma \end{vmatrix} = 1. \tag{7.2.16}$$

The same is valid for the reverse transformation

$$\frac{\partial(x^0, y^0, z^0, ct^0)}{\partial(x, y, z, ct)} = 1.$$

The position vectors $r(x, y, z)$ and $r^0(x^0, y^0, z^0)$, on using (7.2.1), follow the invariant relation

$$|r|^2 - (ct)^2 = r^{0^2} - (ct^0)^2. \tag{7.2.17}$$

Equation (7.2.17) being independent of the relative velocity v, shows the constancy of $|r|^2 - (ct)^2$ when subjected to an LT. The square root $\left(|r|^2 - (ct)^2\right)^{1/2}$ is the length of a 4-vector in a pseudocartesian 4-space representing a physical event.

For small relative velocities, $v \ll c$, $\gamma = 1$, and these reduce to the simple relations

$$x = x^0, \quad y = y^0, \quad z = z^0 - vt^0, \quad ct = ct^0 - \beta z^0, \tag{7.2.18}$$

$$J = 1 - \beta^2 \approx 1, \tag{7.2.19}$$

which are not identical to the rules of Galilean Newtonian transformation for absolute space and absolute time.

The 4-volume element $d\mathbf{x}(= dxdydzcdt)$ transforms as

$$d\mathbf{x} = Jd\mathbf{x}^0 = d\mathbf{x}^0. \tag{7.2.20}$$

If J_0 is the Jacobian of transformation in the three dimensional position space, then

$$J_0 = \frac{\partial(x, y, z)}{\partial(x^0, y^0, z^0)} = \begin{vmatrix} 1 & 0 & 0 \\ 0 & 1 & 0 \\ 0 & 0 & \gamma \end{vmatrix} = \gamma \tag{7.2.21}$$

and the spatial volume element dV transforms as

$$dv = J_0 dV^0 = \gamma dV^0. \tag{7.2.22}$$

7.2.2 The velocity 4-vector

The proper time subinterval dT is defined as

$$dS^2 = -c^2 dT^2, \tag{7.2.23}$$

so that

$$dT^2 = \left(1 - \frac{u^2}{c^2}\right) dt^2 = \left(1 - \frac{u^{0^2}}{c^2}\right) dt^{0^2}, \tag{7.2.24}$$

where the velocity vector u is given by

$$u = \left(\frac{dx}{dt}, \frac{dy}{dt}, \frac{dz}{dt}\right) \tag{7.2.25}$$

and the derivative with respect to t is denoted by dot; u is the velocity of a particle at the point $P(x, y, z, t)$ in the inertial frame K, its fourth component is c. The proper time T, a measure of the length of the time track, is determined by a standard clock which follows the motion of the particle. Like ds, the proper time subinterval dT is an invariant.

The relativistic factor γ of (2.7.10) is a constant because v is a constant velocity. Another relativistic factor, defined by

$$\Gamma = \frac{1}{\sqrt{1 - u^2/c^2}} \tag{7.2.26}$$

is not a constant because u is a variable velocity. The Minkowski 4-velocity, V_μ is defined by

$$V_\mu = \left(\frac{dx_1}{dT}, \frac{dx_2}{dT}, \frac{dx_3}{dT}, \frac{dx_4}{dT}\right) = \frac{(\dot{x}, \dot{y}, \dot{z}, c)}{\sqrt{1 - u^2/c^2}} = \Gamma(\dot{x}, \dot{y}, \dot{z}, c). \tag{7.2.27}$$

$$V_\mu V^\mu = \Gamma^2(\dot{x}^2 + \dot{y}^2 + \dot{z}^2 - c^2) = -c^2 < 0. \tag{7.2.28}$$

It is time-like. The relations (7.2.27) show that

$$\frac{V^2}{c^2} = \frac{u^2/c^2}{1 - u^2/c^2} = \frac{1}{1 - u^2/c^2} - 1. \tag{7.2.29}$$

$$\Gamma = \left(1 + \frac{v^2}{c^2}\right)^{1/2}. \tag{7.2.30}$$

7.2.3 The addition theorem of Einstein for velocity

The transformation of the 4-velocity $U_\mu(= dx_\mu/dt)$ is

$$U_x^0 = \frac{U_x}{\gamma(1+\beta u_z/c)}, \quad U_y^0 = \frac{U_y}{\gamma(1+\beta u_z/c)}, \quad U_z^0 = \frac{U_z}{\gamma(1+\beta u_z/c)},$$

$$\frac{dt^0}{dt} = \gamma\left(1+\left(\beta\frac{U_z}{c}\right)\right). \tag{7.2.31}$$

The corresponding relations for the Minkowski 4-velocity are

$$V_x^0 = V_x, \;\; V_y^0 = V_y, \;\; V_z^0 = \gamma\left[V_z + \beta c\left(1+\frac{V^2}{c^2}\right)^{1/2}\right], \quad \frac{dt^0}{dt} = \left(1+\frac{V^2}{c^2}\right)^{1/2} \tag{7.2.32}$$

because

$$U^2 = \frac{V^2}{1+V^2/c^2}. \tag{7.2.33}$$

The transformations of the components parallel and perpendicular to the velocity of translation v, are

$$U_\parallel^0 = \frac{U_\parallel + v}{1+(\beta U_\parallel/c)}, \quad U_\perp^0 = \frac{U_\perp}{\gamma(1+(\beta U_\parallel/c))}. \tag{7.2.34}$$

$$V_\parallel^0 = \gamma\left[V_\parallel + \beta c\left(1+\frac{V^2}{c^2}\right)^{1/2}\right], \quad V_\perp^0 = V_\perp. \tag{7.2.35}$$

The vector forms of (7.2.34) and (7.2.35), respectively, are

$$u^0 = \frac{u + v\{\gamma - (1-\gamma)(u\cdot v)/v^2\}}{\gamma\{1+(\beta\cdot u)/c\}}, \tag{7.2.36}$$

$$V^0 = V + \frac{1}{v^2}(r-1)v(v.V) + \gamma v\left(1+\frac{V^2}{c^2}\right)^{1/2}. \tag{7.2.37}$$

The velocity vectors u (or V) and v, when non-parallel, enter the velocity transformation relations unsymmetrically. This is the property of non-commutativity of the LTs (see section 7.8).

We next square the components of u^0, add, and obtain

$$u^{0^2} = \frac{u^2 + 2u_z v + v^2 - \beta^2(U_x^2 + U_y^2)}{(1+(U_z\beta/c))^2} \tag{7.2.38}$$

$$= \frac{u^2 + 2uv\cos\theta + v^2 - u^2\beta^2\sin^2\theta}{(1+(\beta u\cos\theta/c))^2} \tag{7.2.39}$$

when

$$u_z = u\cos\theta, \quad u_x^2 + u_y^2 = u_\perp^2 = u^2\sin^2\theta. \tag{7.2.40}$$

Similarly, for the Minkowski velocity v we find that

$$V^{0^2} = \gamma^2\left\{\left(\sqrt{c^2+V^2}\right) + \beta V\cos\phi\right]^2 - c^2$$

$$= \gamma^2\left\{\left(\sqrt{c^2+V^2}\right) + \beta V_z\right\}^2 - c^2, \tag{7.2.41}$$

where

$$V_x^2 + V_y^2 = V^2 \sin^2 \phi, \quad V_z = V \cos \phi. \tag{7.2.42}$$

The relation (7.2.38) can be written as

$$U^{0^2} = \frac{U^2 - U_z^2 + \gamma^2(U_z + v)^2}{\gamma^2(1 + (\beta U_z/c))^2}. \tag{7.2.43}$$

7.3 Group Behaviour of LTs

To show the group behaviour (the LT of an LT is again and LT) consider three co-ordinate systems, K^0, K, K^1, such that the velocity of K with respect to K^0 is v, v' is the velocity of K^1 with respect to K, v'' is the velocity of K^1, with respect to k^0, v, is the velocity of k^0 with respect to K^1 and v^0 is the velocity of K^0 with respect to K; that is,

$$K \xrightarrow{v'} K', \ K^0 \xrightarrow{v} K', \ K \xrightarrow{v^0} K^0, \ K' \xrightarrow{v_1} K^0, \ K^0 \xrightarrow{v''} K', \tag{7.3.1}$$

$$v'' = -v_1, \quad v = -v^0, \tag{7.3.2}$$

The LTs from the K^0 frame to K, and from K^1 to K, therefore, are

$$Z = \gamma\left(z^0 - vt^0\right), \quad ct = \gamma\left(ct^0 - \beta z^0\right), \tag{7.3.3}$$

$$Z' = \gamma'\left(z - v't\right), \quad ct' = \gamma'\left(ct - \beta' z\right), \tag{7.3.4}$$

where

$$\gamma' = \left(1 - \beta'^2\right)^{-1/2}, \quad \gamma'' = \left(1 - \beta''^2\right)^{-1/2}, \quad \text{etc.}, \tag{7.3.5}$$

$\beta' = v'/c$, $\beta'' = v''/c$. Using the value of Z and t from (7.3.3) in (7.3.4) and the addition theorem (7.2.31) of velocities, we obtain

$$z' = \gamma\gamma'\left\{z^0 - vt^0 - v'\left(t^0 - \frac{1}{c}\beta z^0\right)\right\} = \frac{z^0(1 + \beta\beta') - (v + v')t^0}{\sqrt{1 - \beta^2 - \beta'^2 + (\beta\beta')^2}}$$

$$= \frac{\left\{z^0 - \dfrac{(v + v')^2 t^0}{(1 + \beta\beta')}\right\}}{\left\{1 - \dfrac{(v + v')^2}{c^2(1 + \beta\beta')^2}\right\}^{1/2}} = \frac{z^0 - v'' t^0}{\sqrt{1 - (v^{0 2}/c^2)}}, \tag{7.3.6}$$

$$t' = \gamma'\left[\gamma\left(t^0 - \frac{\beta}{c} z_0\right) - \frac{\beta'}{c}\gamma(z^0 - vt^0)\right]$$

$$= \gamma\gamma'\left[t^0\left(1 + \frac{v}{c}\beta'\right) - z^0\left(\frac{\beta}{c} + \frac{\beta'}{c}\right)\right] = \frac{t^0 - \dfrac{v''}{c^2} z^0}{\sqrt{1 - \dfrac{v''^2}{c^2}}}. \tag{7.3.7}$$

Hence LTs form a group.

7.4 Transformation of Velocity and Acceleration Vectors

Differentiating (7.2.30) with respect to T gives

$$F_\mu = \frac{dV_\mu}{dT}. \tag{7.4.1}$$

$$F_\mu V^\mu = F^\mu V_\mu = 0. \tag{7.4.2}$$

Hence the 4-acceleration F^μ is orthogonal to the 4-velocity V_μ. And since V_μ is time-like, F^μ is space-like; so

$$F^\mu F^\nu \eta_{\mu\nu} > 0. \tag{7.4.3}$$

Also

$$\begin{aligned}(F.V) &= \left\{\frac{d}{dT}\left(\frac{dx}{dT}\right)\right\}\frac{dx}{dT} + \left\{\frac{d}{dT}\left(\frac{dy}{dT}\right)\right\}\frac{dy}{dT} + \left\{\frac{d}{dT}\left(\frac{dz}{dT}\right)\right\}\frac{dz}{dT} \\ &= \frac{1}{2}\frac{d}{dT}\left[\left(\frac{dx}{dT}\right)^2 + \left(\frac{dy}{dT}\right)^2 + \left(\frac{dz}{dT}\right)^2\right] \\ &= \frac{1}{2}\frac{d}{dT}\left[-c^2 + \frac{c^2}{1-u^2/c^2}\right] = c^2\Gamma^2\dot{\Gamma},\end{aligned} \tag{7.4.4}$$

where

$$\dot{\Gamma} = \frac{d\Gamma}{dt} = \frac{\Gamma^3}{c^2}(u\cdot f), \tag{7.4.5}$$

$$f_\mu = \frac{d^2x_\mu}{dt^2} = \frac{dU_\mu}{dt}. \tag{7.4.6}$$

And since

$$F_4 = c\frac{d}{dT}\left(\frac{dt}{dT}\right)c\Gamma\dot{\Gamma} = \frac{\Gamma^4}{c}(u\cdot f) \tag{7.4.7}$$

we find that

$$F = \Gamma^2 f + \frac{1}{c}F_4 u = \Gamma^2 f + \frac{1}{c^2}\Gamma^4(u\cdot f)u = \Gamma^4\left[f + \frac{1}{c^2}\left\{u\times(u\times f)\right\}\right], \tag{7.4.8}$$

$$F_1^2 + F_2^2 + F_3^2 - F_4^2 = F^2 - F_4^2 = \Gamma^4 f^2 + \frac{1}{c^2}\Gamma^6(u\cdot f)^2 = \Gamma^6\left[f^2 - \frac{1}{c^2}(u\times f)^2\right]. \tag{7.4.9}$$

So, the inequality of (7.4.3) is verified. Taking vector product of the three-vector V with the three-vector F, and using the relations (7.4.8) for F we obtain

$$(V\times F) = \Gamma^3(u\times f). \tag{7.4.10}$$

Differentiation of (7.6.3) with respect to t gives

$$\frac{dr}{dt} = u = \frac{dr}{dt^0}\bigg/\frac{dt}{dt^0}, \tag{7.4.11}$$

$$\frac{dr}{dt^0} = u^0 + v\left\{\frac{1}{v^2}(\gamma-1)(u^0\cdot v) - \gamma\right\}, \tag{7.4.12}$$

$$\frac{dt}{dt^0} = \gamma\left\{1 - \frac{(u^0\cdot v)}{c^2}\right\}, \tag{7.4.13}$$

$$\frac{dt^0}{dt} = \gamma\left\{1 + \frac{1}{c^2}(u\cdot v)\right\}, \tag{7.4.14}$$

$$u = \frac{\frac{1}{\gamma}u^0 + v\left\{\frac{1}{\gamma v^2}(\gamma-1)(u^0\cdot v) - 1\right\}}{1 - \frac{(u^0\cdot v)}{c^2}} \tag{7.4.15}$$

for the velocity transformation. Similarly, for the reverse results (7.6.4) gives

$$u^0 = \frac{dr^0}{dt}\bigg/\frac{dt^0}{dt} = \frac{\frac{1}{\gamma}u + v\left\{\frac{1}{\gamma v^2}(\gamma-1)(v\cdot u) + 1\right\}}{1 + \frac{(u\cdot v)}{c^2}}. \tag{7.4.16}$$

The transformation relations of the acceleration vector are

$$f = \frac{d^2r}{dt^2} = \left[\frac{d}{dt^0}\left(\frac{dr}{dt}\right)\right]\frac{dt^0}{dt} = \left(\frac{dt^0}{dt}\right)\left[\frac{d}{dt^0}\left\{\frac{dr}{dt^0}\left(\frac{dt^0}{dt}\right)\right\}\right]$$

$$= \left(\frac{dt^0}{dt}\right)\left\{\frac{d}{dt^0}\left[\frac{1}{\gamma}u^0 + v\left\{\frac{1}{\gamma v^2}(\gamma-1)(u^0\cdot v) - 1\right\}\right]\right\} \tag{7.4.17}$$

$$= \left(\frac{dt^0}{dt}\right)\left[\frac{\frac{1}{\gamma}f^0 + \frac{1}{\gamma v^2}(\gamma-1)v(f^0\cdot v)}{\left\{1 - \frac{(u^0\cdot v)}{c^2}\right\}} + \frac{(f^0\cdot v)\left\{\frac{1}{\gamma}u^0 + \frac{(\gamma-1)(u^0\cdot v)v}{\gamma v^2} - v\right\}}{c^2\left\{1 - \frac{(u^0\cdot v)}{c^2}\right\}^2}\right] \tag{7.4.18}$$

$$= \frac{\frac{1}{\gamma}f\left\{1 - \frac{1}{c^2}(u^0\cdot v)\right\} + (f^0\cdot v)\left\{\frac{1}{\gamma v^2}(\gamma-1)v + \frac{1}{\gamma c^2}u^0 - \frac{1}{c^2}v\right\}}{\gamma\left\{1 - \frac{(u^0\cdot v)}{c^2}\right\}^3} \tag{7.4.19}$$

$$f^0 = \frac{\frac{1}{\gamma}f\left\{1 + \frac{1}{c^2}(u\cdot v)\right\} + (f\cdot v)\left\{\frac{1}{\gamma v^2}(\gamma-1)v - \frac{1}{\gamma c^2}v - \frac{1}{c^2}v\right\}}{\gamma\left\{1 + \frac{(u\cdot v)}{c^2}\right\}^3}. \tag{7.4.20}$$

Since the fourth component of the 4-velocity u_μ is c, the fourth component of the 4-acceleration f_μ is zero.

If v is an infinitesimal translational velocity, then

$$u = u^0\left\{1 + \frac{(u^0\cdot v)}{c^2}\right\} - v, \tag{7.4.21}$$

$$f = f^0\left\{1 + \frac{2}{c^2}(u^0\cdot v)\right\} + \frac{1}{c^2}(f^0\cdot v)u^0. \tag{7.4.22}$$

Relations (7.4.14) and (7.4.15), for velocity, show that the three velocity vectors u, u^0, v are coplanar. It is evident that if the velocity is constant in one system of co-ordinates, the velocity is constant in the other system also. But, if the acceleration is constant in one system, it is not constant in another system.

Vector product of u with f gives

$$(u \times f) = \frac{(u^0 \times f^0)}{\gamma^2\{1-(u^0\cdot v)/c^2\}^3} + \frac{(u^0 \times v)(f^0\cdot v)(\gamma - 1)}{\gamma^3 v^2\{1-(u^0\cdot v)/c^2\}^3} + \frac{(v \times f^0)}{\gamma^2\{1-(u^0\cdot v)/c^2\}^3} \times \left\{\frac{1}{\gamma v^2}(\gamma-1)(u^0\cdot v) - 1\right\}$$

$$= \frac{[\{\beta \times (u^0 \times \beta) \times f^0]}{\gamma^3\beta^2\{1-(u^0\cdot v)/c^2\}^3} - \frac{\{\gamma v^2 - (u^0\cdot v)\}(v \times f^0)}{\gamma^3 v^2\{1-(u^0\cdot v)/c^2\}^3} + \frac{(\gamma-1)(f^0\cdot \beta)(u^0 \times \beta)}{\gamma^3\beta^2\{1-(u^0\cdot v)/c^2\}^3}. \quad (7.4.23)$$

Its component parallel to v is given by

$$((u \times f)\cdot v) = \frac{((u^0 \times f^0)\cdot v)}{\gamma^3\{1-(u^0\cdot v)/c^2\}^3}. \quad (7.4.24)$$

If v is an infinitesimal velocity, then

$$(u \times f) = (u^0 \times f^0)\left\{1 + \frac{3}{c^2}(u^0\cdot v)\right\} - (v \times f^0). \quad (7.4.25)$$

$$((u \times f)\cdot v) = ((u^0 \times f^0)\cdot v). \quad (7.4.26)$$

The vector $(u \times f)$ is proportional to the Thomas precession effect, sections 7.11–7.13.

▶ **Example 7.4.1** *If the co-ordinate system K moves relative to the K^0 system parallel to the Z-axis, with the velocity v, and the two co-ordinate systems coincide at time $t^0(=t)=0$, then show from the relation* (7.4.20) *that*

$$f_x^0 = \frac{f_x}{\gamma^2(1+u_z v/c^2)} - \frac{u_x v f_z}{\gamma^2 c^2(1+u_z v/c^2)^3},$$

$$f_y^0 = \frac{f_y}{\gamma^2(1+u_z v/c^2)^2} - \frac{u_y v f_z}{\gamma^2 c^2(1+u_z v/c^2)},$$

$$f_z^0 = \frac{f_z}{\gamma^3(1+u_z v/c^2)^3}.$$

7.5 Transformation Matrix of the Homogeneous LT

The LT relations of (7.2.11), or (7.2.13) can be expressed as

$$Ct = \gamma(ct^0 - (\beta\cdot r_0)), \quad (7.5.1)$$

$$r = \alpha\cdot r^0 - \gamma\beta ct^0, \quad (7.5.2)$$

where

$$\boldsymbol{\alpha} = \boldsymbol{I} + (\gamma - 1)\frac{\boldsymbol{\beta\beta}}{\beta^2} \tag{7.5.3}$$

and $\boldsymbol{I}$ is the unit 3×3 matrix. The 3×3 dyad $\boldsymbol{\alpha}$ is given by

$$\boldsymbol{\alpha} = \begin{vmatrix} 1 + \frac{1}{\beta^2}(\gamma - 1)\beta_x\beta_x & \frac{1}{\beta^2}(\gamma - 1)\beta_x\beta_y & \frac{1}{\beta^2}(\gamma - 1)\beta_x\beta_z \\ \frac{1}{\beta^2}(\gamma - 1)\beta_y\beta_x & 1 + \frac{1}{\beta^2}(\gamma - 1)\beta_y\beta_y & \frac{1}{\beta^2}(\gamma - 1)\beta_y\beta_z \\ \frac{1}{\beta^2}(\gamma - 1)\beta_z\beta_x & \frac{1}{\beta^2}(\gamma - 1)\beta_z\beta_y & 1 + \frac{1}{\beta^2}(\gamma - 1)\beta_z\beta_z \end{vmatrix}. \tag{7.5.4}$$

This matrix being symmetric,

$$\boldsymbol{\alpha} = \boldsymbol{\alpha}^T, \tag{7.5.5}$$

$$\boldsymbol{\alpha}^{-1} = \boldsymbol{I} + \left(\frac{1}{\gamma} - 1\right)\frac{1}{\beta^2}(\beta\beta) = \boldsymbol{\alpha} - \gamma\boldsymbol{\beta\beta}, \tag{7.5.6}$$

where $\boldsymbol{\alpha}^T$ is the transpose and $\boldsymbol{\alpha}^{-1}$ is the inverse of $\boldsymbol{\alpha}$. It is easy to see that

$$\boldsymbol{\alpha}^2 = \boldsymbol{I} + \gamma^2\boldsymbol{\beta\beta}, \quad (\boldsymbol{\alpha}^{-1})^2 = \boldsymbol{I} - \boldsymbol{\beta\beta} \tag{7.5.7}$$

$$(\boldsymbol{\beta}\cdot\boldsymbol{\alpha}) = (\boldsymbol{\alpha}\cdot\boldsymbol{\beta}) = \gamma\boldsymbol{\beta} \tag{7.5.8}$$

$$(\boldsymbol{\beta}\cdot\boldsymbol{\alpha}^{-1}) = (\boldsymbol{\alpha}^{-1}\cdot\boldsymbol{\beta}) = \frac{1}{\gamma}\boldsymbol{\beta}. \tag{7.5.9}$$

Since any vector $\mathbf{P}$ satisfies the relation

$$|\alpha.P|^2 = |P|^2 + \gamma^2(\beta\cdot P)^2, \tag{7.5.10}$$

a 3×3 matrix $\bar{\boldsymbol{\beta}}$ can be defined as

$$(\bar{\boldsymbol{\beta}}\cdot\mathbf{P}) = (\beta \times P), \tag{7.5.11}$$

where, again, P is any vector. The anti-symmetric 3×3 matrix

$$\bar{\boldsymbol{\beta}} = \begin{bmatrix} 0 & -\beta_z & \beta_y \\ \beta_z & 0 & -\beta_x \\ -\beta_y & \beta_x & 0 \end{bmatrix} \tag{7.5.12}$$

satisfies the relations

$$(\alpha\cdot\bar{\boldsymbol{\beta}}) = (\bar{\boldsymbol{\beta}}\cdot\alpha) = (\alpha^{-1}\cdot\bar{\boldsymbol{\beta}}) = (\bar{\boldsymbol{\beta}}\cdot\alpha^{-1}) = \bar{\boldsymbol{\beta}}, \tag{7.5.13}$$

$$\bar{\boldsymbol{\beta}}^2 = \bar{\boldsymbol{\beta}}\bar{\boldsymbol{\beta}} - \bar{\boldsymbol{\beta}}^2\boldsymbol{I}, \tag{7.5.14}$$

where we have used relations of the type

$$(\bar{\boldsymbol{\beta}}^2.P) = \beta \times (\beta \times P) = \beta\beta.P - \beta^2 P. \tag{7.5.15}$$

For a small translational velocity v of an inertial frame, the LT of the relation (7.2.9) reduces to (7.2.11). The 3×3 spatial transformation matrix of (7.5.4) becomes the unit diagonal matrix.

The LTs of the magnetic field H^0 and the electric displacement vector D^0 of the frame K^0 to H and D of the K frame, section 2.12, are

$$H = \gamma[H^0 - (\beta \times D^0)] - \frac{\gamma^2}{\gamma+1}\beta(\beta.H^0), \tag{7.5.16}$$

$$D = \gamma[D^0 + (\beta \times H^0)] - \frac{\gamma^2}{\gamma+1}\beta(\beta.D^0). \tag{7.5.17}$$

The matrix form of these two relations is

$$\begin{bmatrix} D \\ H \end{bmatrix} = \boldsymbol{Q}\cdot \begin{bmatrix} D^0 \\ H^0 \end{bmatrix}, \tag{7.5.18}$$

where Q is the 6×6 transformation matrix given by

$$\boldsymbol{Q} = \begin{bmatrix} \gamma - \frac{\gamma^2\beta_x^2}{\gamma+1} & -\frac{\gamma^2\beta_x\beta_y}{\gamma+1} & -\frac{\gamma^2\beta_x\beta_z}{\gamma+1} & 0 & -\gamma\beta_z & \gamma\beta_y \\ -\frac{\gamma^2\beta_y\beta_x}{\gamma+1} & \gamma - \frac{\gamma^2\beta_y^2}{\gamma+1} & -\frac{\gamma^2\beta_y\beta_z}{\gamma+1} & \gamma\beta_x & 0 & -\gamma\beta_y \\ -\frac{\gamma^2\beta_z\beta_x}{\gamma+1} & -\frac{\gamma^2\beta_z\beta_y}{\gamma+1} & \gamma - \frac{\gamma^2\beta_z^2}{\gamma+1} & -\gamma\beta_y & \gamma\beta_x & 0 \\ 0 & \gamma\beta_z & -\gamma\beta_y & \gamma - \frac{\gamma^2\beta_x^2}{\gamma+1} & -\frac{\gamma^2\beta_x\beta_y}{\gamma+1} & -\frac{\gamma^2\beta_x\beta_z}{\gamma+1} \\ -\gamma\beta_z & 0 & -\gamma\beta_x & -\frac{\gamma^2\beta_y\beta_x}{\gamma+1} & \gamma - \frac{\gamma^2\beta_y^2}{\gamma+1} & -\frac{\gamma^2\beta_y\beta_z}{\gamma+1} \\ -\gamma\beta_y & \gamma\beta_x & 0 & -\frac{\gamma^2\beta_z\beta_x}{\gamma+1} & -\frac{\gamma^2\beta_z\beta_y}{\gamma+1} & \gamma - \frac{\gamma^2\beta_z^2}{\gamma+1} \end{bmatrix}. \tag{7.5.19}$$

When v is parallel to the Z-axis, this relation simplifies to

$$\boldsymbol{Q} = \begin{bmatrix} \gamma & 0 & 0 & 0 & -\beta\gamma & 0 \\ 0 & \gamma & 0 & \beta\gamma & 0 & 0 \\ 0 & 0 & 1 & 0 & 0 & 0 \\ 0 & \beta\gamma & 0 & \gamma & 0 & 0 \\ -\beta\gamma & 0 & 0 & 0 & \gamma & 0 \\ 0 & 0 & 0 & 0 & 0 & 1 \end{bmatrix}. \tag{7.5.20}$$

With the help of (7.5.6) and (7.5.12) equation (7.5.19) is written as

$$\boldsymbol{Q} = \gamma \begin{bmatrix} \boldsymbol{\alpha}^{-1} & -\bar{\boldsymbol{\beta}} \\ \bar{\boldsymbol{\beta}} & \boldsymbol{\alpha}^{-1} \end{bmatrix}, \tag{7.5.21}$$

where α^{-1} is the inverse of α of (7.5.6).

7.6 General Lorentz Transformation

In the general LT, the direction of translation at the uniform rate v is inclined at an angle θ to the Z-axis of the K system. The relations of (7.2.13) are then replaced by the relations

$$r = r^0 + \beta\left\{\frac{1}{\beta^2}(r^0\cdot\beta)(\gamma-1) - \gamma ct^0\right\}, \quad ct = \gamma\{ct^0 - (r^0\cdot\beta)\}, \tag{7.6.1}$$

where $\mathbf{r}$ is the position vector of a point P in the K system, and $\mathbf{r}^0$ is the same of P in K^0. Writing

$$x_4 = ict, \quad x_4^0 = ict^0, \tag{7.6.2}$$

equations (7.6.1) become

$$r = r^0 + \beta(\beta\cdot r^0)\frac{1}{\beta^2}(\gamma-1) + i\gamma\beta x_4^0, \quad x_4 = \gamma\{\gamma_4^0 - i(\beta\cdot r^0)\}. \tag{7.6.3}$$

The reverse relations of transformation, obtained by changing the sign of v, are

$$r^0 = r + \beta(\beta\cdot r)\frac{1}{\beta^2}(\gamma-1) - i\gamma\beta x_4, \quad x_4^0 = \gamma\{x_4 + i(\beta\cdot r)\}. \tag{7.6.4}$$

Hence the 4-linear relations between the x^μs and x^{0^μ}s are

$$x^1 = \left\{1 + (\gamma-1)\frac{\beta_x^2}{\beta^2}\right\}x^{0^1} + (\gamma+1)\frac{\beta_x\beta_y}{\beta^2}x^{0^2} + (\gamma-1)\frac{\beta_z\beta_x}{\beta^2}x^{0^3} + i\gamma\beta_x x^{0^4} \tag{7.6.5}$$

$$x^2 = (\gamma-1)\frac{\beta_y\beta_x}{\beta^2}x^{0^1} + \left\{1 + (\gamma-1)\frac{\beta_y^2}{\beta^2}\right\}x^{0^2} + (\gamma-1)\frac{\beta_y\beta_z}{\beta^2}x^{0^3} + i\gamma\beta_y x^{0^4} \tag{7.6.6}$$

$$x^3 = (\gamma-1)\frac{\beta_z\beta_x}{\beta^2}x^{0^1} + (\gamma-1)\frac{\beta_y\beta_z}{\beta^2}x^{0^2} + \left\{1 + (\gamma-1)\frac{\beta_z^2}{\beta^2}\right\}x^{0^3} + i\gamma\beta_z x^{0^4} \tag{7.6.7}$$

$$x^4 = -i\gamma\beta_x x^{0^1} - i\gamma\beta_y x^{0^2} - i\gamma\beta_z x^{0^3} + \gamma x^{0^4}. \tag{7.6.8}$$

These relations are of the type

$$x^\mu = \Lambda^\mu_{0_\nu} x^{0\nu}, \quad \text{or,} \quad x^{0^\mu} = \Lambda^{0_\nu}_\nu x^\nu, \tag{7.6.9}$$

$$\Lambda^\mu_{0\nu} = \begin{bmatrix} 1+(\gamma-1)\frac{\beta_x^2}{\beta^2} & (\gamma-1)\frac{\beta_x\beta_y}{\beta^2} & (\gamma-1)\frac{\beta_x\beta_z}{\beta^2} & i\gamma\beta_x \\ (\gamma-1)\frac{\beta_y\beta_x}{\beta^2} & 1+(\gamma-1)\frac{\beta_y^2}{\beta^2} & (\gamma-1)\frac{\beta_y\beta_z}{\beta^2} & i\gamma\beta_z \\ (\gamma-1)\frac{\beta_z\beta_x}{\beta^2} & (\gamma-1)\frac{\beta_z\beta_y}{\beta^2} & 1+(\gamma-1)\frac{\beta_z^2}{\beta^2} & i\gamma\beta_z \\ -i\gamma\beta_x & -i\gamma\beta_y & -i\gamma\beta_z & \gamma \end{bmatrix}, \tag{7.6.10}$$

$$\Lambda^\mu_{0\nu}(-\beta) = \Lambda^{\mu-1}_{0\nu}(\beta) = \Lambda^{0\mu}_\nu, \tag{7.6.11}$$

where $\Lambda^{\mu-1}_{0\nu}$ (or $\Lambda^{0\mu}_\nu$) is the inverse of the matrix $\Lambda^\mu_{0\nu}(\beta)$ of the general LT. Both $\Lambda^\mu_{0\nu}$ and $\Lambda^{0\mu}_\nu$ are essentially non-singular matrices.

Comparison of (7.2.11) with the system of equations (7.6.5)–(7.6.8) show that any shift in the origin of the co-ordinate systems for a spatial translation, plus a change in the zero of time, should be taken into account for obtaining the latter system. Specifically, the inhomogeneous LT of (7.6.5)–(7.6.8), which preserves the vacuum velocity of light, is of the form (7.7.1), or

$$x = \wedge_0 \cdot x^0 - x^0, \tag{7.6.12}$$

where $\mathbf{x}^0$ is an arbitrary translation vector of the origin, and $\wedge_0$ is the orthogonal matrix of (7.6.10). Equations (7.6.5)–(7.6.8) can be written as

$$\mathbf{x} = \wedge_0 \cdot \mathbf{x}^0, \tag{7.6.13}$$

where

$$\wedge_{0j}^{i} = \delta_j^i + \frac{1}{\beta^2}\beta_i\beta_j(\gamma - 1), \quad \wedge_{0j}^{4} = -i\beta_j\gamma, \quad \wedge_{04}^{0} = \gamma \tag{7.6.14}$$

and $(i, j) = (1, 2, 3)$. If $\wedge_0^T$ is the transpose of $\wedge_0$, then we can write

$$\wedge_0 \cdot \wedge_0^T = A_0, \tag{7.6.15}$$

where A_0 is a 4×4 matrix to be specified in the following lines. Since the LT is an orthogonality transformation in Euclidean 4-space, the relations

$$\wedge_0 \cdot \wedge_0^T = I, \tag{7.6.16}$$

$$|\wedge_0|^2 = I \tag{7.6.17}$$

hold good, where **I** is the 4×4 unit matrix, and $|\wedge_0|$ is the determinant of the 4×4 transformation matrix $\wedge_0$. Equations (7.6.15) and (7.6.16) give the 10 independent relations,

$$\begin{aligned} \wedge_{0j}^{i} &= 0 \text{ if } i \neq j, (i, j) = (1, 2, 3) \\ \wedge_{0\mu}^{\mu} &= 1 \text{ with } \mu = (1, 2, 3, 4). \end{aligned} \tag{7.6.18}$$

Since $\wedge_{0\mu}^{\nu}$ has 16 elements, in view of (7.6.18), the homogeneous LT has only 6 independent elements. The Lorentz group, therefore, is a 6-parameter group, out of which three specify the relative orientation of the co-ordinate axes (for example, the three Eulerian angles) and the remaining three specify the relative velocity of the two inertial frames of reference (for example, the three components of the relative velocity v of translation). However, since x_0 of (7.6.12) has four independent values, the inhomogeneous LT has ten independent element. Relations (7.6.10) or (7.6.14) show that the determinant $|\wedge_{0\mu}^{\nu}|$ is real. For proper LTs, only those matrices $\wedge_{0\mu}^{\nu}$ are considered, which give

$$|\wedge_{0\mu}^{\nu}| = 1 \tag{7.6.19}$$

out of the two square roots ±1 of (7.6.17), because these lead continuously to the identity transformation.

The element $\wedge_{04}^{4}$ of (7.6.18) is given by

$$\wedge_{04}^{4} = \wedge_{04}^{1}\wedge_{01}^{4} + \wedge_{04}^{2}\wedge_{02}^{4} + \wedge_{04}^{3}\wedge_{03}^{4} + \wedge_{04}^{4}\wedge_{04}^{4}. \tag{7.6.20}$$

Since $\wedge_{04}^{i}$ (for $i = 1, 2, 3$) are imaginary quantities, evidently

$$\wedge_{04}^{4^2} \geq 1. \tag{7.6.21}$$

This inequality ensures the validity of the inequality (7.7.14).

Consider the four co-ordinates x_0^μ of the origin O of the system K, relative to the system K^0 having origin O^0, at time t. Equations (7.6.5)–(7.6.9) then give

$$x_0^i = i\beta_j x^{04}, \quad x_0^4 = \wedge_{04}^4 x^{04}. \tag{7.6.22}$$

These can be expressed as

$$x_0^j = \wedge_{04}^j x^{04}, \quad x_0^4 = \wedge_{04}^4 x^{04}. \tag{7.6.23}$$

The components of the relative velocity v of 0 being

$$\beta_j = \frac{i\wedge_{04}^j x^{04}}{\wedge_{04}^4 x^{04}} = \frac{i\wedge_{04}^j}{\wedge_{04}^4}, \tag{7.6.24}$$

the relations (7.6.22) show that

$$\left(\wedge_{04}^4\right)^2 [1 - \beta_1^2 - \beta_2^2 - \beta_3^2 - \beta_4^2] = 1, \tag{7.6.25}$$

and equations (7.6.20) reduce to

$$\left(\wedge_{04}^4\right)^2 [1 - \beta_1^2 - \beta_2^2 - \beta_3^2 - \beta_4^2] = 1. \tag{7.6.26}$$

Hence

$$\wedge_{04}^4 = \frac{1}{\sqrt{1-\beta^2}} = \gamma. \tag{7.6.27}$$

► **Example 7.6.1** *Find the LT for motion from an unprimed frame K to a primed frame K^1 correct up to the squares of the relative velocity.*

Answer. Consider the motion of a system of n particles. The spatial co-ordinates of the ith particle are known *functions of time* in the frame K. So,

$$\mathbf{r}_i \equiv \mathbf{r}_i(t); \tag{1}$$

and in the frame K^1,

$$r_i' = r_i - vt + \frac{1}{v^2}(\gamma - 1)\{(v \cdot r_i) - v^2 t\}v, \quad t' = \gamma t - \frac{1}{c^2}\gamma(v \cdot r_i). \tag{2}$$

Solving for t up to the order of v^2/c^2 from the second relation of (2) we obtain

$$t = t' + \frac{1}{c^2}(v \cdot r_i(t')) - \frac{v^2}{2c^2}t'. \tag{3}$$

Hence, writing $t = t' + \delta t'$, equation (3) gives

$$\delta t' = \frac{1}{c^2}(v \cdot r_i(t)) - \frac{v^2}{2c^2}t', \tag{4}$$

$$r_i'(t') = r_i(t) - tv + \frac{1}{2c^2}\{(v\cdot r_i(t))\}v - \frac{v^2}{2c^2}tv$$

$$= r_i(t' + \delta t') - (t + \delta t')v + \frac{1}{2c^2}(v\cdot r_i(t'))v - \frac{v^2}{2c^2}t'v. \tag{5}$$

Since, in (5) the last two terms contain the factor $1/c^2$, t is replaced in these by t', and since the first two terms are free from $1/c^2$, t is therefore replaced by $t' + \delta t'$. The Taylor series expansion of the first term shows that

$$r_i(t' + \delta t') = r_i(t') + \delta t' \cdot \frac{\partial r_i(t')}{\partial t'} = r_i(t') + u_i(t')\left\{\frac{1}{c^2}(v\cdot r_i(t'))\right\} - \frac{1}{2c^2}v^2 t' u_i(t'), \tag{6}$$

where

$$u_i = \left[\frac{dr_i(t)}{dt}\right]_{t=t'} \tag{7}$$

and (4) has been used for $\delta t'$. Using moreover this $\delta t'$ in the second term of (5) we find that

$$r_i'(t') = r_i(t') + \frac{1}{c^2}(v\cdot r_i(t'))u_i(t') - \frac{1}{2c^2}v^2 t' u_i(t') - t'v - \frac{1}{2c^2}(v\cdot r_i(t'))v. \tag{8}$$

Differentiating (8) with respect to t' we obtain

$$v_i(t') = v(t') + \frac{1}{c^2}(u_i\cdot r_i(t')\dot{u}_i(t') - \frac{1}{2c^2}v^2 t' \dot{u}_i(t') + \frac{1}{c^2}(v\cdot u_i(t'))u_i(t')$$

$$- \frac{1}{2c^2}v^2 u_i(t') - v - \frac{1}{2c^2}(v\cdot r_i)v. \tag{9}$$

Replacing t' by t, the equations (8) and (9) become

$$r_i' = r_i - vt + \frac{1}{c}(v\cdot r_i)u_i - \frac{1}{2c}v^2 t u_i(t) - \frac{1}{2c^2}(v\cdot r_i)v. \tag{10}$$

$$u_i' = u_i - v + \frac{1}{c^2}(v\cdot u_i)u_i - \frac{1}{2c^2}v^2 t u_i - \frac{1}{2c^2}(v\cdot u_i)v + \frac{1}{c^2}(v\cdot r_i)\dot{u}_i - \frac{1}{2c^2}v^2 t\dot{u}_i. \tag{11}$$

7.7 Some General Features of LT

If (l, m, n) are the direction cosines of v, then

$$\frac{1}{v}(v_x, v_y, v_z) = (l, m, n), \quad l^2 + m^2 + n^2 = 1, \tag{7.7.1}$$

and the transformation matrix of (7.6.10) becomes

$$\wedge^{\mu}_{0\nu} = \begin{bmatrix} 1 + (\gamma - 1)l^2 & (\gamma - 1)lm & (\gamma - 1)nl & i\gamma\beta_x \\ (\gamma - 1)lm & 1 + (\gamma - 1)m^2 & (\gamma - 1)ml & i\gamma\beta_y \\ (\gamma - 1)nl & (\gamma - 1)mn & 1 + (\gamma - 1)n^2 & i\gamma\beta_z \\ -i\gamma\beta_x & -i\gamma\beta_y & -i\gamma\beta_z & \gamma \end{bmatrix}. \tag{7.7.2}$$

Denoting by Λ_1 and Λ_2 the LTs for the velocities v_1 and v_2, respectively, we find that

$$\Lambda_1\Lambda_2 = \begin{bmatrix} 1+(\gamma_1-1)l_1^2 & (\gamma_1-1)l_1m_1 & (\gamma_1-1)n_1l_1 & i\gamma_1\beta_{1x} \\ (\gamma_1-1)l_1m_1 & 1+(\gamma_1-1)m_1^2 & (\gamma_1-1)m_1n_1 & i\gamma_1\beta_{1y} \\ (\gamma_1-1)n_1l_1 & (\gamma_1-1)m_1n_1 & 1+(\gamma_1-1)n_1^2 & i\gamma_1\beta_{1z} \\ -i\gamma_1\beta_{1x} & -i\gamma_1\beta_{1y} & -i\gamma_1\beta_{1z} & \gamma_1 \end{bmatrix} \times$$

$$\begin{bmatrix} 1+(\gamma_2-1)l_2^2 & (\gamma_2-1)l_2m_2 & (\gamma_2-1)n_2l_2 & i\gamma_2\beta_{2x} \\ (\gamma_2-1)l_2m_2 & 1+(\gamma_2-1)m_2^2 & (\gamma_2-1)m_2n_2 & i\gamma_2\beta_{2y} \\ (\gamma_2-1)n_2l_2 & (\gamma_2-1)m_2n_2 & 1+(\gamma_2-1)n_2^2 & i\gamma_2\beta_{2z} \\ -i\gamma_2\beta_{2x} & -i\gamma_2\beta_{2y} & -i\gamma_2\beta_{2z} & \gamma_2 \end{bmatrix}. \tag{7.7.3}$$

Now writing $\Lambda_1\Lambda_2 = [a_{ij}]$ and $\Lambda_2\Lambda_1 = [b_{ij}]$, it is evident that

$$\begin{aligned} a_{12} &= (\gamma_1-1)l_1m_1 + (\gamma_2-1)l_2m_2 + pl_1m_2 + \gamma_1\gamma_2\beta_{1x}\beta_{2y}, \\ b_{12} &= (\gamma_1-1)l_1m_1 + (\gamma_2-1)l_2m_2 + pl_2m_1 + \gamma_1\gamma_2\beta_{1y}\beta_{2x}, \\ a_{12} - b_{12} &= p(l_1m_2 - l_2m_1) + \gamma_1\gamma_2(\beta_1 \times \beta_2) = -(a_{21} - b_{21}), \text{ etc.,} \end{aligned}$$

where $p = (\vee_1 - 1)(\vee_2 - 1)\cos\theta$, $\cos\theta = l_1l_2 + m_1m_2 + n_1n_2$, and θ is the angle between v_1 and v_2. When v_1 and v_2 are not parallel, at lest one component of $(\beta_1 \times \beta_2)$ is non-zero; so, at least one of $a_{ij} - b_{ij}$, with $i \neq j$, does not vanish. Then, from the general LT rule also the relation $\Lambda_1\Lambda_2 \neq \Lambda_2\Lambda_1$, established for non-commutativity in (7.8.11), is obtained. In fact, the transformation matrix Λ_3, given by $\Lambda_3 = \Lambda_1\Lambda_2$, will become an LT if the time axes of $\Lambda_1, \Lambda_2, \Lambda_3$ are coplanar.

The equations (7.6.5)–(7.6.8) can be written as

$$x_\mu = x_\mu^0 + \sum_\nu \varepsilon_{\mu\nu} x_\nu^0, \tag{7.7.4}$$

and more elegantly as

$$x_\mu = \sum_\nu (\delta_{\mu\nu} + \varepsilon_{\mu\nu}) x_\nu^0, \tag{7.7.5}$$

where $\delta_{\mu\nu}$ is the kronecker delta, Appendix 4. Matrix representation of this relation is

$$\boldsymbol{X} = (\boldsymbol{I} + \varepsilon)\boldsymbol{X}^0 = \boldsymbol{A}\boldsymbol{X}^0, \tag{7.7.6}$$

where $\boldsymbol{I}$ is the unit diagonal matrix and $\boldsymbol{A} = \boldsymbol{I} + \varepsilon$. Hence

$$\varepsilon_{\mu\nu} = \begin{bmatrix} (\gamma-1)l^2 & (\gamma-1)lm & (\gamma-1)nl & i\gamma\beta_x \\ (\gamma-1)lm & (\gamma-1)m^2 & (\gamma-1)mn & i\gamma\beta_y \\ (\gamma-1)nl & (\gamma-1)mn & (\gamma-1)n^2 & i\gamma\beta_z \\ -i\gamma\beta_x & -i\gamma\beta_y & -i\gamma\beta_z & \gamma-1 \end{bmatrix} \tag{7.7.7}$$

$$\det \varepsilon_{\mu\nu} = (\gamma-1)^3 lmn = \begin{bmatrix} l & m & n & i\gamma\beta/(\gamma-1) \\ l & m & n & i\gamma\beta/(\gamma-1) \\ l & m & n & i\gamma\beta/(\gamma-1) \\ -i\gamma\beta l & -i\gamma\beta m & -i\gamma\beta n & (\gamma-1) \end{bmatrix} \tag{7.7.8}$$

because $\beta_x = l\beta$, etc. So,

$$\det \varepsilon_{\mu\nu} = -i\gamma(\gamma-1)^3 l^2 m^2 n^2 \begin{vmatrix} 1 & 1 & 1 & i\gamma\beta/(\gamma-1) \\ 1 & 1 & 1 & i\gamma\beta/(\gamma-1) \\ 1 & 1 & 1 & i\gamma\beta/(\gamma-1) \\ 1 & 1 & 1 & i(\gamma-1)/\gamma \end{vmatrix}. \tag{7.7.9}$$

The matrix $\varepsilon_{\mu\nu}$ is therefore singular. Since

$$v = -v^0. \tag{7.7.10}$$

The inverse of (7.6.4) also obtains from (7.6.10) by interchanging r_0, t_0 with r, t, and replacing v by $-v$ respectively.

When the origins O and O^0 of the co-ordinate systems K and K^0, respectively, coincide initially (that is, $t = t^0 = 0$), then the LTs (homogeneous) are four homogeneous linear algebraic equations in r^0, t^0, and one of the four quantities r, t, or, conversely, in r, t and one of the four quantities r^0, t^0. If the origins do not coincide initially, then the LT (inhomogeneous), as in (7.6.12), can be written as

$$x^\mu = \Lambda^\mu_{0\nu} x^{0\nu} - x^\mu_0. \tag{7.7.11}$$

It reduces to the special LT when $x^\mu_0 = 0$. This inhomogeneous transformation indicates that the distinction between past and future is preserved, so it is known as the *orthochronous Poincare transformation.*

The space inversion transformation,

$$r = -r^0, \quad t = t^0, \tag{7.7.12}$$

which is not included in the restricted Lorentz group (the special LT), takes place if K^0 is a right handed system of co-ordinates, and K is left handed, or vice-versa. The condition for avoiding such a reflection is that the determinant of the LT equals + 1. This property cannot be changed if the direction of motion between K^0 and K is changed through rotation of the frame K relative to K^0.

The mere invariance of ds^2 yields a group which is wider than the group of special LT; which is free from any rotation of space axes. The transformation

$$r = r^0, \quad t = -t^0 \tag{7.7.13}$$

leaves the relations (7.2.5) unchanged, but these interchange past and future. If this is to be avoided, then always we must have

$$\Lambda^4_{04} > 0, \tag{7.7.14}$$

where Λ^4_{04} is given by (7.6.10). The reverse condition will always involve time inversion.

The most general boost obtains from (7.2.8) when the spatial co-ordinate axes of both the co-ordinate systems K and K^0 are given the same rotation or reflection. A general LT, like (7.6.1) or (7.6.3), can be decomposed into a boost like (7.2.10), followed by the time reversal transformation (7.7.13), and (or) by the transformation

$$r = \boldsymbol{A}r^0 + \alpha, \quad t = t^0 + a, \tag{7.7.15}$$

where $\boldsymbol{A}$ is a 3×3 orthogonal matrix, α is a 3×1 column matrix, and a is a scalar constant. So, a combination of two boosts, when performed in succession, is a boost only when the two boosts are parallel.

With the help of a rotation operator R, the first equation of (7.6.4) can be written as

$$R\mathbf{r} = r^0 + v^0\left\{\frac{(\gamma^0 - 1)(r^0 \cdot v^0)}{v^{0^2}} - \gamma^0 t^0\right\}. \tag{7.7.16}$$

Thus the operator R transforms r into (Rr), corresponding to the LT without rotation. In general, R is different from unity. The inverse operator R^{-1} represents the rotation of the cartesian axes in the K system, giving these axes the same orientation as the axes in the K^0 system. Relation (7.7.10) can, therefore, be replaced by

$$Rv = -v^0. \tag{7.7.17}$$

Hence the components of v are not equal to the components of $-v^0$. Applying the operator R^{-1} on all the terms of (7.7.16), and using (7.7.17) one gets

$$r = R^{-1}r^0 - v^0\left\{\frac{1}{v^{0^2}}(\gamma^0 - 1)(r^0 \cdot v) - \gamma^0 t^0\right\}, \quad t = \gamma^0\left\{t^0 - \frac{1}{c^2}(v^0 \cdot r^0)\right\}. \tag{7.7.18}$$

The relations inverse to (7.7.18) are

$$r^0 = R^{-1}r - v^0\left\{\frac{1}{v^2}(\gamma^0 - 1)(r \cdot v) - \gamma^0 t\right\}, \quad t^0 = \gamma^0\left(t - \frac{1}{c^2}(v \cdot r)\right) \tag{7.7.19}$$

in which

$$v = |v| = |v^0| = v^0. \tag{7.7.20}$$

7.7.1 Some properties of homogeneous LTs

When the origins 0^0 and 0 coincide initially ($t^0 = t = 0$), the resulting invariance relations

$$S^2 = x_1^{0^2} + x_2^{0^2} + x_3^{0^2} + x_4^{0^4} = x_1^2 + x_2^2 + x_3^2 + x_4^2 \tag{7.7.21}$$

satisfy the linear relations

$$x^\mu = \Lambda^\mu_{0\nu} x^{0\nu}, \tag{7.7.22}$$

where the LT matrix $\Lambda^\mu_{0\nu}$ is given in (7.6.10). Then

$$S^2 = \sum_\mu \left(x^{0\mu^2}\right) = \sum_\mu x^{\mu^2} = \sum_\mu \left(\sum_\nu \Lambda_{0\nu} x^{0\nu}\right)^2. \tag{7.7.23}$$

The differences $(x^{0^{\mu}} - x^{0^{\mu}})$ between the co-ordinates of two points in K^0 also transform as (7.7.23). Hence

$$\sum_{\mu} (x^{0\mu} - \bar{x}^{0\mu})^2 = \sum_{\mu} (x^{\mu} - \bar{x}^{\mu})^2, \tag{7.7.24}$$

where x^{μ} and $\bar{x}^{\mu}$ are the co-ordinates of the same two points in the K system. Since both $\Sigma x_{\mu}^{0^2}$ and $\Sigma \bar{x}_{\mu}^{0^2}$ are invariants, according to (7.7.21), it is obvious that

$$\sum_{\mu} (\bar{x}_{\mu}^0 \bar{x}_{\mu}^0) = \sum_{\mu} (x_{\mu} \bar{x}_{\mu}). \tag{7.7.25}$$

Putting (7.7.4) in (7.7.25), neglecting squares of $\varepsilon_{\mu\nu}$ and using the Einstein summation convention, one gets

$$x_{\mu}^0 \bar{x}_{\mu}^0 = (x_{\mu}^0 + \varepsilon_{\mu\nu} x^{0\nu})(\bar{x}_{\mu}^0 + \varepsilon_{\mu\nu} \bar{x}^{0\nu}) = x_{\mu}^0 \bar{x}_{\mu}^0 + \bar{x}_{\mu}^0 x_{\nu}^0 + x_{\mu}^0 \varepsilon_{\mu\nu} \bar{x}^{0\nu}; \tag{7.7.26}$$

so

$$x_{\mu}^0 \bar{x}_{\mu}^0 = x_{\mu}^0 \bar{x}_{\mu}^0 + 2x_{\mu}^0 \bar{x}^{0\mu} \varepsilon_{\mu\nu} + (\bar{x}_{\mu}^0 x^{0\nu} + x_{\mu}^0 \bar{x}^{0\nu})(\varepsilon_{\mu\nu} + \varepsilon_{\nu\mu}). \tag{7.7.27}$$

The two sides of (7.7.27) become identical when the conditions

$$\varepsilon_{\mu\nu} = -\varepsilon_{\nu\mu}, \quad \varepsilon_{\mu\mu} = 0 \tag{7.7.28}$$

of antisymmetry are imposed.

When (7.7.22) is imposed for x_{μ} and $\bar{x}_{\mu}$ in the right side of (7.2.8), we obtain

$$\sum_{\mu} \left\{ \left(\sum_{\nu} \Lambda_{0\nu}^{\mu} x^{\nu} \right) \left(\sum_{\lambda} \Lambda_{0\lambda}^{\mu} \bar{x}^{\lambda} \right) \right\} = \sum_{\nu,\lambda} \left(\sum_{\mu} \Lambda_{0\nu}^{\mu} \Lambda_{0\lambda}^{\mu} \right) x^{\nu} \bar{x}^{\lambda}. \tag{7.7.29}$$

Comparison shows that

$$\sum_{\mu} (\Lambda_{0\nu}^{\mu} \Lambda_{0\lambda}^{\mu}) = \delta_{\lambda}^{\nu}, \tag{7.7.30}$$

where δ_{λ}^{ν} is the Kronecker delta. This relation and (7.7.22) give

$$\sum_{\nu} (x_{\mu} \Lambda_{0\nu}^{\mu}) = \sum_{\nu,\lambda} (\Lambda_{0\lambda}^{\nu} x_{0\nu} \Lambda_{0\mu}^{\nu}) = \sum_{\lambda} \left\{ x_{0\lambda} \left(\sum_{\nu} \Lambda_{0\lambda}^{\nu} \Lambda_{0\lambda}^{\nu} \right) \right\}$$

$$= \sum_{\nu} (x_{0\nu} \delta_{\mu}^{\nu}) = x_{0\mu}, \tag{7.7.31}$$

and the inverse relation

$$x_{0\mu} = \sum_{\nu} (x_{\nu} \Lambda_{0\nu}^{\mu}). \tag{7.7.32}$$

Also put (7.7.32) in the left hand side of (7.7.25), and obtain

$$\sum_{\nu} (\Lambda_{0\nu}^{\mu} \Lambda_{0\nu}^{\lambda}) = \delta_{\lambda}^{\mu}. \tag{7.7.33}$$

Since (7.7.30) and (7.7.33) are orthogonality relations for the coefficients $\Lambda_{0\nu}^{\mu}$, the homogeneous LT represents rotation of the co-ordinate system in 4-space. The antisymmetric conditions (7.7.28) for infinitesimal LT are equivalent to the orthogonality condition (7.7.30) or (7.7.33) for a finite LT.

7.8 Non-commutativity of General LTs

Using the imaginary factor $i(=\sqrt{-1})$ with time, as in (7.6.2), the special LT relations (7.2.11) and (7.2.12) become

$$x_1 = x_{01}, \ \ x_2 = x_{02}, \ \ x_3 = \gamma(x_{03} + i\beta x_{04}), \ \ x_4 = \gamma(x_{04} - i\beta x_{03}). \tag{7.8.1}$$

$$x_{01} = x_1, \ \ x_{02} = x_2, \ \ x_{03} = \gamma(x_3 - i\beta x_4), \ \ x_{04} = \gamma(x_4 + i\beta x_3). \tag{7.8.2}$$

The non-singular LT matrix of (7.8.1) being

$$\Lambda^{\mu}_{0\nu} = \begin{matrix} & \rightarrow \mu \\ \downarrow \nu & \end{matrix} \begin{bmatrix} 1 & 0 & 0 & 0 \\ 0 & 1 & 0 & 0 \\ 0 & 0 & \gamma & -i\gamma\beta \\ 0 & 0 & i\gamma\beta & \gamma \end{bmatrix} \tag{7.8.3}$$

the inverse matrix $\Lambda^{-1\mu}_{0\nu}(\beta)$ is given by

$$\Lambda^{\mu}_{0\nu}(-\beta) = \Lambda^{-1\mu}_{0\nu}(\beta). \tag{7.8.4}$$

For three initially coinciding co-ordinate systems, K^0, K, K', this subgroup of LTs being commutative, we have

$$\Lambda(v)\,\Lambda\,(v') = \Lambda\left(\frac{v+v'}{1+(vv'/c^2)}\right). \tag{7.8.5}$$

However, for any angular inclination between the space directions of (say) K and K^0, and a relative velocity v, in an arbitrary direction, a much larger group is obtained, the elements of which require six parameters for their specification. Three of these parameters may be the Eulerian angles for rotation of the space axes of the K system, relative to K^0, and three other parameters form the components of v of the motion of K relative to K^0. This full Lorentz group is not commutative, because rotations of co-ordinate axes are not commutative. If R_1 and R_2 are the matrices of the two rotations, then the order in which these rotations are performed is relevant; so

$$R_1 R_2 \neq R_2 R_1. \tag{7.8.6}$$

For a boost v_1 in the x_1 direction, followed by a boost v_2 in the x_2 direction, the transformation matrixes Λ_1 and Λ_2, respectively, are

$$\Lambda_1(\beta_1) = \begin{bmatrix} \gamma_1 & 0 & 0 & -i\gamma_1\beta_1 \\ 0 & 1 & 0 & 0 \\ 0 & 0 & 1 & 0 \\ i\gamma_1\beta_1 & 0 & 0 & \gamma_1 \end{bmatrix}, \ \Lambda_2(\beta_2) = \begin{bmatrix} 1 & 0 & 0 & 0 \\ 0 & \gamma_2 & 0 & -i\gamma_2\beta_2 \\ 0 & 0 & 1 & 0 \\ 0 & i\gamma_2\beta_2 & 0 & \gamma_2 \end{bmatrix}, \tag{7.8.7}$$

where

$$\beta_i = \frac{v_i}{c}, \quad \gamma_i = \frac{1}{\sqrt{1-\beta_i^2}}. \tag{7.8.8}$$

Their product,

$$\Lambda_1(\beta_1)\,\Lambda_2\,(\beta_2) = \begin{bmatrix} \gamma_1 & \gamma_1\gamma_2\beta_1\beta_2 & 0 & -i\gamma_1\gamma_2\beta_1 \\ 0 & \gamma_2 & 0 & -i\gamma_2\beta_2 \\ 0 & 0 & 1 & 0 \\ i\gamma_1\beta_1 & i\gamma_1\gamma_2\beta_2 & 0 & \gamma_1\gamma_2 \end{bmatrix} \tag{7.8.9}$$

does not take the form of the general LT, (7.6.10), even though $\Lambda_1(\beta_1)$ and $\Lambda_2(\beta_2)$ are of that form. In fact, the product $\Lambda_1(\beta_1)\,\Lambda_2\,(\beta_2)$ is of the form (7.6.10) only if the three time axes of the frames K^0, K_1, K_2 are in the same plane. Similarly, the boost v_2 in the X_2 direction first, followed by the boost v_1 in the x_1 direction gives

$$\Lambda_2(\beta_2)\,\Lambda_1\,(\beta_1) = \begin{bmatrix} \gamma_1 & 0 & 0 & -i\gamma_1\beta_1 \\ \gamma_1\gamma_2\beta_1\beta_2 & \gamma_2 & 0 & -i\gamma_1\gamma_2\beta_2 \\ 0 & 0 & 1 & 0 \\ i\gamma_1\gamma_2\beta_1 & i\gamma_2\beta_2 & 0 & \gamma_1\gamma_2 \end{bmatrix}. \tag{7.8.10}$$

Comparison of (7.8.9) and (7.8.10) shows that

$$\Lambda_1(\beta_1)\,\Lambda_2\,(\beta_2) \neq \Lambda_2(\beta_2)\,\Lambda_1\,(\beta_1). \tag{7.8.11}$$

Since successive application of LTs along two directions at a right angle means a rotation, it is evident that the rotations in two opposite senses are not same; so the LTs are not commutative. Galilean transformations (GTs) are commutative, because these do not depend upon the order in which the operations are performed.

The boost parameter (or the rapidity parameter) θ is defined by $\beta = \tanh\theta$; then

$$\gamma = \cosh\theta, \quad \gamma\beta = \sinh\theta. \tag{7.8.12}$$

Evidently, $\theta \leq \beta \leq 1$, $1 \leq \gamma \leq \infty$. The last two relations of (7.2.11) then reduce to rotation type transformations involving hyperbolic functions of θ :

$$x_3 = -x_{0_4}\sinh\theta + x_{0_3}\cosh\theta, \quad x_4 = x_{0_4}\cosh\theta - x_{0_3}\sinh\theta. \tag{7.8.13}$$

These appear in LTs, because the diagonal elements of the matrix tensor $\eta_{\mu\nu}$ of (7.2.7) are not of the same sign. If all the diagonal elements of $\eta_{\mu\nu}$ were of the same sign, and equal, then circular functions would appear in (7.8.13). So, these are said to produce hyperbolic rotation. For real rotation the hyperbolic functions are replaced by circular functions $\cos\theta$ and $\sin\theta$.

7.9 Infinitesimal LT

The elements of the fourth row and column of the matrix of (7.7.7) are linear in the components of v, apart from the factor γ. The other terms are of higher powers in the components of v. So, in the infinitesimal LT, many terms vanish.

Denoting infinitesimal velocity components $\delta v_x, \delta v_y, \delta v_z$, in this section, by v'_x, v'_y, v'_z, we neglect higher powers of these quantities, denote the matrix of the infinitesimal LT by $\delta\Lambda^{\mu}_{0\nu}$ and of ε^{μ}_{ν} by $\delta\varepsilon^{\mu}_{\nu}$, and from relations (7.6.10) and (7.7.7) obtain

$$\delta\Lambda^{\mu}_{0\nu} = \begin{bmatrix} 1 & 0 & 0 & i\beta'_x \\ 0 & 1 & 0 & i\beta'_y \\ 0 & 0 & 1 & i\beta'_z \\ -i\beta'_x & -i\beta'_y & -i\beta'_z & 0 \end{bmatrix}. \tag{7.9.1}$$

$$\delta\varepsilon^{\mu}_{\nu} = \begin{bmatrix} 0 & 0 & 0 & i\beta'_x \\ 0 & 0 & 0 & i\beta'_y \\ 0 & 0 & 0 & i\beta'_z \\ -i\beta'_x & -i\beta'_y & -i\beta'_z & 0 \end{bmatrix}, \tag{7.9.2}$$

where $\beta'_x = v'_x/c$, etc. These relations can be combined as

$$\delta\Lambda^{\mu}_{0\nu} = \delta^{\mu}_{\nu} + \delta\varepsilon^{\mu}_{\nu}, \tag{7.9.3}$$

where δ^{μ}_{ν} is the kronecker delta, and

$$\delta\varepsilon^{i}_{j} = 0,\ \ \delta\varepsilon^{i}_{4} = -\beta'_i,\ \ \delta\varepsilon^{4}_{4} = 0,\ \ (i,j) = (1,2,3). \tag{7.9.4}$$

For two infinitesimal LTs $\delta\Lambda_1$ and $\delta\Lambda_2$, at the velocities $v'_1(=\delta v_1)$ and $v'_2(=\delta v_2)$, the product is

$$\delta\Lambda_1\,\delta\Lambda_2 = \begin{bmatrix} 1+\beta'_{1x}\beta'_{2x} & \beta'_{1x}\beta'_{2y} & \beta'_{1x}\beta'_{2z} & i(\beta'_{1x}+\beta'_{2x}) \\ \beta'_{1y}\beta'_{2x} & 1+\beta'_{1y}\beta'_{2y} & \beta'_{1y}\beta'_{2z} & i(\beta'_{1y}+\beta'_{2y}) \\ \beta'_{1z}\beta'_{2x} & \beta'_{1z}\beta'_{2y} & 1+\beta'_{1z}\beta'_{2z} & i(\beta'_{1z}+\beta'_{2z}) \\ -i(\beta'_{1x}+\beta'_{2x}) & -i(\beta'_{1y}+\beta'_{2y}) & -i(\beta'_{1z}+\beta'_{2z}) & 1+(\beta'_1\beta'_2) \end{bmatrix}. \tag{7.9.5}$$

Similarly determining $\delta\Lambda_2\,\delta\Lambda_1$ we find that

$$\delta\Lambda_1\,\delta\Lambda_2 \neq \delta\Lambda_2\,\delta\Lambda_1\,. \tag{7.9.6}$$

Since

$$\delta\Lambda^{-1\,\mu}_{\ \ \ \nu}(v) = \delta\Lambda^{\mu}_{\nu}(-v) \tag{7.9.7}$$

the two systems K and K^0 are physically indistinguishable.

The Galilean transformation to the frame K, moving with velocity v relative to K^0, is

$$x = x^0 - v_x t^0,\ \ y = y^0 - v_y t^0,\ \ z = z^0 - v_z t^0,\ \ t = t^0; \tag{7.9.8}$$

and the correct LT for small v/c is

$$x = x^0 - \beta_x ct^0,\ \ y = y^0 - \beta_y ct^0,\ \ z = z^0 - \beta_z ct^0, \tag{7.9.9}$$

$$ct = ct^0 - \beta_x x^0 - \beta_y y^0 - \beta_z z^0. \tag{7.9.10}$$

These differ from the GT of (7.9.8) only by the inclusion of the relativity of simultaneity in (7.9.10), which is the only relativistic effect that appears in first order in v/c, and the only one that needs derivation from physical considerations. All other effects are derived mathematically, by iteration of the infinitesimal LT.

7.10 Some Results

7.10.1 Resultant of two successive LTs

Consider the scheme (7.3.1) and the relations (7.3.2) for three inertial reference frames. If the origins of the systems K^0, K, K' coincide at the initial time, $t = t^0 = t^1 = 0$, then eight linear homogeneous algebraic equations are obtained, four of which are for transformation of the co-ordinates (r^0, t^0), to (r, t) and the other four are for transformation of (r, t) to (r', t'). From these the relations between the co-ordinates r^0, t^0 and r', t' are easily obtained. Equations (7.6.1) lead to the relations

$$r' = r + v'\left\{\frac{(r\cdot v')(\gamma' - 1)}{v'^2} - \gamma' t\right\}, \quad t' = \gamma'\left\{t - \frac{(v'\cdot r)}{c^2}\right\}, \tag{7.10.1}$$

$$r = r^0 + v\left\{\frac{(r^0\cdot v)(\gamma - 1)}{v^2} - \gamma t^0\right\}, \quad t = \gamma\left\{t^0 - \frac{(v\cdot r^0)}{c^2}\right\}. \tag{7.10.2}$$

The velocity v'' is obtained with the help of (7.4.15) :

$$v'' = \frac{\left[\dfrac{v'}{\gamma} + v\left\{\dfrac{(\gamma - 1)(v'\cdot v)}{\gamma v^2} + 1\right\}\right]}{1 + \dfrac{(v\cdot v')}{c^2}}. \tag{7.10.3}$$

Similarly,

$$v_1 = -\frac{\left[\dfrac{v}{\gamma} + v'\left\{\dfrac{(\gamma' - 1)(v'\cdot v)}{\gamma' v^2} + 1\right\}\right]}{1 + \dfrac{(v'\cdot v)}{c^2}}. \tag{7.10.4}$$

The transformation from K^1 to the K system is assumed to follow (7.7.18). So

$$v_1 = -R^{-1}v'', \quad Rv'' = -v_1, \tag{7.10.5}$$

where the second relation follows from (7.3.2). Then the transformation from (r', t') to (r^0, t^0) is

$$r' = R^{-1}r^0 - v_1\left\{\frac{(r^0\cdot v'')(\gamma'' - 1)}{v''2} - \gamma'' t^0\right\}, \quad t_2 = \gamma''\left\{t'' - \frac{(v''\cdot r^0)}{c^2}\right\}, \tag{7.10.6}$$

where obviously

$$\gamma''^2 = \frac{1}{1 - \beta''^2}, \quad \beta'' = \frac{v''}{c}. \tag{7.10.7}$$

7.10.2 The general LT using the metric tensor $\eta_{\mu\nu}$

The general LT from the frame K^0 to the frame K, moving with relative velocity v, is

$$x = \left\{1 + (\gamma - 1)\frac{\beta_x^2}{\beta^2}\right\}x^0 + (\gamma - 1)\frac{\beta_x\beta_y}{\beta^2}y^0 + (\gamma - 1)\frac{\beta_x\beta_z}{\beta^2}z^0 - \gamma\beta_x\cdot ct^0, \tag{7.10.8}$$

$$y = (\gamma - 1)\frac{\beta_y\beta_x}{\beta^2}x^0 + \left\{1 + (\gamma - 1)\frac{\beta_y^2}{\beta^2}\right\}y^0 + (\gamma - 1)\frac{\beta_y\beta_z}{\beta^2}z^0 - \gamma\beta_y\cdot ct^0, \tag{7.10.9}$$

$$z = (\gamma - 1)\frac{\beta_z \beta_x}{\beta^2} x^0 + (\gamma - 1)\frac{\beta_y \beta_z}{\beta^2} y^0 + \left\{1 + (\gamma - 1)\frac{\beta_z^2}{\beta^2}\right\} z^0 - \gamma\beta_z \cdot ct^0 \tag{7.10.10}$$

$$ct = -\beta_x \gamma x^0 - \beta_y \gamma y^0 - \beta_z \gamma z^0 + ct^0. \tag{7.10.11}$$

The LT from the frame K to K^0 is obtained by changing the sign of the components of v. The matrix of the coefficients of transformation, $\alpha^\mu_{\nu 0}$, defined in (11.2.25) and $\alpha^{\mu 0}_\nu$, defined in (11.2.27), can be easily evaluated in this case of the general LT. For example,

$$\alpha^\mu_{\nu 0} = \begin{bmatrix} 1 + (\gamma - 1)\frac{\beta_x^2}{\beta^2} & (\gamma - 1)\frac{\beta_x \beta_y}{\beta^2} & (\gamma - 1)\frac{\beta_x \beta_z}{\beta^2} & -\gamma\beta_x \\ (\gamma - 1)\frac{\beta_y \beta_x}{\beta^2} & 1 + (\gamma - 1)\frac{\beta_y^2}{\beta^2} & (\gamma - 1)\frac{\beta_y \beta_z}{\beta^2} & -\gamma\beta_y \\ (\gamma - 1)\frac{\beta_z \beta_x}{\beta^2} & (\gamma - 1)\frac{\beta_y \beta_z}{\beta^2} & 1 + (\gamma - 1)\frac{\beta_z^2}{\beta^2} & -\gamma\beta_z \\ -\gamma\beta_x & -\gamma\beta_y & -\gamma\beta_z & \gamma \end{bmatrix}. \tag{7.10.12}$$

We can assume that

$$\beta_x = 0 = \beta_y, \quad \beta_z = \beta. \tag{7.10.13}$$

Then (7.10.8)–(7.10.11) reduce to an LT and (7.10.12) becomes

$$\alpha^\mu_{\nu 0} = \begin{bmatrix} 1 & 0 & 0 & 0 \\ 0 & 1 & 0 & 0 \\ 0 & 0 & \gamma & -\beta\gamma \\ 0 & 0 & -\beta\gamma & \gamma \end{bmatrix} \tag{7.10.14}$$

with

$$\det \alpha^\mu_{\nu 0} = 1 \tag{7.10.15}$$

and

$$\det \alpha^{\mu 0}_\nu = 1. \tag{7.10.16}$$

The corresponding matrix of the coefficients of reverse relations of transformation are

$$\alpha^{\mu 0}_\nu = \begin{bmatrix} 1 & 0 & 0 & 0 \\ 0 & 1 & 0 & 0 \\ 0 & 0 & \gamma & \beta\gamma \\ 0 & 0 & \beta\gamma & \gamma \end{bmatrix} \tag{7.10.17}$$

We find that

$$\det \alpha^\mu_{\nu 0} = \det \alpha^{\mu 0}_\nu = 1. \tag{7.10.18}$$

7.10.3 The infinitesimal LT using the metric tensor $\eta_{\mu\nu}$

For the infinitesimal LT we ignore powers of components of E greater than unity, take $\gamma = 1$ and find that (7.10.12) reduces to

$$\alpha^{\mu}_{\nu 0} = \begin{bmatrix} 1 & 0 & 0 & -\beta_x \\ 0 & 1 & 0 & -\beta_y \\ 0 & 0 & 1 & -\beta_z \\ -\beta_x & -\beta_y & -\beta_z & 1 \end{bmatrix}. \tag{7.10.19}$$

The reverse transformation matrix is

$$\alpha^{\mu 0}_{\nu} = \begin{bmatrix} 1 & 0 & 0 & -\beta_x \\ 0 & 1 & 0 & -\beta_y \\ 0 & 0 & 1 & -\beta_z \\ \beta_x & \beta_y & \beta_z & 1 \end{bmatrix} \tag{7.10.20}$$

with

$$\det \alpha^{\mu}_{\nu 0} = \det \alpha^{\mu 0}_{\nu} = 1 - \beta^2 = 1. \tag{7.10.21}$$

7.11 Infinitesimal Relative Velocity of Two Systems and Thomas Precession

For an infinitesimal LT, from the K frame of reference to the K^1 frame, the scheme of (7.3.2) gives

$$v'' - v = v' = dv. \tag{7.11.1}$$

Then, retaining terms only up to the first power of v' (or dv), equations (7.10.1) reduce to

$$r' = r - v't, \quad t' = t - \frac{(r \cdot v')}{c^2}; \tag{7.11.2}$$

so equation (7.10.3) becomes

$$v'' = \frac{1}{\gamma}v' + \frac{1}{\gamma v^2}(\gamma - 1)(v' \cdot v)v + v - \frac{1}{c^2}(v' \cdot v)v = v + \frac{1}{\gamma}v' + \frac{(1-\gamma)}{\gamma^2 v^2}(v' \cdot v)v. \tag{7.11.3}$$

Similarly (7.10.4) simplifies to

$$v_1 = v + v' - \frac{1}{c^2}(v' \cdot v)v. \tag{7.11.4}$$

Squaring (7.11.3) and (7.7.24) gives

$$v''^2 = v^2 + \frac{2}{\gamma}(v \cdot v') + \frac{(1-\gamma^2)}{\gamma^4 v^2}(v \cdot v') + \frac{1}{\gamma^2}v'^2 = v^2 + \frac{2}{\gamma^2}(v \cdot v'), \tag{7.11.5}$$

$$v_1^2 = v^2 + \frac{2}{\gamma^2}(v \cdot v'). \tag{7.11.6}$$

Hence, up to the first power of v',

$$v_1^2 = v''^2 = v^2 + \frac{2}{\gamma^2}(v \cdot v'). \tag{7.11.7}$$

Put the value of v_1 from (7.11.4) in (7.10.6) and obtain

$$r' = R^{-1}r^0 + \left\{v + v' - \frac{1}{c^2}(v \cdot v')v\right\}\left\{\frac{1}{v''^2}(r^0 \cdot v^0)(\gamma'' - 1) - \gamma'' t^0\right\}, \tag{7.11.8}$$

where γ'' is given by (7.10.7). Next, put the value of r and t from (7.10.2) in (7.11.2) and obtain

$$\begin{aligned} r' &= r^0 + v\left\{\frac{1}{v^2}(r^0 \cdot v)(\gamma - 1) - \gamma t^0\right\} - \gamma v'\left\{t^0 - \frac{1}{c^2}(r^0 \cdot v)\right\} \\ &= r^0 + (r^0 \cdot v)(\gamma - 1)\frac{v}{v^2} + (r^0 \cdot v)\frac{\gamma v'}{c^2} - \gamma(v + v')t^0. \end{aligned} \tag{7.11.9}$$

When (7.11.5) is used, we find that

$$\gamma'' = \gamma\left\{1 + \frac{1}{c^2}(v \cdot v')\right\}. \tag{7.11.10}$$

Hence the coefficient of t^0 in (7.11.8)

$$= -\gamma\left\{v + v' - \frac{1}{c^2}(v \cdot v')v\right\}\left\{1 + \frac{1}{c^2}(v \cdot v')\right\} = -\gamma(v + v'). \tag{7.11.11}$$

And since this is also the coefficient of t^0 in (7.11.9), when (7.10.8) and (7.10.9) are equated, their temporal parts cancel and the result is

$$\begin{aligned} R^{-1}r^0 &= r^0 + (r^0 \cdot v)(\gamma - 1)\frac{v}{v^2} + (r^0 \cdot v)\frac{\gamma v'}{c^2} - (\gamma'' - 1)\frac{(r^0 \cdot v'')}{v''^2}\left\{v + v' - \frac{1}{c^2}(v' \cdot v)v\right\} \\ &= r^0 + \{v(\gamma - 1) + \beta^2\gamma v'\}\frac{(r^0 \cdot v)}{v^2} - \frac{(r^0 \cdot v'')}{v^2\left\{1 + \frac{2}{\gamma^2 v^2}(v \cdot v')\right\}}\left\{\gamma + \frac{\gamma}{c^2}(v \cdot v') - 1\right\} \\ &\quad \times \left\{v + v' - \frac{1}{c^2}(v \cdot v')v\right\}, \end{aligned} \tag{7.11.12}$$

where the value of v''^2 has been taken from (7.11.10).

Equation (7.11.12) reduces to

$$\begin{aligned} R^{-1}r^0 &= r^0 + \frac{1}{v^2}(\gamma - 1)\left[(r^0 \cdot v)v - \frac{(r^0 \cdot v)}{1 + \frac{2}{\gamma^2 v^2}(v \cdot v')}\left\{v + v' - \frac{v}{c^2}(v \cdot v')\right\}\right] \\ &\quad + \frac{\gamma}{c^2}(r^0 \cdot v)v' - \frac{\gamma(r^0 \cdot v'')}{v^2 c^2}(v' \cdot v)\frac{\left\{v + v' - \frac{v}{c^2}(v \cdot v')\right\}}{1 + \frac{2}{\gamma^2 v^2}(v \cdot v')} \\ &= r^0 + A(B + D) + E(F + G), \end{aligned} \tag{7.11.13}$$

in which

$$\left.\begin{aligned} A &= \frac{(\gamma-1)}{v^2}, \quad B = v(r^0\cdot v), \\ E &= \frac{1}{v^2}, \quad F = \beta^2\gamma(r^0\cdot v)v', \end{aligned}\right\} \tag{7.11.14}$$

$$D = -\frac{(r^0\cdot v'')}{1+\dfrac{2}{\gamma^2 v^2}(v\cdot v')}\left\{v+v'-\frac{v}{c^2}(v\cdot v')\right\}, \tag{7.11.15}$$

$$G = -\frac{\gamma}{c^2}(r^0\cdot v'')(v\cdot v')\frac{\left\{v+v'-\dfrac{1}{c^2}(v\cdot v')v\right\}}{1+\dfrac{2}{\gamma^2 v^2}(v\cdot v')} \tag{7.11.16}$$

Since v' is infinitesimal,

$$D = -(r^0\cdot v'')v - (r^0\cdot v)\left\{v' - \frac{1}{c^2}(v\cdot v')(2-\beta^2)v\right\}, \tag{7.11.17}$$

where $(r^0\cdot v'')$ has been replaced by $(r^0\cdot v)$ in the terms which already contain the first power of infinitesimal velocity v'.

Equation (7.11.13) can therefore be replaced by

$$\begin{aligned} R^{-1}r^0 &= r^0 + \frac{(\gamma-1)}{v^2}\left\{(r^0\cdot v)-(r^0\cdot v'')\right\}v + \frac{1}{v^2}(r^0\cdot v) \\ &\quad\times\left[(\gamma-1)\left\{-v'+\frac{1}{v^2}(v\cdot v')(2-\beta^2)v\right\}+\beta^2\gamma v' - \frac{\beta^2}{v^2}\gamma(v\cdot v')\frac{\left\{v+v'-\dfrac{1}{c^2}(v\cdot v')v\right\}}{1+\dfrac{2}{\gamma^2 v^2}(v\cdot v')}\right] \\ &= r^0 + \frac{(\gamma-1)}{v^2}\{r^0\cdot(v-v'')\}v + \frac{1}{v^2}(r^0\cdot v)Q, \end{aligned} \tag{7.11.18}$$

where

$$\begin{aligned} Q &= \beta^2\gamma v' - (\gamma-1)\left\{v'-\frac{1}{v^2}v(v\cdot v')(2-\beta^2)\right\} - \frac{\beta^2}{v^2}\gamma(v\cdot v')\frac{\left\{v+v'-\dfrac{(v\cdot v')}{c^2}v\right\}}{\left\{1+\dfrac{2}{\gamma^2 v^2}(v\cdot v')\right\}} \\ &= \beta^2\gamma v' - (\gamma-1)\left[v' - \frac{(v\cdot v')}{v^2}v\left\{(\gamma-1)(2-\beta^2)-\gamma\beta^2\right\}\right] \\ &= \left(1-\frac{1}{\gamma}\right)\left\{v' + \frac{(v\cdot v')}{v^2}\left(\frac{1}{\gamma}-1\right)v\right\}. \end{aligned} \tag{7.11.19}$$

Equation (7.11.3) gives

$$v'' - v = \frac{1}{\gamma}\left\{v' + \frac{1}{\gamma v^2}(v\cdot v')(1-\gamma)v\right\}. \tag{7.11.20}$$

So (7.11.19) reduces to

$$Q = \gamma\left(1-\frac{1}{\gamma}\right)(v''-v), \tag{7.11.21}$$

and (7.11.18) becomes

$$\begin{aligned} R^{-1}r^0 &= r^0 + \frac{(\gamma-1)}{v^2}\{(r^0\cdot v) - (r^0\cdot v'')\}\, v + \frac{\gamma}{v^2}(r^0\cdot v)\left(1-\frac{1}{\gamma}\right)(v''-v) \\ &= r^0 + \frac{1}{v^2}(\gamma-1)\{v\times(v''-v)\}\times r^0. \end{aligned} \tag{7.11.22}$$

Hence (7.11.1) can be written as

$$R^{-1}r^0 = r^0 + \frac{(\gamma-1)}{v^2}(v\times dv). \tag{7.11.23}$$

The corresponding inverse relation is

$$Rr^0 = r^0 + d\theta\times r^0, \tag{7.11.24}$$

where

$$d\theta = -\frac{(\gamma-1)}{v^2}(v\times dv). \tag{7.11.25}$$

Equation (7.11.24) shows that the rotation operator R turns the vector r^0 through the angle $|d\theta|$ about the direction perpendicular to the plane of v and $|d\theta|$ determines in (7.11.25) the infinitesimal change dv in the velocity of the reference frame K^1 relative to K. Dividing $d\theta$ by dt the expression for the angular velocity vector Ω is obtained

$$\Omega = -\frac{(\gamma-1)}{v^2}\left(v\times\frac{dv}{dt}\right). \tag{7.11.26}$$

Here dv/dt is the acceleration as seen from the K^0 system of reference. For low velocities,

$$\Omega = -\frac{1}{2c^2}\left(v\times\frac{dv}{dt}\right). \tag{7.11.27}$$

This Ω is the Thomas precession frequency. It is further discussed in sections 7.12 and 7.13.

► **Example 7.11.1** *Find the Thomas precession angle for uniform circular motion at the rate w about the the line through its center and perpendicular to its plane.*

Answer. In cylindrical polar co-ordinates the velocity and acceleration are $r\dot{\omega}\hat{\theta}$ and $-r\omega^2\hat{r}$ at the point (r,θ,ϕ); $\hat{r},\hat{\theta},\hat{z}$ are the unit vectors along the positive directions of the radius, the direction of increase of θ and the z direction. So the vector product in the right hand side of (7.11.26) becomes

$$\left(v\times\frac{dv}{dt}\right) = r^2\omega^3\hat{z} \tag{1}$$

and the angular frequency Ω is given by

$$\Omega = -\frac{(\gamma-1)}{v^2}r^2\omega^3\hat{z}, \tag{2}$$

where $v = r\omega$. Integrating Ω with respect to time over the period $2\pi/\omega$ we obtain

$$\theta = \int_0^{2\pi/\omega}\Omega dt = -\frac{(\gamma-1)}{v^2}r^2\omega^2\hat{z}\frac{2\pi}{\omega} \tag{3}$$

because the integrand is independent of t. Hence the angle of precession, through which the initial position must be turned to reach the final position, is given by

$$\theta = -2\pi\left[\frac{1}{\sqrt{1-\gamma^2\omega^2/c^2}} - 1\right] = \frac{\pi\gamma^2\omega^2}{c^2}. \tag{4}$$

► **Example 7.11.2** *For uniform motion round a regular polygon of n sides show that the angle of Thomas precession θ is given by*

$$\theta = -(n-1)\left[\frac{1}{\sqrt{1-\beta^2}} - 1\right]\sin\frac{2\pi}{n}.$$

Moreover, show that the limiting value of this angle, when $n \to \infty$, coincides with the result for uniform circular motion of exercise 7.11.1.

Answer. Let AB and BC be two consecutive sides of the polygon, as in the fig. 7.11.1, and the motion be in the senses AB and BC. Let 2γ be the angle ABC. Then

$$2\pi + 2\gamma n = 2\pi n, \tag{1}$$

So

$$2\gamma = 2\pi\left(1 - \frac{1}{n}\right). \tag{2}$$

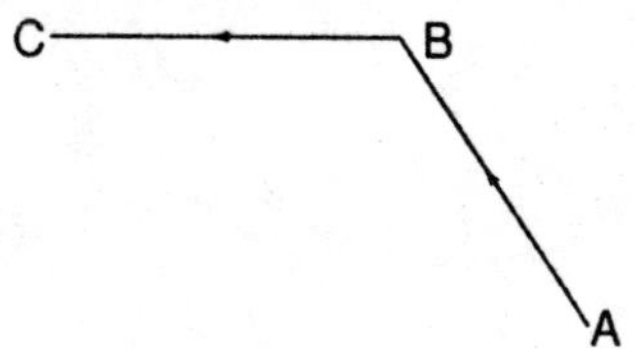

Fig. 7.11.1 It shows the two consecutives sides of a regular polygon.

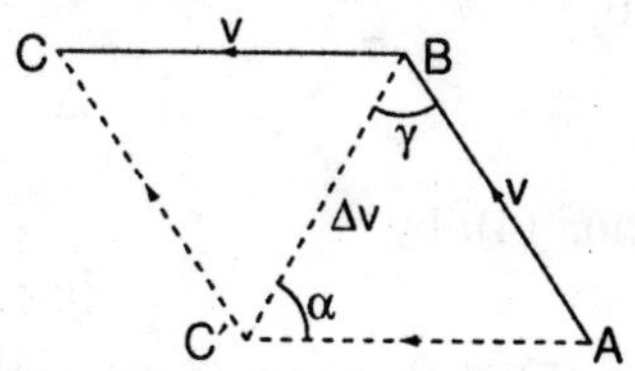

Fig. 7.11.2 It shows the velocities of the motion.

At each corner the velocity alters by the amount

$$\Delta v = 2v\cos\gamma, \tag{3}$$

because, vectorially, $v_{BC} - v_{AB} = \Delta v$; so fig. 7.11.2 gives $\Delta v = v_{BC} + v_{BA} = v_{BC'}$. The angle γ between BC' and BC (or AB) is, obviously, also the angle between Δv and v. Hence

$$|v \times \Delta v| = 2v^2\cos\gamma\sin\gamma = v^2\sin 2\gamma. \tag{4}$$

So, at the corner B the precession angle, due to change in velocity, is

$$-\frac{(\gamma-1)}{v^2}v^2|\sin 2\gamma|.$$

Since, for the polygon of n sides to reach the initial point again, change of direction at only $n-1$ corners will occur, so the Thomas precession angle

$$\theta = -(n-1)(\gamma-1)|\sin 2\gamma| = -(n-1)(\gamma-1)\sin\frac{2\pi}{n} \text{ for } n \geq 4.$$

For $n = 3$, we find that

$$\theta = -2(\gamma-1)\sin\frac{2\pi}{3}. \tag{5}$$

And since

$$\lim_{n\to\infty} \frac{\sin(2\pi/n)}{2\pi/(n-1)} = 1$$

equation (5) can be written as

$$\theta = -2\pi(\gamma - 1)\frac{\sin(2\pi/n)}{2\pi/(n-1)} \to -2\pi(\gamma - 1) \text{ as } n \to \infty.$$

This is the formula (4) of exercise 7.11.1.

▶ **Example 7.11.3** *Find the Thomas precession for uniform motion round an irregular polygon of n sides.*

Answer. From exercise 7.11.2 replace equation (1) by

$$2\pi + \sum_{i=1}^{n} \theta_i = 2\pi n;$$

equation (2) by

$$\sum_{i=1}^{n} \theta_i = 2\pi(n-1); \tag{1}$$

equation (3) by

$$\Delta v_i = 2v\cos\left(\frac{\theta_1}{2}\right); \tag{2}$$

and so, equation (4), by

$$|v \times \Delta v_i| = v^2 \sin\theta_i. \tag{3}$$

Then equation (5) of that exercise should be replaced by

$$\theta = -(\gamma - 1)\sum_{i=1}^{n-1} |\sin\theta_i| = -(\gamma - 1)\left[|\sin\theta_1| + |\sin\theta_2| + \cdots + |\sin\theta_{n+1}|\right] \tag{4}$$

for the precession angle, where $\theta_1 + \theta_2 + \cdots + \theta_{n-1} = 2\pi(n-1) - \theta_n$.

7.12 More about Thomas Precession

Consider the motion of a rigid body A, in the inertial frame K^0. A point $P(x, y, z)$ in the body moves arbitrarily. The inertial frame K_p^0 is fixed to the body at P. At the instant t the point P can be regarded as fixed instantaneously, and the system K_p^0 as travelling with velocity v (say) with respect to K^0. At the latter instant $t + dt$ the point P has velocity $v + dv$ in the K^0 system, which should therefore have the angular velocity Ω, given by (7.11.26).

When the body A moves without change of direction, with respect to the reference system in which it is instantaneously at rest, A is travelling parallel to itself with proper translation. If the point P of A returns to the same place after travelling round a closed path, the solid does return to its original position because the initial and final positions of the axes of the K_p^0 system are different. Then, to recover the original orientation, the body must be turned through the Thomas precession angle.

This kinematic effect depends on the motion of an object, and so, on the relativistic meaning of simultaneity in different frames of reference. It is due to the acceleration (relative to the laboratory frame) experienced by a body orbiting perpendicular to the direction of its own motion. The reference frame fixed with the center of gravity of the moving body is a sequence of inertial frames whose successive origins move at each instant with the non-uniform velocity of the *cg* of the body. So, rigorously speaking, Thomas precession should be studied on the basis of the general theory of relativity. For a circular path, using the transformations for Lorentz rotation, this effect has been studied, Arzelies (1966, 1972).

Macroscopic bodies, spinning about sun and the earth, experience Thomas precession. This has led to proposals for an experiment to test the general theory of relativity, Schiff (1966).

For rotation, the rule of addition of velocities is not decided by Euclidean plane geometry; the Lobachevskian hyperbolic trigonometry determines it. The rotation of the axes in successive transformation, from transverse acceleration, makes the spin turn. For a spinning electron, with a sidewise acceleration due to its orbital motion about the nucleus, the rest frame is different at every moment; so the axis of its spin in the laboratory frame changes direction even in the absence of torque, giving rise to this precession. Electrons accelerated by electric field in atoms, neutrons and photons, accelerated by nuclear forces, and earth satellites accelerated by gravitational forces, experience this effect.

A spinning orbital electron moves in an orbit in the axially symmetric electric field of rest frame of the atom. For an observer "riding with" the electron, the other changes appear as in motion, and this motion produces the effect of a magnetic field. This field, acting on the magnetic moment and on the electrons, gives rise to precession of the spinning particles, and splitting of the spectral lines. The precessional angular velocity is opposite in direction to the angular velocity of precession of the spin in the virtual magnetic field, and just one-half as large. This treatment gave birth to the idea of spinning electron and the involvement of special theory of relativity in quantum mechanics.

7.13 An Alternative Derivation of the Thomas Precession Effect

Since successive application of two LTs is equivalent to a single LT, plus a rotation, LTs are described in terms of non-commutative matrices, and the results of this application depend upon the order in which it is performed.

For an electron, if $v(t)$ is its velocity at time t, and $v'(= v + \delta v)$ is the velocity at time $t'(= t + \delta t)$, then let the comoving reference frames, at these times, be denoted by K and K' respectively. The co-ordinates (r, t) of a point P in the K frame and the co-ordinates (r^0, t^0) in the K^0 frame are connected by the LT. So, if $\mathbf{x} \equiv (r, t)$, and $\mathbf{x}^0 \equiv (r^0, t^0)$, then

$$\mathbf{x} = \wedge(v)\mathbf{x}^0. \tag{7.13.1}$$

Similarly, $\mathbf{x}' \equiv (r', t')$ are the co-ordinates of P in the K' frame,

$$\mathbf{x}' = \wedge(v')\mathbf{x}^0. \tag{7.13.2}$$

Then the matrix $\wedge^R$ of transformation from K to K' is obtained, if we write

$$\mathbf{x}' = \wedge^R\mathbf{x}, \tag{7.13.3}$$

$\wedge^R$ is called the *restricted LT* because it may not be a pure LT like $\wedge(v)$ and $\wedge(v')$ which obtain from (7.6.10), Goldstein (1980, p. 284). Then

$$\wedge^R = \wedge(v')\wedge^{-1}(v) = \wedge(v')\wedge(-v) \tag{7.13.4}$$

because $\wedge^{-1}(v) = \wedge(-v)$, where $\wedge^{-1}(v)$ is the inverse matrix of $\wedge(v)$. Without any loss of generality, we take $v_1 = 0, v_2 = 0, v_3 = v, v_1' = 0, v_2' \neq 0, v_3' \neq 0$, where the components of v and v' are denoted by (v_1, v_2, v_3) and (v_1', v_2', v_3'), respectively, and obtain

$$\wedge^{(-v)} = \begin{bmatrix} 1 & 0 & 0 & 0 \\ 0 & 1 & 0 & 0 \\ 0 & 0 & \gamma & -i\gamma\beta \\ 0 & 0 & i\gamma\beta & \gamma \end{bmatrix} \tag{7.13.5}$$

$$\wedge(v') = \begin{bmatrix} 1 & 0 & 0 & 0 \\ 0 & 1+\frac{\beta_2'^2}{\beta'^2}(\gamma'-1) & \frac{1}{\beta'^2}(\gamma'-1)\beta_2'\beta_3' & i\gamma'\beta_2' \\ 0 & \frac{1}{\beta'^2}(\gamma'-1)\beta_2'\beta_3' & 1+\frac{\beta_3'^2}{\beta'^2}(\gamma'-1) & i\gamma'\beta_3' \\ 0 & -i\gamma'\beta_2' & -i\gamma'\beta_3' & \gamma' \end{bmatrix}. \tag{7.13.6}$$

Now writing $v_3' = v + \delta v_3, v_2' = \delta v_2$, and keeping only terms of first order in δv, equation (7.13.6) becomes

$$\wedge(v') = \wedge(v+\delta v) = \begin{bmatrix} 1 & 0 & 0 & 0 \\ 0 & 1 & \frac{1}{\beta}(\gamma-1)\delta\beta_2 & i\gamma\delta\beta_2 \\ 0 & \frac{1}{\beta}(\gamma-1)\delta\beta_2 & (\gamma+\gamma^3\beta\delta\beta_3) & i\gamma(\beta+\gamma^2\delta\beta_3) \\ 0 & -i\gamma\delta\beta_2 & -i\gamma(\beta+\gamma^2\delta\beta_3) & (\gamma+\gamma^3\beta\delta\beta_3) \end{bmatrix} \tag{7.13.7}$$

The transformation matrix $\wedge^R$ of (7.13.4) is then easily evaluated :

$$\wedge^R = \wedge(v')\wedge(-v) = \begin{bmatrix} 1 & 0 & 0 & 0 \\ 0 & 1 & \frac{1}{\beta}(\gamma-1)\delta\beta_2 & i\gamma\delta\beta_2 \\ 0 & \frac{1}{\beta^2}(\gamma-1)\delta\beta_2 & (\gamma+\gamma^3\beta\delta\beta_3) & i\gamma(\beta+\gamma^2\delta\beta_3) \\ 0 & -i\gamma\delta\beta_2 & -i\gamma(\beta+\gamma^2\delta\beta_3) & (\gamma+\gamma^3\beta\delta\beta_3) \end{bmatrix} \times$$

$$\begin{bmatrix} 1 & 0 & 0 & 0 \\ 0 & 1 & 0 & 0 \\ 0 & 0 & \gamma & -i\gamma\beta \\ 0 & 0 & i\gamma\beta & \gamma \end{bmatrix}$$

$$= \begin{bmatrix} 1 & 0 & 0 & 0 \\ 0 & 1 & -\frac{1}{\beta}(\gamma-1)\delta\beta_2 & i\gamma\delta\beta_2 \\ 0 & \frac{1}{\beta}(\gamma-1)\delta\beta_2 & 1 & i\gamma^2\delta\beta_3 \\ 0 & -i\gamma\delta\beta_2 & -i\gamma^2\delta\beta_3 & 1 \end{bmatrix}. \tag{7.13.8}$$

This infinitesimal transformation can be written as the sum,

$$\wedge^R = \begin{bmatrix} 1 & 0 & 0 & 0 \\ 0 & 1 & 0 & i\gamma\delta\beta_2 \\ 0 & 0 & 1 & i\gamma^2\delta\beta_3 \\ 0 & -i\gamma\delta\beta_2 & -i\gamma^2\delta\beta_3 & 1 \end{bmatrix} + \begin{bmatrix} 0 & 0 & 0 & 0 \\ 0 & 0 & -\frac{1}{\beta}(\gamma-1)\delta\beta_2 & 0 \\ 0 & \frac{1}{\beta}(\gamma-1)\delta\beta_2 & 0 & 0 \\ 0 & 0 & 0 & 0 \end{bmatrix} \tag{7.13.9}$$

of an LT for the infinitesimal velocity $\delta v \equiv (0, \gamma\delta\beta_2, \gamma^2\delta\beta_3)$, together with an infinitesimal rotation

$$\mathbf{d}\theta = \left(\frac{\gamma-1}{\beta}\delta\beta_2, 0, 0\right),$$

as may be seen from the relations (15)–(17) of exercise (7.13.1). In vector notation, the infinitesimal velocity vector is

$$\gamma\left\{\delta v + (\gamma-1)(v\cdot\delta v)\frac{v}{v^2}\right\}$$

and the rotation angle is

$$(\gamma-1)\frac{(v\times\delta v)}{v^2}$$

in time δt. So, the Thomas precession formula (7.11.26) is verified.

▸ **Example 7.13.1** *Obtain the Thomas precession frequency formula assuming that v' is an infinitesimal velocity.*

Answer. Since v' is an infinitesimal velocity, the infinitesimal LT, $\delta\wedge^{\mu}_{0\nu}$ of (7.9.1) is used for v'.

$$\delta\wedge^{\mu}_{\nu'}(v') = \begin{bmatrix} 1 & 0 & 0 & 0 \\ 0 & 1 & 0 & i\beta'_2 \\ 0 & 0 & 1 & i\beta'_3 \\ 0 & -i\beta'_2 & -i\beta'_3 & 1 \end{bmatrix}, \tag{1}$$

where, without any loss of generality, v' is taken to be lying in the YZ plane. Then, using (7.13.5), we find that

$$\wedge'' = \delta\wedge'(\beta')\wedge(\beta) = \begin{bmatrix} 1 & 0 & 0 & 0 \\ 0 & 1 & \gamma\beta\beta'_2 & i\gamma\beta'_2 \\ 0 & 0 & \gamma+\gamma\beta\beta'_3 & i\gamma\beta+i\gamma\beta'_3 \\ 0 & -i\beta'_2 & -i\gamma\beta'_3-i\gamma\beta & \gamma+\gamma\beta\beta'_3 \end{bmatrix}. \tag{2}$$

Retaining only the largest terms this relative gives

$$\wedge'' = \begin{bmatrix} 1 & 0 & 0 & 0 \\ 0 & 1 & \gamma\beta\beta'_2 & i\gamma\beta'_2 \\ 0 & 0 & \gamma & i\gamma\beta \\ 0 & -i\beta'_2 & -i\gamma\beta & \gamma \end{bmatrix}. \tag{3}$$

Since $\wedge''_{jk}$ are not symmetric, $\wedge''$ cannot be a pure LT (the boost transformation). Like $\wedge^R$ of (7.13.3), it is a restricted LT. If v'' be the effective velocity between the origins O^0 and O' of the two systems K^0 and K', respectively, then the relations (7.6.24) give

$$\beta''_1 = 0, \quad \beta''_2 = \frac{1}{\gamma}\beta'_2, \quad \beta''_3 = \beta. \tag{4}$$

Since β'_2 is infinitesimal, the relativistic factor γ'' approximately equals γ^2. So,

$$\gamma'' \approx \gamma, \beta'' \approx \beta. \tag{5}$$

The pure LT, for the relative velocity $-v''$, denoted by $\wedge(-\beta'')$, is obtained from (7.6.10). Retaining only the largest terms in $\wedge(-\beta'')$, and using (4) and (5) we obtain

$$\wedge(-\beta'') = \begin{bmatrix} 1 & 0 & 0 & 0 \\ 0 & 1 & \frac{1}{\gamma\beta}(\gamma-1)\beta'_2 & -i\beta'_2 \\ 0 & \frac{1}{\gamma\beta}(\gamma-1)\beta'_2 & \gamma & -i\gamma\beta \\ 0 & i\beta'_2 & i\gamma\beta & \gamma \end{bmatrix}. \tag{6}$$

If a pure LT is denoted by $\wedge^P$ and a restricted LT by $\wedge^R$, then the relation

$$\wedge^R = R\wedge^P \tag{7}$$

defines the rotation matrix R. Taking $\wedge''$ of (3) for the $\wedge^R$ and $\wedge^{(\beta'')}$ of (6) for $\wedge^P$, this relation gives

$$R = \wedge^R\, \wedge^{-P}\, (\beta''), \tag{8}$$

where

$$R = \begin{bmatrix} 1 & 0 & 0 & 0 \\ 0 & 1-\beta'^2_2 & \frac{1}{\gamma\beta}(\gamma-1)\beta'_2 & 0 \\ 0 & -\frac{1}{\beta\gamma}(\gamma-1)\beta'_2 & 1 & 0 \\ 0 & 0 & 0 & 1 \end{bmatrix} \tag{9}$$

and $\wedge^{-P}$ is the inverse of $\wedge^P$, and the relation

$$\wedge^{-P}(\beta'') = \wedge^P(-\beta'') \tag{10}$$

has been used. Neglecting squares of β''_2, equation (9) gives

$$R = \begin{bmatrix} 1 & 0 & 0 & 0 \\ 0 & 1 & \frac{1}{\beta\gamma}(\gamma-1)\beta'_2 & 0 \\ 0 & -\frac{1}{\beta\gamma}(\gamma-1)\beta'_2 & 1 & 0 \\ 0 & 0 & 0 & 1 \end{bmatrix}. \tag{11}$$

Its space part is

$$R_S = \begin{bmatrix} 1 & 0 & 0 \\ 0 & 1 & \frac{1}{\beta\gamma}(\gamma-1)\beta_2' \\ 0 & -\frac{1}{\beta\gamma}(\gamma-1)\beta_2' & 1 \end{bmatrix}. \tag{12}$$

The elements of the antisymmetric rotation matrix

$$\varepsilon = \begin{bmatrix} 0 & d\theta_3 & -d\theta_2 \\ -d\theta_3 & 0 & d\theta_1 \\ d\theta_2 & -d\theta_1 & 0 \end{bmatrix}, \tag{13}$$

where the three independent parameters specifying the rotation, $d\theta_1, d\theta_2, d\theta_3$, Goldstein (1980, p. 169), form the components of an infinitesimal vector. The change in the components of a vector, under the infinitesimal transformation of the co-ordinate system, is then expressed in the matrix form

$$\boldsymbol{X} - \boldsymbol{X}^0 = d\boldsymbol{x} = \boldsymbol{\varepsilon X}, \tag{14}$$

which can be written as

$$\left.\begin{aligned} dx_1 &= x_2 d\theta_3 - x_3 d\theta_3; \\ dx_2 &= x_3 d\theta_1 - x_1 d\theta_3; \\ dx_3 &= x_1 d\theta_2 - x_2 d\theta_1 \end{aligned}\right\} \tag{15}$$

and, more compactly as

$$dr = (r \times d\theta). \tag{16}$$

Comparison of R_S of equation (12) with $\boldsymbol{I} + \boldsymbol{\varepsilon}$, where $\boldsymbol{I}$ is the 3×3 unit matrix and $\boldsymbol{\varepsilon}$ is given by (13), shows that R implies rotation of the co-ordinate system K' with respect to the K^0 system, about the x axis, through an infinitesimal angle $d\theta$, given by

$$d\theta = \frac{1}{\gamma\beta}\beta_2'(\gamma - 1) = \beta_2''\beta\frac{(\gamma-1)}{\beta^2}. \tag{17}$$

Equations (4) show that the laboratory observer will see a change dv in the velocity of the particle in the time dt apart. The infinitesimal velocity dv has the only non-zero component v_2'' parallel to the Y-axis. Since the initial Z axis has been chosen along the direction of ∴ the infinitesimal rotation vector $d\theta$ is written as

$$d\theta = -\frac{1}{v^2}(\gamma - 1)(v \times dv). \tag{18}$$

The corresponding precession frequency is the Ω of (7.11.26).

7.14 Limiting Sum of n Identical Infinitesimal LTs

We now consider a series of n infinitesimal LTs, such that all of them have the same magnitude and direction. So $\delta v_1 = \delta v_2 = \delta v_n = \delta v$ (say). As $n \to \infty$, we find that

$$\frac{\delta v}{c} \to 0 \quad \text{and} \quad \left| n\frac{\delta v}{c} \right| \to \theta, \tag{7.14.1}$$

where θ is a finite quantity. It will be seen that the limiting result is the LT of (7.6.10).

We write (7.9.1) as

$$\delta\wedge = \boldsymbol{I} + i\delta\beta[\boldsymbol{l}], \tag{7.14.2}$$

where

$$[\boldsymbol{l}] = \begin{bmatrix} 0 & 0 & 0 & l_x \\ 0 & 0 & 0 & l_y \\ 0 & 0 & 0 & l_z \\ -l_x & -l_y & -l_z & 0 \end{bmatrix} \tag{7.14.3}$$

l_x, l_y, l_z are the direction cosines of δv, and $\boldsymbol{I}$ is the unit tensor. Raising δA to the power of the positive integral number n, we find from (7.14.2) that

$$\begin{aligned}\delta\wedge^n &= (\boldsymbol{I} + i\delta\beta[\boldsymbol{l}])^n \\ &= \boldsymbol{I} + n_i\delta\beta[\boldsymbol{l}] + \frac{n(n-1)}{2!}(i\delta\beta)^2[\boldsymbol{l}]^2 + \frac{1}{3!}n(n-1)(n-2)(i\delta\beta)^3[\boldsymbol{l}]^3 + \cdots,\end{aligned} \tag{7.14.4}$$

where the binomial expansion of $(\boldsymbol{I} + i\delta\beta[\boldsymbol{l}])^n$ and the relation $I^n = I$, for the unit matrix I have been used. Squaring $[\boldsymbol{l}]$ we obtain

$$[\boldsymbol{l}]^2 = \begin{bmatrix} l_x^2 & l_x l_y & l_x l_z & 0 \\ l_y l_x & l_y^2 & l_y l_z & 0 \\ l_z l_x & l_z l_y & l_z^2 & 0 \\ 0 & 0 & 0 & 1 \end{bmatrix}. \tag{7.14.5}$$

Also, since $l_x^2 + l_y^2 + l_z^2 = 1$, we find that

$$[\boldsymbol{l}]^3 = -[\boldsymbol{l}], \quad [\boldsymbol{l}]^4 = -[\boldsymbol{l}]^2. \tag{7.14.6}$$

Hence, using (7.14.1) for $n \to \infty$, we easily find that

$$\lim_{n\to\infty} (\delta\wedge)^n = I + i\left(\theta + \frac{\theta^3}{3!} + \frac{\theta^5}{5!} + \cdots\right)[\boldsymbol{l}] - \left(\frac{\theta^2}{2!} + \frac{\theta^4}{4!} + \cdots\right)[\boldsymbol{l}]^2. \tag{7.14.7}$$

Putting the expressions for the matrices $[\boldsymbol{l}]$ and $[\boldsymbol{l}]^2$ from (7.14.3) and (7.14.5), and of $\boldsymbol{I}$ in this equation we obtain

$$\lim_{n\to\infty} (\delta \boldsymbol{L})^n = \begin{vmatrix} 1 + (\cosh\theta - 1)l_x^2 & (\cosh\theta - 1)l_x l_y & (\cosh\theta - 1)l_x l_z & i\sinh\theta l_x \\ (\cosh\theta - 1)l_y l_x & 1 + (\cosh\theta - 1)l_y^2 & (\cosh\theta - 1)l_y l_z & i\sinh\theta l_y \\ (\cosh\theta - 1)l_z l_y & (\cosh\theta - 1)l_y l_z & 1 + (\cosh\theta - 1)l_z^2 & i\sinh\theta l_z \\ -i\sinh\theta l_x & -i\sinh\theta l_y & -i\sinh\theta l_z & \cosh\theta \end{vmatrix}, \tag{7.14.8}$$

where we have used the hyperbolic formulas

$$\cosh\theta = 1 + \frac{\theta^2}{2!} + \frac{\theta^4}{4!} + \cdots, \quad \sinh\theta = \theta + \frac{\theta^3}{3!} + \frac{\theta^5}{5!} + \cdots \tag{7.14.9}$$

The origin of the K^0 system has the co-ordinates $(0, 0, 0, t^0)$ in the K^0 system and (x, y, z, t) in the K system at time t^0 or t. The transformation relations (7.14.8), at this point,

$$x = i \sinh \theta ict_0 l_x, \quad y = i \sinh \theta ict_0 l_y, \quad z = i \sinh \theta ict_0 l_z, \quad ict = \cosh \theta ict \tag{7.14.10}$$

show that

$$x = -l_x ct \tanh \theta, \quad y = -l_y ct \tanh \theta, \quad z = -l_z ct \tanh \theta. \tag{7.14.11}$$

Hence, the velocity v of the K frame is given by

$$v = \frac{1}{t}(-x, y, -z) = lc \tanh \theta. \tag{7.14.12}$$

Squaring this relation we find that

$$\tanh \theta = \beta, \quad \cosh \theta = \gamma, \quad \sinh \theta = \gamma\beta \tag{7.14.13}$$

because

$$l = \frac{\boldsymbol{\beta}}{\beta}. \tag{7.14.14}$$

Chapter 8

Physics of Rotation

8.1 Introduction

A general motion is composed of translation and rotation. Also translation is the limit of rotation about axis at infinity. The direction of rotation is that of a right handed screw, driven along the direction of the angular velocity vector Ω. The centrifugal force of rotation is directed radially outward, and is proportional to the square of the angular velocity. Like the electrostatic force, it acts even when the particle is stationary in the moving system. The Coriolis force is proportional to the vector product of angular velocity and particle velocity. Like the Lorentz force, it acts only on moving particles and media. An elevator, a moving train, or rotating platforms, or the frame attached to the inside of an artificial satellite are examples of accelerated frames of reference.

According to the principle of equivalence of the general theory of relativity (GTR) all accelerations, including rotation, are non-permanent gravitational fields, pervading all space. Rotation, like gravitational fields, makes the space-time continuum curved. The metric tensor for any specified rotation is not unique; and huge calculations are necessary for finding the sophisticated aspects of rotation in non-Euclidean 4-space. In the linearized approximation of the equivalent formulas for rotation, other accelerations and gravitation, superposition is valid, so that the fields for bodies with more complex structures can be built up from superposition of the parts. A simple case of field of rotation (see section 8.12) of an earth satellite is considered for illustration. Numerical results show that some consequences are too small for laboratory detection, but these could be significant for Celestial bodies.

All rigid bodies are essentially elastic in nature. So, these are deformable by application of stress, which may be due to rotation. Although a system in rotation settles down to a steady state, this will not necessarily be the unstressed original state (the ground state), which is free from all stresses in which any induced vibrations have died away. To avoid the effects of stress due to rotation, only small angular velocities of rotation are considered here.

Of the simple systems, for instance, the system with constant linear velocity, linear acceleration and angular rotation, only angular rotation does not permit the introduction of an instantaneous rest frame for the entire system, and the metric cannot be placed in diagonal form. For this reason, only effects of the first order of rotation velocity, and not higher, are considered in rotating frames of reference.

Lorentz rotation is a rigid rotation with local time dilation and length contraction along arcs of circles of given radii, both of which follow from the LT. After finding the related metric for small rotational velocities, we proceed with the general problem of Lorentz rotation without this approximation and also consider transformation from GN rotation to this and vice-versa.

All differentials of Lorentz rotation are not exact, that is, these differentials are not perfect or holonomic. These non-holonomic co-ordinate differentials are artificially made perfect or holonomic following *Trocheris* (1949) by adding some terms. Some auxiliary relations and consistency conditions are necessary to make the Trocheris rotation valid for all values of the uniform angular velocity Ω. These conditions give rise to a new metric of space for non-rigid and axial rotation at a uniform rate. It can be expanded in powers of $r\Omega/c(=\beta)$ and the first few terms of this expansion give the metric of other rotations which generate it.

Kinetics and thermodynamics are not developed properly for rotation and Lorentz transformation for translation at uniform rates. On the problems of thermodynamics of rotating bodies, *Ter Haar and Wergeland* (1973), remarked : *One will not need a theory of moving system in a wider sense for situations when it is necessary to consider the total (linear and angular) momentum as thermodynamical variables in the same sense as the energy. Landau and Lifshitz's* (1969) thermodynamics of rotating bodies is a case in point and other examples are conceivable though not a common occurrence". Some key-words and expressions for physics of rotation are : Lorentz rotation, co-ordinate free rotation, metric-independent field equations in rotating frames, differential or elliptic rotation; Galilean-Newtonian (GN) rules of transformation for rotation; holonomic differentials or frames of rotation; anholonomic (non-integrable) differentials or frames of rotation; physics, (kinetics, kinematics, mechanics, electrodynamics) in rotating frames; natural motion of both micro bodies and macro bodies of different dimensions and constitutions in rotating frames.

8.2 Limitations of Relativistic Development

Since relativistically no body is rigid, the rigid body dynamics has no relativistic generalization. Covariant formulation is not always possible of a given problem because not all types of forces are available in a covariant form. For example, the ordinary force of gravitation, which propagates at infinite speed, has an action-at-a distance effect, and so cannot be expressed in a covariant manner. Sometimes, expressions for the field potentials are determined by considerations other than those of mechanics. Only when these extra-mechanical theories are brought under a covariant formulation, for their transformation behaviour, can one find a covariant Lagrangian or energy.

The relativistic formulations of statics and thermodynamics are not complete for the relativistic generalization; one reason being that thermodynamical equilibrium, per definition is a state, where all relative motions between parts of a system have come to rest; so, whenever such motions are present they imply some dissipation of energy, which means the existence of non-equilibrium. In certain systems, with internal motions, it is possible to assign local values of temperature and other thermodynamical variables, but then equilibrium between adjacent parts, which are moving with finite relative velocities, is not possible. So, a theory of moving systems in a wider sense is necessary for situations, where the total (linear and angular) momentum is considered as a thermodynamic variable in the same sense as the energy.

8.3 Kinematics and Kinetics of Motion

Kinetics and kinematics are the two divisions of mechanics. Kinematics deals with the shape of the trajectories followed by a point in motion, as well as the time necessary for the particles to traverse a certain length of these respective trajectories. This is done apart from all considerations of force, mass and energy. So, kinematics requires the utilization of two fundamental concepts, length and time, and is not concerned with the physical realization of the motions.

Bodies following kinematically identical motions move along geometrically identical paths in such a fashion that after identical time intervals, they occupy identical positions on those paths. Topics of kinematical aspects of rotation are infinitesimal rotation, transformation features of rotation and differentials of independent variables of rotation.

Kinetics, on the other hand, is concerned with the effects of force on the motion of bodies. So, under kinetics, transition of a system to a state of equilibrium, called *kinetic equilibrium*, is studied, be it static or dynamic. So, topics like the constants of motion, motion, entropy, statistical state of a body, equilibrium configurations, transformation of energy, thermodynamics of rotating bodies come under kinetics of motion.

These divisions have lost their importance due to availibility of some formalisms of broad generalization for unification of mechanics. *Goldstein* (1980) has avoided division of mechanics into kinetics and kinematics. However, in relativistic mechanics, it becomes convenient to specify the kinetical and kinematical aspects of mechanics, and the inherent limitations for achieving a thorough development of these aspects. *Arzelies* (1966) has produced a monograph on the kinematics of motion under the STR. But no monograph is known to exit on relativistic kinetics.

Relativistic kinetics of rotation, even for Newtonian limits of absolute time and space, have not been developed properly. The difficulties of development of relativistic statics and thermodynamics have been discussed in the literature, *Gron* (1973), *Yuen* (1970), *Newburgh* (1979), and *Landau and Lifshitz* (1969), Topics of kinetics of rotation including the Boltzmann distribution function in rotating frames, have been discussed, *Chapman and Cowling* (1970), *Landau and Lifshitz* (1969).

In pre-relativistic kinematics the time is a universal variable, independent of the reference system chosen, and the formulas for change of axes are valid, whatever co-ordinate system may have been chosen. For example, one of these might be fixed to earth, and the other might rotate with respect to earth.

8.4 Forces and Fields

Forces appear when bodies interact on each other. If there are no bodies to interact, there is no force. Bodies may *interact* either when they are separated in space or in contact. If the bodies are thus separated the evidence of their interaction is a change with time of their distance of separation, and if they are in contact, the evidence of interaction is relative deformation, small or large.

Inertia implies the tendency of a body to resist any change in its state of uniform motion. Force is needed not for maintaining a given state of uniform motion, but for changing it. In other words, the external force is related not to velocity, but to acceleration which is the rate of change of velocity. To produce acceleration, (change of the velocity) of the body, it becomes necessary to take account of the inertia. Mass is a quantitative expression of the inertia of a body.

For validity of the concept of action-at-a distance, forces must be transmitted instantaneously, that is, at infinite speeds; but, relativistically speaking, these transmit at speeds less than the vacuum speed c of light. So this concept has no sense in the relativity theory. However, EM forces propagate at phase velocities of the order of e or more; so electrodynamics automatically satisfies this requirement of relativity theory.

8.5 Newtonian Laws of Motion

Newton's first law states : A body, on which no forces act, continues in a state of rest or of uniform motion in a straight line. This law is not universal because if an observer A determines that the motion of forceless bodies is uniform and rectilinear, another observer B, who is accelerating relative to A, would conclude that Newton's first law is not true. Actually, this law is not true for all observers or all reference frames. It is true for infinitely many *privileged* observers and *privileged* frames; these are the so-called inertial frames of reference. So, it is an affirmation of the existence of inertial frames in which a body moving with constant velocity is acted upon by zero net force. In these frames, this is the law of motion of the body :

$$\frac{dp}{dt} = F, \tag{8.5.1}$$

where the force vector F is free from certain additive terms, which are typical of accelerated frames of reference, and where p is the momentum vector of the body at time t. For relativistic motion, obviously,

$$p = \frac{mu}{\sqrt{1 - u^2/c^2}} \equiv m\gamma u, \tag{8.5.2}$$

where m is the rest mass and the relativistic factor

$$\gamma = \frac{1}{\sqrt{1 - u^2/c^2}}. \tag{8.5.3}$$

8.6 Absolute Space and Absolute Time

Since all observers are not alike, Newton rationalized the situation by postulating the existence of *absolute space*. Only when measured against the reference frame of the absolute space do Newton's laws of motion have their familiar form. If A is at rest in the absolute space and B accelerates relative to that space, the laws of Newton, as they stand, will not hold in the rest frame of B. Here Newton introduced the concept of an inertial force. Then, in B, these laws can be reinstated by postulating the existence of an equal and opposite force which cancels the force of acceleration of B.

If a system of co-ordinates exist in which the equation of motion (8.5.1) holds good, then, obviously, there exists an infinite number of systems of co-ordinate axes having the same property and these systems move with respect to each other with uniform velocity of translation. The validity of this statement can be seen from the transformation relations

$$r = r^0 - vt^0, \quad t = t^0, \tag{8.6.1}$$

for uniform translation at the rate v, and from the fact that, for these

$$\ddot{r} = \ddot{r}^0. \tag{8.6.2}$$

So acceleration is then same for the two systems K and K^0. Here v is velocity of K relative to K^0. Hence the force is a physical reality independent of the choice of co-ordinate axes, and the equation

$$\text{mass} \times \text{acceleration} = \text{force} \tag{8.6.3}$$

remains the same in all inertial systems of co-ordinates axes. The equations of dynamics are therefore invariants of the rules of GN transformation for translation at uniform rates, and every particular set of co-ordinate axes of it can be considered as a system with the same right as any other set of rectangular cartesian axes of the system, and the term absolute rest looses its importance in mechanics.

The second relation of (8.6.1) gives the postulate of universal time. Thus it is impossible to detect uniform motion of translation from equation (8.5.1) for motion, and consequently the inertial system cannot be determined by experiments in classical mechanics.

8.7 Mass Non-conservation

Relativistically, the energy of a body at rest contains, in addition to the rest energy of its constitutive elements, the kinetic energy of the particles and the energy of their mutual interaction. In other words,

$$mc^2 \neq \sum_n m_n c^2, \tag{8.7.1}$$

where m is the total mass of the particles of the system and m_n is the mass of the nth particle. So

$$m \neq \sum_n m_n. \tag{8.7.2}$$

Thus, in relativistic mechanics, the law of conservation of mass does not hold good. Instead, only the law of conservation of energy is valid when the rest energies of the particles are included.

8.8 Familiar Theory of Rotation

8.8.1 Newtonian rotation, and the fixed star reference frame

The Newtonian notion of *absolute space* relates to a system of reference relative to which every motion should be measured. A body can move with uniform and constant velocity v relative to the fixed stars, which are almost at rest relative to the *absolute space.* The motions of stars are recorded relative to earth. So, a rate of precession, and mutation of the earth, a velocity of the solar system, and a rate of rotation of the galaxy, exist. Taken together these make the corrected star motion relative to earth average to zero, *Sard* (1970, p. 14). This is the basis of the fixed star reference frame. The velocity of stars near the sun, with respect to the galactic center, is about 200 km/sec, on a radius of about 2.5×10^{22} cm; it corresponds to a very small centripetal acceleration 10^{-8} cm/sec^2. Therefore, the fixed star frame does not rotate with respect to the terrestrial inertial frames. The inertial force due to the translational acceleration of a non-inertial frame is like the force due to homogeneous gravitational field, produced by bodies at distances large compared with the dimensions of the region in which the particle moves; there is no known means to distinguish between these two forces.

8.8.2 The basic relations

Consider two frames of reference $K^0(OX^0Y^0Z^0t^0)$ and $K(OXYZt)$, where K rotates relative to K^0 at the rate Ω about OZ (or OZ^0). The inertial frame K^0 is free from any acceleration relative to the fixed stars. The rotation velocity v of K relative to K^0 is

$$V = (\Omega \times r). \tag{8.8.1}$$

When Ω is constant, this rotation is rigid. Rigid bodies are deformable under stress, which may be due to rotation; although a system in rotation settles down, due to a steady state, this will not necessarily be the original state (the ground state). So, in a finite region of space, we assume the inequality

$$r\Omega \ll c, \tag{8.8.2}$$

where r is the radial distance from the axis of rotation.

The Galilean-Newtonian (GN) rules of transformation involving the absolute time

$$r^0 = r + vt, \quad t^0 = t, \tag{8.8.3}$$

are valid for translation at uniform rates v from the frame K^0 to the frame K. Assumption of absolute time is that any two inertial systems are connected by a Galilean transformation; so, the velocity of all materials and light would be different in two systems moving with uniform relative velocity v. The equation of motion of a freely falling body,

$$m\ddot{r} = 0, \tag{8.8.4}$$

remains invariant under these Galilean rules.

For rotation, of a particle valid in a finite region, these rules are

$$x = x^0 \cos\Omega t^0 + y^0 \sin\Omega t^0, \quad y = -x^0 \sin\Omega t^0 + y^0 \cos\Omega t^0,$$

$$z = z^0, \quad t = t^0, \quad x^2 + y^2 + z^2 = r^2 = r^{0^2} = x^{0^2} + y^{0^2} + z^{0^2}. \tag{8.8.5}$$

In polar co-ordinates these relations become

$$r = r^0, \quad \theta = \theta^0 - \Omega t^0, \quad z = z^0, \quad t = t^0, \tag{8.8.6}$$

where the angle in polar co-ordinates is θ in the K system and θ^0 in the K^0 system. Equation of motion (8.8.4), does not remain invariant for these transformations. If the axis of rotation is at infinity and $\Omega \to 0$ such that $r\Omega \to v$, then the rotation velocity becomes a uniform translational velocity, and the concept of absolute time is preserved. Since θ^0 and t^0 appear in a linear combination in (8.8.6), only effects of the order of $r\Omega/c$, and not higher, are obtained from this theory. The reverse relations are obtained by changing the sign of Ω.

In fluids, Newtonian rotation induces a component of vorticity in the direction of the vector Ω, so the stream lines become closely wound spirals with motions mainly confined to the family of parallel planes which are perpendicular to Ω.

Conventionally, θ increases in the anticlockwise sense; so Ω is positive for anticlockwise rotation about OZ^0, when the observer looks antiparallel to OZ^0 in the K^0 frame. And Newton's laws of motion assume the existence of a rigid inertial co-ordinate system K, precessing with angular velocity $\Omega(t)$ around the origin O fixed in it. In the K system the velocity u of a particle is given by

$$u = \dot{r}. \tag{8.8.7}$$

8.8.3 Rules of transformation

In the inertial frame K^0, the unit vectors do not change direction. But the unit vectors fixed in the K frame undergo rotation. So, for differential operators, the rules of transformation are

$$\left.\begin{aligned} \frac{\partial}{\partial x^0} &= \frac{\partial x}{\partial x^0}\frac{\partial}{\partial x} + \frac{\partial y}{\partial x^0}\frac{\partial}{\partial y} = \cos\Omega t^0\frac{\partial}{\partial x} - \sin\Omega t^0\frac{\partial}{\partial y}, \\ \frac{\partial}{\partial y^0} &= \frac{\partial x}{\partial y^0}\frac{\partial}{\partial x} + \frac{\partial y}{\partial y^0}\frac{\partial}{\partial y} = \sin\Omega t^0\frac{\partial}{\partial x} + \cos\Omega t^0\frac{\partial}{\partial y}, \\ \frac{\partial}{\partial z^0} &= \frac{\partial}{\partial z}, \quad \frac{\partial}{\partial t} = \frac{\partial}{\partial t^0} \end{aligned}\right\}. \tag{8.8.8}$$

$$\nabla^0 = \hat{x}^0\frac{\partial}{\partial x^0} + \hat{y}^0\frac{\partial}{\partial y^0} + \hat{z}^0\frac{\partial}{\partial z^0} = \hat{x}\frac{\partial}{\partial x} + \hat{y}\frac{\partial}{\partial y} + \hat{z}\frac{\partial}{\partial z} = \nabla, \tag{8.8.9}$$

where $\hat{x}^0$, $\hat{y}^0$, $\hat{z}^0$ are the unit vectors parallel to OX^0, OY^0, OZ^0 and $\hat{x}$, $\hat{y}$, $\hat{z}$ are the same parallel to OX, OY, OZ.

If A is a vector field, and the gradient operators for these two systems are ∇^0 and ∇, then

$$\nabla^0 = \nabla, \quad \nabla^0 \times A = \nabla \times A, \quad \nabla^0 . A = \nabla . A. \tag{8.8.10}$$

So ∇ is invariant for the transformation, and

$$\nabla . v = \nabla^0 \times v = \nabla^0 . [\hat{x}^0(\Omega_2 z^0 - \Omega_3 y^0) + \hat{y}^0(\Omega_3 x^0 - \Omega_1 z^0) + \hat{z}^0(\Omega_1 y^0 - \Omega_2 x^0)] = 0 \tag{8.8.11}$$

$$\nabla \times v = \nabla^0 \times v = \nabla^0 \times (\Omega \times r^0) = 2\Omega. \tag{8.8.12}$$

$$(A . \nabla)v = (A . \nabla^0)(\Omega \times r^0) = \Omega \times A, \tag{8.8.13}$$

$$\begin{aligned} (\nabla(v . A)) &= \nabla^0(v . A) = (v . \nabla^0)A + (\Omega \times A) + v \times (\nabla^0 \times A) + 2(A \times \Omega) \\ &= (v . \nabla^0)A - (\Omega \times A) + v \times (\nabla^0 \times A). \end{aligned} \tag{8.8.14}$$

The time rate of change of A is expressed as

$$\frac{dA}{dt} = \left(\frac{dA}{dt}\right) + (\Omega \times A), \tag{8.8.15}$$

where in (dA/dt) the differentiation of A is carried out in the rotating frame K, assuming that the unit vectors $\hat{x}$ and $\hat{y}$ are fixed (*Goldstein* (1980, p. 174)).

If the relative velocity $v_m(x, y, z, t)$ between the inertial frame K^0 and the frame K is non-uniform, then the relation

$$\left(\frac{dA}{dt^0}\right)_{k^0} = \left(\frac{dA}{dt}\right)_k + (\Omega \times A) \tag{8.8.16}$$

equates the time derivative of A in the two reference frames K^0 and K. In terms of partial derivatives the total time derivatives of A in the two frames, following Shiozawa (1973), are

$$\left(\frac{dA}{dt^0}\right)_k = \frac{dA}{dt^0} + (v_m \cdot \nabla)A, \tag{8.8.17}$$

$$\left(\frac{dA}{dt}\right)_k = \frac{dA}{dt} + ((v_m - v)\cdot \nabla)A. \tag{8.8.18}$$

Hence

$$\frac{\partial A}{\partial t^0} = \frac{\partial A}{\partial t} - (v\cdot \nabla)A + (\Omega \times r). \tag{8.8.19}$$

For a scalar variable ϕ the corresponding relation is

$$\frac{\partial \phi}{\partial t^0} = \frac{\partial \phi}{\partial t^0} - (v\cdot \nabla^0)\phi = \frac{d\phi}{dt} - (v\cdot \nabla^0)\phi = \frac{\partial \phi}{\partial t} - (v\cdot \nabla^0)\phi. \tag{8.8.20}$$

Putting the value of $(\Omega \times A)$ from (8.8.13) in (8.8.19), we get

$$\frac{\partial A}{\partial t^0} = \frac{\partial A}{\partial t} - (v\cdot \nabla)A + (A\cdot \nabla)v. \tag{8.8.21}$$

Again, since

$$\nabla^0 \times (v \times A) = \nabla \times (v \times A) = (A\cdot \nabla)v - (v\cdot \nabla)A + v(\nabla\cdot A) - A(\nabla\cdot v) \tag{8.8.22}$$

equation (8.8.21) becomes

$$\frac{\partial A}{\partial t^0} = \frac{\partial A}{\partial t} + \nabla \times (v \times A) - v(\nabla\cdot A) + A(\nabla\cdot v). \tag{8.8.23}$$

Using (8.8.11) for $(\nabla\cdot v)$ it can be written as

$$\frac{\partial A}{\partial t^0} + v(\nabla\cdot A) = \frac{\partial A}{\partial t} + \nabla \times (v \times A). \tag{8.8.24}$$

8.9 Mass Motion in Inertial Frames

We consider two masses m_1 at P_1 and m_2 at P_2, of fig. 8.9.1; r_1 and r_2 are the vectors from P_1 and P_2 respectively to the volume element dV at Q. Then

$$M = m_1 + m_2 + \iiint\limits_V \rho^m dV, \quad r_1 = R + r_2, \tag{8.9.1}$$

where ρ^m is the mass density, and M is the total mass of the universe. The momentum of the Universe relative to the observers P_1 and P_2, respectively, are

$$p_1 = m_2\dot{R} + \iiint \rho \dot{r}_1 dV, \tag{8.9.2}$$

$$p_2 = -m_1\dot{R} + \iiint \rho \dot{r}_2 dV = -m_1\dot{R} + \int \rho \dot{r}_1 dV - \dot{R}(M - m_1 - m_2)$$

$$= (m_2 - M)\dot{R} + \iiint \rho \dot{r}_1 dV, \tag{8.9.3}$$

where $R = P_1P_2 = r_1 - r_2$. Hence we obtain

$$p_1 = p_2 + M\dot{R}; \tag{8.9.4}$$

so that if relative to P_1 the momentum p_1 of the universe is constant, this momentum will also be constant for any other observer who is moving with constant velocity $\dot{R}$ relative to P_1; and, for inertial observers, the total momentums of the universe is constant.

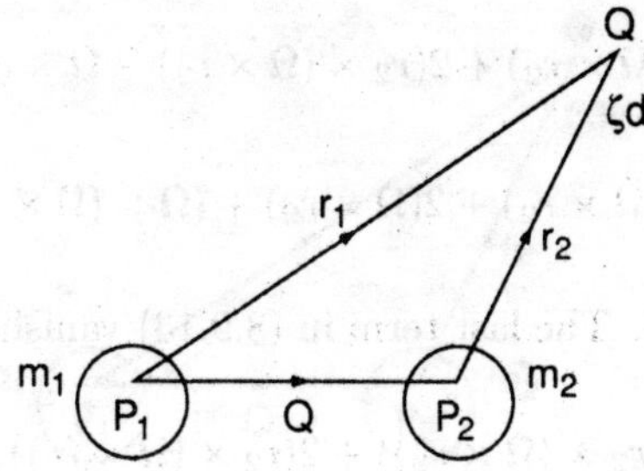

Fig. 8.9.1 The vectors joining the points P_1 and P_2 to the point Q where the mass element pdV in the volume element dV is located.

If the acceleration of m_2 relative to m_1 is $\ddot{R}$, the rate of change of momentum of the universe in the eyes of P_2 is

$$\dot{p}_2 = -M\ddot{R}. \tag{8.9.5}$$

Hence all the mass of the universe appears to him to be concentrated at the position of the inertial observer P_1 and moving with him; and, according to P_2, the center of mass of the rest of the universe moves with acceleration $-\ddot{R}$, as if the universe were pervaded with gravitational field of this magnitude and direction. When accelerated from electromagnetic or other non-gravitational forces, P_2 will observe such a gravitational field, in the direction opposite to the acceleration $\ddot{R}$ that he has acquired with respect to the system of inertial observers.

If L_1 and L_2 are the orbital angular momentums of the universe about P_1 and P_2 respectively, then neglecting intrinsic spin, it follows that

$$L_1 = (R \times m_1\dot{R}) + \iiint (r_1 \times \rho\dot{r}_1)\, dV, \tag{8.9.6}$$

$$L_2 = (R \times m_1\dot{R}) + \iiint (r_2 \times \rho\dot{r}_1)\, dV. \tag{8.9.7}$$

When both P_1 and P_2 are located at the same point, but P_2 is rotating with angular velocity Ω relative to P_1,

$$R = 0, \quad r_1 = r_2, \quad \dot{r}_1 = \dot{r}_2 + (\Omega \times r_2), \tag{8.9.8}$$

$$L_1 = \iiint (r_1 \times \rho\dot{r}_1)\, dV = L_2 + \iiint \rho(r_2 \times (\Omega \times r_2))\, dV, \tag{8.9.9}$$

$$L_2 = \iiint (r_2 \times \rho\dot{r}_2)\, dV. \tag{8.9.10}$$

For a given L_2 we may choose Ω so that

$$L_1 = 0. \tag{8.9.11}$$

Then P_1 is an inertial observer, Ω is the angular velocity of P_2 relative to P_1, and

$$L_2 = -\iiint \rho(r_2 \times (\Omega \times r_2))\, dV. \tag{8.9.12}$$

If the density of the universe does not change appreciably during the period $2\pi/\Omega$ of rotation, then

$$L_2 = -\iiint \rho dV \left[(r_2 \times (\dot{\Omega} \times r_2) + 2(r_2 \times (\Omega \times r_2) - \Omega \times (r_2 \times \dot{r}_2)\right]$$

$$= -\iiint \rho dV \left[r_2 \times \{(\dot{\Omega} \times r_2) + 2(\Omega \times \dot{r}_2) + (\Omega \times (\Omega \times r_2) - (\Omega_2 \times (r_1 \times \dot{r}_1))\right], \tag{8.9.13}$$

where (8.9.8) has been used. The last term in (8.9.13) vanishes due to (8.9.11), so we can write

$$L_2 = \iiint \rho dV\, [(r_2 \times (\Omega \times r_2)) + 2(r_2 \times (\Omega \times \dot{r}_2)) + (r_2 \times (\Omega \times (\Omega \times r_2)))] \tag{8.9.14}$$

$$= \iiint (r_2 \times \rho f_2)\, dV. \tag{8.9.15}$$

with

$$f_2 = -(\dot{\Omega} \times r_2) - 2(\Omega \times \dot{r}_2) - (\Omega \times (\Omega \times r_2)) \tag{8.9.16}$$

f_2 is, evidently, the sum of the Coriolis force, the centrifugal force and a force proportional to $\dot{\Omega}$. Hence, to the rotating observer P_2 the angular momentum is changing as if the universe was under the influence of a torque which arises from the force per unit mass f_2, given by (8.9.16). This is, as if, throughout the universe there is a gravitational field f_2 per unit mass so that a particle of mass m moving relative to P_2 under an extra force F would experience the motion from

$$m\ddot{r}_2 = F + mf_2. \tag{8.9.17}$$

Thus, in presence of matter, outside P_2, this observer can single out a special set of co-ordinate system in which the laws of physics assume a particularly simple form. The extra forces experienced in the rotating frame of P_2 are on the rest of the matter in the universe, because equation (8.9.15) is meaningless when the density $\rho^m = 0$. This conclusion supports the principle of Mach[1] that the origin of the three forces of (8.9.16) lies in the distant parts of the universe. But, within the limits of this Newtonian theory of rotation, these inertial forces are of kinematical origin in the distant matter in the sense that the gravitational field of the earth is generated by the matter of the earth.

8.10 Rotation of a Rigid Body about a Fixed Point

Consider rotation of a rigid body about a fixed point. The co-ordinates of a point in the body are $r' = (x'_1, x'_2, x'_3)$ with respect to the rectangular cartesian axes OX'_1, OX'_2, OX'_3 in the body, and the co-ordinates are $r = (x_1, x_2, x_3)$ with respect to rectangular cartesian co-ordinate axes

[1]Mach's hypothesis : The inertia of one body is due to the presence of all other bodies in the universe. Mach felt that the bulk material of the universe should some how determine the inertial, and hence the gravitational, properties of individual bodies. Matter is assumed to be distributed homogeneously in the universe, and the geometry of the space is determined by the matter distribution.

(OX_1, OX_2, OX_3), fixed in space, that is, axes fixed in an inertial system of co-ordinates. Since one point of the body is fixed in space, the origin O for both the co-ordinate systems is taken at that point. The linear relations between the co-ordinates of the two systems are given by

$$X_i^1 = A_{ij}X_j \tag{8.10.1}$$

or by

$$r' = (A \cdot r), \tag{8.10.2}$$

where A_{ij} is the transformation matrix. Since A is orthogonal,

$$r = (A^{-1} \cdot r') = (A^T \cdot r'), \tag{8.10.3}$$

where A^T is the transpose of A and A^{-1} is the inverse. So det $A \neq 0$ and

$$A_{ij}^T = A_{ji}, \quad (A \cdot A^T) = I, \tag{8.10.4}$$

where I is the unit diagonal matrix.

Since the origin and the co-ordinate axes OX_1', OX_2', OX_3' are fixed in the body

$$\dot{r}' = 0 \tag{8.10.5}$$

so that from (8.10.2), after differentiation with respect to time, we find that

$$(\mathbf{A} \cdot \dot{r}) = -(\dot{\mathbf{A}} \cdot r). \tag{8.10.6}$$

Multiplying both sides by the inverse matrix $\mathbf{A}^{-1}$ it follows that

$$\dot{r} = -(\mathbf{B} \cdot r), \tag{8.10.7}$$

where

$$\mathbf{B} = (\mathbf{A}^{-1} \cdot \dot{\mathbf{A}}) = -(\mathbf{A}^{-1} \cdot \mathbf{A}) = -(\mathbf{A}^T \cdot \mathbf{A}) = -\mathbf{B}^T. \tag{8.10.8}$$

Hence $\mathbf{B}$ is antisymmetric; so

$$\Omega \equiv (B_{23}, B_{31}, B_{12}), \tag{8.10.9}$$

$$\dot{r} \equiv (\Omega \times r) \tag{8.10.10}$$

and r is rotating with the angular velocity Ω about axes fixed in space.

If r' denotes the position of a particle which is not necessarily fixed in the body, equation (8.10.1) still holds good; moreover $\dot{r}' \neq 0$, and differentiation of (8.10.6) with respect to time gives

$$\dot{r}' = (\mathbf{A} \cdot \dot{r}) + (\dot{\mathbf{A}} \cdot r), \tag{8.10.11}$$

$$\ddot{r}' = (\mathbf{A} \cdot \ddot{r}) + 2(\dot{\mathbf{A}} \cdot \dot{r}) + (\ddot{\mathbf{A}} \cdot r). \tag{8.10.12}$$

Multiplying both sides of these two equations by $\mathbf{A}^{-1}$ gives

$$(\mathbf{A}^{-1} \cdot \dot{r}') = \dot{r} + (\mathbf{B} \cdot r), \tag{8.10.13}$$

$$(\mathbf{A}^{-1} \cdot \ddot{r}') = \ddot{r} + (\mathbf{B} \cdot \dot{r}) + ((\dot{\mathbf{B}} + \mathbf{B}^2) \cdot r) \tag{8.10.14}$$

because

$$\dot{\mathbf{B}} = (\dot{\mathbf{A}}^{-1} \cdot \dot{\mathbf{A}}) + (\mathbf{A}^{-1} \cdot \ddot{\mathbf{A}}), \tag{8.10.15}$$

$$B^2 = (-\mathbf{A}^{-1} \cdot \mathbf{A})(\mathbf{A}^{-1} \cdot \dot{\mathbf{A}}) = -(\dot{\mathbf{A}}^{-1} \cdot \dot{\mathbf{A}}). \tag{8.10.16}$$

Instantaneously, when the two systems of co-ordinate coincide with each other, we have $A = I$ and equations (8.10.13) and (8.10.14) reduce to

$$\dot{r}' = \dot{r} - (\Omega \times r), \tag{8.10.17}$$

$$\ddot{r}' = \ddot{r} - 2(\Omega \times \dot{r}) - (\dot{\Omega} \times r) + (\Omega \times (\Omega \times)). \tag{8.10.18}$$

These two equations give

$$\ddot{r} = \ddot{r}' + 2(\Omega \times r') + (\dot{\Omega} \times r') + (\Omega \times (\Omega \times r')). \tag{8.10.19}$$

Since the left hand side is the acceleration of the particle relative to the inertial co-ordinate system $OX_1X_2X_3$, we obtain

$$m\ddot{r} = F, \tag{8.10.20}$$

$$m\ddot{r}' = mf + F, \tag{8.10.21}$$

where f is the f_2 of (8.9.31). The angular momentum

$$\mathbf{S} = \iiint \rho(\Omega \times \dot{r})\, dV \tag{8.10.22}$$

$$= \iiint \rho(r \times (\Omega \times r))\, dV = (\mathbf{I}^0 \cdot \Omega), \tag{8.10.23}$$

where (8.10.10) has been used and I^0 is the moment of inertia tensor

$$I_{ij} = \iiint \rho dV (r^2 \delta_{ij} - x_i x_j). \tag{8.10.24}$$

$$S_i = I^0_{ij} \Omega_j. \tag{8.10.25}$$

Here Einstein summation convention is meant for repeated index and δ_{ij} is the Kronecker delta. Also, putting in the kinetic energy expression

$$\mathcal{E} = \frac{1}{2} \iiint \rho \dot{r}^2 dV, \tag{8.10.26}$$

the relation (8.10.10), it follows that

$$\mathcal{E} = \frac{1}{2}(\Omega \cdot S). \tag{8.10.27}$$

Let the co-ordinate axes $OX_1'X_2'X_3'$ coincide with the principal axes of inertia of the body. These axes being fixed in the body, the principal moments of inertia $I^{0'}_{11}, I^{0'}_{22}, I^{0'}_{33}$ are constants for rigid bodies; so their time derivatives vanish. Differentiating the transformation relation

$$I' = \mathbf{A} I^0 \mathbf{A}^{-1} \tag{8.10.28}$$

with respect to time we obtain

$$\mathbf{A} I^0 \dot{\mathbf{A}}^{-1} + \mathbf{A} \dot{I}^0 \mathbf{A}^{-1} + \dot{\mathbf{A}} I^0 \mathbf{A}^{-1} = 0. \tag{8.10.29}$$

Solving for $\dot{I}^0$ from this relation gives

$$\dot{I}^0 = -I^0 \dot{\mathbf{A}}^{-1} \mathbf{A} - \mathbf{A}^{-1} \dot{\mathbf{A}} I^0. \tag{8.10.30}$$

Using (8.10.8) in this relation we have

$$\dot{I}^0 = I^0\mathbf{B} - \mathbf{B}I^0. \tag{8.10.31}$$

Differentiating now (8.10.23) with respect to time and using the result in this relation gives

$$\dot{S} = (I^0\mathbf{B} - \mathbf{B}I^0)\Omega + I^0\dot{\Omega}. \tag{8.10.32}$$

Using again (8.10.23) this relation reduces to

$$\dot{S} = I^0\mathbf{B}\Omega - (\mathbf{B}\cdot\mathbf{S}) + I^0\dot{\Omega}. \tag{8.10.33}$$

Also, since $\mathbf{B}$ is antisymmetric, the relations (8.10.9) show that

$$(\mathbf{B}\cdot\Omega) = 0. \tag{8.10.34}$$

Hence (8.10.33) reduces to

$$\dot{S} = I^0\dot{\Omega} - (\mathbf{B}\cdot S). \tag{8.10.35}$$

Next, equating (8.10.7) with (8.10.10) and replacing in the result, r by S we find that (8.10.35) reduces to

$$\dot{S} = I^0\dot{\Omega} + (\Omega \times S). \tag{8.10.36}$$

For the moment of the forces we equate the applied torque N to the rate of change of angular momentum. When the two systems OX_1X_2, X_3 and $OX_1'X_2'X_3'$ coincide, instantaneously, the Euler equations

$$N' = I^0\dot{\Omega}' + (\Omega' \times S') = I^0\dot{\Omega}' + (\Omega' \times (I^0\Omega')) \tag{8.10.37}$$

follow. Their component equations are

$$N_A' = I_A^0\dot{\Omega}_x' + (I_C^{0'} - I_B^{0'})\Omega_y'\Omega_z', \tag{8.10.38}$$

$$N_B' = I_B^0\dot{\Omega}_y' + (I_A^{0'} - I_C^{0'})\Omega_z'\Omega_x', \tag{8.10.39}$$

$$N_C' = I_A^0\dot{\Omega}_z' + (I_B^{0'} - I_A^{0'})\Omega_x'\Omega_y'. \tag{8.10.40}$$

8.11 Differential Rotation

Differential rotation (also called *non-rigid* or *non-uniform rotation*) is an incompressible velocity field symmetric about the axis of rotation. This rotation depends on the structure of the body and of the major external influence. The relation between the time period of rotation and the position vector $\mathbf{r}$ is determined from observations on the change of Ω with r. The condition for rigid (uniform) rotation is $\nabla\Omega = 0$, where Ω is the angular rate of rotation; for non-rigid rotation $\nabla\Omega \neq 0$. In cylindrical polar co-ordinates, the velocity u for non-rigid rotation is given by $u = \Omega(r, z)\hat{z} \times r$, where r is the vector radial distance from the axis and $\hat{Z}$ is the unit vector along the Z axis. If the stress lines form ellipses with centers in OZ, and even ellipses with some amount of azimuthal and axial distortions, the rotation is mathematically difficult to treat.

Observations of planets and stars show that their Ω rises to its peak value near the axis, and gradually decays away from there. So, all points of an axially symmetric surface of constant

r rotate at the same velocity. But the neighbourhood points, with slightly different r, have different velocities. Moreover, friction from tangential stresses and strains of twist, due to contact between particles, develops on surfaces of slightly different **r**, moving with continuously changing velocities.

Observations of the motion of sunspots and other markings of the photosphere indicate that different latitudes rotate with different angular velocities. The sun rotates faster at the equator than at high latitudes, the period of rotation is 25 days near the equator, and 27 days at latitude 35^0. Variation of the angular velocity with the latitude ϕ, indicated also by Doppler effect measurements, is probably of the type

$$\Omega = \Omega_a + \Omega' \cos^2 \phi = \Omega_e - \Omega' \sin^2 \phi,$$

where $\Omega_e = 14.4^0/\text{day}$, $\Omega_a = 11.6^0/\text{day}$, $\Omega' = 2.8^0/\text{day}$.

The evolution of magnetic field in non-uniform rotating cosmic bodies is in conformity with Ferraro law of isorotation (Exercise 4.4.8). This law gives the average state of rotation, provided that the shape of the magnetic lines of force is known, because every point of such a line must rotate with the same average angular velocity; otherwise very strong electric currents would be induced. Relativistically, all rotations are non-rigid and the introduction of an instantaneous rest frame for the entire system is not possible because the space-time metric cannot be reduced then to the diagonal form. So, for simplicity, effects of rotation of the order of v/c only, following from Newtonian and Galilean concepts of motion, are studied. The relativistic non-rigidity of rotation is not related is the non-rigid rotation discussed here. Moreover, the relativistic non-rigidity is very weak but the other non-rigidity is not so weak.

8.12 Fields of Rotation of an Earth Satellite

Neglecting the effects of Sun, Moon and other nearby objects, on motion of a free satellite P, rotating at the constant rate Ω, its dynamics is studied due to the gravitational force of earth. For this purpose we consider a stationary Cartesian co-ordinate system K, with origin at the center of the earth, and having (x^0, y^0, z^0) for co-ordinates of P. Then the only force, acting on P, is the inverse square of distance attraction, from the center of the earth (radius R_0). At the earth surface, the acceleration is $g = 32.2$ feet/sec^2 $= 981$ cm/sec^2; at any radius R, this acceleration is

$$f = g\frac{R_0^2}{R^2}. \tag{8.12.1}$$

The Kepler law for motion of P is

$$\ddot{R} = -Rg\frac{R_0^2}{R^3}. \tag{8.12.2}$$

A second co-ordinate system K is assumed to rotate with angular velocity Ω about OZ (which coincides with OZ^0); the co-ordinates of P in K are (x, y, z). Then, the equations of motion are

$$\ddot{x} = 2\Omega\dot{y} + x\left(\Omega^2 - \frac{q}{R^3}R_0^2\right). \tag{8.12.3}$$

$$\ddot{y} = -2\Omega\dot{x} + y\left(\Omega^2 - \frac{q}{R^3}R_0^2\right). \tag{8.12.4}$$

$$\ddot{z} = \frac{2q}{R^3}R_e^2. \tag{8.12.5}$$

One solution of these equations is

$$X = 0, \quad Y = Re, \quad Z = 0, \quad \dot{X} = 0, \quad \dot{Y} = 0, \quad \dot{Y} = 0, \tag{8.12.6}$$

where $Re = (gR_0^2/\Omega^2)^{1/3}$. Hence, a satellite, describing a circular orbit with constant angular velocity Ω, remains stationary in the system K. If (x', y', z') is a point in the immediate neighbourhood of the satellite, in a third co-ordinate system K', centered on P, then the transformation relations

$$x = x', \quad y = y' + Re, \quad z = z' \tag{8.12.7}$$

and the approximate relations

$$R = R_e + y', \quad g\frac{R_0^2}{R^3} = \Omega^2\left(1 - \frac{3y'}{R_e}\right) \tag{8.12.8}$$

are valid. The equations of motion,

$$\ddot{x}' = 2\Omega\dot{y}', \quad \ddot{y}' = -2\Omega x' + 3\Omega^2 y', \quad \ddot{z} = -\Omega^2 z', \tag{8.12.9}$$

are these of an object near the satellite, travelling on a circular orbit in a reference co-ordinate system that is centered on it and rotating with it. In vector form these become

$$\ddot{R}' = \nabla\phi + 2(\dot{R}' \times \Omega), \tag{8.12.10}$$

with

$$\frac{\partial\phi}{\partial x'} = 0, \quad \frac{\partial\phi}{\partial y'} = 3\Omega^2 y', \quad \frac{\partial\phi}{\partial z'} = -\Omega^2 z' \tag{8.12.11}$$

$R' = (x', y', z')$, and the potential ϕ is given by

$$\phi(x', y', z') = \frac{\Omega^2}{2}(3y'^2 - z'^2). \tag{8.12.12}$$

The negative Y axis points to the center of the earth, and the positive X axis is along the direction of motion of the particle. The apparent acceleration has two parts. The potential function of the position co-ordinates is the 'gravostatic' part of the potentials. The other is the 'gravo magnetic' (or Coriolis) component; it is a function of the velocity and rotation frequency. This interpretation of the velocity dependent Coriolis force of gravitation arising in satellite flights round the earth, resembles that of motion of an electron in a field of electrostatic potential ϕ and magnetic induction vector $2mc\Omega/q$. The equivalent magnetic field is uniform and parallel to the axis of rotation.

In the EM theory, the magnetic field component, acting on a charge, is produced by other charges rotating in the wires of a solenoidal coil. Then the Coriolis force is produced by the balance of the Universe rotating about the satellite, and a moving body induces gravity waves, of a Coriolis field, like the EM waves of a moving charge, because, otherwise, the gravity waves would propagate at infinite speed. The equivalent gravitational scalar potential ϕ and vector potential A give effects of accuracy sufficient to determine whether they have any significance.

8.13 Non-inertial Co-ordinate Systems in Absolute Space and Time

Dynamics of particles and fluids is studied only in inertial systems of co-ordinates. Even expressions for energy and energy-like quantities in accelerated co-ordinate systems are not available in text books. In this respect *Wells* (1967,, section 29(4), p. 25) is an exception.

Consider the kinetic energy expression

$$T = \frac{1}{2}\sum_i (m_i v_i^2) \tag{8.13.1}$$

in an inertial frame, where v_i is measured relative to an inertial frame of co-ordinates. In an accelerated co-ordinate system, due to the relative motion of the system, some or all the chosen co-ordinates are non-inertial. The velocity is obtained from the equations of transformation to the inertial co-ordinate axes in terms of the co-ordinates of the accelerated frame. So, the velocity is not that relative to the non-inertial reference axes.

Consider the orthogonal co-ordinate axes O_2X_2, O_2Y_2, O_2Z_2 of the frame 2 in motion with constant acceleration a, relative to the inertial (laboratory) orthogonal co-ordinate axes (O_1X_1, O_1Y_1, O_1Z_1), fig. 8.13.1 of frame 1.

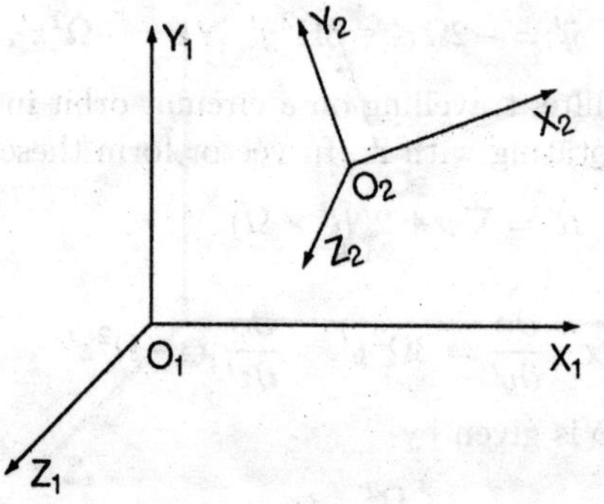

Fig. 8.13.1 Shows two coordinate systems 1 and 2. System 1 is $O_1X_1Y_1Z_1$ and system 2 is $O_2X_2Y_2Z_2$. System 1 is the laboratory inertial frame, and system 2 is accelerating, so that it has rotation as well as translation relative to the system 1.

Then the transformation relations for the co-ordinates are

$$\left.\begin{aligned} x_1 &= x_0 + l_1x_2 + l_2y_2 + l_3z_2 \\ y_1 &= y_0 + m_1x_2 + m_2y_2 + m_3z_2 \\ z_1 &= z_0 + n_1x_2 + n_2y_2 + n_3z_2 \end{aligned}\right\}, \tag{8.13.2}$$

where l_1, m_1, n_1 are the direction cosines of O_2x_2; (l_2, m_2, n_2) and (l_3, m_3, n_3) are those of O_2y_2 and O_2z_2. For a known motion, the values of (x_0, y_0, z_0) and the direction cosines are known functions of time t. Differentiation with respect to t gives

$$\dot{x}_1 = \dot{x}_0 + l_1\dot{x}_2 + l_2\dot{y}_2 + l_3\dot{z}_2 + \dot{l}_1x_2 + \dot{l}_2y_2 + \dot{l}_3z_2, \tag{8.13.3}$$

etc. In terms of the inertial velocity v and uniform acceleration a of the frame $O_2x_2y_2z_2$ relative to the frame $O_1x_1y_1z_1$, the vector form transformation relation is

$$r_1 = vt + \frac{1}{2}at^2 + r_2 \tag{8.13.4}$$

for the position vector. Hence the velocity transformation relation is

$$\dot{r}_1 = v + at + \dot{r}_2. \tag{8.13.5}$$

Then the kinetic energy of a particle of mass m at a point P in the system 2 is

$$T = \frac{1}{2}m\left[(v_x + a_x t + \dot{x}_2)^2 + (v_y + a_y t + \dot{y}_2)^2 + (v_z + a_z t + \dot{z}_1)^2\right]. \tag{8.13.6}$$

This expansion for the kinetic energy in an accelerated reference frame is an explicit function of time t.

For two dimensional motion in the xy plane the relations (8.13.2) simplify to

$$x_1 = x_0 + x_2 \cos\theta - y_2 \sin\theta. \tag{8.13.7}$$

$$y_1 = y_0 + x_2 \sin\theta + y_2 \cos\theta. \tag{8.13.8}$$

It is evident that x_1 and y_1 are functions of both x_2 and y_2. If O_2 is moving with velocity v, relative to O_1, and at the same time the co-ordinate system 2 is rotating with constant angular velocity Ω about $O_2 z_2$ (which is parallel to $O_1 z_1$), so that $\theta = \Omega t$, for finding the kinetic energy and other expressions for motion in the rotating frame, we obtain

$$x_1 = v_x t + x_2 \cos\Omega t - y_2 \sin\Omega t. \tag{8.13.9}$$

$$y_1 = v_y t + x_2 \sin\Omega t + x_2 \cos\Omega t. \tag{8.13.10}$$

8.14 The Metric Tensor for GN Rotation

Consider the frame of reference K rotating uniformly relative to the laboratory inertial frame K^0. For a point P the cylindrical polar co-ordinates r^0, θ^0, z^0, t^0 are chosen in K^0 and (r, θ, z, t) in K. Then, in the relations for transformation of these co-ordinates from one system to another, the dimensions of all the co-ordinates are not the same; θ, for instance, is dimensionless and r has the dimension of length.

For the K^0 frame we write

$$x^{1^0} = r^0, \quad x^{2^0} = \theta^0, \quad x^{3^0} = z^0, \quad x^{4^0} = ct^0, \tag{8.14.1}$$

for the K frame, we similarly have

$$x^1 = r, \quad x^2 = \theta, \quad x^3 = z, \quad x^4 = ct. \tag{8.14.2}$$

Then

$$x^{1^0} = x', \quad x^{2^0} = \theta^0 = x^2 + \frac{\Omega}{c}x^4, \quad x^{3^0} = x^3, \quad x^{4^0} = x^4, \tag{8.14.3}$$

$$x^1 = x^{1^0}, \quad x^2 = x^{2^0} - \frac{\Omega}{c}x^{4^0} = \theta^0 - \Omega t^0, \quad x^3 = x^{3^0}, \quad x^4 = x^{4^0}. \tag{8.14.4}$$

The non-zero coefficients of transformation. $\alpha_\nu^{\mu^0}$ and $\alpha_{\nu^0}^\mu$, obtained from (11.2.19) and (11.2.21), are

$$\alpha_{\mu^0}^\mu = 1, \quad \alpha_{4^0}^2 = -\frac{\Omega}{c}, \quad \alpha_\mu^{\mu^0} = 1, \quad \alpha_4^{2^0} = \frac{\Omega}{c}, \tag{8.14.5}$$

where the repeated index convention does not mean summation over that index. Then

$$\left.\begin{aligned}
\alpha_1^{1^0} &= \cos\Omega t,\ \alpha_2^{1^0} = -\sin\Omega t,\ \alpha_3^{1^0} = 0,\\
\alpha_4^{1^0} &= -\frac{\Omega}{c}(x\sin\Omega t + y\cos\Omega t) = -\frac{\Omega}{c}y^0,\\
\alpha_1^{2^0} &= \sin\Omega t,\ \alpha_2^{2^0} = \cos\Omega t,\ \alpha_3^{2^0} = 0,\\
\alpha_4^{2^0} &= \frac{\Omega}{c}(x\cos\Omega t - y\sin\Omega t) = -\frac{\Omega}{c}x^0,\\
\alpha_1^{3^0} &= 0,\ \alpha_2^{3^0} = 0,\ \alpha_3^{3^0} = 1,\ \alpha_4^{3^0} = 0,\\
\alpha_1^{4^0} &= 0,\ \alpha_2^{4^0} = 0,\ \alpha_3^{4^0} = 0,\ \alpha_4^{4^0} = 1,
\end{aligned}\right\}. \tag{8.14.6}$$

$$\det|\alpha^\mu_{\nu^0}| = \begin{vmatrix} 1 & 0 & 0 & 0\\ 0 & 1 & 0 & -\Omega/c\\ 0 & 0 & 1 & 0\\ 0 & 0 & 0 & 1\end{vmatrix}. \tag{8.14.7}$$

Equations (8.14.6) give

$$\det|\alpha^{\mu^0}_\nu| = \begin{vmatrix} \cos\Omega t & -\sin\Omega t & 0 & -\frac{\Omega}{c}y^0\\ \sin\Omega t & \cos\Omega t & 0 & \frac{\Omega}{c}x^0\\ 0 & 0 & 1 & 0\\ 0 & 0 & 0 & 1\end{vmatrix} = 1. \tag{8.14.8}$$

Similarly, $\det|\alpha^\mu_{\nu^0}|$ is determined by changing the sign of Ω in (8.14.8) or directly. The invariant 4-interval ds is given by

$$ds^2 = dx^{0^2} + dy^{0^2} + dz^{0^2} - (cdt^0)^2 = g_{\mu^0\nu^0}dx^{\mu^0}dx^{\nu^0} \tag{8.14.9}$$

$$= dx^2 + dy^2 + dz^2 - (1-\beta^2)(cdt)^2 + 2\left(-\frac{y}{c}\Omega\right)dx\cdot cdt + 2\left(\frac{x\Omega}{c}\right)dy\cdot cdt \tag{8.14.10}$$

$$= dr^{0^2} + (r^0 d\theta^0)^2 + dz^{0^2} - (cdt^0)^2 \tag{8.14.11}$$

$$= g_{\mu\nu}dx^\mu dx^\nu = -c^2dT^2 \tag{8.14.12}$$

$$= dr^2 + (rd\theta)^2 + dz^2 + 2\beta\cdot rd\theta\cdot cdt - (1-\beta)^2(cdt)^2, \tag{8.14.13}$$

where

$$\beta = \frac{v}{c} = \frac{r\Omega}{c} = \frac{r^0\Omega}{c},\quad \gamma^2 = \frac{1}{1-\beta^2} \tag{8.14.14}$$

and dT is the proper time sub-interval. In the k frame the metric elements being symmetric, $g_{\mu\nu} = g_{\nu\mu}$, and for the Cartesian co-ordinates,

$$g_{\mu\nu} = \begin{bmatrix} 1 & 0 & 0 & -\dfrac{y\Omega}{c} \\ 0 & 1 & 0 & \dfrac{x\Omega}{c} \\ 0 & 0 & 1 & 0 \\ -\dfrac{y\Omega}{c} & \dfrac{x\Omega}{c} & 0 & -(1-\beta^2) \end{bmatrix}. \tag{8.14.15}$$

$$g^{\mu\nu} = \begin{bmatrix} 1-\dfrac{y^2\Omega^2}{c^2} & \dfrac{xy\Omega^2}{c^2} & 0 & -\dfrac{y\Omega}{c} \\ \dfrac{xy\Omega^2}{c^2} & 1-\dfrac{x^2\Omega^2}{c^2} & 0 & \dfrac{x\Omega}{c} \\ 0 & 0 & 1 & 0 \\ -\dfrac{y\Omega}{c} & \dfrac{x\Omega}{c} & 0 & -1 \end{bmatrix} \tag{8.14.16}$$

and

$$\det g_{\mu\nu} = g = -1 = \det g^{\mu\nu}. \tag{8.14.17}$$

Equation (8.14.10) can be written as

$$ds^2 = -\left[1 - \left\{\frac{u + (\Omega \times r))}{c}\right\}^2\right](cdt)^2 = -\left(1 - \frac{u^{0^2}}{c^2}\right)(cdt)^2, \tag{8.14.18}$$

where

$$u^0 = u + (\Omega \times r), \tag{8.14.19}$$

$$u = \frac{dr}{dt}, \quad u^0 = \frac{dr^0}{dt^0} = \frac{dr}{dt} + (\Omega \times r). \tag{8.14.20}$$

So, u^0 is the ordinary velocity of the particle in the K^0 frame, and u is the same in the K frame; the two radii vectors r and r^0 are obviously identical. We see that when

$$g = \frac{1}{c}(\Omega \times r), \tag{8.14.21}$$

we evidently get

$$g_i = \frac{g_{i4}}{g_{44}}. \tag{8.14.22}$$

The formal compatibility between (8.14.15) and (8.14.19) may be purely accidental, and does not imply that the relativistic equations of motion can be taken over from non-relativistic mechanics without any modification.

In cylindrical co-ordinates, equations (8.14.9) and (8.14.11) give

$$g_{\mu^0\nu^0} = \begin{bmatrix} 1 & 0 & 0 & 0 \\ 0 & r^{0^2} & 0 & 0 \\ 0 & 0 & 1 & 0 \\ 0 & 0 & 0 & -1 \end{bmatrix}; \tag{8.14.23}$$

so,

$$g^{\mu^0\nu^0} = \begin{bmatrix} 1 & 0 & 0 & 0 \\ 0 & 1/r^{0^2} & 0 & 0 \\ 0 & 0 & 1 & 0 \\ 0 & 0 & 0 & -1 \end{bmatrix}, \tag{8.14.24}$$

$$\det g_{\mu^0\nu^0} = -r^{0^2} = g^0, \quad \det g^{\mu^0\nu^0} = -\frac{1}{r^{0^2}} = \frac{1}{g^0}. \tag{8.14.25}$$

Similarly, equation (8.14.13) gives

$$g_{\mu\nu} = \begin{bmatrix} 1 & 0 & 0 & 0 \\ 0 & r^2 & 0 & \beta r \\ 0 & 0 & 1 & 0 \\ 0 & \beta r & 0 & -(1-\beta^2) \end{bmatrix}. \tag{8.14.26}$$

The inverse of this matrix is

$$g^{\mu\nu} = \begin{bmatrix} 1 & 0 & 0 & 1 \\ 0 & 1/2r^2 & 0 & 0 \\ 0 & 0 & 1 & 0 \\ 1 & 0 & 0 & -1 \end{bmatrix}. \tag{8.14.27}$$

Then evidently

$$\det g_{\mu\nu} = g = -r^2, \quad \det g^{\mu\nu} = -\frac{1}{r^2} = \frac{1}{g}. \tag{8.14.28}$$

Equation (8.14.13) can be written as

$$ds^2 = dr^2 + \gamma^2 (rd\theta)^2 + (dz)^2 - \left(\beta\gamma r d\theta - \frac{1}{\gamma} cdt\right)^2. \tag{8.14.29}$$

Then the spatial metric tensor γ_{ij}, defined (11.3.20), becomes

$$\gamma_{ij} = \begin{bmatrix} 1 & 0 & 0 \\ 0 & r^2\gamma^2 & 0 \\ 0 & 0 & 1 \end{bmatrix}. \tag{8.14.30}$$

Therefore,

$$\det \gamma_{ij} = |\gamma_{ij}| = \hat{\gamma} = r^2\gamma^2, \tag{8.14.31}$$

$$a_1 = 0, \quad a_2 = r\beta\gamma, \quad a_3 = 0, \tag{8.14.32}$$

$$dl^2 = dr^2 + \gamma^2(rd\theta)^2 + (dz)^2, \tag{8.14.33}$$

$$\frac{1}{\Gamma^2} = \left(\frac{1}{\gamma} - \frac{\beta}{c}\gamma r\dot{\theta}\right)^2 - \frac{1}{c^2}\left(\dot{r}^2 + \gamma^{2'} r^2\dot{\theta}^2 + \dot{z}^2\right)$$

$$= 1 - \frac{1}{c^2}(r\dot{\theta} + \beta c)^2 - \frac{1}{c^2}(\dot{r}^2 + \dot{z}^2) = \left(\frac{dT}{dt}\right)^2, \tag{8.14.34}$$

where dot denotes derivative with respect to t, and the quantities a_i, dl and Γ are defined in the relations (11.3.2)–(11.3.4). This transformation is called an *orthodox approach to rotation*, *Corum* (1977). It preserves the absolute character of time.

Putting

$$W = u - v, \tag{8.14.35}$$

we find that

$$\frac{1}{\Gamma^2} = 1 - \frac{W^2}{C^2} \tag{8.14.36}$$

because $v \equiv (0, r\Omega, 0)$ and so $W \equiv (\dot{r}, r(\dot{\theta} - \Omega), \dot{z})$.

We can write (8.14.29) as

$$ds^2 = dl^2 - (cdt^*)^2, \tag{8.14.37}$$

where

$$cdt^* = \frac{c}{\gamma}dt - \beta\gamma \cdot rd\theta \tag{8.14.38}$$

and dl is given in (8.14.33).

The expression for the velocity of light c^* in the rotating frame K' in this theory is obtained by putting $ds = 0$ in the relation (8.14.37). Thus

$$c^* = \frac{dl}{dt} = \frac{c}{\gamma} - \beta\gamma u_\theta, \tag{8.14.39}$$

where

$$u_\theta = r\frac{d\theta}{dt}. \tag{8.14.40}$$

The non-zero components of the Christoffel symbol

$$\Gamma^\lambda_{\eta\nu} = \frac{1}{2}g^{\lambda\mu}\left(\frac{\partial g_{\mu\nu}}{\partial x^\eta} + \frac{\partial g_{\mu\eta}}{\partial x^\nu} + \frac{\partial g_{\eta\nu}}{\partial x^\mu}\right), \tag{8.14.41}$$

for this rotation, are

$$\Gamma'_{24} = \Gamma'_{42} = -\frac{\Omega}{C}, \quad \Gamma'_{44} = -\frac{x\Omega^2}{C^2}, \tag{8.14.42}$$

$$\Gamma^2_{14} = \Gamma^2_{41} = \frac{\Omega}{c}, \quad \Gamma^2_{44} = -\frac{y\Omega^2}{C^2}. \tag{8.14.43}$$

Equations (8.14.13) and (8.14.37) have been briefly discussed in text books, actually these are equations (89.2) of *Landau and Lifshitz* (1975), equation (8.77) of *Moller* (1972) and equations (4.84) and (4.90) of Adder, *Bazin and Schiffer* (1975).

8.15 Theory of Lorentz Rotation

Consider the transformation relations

$$r = r', \quad \theta = \theta', \quad z = z', \quad t = \gamma' t' \tag{8.15.1}$$

from the K frame to another frame K' which is characterized by a time dilation of the Lorentz type, where

$$\gamma' = \frac{1}{\sqrt{1-\beta'^2}} = \gamma, \quad \beta' = \frac{r'\Omega}{c} = \beta \tag{8.15.2}$$

because $r = r'$. Then, evidently, the relations of transformation from the K^0 frame to the frame K' are

$$r' = r^0, \quad \theta' = \theta^0 - \Omega t^0, \quad z' = z^0, \quad t' = \frac{t^0}{\gamma}. \tag{8.15.3}$$

The relations of reverse transformation, are

$$r^0 = r', \quad \theta^0 = \theta' + \Omega\gamma t', \quad z^0 = z', \quad t^0 = \gamma t'. \tag{8.15.4}$$

It may be noted that the relations of (8.15.4) do not follow from those of (8.15.3) by changing the sign of Ω and interchanging the positions of the prime and the superscript 0 in these relations. The reason is that the transformation from an inertial frame of reference to a non-inertial one is different from the reverse transformation. This is because it is possible to have the same matrix for transformation from several non-inertial frame to a non-inertial frame without choosing the geometry of the problem.

Equation (8.14.13), after transformation to the rotating frame K', which means using (8.15.1), reads

$$\begin{aligned} ds^2 = dr^2(1-\beta^2\gamma^4\Omega^2 t'^2) + (rd\theta)^2 + dz^2 - (cdt')^2 + 2\beta^2\gamma^3\Omega t'\cdot dr\cdot rd\theta \\ - 2\beta\gamma^2\Omega t'\cdot cdt'\cdot dr + 2\beta\gamma rd\theta\cdot cdt', \end{aligned} \tag{8.15.5}$$

where the unnecessary primes over r, θ, z have been dropped. This relation can be reduced to the form convenient for interpretation,

$$ds^2 = dr^2 + \frac{(rd\theta)^2}{1-\beta^2} + dz^2 - (cdt^*)^2, \tag{8.15.6}$$

where

$$cdt^* = cdt' - \beta\gamma\cdot rd\theta + \beta\gamma^2\Omega t'\cdot dr. \tag{8.15.7}$$

Writing (8.15.6) as

$$ds^2 = \gamma_{ij}dx^i dx^j - (cdt' - \beta\gamma rd\theta + \beta\gamma^2\Omega t'.dr) \tag{8.15.8}$$

we obtain

$$\gamma_{ij} = \begin{bmatrix} 1 & 0 & 0 \\ 0 & r^2\gamma^2 & 0 \\ 0 & 0 & 1 \end{bmatrix}, \tag{8.15.9}$$

where

$$dx^i \equiv (dr, d\theta, dz). \tag{8.15.10}$$

Since the elements of γ_{ij} are independent of time, the Galilean rotation is essentially rigid. It is then easy to see that

$$\frac{dT}{dt'} = \frac{1}{\Gamma} = \left[\left(\frac{dt^*}{dt'}\right)^2 - \left(\frac{dr}{cdt'}\right)^2 - \frac{r^2}{1-\beta^2}\left(\frac{d\theta}{dt'}\right)^2 - \left(\frac{dz}{cdt'}\right)^2\right]^{1/2}$$

$$= \left[\left\{1 - \frac{\beta\gamma r}{c}\left(\frac{d\theta}{dt'}\right) + \frac{\beta\gamma^2\Omega t'}{c}\left(\frac{dr}{dt'}\right)\right\}^2\right.$$

$$\left.-\frac{1}{c^2}\left\{\left(\frac{dr}{dt'}\right)^2 + \gamma^2\left(r\frac{d\theta}{dt'}\right)^2 + \left(\frac{dz}{dt'}\right)\right\}\right]^{1/2}$$

$$= \left[1 - \frac{2}{c}\beta\gamma\left(u'_\theta - \gamma\Omega t' u'_r\right) - \frac{2\beta^2\gamma^3}{c^2}\Omega t' \cdot u_{\theta'}u_{r'}\right.$$

$$\left.-\frac{1}{c^2}\left\{(1-\beta^2\gamma^4\Omega^2 t'^2)u_{r'}^2 + u_{\theta'}^2 + u_{z'}^2\right\}\right]^{1/2}, \qquad (8.15.11)$$

where

$$u_{r'} = \frac{dr}{dt'}, \quad u_{\theta'} = \frac{rd\theta}{dt'}, \quad u_{z'} = \frac{dz}{dt'} \qquad (8.15.12)$$

and dT is the proper time sub-interval defined in (11.2.41).

Putting $ds = 0$, $dr = 0$ and $dz = 0$ in (8.15.6) we obtain the condition for free propagation of light rays along circular paths at a fixed radius :

$$\gamma^2(rd\theta)^2 - c^2\left(dt' - \frac{\beta\gamma}{c}rd\theta\right)^2 = 0. \qquad (8.15.13)$$

Solving for dt' we have two values :

$$cdt' = \beta\gamma rd\theta \pm \gamma rd\theta. \qquad (8.15.14)$$

These give the velocity of light along the arc of the circle of radius r propagating in the clockwise (CW) and counter Clockwise (CCW) senses. The corresponding circulation times, denoted by t'_1 and t'_2, are obtained by integrating θ between 0 and 2π and between 0 and -2π. So,

$$t'_1 = \frac{2\pi}{c}r\gamma(1+\beta), \quad t'_2 = \frac{2\pi}{c}r\gamma(1-\beta). \qquad (8.15.15)$$

The time difference,

$$t'_1 - t'_2 = \frac{4\pi}{c}r\beta\gamma, \qquad (8.15.16)$$

is used in the study of Sagnac effect, and also in the ring laser effects, *Post* (1967).

8.16 Trocheris Rotation

We now consider transformation from the frame, K^0, to the frame K, rotating about the Z axis, at the constant rate Ω. For this purpose, we assume the validity of local length contraction and time dilation of the type provided by LTs along intervals of arcs of circles of constant radii. This transformation is

$$dr^0 = dr, \quad r^0 d\theta^0 = \gamma(rd\theta + \beta cdt), \quad dz^0 = dz,$$
$$cdt^0 = \gamma(cdt + \beta rd\theta), \tag{8.16.1}$$

where

$$\beta = \frac{v}{c} = \frac{1}{c}(\Omega \times r), \quad \gamma = \frac{1}{(1-\beta^2)^{1/2}}. \tag{8.16.2}$$

The matrix of this transformation, in which the $\mu\gamma$eth element is denoted by $\alpha_\nu^{\mu^0} (= \partial x^{\mu^0}/\partial x^\nu)$ is

$$\alpha_\nu^{\mu^0} = \begin{bmatrix} 1 & 0 & 0 & 0 \\ 0 & \gamma & 0 & \beta\gamma \\ 0 & 0 & 1 & 0 \\ 0 & \beta\gamma & 0 & \gamma \end{bmatrix}. \tag{8.16.3}$$

The reverse relations of transformation obtain by changing the sign of Ω in (8.16.1),

$$dr = dr^0, \quad rd\theta = \gamma(r^0 d\theta^0 - \beta cdt^0), \quad dz = dz^0,$$
$$cdt = \gamma(cdt^0 - \beta r^0 d\theta^0). \tag{8.16.4}$$

Trocheris (1949) noted that the GN relations of transformation (Section 8.14) are valid in finite regions of space and that if the axis of rotation goes to infinity while Ω tends to zero in such a way that $r\Omega$ tends to an finite value, then the uniform rotation tends to translation of the GN type at the uniform rate v. To find a transformation scheme which, in this limit, tends to an LT, we considered the Lorentz type relations (8.16.1) and (8.16.4) for small rotational velocities ($\gamma \approx 1$) and obtained the simpler set of transformation relations

$$dr^0 = dr, \quad r^0 d\theta^0 = rd(\theta + \Omega t), \quad dz^0 = dz.$$
$$dt^0 = dt + \frac{1}{c}\beta rd\theta. \tag{8.16.5}$$

Since $\beta (= r\Omega/c)$ depends on r, the time differential dt^0 is not a perfect differential; so this transformation is non-integrable or anholonomic. This means that two such infinitesimal transformations would contradict each other in their common range between r and $r + dr$ when taken for distances r and $r + dr$ from the axis of rotation. The differentials $d\theta^0$ and dt^0 of (8.16.1) and $d\theta$ and dt of (8.16.4) are also anholonomic. Here we do not study the problem involving the non-integrable differentials, as it is, but following *Trocheris* (1949), make these exact artificially by adding some extra terms, to see the consequences. There is no ambiguity in the choices of extra terms added, because a non-integrable differential cannot be made exact in more than one way, at least in the cases we are aware of. This exactization leads to some interesting consequences which we consider here.

Using consistency checks in the final results of this theoretical artifice[2] we obtain a set of auxiliary relations of transformation for rotation, which contain the most concentrated

[2] The consistency checks for identification of the auxiliary metric of rotation with that of a real rotation are : (1) The relations of transformation which obtain from the auxiliary metric for rotation are valid for all real constant values of the angular velocity Ω;

development of relativistic space time distortions of the Lorentz type from rotation. So, this process of making the differentials exact for rotation seems to have physical significance, which should be investigated.

As the expression for cdt^0 in (8.16.5) contains the non-integrable term $r^2\Omega d\theta/c$, we write

$$ct^0 = ct + \frac{r^2\Omega}{c}(\theta - \theta_1), \tag{8.16.6}$$

where θ_1 is the constant of integration of the exact differential which on integration gives (8.16.6). Thus the term cdt becomes a total differential of (8.16.6). Co-ordinates in both the frames K^0 and K of reference can be chosen such that θ_1 vanishes, giving

$$r^0 = r, \quad \theta^0 = \theta + \Omega t, \quad ct^0 = ct + \frac{r^2\Omega}{c}\theta, \quad z^0 = z. \tag{8.16.7}$$

For the small velocity approximation, we write $\gamma \approx 1$ and recover the similar reverse relations of transformation,

$$r = r^0, \quad \theta = \theta^0 - \Omega t^0, \quad ct = ct^2 - \frac{r^2\Omega}{c}\theta^0, \quad z = z^0. \tag{8.16.8}$$

However, solving for θ and t from the relations (8.16.7), in place of approximate values of (8.16.8), we actually obtain

$$\theta = \frac{\theta^2 - \Omega t^0}{1-\beta^2}, \quad t = \frac{t^0 - (r^2\Omega/c^2)\theta^0}{1-\beta^2}. \tag{8.16.9}$$

The non-zero elements of the transformation matrices α^{μ^0} and α^{μ} obtain from (8.16.7) :

$$\alpha^{\mu^0}_{\mu} = \frac{\partial x^{\mu^0}}{\partial x^{\mu}} = 1, \quad \alpha^{2^0}_{4} = \frac{\partial x^{2^0}}{\partial x^4} = \frac{\Omega}{c},$$

$$\alpha^{4^0}_{1} = \frac{\partial x^{4^0}}{\partial x^1} = 2\beta\Omega, \quad \alpha^{4^0}_{2} = \frac{\partial x^{4^0}}{\partial x^2} = r\beta, \tag{8.16.10}$$

$$\alpha^{\mu}_{\mu^0} = \frac{\partial x^{\mu}}{\partial x^{\mu^0}} = 1, \quad \alpha^{2}_{4^0} = \frac{\partial x^2}{\partial x^{4^0}} = -\frac{\Omega}{c},$$

$$\alpha^{4}_{2^0} = \frac{\partial x^4}{\partial x^{2^0}} = -4\beta, \quad \alpha^{4}_{1^0} = \frac{\partial x^4}{\partial x'^0} = -2\beta\theta^0 = -2\beta(\theta + \Omega t). \tag{8.16.11}$$

Here $\beta = r\Omega/c$, and repeated index does not mean summation over that index. The velocity transformation relations are

$$\frac{dr^0}{dt^0} = u_{r^0} = \frac{u_r}{dt^0/dt} = \frac{u^r}{1 + \frac{\beta}{c}u_\theta + \frac{2}{c}\beta\theta u_r}, \tag{8.16.12}$$

---

(2) The group property of classical rotation at uniform rates about a fixed axis is satisfied by this rotation;
(3) The velocity of this rotation has the upper limit c. See section 8.17.

The conditions that the Trocheris rotation is an one parameter (namely the angular velocity Ω) group of infinitesimal transformation in r, θ, z, t of the relations of (8.17.1) or (8.17.2), yield two differential equations which we call the auxiliary relations. These are new type of transformation relations of the relativity theory for finding the 4-space sub-interval ds of a new rotation metric.

$$r^0\frac{d\theta^0}{dt^0} = u_{\theta^0} = \frac{r\dot{\theta} + \beta c}{dt^0/dt} = \frac{u_\theta + \beta c}{1 + \frac{\beta}{c}u_\theta + \frac{2}{c}\beta\theta u_r}$$

$$= \frac{[u + (\Omega \times r)]_\theta}{1 + \frac{1}{c^2}(u\cdot(\Omega \times r)) + \frac{2\Omega}{c^2}\theta(r\cdot u)}, \tag{8.16.13}$$

$$\frac{dz^0}{dt^0} = u_{z^0} = \frac{u_z}{1 + \frac{\beta}{c}u_\theta + \frac{2}{c}\beta\theta u_r}, \tag{8.16.14}$$

where

$$\frac{dt^0}{dt} = 1 + \frac{\beta}{c}u_\theta + \frac{2}{c}\beta\theta u_r. \tag{8.16.15}$$

The differential relations, obtained from (8.16.7), are

$$dr^0 = dr, \quad d\theta^0 = d\theta + \Omega dt, \quad dz^0 = dz,$$

$$cdt^0 = cdt + \frac{r^2}{c}\Omega d\theta + \frac{2}{c}r\Omega\theta dr. \tag{8.16.16}$$

The reverse relations of transformation are

$$dr = dr^0, \quad d\theta = d\theta^0 - \Omega dt^0, \quad dz = dz^0, \tag{8.16.17}$$

$$cdt = cdt^0 - \frac{r^0\Omega}{c}(2\theta^0 dr^0 + r^0 d\theta^0).$$

Equations (8.16.17) can be combined vectorially to give

$$dr = dr^0 - (\Omega \times r^0)dt^0,$$

$$dt = dt^0 - \frac{1}{c^2}(\Omega \times r^0)\cdot dr^0 - \frac{2\Omega}{c^2}\theta^0(r^0\cdot dr^0). \tag{8.16.18}$$

The interval ds reads

$$ds^2 = (1 - 4\beta^2\theta^2)dr^2 + \frac{1}{\gamma^2}(rd\theta)^2 + dz^2 - \frac{1}{\gamma^2}(cdt)^2 - 2.2\beta\theta\cdot dr\cdot cdt. \tag{8.16.19}$$

It can be written as

$$ds^2 = dr^2 + \frac{r^2}{\gamma^2}\left(d\theta - \frac{2}{r}\beta^2\gamma^2\theta dr\right)^2 + dz^2 - \frac{c^2}{\gamma^2}\left(dt + \frac{2}{c}\beta\gamma^2\theta dr\right)^2$$

$$= dr^2 + \frac{1}{\gamma^2}(rd\theta^*)^2 + dz^2 - \frac{1}{\gamma^2}(cdt^*)^2, \tag{8.16.20}$$

where

$$d\theta^* = d\theta - \frac{2}{r}\beta^2\gamma^2\theta dr, \tag{8.16.21}$$

$$dt^* = dt + \frac{2}{c}\beta\gamma^2\theta dr \tag{8.16.22}$$

$d\theta^*$ is an angle like displacement and dt^* is a time like sub-interval. The metric tensor $g_{\mu\nu}$, for Trocheris rotation, is

$$g_{\mu\nu} = \begin{bmatrix} 1-4\beta^2\theta^2-2\beta^2\theta r & 0 & -2\beta\theta \\ -2\beta^2\theta r & \dfrac{r^2}{\gamma^2} & 0 & 0 \\ 0 & 0 & 1 & 0 \\ -2\beta\theta & 0 & 0 & -\dfrac{1}{\gamma^2} \end{bmatrix}. \tag{8.16.23}$$

The corresponding contravariant metric tensor $g^{\mu\nu}$ is evaluated as the inverse of the matrix of $g_{\mu\nu}$ from (8.16.23)

$$g^{\mu\nu} = \begin{bmatrix} 1 & \dfrac{2}{r}\beta^2\gamma^2\theta & 0 & -2\beta\gamma^2\theta \\ \dfrac{2}{r}\beta^2\gamma^2\theta & \dfrac{\gamma^2}{r^2}(1+4\beta^4\gamma^2\theta^2) & 0 & -\dfrac{4}{r}\beta^3\gamma^4\theta^2 \\ 0 & 0 & 1 & 0 \\ -2\beta\gamma^2\theta & -\dfrac{4}{r}\beta^3\gamma^4\theta^2 & 0 & -\gamma^2(1-4\beta^2\gamma^2\theta^2) \end{bmatrix}. \tag{8.16.24}$$

The determinants of (8.16.23) and (8.16.24) are

$$g = \det g_{\mu\nu} = -\frac{r^2}{r^4}, \quad \det g^{\mu\nu} = -\frac{\gamma^4}{r^2} = \frac{1}{g}. \tag{8.16.25}$$

Equations (8.16.20) show that the length element dl, in position space, is given by

$$dl^2 = dr^2 + \frac{r^2}{\gamma^2}\left(d\theta - \frac{2}{r}\beta^2\gamma^2\theta dr\right)^2 + dz^2 = \gamma_{ij}dx^i dx^j, \tag{8.16.26}$$

where

$$\gamma_{ij} = \begin{bmatrix} 1+4\beta^4\gamma^2\theta^2 & -2\beta^2\theta r & 0 \\ -2\beta^2\theta r & \dfrac{r^2}{\gamma^2} & 0 \\ 0 & 0 & 1 \end{bmatrix}. \tag{8.16.27}$$

Then

$$\det \gamma_{ij} = \hat{\gamma} = \frac{r^2}{\gamma^2}. \tag{8.16.28}$$

Since the elements of γ_{ij} are independent of time t, the Trocheris rotation is essentially of rigid nature.

Putting $ds^2 = -c^2dT^2$ in (8.16.20) we evaluate the quantity dT/dt :

$$\frac{1}{\Gamma} = \frac{dT}{dt} = \left[\frac{1}{\gamma^2}\left(1+\frac{2\beta}{c}\theta\dot{r}\gamma^2\right)^2 - \frac{1}{c^2}\left\{\frac{r^2}{\gamma^2}\left(\dot{\theta} - \frac{2\beta^2}{r}\theta\dot{r}\gamma^2\right)^2 - \dot{z}^2\right\}\right]^{1/2}. \tag{8.16.29}$$

The Christofell symbol $\Gamma^{\lambda}_{\mu\nu}$ is defined as

$$\gamma^{\lambda}_{\mu\nu} = \frac{g^{\lambda\alpha}}{2}\left(\frac{\partial g_{\alpha\mu}}{\partial x^{\nu}} + \frac{\partial g_{\alpha\nu}}{\partial x^{\mu}} - \frac{\partial g_{\mu\nu}}{\partial x^{\alpha}}\right). \tag{8.16.30}$$

Since $g_{33} = 1, g_{3\mu} = 0$, for $\mu \neq 3$ it is easy to see that

$$\Gamma^{3}_{\mu\nu} = 0, \quad \Gamma^{\lambda}_{3\nu} = 0, \quad \Gamma^{\lambda}_{\mu 3} = 0, \tag{8.16.31}$$

and $\Gamma'_{11}, \Gamma'_{12}$ and Γ'_{13} vanish. Moreover, $\Gamma_{\mu\nu}$ is symmetrical with respect to the indices μ and ν; so

$$\Gamma^{\lambda}_{\mu\nu} = \Gamma^{\lambda}_{\nu\mu}. \tag{8.16.32}$$

The non-zero elements of $\Gamma^{\lambda}_{\mu\nu}$ are

$$\begin{aligned}
&\Gamma'_{24} = \Gamma'_{41} = -\beta, \quad \Gamma'_{44} = -\frac{\beta^2}{2r}, \quad \Gamma^{2}_{11} = -\frac{2\theta}{c^2}\Omega^2\gamma^2,\\
&\Gamma^{2}_{12} = \Gamma^{2}_{21} = \frac{1}{r}(1-\beta^2\gamma^2), \quad \Gamma^{2}_{14} = \Gamma^{2}_{41} = \frac{\gamma^2\Omega}{cr}, \quad \Gamma^{2}_{22} = -2\beta^2\gamma^2\theta,\\
&\Gamma^{2}_{24} = \Gamma^{2}_{42} = -\frac{2}{r}\beta^3\gamma^2\theta, \quad \Gamma^{2}_{44} = -\frac{2\Omega^2}{c^2}\beta^2\gamma^2\theta,\\
&\Gamma^{4}_{11} = \frac{2}{r}\beta\gamma^2\theta, \quad \Gamma^{4}_{12} = \Gamma^{4}_{21} = \beta\gamma^2, \quad \Gamma^{4}_{14} = \Gamma^{4}_{41} = -\frac{\beta^2\gamma^2}{r},\\
&\Gamma^{4}_{22} = 2\beta\gamma^2 r\theta, \quad \Gamma^{4}_{24} = \Gamma^{4}_{24} = 2\beta^2\theta\gamma^2, \quad \Gamma^{4}_{44} = \frac{2\theta}{r}\beta^3\gamma^2.
\end{aligned}$$

The expressions for the 4-velocity tensor $V^{\mu}(= dx^{\mu}/dT)$ and the 4-acceleration tensor $F^{\mu}(= dV^{\mu}/dT)$ in the K frame are

$$V_r = \frac{dr}{dT} = \Gamma\frac{dr}{dt} = \Gamma u_r, \quad V_\theta = \Gamma u_\theta, V_z = \Gamma u_z, \quad V_4 = \Gamma c \text{ and}$$

$$F_r = \frac{dV_r}{dT} = \Gamma\frac{d}{dt}(\Gamma u_r), \quad F_\theta \frac{d}{dt}(\Gamma u_\theta), \quad F_z = \Gamma\frac{d}{dt}(\Gamma u_2), \quad F_4 = \Gamma\frac{d}{dt}(\Gamma c),$$

where $u_r = dr/dt, u_\theta = rd\theta/dt, u_z = dz/dt, u_4 = c$.

8.17 Further Study of Trocheris Rotation

Some consistency checks and other checks are possible for the development of Trocheris rotation on a realistic basis. Thus we obtain a set of auxiliary relations which are relations of transformation for a metric of rotation. The first few terms of expansions of this metric in positive integral powers of $(r\Omega/c)$ yield the metric of GN rotation first and then the metric of the Lorentz rotation. Auxiliary relations of any type do not appears in the analysis of GN rotation and Lorentz rotation.

One consistency check is to see whether the transformations considered in section (8.16) are valid for all constant values of the angular velocity Ω. Another is to see whether the group property of classical uniform rotation about an axis is retained in this rotation, and still another is to impose that the rotation velocity has the upper limit c. So, if two frames of reference are in uniform rotation about the same axis, they must be in uniform rotation with respect to each other. *Trocheris* (1949) noted that there is only one transformation satisfying this requirement which reduces to the transformation of section 8.16 in the first approximation in Ω. It is determined as the one parameter group of transformations in r, θ, z, t, corresponding to the infinitesimal transformation

$$\delta r = 0, \quad \delta\theta = t\delta\Omega, \quad \delta z = 0, \quad c\delta t = \frac{r}{c}\delta\Omega \cdot r\theta. \tag{8.17.1}$$

In the K^0 frame the corresponding relations are

$$\delta r^0 = 0, \quad \delta\theta^0 = t^0\delta\Omega, \quad \delta z^0 = 0, \quad c\delta t^0 = \frac{r^0}{c}\delta\Omega \cdot r^0\theta^0. \tag{8.17.2}$$

Equations (8.17.1) yield the differential equations

$$\frac{d^2\theta}{d\Omega^2} = \frac{r^2\theta}{c^2}, \quad \frac{d^2t}{d\Omega^2} = \frac{r^2t}{c^2}. \tag{8.17.3}$$

The solutions being

$$\theta = P\cosh\beta + Q\sinh\beta, \quad t = R\cosh\beta + L\sinh\beta, \tag{8.17.4}$$

where the quantities P, Q, R, L are the four constants of integration of (8.17.3) and $\beta = r\Omega/c$. Some conditions to find these are

$$\theta = \theta^0 \quad \text{and} \quad t = t^0 \text{ when } \Omega = 0. \tag{8.17.5}$$

These give $P = \theta^0, R = t^0$. Differentiating θ of (8.17.4) with respect to Ω and using (8.17.1) we obtain the relations

$$\frac{d\theta}{d\Omega} = \frac{r}{c}(\theta^0\sinh\beta - Q\cosh\beta) \tag{8.17.6}$$

$$= t = t^0\cosh\beta + L\sinh\beta. \tag{8.17.7}$$

Comparison of coefficients of $\cosh\beta$ and $\sinh\beta$ from both sides of these relations gives the other requisite conditions :

$$L = \frac{r}{c}\theta^0, \quad Q = \frac{c}{r}t^0. \tag{8.17.8}$$

Hence equation (8.17.4) finally become

$$\theta = \theta^0\cosh\beta^0 - \frac{ct^0}{r^0}\sinh\beta^0, \tag{8.17.9}$$

$$t = t^0\cosh\beta^0 - \frac{r^0\theta^0}{c}\sinh\beta^0, \tag{8.17.10}$$

where $\beta^0 = r^0\Omega/c$. The reverse set of such relations is

$$\theta^0 = \theta\cosh\beta + \frac{ct}{r}\sinh\beta, \tag{8.17.11}$$

$$t^0 = t\cosh\beta + \frac{r\theta}{c}\sinh\beta. \tag{8.17.12}$$

We call these an auxiliary set of relations of transformation for rotation, and study its metric tensor in section 8.18.

For variation of the parameter Ω, in equations (8.17.11) and (8.17.12), in place of (8.17.8) we obtain

$$\delta\theta^0 = \frac{r}{c}\left(\theta \sinh\beta + \frac{ct}{r}\cosh\beta\right)\delta\Omega = t^0\delta\Omega. \tag{8.17.13}$$

$$\delta t^0 = \frac{r}{c}\left(t\sinh\beta + \frac{r\theta}{c}\cosh\beta\right)\delta\Omega = \frac{r^2}{c^2}\left(\theta\cosh\beta + \frac{ct}{r}\sinh\beta\right)\delta\Omega = \frac{r^2\theta^2}{c^2}\delta\Omega. \tag{8.17.14}$$

Since these relations are identical to the corresponding relations of (8.17.2), the corresponding second order differential equations for θ^0 and t^0 will be same as those of (8.17.3), and obviously the solutions will also be same, namely (8.17.11) and (8.17.12). So the transformation relations (8.17.11) and (8.17.13) are recovered for variations of Ω.

8.17.1 Summation law for angular velocities

To show the validity of the summation law, for angular velocities, consider three rotating frames, K, K_1 and K^1, such that the K_1 frame rotates relative to the K-frame at the rate Ω^0, the K^1 frame at the rate Ω_1 relative to the K_1 frame, and the K^1 frame at the rate Ω' relative to the frame K. To show that the sum rule

$$\Omega' = \Omega^0 + \Omega_1 \tag{8.17.15}$$

holds good, write

$$\beta^0 = \frac{r\Omega^0}{c},\quad \beta_1 = \frac{r\Omega_1}{c},\quad \beta' = \frac{r\Omega^1}{c}. \tag{8.17.16}$$

Then the twelve relations, like those of (8.17.9)–(8.17.12) involving Ω_0, Ω_1 and Ω^1 are

$$\theta_1 = \theta' \cos\beta_1 + \frac{ct'}{r'}\sin\beta_1, \tag{8.17.17}$$

$$t_1 = t'\cosh\beta_1 + \frac{r'\theta'}{c}\sinh\beta_1, \tag{8.17.18}$$

$$\theta = \theta_1\cosh\beta^0 + \frac{ct_1}{c}\sinh\beta^0, \tag{8.17.19}$$

$$t = t_1\cosh\beta^0 + \frac{r\theta_1}{c}\sinh\beta^0, \tag{8.17.20}$$

$$\theta' = \theta\cosh\beta' - \frac{ct}{r}\sinh\beta', \tag{8.17.21}$$

$$t' = t\cosh\beta' - \frac{r\theta}{c}\sinh\beta', \tag{8.17.22}$$

$$\theta = \theta'\cosh\beta' + \frac{ct'}{r}\sinh\beta', \tag{8.17.23}$$

$$t = t'\cosh\beta' + \frac{r\theta'}{c}\sinh\beta', \tag{8.17.24}$$

$$\theta_1 = \theta\cosh\beta^0 - \frac{ct}{r}\sinh\beta^0, \tag{8.17.25}$$

$$t_1 = t\cosh\beta^0 - \frac{r\theta}{c}\sinh\beta^0, \tag{8.17.26}$$

$$\theta' = \theta_1\cosh\beta_1 - \frac{ct_1}{r_1}\sinh\beta_1, \tag{8.17.27}$$

$$t' = t_1\cosh\beta_1 - \frac{r_1\theta_1}{c}\sinh\beta_1. \tag{8.17.28}$$

Putting the value of θ and t^0 from (8.17.33) and (8.17.24) in (8.17.9) we obtain

$$\theta = \left(\theta'\cosh\beta_1 + \frac{ct'}{r}\sinh\beta_1\right)\cosh\beta^0 + \frac{c}{r^0}\left(t'\cosh\beta_1 + \frac{r\theta'}{c}\sinh\beta_1\right)\sinh\beta^0$$

$$= \theta'\cosh(\beta_1+\beta^0) + \frac{ct'}{r}\sinh(\beta_1+\beta^0). \tag{8.17.29}$$

Thus (8.17.19) is verified with the help of (8.17.15).

In the theory of transformation for GN rotation, or Lorentz rotation, it is not possible to find relations analogous to (8.17.2) and (8.17.3), because the first order differential equations, obtained from (8.15.4), are

$$\frac{d\theta}{d\Omega} = t, \quad \frac{dt}{d\Omega} = \frac{r^2}{c^2}\Omega t. \tag{8.17.30}$$

Of these only the second relation can be integrated to give

$$t = t^0\exp\left(\frac{r^2\Omega^2}{2c^2}\right) \tag{8.17.31}$$

and the similar indefinite integral of the first relation cannot be obtained because that partial differential equation is essentially non-holonomic in character[3].

8.17.2 Upper limit of rotation velocity

If the rotation velocity has an upper limit c, then, for the fixed observer, a fixed point of the rotating system never has a velocity greater than c. To show this consider in the rotating system K, the point P, with co-ordinates r, θ, z, t to stationary. In the system K^0, the point P moves round a circle $r^0 = r$; the co-ordinates of P in K are given by the relations (8.17.11) and (8.17.12), where r, θ, z are constants and only t varies. We can then put $\theta = 0$ and, from (8.17.11), obtain

$$\theta^0 = \frac{ct^0}{r^0}\tanh\beta. \tag{8.17.32}$$

Hence

$$v^0 = r^0\frac{d\theta^0}{dt^0} = c\tanh\beta. \tag{8.17.33}$$

Evidently $v^0 < c$, and as $r^0 \to \infty$, $v^0 \to c$. Equation (8.17.11) can now be written as

$$r^0\theta^0 = \frac{r\theta}{\operatorname{sech}\beta} + \frac{ct\tanh\beta}{\operatorname{sech}\beta} = \gamma(r\theta + vt). \tag{8.17.34}$$

[3] As earlier, we can make the first equation of (8.17.30) exact by adding the term $\Omega dt/d\Omega$ with the first term on its right side, and obtain after integration the relation $\theta - \Omega t = \theta_0$, where $\theta = \theta_0$ and $t = t_0$ when $\Omega = 0$. Then, following the deduction of equations (8.17.13) and (8.17.14) we obtain $\delta\theta_0 = -t\delta\Omega$ and $\delta t_0 = -(r^2\Omega^2/c^2)\Omega t_0\delta\Omega$. These give $\delta\theta_0 = -t_0\exp(r^2/2c^2)\delta\Omega$ and $\delta t_0 = -(r^2/c^2)\Omega t_0\partial\Omega$. We therefore, find that the exactization in this case does not gives relations corresponding to (8.17.13) and (8.17.14) for (8.17.1).

with

$$v = c \tanh \beta \quad \text{and} \quad \gamma = \frac{1}{\sqrt{1 - v^2/c^2}}. \tag{8.17.35}$$

Equation (8.17.12) similarly gives

$$t^0 = \gamma \left(t + \frac{v}{c^2} r\theta \right). \tag{8.17.36}$$

When $r\theta$ is replaced by x, $r^0\theta^0$ by x^0, $r\Omega$ by c and v is regarded as a uniform translation, parallel to the X axis, these two equations reduce to the LT relations.

$$x^0 = \gamma(x + vt), \quad t^0 = \gamma \left(t + \frac{xv}{c^2} \right). \tag{8.17.37}$$

Hence the transformation relations for the Trocheris rotation tend to those of a LT in a finite region when the axis of rotation goes away to infinity, and the velocity $r\Omega$ tends to a finite constant value. In this rotation study the velocity for the LT is not the limit of $r\Omega$, but the limit of the function of $c \tanh(r\Omega/c)$, the first term of expansion of which is $r\Omega$. Hence transformations like those of (8.17.11) and (8.17.12) contain the most saturated contribution to the distortions in space and time due to rotation.

8.17.3 Time taken by light to transmit along various paths

Consider light rays going round a *parallel* r = constant, z = constant with the help of suitable mirror devices. On such a circle the relations (8.17.34) and (8.17.35) are identical with a LT on r, θ, t; so the composition of velocities around such a circle is the same as that of linear velocities, and, in the rotating system, the light goes round such a circle, or a part of it, with velocity c in all directions. Then two clocks A and B, on the same parallel, can be adjusted in the following manner : A light signal is transmitted from A at time t_1, goes round the circle, is reflected by B, and is recovered back in A at time t_2. The two clocks are adjusted if the reflection occurs on the clock B when it shows the time $(t_1 + t_2)/2$. The point events times t at A and B in the rotating system have, for the fixed observer, time co-ordinates which differ by the amount

$$(\theta_B - \theta_A) \frac{r}{c} \sinh \left(\frac{r\Omega}{c} \right). \tag{8.17.38}$$

This difference depends upon the exact value chosen for θ_A and θ_B. For example, consider two clocks A and B at the same point in the rotating frame. When B coincides again with A, the rotating observer will still read the same time by both clocks, but, for the fixed observer, B will be slow with respect to A by the amount

$$\delta t' = 2\pi \cdot \frac{r}{c} \cdot \sinh \beta. \tag{8.17.39}$$

The definition of the exact value of θ for a clock in the rotating system is equivalent to defining the time shown by this clock for the fixed observer. Actually, a definite value of θ cannot be ascribed to an abstract point in the rotating system, but only to a material support whose co-ordinate θ is found to vary continuously in its displacement. The same can be said about the value of θ^0 for a clock in the inertial system K^0 and the time it shows for the rotating observer.

8.18 Rotation from the Auxiliary Relations of the Trocheris Rotation

The auxiliary relations of transformations (8.17.9) and (8.17.10), or the reverse relations (8.17.11) and (8.17.12), depict rotation about a fixed axis at the uniform rate Ω, and are remarkable for their group behaviour. Moreover, the differentials contain hyperbolic trigonometrical ratios of $r\Omega/c$, and not just these quantities or the factor

$$\gamma\left(=1/\sqrt{1-r^2\Omega^2/c^2}\right)$$

as coefficients. So, they provide a more concentrated contribution of the rotational velocity than the other known or conceivable transformation relations which contain only one or two terms of expansion of these relations, as will be evident from a scrutiny of the transformation relations of the preceding sections. The importance of the relations (8.17.9) and (8.17.10) is due to this speciality.

Differentials of (8.17.11) and (8.17.12) are

$$d\theta^0 = d\theta\cosh\beta + \left\{\left(\beta\theta - \frac{ct}{r}\right)\sinh\beta + \Omega t\cosh\beta\right\}\frac{dr}{r} + \frac{c}{r}dt\sinh\beta. \tag{8.18.1}$$

$$dt^0 = dt\cosh\beta + \{(\Omega t + \theta)\sinh\beta + \beta\theta\cosh\beta\}\frac{dr}{c} + \frac{r}{c}d\theta\sinh\beta. \tag{8.18.2}$$

For these the expression for the interval ds becomes

$$\begin{aligned} ds^2 &= (dr)^2\Bigg[1 - (\beta\theta)^2 + (\Omega t)^2 - \frac{1}{2}\left(\frac{ct}{r}\right)^2 + \frac{\theta^2}{2} + 2\theta\Omega + \frac{1}{2}\left(\frac{c^2t^2}{r^2} - \theta^2 - 4\theta\Omega t\right)\cosh(2\beta) \\ &\quad - \left(\beta\theta^2 + \frac{\Omega}{r}ct^2\right)\sinh(2\beta)\Bigg] + (rd\theta)^2 + (dz)^2 - (cdt)^2 \\ &\quad -2\cdot rd\theta\cdot dr\cdot\left[\frac{ct}{2r}\sinh(2\beta) - \Omega t - \frac{\theta}{2} + \frac{\theta}{2}\cosh(2b)\right] \\ &\quad -2dr\cdot cdt\cdot\left[\beta\theta - \frac{ct}{2r} + \frac{ct}{2r}\cosh(2\beta) + \frac{\theta}{2}\sinh(2\beta)\right] \\ &= (dr)^2 + (rd\theta^*)^2 + (dz)^2 - (cdt^*)^2 \end{aligned} \tag{8.18.3}$$

with

$$d\theta^* = d\theta - \frac{1}{r}\left[\frac{ct}{2r}\sinh(2\beta) - \Omega t - \frac{\theta}{2} + \frac{\theta}{2}\cosh(2\beta)\right]dr, \tag{8.18.4}$$

$$dt^* = dt + \frac{1}{c}\left[\beta\theta - \frac{ct}{2r} + \frac{ct}{2r}\cosh(2\beta) + \frac{\theta}{2}\sinh(2\beta)\right]dr. \tag{8.18.5}$$

The Trocheris theory of rotation, and also its relativistic generalization do not provide any expression for the difference in the times of paths of light from the same source, due to the circular motion in the CW and CCW senses when the co-ordinate system K is rotating in one sense. This is seen from (8.16.20) by putting in that $ds = 0, dr = 0, dz = 0$, solving for dt, and then integrating as in the end of section 8.15. But *Post* (1967) noted that the relation for t_0

in (8.16.7) can be applied to the periphery of a circular optical path yielding the differential relation

$$dt^0 = dt + \frac{\Omega}{c^2} r^2 d\theta \tag{8.18.6}$$

which implies local time non-uniformity. Integration of $d(t^0 - t)$ over the angles $\pm 2\pi$ for the CW and CCW beams, and the difference of the corresponding results give the value

$$\frac{4\pi r^2}{c^2}, \tag{8.18.7}$$

which is (8.15.16) for $\gamma = 1$.

Equation (8.18.3) can be written as

$$ds^2 = dl^2 - (cdt^*)^2, \tag{8.18.8}$$

$$\begin{aligned}
\text{where} \quad dl^2 &= (dr)^2 + (rd\theta^*)^2 + (dz)^2 \\
&= (dr)^2 + \left[rd\theta - \left\{\frac{ct}{2r}\sinh(2\beta) - \Omega t - \frac{\theta}{2} + \frac{\theta}{2}\cosh(2\beta)\right\} dr\right]^2 + (dz)^2 \\
&= \left[1 + \left\{\frac{ct}{2r}\sinh(2\beta) - \Omega t - \frac{\theta}{2} + \frac{\theta}{2}\cosh(2\beta)\right\}^2\right](dr)^2 \\
&\quad -2r\left\{\frac{ct}{2r}\sinh(2\beta) - \Omega t - \frac{\theta}{2} + \frac{\theta}{2}\cosh(2\beta)\right\} drd\theta + (rd\theta)^2 + (dz)^2 \\
&= \gamma_{ij} dx^i dx^j
\end{aligned} \tag{8.18.9}$$

with

$$\gamma_{ij} = \begin{bmatrix}
1 + \left\{\frac{ct}{2r}\sinh(2\beta) - \Omega t - \frac{\theta}{2} + \frac{\theta}{2}\cosh(2\beta)\right\}^2 & -r\left\{\frac{ct}{2r}\sinh(2\beta) - \Omega t - \frac{\theta}{2} + \frac{\theta}{2}\cosh 2\beta\right\} & 0 \\
-r\left\{\frac{ct}{2r}\sinh(2\beta) - \Omega t - \frac{\theta}{2} + \frac{\theta}{2}\cosh(2\beta)\right\} & r^2 & 0 \\
0 & 0 & 1
\end{bmatrix}. \tag{8.18.10}$$

Since the elements of γ_{ij} depend on time, this rotation is non-rigid, *Moller* ((1972), p. 417, after equation (10.284)).

8.19 Critical Comments and Clarification

8.19.1 Anholonomic differentials of rotation

Non-integrable constraints are found to appear in some problems of rigid body mechanics. For instance these exist in the dynamics of motion of billiard balls and in rolling motion. These constraints also appear in problems of the relativistic rotation theory.

An important feature of rotating frames is their anholonomity or non-integrability, in which any inherent non-integrability of the differentials of transformation is not disturbed artificially. In this regard, wrote *Corum* (1977). *A consistent theory of relativistic rotation, satisfying the principle of general covariance, inherently requires the presence of the object of anholonomity.* This theory provides a single simple framework for discussing and even clarifying the description of this rotation. Anholonomical co-ordinates eliminate superfluous equations; so the interpretation of physical results is less complicated when all these are referred to orthogonal reference frames. The anholonomic approach gives a consistent theory of relativistic rotation and can distinguish between the physically different operations of source rotation and observer rotation in a flat space-time.

Anolonomic co-ordinate differentials are the logical extension of differentials of STR to non-inertial frames of GTR. These differentials ensure orthogonality of reference frames and retain unique meaning of the components of the field tensors in all frames of reference. So these save much conceptual simplicity. The covariant nature of the Maxwell electrodynamics, under relativistic rotation, is actually attainable only with the inclusion of the object of anholonomity.

The method of instantaneous local reference frames, ensuring the holonomic character of transformation of co-ordinate differentials, is not free from ambiguities, due to their oblique character. Actually, the obliqueness of tetrad frames is due to the existence of some non-diagonal elements (of the type $g_{\mu\gamma} \neq 0$ for $\mu \neq \gamma$) in the metric tensor. Then EM forces proportional to the particle velocities appear, even when a charged particle moves in a purely electric field. Both for holonomic differentials for instantaneous reference frames, and for GN frames for rotation, the universe is described from an oblique frame of reference, skewed basis vectors are employed, and EM forces proportional to charge velocities appear even when the charges move in the purely electrostatic field, the reason being that the references are not orthogonal.

8.19.2 Tetrad formalism

Method for study of physics in rotating frames should depend on the nature of the problem considered. On the confusion, *Pirani* (1957) writes, *Many of the difficulties of interpretation of general relativity theory arise from the use of co-ordinate systems to which no exact physical significance is ascribed. One way of avoiding such difficulties is to employ the tetrad formalism.* The method of local inertial tetrads is not free from objections, and that it can be made equivalent to the method of *global* non-inertial reference at least locally, *Post and Bahulikar* (1971). It should be ascertained that it is a suitable mathematical expedient that enhances physical clarity, as claimed by its proponents, *Mo* (1970) and *Corum* (1980).

The anholonomic approach should be developed for relativistic rotation. Exercises would be helpful on transformation for some values of the angular velocity and on the kinematics in the anholonomic rotating systems, like the holonomic Trocheris rotation exercises, *Trocheris* (1949), *Chakraborty and Sarkar* (1985).

8.20 Relative Angular Advancement of Frames

For rotation, an observer may carry a reference axis along a circular path. After proceeding over such a closed path, the observer finds that his reference axis is not at the starting orientation; the net gap, or angular defect $\delta\theta$, or the corresponding precession frequency $8\theta(2\pi/\Omega)$, can be evaluated from the space time metric of that rotation. If one carries a given axis in a given frame of reference, along a spatially closed curve, with an acceleration not parallel to his velocity, then, as seen by an inertial observer, the reference axis will have rotated through the small angle $\delta\theta^0$, which is the Thomas precession effect. It is purely a kinematic effect, depending on the inherent anholonomicity of the transformation, for relativistic rotation. It is also independent of the cause of acceleration, Corum (1977). It arises even in inertial frames of reference, because under successive LTs, a orientation changes by an amount equivalent to that obtained by a LT plus a spatial rotation.

The quantity $\theta - \theta_0$ is the angular advancement (rotation) of the K frame relative to the K^0 frame, and $\theta^0 - \theta$ is the same of the K^0 frame relative to the K frame. The rates of changes of these relative angular displacements, as seen by observers fixed in the two frames, are

$$\frac{d}{dt}(\theta^0 - \theta) \quad \text{and} \quad \frac{d}{dt^0}(\theta^0 - \theta). \tag{8.20.1}$$

If Ω_T is their difference, then

$$\Omega_T = \frac{d}{dt^0}(\theta^0 - \theta) - \frac{d}{dt}(\theta^0 - \theta) = \left(\frac{dt}{dt^0} - 1\right)\left\{\frac{d}{dt}(\theta^0 - \theta)\right\}. \tag{8.20.2}$$

This Ω_T corresponds to a change in time *measure* that is associated with a gravitational potential. If the time scales for the two frames K and K^0 are different, then Ω_T is non-zero and gives the Thomas precession rate of rotating frames, *Post* (1967) and *Chakraborty* and *Sarkar* (1985).

8.21 Lagrangian and Hamiltonian for GN Rotation

If the force F, acting on a particle of a system, has a non-conservative part F^0, and the conservative part is $-\nabla V$, then

$$F = -\nabla V + F^0 \tag{8.21.1}$$

and the Euler-Lagrangian equation, for the Lagrangian density $\mathcal{L}$, reads

$$\frac{d}{dt}\left(\frac{\partial \mathcal{L}}{\partial \dot{x}_i}\right) - \frac{\partial \mathcal{L}}{\partial x_i} = F_i^0, \tag{8.21.2}$$

where, obviously,

$$\mathcal{L} = T - V \tag{8.21.3}$$

T is the kinetic energy density and V is the potential energy per unit volume of the body at the point r; dot denotes time derivate.

Consider an infinitesimal variation of the Lagrangian $\mathcal{L}$. Since $\mathcal{L}$ is a function of the position vector r_i, of the ith particle, and u_i is its velocity, we can write

$$\delta\mathcal{L} = \sum\left[\left(\frac{\partial \mathcal{L}}{\partial r_i}\cdot \delta r_i\right) + \left(\frac{\partial \mathcal{L}}{\partial u_i}\cdot \delta u_i\right)\right]. \tag{8.21.4}$$

For an infinitesimal rotation $\delta\Omega$ of the system of co-ordinates K, the increment δr_i in the radius vector, from the origin, on the axis of rotation, to the ith particle, is given by

$$\delta \mathbf{r}_i = (\delta\Omega \times \mathbf{r}_i). \tag{8.21.5}$$

The corresponding velocity increment is

$$\delta \mathbf{u}_i = (\delta \boldsymbol{\Omega} \times \mathbf{u}_i). \tag{8.21.6}$$

Hence equation (8.21.4) reads

$$\delta \mathcal{L} = \sum_i \left[\left\{ \frac{\partial \mathcal{L}}{\partial r_i} . (\delta \Omega \times r_i) \right\} + \left\{ \frac{\partial \mathcal{L}}{\partial u_i} . (\delta \Omega \times u_i) \right\} \right]. \tag{8.21.7}$$

Now we put

$$\frac{\partial \mathcal{L}}{\partial r_i} = \dot{p}_i, \tag{8.21.8}$$

$$\frac{\partial \mathcal{L}}{\partial u_i} = p_i \tag{8.21.9}$$

in this equation, and obtain

$$\delta \mathcal{L} = \sum_i \left[\{ \dot{p}_i \cdot (\delta \Omega \times r_i) \} + \{ p_i \cdot (\delta \Omega \times u_i) \} \right]. \tag{8.21.10}$$

Readjusting the vectors of both the scalar triple products, this relation becomes

$$\begin{aligned} \delta \mathcal{L} &= \left[\delta \Omega \cdot \left\{ \sum_i ((r_i \times \dot{p}_i) + (u_i \times p_i)) \right\} \right] \\ &= \left[\delta \Omega \cdot \frac{d}{dt} \left\{ \sum_i (r_i \times p_i) \right\} \right] = \left(\delta \Omega \cdot \frac{dL}{dt} \right), \end{aligned} \tag{8.21.11}$$

where

$$L = \sum_i (r_i \times p_i) \tag{8.21.12}$$

L is the angular momentum vector of the system. L like the linear momentum P, is also additive even when the particles interact mutually, and depends on the choice of origin, because if we put $r_i = r_i^0 + r^0$, we find that

$$L = \sum_i (r_i^0 \times p_i) + r^0 \times \left(\sum_i p_i \right) = L_0 + (r^0 \times p). \tag{8.21.13}$$

Here L^0 is the angular momentum when the origin is shifted to $\mathbf{r}^0$ and $\mathbf{p}$ is the vector sum of linear momentums of the particles of the system.

If v is the velocity of the frame K relative to the frame K^0, u_i and u_i^0 are the velocities of the ith particle in the two frames, then

$$u_i^0 = u_i + v, \quad v = (\Omega \times r). \tag{8.21.14}$$

Equation (8.21.12) shows that the angular momentums L and L^0, in the two frames, are given by

$$L^0 = L + \mu(R \times v), \tag{8.21.15}$$

$$L^0 = \sum_i (r_i^0 \times p_i^0), \tag{8.21.16}$$

with

$$R = \frac{1}{\mu} \sum (r_i^0 \times p_i^0), \tag{8.21.17}$$

$$\mu = \sum_i m_i. \tag{8.21.18}$$

The quantity μ is the total mass of the system. Equation (8.21.17) is the law of transformation of angular momentum; the same, for the ordinary momentum p, being

$$p^0 = p + \mu v. \tag{8.21.19}$$

Energy of a mechanical system, which is at rest as a whole, is usually called its internal energy $\mathcal{E}_i$. This includes the kinetic energy of relative motion of the particles in the system, and the potential energy of their interaction. If $\mathcal{E}$ and $\mathcal{E}_0$ are the energies of the mechanical system in the two frames, then

$$\mathcal{E} = \frac{1}{2}\left(\sum_i m_i u_i^2\right) + V = \frac{1}{2}\left[\sum m_i (u_i^0 - v)^2\right] + V$$

$$= \mathcal{E}^0 - (v \cdot p^0) + \frac{\mu}{2} v^2 \tag{8.21.20}$$

with

$$\mathcal{E}^0 = \frac{1}{2}\left(\sum_i m_i u_i^2\right) + V. \tag{8.21.21}$$

Denoting the Lagrangian in the two frames, by $\mathcal{L}$ and L^0, with the help of (8.21.14), we obtain

$$\mathcal{L} = L^0 - (v \cdot p^0) + \frac{\mu}{2} v^2. \tag{8.21.22}$$

Transformation of the action S (the integral of $\mathcal{L}$ with respect to time, over a fixed interval) is found to be

$$S = S^0 - \mu (v \cdot R^0) + \frac{\mu}{2} v^2 t, \tag{8.21.23}$$

where the second term in the right hand side is obtained from the fact that

$$\int p^0 \, dt = \sum_i (m_i r_i^0) = \mu R^0. \tag{8.21.24}$$

This relations defines μR^0 for (8.21.23). For transformation to cylindrical co-ordinates, in the same reference frame K, we put

$$x_i = r_i \cos \theta_i, \quad y_i = r_i \sin \theta_i,$$

in (8.21.3), and obtain

$$\mathcal{L} = \frac{1}{2} \sum_i \left\{ m_i (\dot{r}_i^2 + r_i^2 \dot{\theta}_i^2 + \dot{z}_i^2) \right\} - V. \tag{8.21.25}$$

$$L_z = \sum_i \{ m_i (x_i \dot{y}_i - \dot{x}_i y_i) \} = \sum_i (m_i r_i^2 \dot{\theta}_i). \tag{8.21.26}$$

Differentiating $\mathcal{L}$ with respect to θ_i we have

$$L_z = \sum_i \left(\frac{\partial \mathcal{L}}{\partial \theta_i}\right), \tag{8.21.27}$$

where, evidently, θ_i is the angle of rotation of the ith particle about the Z axis.

In quasistatic problems both linear and angular momentums are conserved; so

$$p = m(u + v) = p^0 = mu^0, \tag{8.21.28}$$

$$L = (r \times p) = L^0(r \times p^0). \tag{8.21.29}$$

In the rotating frame K, the Lagrangian $\mathcal{L}_f$ of motion of a non-relativistic particle is

$$\mathcal{L}_F = -mc^2 \left[1 - \frac{(u + v)^2}{2c^2}\right]. \tag{8.21.30}$$

The total differential $d\mathcal{L}_F$ of this Lagrangian, Landau and Lifshitz (1978, p. 126–128), is

$$\begin{aligned} d\mathcal{L}_F &= m[(u \cdot du) + (du \cdot v) + \{(u \cdot (\Omega \times dr) + (\Omega \times r) \cdot (\Omega \times dr)\}] \\ &= m[\{(u + v) \cdot du\} + \{dr \cdot (u \times \Omega)\} + \{dr \cdot ((\Omega \times r) \times \Omega)\}] \\ &= m(u + v) \cdot dr + \Omega \cdot (dr \times (u + v)). \end{aligned} \tag{8.21.31}$$

So, evidently

$$\frac{\partial \mathcal{L}_F}{\partial u} = m(u + v). \tag{8.21.32}$$

$$\frac{\partial \mathcal{L}_F}{\partial r} = -m[(u \times \Omega) + (v \times \Omega)]. \tag{8.21.33}$$

Consequently, it follows that

$$\frac{d}{dt}\left(\frac{\partial \mathcal{L}_F}{\partial u_i}\right) - \frac{\partial \mathcal{L}_F}{\partial r} = \frac{d}{dt}(mu) + 2m(\Omega \times u) - m(v \times \Omega) \tag{8.21.34}$$

and the Lagrangian equation (8.21.2) yields the equation of motion involving the Coriolis force $2m(\Omega \times u)$ and the centrifugal force $m(v \times \Omega)$. For the Hamiltonian $\mathcal{H}_m$ we have

$$\mathcal{H}_m = (P \cdot u) - \mathcal{L}_m, \tag{8.21.35}$$

$$P = \frac{\partial \mathcal{L}_m}{\partial u} \tag{8.21.36}$$

$\mathcal{L}_m$ being the Lagrangian of the particle, for both motion and other interactions, and P is the generalized momentum. Then

$$\nabla \cdot \mathcal{H}_m = -\frac{dP}{dt} = -\frac{d}{dt}\left(\frac{\partial \mathcal{L}_m}{\partial u}\right). \tag{8.21.37}$$

Since $\mathcal{H}_m$ is also energy of the particle,

$$\mathcal{E} = \mathcal{H}_m = m\left[c^2 + \frac{1}{2}(u^2 - v^2)\right] \tag{8.21.38}$$

$\mathcal{H}_m$ does not contain any term linear in the velocity u. Rotation of the frame of reference adds a term depending on the co-ordinates of the particle and proportional to the square of its angular velocity. The centrifugal potential energy term $-mv^2/2$ is due to the centrifugal force of rotation.

8.22 Angular Momentum Analysis

The generalized angular momentum tensor in 4-space is defined as

$$L^{\mu\nu} = \sum (x^\mu p^\nu - x^\nu p^\mu). \tag{8.22.1}$$

The spatial components of this anti-symmetric tensor are the components of the vectorial angular momentum,

$$L = \sum (r \times p). \tag{8.22.2}$$

The components L^{14}, L^{24}, L^{34} form the vector

$$\sum \left(ctp - \frac{\mathcal{E}}{c} r \right). \tag{8.22.3}$$

Thus, we can take $L^{\mu\nu}$ in the form

$$L^{\mu\nu} = \left[L, \sum \left(ctp - \frac{\mathcal{E}}{c} r \right) \right]. \tag{8.22.4}$$

For a closed system $L^{\mu\nu}$ are constants; so, in particular, we write

$$\sum \left(ctp - \frac{\mathcal{E}}{c} r \right) = \text{constant}. \tag{8.22.5}$$

Moreover, since the total energy $\Sigma\mathcal{E}$ is also conserved, this relation can be expressed as

$$\frac{(\Sigma\mathcal{E}r)}{(\Sigma\mathcal{E})} - \frac{c^2 t(\Sigma p)}{(\Sigma\mathcal{E})} = \text{constant}. \tag{8.22.6}$$

Hence the point with radius vector R, given by

$$V = \frac{c^2}{(\Sigma\mathcal{E})}(\Sigma p), \tag{8.22.7}$$

moves with the uniform velocity

$$V = \frac{c^2}{(\Sigma\mathcal{E})}(\Sigma p). \tag{8.22.8}$$

Relation (8.22.7) is valid only if interaction between the particles is neglected, though classically, the formula for the centre of inertia is applied to both interacting and non-interacting particles. The components of R, thus defined, do not constitute the space part of any 4-vector; so, they do not transform, from one reference frame to another, like the co-ordinates of a point. Thus, we get different points for the centre of inertia of a given system with respeçt to different frames of reference.

Consider an infinitesimal rotation in 4-space. The co-ordinates x^μ of one particle of a system take on new values $x^{\mu'}$ such that the differences $x^{\mu'} - x^\mu$ are linear functions of x^μ; so we write

$$x^{\mu'} - x^\mu = x_\mu \delta\Omega^{\mu\nu} \tag{8.22.9}$$

with infinitesimal coefficients $\delta\Omega^{\mu\nu}$. Since $x^\mu x_\mu$ is an invariant, where the Einstein convention is implied for repeated index, we have

$$x^{\mu'} x_{\mu'} = x_\mu x^\mu. \tag{8.22.10}$$

Substituting for $x^{\mu'}$ from (8.22.9) in this relation and dropping the term quadratic in $\delta\Omega_\mu$, we obtain

$$x^\mu x^\nu \delta\Omega_{\mu\nu} = 0. \tag{8.22.11}$$

Since $x^\mu x^\nu$ is a symmetric tensor, $\delta\Omega_{\mu\nu}$ must be anti-symmetric for the product to be zero. So

$$\delta\Omega_{\mu\nu} = -\delta\Omega_{\nu\mu}. \tag{8.22.12}$$

For the vectorial displacement along the arc of a circle of radius r, due to rotation, we can write

$$\delta r = (\Omega\delta t) \times r + (\delta\Omega t) \times r = (\Omega\delta t + \delta\Omega t) \times r. \tag{8.22.13}$$

If we write the corresponding tensor relation as

$$\delta x_\mu = \delta\Omega_{\mu\nu} x^\nu, \tag{8.22.14}$$

then we have

$$\left.\begin{aligned}\delta\Omega_{12} &= -t\delta\Omega_3 - \Omega_3\delta t\\ \delta\Omega_{13} &= t\delta\Omega_3 - \Omega_2\delta t\end{aligned}\right\}. \tag{8.22.15}$$

For a system of non-interacting particles, Chakraborty (1987, relation (9.16)), we have

$$\delta S = -\sum(p^\mu \delta x_\mu), \tag{8.22.16}$$

where S is the action integral. Putting the value of δx_μ from (8.22.14) in this relation we obtain

$$\delta S = -\delta\Omega_{\mu\nu}(\Sigma p^\mu x^\nu). \tag{8.22.17}$$

Now resolve the tensor $(\Sigma p^\mu x^\nu)$ into its symmetric and anti-symmetric parts. Then, since the product of a symmetric and anti-symmetric tensor is zero, the relation (8.22.7) becomes

$$\delta S = -\frac{1}{2}\delta\Omega_{\mu\nu}\left\{\Sigma(p^\mu x^\nu - p^\nu x^\mu)\right\}. \tag{8.22.18}$$

It can be written as

$$\frac{\partial S}{\partial\Omega_{\mu\nu}} = -\frac{1}{2}\{\Sigma(p^\mu x^\nu - p^\nu x^\mu)\}. \tag{8.22.19}$$

For a closed system, owing to the isotropy of space time, the Lagrangian does not change under a rotation in 4-space, the parameters $\delta\Omega_{\mu\nu}$ are cyclic co-ordinates, and, as already mentioned, the components of the angular momentum tensor are conserved.

8.23 Relativistic Spin and Angular Momentum

Relativistically, the spin S of a system is defined as its angular momentum in the inertial frame K^0, in which its total momentum is zeró, that is, the rest frame of the system, as a whole. The angular momentum, in this frame, in general, is not given by (8.22.2), because many of the known particles have intrinsic angular momentum that must be combined with the orbital $(r \times p)$ terms. Rather, it is defined by the law of motion, in which the rate of change of angular momentum is equated with the net torque N given by

$$N = \sum(r \times F) = \sum\left(r \times \frac{dp}{dt}\right) = \frac{d}{dt}\sum(r \times p) = \frac{dL}{dt}, \tag{8.23.1}$$

where F is the force acting on a particle of the system. In the absence of any torque, the angular momentum is a constant of motion.

On microscopic systems torques arise from interaction of their electric or magnetic moments with applied EM fields. The magnetic dipole moment density M gives rise to the torque

$$N = (M \times B), \tag{8.23.2}$$

which tends to make the moment point along the field, and a force

$$F = (M \cdot \nabla)B \tag{8.23.3}$$

tending to drive the system to regions of stronger fields. This force vanishes in a homogeneous magnetic field, whose only effect is then to produce a torque on the system. In the absence of magnetic field, the angular momentum is conserved.

Particles, such as the electron, have spin. The spin S is a pseudo vector, whose rate of change with time equals the applied torque N,

$$N = \frac{dS}{dt}. \tag{8.23.4}$$

The law of conservation of angular momentum of an isolated system, is equivalent to the isotropy of inertial frames. The spin of a space craft is its angular momentum L in its centre of mass frame. The spin of an atom is its spin plus orbit, in its centre of mass frame. The spin of one of its electrons, or the nucleus, is its angular momentum in its rest frame.

Unlike the momentum, which has definite components in each reference frame, angular momentum is defined only in one particular frame of reference. It does not transform. Any statement about it refers to the rest frame as of that instant. If we say that in the laboratory the spin of an electron makes the angle θ with its velocity, we mean that in the rest frame of the electron the spin makes this angle with the line of motion of the laboratory, Corben (1968), Sard (1970).

8.23.1 Angular velocity and deformation analysis

Using the metric tensor and Christoffel symbol $\Gamma^{\lambda}_{\mu\nu}$ the angular momentum tensor Ω^{ij} is defined as

$$\Omega^{ij} = -\frac{c}{\sqrt{-g_{44}}}\left[\frac{1}{2}\frac{\partial g^{ij}}{\partial x^4} + g^{i\mu}\Gamma^{i}_{\mu 4}\right]. \tag{8.23.5}$$

The corresponding angular velocity vector Ω_i is

$$\Omega_i = \frac{1}{2}\sqrt{\bar{\gamma}}e_{ijk}\Omega^{jk}, \tag{8.23.6}$$

where e_{ijk} is the Levi Civita symbol of third order. The scalar quantity

$$\Omega = \sqrt{\Omega_i\Omega^i} = \sqrt{\gamma^{ij}\Omega_i\Omega_j} \tag{8.23.7}$$

is the angle at a given point through which the reference frame K rotates in unit proper time. The quantity

$$D_{ij} = \frac{c}{2\sqrt{-g_{44}}}\frac{\partial \gamma_{ij}}{\partial x^4} \tag{8.23.8}$$

is the deformation tensor, Zeldovich and Novikov (1978, p. 21). The corresponding scalar quantity

$$D = D_{ij}\gamma^{ij} \tag{8.23.9}$$

is the relative part of expansion of the volume element dV; so

$$D = \frac{1}{dV}\frac{d}{dT}(dV), \tag{8.23.10}$$

where dT is the proper time sub-interval.

8.24 Basics of Lagrangian Formalism in Non-inertial Frames

8.24.1 The action integral and the Lagrangian

Consider the invariant action integral

$$S = -mc\int_a^b \sqrt{-ds^2} \tag{8.24.1}$$

for motion of a particle of mass m between two given world points 'a' and 'b', which are the events of its arrival at the inertial position and the departure at the final position at definite times t_1 and t_2. It can also be written as

$$S = \int_{t_1}^{t_2} Ldt. \tag{8.24.2}$$

Then, obviously,

$$L = -mc\sqrt{-g_{\mu\nu}\dot{x}^\mu\dot{x}^\nu}, \tag{8.24.3}$$

where dot denotes derivative with respect to time t. For inertial frames

$$L = -mc\sqrt{1-\frac{u^2}{c^2}}, \tag{8.24.4}$$

where $u^\mu = dx^\mu/dt$ and $u = dr/dt$. This L, being the Lagrangian of motion, the L of (8.24.3) is also called the same for non-inertial frames, $2L$ has the dimension of kinetic energy of a particle. If m is replaced by the mass density, then the Lagrangian is called the *Lagrangian density,* and is denoted by $\mathcal{L}$ in place of L. The factor $-mc$ is incorporated in S and L to recover the standard form of the generalized momentum P^μ from (8.24.34).

For the Lagrangian L of motion and action, the action principle

$$\delta\int_{P_1}^{P_2} \sqrt{-ds^2} = 0 \tag{8.24.5}$$

can be expressed as

$$\delta\int_{P_1}^{P_2} L(x^\mu, \dot{x}^\mu)dp = 0, \tag{8.24.6}$$

where dot is a derivative with respect to the parameter P (P being an arbitrary parameter on which the co-ordinates x^μ depend) and P_1 and P_2 are two given points in space. It is possible to specify the limits such that the world point P_2 is in the region of absolute future with respect to P_1. There are all possible paths connecting P_2 with P_1, but the actions takes on an extreme value only on the worldline along which the particle actually moves under the action of the given forces. Though the terminal points P_1 and P_2 are fixed, the path can be deformed by giving the intermediate points arbitrary infinitesimal displacements.

The relation (8.24.6) is valid for all variations δx^μ satisfying the boundary conditions

$$\delta x^\mu(p_1) = 0, \quad \delta x^\mu(p_2) = 0. \tag{8.24.7}$$

Equation (8.24.6) can be written as

$$\delta \int_{p_1}^{p_2} L dp = \int_{p_1}^{p_2} \left[\frac{\partial L}{\partial x^\mu}\delta x^\mu(p) + \frac{\partial L}{\partial \dot{x}^\mu}\delta \dot{x}^\mu(p)\right] dp = 0. \tag{8.24.8}$$

Remembering that

$$\delta \dot{x}^\mu = \frac{d}{dp}(\delta x^\mu), \tag{8.24.9}$$

the second term of the integral in (8.24.8) is integrated partially; then using the terminal conditions (8.24.9) that equation becomes

$$\delta \int_{p_1}^{p_2} L dp = \int_{p_1}^{p_2} \left[\frac{\partial L}{\partial x^\mu} - \frac{d}{\partial p}\left(\frac{\partial L}{\partial \dot{x}^\mu}\right)\right] \delta x^\mu(p) dp = 0, \tag{8.24.10}$$

giving the Euler-Lagrange equation (8.21.2) for $F_i^0 = 0$ or (8.24.19). This relation is not unique, because when L satisfies it, the quantity $aL + b$ also satisfies it for all constants a and b.

If L is a homogeneous function of nth degree, in the variables $\dot{x}^\mu$, then, by Euler's theorem of differential calculus,

$$\dot{x}^\mu \frac{\partial L}{\partial \dot{x}^\mu} = nL \tag{8.24.11}$$

and equation (8.24.28) gives

$$\begin{aligned}\frac{\partial L}{\partial p} &= \frac{\partial L}{\partial \dot{x}^\mu}\dot{x}^\mu + \frac{\partial L}{\partial \dot{x}^\mu}\frac{d\dot{x}^\mu}{dp} = \frac{d}{dp}\left(\frac{\partial L}{\partial \dot{x}^\mu}\right)\dot{x}^\mu + \frac{\partial L}{\partial \dot{x}^\mu}\frac{d\dot{x}^\mu}{dp} \\ &= \frac{d}{dp}\left(\frac{\partial L}{\partial \dot{x}^\mu}\dot{x}^\mu\right) = \frac{d}{dp}(nL) = n\frac{dL}{dp}.\end{aligned} \tag{8.24.12}$$

Hence (8.24.28) reduces to

$$(n-1)\frac{\partial L}{\partial p} = 0. \tag{8.24.13}$$

If $n \neq 1$, this equation gives

$$L(x^\mu, \dot{x}^\mu) = \text{constant}. \tag{8.24.14}$$

The constancy of the Lagrangian is an integral of motion. The condition (8.24.11) shows that L can be written, as in (8.24.21) because $g_{\mu\nu}$ are only functions of the co-ordinates x^μ. The path of the particle is space-like when the constant is positive, time like when negative and a null geodetic when it is zero.

With the help of the relation (8.24.3) equation (8.24.2) can be expressed as

$$S = -mc\int_a^b \sqrt{-U_\mu U^\mu} dT = -mc\int_{t_1}^{t_2} \sqrt{-g_{\mu\nu}\frac{dx^\mu}{dt}\frac{dx^\nu}{dt}}\, dt$$

$$= -mc\int_{p_1}^{p_2} \sqrt{-g_{\mu\nu}\frac{dx^\mu}{dp}\frac{dx^\nu}{dp}}\, dp. \tag{8.24.15}$$

Equation (8.24.3) can be written as

$$L = -\frac{m}{\Gamma}c^2 = -mc\sqrt{\left\{\left(a_i u^i - c\sqrt{-g_{44}}\right)^2 - \imath^2\right\}}, \tag{8.24.16}$$

where

$$\frac{1}{\Gamma} = \frac{dT}{dt}, \tag{8.24.17}$$

$$\imath^2 = \gamma_{ij}\frac{dx^i}{dt}\frac{dx^j}{dt}, \tag{8.24.18}$$

$$a_i = \frac{g_{4i}}{\sqrt{-g_{44}}}. \tag{8.24.19}$$

In view of (8.24.15), an alternative form for L, in terms of p, is

$$L = -mc\sqrt{-g_{\mu\nu}\frac{dx^\mu}{dp}\frac{dx^\nu}{dp}}. \tag{8.24.20}$$

For many purposes, the simpler expression

$$L = g_{\mu\nu}\frac{dx^\mu}{dp}\frac{dx^\nu}{dp}, \tag{8.24.21}$$

which follows from the action relation

$$\delta\int_{p_1}^{p_2} g_{\mu\nu}\frac{dx^\mu}{dp}\frac{dx^\nu}{dp}dp = 0, \tag{8.24.22}$$

would do.

8.24.2 The canonical momentum and the Hamilton equations

The canonical momentum P in 3-space is defined as

$$P_i = \frac{\partial L}{\partial u_i} = mc\frac{\left(a_i u^i - c\sqrt{-g_{44}}\right)a_i + \gamma_{ij}u^j}{(a_i u^i - c\sqrt{-g_{44}})^2 - \imath^2}$$

$$= m\Gamma u_i - m\Gamma\left(a_i u^i - c\sqrt{-g_{44}}\right)a_i. \tag{8.24.23}$$

Define the kinetic energy $\mathcal{E}$ by

$$\mathcal{E} = -m\Gamma c\left(a_i u^i - c\sqrt{-g_{44}}\right) \tag{8.24.24}$$

and the particle 3-momentum P_i by

$$p_i = m\Gamma u_i. \tag{8.24.25}$$

Then equation (8.24.23) becomes

$$\frac{\partial L}{\partial u_i} = P_i \tag{8.24.26}$$

$$= p_i + \frac{\mathcal{E}}{c} a_i. \tag{8.24.27}$$

Then the Euler-Lagrange equation,

$$\frac{d}{dt}\left(\frac{\partial L}{\partial \dot{x}^\mu}\right) - \frac{\partial L}{\partial x^\mu} = 0, \tag{8.24.28}$$

gives

$$\frac{dP_i}{dt} = \frac{\partial L}{\partial x^i}. \tag{8.24.29}$$

With the help of (8.24.17), equation (8.24.16) becomes

$$L = -m\Gamma\left[\left(a_i u^i - c\sqrt{-g_{44}}\right)^2 - l^2\right]. \tag{8.24.30}$$

The Hamiltonian $\mathcal{H}$ is given by

$$\mathcal{H} = P_i u^i - L \tag{8.24.31}$$

$$= m\Gamma\left[u^2 + \left(c\sqrt{-g_{44}} - a_i u^i\right)\left(a_i u^i\right) + \left\{\left(c\sqrt{-g_{44}} - a_i u^i\right)^2 - l^2\right\}\right]$$

$$= m\Gamma c\sqrt{-g_{44}}\left(c\sqrt{-g_{44}} - a_i u^i\right) = \mathcal{E}\sqrt{-g_{44}} = -cP_4, \tag{8.24.32}$$

where P_4 is P_μ for $\mu = 4$ with

$$P_\mu = \frac{\partial L}{\partial u^\mu}. \tag{8.24.33}$$

The Hamiltonian $\mathcal{H}$ (or $-CP_4$) is the total energy of the particle in the gravitational field. The $E - L$ equation (8.24.28) and the definition of $\mathcal{H}$ in (8.24.31) give the equation

$$\frac{d\mathcal{H}}{dt} = -\frac{\partial L}{\partial t}. \tag{8.24.34}$$

In 4-space, the two relations (8.24.29) and (8.24.30), together with (8.24.32), give the relation

$$\frac{dP_\mu}{dt} = -\frac{\partial L}{\partial x^\mu}. \tag{8.24.35}$$

Regarding $\mathcal{H}$ as a function of the canonically conjugate variables P_μ, x^μ and t, the equations of motion may be written in the Hamiltonian form

$$\frac{dP_\mu}{dt} = -\frac{\partial \mathcal{H}}{\partial x^\mu}, \tag{8.24.36}$$

$$\frac{dx^\mu}{dt} = \frac{\partial \mathcal{H}}{\partial P_\mu}. \tag{8.24.37}$$

8.24.3 Lagrangian density concept

The variational principle depends on the variation of a time integral of the term (8.24.2). The quantity S can be regarded as a legitimate scalar in 4-space, and the Lagrangian L is identified as the difference between two quantities which are a kinetic energy like quantity and a potential energy like quantity for the system described by this L. This Lagrangian is a function of x^μ and of the first time derivatives of the three position co-ordinates x^i. Actually, for discrete systems, the equations of motion can be obtained from (8.24.2) with the help of the variational technique, in a form which is valid for arbitrary curvilinear co-ordinates.

The variational procedure for discrete systems can be generalized by replacing it by a continuous system, for example, in the case of free-space electrodynamics. Then the kinetic energy is replaced by magnetic energy, and potential energy by electrical energy, for transition from the discrete to the corresponding continuous systems. For the Lagrangian L for a continuous system, the explicit form of the constitutive relations of electrodynamics is necessary. So, the Lagrangian L is replaced by a single triple integral

$$L = \int \mathcal{L}_0 dV \tag{8.24.38}$$

in 3-space of position co-ordinates, where dV is a volume element of volume V, $\mathcal{L}_0$ is the locally defined Lagrangian density, and is not dependent on what exists in other points of space. The action S of (8.24.2) therefore becomes

$$S_0 = \int_{t_1}^{t_2} \iiint_{\tau} \mathcal{L}_0 dV \cdot cdt, \tag{8.24.39}$$

where $S_0 = cS$.

In curved spaces, an invariant function $\mathcal{L}_0(x^\mu)$ of the points in a region R in 4-space gives a scalar quantity under an arbitrary co-ordinate transformation. Hence

$$\mathcal{L}_0(x^\mu) = \alpha_0'(x^{\mu'}) \tag{8.24.40}$$

and the transformation of the 4-volume element $d\mathbf{x}$ is ·

$$\sqrt{-g}\, d\mathbf{x} = \sqrt{-g'}\, d\mathbf{x}'. \tag{8.24.41}$$

Moller (1972), equation (9.63), or Adler et al (1975), p. 78.

$$S_0 = \iiiint_R \mathcal{L}_0(x^\mu)\sqrt{-g}\, d\mathbf{x}. \tag{8.24.42}$$

where $d\mathbf{x}(= dV \cdot cdt)$, is an invariant according to Hamilton's principle of least action, provided that the 4-space integration is taken from the lowest to the hightest values of the co-ordinates that are compatible with the region of integration R. Equation (8.24.42) can be written as

$$S_0 = \iiiint_R \mathcal{L}_0(x^\mu)\sqrt{-g}\, d\mathbf{x} = \iiiint_R \mathcal{L}_0(x^{\mu'})\sqrt{-g}\frac{\partial(x^1, x^2, x^3, x^4)}{\partial(x^{1'}, x^{2'}, x^{3'}, x^{4'})} d\mathbf{x}', \tag{8.24.43}$$

where the Jacobian of transformation is

$$J = \frac{\partial(x^1, x^2, x^3, x^4)}{\partial(x^{1'}, x^{2'}, x^{3'}, x^{4'})}. \tag{8.24.44}$$

Relations (8.24.41) and (8.24.43) give

$$S_0 = \iiiint_R \mathcal{L}_0'(x^{\mu'})\sqrt{-g'}\,d\mathbf{x}'. \tag{8.24.45}$$

Hence, writing

$$\mathcal{L}(x^{\mu}) = \sqrt{-g}\,\mathcal{L}_0(x^{\mu}), \tag{8.24.46}$$

we find that the invariant action integral (8.24.43) takes the form

$$S_0 = \iiiint_R \mathcal{L}(x^{\mu})d\mathbf{x}. \tag{8.24.47}$$

The Lagrangian $\mathcal{L}(x^{\mu})$, defined in (8.24.46) is, therefore, a scalar density of weight +1 (or a tensor density of zero order). Adler et al (1975, section 3.5). The transformation law for $\alpha(x^{\mu})$, therefore, is

$$\mathcal{L}'(x^{\mu'}) = \left|\alpha_{\nu}^{\mu'}\right| \mathcal{L}(x^{\mu}) = \frac{1}{|\alpha_{\nu'}^{\mu}|}\mathcal{L}(x^{\mu}), \tag{8.24.48}$$

where $|\alpha_{\nu}^{\mu'}|$ and $|\alpha_{\nu'}^{\mu}|$ are the determinants of the transformation coefficients. Hence $\mathcal{L}(x^{\mu})$ and the 4-volume element $d\mathbf{x}$ transform as

$$\mathcal{L}'(x^{\mu'}) = J\mathcal{L}(x^{\mu}) = \sqrt{-g}\mathcal{L}(x^{\mu}). \tag{8.24.49}$$

$$d\mathbf{x}' = d\tau'\cdot cdt' = c\cdot j^{-1}\cdot d\tau\cdot cdt = \sqrt{-g}d\mathbf{x}, \tag{8.24.50}$$

where the Jacobian J is defined in (8.24.44).

Equation (8.24.47) shows that, dimensionally speaking,

$$[\mathcal{L}] = [S_0][l^{-3}t^{-1}], \tag{8.24.51}$$

where $[x]$ means the dimension of x, the unit of action of $[S]$ is

$$[S] = \left[\frac{ml^2}{t}\right]. \tag{8.24.52}$$

So

$$[S_0] = \left[\frac{ml^3}{t^2}\right]. \tag{8.24.53}$$

The factor $l^{-3}t^{-1}$ of (8.24.51) characterizes a density of weight +1 in 4-space, Post (1962, Chapter II). Hence also, $\mathcal{L}$ is a scalar density of weight +1 in 4-space.

8.25 Lagrangian of a Free Particle in GN Rotation

For GN rotation take ds^2 from (8.14.13) and $\mathcal{L}$ from (8.24.21), then denoting the derivative with respect to p by a dot, in cylindrical polar co-ordinates, we obtain

$$L = \dot{r}^2 + r^2\dot{\theta}^2 + \dot{z}^2 + 2\beta\cdot r\dot{\theta}c\dot{t} - (1-\beta^2)c^2\dot{t}^2, \tag{8.25.1}$$

where dot means the derivative with respect to the parameter p. In place of (8.24.21), if (8.24.3) is used, then

$$L = -mc\left[c^2 - \dot{r}^2 - r^2\dot{\theta}^2 - \dot{z}^2 - 2\beta\cdot r\dot{\theta}\cdot c\dot{t} + (1-\beta^2)c^2\dot{t}^2\right]^{1/2}, \tag{8.25.2}$$

where also dot means derivative with respect to p. The equations of motion for a free particle are determined from (8.24.28) and (8.25.1) or (8.25.2). Since the right hand side of (8.24.28) is zero, we can utilize any of these two expressions for L. The expression (8.25.1) being simpler, for a free particle we use that and obtain

$$\frac{\partial L}{\partial \dot{r}} = 2\dot{r}, \quad \frac{\partial L}{\partial \dot{z}} = 2\dot{z}, \tag{8.25.3}$$

$$\frac{\partial L}{\partial \dot{\theta}} = 2r^2\dot{\theta} - 2rc\dot{t}\beta, \tag{8.25.4}$$

$$\frac{\partial L}{\partial \dot{t}} = -2c^2(1-\beta^2)\dot{t} - 2r\dot{\theta}\cdot c\beta, \tag{8.25.5}$$

$$\frac{\partial L}{\partial r} = 2r\dot{\theta}^2 - 4r\dot{\theta}\cdot\Omega\dot{t} + 2r\Omega^2\dot{t}^2, \tag{8.25.6}$$

$$\frac{\partial L}{\partial \theta} = 0, \quad \frac{\partial L}{\partial z} = 0, \quad \frac{\partial L}{\partial t} = 0, \tag{8.25.7}$$

$$\frac{d}{dt}\left(\frac{\partial L}{\partial \dot{r}}\right) = 2\ddot{r}, \quad \frac{d}{d\beta}\left(\frac{\partial L}{\partial \dot{z}}\right) = 2\ddot{z}. \tag{8.25.8}$$

Hence the EL equations for $x^\mu = r,\ \theta,\ z,\ t$ are

$$\ddot{r} = r\dot{\theta}^2 - 2r\dot{\theta}\cdot\Omega\dot{t} + \dot{t}^2\cdot r\Omega^2, \tag{8.25.9}$$

$$\frac{d}{dp}(r^2\dot{\theta} - r^2\Omega\ddot{t}) = 0, \tag{8.25.10}$$

$$\frac{d}{dt}\left[-\frac{c^2}{\gamma^2}\dot{t} - r^2\Omega\dot{\theta}\right] = 0. \tag{8.25.11}$$

$$\ddot{z} = 0. \tag{8.25.12}$$

Denoting the derivative with respect to the proper time T by a prime, these relations reduce to

$$r'' - r\theta'^2 - 2r\theta'\cdot\Omega t' + t'^2 r\Omega^2 = 0, \tag{8.25.13}$$

$$\frac{d}{dT}(r^2\theta' - r^2\Omega t') = 0, \tag{8.25.14}$$

$$\frac{d}{dT}\left(-\frac{c^2}{\gamma^2}t' - r^2\Omega\theta'\right) = 0, \tag{8.25.15}$$

$$z'' = 0. \tag{8.25.16}$$

Equations (8.25.14) and (8.25.15) give the conservation conditions

$$r^2\theta' - r^2\Omega t' = h, \tag{8.25.17}$$

$$\frac{c^2}{\gamma^2}t' + r^2\Omega\theta' = k, \tag{8.25.18}$$

where h and k are constants of the motion. Putting the value of $r^2\theta'$ from (8.25.17) into (8.25.18) we find that

$$t' = \text{constant}. \tag{8.25.19}$$

The time scale can be adjusted such that this constant equals unity, then

$$t' = 1. \tag{8.25.20}$$

Differentiation of (8.25.17) with respect to T gives

$$2r'\theta' + r\theta'' - 2r'\Omega = 0. \tag{8.25.21}$$

Since $r \neq 0$, this equation can be written as

$$\frac{1}{r}\frac{d}{dT}(r^2\theta') = 2r'\Omega. \tag{8.25.22}$$

Also using (8.25.26) equation (8.25.13) becomes

$$r'' - r\theta'^2 = -2r\Omega\theta' + r\Omega^2. \tag{8.25.23}$$

Equations (8.25.22) and (8.25.23) simplify to give

$$\frac{d}{dT}\{r^2(\theta' + \Omega)\} = 0, \tag{8.25.24}$$

$$r'' - r(\theta' + \Omega) = 0. \tag{8.25.25}$$

Since the components of the Coriolis force $2(\Omega \times u)$ are $(-2r\theta'\Omega, 2r'\Omega, 0)$, and components of the centrifugal force $\Omega \times (\Omega \times r)$ are $(-r\Omega^2, 0, 0)$, it is evident that the right hand side of (8.25.22) and (8.25.23) contain only the radial and azimuthal components of these forces of rotation.

8.26 Lagrangian of a Free Particle in Lorentz Rotation

For the Lorentz equations (8.14.13) and (8.24.21) show that in place of (8.25.1) and (8.25.2) for GN rotation, for Lorentz rotation we have

$$L = (1 - \beta^2\gamma^4\Omega^2t^2)\dot{r}^2 + r^2\dot{\theta}^2 + \dot{z}^2 - c^2\dot{t}^2 + 2\beta^2\gamma^2\Omega t\cdot \dot{r}\cdot r\dot{\theta}$$
$$- 2\beta\gamma^2\cdot \Omega t\cdot c\dot{t}\cdot \dot{r} + 2\beta\gamma\cdot r\dot{\theta}\cdot c\dot{t}, \tag{8.26.1}$$

or

$$L = -mc[c^2 - (1 - \beta^2\gamma^4\Omega^2t^2)\dot{r}^2 - r^2\dot{\theta}^2 - \dot{z}^2 + c^2\dot{t}^2 - 2\beta^2\gamma^2\Omega t\cdot \dot{r}\cdot r\dot{\theta}$$
$$- 2\beta\gamma^2\Omega t\cdot c\dot{t}\cdot \dot{r} + 2\beta\gamma r\dot{\theta}\cdot c\dot{t}]. \tag{8.26.2}$$

The components of the canonical (generalized) 3-momentum P_i are

$$P_r = \frac{\partial L}{\partial \dot{r}} = m\Gamma[\dot{r} - (c - \beta\gamma r\dot{\theta} + \beta\gamma^2\Omega t\cdot \dot{r})\beta\gamma^2\Omega t], \tag{8.26.3}$$

where

$$\frac{1}{\Gamma^2} = \frac{1}{c^2}[c^2 - (\beta\gamma r\dot{\theta} + \beta\gamma^2\Omega t\cdot \dot{r})^2 - \dot{r}^2 - \gamma^2(r\dot{\theta})^2 - \dot{z}^2]. \tag{8.26.4}$$

The expression for the kinetic energy $\mathcal{E}$, equation (8.24.25), is

$$\mathcal{E} = m\Gamma c(c - \beta\gamma r\dot{\theta} + \beta\gamma^2\Omega t\cdot \dot{r}). \tag{8.26.5}$$

Then (8.26.3) becomes

$$P_r = m\Gamma[\dot{r} + (c - \beta\gamma r\dot{\theta} + \beta\gamma^2\Omega t\cdot \dot{r})\gamma^2\beta\Omega t] = p_r + \frac{1}{c}\mathcal{E}\gamma^2\beta\Omega t, \tag{8.26.6}$$

where the particle momentum 3-vector is **p**, with components

$$(p_r, p_\theta, p_z) = m\Gamma(\dot{r}, r\dot{\theta}, \dot{z}). \tag{8.26.7}$$

The other two components, P_θ and P_z, of the canonical momentum P can be evaluated easily.

The EL equations for the Lagrangian of (8.26.2) would involve secularly growing terms in time t, because the frame of reference is accelerated. For this reason, confusion prevails in the interpretation of these equations. The conservation equation, corresponding to (8.25.17) for GN rotation, in this case, is

$$r^2\dot{\theta} - r\beta^2\gamma^3\cdot \Omega t\cdot \dot{r} - r^2\gamma\Omega t = h. \tag{8.26.8}$$

It provides a constant of the motion.

8.27 Constants of Motion

The word 'isotropy' means the same in all directions. So, by the conservation laws in isotropic space, are meant the invariance of the mechanical properties of a closed system when it is rotated as a whole in any manner in space. For example, an infinitesimal rotation of the system obtains the condition for the Lagrangian to remain constant. This condition is of the type of conservations for homogeneity of space. These laws ensure that the mechanical properties of a closed system are unchanged by any parallel displacement of the entire system in space. So, for an infinitesimal displacement $\delta\mathbf{r}$, from the position $\mathbf{r}$ to the position $\mathbf{r} + \delta\mathbf{r}$, the Lagrangian $\mathcal{L}$ remains unchanged in the motion of a closed system.

The change in $\mathcal{L}$ resulting from an infinitesimal change in the co-ordinates and velocities is given by (8.21.4)

$$\delta\mathcal{L} = \sum_i \left(\frac{\partial\mathcal{L}}{\partial r_i}\cdot \delta r_i + \frac{\partial\mathcal{L}}{\partial u_i}\cdot \delta u_i\right). \tag{8.27.1}$$

Using the form (8.21.11) for $\mathcal{L}$ we find that

$$L = \sum_i (r_i \times p_i) = \text{constant}, \tag{8.27.2}$$

where the summation in i means running over all the particles forming the system. This is an integral of motion. Like the linear momentum, L is additive. There are no other additive integrals of motion.

This law of conservation of angular momentum is valid for closed systems. It is also valid in some cases of a system in motion by an external field; for example, the component along the axis of uniform rotation of a system which is symmetric about that axis, is conserved because such a rotation does not alter the mechanical properties of the system. This conservation law is a consequence of the fact that the Lagrangian of a closed system does not change under a rotation of the system as a whole due to isotropy of space.

When the velocities of the particles are fixed and the change $\delta\mathcal{L}$ results from an infinitesimal parallel change in the position vector of every particle, by the same amount $\delta\mathbf{r}$, we find that

$$\delta\mathcal{L} = \sum_i \left(\frac{\partial \mathcal{L}}{\partial \mathbf{r}_i} \cdot \delta\mathbf{r}_i \right) = \left(\delta\mathbf{r} \cdot \sum_i \frac{\partial \mathcal{L}}{\partial \mathbf{r}_i} \right) = 0, \tag{8.27.3}$$

where the summation is over all the particles of the system. Since $\delta\mathbf{r}$ is arbitrary,

$$\sum_i \frac{\partial \mathcal{L}}{\partial r_i} = 0. \tag{8.27.4}$$

Putting this relation in (8.21.2) for $F_i^0 = 0$ we obtain

$$\sum_i \frac{d}{dt} \left(\frac{\partial \mathcal{L}}{\partial \mathbf{u}_i} \right) = \frac{d}{dt} \left(\sum_i \frac{\partial \mathcal{L}}{\partial \mathbf{u}_i} \right) = 0 \tag{8.27.5}$$

and from (8.21.9) we find that

$$\sum_i (m_i \mathbf{u}_i) = p = \sum_i \left(\frac{\partial \mathcal{L}}{\partial \mathbf{u}_i} \right) = \text{constant}, \tag{8.27.6}$$

where p is the momentum of the systems. Momentum is additive even when mutual interaction between particles of a closed system cannot be neglected. Conservation of momentum follows from the homogeneity of space which allows the mechanical properties of a closed system to remain unchanged by any parallel displacement of the entire system in space.

By the principle of homogeneity of time the Lagrangian of a closed system does not depend explicitly on time. So the total derivative of $\mathcal{L}$ can be written as

$$\frac{d\mathcal{L}}{dt} = \sum_i \left(\frac{\partial \mathcal{L}}{\partial q_i} \dot{q}_i \right) + \sum_i \left(\frac{\partial \mathcal{L}}{\partial \dot{q}_i} \ddot{q}_i \right). \tag{8.27.7}$$

Using the Euler-Lagrange equation

$$\frac{d}{dt} \left(\frac{\partial \mathcal{L}}{\partial \dot{q}_k} \right) - \frac{\partial \mathcal{L}}{\partial q_k} = 0 \tag{8.27.8}$$

to eliminate $\partial\mathcal{L}/\partial q_i$ we get

$$\begin{aligned} \frac{d\mathcal{L}}{dt} &= \sum_i \left(\frac{d}{dt} \frac{\partial \mathcal{L}}{\partial \dot{q}_i} \dot{q}_i + \frac{\partial \mathcal{L}}{\partial \dot{q}_i} \ddot{q}_i \right) \\ &= \sum_i \left(\dot{q}_i \frac{d}{dt} \frac{\partial \mathcal{L}}{\partial \dot{q}_i} + \ddot{q}_i \frac{\partial \mathcal{L}}{\partial \dot{q}_i} \right) = \sum_i \frac{d}{dt} \left(\dot{q}_i \frac{\partial \mathcal{L}}{\partial \dot{q}_i} \right). \end{aligned} \tag{8.27.9}$$

Hence we can write

$$\frac{d}{dt}\left[\mathcal{L} - \sum_i \dot{q}_i \frac{\partial \mathcal{L}}{\partial \dot{q}_i}\right] = 0, \tag{8.27.10}$$

and consequently obtain the energy conservation law

$$\mathcal{E} = \sum_i \left(\dot{q}_i \frac{\partial \mathcal{L}}{\partial \dot{q}_i}\right) - \mathcal{L} = \text{constant}. \tag{8.27.11}$$

So, for a closed system of n particles, even when there are forces of mutual interaction among these particles and not with anything outside the system, the single condition (8.27.11) and three conditions, in (8.27.6) and (8.27.2), give a total of seven conditions of conservation.

8.28 Macroscopic Motion and Entropy

In macroscopic motion of a body the separate microscopic sub-parts take part as a whole; the body is actually divided into a large number of small (but sufficiently large to be regarded as macroscopic) sub-parts. The mass, energy and momentum of the ith sub-part are denoted by M_i, $\mathcal{E}_i$, p_i, respectively. The entropy[4] S_i of this part is a function of its internal energy which is the difference between its total energy $\mathcal{E}_i$ and its kinetic energy of macroscopic motion, namely $P_i^2/2M_i$. Hence, the total energy S of the body can be written in the form

$$S = \sum_i S_i, \tag{8.28.1}$$

where S_i is a function of the internal energy $\mathcal{E}_i - p_i^2/2M_i$. If the body is a closed system its total momentum, angular momentum and energy, are conserved. So we have

$$\sum \mathbf{p}_i = \text{constant}, \quad \sum_i (\mathbf{r}_i \times \mathbf{p}_i) = \text{constant}, \tag{8.28.2}$$

where $\mathbf{r}_i$ is the position vector of the ith sub-part of the body.

In the equilibrium state the entropy of the body, expressed as a function of the moment p_i, is a maximum when subjected to these additional conditions. Hence, by Lagrange's method of undetermined multipliers, the derivative with respect to p_i of the sum

$$\sum_i [S_i + (A \cdot p_i) + (B \cdot (r_i \times p_i))] \tag{8.28.3}$$

is to be equated to zero for arbitrary constant values of A and B.

The velocity v_i of the ith part of the body is defined as

$$v_i = \frac{p_i}{M_i} \tag{8.28.4}$$

[4]Entropy σ, a dimensionless quantity, is a consequence of purely statistical regularities alone and has no meaning in non-macroscopic systems. The entropy $S = k\sigma$ has the dimension of k_B, where k_B is the Boltzmann constant. Entropy is defined such that it cannot be negative. It has the dimension of a logarithm of action. This means that a change in the units of action would mean a change of the entropy by an additive constant. Hence, classically, entropy is a quantity uncertain to the extent of an additive constant depending on the choice of units. The difference in entropy, that is, the change in entropy induced by a process, does not depend on the choice of units.

and the temperature T by

$$\frac{dS}{d\mathcal{E}} = T. \tag{8.28.5}$$

Using these definitions the derivatives of S_i, with respect to the components of p_i, are given by

$$\frac{\partial S_i}{\partial p_i} = \frac{\partial S_i}{\partial Q_i}\frac{dQ_i}{dp_i} = -\frac{p_i}{M_i}\frac{1}{T} = -\frac{v_i}{T} \tag{8.28.6}$$

because

$$Q_i = \mathcal{E} - \frac{p_i^2}{2M_i}. \tag{8.28.7}$$

Here Q_i is the internal energy of the ith part of the body. Hence differentiating (8.28.3) with resepect to p_i, and equating that expression to zero, we obtain

$$-\frac{v_i}{T} + A + (B \times \mathbf{r}_i) = 0, \quad \text{or,} \quad \mathbf{v}_i = \mathbf{u} + (\Omega \times \mathbf{r}_i), \tag{8.28.8}$$

where $\mathbf{u} = TA$, $\Omega = TB$ and $\mathbf{u}$ and Ω are constant vectors. Therefore, if the velocities of all parts of the body are given by (8.28.8), then the body translates as a whole with constant angular velocity Ω. So a closed system, in thermodynamic equilibrium, can perform only uniform translation and a rotation as a whole and internal macroscopic motion is ruled out. When the bodies are at rest, $Q_i = \mathcal{E}_i$ and so $\mathcal{E}(= \Sigma_i \mathcal{E}_i)$ is the internal energy of the body. Maximum of the entropy requires the second derivative with respect to p_i to be negative and so $T > 0$, that is, the temperature can only be positive.

8.29 Statistical State of a Body

Liouville theorem (section 3.14) shows that the distribution function f is constant along a phase trajectory of a sub-system. For quasi-closed systems this result is valid only for not too long intervals of time, during which the sub-system behaves with sufficient accuracy like a closed one. Hence the distribution function f can depend only on those combinations of the variables which remain constant while the sub-system moves like a closed system. These being the first integrals of motion, the distribution function is itself an integral of motion, and the logarithm of that equation is an additive integral of motion.

The seven constants of motion are the only seven independent integrals of motion for a closed system. If $\mathcal{E}_i(p,q), p_i(p,q), M_i(p,q)$ are these seven quantities, for the ith sub-system, then $\log f_i$ must be a linear combination of these quantities. So

$$\log f_i = \alpha_i + \beta\mathcal{E}_i(p,q) + (A_i \cdot p_i(p,q)) + (B \cdot M_i(p,q)), \tag{8.29.1}$$

where α_i, β, A, B are constant coefficients, α_i is obtained from the normalizing condition

$$\int f_i dp_i dq_i = 1, \tag{8.29.2}$$

and the other seven constants β, A, B, from the seven constant values of the additive integrals of motion for the whole closed system.

This distribution function should be constant for all points of the phase space, where

$$\mathcal{E} = \mathcal{E}_0, \quad p = p_0, \quad L = L_0 \tag{8.29.3}$$

$\mathcal{E}_0$, p_0, M_0 are given constants and f should vanish for all other points. This f would remain constant along any phase space trajectory and so should satisfy the Liouville theorem. The distribution function can therefore be written as

$$f = \text{constant } \delta(\mathcal{E} - \mathcal{E}_0)\delta(p - p_0)\delta(M - M_0), \tag{8.29.4}$$

where $\delta(x)$ is the delta function of Dirac. At all points of phase space for which $\mathcal{E}$, p, M differ from $\mathcal{E}_0$, p_0, M_0, the distribution function will vanish, f will become infinite and the integral

$$\int f dp dq \neq 0 \tag{8.29.5}$$

is non-zero. This is valid for the microcanonical distribution.

The seven integrals of motion completely determine the statistical properties of a closed system; so the large number of inertial conditions required of statistical mechanics are replaced by these integrals. Thus, depending on these integrals, a simple distribution function, satisfying the Liouville theorem, can be used. Since momentum and angular momentum of a closed system are connected with its motion of uniform translation and rotation, the statistical state of a body, performing a given motion, depends only upon its energy.

8.30 Transformation of Energy

The Lagrangian $\mathcal{L}$ in non-relativistic motion is given by

$$\mathcal{L} = \sum \left[-mc^2 + \frac{m}{2}u^2 - m(u\cdot(\Omega \times r)) + \frac{m}{2}(\Omega \times r)^2 - q\left(\phi - \frac{1}{c}(u\cdot A)\right)\right] - V, \tag{8.30.1}$$

where the summation is over all the macroscopic sub-parts of the body, m is its rest mass and V is the potential of the sum of the forces of non-electrical origin acting on it. The momentum $\mathbf{p}$ of a single particle in the rotating K frame, is given by

$$\mathbf{p} = \frac{\partial \mathcal{L}}{\partial \mathbf{u}} = m\mathbf{u} - m(\Omega \times \mathbf{r}) + \frac{q}{c}A. \tag{8.30.2}$$

The mechanical and thermodynamic properties of a body are not affected by uniform translation, only the kinetic energy is added to its energy. The energy $\mathcal{E}$, being the Hamiltonian $\mathcal{H}$, in the K system, is

$$\mathcal{E} = \sum (p\cdot u) - \mathcal{L} = \sum \left[\frac{m}{2}\{u^2 - (\Omega \times r)^2\} + q\phi + mc^2\right] + V. \tag{8.30.3}$$

If $\mathbf{u}_0, \mathbf{p}_0, \mathcal{L}_0, \phi_0, A_0$ are the particle velocity, momentum, the Lagrangian, the scalar potential and the vector potential, respectively, in the K^0 frame, then

$$\mathbf{u}_0 = \mathbf{u} - (\Omega \times \mathbf{r}), \quad \mathbf{p}_0 = \frac{\partial \mathcal{L}_0}{\partial \mathbf{u}_0} = m_0\mathbf{u}_0 + \frac{q}{c}A_0, \tag{8.30.4}$$

$$\mathcal{L} = \sum \left[\frac{m}{2}u_0^2 - mc^2 - q\left(\phi_0 - \frac{1}{c}(u_0\cdot A_0)\right)\right] - V \tag{8.30.5}$$

for the mass m and charge q of the sub-part. The relations of (8.30.2) and (8.30.4) show that

$$\mathbf{p} = \mathbf{p}_0 + \frac{1}{c}(A_0 - A), \tag{8.30.6}$$

$$L = (\mathbf{r} \times \mathbf{p}) = L_0 + \frac{q}{c} r \times (A_0 - A), \quad L_0 = (\mathbf{r} \times \mathbf{p}_0). \tag{8.30.7}$$

The energy $\mathcal{E}_0$ of the macroscopic system in the K^0 frame is then given by

$$\mathcal{E}_0 = \sum \left[mc^2 + \frac{m}{2} u_0^2 + q\phi_0 \right] + V. \tag{8.30.8}$$

Using this value of $\mathcal{E}_0$ in (8.30.3) and (8.30.4) we obtain

$$\mathcal{E} = \mathcal{E}_0 + \sum [(M_0 \cdot \Omega) + q(\phi - \phi_0)]. \tag{8.30.9}$$

This is the law of transformation of energy in a uniformly rotating frame of reference. Since the energies in the two frames can be written as

$$\frac{p^2}{2m} + (\Omega \cdot M) = \frac{m}{2} \mathbf{u}^2 - \frac{m}{2} (\Omega \times \mathbf{r})^2 + \frac{q}{c} (\mathbf{u} \cdot A) + \frac{q^2 A^2}{2mc^2}, \tag{8.30.10}$$

$$\frac{p_0^2}{2m} = \frac{m}{2} u_0^2 + \frac{q}{c} (u_0 \cdot A_0) + \frac{q^2 A_0^2}{2mc^2}, \tag{8.30.11}$$

we then find that

$$\mathcal{E} = \sum \left[\frac{p^2}{2m} + (\Omega \cdot M) + mc^2 + q \left(\phi - \frac{1}{c} (u \cdot A) \right) - \frac{q^2 A^2}{2mc^2} \right] + V. \tag{8.30.12}$$

$$\mathcal{E}_0 = \sum \left[\frac{p_0^2}{2m} + mc^2 + q \left(\phi_0 - \frac{1}{c} (u_0 \cdot A_0) \right) - \frac{q^2 A_0^2}{2mc^2} \right] + V. \tag{8.30.13}$$

In the absence of external fields $\phi = 0$, $A = 0$,

$$\mathcal{E} = \mathcal{E}_0 + (L_t^0 \cdot \Omega), \tag{8.30.14}$$

$$\mathbf{p} = \mathbf{p}_0, \quad L = L_0, \tag{8.30.15}$$

$$\mathcal{E} = \sum \left[\frac{p^2}{2m} - \frac{m}{2} (\Omega \times \mathbf{r})^2 \right] + V, \tag{8.30.16}$$

where L_t^0 is the total angular momentum of the system in the K^0 frame and L_t is the same in the K system.

8.31 Thermodynamics of Rotating Bodies

Dropping the subscript t from L in (8.30.14) we find that

$$\mathcal{E}(p, q) = \mathcal{E}_0(p, q) + (\Omega \cdot L^0(p, q)) = \mathcal{E}_0 + (\Omega \cdot L), \tag{8.31.1}$$

where p's are the momentums and q's are the co-ordinates of particles of the body. Hence

$$\frac{\partial \mathcal{E}(p, q)}{\partial \Omega} = L \tag{8.31.2}$$

because $\mathcal{E}$ depends on the angular velocity Ω as a parameter. Averaging this equation over the statistical distribution we obtain

$$\left(\frac{\partial \mathcal{E}}{\partial \Omega} \right)_S = L \tag{8.31.3}$$

for constant entropy S (the entropy S remains constant in adiabatic processes), where $\mathcal{E} = \langle \mathcal{E}(p,q) \rangle$, $L = L(p,q)$ and these are the mean thermodynamic energy and angular momentum of the system.

The differential of energy in the K system is given by

$$d\mathcal{E} = Tds + (L \cdot d\Omega), \tag{8.31.4}$$

and the free energy is

$$F = \mathcal{E} - TS. \tag{8.31.5}$$

For the differential of F we have

$$dF = -SdT + (L \cdot d\Omega). \tag{8.31.6}$$

Averaging (8.31.1) we get

$$\mathcal{E} = \mathcal{E}_0 + (L \cdot d\Omega). \tag{8.31.7}$$

Differentiating this relation and using (8.31.4) we obtain

$$d\mathcal{E}_0 = TdS - (\Omega \cdot dL) \tag{8.31.8}$$

and for dF_0 in the K^0 system we have

$$dF_0 = -SdT - (\Omega \cdot dL) \tag{8.31.9}$$

Here L, and not Ω, is the independent variable, and

$$\Omega = \left(\frac{\partial \mathcal{E}_0}{\partial L}\right)_S = \left(\frac{\partial F_0}{\partial L}\right)_T, \tag{8.31.10}$$

where $(\partial F_0/\partial L)_T$ is the derivative of F_0 for constant temperature.

To find the influence of the centrifugal and Coriolis forces of a uniformly rotating body, on its thermodynamic properties, we must note that the former force is proportional to the dimensions of the body (because it is proportional to the axial distance) and the latter is completely independent of the size of the body. So the influence of the centrifugal force is much more than the Coriolis force; for this reason the latter is completely neglected in the study of kinetics of rotation. Therefore, for thermal equilibrium,

$$\mu_0(p,T) - \frac{m}{2}\Omega^2 r^2 = \text{constant}, \tag{8.31.11}$$

where μ_0 is the chemical potential of the stationary body, m is its mass, r is its distance from the axis of rotation, and $mr^2\Omega^2/2$ is the centrifugal energy of the body. Moreover,

$$\mathcal{E} = Q_i + \frac{L^2}{2I}, \tag{8.31.12}$$

where $\mathcal{E}$ is the total energy of the body, Q_i is its internal energy, I is the moment of inertia about the axis of rotation and L is the angular momentum. Since rotation changes the distribution of mass of the body, its I and Q_i would be independent of Ω (or L) only when Ω is sufficiently small. Since entropy of an isolated body is a function of Q_i, using (8.28.3) we have

$$S \equiv S(Q) \equiv S\left(\mathcal{E} - \frac{L^2}{2I}\right). \tag{8.31.13}$$

Hence the total energy and angular momentum of a body are conserved and S is a maximum when $\mathcal{E}$ and L are given, and for its equilibrium the axis of rotation must be one of its principle axis.

8.32 Angular Velocity Distribution of a Rotating Body

The kinetic energy of rotation of a solid body is given by

$$\mathcal{E}_{\text{rot}} = \frac{1}{2}(I_1\Omega_1^2 + I_2\Omega_2^2 + I_3\Omega_3^2) = \frac{1}{2}\left(\frac{L_1^2}{I_1} + \frac{L_2^2}{I_2} + \frac{L_3^2}{I_3}\right), \tag{8.32.1}$$

where I_1, I_2, I_3 are the principal moments of inertia, Ω_1, Ω_2, Ω_3 are the components of Ω along the principal axes of inertia and

$$L_i = I_i\Omega_i^2 \tag{8.32.2}$$

L_1, L_2, L_3 are the components of angular momentum; these can be regarded as the generalized momentum components for the velocities. Hence the normalized probability distribution of angular momentum is given by

$$df_M = \frac{dL_1 dL_2 dL_3}{(2\pi kT)^{3/2}(I_1 I_2 I_3)^{1/2}} \exp\left[-\frac{1}{2kT}\left(\frac{L_1^2}{I_1} + \frac{L_2^2}{I_2} + \frac{L_3^2}{I_3}\right)\right] \tag{8.32.3}$$

and of angular velocity by

$$df_\Omega = \frac{(I_1 I_2 I_3)^{1/2} d\Omega_1 d\Omega_2 d\Omega_3}{(2\pi kT)^{3/2}} \exp\left[-\frac{1}{2kT}(I_1\Omega_1^2 + I_2\Omega_2^2 + I_3\Omega_3^2)\right]. \tag{8.32.4}$$

The mean square of the angular velocity and angular momentum of a body therefore are,

$$\bar{\Omega}^2 = kT\left(\frac{1}{I_1} + \frac{1}{I_2} + \frac{1}{I_3}\right), \tag{8.32.5}$$

$$\bar{L}^2 = kT(I_1 + I_2 + I_3). \tag{8.32.6}$$

8.33 Finite Rotation

Consider rotation of a cube, whose three concurrent edges at the point O are OX, OY, OZ, by the angle $\pi/2$ about OZ, followed by the same about OX, both in the anticlockwise sense. This gives a position to the cube which is different from that obtained by rotating first about OX and then about OZ through angles $\pi/2$. So, a rotation A_1 first and then A_2 of a rigid body is, in general, different from the rotation A_2 first and then A_1. Mathematically, if A_1 and A_2 are the two matrices of the two rotations, then the result of the first operation is represented by the matrix product A_1A_2, and the same of the second operation by A_2A_1. Since the matrices A_1A_2 and A_2A_1 are, in general, different, the two operations give different results. It is seen that the first rotation, given by the matrix A_1A_2, is equivalent to the rotation by the angle $4\pi/3$ about the line through O and the centre of the cube, that is, about a line which makes equal angles with the edges OX, OY, OZ.

For rotation through finite angle ϕ_1 about the Z axis, we have the transformation relations

$$\left.\begin{aligned} x_1 &= x\cos\phi_1 + y\sin\phi_1 \\ y_1 &= -x\sin\phi_1 + y\cos\phi_1 \\ z_1 &= z \end{aligned}\right\}. \tag{8.33.1}$$

etc. Then the square matrices A_1 and A_2 of the two rotations are

$$A_1 = \begin{bmatrix} \cos\phi_1 & \sin\phi_1 & 0 \\ -\sin\phi_1 & \cos\phi_1 & 0 \\ 0 & 0 & 1 \end{bmatrix}. \tag{8.33.2}$$

$$A_2 = \begin{bmatrix} \cos\phi_2 & \sin\phi_2 & 0 \\ -\sin\phi_2 & \cos\phi_2 & 0 \\ 0 & 0 & 1 \end{bmatrix}. \tag{8.33.3}$$

Since

$$A_1 A_2 \neq A_2 A_1, \tag{8.33.4}$$

finite rotations do not satisfy the requirements of vectors. So any finite rotation cannot be represented by a simple vector.

For rotation through angle ϕ, the operator relation

$$A = \cos\frac{\phi}{2} - i\sin\frac{\phi}{2}(\sigma_x\cos\alpha + \sigma_y\cos\beta + \sigma_z\cos\gamma), \tag{8.33.5}$$

is used, where α, β, γ are the angles the axis of rotation make with the co-ordinate axes,

$$\sigma_x = \begin{bmatrix} 0 & 1 \\ 1 & 0 \end{bmatrix} = ii_0, \tag{8.33.6}$$

$$\sigma_y = \begin{bmatrix} 0 & -i \\ i & 0 \end{bmatrix} = ij_0, \tag{8.33.7}$$

$$\sigma_z = \begin{bmatrix} 1 & 0 \\ 0 & -1 \end{bmatrix} = ik_0, \tag{8.33.8}$$

where

$$i = \sqrt{-1}, \quad i_0^2 = j_0^2 = k_0^2 = -1 \tag{8.33.9}$$

$$\sigma_x^2 = \sigma_y^2 = \sigma_z^2 = 1 \tag{8.33.10}$$

$$\sigma_x\sigma_y = -\sigma_y\sigma_x = i\sigma_x \tag{8.33.11}$$

and cyclic permutations can be used.

The quantity A of (8.33.5) is called *Hamilton's quaternions,* spin matrix or spinor transformation. For more information on these rotation operators see Misner et al (1973, Chapter 41), Pars (1965, Chapter VIII).

8.34 Infinitesimal Rotation

Let $\mathbf{r}$ be the position vector relative to the point P of a particle before the displacement, and $\mathbf{r} + \delta\mathbf{r}$ be the same after the displacement. If it is an infinitesimal rotational displacement then

$$\delta\mathbf{r} = ((\Omega\delta t) \times \mathbf{r}) + ((\delta\Omega t) \times \mathbf{r}) = (\delta\mathbf{n} \times \mathbf{r}), \tag{8.34.1}$$

where Ω is the angular velocity of rotation, and

$$\delta\mathbf{n} = \mathbf{\Omega}\delta t + \delta\mathbf{\Omega}t. \tag{8.34.2}$$

The point P moves perpendicular to the plane of the vectors $\mathbf{r}$ and $\delta\mathbf{n}$ in the same way as $\delta\mathbf{n} \times \mathbf{r}$.

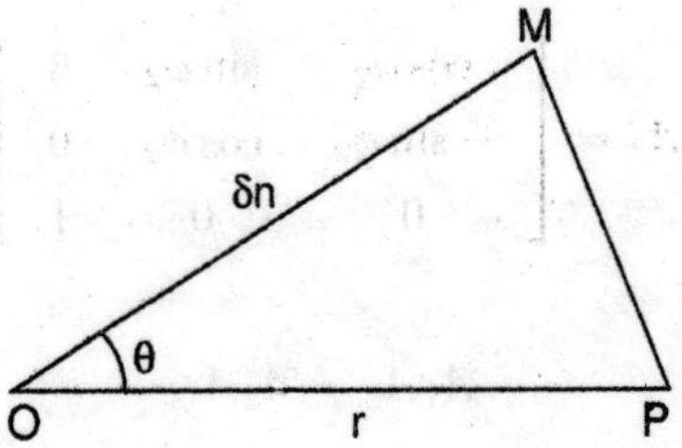

Fig. 8.34.1 Shows the displacement due to the infinitesimal rotation δn.

Let PM of fig. 8.34.1 be the perpendicular from P to the line through O parallel to δn. Then

$$|\delta r| = PM\delta n = r\sin\theta\delta n \tag{8.34.3}$$

because $PM = r\sin\theta$ if θ is the angle between $\mathbf{r}$ and $\delta\mathbf{n}$. If the second infinitesimal rotation puts the particle at the point R, then

$$OR = \mathbf{r} + (\delta\mathbf{n}\times\mathbf{r}) + \delta\mathbf{n}'\times(\mathbf{r} + (\delta\mathbf{n}\times\mathbf{r})), \tag{8.34.4}$$

where $\delta n'$ is the second infinitesimal rotation vector. Next, neglecting infinitesimals of the second order, the resultant displacement is given by

$$\mathbf{OR} - \mathbf{OP} = ((\delta\mathbf{n} + \delta\mathbf{n}')\times\mathbf{r}) = ((\delta\mathbf{n}' + \delta\mathbf{n})\times\mathbf{r}). \tag{8.34.5}$$

This is the displacement due to a single rotation of amount $\delta\mathbf{n} + \delta\mathbf{n}'$. Hence, the resultant of two infinitesimal rotations about the same point O is the vector sum of these rotations.

For rotations through an infinitesimal angle $d\phi$, correct up to the first order of infinitesimals, the relations of the type (8.33.2) are written as

$$I + \varepsilon = \begin{bmatrix} 1 & d\phi & 0 \\ -d\phi & 1 & 0 \\ 0 & 0 & 1 \end{bmatrix}, \tag{8.34.6}$$

where

$$I = \begin{bmatrix} 1 & 0 & 0 \\ 0 & 1 & 0 \\ 0 & 0 & 1 \end{bmatrix} \tag{8.34.7}$$

The elements of the unit matrix I are given by the Kronecker delta δ_{ij}. Separating ε from (8.34.5), and using (8.34.7), for this infinitesimal matrix, we obtain

$$\varepsilon = \begin{bmatrix} 0 & d\phi & 0 \\ -d\phi & 0 & 0 \\ 0 & 0 & 0 \end{bmatrix} = d\phi\begin{bmatrix} 0 & 1 & 0 \\ -1 & 0 & 0 \\ 0 & 0 & 0 \end{bmatrix}. \tag{8.34.8}$$

It is non-symmetric though the original matrix A_1 of (8.33.2) is not so. The relation of (8.34.8) indicate that the more general non-symmetric matrix, for the general three dimensional infinitesimal rotation, Goldstein (1980, section 4.8, p. 168–170), is given by

$$\varepsilon = \begin{bmatrix} 0 & d\phi_3 & -d\phi_2 \\ -d\phi_3 & 0 & d\phi_1 \\ d\phi_2 & -d\phi_1 & 0 \end{bmatrix}. \tag{8.34.9}$$

The three infinitesimal angles $d\phi_1, d\phi_2, d\phi_3$ are identified as the three independent parameters specifying rotation.
Up to the first order of infinitesimals, the determinant of $I + \varepsilon$ of (8.34.6) is

$$\det(I + \varepsilon) = +1. \tag{8.34.10}$$

Hence, infinitesimal orthogonal transformations of co-ordinate axes, correspond to proper infinitesimal rotations. The difference between the components, along the same directions, of a vector, in two sets of co-ordinate axes, is an infinitesimal. In general, this infinitesimal should be of the type

$$x_1^0 - x_1 = \varepsilon_{11}x_1 + \varepsilon_{12}x_2 + \varepsilon_{13}x_3, \tag{8.34.11}$$

where the matrix elements ε_{ij} are infinitesimal quantities, as may be seen from (8.34.5). Equation (8.34.11) can be written in general form as

$$x_i^0 = x_i + \varepsilon_{ij}x_j \tag{8.34.12}$$

and, more elegantly as

$$x_i^0 = (\delta_{ij} + \varepsilon_{ij})x_j, \tag{8.34.13}$$

where repeated index means summation over that index and δ_{ij} is the Kronecker delta. Hence, we can write

$$x^0 = (I + \varepsilon)x = Ax, \tag{8.34.14}$$

where I is the unit matrix of (8.34.7) and

$$A = I + \varepsilon \tag{8.34.15}$$

The inverse matrix A^{-1} is given by

$$A^{-1} = I - \varepsilon, \quad A\cdot A^{-1} = I. \tag{8.34.16}$$

If A^T and ε^T are the transposes of A and ε, respectively, then

$$A^T = I + \varepsilon^T. \tag{8.34.17}$$

Since, for orthogonal transformations,

$$\varepsilon^{-1} = \varepsilon^T,$$

we have

$$\varepsilon^T = -\varepsilon, \quad \text{or,} \quad \varepsilon_{ij} = -\varepsilon_{ji}. \tag{8.34.18}$$

This shows that the matrix ε is non-symmetrical.

If $I + \varepsilon_1$ and $I + \varepsilon_2$ are the matrices of two infinitesimal transformations, then neglecting higher order infinitesimals like the matrix product $\varepsilon_1 \cdot \varepsilon_2$ we have

$$A_1 A_2 = I + \varepsilon_1 + \varepsilon_2. \tag{8.34.19}$$

Hence transformations of infinitesimal rotations allow the cumulative interchange of ε_1 and ε_2; so rotation in reverse order leads to the same result.

The transformation relation (8.34.12) is written as

$$x^0 - x = d\mathbf{x} = \varepsilon x, \tag{8.34.20}$$

where $d\mathbf{x}$ is the change in the components of the vector x under infinitesimal transformation. Using (8.34.14) for ε, the expanded form, obtained with the help of the relation (8.34.9), is

$$d\mathbf{x}_1 = x_2 d\phi_3 - x_3 d\phi_2, \tag{8.34.21}$$

$$d\mathbf{x}_2 = x_3 d\phi_1 - x_1 d\phi_3, \tag{8.34.22}$$

$$d\mathbf{x}_3 \quad x_1 d\phi_2 - x_2 d\phi_1. \tag{8.34.23}$$

The vector form of these relations is

$$dr = (r \times d\phi), \tag{8.34.24}$$

where the components of the vector $d\boldsymbol{\phi}$ are $d\phi_1, d\phi_2, d\phi_3$.

The increment $\delta\mathbf{r}$, in the radius vector $\mathbf{r}$, fig. 8.34.2, from an origin on the axis to a particle of the rotating system is due to $d\Omega$. For rigid rotation the linear displacement of the tip of the radius vector is given by

$$|\delta r| = r \sin\theta \delta\Omega \tag{8.34.25}$$

and $\delta\mathbf{r}$ is perpendicular to the plane of $\mathbf{r}$ and $\delta\boldsymbol{\Omega}$.

So

$$\delta\mathbf{r} = (\delta\boldsymbol{\Omega} \times \mathbf{r}) \tag{8.34.26}$$

and

$$\delta\mathbf{v} = (d\boldsymbol{\Omega} \times \mathbf{v}), \tag{8.34.27}$$

where $\delta\mathbf{v}$ is the velocity increment corresponding to the displacement $\delta\mathbf{r}$. Sec de Groot and Suttorp (1972, equation (12), p. 411, and pages 439–443).

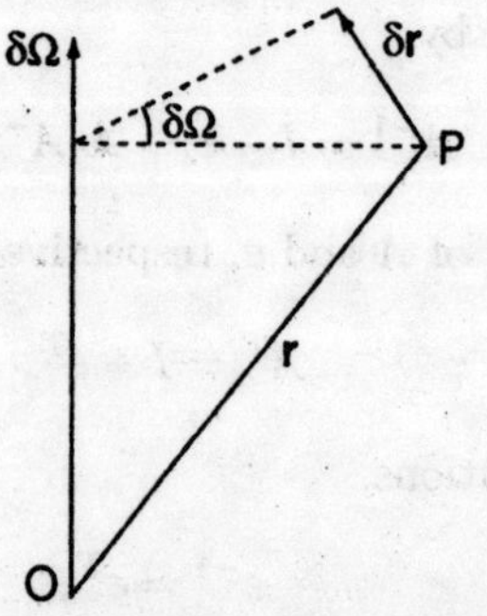

Fig. 8.34.2 Shows the increment in the radius vector from origin on the axis to a particle in the rotating system for rigid rotation.

8.35 The Pseudo Vector Character of Infinitesimal Rotation

The infinitesimal rotations, in general, behave like axial vectors. To show that consider the transformation of the infinitesimal matrix ε of (8.34.20)

$$\varepsilon^0 = B\varepsilon B^{-1}, \tag{8.35.1}$$

where B^{-1} is the inverse of the matrix B. In the K^0 frame, equation (8.34.20) is

$$dx^0 = \varepsilon^0 x^0. \tag{8.35.2}$$

Since B is orthogonal,

$$B^{-1} = B^T, \tag{8.35.3}$$

where, as in (8.34.17), the matrix B^T is the transpose of B. Hence relation (8.35.1) gives

$$\varepsilon^0_{jk} = B_{jm}\varepsilon_{mn}B_{kn}, \tag{8.35.4}$$

where repeated index means summation over that index.

If a matrix is antisymmetric, it remains so under any orthogonal transformation. So, (8.35.2) can be written in the same form as (8.34.21)–(8.34.23) in the K^0 system

$$\left.\begin{aligned} dx^0_1 &= x^0_2 d\phi^0_3 - x^0_3 d\phi^0_2, \\ dx^0_2 &= x^0_3 d\phi^0_1 - x^0_1 d\phi^0_3, \\ dx^0_3 &= x^0_1 d\phi^0_2 - x^0_2 d\phi^0_1, \end{aligned}\right\} \tag{8.35.5}$$

and the form of (8.34.2) is preserved

$$d\mathbf{r}^0 = (\mathbf{r}^0 \times d\boldsymbol{\phi}^0). \tag{8.35.6}$$

The relation between $d\phi$ and ε, obtained from (8.34.9), is

$$d\phi_i = \frac{1}{2}e_{ijk}\varepsilon_{jk}, \tag{8.35.7}$$

where e_{ijk} is the Levi Civita symbol (Appendix 4). In the K^0 frame, similarly we obtain

$$d\phi^0_i = \frac{1}{2}e_{ijk}\varepsilon^0_{jk}. \tag{8.35.8}$$

The converse transformation relation is

$$\varepsilon_{jk} = e_{ijk}d\phi_i. \tag{8.35.9}$$

Hence, using (8.35.4) from (8.35.8), we obtain

$$d\phi^0_i = \frac{1}{2}e_{ijk}B_{jm}B_{km}. \tag{8.35.10}$$

Substitution of the value of ε_{jkl} from (8.35.9) in (8.35.10) gives

$$d\phi^0_i = \frac{1}{2}e_{ijk}e_{lmn}B_{jm}B_{kn}d\phi_l. \tag{8.35.11}$$

Since B is an orthogonal matrix,

$$B_{\mu\nu} B_{\lambda\nu} = \delta^{\mu}_{\lambda}. \tag{8.35.12}$$

Multiplying both sides of (11.2.20) by $B_{il} B_{il}$ and using (8.35.12) we obtain

$$\sum_l B_{il} \cdot \frac{1}{2} \sum [e_{ijk} e_{lmn} B_{jm} B_{kn}] = \sum_l [B_{il} (B_{il} \det B)] \,. \tag{8.35.13}$$

This equation can be written as

$$\sum_l B_{il} \left\{ \frac{1}{2} \sum_{j,k,m,n} e_{ijk} e_{lmn} B_{jm} B_{kn} - B_{il} \det B \right\} = 0. \tag{8.35.14}$$

Since B_{il} are elements of any arbitrary orthogonal matrix, we can write

$$\frac{1}{2} \sum_{j,k,m,n} e_{ijk} e_{lmn} B_{jm} B_{kn} = B_{il} \det B. \tag{8.35.15}$$

Equation (8.35.11), with the help of (8.35.15), for the transformation of $d\phi_i$ from the K frame to the K^0 frame, becomes

$$d\phi^0_i = (\det B) \left(\sum_l B_{il} d\phi_l \right) . \tag{8.35.16}$$

For proper rotation $\det B = 1$, and for improper rotation $\det B = -1$, and $d\Omega$ is not a polar vector.

8.36 General Infinitesimal Displacement

A general, infinitesimal displacement consists of the vector sum of an infinitesimal translational displacement of the base point O of Fig. 8.34.1 and an infinitesimal rotational (rigid) displacement with respect to O. Following Synge and Griffith (1949, section 10.5), we consider these displacements here.

Displacement of a point P, of a rigid body, having position vector r, depends on r and is denoted by $A(r)$. A further displacement is similarly denoted by $B(r + A(r))$ because it depends on the position of the particle which is now $A(r)$. Up to the first order of displacements $B(r + A(r)) = B(r)$. So, correct up to the first order of terms the resultant displacement of the particle is

$$A(r) + B(r). \tag{8.36.1}$$

The order of these infinitesimal displacements can be changed, because

$$A(r) + B(r) = B(r) + A(r).$$

This is for a sequence of displacements, as if they are applied separately to the body in its origin position.

The general infinitesimal displacements δl of any particle of a body, should depend on the position vector $\mathbf{r}$ of the particle at P relative to a base point O, the translational displacement $\delta\mathbf{S}$ of 0 and rotation $\delta\mathbf{n}$ of the body. Hence

$$\delta\mathbf{l} = \delta\mathbf{S} + (\delta\mathbf{n} \times \mathbf{r}). \tag{8.36.2}$$

Changing the base point O to O', let $\delta s'$ be the translational displacement, and $\delta n'$ be the rotation with respect to O'. If $OO' = a$, the displacement $\delta s'$ of O' is given by

$$\delta \mathbf{S}' = \delta S + (\delta n \times a). \tag{8.36.3}$$

The displacement of the particle at P has now the equivalent expressions, according as the base point is O or O'. Equating them we get

$$\boldsymbol{\delta} + (\delta \mathbf{n} \times (\mathbf{r}' + \mathbf{a})) = \delta \mathbf{s}' + (\delta \mathbf{n}' \times \mathbf{r}'), \tag{8.36.4}$$

where $\mathbf{r}'(= \mathbf{O}'P)$ is the position vector of P relative to O'. Combining (8.36.4) with (8.36.3) we find that

$$(\delta \mathbf{n} \times \mathbf{r}') = (\delta \mathbf{n}' \times \mathbf{r}'). \tag{8.36.5}$$

Since this relation is valid for all values of r', we write

$$\delta \mathbf{n} = \delta \mathbf{n}'. \tag{8.36.6}$$

So for infinitesimal, rotation is independent of the base point. Combination of translation and rotation is a screw displacement which consists of a translation and rotation about an axis parallel to the direction of translation δs and rotation, $\delta \mathbf{n}$. Then changing the base point to O', where $OO' = r$ we can write

$$\delta \mathbf{s}' = \delta \mathbf{s} + (\delta \mathbf{n} \times \mathbf{r}). \tag{8.36.7}$$

If $\delta \mathbf{s}' = p\delta \mathbf{n}'$, where p is the pitch of the screw which is a scalar quantity, then using (8.36.6) we find that

$$p\delta \mathbf{n} = \delta \mathbf{s} + (\delta \mathbf{n} \times \mathbf{r}) \tag{8.36.8}$$

and the displacement is reduced to that of a screw type. The scalar product of δs with δn gives the expression for p,

$$p = \frac{(\delta \mathbf{n} \cdot \delta \mathbf{s})}{(\delta n)^2}. \tag{8.36.9}$$

The point O' is anywhere in a line which is the axis of the screw because then (8.36.8) remains unchanged.

8.37 Co-ordinate Free Rotation

Co-ordinate free rotation formalism requires no particular set of axes for use. Vectors determine these rotations in terms of the unit vector along the axis of rotation and the angle of rotation, Mathews (1976).

Consider an arbitrary vector $\mathbf{v}$ which is to undergo a rotation through an angle θ about an axis along which the unit vector is $\mathbf{i}$. Let $\mathbf{v}'$ be the replacement of $\mathbf{v}$ after the rotation, fig. 8.37.1.

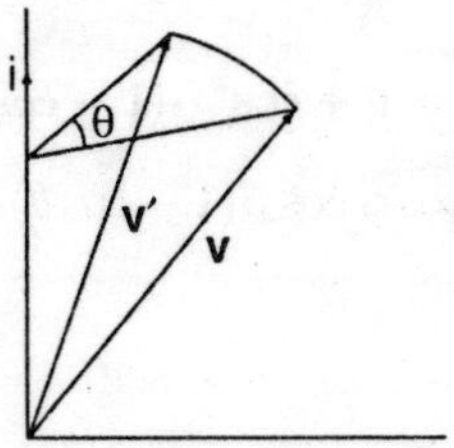

Fig. 8.37.1 Rotation of the vector **v** about the axis i, through the angle θ, to produce the new vector **v**′.

The vector $\mathbf{v}$ is decomposed into a component v_{11}, parallel to i, and a component $v_\perp$ perpendicular to i. From Fig. 8.37.1 it is clear that i points upward out of the plane of the figure. Hence

$$v_{11} = i(i \cdot v)$$

$$v_\perp = v - v_{11} = v - i(i \cdot v)$$

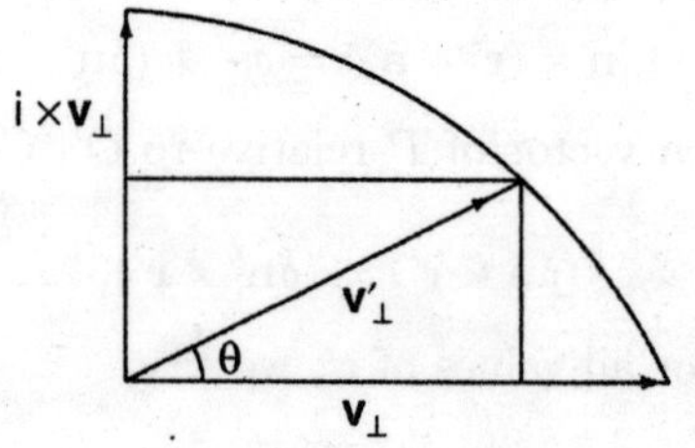

Fig. 8.37.2 View of the plane perpendicular to i, containing $\mathbf{v}_\perp$ and $\mathbf{v}'_\perp$.

Since rotation leaves v_{11} unaffected, we can write $v'_{11} = v_{11}$. But rotation affects $v_\perp$ because, fig. 8.37.2,

$$v'_\perp = v_\perp \cos\theta + i \times v_\perp \sin\theta. \quad (8.37.1)$$

Hence

$$\begin{aligned} v' = v'_{11} + v'_\perp &= v_{11} + v_\perp \cos\theta + (i \times v_\perp)\sin\theta \\ &= (v \cdot i)i + (v - (v \cdot i)i)\cos\theta + (i \times v)\sin\theta \\ &= v_\perp \cos\theta + (v \cdot i)i(1 - \cos\theta) + (i \times v)\sin\theta. \end{aligned}$$

We now assume that a right handed triad of unit vectors e_1, e_2, e_3 is rotated about the axis $\mathbf{i}$ through an angle θ. Then â new triad e'_1, e'_2, e'_3 represents the direction of $\mathbf{v}'$. The direction cosines are $\lambda_{ij} = (e'_i \cdot e_j)$. Then $e'_i = e_i \cos\theta + (e_i \cdot i)i(1 - \cos\theta) + (i \times e_i)\sin\theta$, so that

$$\lambda_{ij} = \delta_{ij}\cos\theta + (i \cdot e_i)(i \cdot e_j)(1 - \cos\theta) + (i \cdot (e_i \times e_j))\sin\theta,$$

where we have used the vector identity $A \cdot (B \times C) = C \cdot (A \times B)$.

Let the angles the rotation axis i make with the original triad e_i, e_2, e_3 be α, β, γ. Then the direction cosines of i are $\cos\alpha, \cos\beta, \cos\gamma$, and we can write

$$\cos\alpha = (i \cdot e_1), \quad \cos\beta = (i \cdot e_2), \quad \cos\gamma = (i \cdot e_3)$$

and find that

$$\begin{aligned} \lambda_{11} &= \cos\theta + \cos^2\alpha(1 - \cos\theta), \\ \lambda_{12} &= \cos\alpha\cos\beta(1 - \cos\theta) + \cos\gamma\sin\theta, \text{ etc.} \end{aligned}$$

Chapter 9

Exterior Differential Forms and Electrodynamics

9.1 Preliminary Remarks

The method of exterior differential forms is a generalization of the vector method, and is helpful in tensor calculus. Its theory has developed in its own manner, and so cannot be identified with vectors or tensors. The various relevant notions have been searched, found out and knitted to yield a pattern which makes vector methods look triffle and drives to other deeper aspects of differential geometry in a very elegant manner. Symmetry principles are not important in classical mechanics, because its mathematical frame work had not been well defined for a long time. The quantitative dynamics, developed by Lagrange, Hamilton, Jacobi and others, does not solve many problems of celestial mechanics. So, new methods and concepts emerged for a qualitative dynamics which makes use of differential forms on manifolds. This theory is useful in the development of electrodynamics in rotating and other accelerated frames.

Both theorems of Stokes and Gauss of vector calculus state that the integral of some types of derivatives of a vector field, over a region of certain dimensions, equals the integral of that field over the boundary of that region. These theorems convert the global equations of electrodynamics of Maxwell in the original integral forms into local differential equations. But vector methods cannot either connect or provide any general result from which the two vector relations.

$$\mathbf{V} \times (\mathbf{V}\phi) = \mathbf{0} \quad \text{and} \quad \mathbf{V} \cdot (\nabla \times A) = \mathbf{0} \tag{9.1.1}$$

follow easily, where ϕ and A are a scalar and a vector field function respectively. But the methods of differential forms unify these theorems and other results. Actually, a more general type of "derivative" operator of theory of exterior differential forms leads to both these equations and establishes an unifying connection between these vector properties. The theorems in vector analysis, including the vector identities and the classical vector analysis results in three dimensions can be derived from this theory. This theory generalizes the properties of transformation of integrands of multiple integrals under co-ordinate transformation. For example, the double integral

$$\iint\limits_R f(x, y)dxdy \tag{9.1.2}$$

transforms for changes to polar co-ordinates (r, θ), with $x = r\cos\theta$ and $y = r\sin\theta$, to the double integral

$$\iint f(r\cos\theta, r\sin\theta) r d\theta dr. \tag{9.1.3}$$

Here the function $f(x, y)$ transforms as a scalar field, and

$$dxdy = J(x, y; r, \theta) dr d\theta, \tag{9.1.4}$$

where the Jacobian of the transformation is

$$J(x, y; r, \theta) = r, \tag{9.1.5}$$

$$dxdy = r dr d\theta \tag{9.1.6}$$

dx and dy are not calculated as functions of dr and $d\theta$ and then multiplied. Actually, the integral of $f(x, y)$ and $dxdy$ must be transformed as such. Hence, an algebra of quantities, like dx and dy, should be constructed, which transform properly under co-ordinate transformations. The algebra is available in the theory of exterior differential forms.

Analytically speaking, differential forms are completely anti-symmetric tensors. Pictorially these are intersecting stacks of surfaces. The mathematical formalism for manipulating differential forms is known as the *exterior calculus*, which is applicable not only to space-time, but also to more general geometrical systems. The concepts of differential forms and exterior derivatives are metric free. Metric appears only in the concept of Hodge duality operation (section 9.2.5), that carries attention away from the Farady F to the dual $^*F$ (section 9.4.5).

Differential forms have simple properties of transformation under mappings or change of co-ordinates. For this reason, they are ideally suited to work on spaces that are defined by means of local co-ordinate systems. Flanders (1963, p. 3) writes "... in many ... situations the use of the exterior calculus, often combined with the method of moving frames of E. Cartan, leads to decisive results in a way which is very difficult with tensors alone. Sometimes, a combination of techniques is in order." Moreover, this method can be used to get rid of superfluous terms which sometimes arise in tensor theory.

The symbols ϕ and ψ generally would represent one-form, as distinct from w whose order is arbitrary. The letters p and q usually denote the order of a form itself of that order. The limits of a summation will be omitted when these are obvious.

9.2 The Hodge Duality and Star Operations

The unique Hodge decomposition theorem is : Every p-form can be decomposed into the sum of the exterior derivatives of a $(p-1)$ form, the co-derivative of a $(p-1)$ form and a harmonic term. This theorem, the relation (9.6.5), is the differential form analogue of the Helmholtz theorem of vector calculus, which shows that a vector field can be uniquely decomposed into the gradient of some scalar field, the curl of some vector field and a vector that satisfies Laplace's equation, the relation (9.12.4).

The Hodge star operator, or dual, associates to every p-form $wa(n-p)$ for $^*w$. This dual is not to be confused with the idea of dual vector space in which sets of vector are related to sets of one forms. But here anti-symmetric tensors of order p are related to anti-symmetric tensors of order $n-p$ in a n-space. So, if $n = 4$, and w is a one-form, then $^*w$ must be a three-form.

In a n-space, the dual of a basis p-form (with $p \leq n$),

$$dx^{i_1} \wedge dx^{i_2} \wedge \cdots \wedge dx^{i_p} \tag{9.2.1}$$

is defined to be the $(n-p)$ form

$$\varepsilon_{i_1 \cdots i_p \, i_{p+1} \cdots i_n} dx^{i_{p+1}} \wedge dx^{i_{p+2}} \wedge \cdots \wedge dx^{i_n}, \tag{9.2.2}$$

where the symbol $\wedge$ is called the *wedge product* or the *exterior product* and the numbers $i_1, \cdots, i_p, i_{p+1}, \cdots, i_n$ are the first n natural numbers, that is, the numbers $i_{p+1}, i_{p+2}, \cdots, i_n$ complete the set $i_1, i_2, \cdots, i_p$. Here $\varepsilon_{i_1 i_2 \cdots i_n}$ is the Levi Civita tensor of nth order. If w is a p-form in the n-space, then

$$^{**}w = (-1)^{p-1} w. \tag{9.2.3}$$

For a two-form, denoted by F, we set $p = 2$ and obtain

$$^{**}F = -F. \tag{9.2.4}$$

In this sense, the star operator has the properties of imaginary numbers, and for two-forms we can write

$$^{**} = -1 \quad \text{or} \quad * = i. \tag{9.2.5}$$

Hence

$$e^{*\alpha} = e^{i\alpha} = \cos\alpha + i \sin\alpha = \cos\alpha + * \sin\alpha. \tag{9.2.6}$$

9.3 The Exterior Algebra of Forms

In an n-dimensional Euclidean space, with rectangular co-ordinates $x^1, x^2, \cdots, x^n$, let a point P be denoted by the vector $\mathbf{x}$, where

$$\mathbf{x} \equiv (x^1, x^2, \cdots, x^n). \tag{9.3.1}$$

At each point $\mathbf{x}$ the basis one forms are dx^i with $i = 1, 2, \cdots, n$. An arbitrary one-form w is a linear combination of basis one forms. So the general basis one forms can be expressed as the sum

$$\omega^j = \sum_{i=1}^{n} \left(a_i^j \, dx^i \right). \tag{9.3.2}$$

The set of one forms, at a point, forms a vector space which is duel to the tangent space at the point. Writing

$$\omega^1 = \sum_i a_i^1 \, dx^i \tag{9.3.3}$$

and

$$W^2 = \sum_i (a_i^2 dx^i), \tag{9.3.4}$$

we have

$$W^1 + W^2 = \sum_i (a_i^1 + a_i^2) dx^i. \tag{9.3.5}$$

A two form F can be defined as the Wedge product

$$F = F_{\mu\nu} dx^\mu \wedge dx^\gamma, \tag{9.3.6}$$

where the quantity $dx^\mu \wedge dx^\gamma$ is called a *bi-vector.* The rules of this product are

$$\text{(i)} \quad dx^i \wedge dx^j = -dx^j \wedge dx^i, \tag{9.3.7}$$

so the exterior product is anti-commutative;

$$\text{(ii)} \quad \left(\sum_i a_i dx^i\right) \wedge dx^j = \sum_i a_i (dx^i \wedge dx^j), \tag{9.3.8}$$

and (iii) this product is distributive. From (i) it follows that for $i = 1$

$$dx^i \wedge dx^i = 0 \tag{9.3.9}$$

and the number of independent two-forms, at a point, is $n(n-1)/2$, or ${}^nC_2$. Two-forms can be obtained from one-forms in the manner shown in the relation (9.5.3).

A one-form, as in the relation (9.3.2), is a number produced from a vector. A two-form, as in the relation (9.3.6), is also a number; but it is produced out of an oriented surface, that is, a surface with a sense of circulation indicated on it.

Two-forms are either open or general. The closed two-forms have the property that as many tubes enter a closed two-form as have emerged from it, so exterior derivative of it is zero and no three-form is derivable from it other than the trivial zero three-form. A general two-form sends across a closed two-surface a non-zero net number of tubes, and therefore permits the definition of a non-trivial three-form as its exterior derivative. A general two-form generalizes the concept of Faraday on tubes of force to 4-spaces of both Euclid and Riemann.

The honey comb structure of tubes in a box is an illustration of a two-form in a 3-space. It yields a number for each choice of smooth element of two-surface slicing through the three-volume. A two-form actually gives the number of tubes sliced through, not the shape of the tubes. Through a specified element of two-surfaces fixed in a three-space, the count of tubes is unchanged. Direction and density of the tubes can be ascertained thus, and not the specific location or orientation for the tube walls. Thus the following representations give one and the same two-form F:

$$F = F_{\mu\nu}(2dx^\mu) \wedge \left(\frac{1}{2}dx^\nu\right) = F_{\mu\nu} d\left(\frac{x^\mu - x^\nu}{\sqrt{2}}\right) \wedge d\left(\frac{x^\mu + x^\nu}{\sqrt{2}}\right)$$

$$= F_{\mu\nu} \frac{(\alpha dx^\mu + \beta dx^\nu)}{(\alpha\delta - \beta\gamma)^{1/2}} \wedge \frac{(\gamma dx^\mu + \delta dx^\nu)}{(\alpha\delta - \beta\gamma)^{1/2}}, \tag{9.3.10}$$

where all the four functions α, β, r, δ depend on the position co-ordinates.

Actually the two-forms like F also count much and any one of the one-forms forming it is not of so much importance always. Hence like lines of force, tubes of force are also not physically negligible concepts for the same structure. The lines of force picture has the advantage of not imposing any specific composition of a two-form into the product of one-forms. However, representation of the two-forms as the exterior product of one-forms is often useful. The conceptually abstract tubes of force can be regarded pictorially as elementary tubes. Any visualization of the walls of these tubes indicates that two-forms can be built out of one-forms, but as indicated in (9.3.10), many combinations of one forms give rise to the same two-forms.

More generally, an arbitrary number of one-forms can be multiplied to get a basis p-form

$$dx^{i_1} \wedge dx^{i_2} \wedge \cdots \wedge dx^{i_p}. \tag{9.3.11}$$

The three-form, called a *tri-vector,* is

$$dx^i \wedge dx^j \wedge dx^k = (dx^i \wedge dx^j) \wedge dx^k = dx^i \wedge (dx^j \wedge dx^k). \tag{9.3.12}$$

In the n dimensional space if $p > n$, then at least one of the dx^i's is duplicated in this product. But, anti-commuting the necessary number of times we obtain in the product a term of the form $dx^i \wedge dx^i$, which is zero. Hence, a p-form is zero for $p > n$, and we need only consider p-forms for which $p \leq n$.

An arbitrary p-form is

$$W = \sum_{i_1 \cdots i_p} a_{i_1 \cdots i_p} dx^{i_1} \wedge dx^{i_2} \wedge \cdots \wedge dx^{i_p}. \tag{9.3.13}$$

Following the argument for ascertaining the number of two-forms, it is evident that the number of independent basis p-forms is

$$n!/p!(n-p)!. \tag{9.3.14}$$

These considerations are valid at a point of the underlying n-space x, and can also be extended by defining forms valid at each point such that they are smoothly connected. Thus a field of one-forms is given by

$$\omega(x) = \sum_i a_i(x) dx^i, \tag{9.3.15}$$

where $a_i(x)$ is a set of differentiable functions rather than a set of numbers. In this manner, a one-form at any given point x_0 in the n-space is defined by

$$W(x_0) = \sum_i a_i(x_0) dx^i. \tag{9.3.16}$$

This is generalized to a field of p-forms in the obvious way, then an arbitrary p-form is given by

$$W(x) = \sum_{i_i \cdots i_p} a_{i_1 \cdots i_p}(x) dx^{i_1} \wedge dx^{i_2} \wedge \cdots \wedge dx^{i}_{p}. \tag{9.3.17}$$

A zero form is a scalar field. It W is a p-form and η is a q-form, then each of the q one-forms of η must be commuted through each of the p one-forms of W, thus introducing a factor $(-1)^{pq}$ in the Wedge product. Hence

$$W \wedge \eta = (-1)^{pq} \eta \wedge \omega. \tag{9.3.18}$$

For arbitrary forms the corresponding result follows by the distribution law. If both ω and η are one-forms, we put $p = 1, q = 1$ and the result of relation (9.3.7) follows.

9.4 Exterior Algebra in Three-dimensions

In a three-dimensional space of position co-ordinates, the basis one-forms at a point are

$$dx^1, dx^2, dx^3. \tag{9.4.1}$$

Therefore, basis two-forms are

$$dx^1 \wedge dx^2, \quad dx^1 \wedge dx^3, \quad dx^2 \wedge dx^3 \tag{9.4.2}$$

and the only one basis three-form is

$$dx^1 \wedge dx^2 \wedge dx^3 . \tag{9.4.3}$$

Thus if ϕ, W and η are the arbitrary one-form, two-form and three-form respectively., in a 3-space, then

$$\phi = a_1 dx^1 + a_2 dx^2 + a_3 dx^3 = \sum_i a_i dx^i , \tag{9.4.4}$$

$$\omega = a_{12} dx^1 \wedge dx^2 + a_{13} dx^1 \wedge dx^3 + a_{23} dx^2 \wedge dx^3 , \tag{9.4.5}$$

$$\eta = a dx^1 \wedge dx^2 \wedge dx^3 . \tag{9.4.6}$$

The Hodge star operation (section 9.2) gives

$$\left.\begin{aligned} *dx^1 &= dx^2 \wedge dx^3 \\ *dx^2 &= dx^3 \wedge dx^1 \\ *dx^3 &= dx^1 \wedge dx^2 \end{aligned}\right\} \tag{9.4.7}$$

$$\left.\begin{aligned} *(dx^1 \wedge dx^2) &= dx^3 \\ *(dx^2 \wedge dx^3) &= dx^1 \\ *(dx^3 \wedge dx^1) &= dx^2 \end{aligned}\right\} \tag{9.4.8}$$

$$*f = f dx^1 \wedge dx^2 \wedge dx^3 , \tag{9.4.9}$$

$$*(dx^1 \wedge dx^2 \wedge dx^3) = 1, \tag{9.4.10}$$

where f is a scalar field function. Obviously, therefore,

$$d^* f = 0. \tag{9.4.11}$$

The duals of (9.4.4)–(9.4.6) are the sums of the duals of their component elements; so the dual is a linear operator. Hence

$$*(*\phi) = \phi, \quad *(*\omega) = \omega, \quad *(*\eta) = \eta. \tag{9.4.12}$$

For $n = 3$, equation (9.5.4) gives

$$d\phi = \left(\frac{\partial a_2}{\partial x^1} - \frac{\partial a_1}{\partial x^2} \right) dx^1 \wedge dx^2 + \left(\frac{\partial a_3}{\partial x^1} - \frac{\partial a_1}{\partial x^3} \right) dx^1 \wedge dx^3 + \left(\frac{\partial a_3}{\partial x^2} - \frac{\partial a_2}{\partial x^3} \right) dx^2 \wedge dx^3 . \tag{9.4.13}$$

This relation is reminiscent of the curl operator. Similarly, for $n = 3$, relation (9.5.2) gives

$$df = ((\nabla f) \cdot dx), \tag{9.4.14}$$

which is reminiscent of the gradient operation. A more general relation of the type (9.4.4) is

$$\phi_i = \sum_{j=1}^{3} a_{ij} dx^j . \tag{9.4.15}$$

Then from (9.4.9) it is evident that

$$\phi_1 \wedge \phi_2 \wedge \phi_3 = \begin{vmatrix} a_{11} & a_{12} & a_{13} \\ a_{21} & a_{22} & a_{23} \\ a_{31} & a_{32} & a_{33} \end{vmatrix} dx^1 \wedge dx^2 \wedge dx^3 \tag{9.4.16}$$

$$= * \begin{vmatrix} a_{11} & a_{12} & a_{13} \\ a_{21} & a_{22} & a_{23} \\ a_{31} & a_{32} & a_{33} \end{vmatrix}$$

and equation (9.5.11), for the transformation properties of triple-integrals, gives

$$\mathbf{dx}^1 \wedge \mathbf{dx}^2 \wedge \mathbf{dx}^3 = J(x, x)\mathbf{dx}^1 \wedge \mathbf{dx}^2 \wedge \mathbf{dx}^3, \tag{9.4.17}$$

where J is the Jacobian of the transformation. Similarly, in terms of differential forms, the double integral of the expression (9.1.2) should now read

$$\iint f(x, y)\mathbf{dx} \wedge \mathbf{dy}. \tag{9.4.18}$$

Its transformation to polar co-ordinates is

$$\iint f(x, y)\mathbf{dx} \wedge \mathbf{dy} = \iint f(r\cos\theta, r\sin\theta)J(x, y)\mathbf{dr} \wedge \boldsymbol{d\theta}. \tag{9.4.19}$$

Thus only with the help of the algebra of forms can one treat $\mathbf{dx}$ and $\mathbf{dy}$ as algebraic objects.

9.5 Exterior Derivatives

For the p-form w, defined in (9.3.17), the exterior derivative

$$dw = \sum_{j, i_1, \cdots, i_p} \frac{\partial a_{i_1 \cdots i_p}}{\partial x^j} dx^j \wedge dx^{i_1} \wedge \cdots \wedge dx^{i_p} \tag{9.5.1}$$

is a $(p+1)$ form. Hence $dw = 0$ if $p = n$. If $p = 0$, then w is the scalar f, and the gradient operator type relation

$$dw = df = \sum_i \frac{\partial f}{\partial x^i} dx^i \tag{9.5.2}$$

is obtained. If $p = 1$, then $\mathbf{w}$ is a one-form, given by (9.3.15) or (9.4.4), so

$$\mathbf{dw} = d\phi = \sum \frac{\partial a_j}{\partial x^j} \mathbf{dx}^j \wedge \mathbf{dx}^i \tag{9.5.3}$$

$$d\phi = \sum_{j<i} \left(\frac{\partial a_i}{\partial x^j} - \frac{\partial a_j}{\partial x^i} \right) \mathbf{dx}^j \wedge \mathbf{dx}^i. \tag{9.5.4}$$

The following Lemma are useful (for their proofs see Flanders (1963) and Deschamps (1970)) :

(i) For arbitrary p-forms and q-forms

$$d(p \wedge q) = dp \wedge q + (-1)^p p \wedge dq. \tag{9.5.5}$$

Particularly, if f and g are scalar functions, then

$$(i_1)\ \ d(f) = gdf + fdg. \tag{9.5.6}$$

$$(i_2)\ \ d(f\phi) = df \times \phi + fd\phi. \tag{9.5.7}$$

$$(i_3)\ \ d(\phi \wedge \psi) = d\phi \wedge \psi - \phi \wedge d\psi. \tag{9.5.8}$$

(ii) If α and β are both p forms, then

$$\alpha \wedge {}^*\beta = \beta \wedge {}^*\alpha. \tag{9.5.9}$$

For transformation from the co-ordinate system k with co-ordinates $\mathbf{x}$ of a point p in n-space to the co-ordinate system $\bar{k}$ with co-ordinates $\bar{\mathbf{x}}(\equiv x^{-1}, x^{-2}, x^{-3}, \cdots, x^{-n})$ for the same point P, we can write

$$\bar{x}^i = f_i(x), \quad i = 1, 2, \cdots, n. \tag{9.5.10}$$

Then

$$d\bar{x}^i = \sum_i \frac{\partial f_i}{\partial x^j} dx^j. \tag{9.5.11}$$

In addition to the notion of exterior derivative dw of w, the notion of the co-derivative δw can be introduced in the following manner :

$$\delta w = (d^* w). \tag{9.5.12}$$

The co-derivative thus maps a p-form into a $(p-1)$ form. The combination of $d\delta + \delta d$ is defined as the generalized Laplace operator. To show that, using scalar operation, we obtain

$$\begin{aligned}(d\delta + \delta d)f &= (d^* d^* + {}^* d^* d)f = *d^* df = *d^* \left(\sum_i \frac{\partial f}{\partial x^i} dx^i\right) \\ &= *d\left(\sum_i e_{ijk} \frac{\partial f}{\partial x^i} dx^j \wedge dx^k\right),\end{aligned} \tag{9.5.13}$$

where e_{ijk} is the third order anti-symmetric Levi Civita unit tensor; so

$$e_{ijk} = 1 \text{ if } (i, j, k) = (1, 2, 3) \tag{9.5.14}$$

under all even permutations of these indices; and $= 0$ if any two or more indices are repeated. Equation (9.5.13) can be written as

$$(d\delta + \delta d)f = *\left(\sum_{i,l} \varepsilon_{ijk} \frac{\partial^2 f}{\partial x^l \partial x^i} dx^l \wedge dx^j \wedge dx^k\right). \tag{9.5.15}$$

For $n = 3$, therefore, we obtain

$$(d\delta + \delta d)f = \sum_{i,l} e_{ijk}e_{ljk}\frac{\partial^2 k}{\partial x^l \partial x^i} = \sum_{i,l} \delta_{il}\frac{\partial^2 f}{\partial x^l \partial x^i}$$

$$= \sum_{i=1}^{3} \frac{\partial^2 f}{\partial x^{i2}}, \tag{9.5.16}$$

where the relation (9.4.11) has been utilized.

9.6 Fundamental Theorems of Differential Forms

9.6.1 Poincare's lemma

If w is an arbitrary p-form, then

$$d^2 w = 0 \tag{9.6.1}$$

because, equation (9.5.1) is recovered with the help of (9.3.13). Further differentiation of (9.5.1) gives

$$d^2\omega = \sum_{k,j,i} \frac{\partial^2}{\partial x^k \partial x^j} a_{i_1} \cdots a_{i_p} d\mathbf{x}^k \wedge d\mathbf{x}^j \wedge d\mathbf{x}^{i_1} \wedge \cdots \wedge d\mathbf{x}^{i_p}$$

$$= \sum_{j,k,i} \frac{\partial^2}{\partial x^j \partial x^k} a_{i_1 \cdots i_p} d\mathbf{x}^j \wedge d\mathbf{x}^k \wedge d\mathbf{x}^{i_1} \wedge \cdots \wedge d\mathbf{x}^{i_p}, \tag{9.6.2}$$

where the dummy indices j and k have been interchanged. With the help of the skew symmetry of $d\mathbf{x}^j \wedge d\mathbf{x}^k$, and the equality of mixed partial derivatives, this relation can be written as

$$d^2 w = -\sum_{j,k,i} \frac{\partial^2 a_i \cdots i_p}{\partial x^k \partial x^j} d\mathbf{x}^k \wedge d\mathbf{x}^j \wedge d\mathbf{x}^{i_1} \wedge \cdots \wedge d\mathbf{x}^{i_p}$$

$$= -d^2 w = 0. \tag{9.6.3}$$

9.6.2 Converse of the Poincare lemma

If $d\lambda = 0$, where λ is a p-form, then subject to certain restrictions on the topological nature of the domain of validity of this lemma, not discussed here, then, there exists a $(p-1)$ form w such that

$$\lambda = dw. \tag{9.6.4}$$

9.6.3 The Hodge decomposition theorem

Any p-form, section 9.2, can be written as the sum

$$\omega = d\alpha + \delta\beta + \gamma, \tag{9.6.5}$$

where α is a $(p-1)$ form, β is a $(p+1)$ form and γ is a harmonic p-form, satisfying Laplace's equation

$$\nabla^2\gamma = 0. \tag{9.6.6}$$

9.6.4 Generalized Stoke's theorem

If ω is a p-form; D is a $(p+1)$ dimensional domain and ∂D is the closed p dimensional boundary of D, the integral of w over the boundary of the region D is equal to the integral of the exterior derivative over the region D, then

$$\int_{\partial D} \omega = \int_{D} dw. \tag{9.6.7}$$

In 3-space, when ω is, respectively, a one-and two-forms, this theorem encompasses the duo of the theorems of Stokes and Gauss. Moreover, when w is a zero-form, this theorem gives the fundamental theorem of calculus.

9.7 Exterior Algebra in Pseudo-Cartesian 4-space

In the pseudo-cartesian (also called the *pseudo-Euclidean*) Minkowskian 4-space (also called the space of 3 + 1 dimensions to distinguish 3-space of position space and 1-space of time) time is the 4th dimension, so the metric has the diagonal elements $(1, 1, 1, -1)$. So, extension is required of the concept of duality to this space; and of the rules of association between vectors and one-forms to an association between higher order tensors and forms.

In 4-space the dual of the one-form (9.3.2) is defined as the 3-form

$$*w = e_{\mu\alpha\beta\gamma} A^{\mu} \mathbf{dx}^{\alpha} \wedge \mathbf{dx}^{\beta} \wedge \mathbf{dx}^{\gamma} \tag{9.7.1}$$

and the two-form (9.3.6) has the dual

$$*F = \frac{1}{2} e_{\alpha\beta\mu\gamma} F^{\mu\gamma} dx^{\alpha} \wedge dx^{\beta}, \tag{9.7.2}$$

which is also a two-form, where $e_{\alpha\beta\mu\gamma}$ is the Levi Civita symbol of fourth order.

9.8 Vector Results from Differential Forms

To every ordinary vector A in 3-space of position co-ordinates, a one-form can be associated. Also to every operation on vectors an operation on forms can be associated. Such a covariant vector A can be written as

$$A = A_1 \hat{\mathbf{x}}_1 + A_2 \hat{\mathbf{x}}_2 + A_3 \hat{\mathbf{x}}_3, \tag{9.8.1}$$

where $\hat{x}_1, \hat{x}_2, \hat{x}_3$ are the unit vectors parallel to the co-ordinate axes Ox_1, Ox_2, Ox_3 respectively. The corresponding one-form is

$$\phi_A = A_1 dx^1 + A_2 dx^2 + A_3 dx^3. \tag{9.8.2}$$

The association of the one-form ϕ_A with A is considered under a co-ordinate transformation. With the function f is thus associated the one-form f.

9.9 Correspondences between Some Vector Relations and Differential Forms

Consider the vector formula

$$(A \times B) = (A_2B_3 - A_3B_2)\hat{\mathbf{x}}_1 + (A_3B_1 - A_1B_3)\hat{\mathbf{x}}_2 + (A_1B_2 - A_2B_1)\hat{\mathbf{x}}_3. \qquad (9.9.1)$$

The appropriate operations on ϕ_A and ϕ_B should yield $\phi_{A\times B}$. For this we note that

$$\phi_A \wedge \phi_B = (A_1 d\mathbf{x}^1 + A_2 d\mathbf{x}^2 + A_3 d\mathbf{x}^3) \wedge (B_1 d\mathbf{x}^1 + B_2 d\mathbf{x}^2 + B_3 d\mathbf{x}^3). \qquad (9.9.2)$$

But this is a two-form. Since the coefficients of the basis forms give (9.9.1), we take the dual of $\phi_A \wedge \phi_B$ and obtain

$$\begin{aligned} *(\phi_A \wedge \phi_B) &= (A_2B_3 - A_3B_2)dx^1 + (A_3B_1 - A_1B_3)dx^2 \\ &\quad + (A_1B_2 - A_2B_1)dx^3. \end{aligned} \qquad (9.9.3)$$

Since the right hand side of (9.9.3) is a one-form, and since the coefficients are appropriate for the formula (9.9.1), we can associate (9.9.3) with the vector operation of (9.9.1) as

$$*(\phi_A \wedge \phi_B) = \phi_C. \qquad (9.9.4)$$

Consider the vector formula for the scalar product,

$$(A\cdot B) = A_1B_1 + A_2B_2 + A_3B_3. \qquad (9.9.5)$$

Since $(A\cdot B)$ is a scalar, it must be a zero form; also it is an algebraic operation without the involvement of derivatives. Since both ϕ_A and ϕ_B are one-forms, the operation $*(\phi_A \wedge \phi_B)$ or $*(\phi_B \wedge \phi_A)$ is a zero-form, and would be appropriate. To see that expand $*(\phi_A \wedge \phi_B)$ and obtain

$$\begin{aligned} *(\phi_A \wedge *\phi_B) &= *[(A_1 d\mathbf{x}^1 + A_2 d\mathbf{x}^2 + A_3 d\mathbf{x}^3) \wedge *(B_1 d\mathbf{x}^1 + B_2 d\mathbf{x}^2 + B_3 d\mathbf{x}^3)] \\ &= *[(A_1 dx^1 + A_2 dx^2 + A_3 dx^3) \wedge (B_1 dx^2 \wedge dx^3 + B_2 dx^3 \wedge dx^1 + B_3 dx^1 \wedge dx^2)] \\ &= *[(A_1B_1 + A_2B_2 + A_3B_3) dx^1 \wedge dx^2 \wedge dx^3] \\ &= (A_1B_1 + A_2B_2 + A_3B_3) \\ &= (A\cdot B) = (B\cdot A) = *(\phi_B \wedge \phi_A). \end{aligned} \qquad (9.9.6)$$

Consider next the vector formula

$$(\nabla \times B) = \left(\frac{\partial B_3}{\partial x^2} - \frac{\partial B_2}{\partial x^3}\right)\hat{x}_1 + \left(\frac{\partial B_1}{\partial x^3} - \frac{\partial B_3}{\partial x^1}\right)\hat{x}_2 + \left(\frac{\partial B_2}{\partial x^1} - \frac{\partial B_1}{\partial x^2}\right)\hat{x}_3. \qquad (9.9.7)$$

Since $d\phi_B$ is a two-form, its dual is a one-form. So, for establishing this formula we consider

$$
\begin{aligned}
*(d\phi_B) &= *\left[d\left(B_1 dx^1 + B_2 dx^2 + B_3 dx^3\right)\right] \\
&= *\left[\left\{\frac{\partial B_1}{\partial x^2}(dx^2 \wedge dx^1) + \frac{\partial B_1}{\partial x^3}(dx^3 \wedge dx^1)\right\} + \left\{\frac{\partial B_2}{\partial x^1}(dx^1 \wedge dx^2) + \frac{\partial B_2}{\partial x^3}(dx^3 \wedge dx^2)\right\}\right. \\
&\quad \left. + \left\{\frac{\partial B_3}{\partial x^1}(dx^1 \wedge dx^3) + \frac{\partial B_3}{\partial x^2}(dx^2 \wedge dx^3)\right\}\right] \\
&= \left(\frac{\partial B_3}{\partial x^2} - \frac{\partial B_2}{\partial x^3}\right) dx^1 + \left(\frac{\partial B_1}{\partial x^3} - \frac{\partial B_3}{\partial x^1}\right) dx^2 + \left(\frac{\partial B_2}{\partial x^1} - \frac{\partial B_1}{\partial x^2}\right) dx^3. \qquad (9.9.8)
\end{aligned}
$$

9.10 Correspondence between Some More Vector Relations and Differential Forms

The tabular form of some of these correspondences is

Vector	Differential form
$A = A_1\hat{x}_1 + A_2\hat{x}_2 + A_3\hat{x}_3$	$\phi_A = A_1 dx^1 + A_2 dx^2 + A_3 dx^3$
f (scalar function)	f (zero form)
$(A \cdot B)$	$*(\phi_A \wedge *\phi_B)$ or $*\ (\phi_B \wedge *\phi_A)$
$(\nabla \cdot B)$	$*[d(*\phi_B)]$
∇f	df
$(A \times B)$	$*(\phi_A \wedge \phi_B)$
$(V \times B)$	$*(d\phi_B)$

Here all the ϕ's are one-forms.

9.11 Vector Identities and Differential Forms

The tabular form of the vector identities and the corresponding statements of differential forms are

Vector identities	Differential form
1. $\boldsymbol{\nabla}(fg) = f\boldsymbol{\nabla}g + g\boldsymbol{\nabla}f$	$d(fg) = fdg + gdf$
2. $\boldsymbol{\nabla}(\mathbf{A}\cdot f) = (\boldsymbol{\nabla}f\cdot\mathbf{A}) + f(\boldsymbol{\nabla}\cdot\mathbf{A})$	$*[d^*(f\phi_A)] = *(df \wedge *\phi_A) + f^*(d^*\phi_A)$
3. $\boldsymbol{\nabla}(\mathbf{A}\times\mathbf{B}) = \mathbf{B}(\boldsymbol{\nabla}\times\mathbf{A}) - \mathbf{A}(\boldsymbol{\nabla}\times\mathbf{B})$	$*[d^{**}(\phi_A \wedge \phi_B)] = *(\phi_B \wedge d\phi) - *(\phi_A \wedge d\phi_B)$
4. $\boldsymbol{\nabla}\times(f\mathbf{A}) = (\boldsymbol{\nabla}f)\times\mathbf{A} + f(\boldsymbol{\nabla}\times\mathbf{A})$	$*[d(f\phi_A)] = *(df \wedge \phi_A) + f * (d\phi_A)$
5. $(\mathbf{A}\cdot(\mathbf{B}\times\mathbf{C})) = (\mathbf{C}\cdot(\mathbf{A}\times\mathbf{B})) = (\mathbf{B}\cdot(\mathbf{C}\times\mathbf{A}))$ $= (\mathbf{B}\cdot(\mathbf{C}\times\mathbf{A}))$	$*(\phi_A \wedge \phi_B \wedge \phi_C) = *(\phi_C \wedge \phi_A \wedge \phi_B)$ $= *(\phi_B \wedge \phi_C \wedge \phi_A)$
6. $(\mathbf{A}\times(\mathbf{B}\times\mathbf{C})) = \mathbf{B}(\mathbf{A}\cdot\mathbf{C}) - \mathbf{C}(\mathbf{A}\cdot\mathbf{B})$	$*[\phi_A \wedge *(\phi_B \wedge \phi_C)]$ $= \phi_B^*[\phi_A \wedge^* \phi_C] - \phi_C^*[\phi_A \wedge \phi_C^*]$
7. $(\mathbf{A}\times\mathbf{B})\cdot(\mathbf{C}\times\mathbf{D})$ $= (\mathbf{A}\cdot\mathbf{C})(\mathbf{B}\cdot\mathbf{D}) - (\mathbf{A}\cdot\mathbf{D})(\mathbf{B}\cdot\mathbf{C})$	$*[*(\phi_A \wedge \phi_B) \wedge (\phi_C \wedge \phi_D)$ $= *(\phi_A \wedge \phi_C) * (\phi_B \wedge^* \phi_D)$ $- * (\phi_A \wedge^* \phi_D) * (\phi_B \wedge^* \phi_C)$
8. $\boldsymbol{\nabla}\times(\boldsymbol{\nabla}\phi) = 0$	$d^2 f = 0$
9. $\boldsymbol{\nabla}\cdot(\boldsymbol{\nabla}\times\mathbf{A}) = 0$	$*(d^2\phi_A) = 0$

The first four of these identities are all implied in the relations (9.5.5)–(9.5.8). The fifth identity follows from the relation (9.3.18). For proofs of differential forms corresponding to the sixth identity, we consider the relation

$$*(\phi_B \wedge \phi_C) = *\left(\sum_{i,j} B_i C_j dx^i \wedge dx^j\right) = \sum_{i,j,k} B_i C_j \varepsilon_{ijk} dx^k. \tag{9.11.1}$$

Consequently, we find that

$$\begin{aligned} *[\phi_A \wedge^* (\phi_B \wedge \phi_C)] &= *[(\Sigma A_i dx^i) \wedge \Sigma(B_i C_j e_{ijk} dx^k) \\ &= *\Sigma A_i B_i C_j dx^i \wedge dx^k e_{ijk} + *\Sigma A_j B_i C_j dx^j \wedge dx^k e_{ijk} \\ &= \Sigma e_{ijk} e_{ikj} A_i B_i C_j dx^j + \Sigma e_{ijk} e_{ijk} A_j B_i C_j dx^i \\ &= -\phi_C^*(\phi_A \wedge^* \phi_B) + \phi_B^*(\phi_A \wedge^* \phi_C), \end{aligned} \tag{9.11.2}$$

because it is easy to see that

$$e_{ijk} e_{ikj} = -1 \quad \text{and} \quad e_{ijk} e_{ijk} = 1, \tag{9.11.3}$$

where i, j, k are different, and otherwise zero.

For the proof of the differential form result corresponding to the seventh identity, we consider the result

$$*(\phi_A \wedge \phi_B) \wedge (\phi_C \wedge \phi_D) = (\Sigma A_i B_j e_{ijk} dx^k) \wedge (\Sigma C_p D_q dx^p \wedge dx^q). \tag{9.11.4}$$

The right hand side of this relation, on expansion, gives

$$\begin{aligned} &[(A_2B_3 - A_3B_2)dx^1 + (A_3B_1 - A_1B_3)dx^3 + (A_1B_2 - A_2B_1)dx^3] \\ &[(C_1D_2 - C_2D_1)dx^1 \wedge dx^2 + (C_2D_3 - C_3D_2)dx^2 \wedge dx^3 + (C_3D_1 - C_1D_3)dx^3 \wedge dx^1] \\ = &[(A_2B_3 - A_3B_1)(C_2D_3 - C_3D_2) + (A_3B_1 - A_1B_3)(C_3D_1 - C_1D_3) \\ &+(A_1B_2 - A_2B_1)(C_1D_2 - C_2D_1)]\mathbf{dx}^1 \wedge \mathbf{dx}^2 \wedge \mathbf{dx}^3. \end{aligned} \tag{9.11.5}$$

In the general form the right hand side of (9.11.4)

$$= \Sigma(A_i B_j C_p D_q e_{ijk} dx^k \wedge dx^p \wedge dx^q) + \Sigma(A_i B_j C_q D_p e_{ijk} dx^k \wedge dx^q \wedge dx^p). \tag{9.11.6}$$

Hence a star operator operating on both sides of (9.11.4), with the help of (9.11.6), gives

$$\begin{aligned} *[*(\phi_A \wedge \phi_B) \wedge (\phi_C \wedge \phi_D)] &= \Sigma(A_i B_j C_p D_q e_{ijk} e_{ipq}) + \Sigma(A_i B_j C_q D_p e_{ijk} e_{ipq}) \\ &= *(\phi_A \wedge^* \phi_C) * (\phi_B \wedge^* \phi_D) - {}^*(\phi_A \wedge {}^*\phi_D) * (\phi_B \wedge^* \phi_C). \end{aligned} \tag{9.11.7}$$

The eighth and ninth identities are obtained by applying the Poincare lemma of section 9.7 to a zero form and to a one-form.

9.12 Vector Theorems

(a) If $(\boldsymbol{\nabla} \times \mathbf{E}) = 0$, then for any scalar function ψ of position co-ordinates, $\mathbf{E} = \nabla\psi$. The corresponding differential form statement is : If ω is a one-form and α is a zero-form, then

$$*(d\mathrm{w}) = d\mathrm{w} = 0 \text{ for } \boldsymbol{\omega} = \boldsymbol{d\alpha}. \tag{9.12.1}$$

This theorem follows from the converse of the Poincare lemma, equation (9.6.4).

(b) If $(\nabla \cdot B) = 0$, then for any vector function A of position co-ordinates,

$$\mathbf{B} = \boldsymbol{\nabla} \times \mathbf{A}.$$

The differential form statement of this vector theorem is : If

$$*(d^*\mathbf{w}) = 0, \tag{9.12.2}$$

then a one-form β exists such that

$$*\mathbf{W} = d\boldsymbol{\beta} \quad \text{or} \quad \mathbf{w} = *(d\boldsymbol{\beta}). \tag{9.12.3}$$

This theorem also follows from the converse, (9.6.4), of Pioncare's lemma.

(c) Any vector A can be written as

$$\mathbf{A} = \boldsymbol{\nabla} f + \boldsymbol{\nabla} \times \mathbf{B} + \mathbf{Q} \tag{9.12.4}$$

with

$$\nabla^2 Q = 0 \tag{9.12.5}$$

f is a scalar function, and **B** and **Q** are vectors. The corresponding differential form statement is : If **w** is a p-form, then the quantities α, β, γ exist such that

$$\omega = d\alpha + \delta\beta + \gamma, \tag{9.12.6}$$

where γ is a harmonic function; so

$$\nabla^2\gamma = 0. \tag{9.12.7}$$

This is the Helmholtz theorem.

(d) Theorem of Gauss :

$$\iiint_{\tau} (\boldsymbol{\nabla} \cdot A) d\tau = \iint_{\Sigma} (\mathbf{A} \cdot d\boldsymbol{\Sigma}). \tag{9.12.8}$$

For this theorem the vector result

$$(\boldsymbol{\nabla} \cdot \mathbf{A}) d\tau = \left(\frac{\partial A_1}{\partial x_1} + \frac{\partial A_2}{\partial x_2} + \frac{\partial A_3}{\partial x_3} \right) dx^1 dx^2 dx^3 \tag{9.12.9}$$

corresponds to the differential form result

$$d(*\phi_A) = \left(\frac{\partial A_1}{\partial x_1} + \frac{\partial A_2}{\partial x_2} + \frac{\partial A_3}{\partial x_3} \right) \mathbf{dx}^1 \wedge \mathbf{dx}^2 \wedge \mathbf{dx}^3, \tag{9.12.10}$$

and the vector result

$$(\mathbf{A} \cdot d\boldsymbol{\Sigma}) = V_1 \mathbf{dx}^2 \mathbf{dx}^3 + V_2 \mathbf{dx}^3 \mathbf{dx}^1 + V_3 \mathbf{dx}^1 \mathbf{dx}^2 \tag{9.12.11}$$

corresponds to the differential form result

$$*\phi_A = A_1 dx^2 \wedge dx^3 + A_2 dx^3 \wedge dx^1 + A_3 dx^1 \wedge dx^2. \tag{9.12.12}$$

If now we choose a two-form $\boldsymbol{w} = *\phi_A$ we get

$$\int_D d(*\phi_A) = \int_{\partial D} *\phi_A. \tag{9.12.13}$$

9.13 Theorem of Stokes

This theorem reads

$$\iint_\Sigma (\nabla \times \mathbf{A}) d\mathbf{\Sigma} = \oint_\Gamma (\mathbf{A} \cdot d\boldsymbol{l}). \tag{9.13.1}$$

Since

$$d\mathbf{\Sigma} = \hat{\mathbf{x}}_1 dx^2 dx^3 + \hat{\mathbf{x}}_2 dx^3 dx^1 + \hat{\mathbf{x}}_3 dx^1 dx^2,$$

we find that

$$((\nabla \times \mathbf{A}) \cdot d\mathbf{\Sigma}) = \left(\frac{\partial A_3}{\partial x^2} - \frac{\partial A_2}{\partial x^3}\right) dx^2 dx^3 + \left(\frac{\partial A_1}{\partial x^3} - \frac{\partial A_3}{\partial x^1}\right) dx^3 dx^1 + \left(\frac{\partial A_2}{\partial x^1} - \frac{\partial A_1}{\partial x^2}\right) dx^1 dx^2. \tag{9.13.2}$$

The corresponding differential form statement is

$$d\phi_A = \left(\frac{\partial A_3}{\partial x^2} - \frac{\partial A_2}{\partial x^3}\right) \mathbf{dx}^2 \wedge \mathbf{dx}^3 + \left(\frac{\partial A_1}{\partial x^3} - \frac{\partial A_3}{\partial x^1}\right) \mathbf{dx}^3 \wedge \mathbf{dx}^1 + \left(\frac{\partial A_2}{\partial x^1} - \frac{\partial A_1}{\partial x^2}\right) \mathbf{dx}^1 \wedge \mathbf{dx}^2. \tag{9.13.3}$$

Also, corresponding to $(\mathbf{A} \cdot d\boldsymbol{l})$ we have ϕ_A. Then if $\mathbf{w}$ is a one-form

$$W = \phi_A \tag{9.13.4}$$

and

$$\int_D d\phi_A = \int_{\partial D} \phi_A. \tag{9.13.5}$$

9.14 Derivation of the Maxwell Equations

The 4-current j_μ is associated with the one-form

$$j_\mu dx^\mu \tag{9.14.1}$$

and the covariant, skew-symmetric EM field tensor $F_{\mu\gamma}$ is associated with the two-form

$$\mathbf{F} = \frac{1}{2} F_{\mu\gamma} \mathbf{dx}^\mu \wedge \mathbf{dx}^\gamma, \tag{9.14.2}$$

where F is called the *Faraday*, the dual $*\mathbf{F}$ of $\mathbf{F}$ is called the *Maxwell*. Evidently

$$F = -\ C(E_x dl \wedge \mathbf{dx} + E_y dt \wedge \mathbf{dy} + E_z \mathbf{dt} \wedge \mathbf{dz}) + B_x \mathbf{dy} \wedge \mathbf{dz} + B_y \mathbf{dz} \wedge \mathbf{dx} + B_z \mathbf{dx} \wedge \mathbf{dy}. \tag{9.14.3}$$

The energy-momentum one-form is of the type

$$\alpha = \alpha_\mu W^\mu \tag{9.14.4}$$

$$\text{or,}\quad p = p_x dx + p_y dy + p_z dz - Edt. \tag{9.14.5}$$

In a n-space, the dual of a basis p-form (with $p \leq n$),

$$\mathbf{dx}^{i_1} \wedge \mathbf{dx}^{i_2} \wedge \cdots \wedge \mathbf{dx}^{i_p}$$

is defined to be the $(n-p)$ form

$$e_{i_1 i_2 \cdots i_p, i_{p+1}, \cdots, i_n} \mathbf{dx}^{i_{p+1}} \wedge \mathbf{dx}^{i_{p+2}} \cdots \wedge \mathbf{dx}^{i_n},$$

where $e_{i_1 i_2 \cdots i_n}$ is the Levi Civita tensor of nth order and the set of numbers $i_1, i_2, \cdots, i_p$, $i_{p+1}, \cdots, i_n$ is the first n natural numbers (that is, the number $i_{p+1}, i_{p+2}, \cdots, i_n$ complete the set $i_1, i_2, \cdots, i_n$).

The Maxwell $*F$ which is the dual of the Faraday F is given by

$$*\mathbf{F} = -\ CB_x \mathbf{dx} \wedge \mathbf{dt} - CB_y \mathbf{dy} \wedge \mathbf{dt} - CB_z \mathbf{dz} \wedge \mathbf{dt} + E_x \mathbf{dy} \wedge \mathbf{dz} + E_y \mathbf{dz} \wedge \mathbf{dx} + E_z \mathbf{dx} \wedge \mathbf{dy}. \tag{9.14.6}$$

So

$$**F = -\mathbf{F}. \tag{9.14.7}$$

For the p-form W, denoted by

$$W = \sum_{i_1, \cdots, i_p} a_{i_1} a_{i_2} \cdots a_{i_p} dx^{i_1} \wedge dx^{i_2} \cdots \wedge dx^{i_p},$$

the exterior derivative dW is a $(p+1)$ form given by

$$dW = \sum_{j_1 i_1, \cdots, i_p} \frac{\partial a_{i_1 i_2 \cdots i_p}}{\partial x^j} \mathbf{dx}^j \wedge \mathbf{dx}^{i_1} \cdots \wedge \mathbf{dx}^{i_p}.$$

So, taking the exterior derivative of F of (9.14.3) in the co-ordinate system appropriate for a local Lorentz frame, we obtain

$$\begin{aligned} d\mathbf{F} &= d[-CE_x \mathbf{dt} \wedge \mathbf{dx} - CE_y \mathbf{dt} \wedge \mathbf{dy} - CE_z \mathbf{dt} \wedge \mathbf{dz} \\ &\quad + B_x dy \wedge dz + B_y dz \wedge dx + B_z dx \wedge dy] \\ &= -\left(\frac{\partial E_x}{\partial t} dt + \frac{\partial E_x}{\partial x} dx + \frac{\partial E_x}{\partial y} dy + \frac{\partial E_z}{\partial t} dz\right) \wedge c\mathbf{dt} \wedge dx. \end{aligned} \tag{9.14.8}$$

Arranging all the terms as a linear combination of the four three-forms $dx \wedge dy \wedge dz$, $dt \wedge dy \wedge dz$, $dt \wedge dz \wedge dx$, $dt \wedge dx \wedge dy$, we obtain

$$dF = \left(\frac{\partial B_x}{\partial x} + \frac{\partial B_y}{\partial y} + \frac{\partial B_z}{\partial z}\right) dx \wedge dy \wedge dz$$
$$+ \left(C\frac{\partial E_z}{\partial y} - C\frac{\partial E_y}{\partial z} + \frac{\partial B_x}{\partial t}\right) dt \wedge dy \wedge dz$$
$$+ \left(C\frac{\partial E_x}{\partial y} - C\frac{\partial E_z}{\partial x} + \frac{\partial B_y}{\partial t}\right) dt \wedge dz \wedge dx$$
$$+ \left(C\frac{\partial E_y}{\partial x} - C\frac{\partial E_x}{\partial y} + \frac{\partial B_z}{\partial t}\right) dt \wedge dx \wedge dy. \tag{9.14.9}$$

The homogeneous pair of Maxwell equations

$$\nabla \times E = -\frac{1}{C}\frac{\partial B}{\partial t}, \quad (\nabla \cdot B) = 0 \tag{9.14.10}$$

in tensor notation is

$$E[\mu\nu, \lambda] = 0 \tag{9.14.11}$$

$[\mu\nu, \lambda]$ denotes a skew symmetry in all the indices, and a comma denotes differentiation with respect to the index following it. The corresponding relation, involving differential forms and valid at all points, in the empty 4-space of Minkowski, is

$$dF = 0. \tag{9.14.12}$$

The Faraday F is therefore, a closed two-form, and the tubes of it nowhere come to an end. Moreover, as many tubes of it enter a closed two-surface as emerge from that. This F is always expressible as

$$\mathbf{F} = d\mathbf{A}, \tag{9.14.13}$$

where A a one-form, is the 4-potential A^{μ}.

The inhomogeneous pair of Maxwell equations

$$\boldsymbol{\nabla} \times \mathbf{H} = \frac{1}{C}\frac{\partial D}{\partial t} + \frac{4\pi}{C}j, \quad (\nabla \cdot D) = 4\pi\rho \tag{9.14.14}$$

in tensor notation appears to be

$$\mathbf{H}^{\mu\nu}_{,\nu} = \frac{4\pi}{c} j\mu. \tag{9.14.15}$$

In vacuum

$$B = H, \quad D = E, \tag{9.14.16}$$

$$\mathbf{H}^{\mu\gamma} = F^{\mu\gamma}. \tag{9.14.17}$$

Then (8.14.15) is replaced by

$$F^{\mu\nu}_{,\nu} = \frac{4\pi}{C} j^{\mu}. \tag{9.14.18}$$

In the language of differential forms this relation reads

$$d * F = \frac{4\pi}{C} * j. \tag{9.14.19}$$

The dual $*j$ is the three-form charge in this formalism.

The exterior derivative of the dual $*F$, given in (9.14.6), is

$$d*F = \left(\frac{\partial E_x}{\partial x} + \frac{\partial E_y}{\partial y} + \frac{\partial E_z}{\partial z}\right) dx \wedge dy \wedge dz$$
$$+ \left(\frac{\partial E_x}{\partial t} - C\frac{\partial B_z}{\partial y} + C\frac{\partial B_y}{\partial z}\right) dt \wedge dy \wedge dz$$
$$+ \left(\frac{\partial E_y}{\partial t} - C\frac{\partial B_x}{\partial z} + C\frac{\partial B_z}{\partial x}\right) dt \wedge dz \wedge dx$$
$$+ \left(\frac{\partial E_z}{\partial t} - C\frac{\partial B_y}{\partial x} + C\frac{\partial B_x}{\partial y}\right) dt \wedge dx \wedge dy \tag{9.14.20}$$
$$= 4\pi[(\rho dx \wedge dy \wedge dz) - J_x dt \wedge dy \wedge dz - j_y dt \wedge dz \wedge dx - j_z dt \wedge dx \wedge dy]. \tag{9.14.21}$$

This equation yields the inhomogeneous pair of Maxwell equations (9.14.14) for free charges in vacuum. In vacuum, free from all charges, $j = 0$; then $*F$, like F, is a closed two-form. Neither of the two equations (9.14.12) and (9.14.19) make any reference to the metric $g_{\mu\gamma}$. In dielectrics we introduce the one-forms

$$W_1 = E_1 dx^1 + E_2 dx^2 + E_3 dx^3 \tag{9.14.22}$$
$$W_2 = B_1 dx^2 dx^3 + B_2 dx^3 dx^1 + B_3 dx^1 dx^2 \tag{9.14.23}$$
$$W_3 = H_1 dx^1 + H_2 dx^2 + H_3 dx^3 \tag{9.14.24}$$
$$W_4 = D_1 dx^2 dx^3 + D_2 dx^3 dx^1 + D_3 dx^1 dx^2 \tag{9.14.25}$$
$$W_5 = j_1 dx^2 dx^3 + j_2 dx^3 dx^1 + j_3 dx^1 dx^2. \tag{9.14.26}$$

These involve differentials of position co-ordinates only. Denote the exterior derivatives with respect to space variables by primes and with respect to time t by dot. Thus

$$\frac{\partial}{\partial t} w_1 = \dot{w}_1 = \dot{E}_1 dx^1 + \dot{E}_2 dx^2 + \dot{E}_3 dx^3 \text{ etc.} \tag{9.14.27}$$

Consider now the equations

$$d^1 W_1 = -\frac{1}{C}\dot{W}_2 \tag{9.14.28}$$
$$d^1 W_3 = \frac{4\pi}{C} W_5 + \frac{1}{C}\dot{W}_4 \tag{9.14.29}$$
$$d^1 W_2 = 0 \tag{9.14.30}$$
$$d^1 W_4 = 4\pi\rho dx^1 dx^2 dx^3. \tag{9.14.31}$$

These yield the Maxwell equations (9.14.10) and (9.14.14) in dielectrics. The Poynting flux

$$\mathbf{S} = \frac{C}{4\pi}(\mathbf{E} \times \mathbf{H}) \tag{9.14.32}$$

is obtained from

$$\frac{C}{4\pi}(W_1 \wedge W_3) = S_1 dx^2 \wedge dx^3 + S_2 dx^3 \wedge dx^1 + S_3 dx^1 \wedge dx^2. \tag{9.14.33}$$

The Poynthing theorem

$$\text{div } S + \frac{1}{4\pi}\left[\left(H.\frac{\partial B}{\partial t}\right) + \left(E.\frac{\partial D}{\partial t}\right)\right] = -(E\cdot j) \tag{9.14.34}$$

follows from

$$\begin{aligned} d'(W_1 \wedge W_3) &= d'W_1 \wedge W_3 - W_1 \wedge d'W_3 \\ &= -\frac{1}{C}\dot{W}_2 \wedge W_3 - W_1 \wedge \left(\frac{4\pi}{c}W_5 + \frac{1}{C}\dot{W}_4\right) \\ &= -\frac{1}{C}\dot{W}_2 \wedge W_3 - \frac{4\pi}{C}W_1 \wedge W_5 - \frac{1}{C}W_1 \wedge \dot{W}_4. \end{aligned} \tag{9.14.35}$$

Like the 2-form **F** representing the EM field E, B, (the electric field intensity vector and the magnetic induction respectively) let the 2-form G represent the EM field (H, D) (namely, the magnetic intensity vector and the electric displacement vector), and let the 3-current j be the 4-current density. Then

$$\begin{aligned} G = &-(H_x dt \wedge dx + H_y dt \wedge dy + H_z dt \wedge dz) \\ &-D_x dy \wedge dz - D_y dz \wedge dx - D_z dx \wedge dy. \end{aligned} \tag{9.14.36}$$

If the symbol d is taken to be the exterior derivative, the two set of Maxwell equations can be succinctly written as

$$d\mathbf{F} = 0, \tag{9.14.37}$$

$$d\mathbf{G} = \frac{4\pi}{c}j, \tag{9.14.38}$$

in place of the relations (9.14.28)–(9.14.31). Since $\mathbf{dd} = 0$, it follows from (9.14.37) and (9.14.38) that

$$dj = 0. \tag{9.14.39}$$

This relation ensures the total conservation of electric charge. Conversely, the existence of the relation (9.14.39) does not imply the existence of a field **G** yielding (9.14.38). The stronger argument that all cyclic integrals of j vanish,

$$\oint j = 0, \tag{9.14.40}$$

may be called the *global conservation law* for electric charge in contrast to the local conservation law (9.14.40), Post (1974). The 3-form j is now called *exact*.

Similarly, the local statement (9.14.37) may be taken as the local conservation of flux law. The fact that (9.14.37) is valid energy, everywhere at all time, does not yet imply the existence of a field A (say, the 4-vector potential A^{μ}) such that

$$d\mathbf{A} = \mathbf{F}. \tag{9.14.41}$$

However, if all cyclic integrals of **F** vanish,

$$\oint \mathbf{F} = 0, \tag{9.14.42}$$

then a field **A** exists, satisfying the condition (9.14.41), Post (1974). Then **F** is called *exact*.

The conservation law of flux and electric charge are global conservation laws. So, the existence of **F** and **j** expresses the flux conservation (specifically, global conservation) as inferred from Faraday's law of EM induction and electric charge conservation. The fields **A** and **G** exist by virtue of the special physical properties attributed to the fields **F** and **j** if the manifold over which **F** is defined as compact and simply connected, Post (1974).

Chapter 10

Physics of Rotating Bodies for Electrodynamics

10.1 Introductory Remarks

We have derived in this chapter the Maxwell equations for Newtonian rotation, including the associated Schiff charge effect, which is numerically negligible, without discussion on the involved basic problems of electrodynamics of rotation. For small rotational velocities, the Newtonian rules of transformation of the Maxwell equations, depending on absolute time, are sufficient. The MHD equations for Newtonian rotation, the system of equations for rotation of fluids of negative and positive charges and the equations for one fluid average of these fluids have been discussed.

The physical clarity of the EM effects is helpfull in the understanding of field aspects of rotation. This clarification is not possible of rotation by the traditional treatments in mechanics. Even the modern field theory of classical mechanics, Goldstein (1980), does not consider these aspects of rotation and other accelerations. The field aspects of rotation can be covered under the heading *A story of rotation, when treated as a field.* For this purpose analogies of the forces of rotation should be determined with the help of classical forces of electric and magnetic fields, acting on charged particles. This new analysis and interpretation of rotation is complementary to the broader topic of relations between field theory and mechanics of particles and media. Such indirect means of study are commendable for understanding the sophisticated aspects of physics of rotation. Since other accelerations in the equations of motion can also be expressed in terms of the 4-potential, it seems that such applications of analogy with EM forces is possible for these as well.

Some of the field aspect of rotation are only qualitatively understood in this manner; some are derived quantitatively from analogy with phenomenologically developed electrodynamics, and some are sophisticated aspects which must be difficult to verify, because of their smallness.

Between the two types of quantities for analogy an isomorphism is found to exist between the force vectors, solenoidal force vectors, vector potentials, source density, current density, moment, capacity of space and permeability of space, Forward (1961).

The concept of field of angular velocity Ω of rotation is that a rotating observer can assign this Ω to every point of position space. When Ω varies slowly with time this observer can measure the Coriolis, centrifugal and other Ω dependent forces, and see the field of changing Ω. When Ω is uniform this field is uniform. The inertial field of Ω, therefore, is just as real in the rotating frame as is the EM field in the Newtonian frame, since both are defined similarly. A charge q, in a Newtonian frame, behaves like a particle of mass m in a rotating frame. The space-time variation of EM field is known from *the experimentally based fundamental laws of Coulomb, Faraday and others.* But the laws guiding the Ω field are not known.

One consequence of the analogy between EM forces and forces of rotation is the Faraday law of induction for rotation, which can be developed both for particles and fluids. For particles this law states that the rate of work done by a particle of unit mass, in moving along a closed circuit, due to the force of rotation, equals the rate of loss of flux of 2Ω, through the surface enclosed by the circuit. For a fluid flowing in a tube, in a rotating frame, this law is : The change in the flux of 2Ω through a fluid current, equals the total impulse, per unit mass, given to the fluid in the tube. The field concepts are inherent and highly developed in the EM theory, but not in the physics of rotation. For instance, sophisticated rules of transformation, for the field of rotation, do not exist like those of Lorentz transformation (LT) for uniform translation, or like those of the EM fields, which follow as a transformation of second order antisymmetric field tensors, from the LT.

The analogy between rotation and the EM effects shows that :

(1) A vector potential can be defined for the Coriolis force;

(2) Problems can be defined for solution by this analogy;

(3) In electromagnetism an operative way of determining B is by measuring the torque on a closed current loop, or the precession of a rotating charged body. In a rotating reference system, to measure Ω we take a top and measure its precession velocity with no applied torque, or the torque necessary to keep its axis fixed.

Investigation is not complete on the why and how of rotation, including its sophisticated aspects. However, Galialean invariance (absoluteness of time) is such an effective theory that it has not yet given rise to any experimental discrepancy for rotation, though discrepancies exist for uniform translation, and these gave rise to the special theory of relativity (STR). So, works on these aspects do not have any possibility of experimental verification, or application in recent times.

The sophisticated EM equivalents of forces of acceleration and their induced fields may become significant in celestial mechanics, owing to the availability of large masses for their effective manifestation. Laboratory experiments cannot detect these weak effects.

10.2 Field of Rotation

Effects of rotation, when treated as a field, are :

(1) A drift due to rotation, independent of the mass and charge of particles, exists like the $E \times B$ drift in EM field studies. For a constant angular velocity Ω, this drift is

$$-\frac{1}{2}(\Omega \times r), \tag{10.2.1}$$

where $r(x, y, z)$ is the position vector of a point. Another drift, which depends on the angular acceleration, and is perpendicular to Ω and to the drift of (10.2.1) is

$$-\frac{1}{2\Omega^2}(\dot{\Omega} \times (\Omega \times r)). \tag{10.2.2}$$

(2) Alfven wave-like transverse waves, due to rotation, propagate parallel to Ω with the phase velocity

$$C_A = \frac{m\Omega}{q(\pi\rho^m)^{1/2}} = \left(\frac{2mc^2}{q}\right)^{1/2}, \tag{10.2.3}$$

which, for an electron, is about 60 cm/sec, where m is the particle mass, q is the charge and ρ^m is the mass density.

(3) Like the division of the electromagnetic energy into an electric energy and a magnetic energy, the rotation energy separates into the following two parts :

$$\frac{m^2}{8\pi q^2}(\Omega^4 + \dot{\Omega}^2)(x^2 + y^2) \tag{10.2.4}$$

$$\frac{4m^2}{8\pi q^2}c^2\Omega^2, \tag{10.2.5}$$

where Ω is parallel to the z-axis. The equivalent Poynting flux is

$$S = \frac{m^2c^2}{2\pi q^2}[\Omega^2(\Omega \times r) \times \{(\dot{\Omega} \times r) \times \Omega\}]. \tag{10.2.6}$$

The Poynting theorem disintegrates into two parts. One part is that the sum of the divergence of S and the time rate of variation of the magnetic energy density—like quantity of (10.2.5) is zero. The other part is that the sum of the rate of work done by the forces of rotation, and the time derivative of the electric energy density like quantity of (10.2.4) vanishes. Like the Maxwell stress tensor of EM theory, the stress tensor elements of rotation, and the Lagrangian of rotation when treated as a field, also exist.

10.3 Equations of Motion of a Rotating MHD Material

The rotating frame identified with the rotation at the constant rate Ω about the z axis is the K frame whose origin is the center of gravity of the material body. The rest frame at the same origin O is the K^0 frame which is identified with the K frame for $\Omega = 0$.

The rotation velocity equals $r\Omega$ at the point P, whose position vector is r. It is much less than c, the phase velocity of light at any point. So, $r\Omega \ll c$. Then the fluid element at P has the velocity components u_x, u_y, u_z in the K frame, and u_x^0, u_y^0, u_z^0 in the K^0 frame, and

$$U_x = U_x^0 + \Omega y, \quad U_y = U_y^0 - \Omega x, \quad U_z = U_z^0, \tag{10.3.1}$$

$$\text{or,} \quad \mathbf{u} = \mathbf{u}^0 - (\Omega \times \mathbf{r}). \tag{10.3.2}$$

Differentiation of (10.3.1) gives

$$\left(\frac{du_x^0}{dt}\right)_0 = \frac{du_x}{dt} - 2\Omega u_y - \Omega^2 x, \tag{10.3.3}$$

$$\left(\frac{du_y^0}{dt}\right)_0 = \frac{du_y}{dt} + 2\Omega u_x - \Omega^2 y, \tag{10.3.4}$$

$$\left(\frac{du_z^0}{dt}\right)_0 = \frac{du_z}{dt}. \tag{10.3.5}$$

These relations are equivalent to the single vector equation

$$\left(\frac{d\mathbf{u}^0}{dt}\right)_0 = \frac{d\mathbf{u}}{dt} + 2(\mathbf{\Omega} \times \mathbf{u}) - \frac{1}{2}\nabla(|\mathbf{\Omega} \times \mathbf{r}|^2), \tag{10.3.6}$$

where $(dx^0/dt)_0$ is the derivative of x^0 with respect to the variables of the K^0 frame. The Coriolis force is $2(\mathbf{\Omega} \times \mathbf{u})$, and the centrifugal force is $-\,\mathrm{grad}(|\mathbf{\Omega} \times \mathbf{r}|^2/2)/2$.

Since the rotating vectors are differentiated only with respect to time, the hydrodynamical equation of motion, (Chandrasekher (1961) sections 21–26) is

$$\frac{du_i}{dt} = \frac{\partial u_i}{\partial t} + u_j\frac{\partial u_i}{\partial x_j} = x_i + \frac{1}{\rho}\frac{\partial P_{ij}}{\partial x_j} + \frac{\partial}{\partial x_i}\left(\frac{1}{2}|\Omega \times r|^2\right) + 2\varepsilon_{ijk}u_j\Omega_k \tag{10.3.7}$$

for $i = x, y, z$ or 1, 2, 3; ε_{ijk} is the alternating unit tensor of (A4.3); x_i is the ith component of the sum of the external forces acting on the fluid and P_{ij} is the stress tensor. For incompressible fluids this equation becomes

$$\frac{\partial u_i}{\partial t} + u_j\frac{\partial u_i}{\partial x_j} = x_i - \frac{\partial}{\partial x_i}\left(\frac{1}{\beta}p - \frac{1}{2}|\Omega \times r|^2\right) + \nu\nabla^2 u_i + 2\varepsilon_{ijk}u_j\Omega_k, \tag{10.3.8}$$

where $\nu\nabla^2 u_i$ is the viscous force per unit mass per unit volume. Since

$$u_j\frac{\partial u_i}{\partial x_j} = \frac{1}{2}\frac{\partial}{\partial x_j}|u|^2 - \varepsilon_{ijk}u_j w_k, \quad \mathbf{w} = \nabla \times \mathbf{u}, \tag{10.3.9}$$

equation (10.3.8) is written as

$$\frac{\partial u_i}{\partial t} - \varepsilon_{ijk}u_j w_k = -\frac{\partial \varpi}{\partial x_i} + 2\varepsilon_{ijk}u_j\Omega_k, \tag{10.3.10}$$

where

$$\varpi = \frac{1}{\rho}p + \frac{1}{2}|u|^2 + V - \frac{1}{2}|\Omega \times r|^2, \quad x_i = -\frac{\partial V}{\partial x_i}. \tag{10.3.11}$$

So w is the vorticity of the fluid. Equation (10.3.10) now reads

$$\frac{\partial \mathbf{u}}{\partial t} - \mathbf{u} \times (\mathbf{w} + 2\mathbf{\Omega}) = -\nabla\varpi. \tag{10.3.12}$$

Taking curl of all the terms of this equation, in the rotating frame, we obtain

$$\frac{\partial \mathbf{w}}{\partial t} - \nabla \times \{\mathbf{u} \times (\mathbf{w} + 2\mathbf{\Omega})\} = 0. \tag{10.3.13}$$

So, for a surface Σ, enclosed by a simple closed curve Γ, following the fluid motions, the conservation condition is

$$\iint (w + 2\Omega)\cdot d\Sigma = \text{constant} \tag{10.3.14}$$

and the vector $\mathbf{w} + 2\mathbf{\Omega}$ plays the role of vorticity.

10.3.1 Some general considerations

We define the dimensionless numbers A_R and R as

$$A_R = \frac{C_A}{l\Omega}, \quad R = \frac{v}{l\Omega},$$

in which l is a typical length, v is a speed relative to the rotating frame, C_A is the Alfven speed and Ω is the average angular velocity of rotation of the system. A_R is called the *Alfven number* for rotational motion. For large rotation velocities the flow is strongly influenced by the Coriolis force in the fluid interiors; and Ω is so large in atmospheres of planets, the sun, pulsars and other stars that $A_R < 1$ and $R \ll 1$. The time scale of the effects which develop is $\Omega(l/C_A)^2$. It is much larger than l/C_A, Acheson and Hide (1973). The Coriolis force modify the hydromagnetic waves. Thus, a plane polarized wave is split into two circularly polarized components with different phase velocities and the group velocity is not exactly parallel to the line of force, Lehnert (1954, 1971).

For incompressible fluids the viscous force $\nu\nabla^2 u$ of (10.3.8) equals $-\nu\nabla\times(\nabla\times u)$. The rate of work done against the viscous force is $-\nu(u\cdot\nabla\times(\nabla\times u))$. If the fluid is in a perfectly rough container of volume V and surface Σ, then the boundary condition is $u = 0$ on Σ. Integrating over V the total rate of dissipation of energy is the integral

$$-\nu\iiint_V (u\cdot\nabla\times(\nabla\times u))\,dV = -\nu\iiint_V (\nabla\times u)^2 dV + \nu\iint_\Sigma u\times(\nabla\times u)\cdot d\Sigma,$$

where the vector formula (A3.11) has been used. On applying the boundary conditions the second term vanishes and the loss can be calculated if u is known from the conditions of a problem.

10.4 Equations of Motion for a Rotating Plasma

We consider here the basic equations for a rotating plasma following Brinkman (1960). The equations of motion for ions, each of mass M, charge q_i and electrons, each of mass m and charge q_e, are

$$N_i M\left\{\frac{\partial u_i}{\partial t} + (u_i\cdot\nabla)u_i\right\} = q_i N_i\left(E + \frac{1}{c}(u_i\times H)\right) - \nabla p_i - q_i^2 N_i N_e \eta(u_i - u_e), \tag{10.4.1}$$

$$N_e M\left\{\frac{\partial u_e}{\partial t} + (u_i\cdot\nabla)u_e\right\} = q_e N_e\left(E + \frac{1}{c}(u_e\times H)\right) - \nabla p_e - q_e^2 N_e N_i \eta(u_e - u_i), \tag{10.4.2}$$

where η is the plasma resistivity. The mean velocity $\mathbf{u}$ and the current density i are defined as

$$\mathbf{u} = \frac{N_i M\mathbf{u}_i + N_e m u_e}{N_i M + N_e m}, \tag{10.4.3}$$

$$j = q_i N_i \mathbf{u}_i + q_e N_e \mathbf{u}_e. \tag{10.4.4}$$

If the material vorticity is ω_m, then

$$w_m = \nabla\times\mathbf{u}, \tag{10.4.5}$$

the electrical vorticity

$$w_E = \frac{\nabla \times j}{(q_i N_i - q_e N_e)^{1/2}} \tag{10.4.6}$$

and the magnetic vorticities w_{B_i} and w_{B_e} are given by

$$w_{B_i} = \frac{q_i B}{Mc} = \frac{q_i}{Mc}\nabla \times A, \quad w_{B_e} = \frac{q_e B}{mc} = \frac{q_e}{mc}\nabla \times A, \tag{10.4.7}$$

where A is the vector potential; so $B = \text{curl } A$. For a neutral plasma,

$$\left|\frac{q_i}{e}N_i\right| = \left|\frac{q_e}{e}N_e\right| = N \text{ (say)}. \tag{10.4.8}$$

Moreover, ions being very heavy compared to the electrons,

$$\left|\frac{q_i}{e}M\right| \gg \left|\frac{q_e}{e}m\right|. \tag{10.4.9}$$

Neglecting, therefore, $(q_e m/e)$ with respect to $(q_i M/e)$, putting $q_e = -e$ and $q_i = e$, $w_{B_i} = 0$, and writing w_B for w_{B_e}, we obtain

$$\mathbf{w}_E = \nabla \times \left(\frac{j}{eN}\right), \quad \mathbf{w}_B = -\frac{eB}{mc}, \tag{10.4.10}$$

$$\mathbf{u} = \mathbf{u}_i + \frac{m}{M}\mathbf{u}_e, \quad \mathbf{j} = eN(\mathbf{u}_i - \mathbf{u}_e). \tag{10.4.11}$$

The transformation inverse to (10.4.11) is

$$\mathbf{u}_e = \mathbf{u} - \mathbf{i}, \quad \mathbf{u}_i = \mathbf{u} + \frac{m}{M}\mathbf{i}, \tag{10.4.12}$$

where $i = j/eN$. Therefore, (10.4.1) is written as

$$\frac{\partial \mathbf{u}_i}{\partial t} + (\mathbf{u}_i \cdot \nabla)\mathbf{u}_i = \frac{e}{M}\left(E + \frac{1}{c}(\mathbf{u}_i \times B)\right) - \frac{1}{\rho^m}\nabla p_i - \frac{N_e^2}{M}\eta(\mathbf{u}_i - \mathbf{u}_e). \tag{10.4.13}$$

Utilizing the vector relation

$$(\mathbf{u}_i \cdot \nabla)\mathbf{u}_i = \nabla\left(\frac{u_i^2}{2}\right) - \mathbf{u}_i \times (\nabla \times \mathbf{u}_i)$$

in this equation, we obtain

$$\frac{\partial u_i}{\partial t} - \mathbf{u}_i \times (\nabla \times \mathbf{u}_i) = \frac{e}{M}\left(E + \frac{1}{c}(\mathbf{u}_i \times B)\right) - \nabla\left(\frac{1}{\rho_i^m}p_i + \frac{1}{2}u_i^2\right) - \frac{Ne^2}{M}\eta(\mathbf{u}_i - \mathbf{u}_e), \tag{10.4.14}$$

where N is the assumed constant; so $(\nabla p_i)/\rho_i^m$ is replaced by $\nabla(p_i/\rho_i)$.

Using (10.4.11) and (10.4.12) in (10.4.14) we obtain

$$\frac{\partial}{\partial t}\left(\mathbf{u} + \frac{m}{M}\mathbf{i}\right) - \left(\mathbf{u} + \frac{m}{M}\mathbf{i}\right) \times \nabla \times \left(\mathbf{u} + \frac{m}{M}\mathbf{i}\right)$$
$$= \frac{e}{M}E - \nabla\left(\frac{1}{\rho_i^m}p_i + \frac{1}{2}u_i^2\right) - \frac{m}{M}\left(\mathbf{u} + \frac{m}{M}\mathbf{i}\right) \times w_B - \frac{e}{M}\eta j. \tag{10.4.15}$$

This equation is also written as

$$\frac{\partial \mathbf{u}}{\partial t} - (\mathbf{u} \times w_m) + \frac{m}{M}\frac{\partial \mathbf{i}}{\partial t} - \frac{m}{M}(u \times w_E) - \frac{m}{M}(i \times w_m) - \frac{m^2}{M^2}(i \times w_B)$$

$$= \frac{e}{M}E - \frac{m}{M}(u \times w_B) - \frac{m^2}{M^2}(i \times w_B) - \nabla\left(\frac{1}{\rho_i^m}p_i + \frac{1}{2}u_i^2\right) - \frac{Ne^2}{M}\eta_i. \quad (10.4.16)$$

Now, taking curl of both sides, and using the Maxwell equation

$$c\nabla \times E = -\frac{\partial B}{\partial t}$$

we get

$$\frac{\partial w_m}{\partial t} - \nabla \times (\mathbf{u} \times w_m) + \frac{m}{M}\frac{\partial w_E}{\partial t} - \frac{m}{M}\nabla \times (\mathbf{u} \times w_E) - \frac{m}{M}\nabla \times (\mathbf{i} \times w_m) - \frac{m^2}{M^2}\nabla \times (\mathbf{i} \times w_E)$$

$$= -\frac{m}{M}\frac{\partial \mathbf{w}_B}{\partial t} - \frac{m}{M}\nabla \times (\mathbf{u} \times w_B) - \frac{m^2}{M^2}\nabla \times (\mathbf{i} \times w_B) - \frac{Ne^2}{M}\eta w_E. \quad (10.4.17)$$

Equation (10.4.1) can similarly be written as

$$\frac{\partial u_e}{\partial t} - \mathbf{u}_e \times (\nabla \times \mathbf{u}_e) = -\frac{e}{m}E + (\mathbf{u}_e \times w_B) - \nabla\left(\frac{1}{\rho_e^m}p_e + \frac{1}{2}u_e^2\right) + \frac{Ne^2}{m}\eta(\mathbf{u}_e - \mathbf{u}_i). \quad (10.4.18)$$

With the help of (10.4.11) and (10.4.12) this equation reduces to

$$\frac{\partial}{\partial t}(\mathbf{u} - \mathbf{i}) - (\mathbf{u} - \mathbf{i}) \times \nabla \times (\mathbf{u} - \mathbf{i}) = -\frac{e}{m}E + (\mathbf{u} - \mathbf{i}) \times \mathbf{w}_B - \nabla\left(\frac{1}{\rho_e^m}p_e^*\right) - \frac{\eta e}{m}\mathbf{j}, \quad (10.4.19)$$

where

$$\frac{p^*}{\rho_e^m} = \frac{p_e}{\rho_e^m} + \frac{1}{2}u_e^2;$$

it can now be written as

$$\frac{\partial u}{\partial t} - \frac{\partial i}{\partial t} - (\mathbf{u} \times \mathbf{w}_m) + (\mathbf{u} \times \mathbf{w}_i) + (\mathbf{i} \times \mathbf{w}_m) - (\mathbf{i} \times \mathbf{w}_E)$$

$$= -\frac{e}{m}E + (\mathbf{u} \times w_B) - (\mathbf{i} \times w_B) - \nabla\left(\frac{1}{\rho_0^m}p_e^*\right) - Ne^2\eta\frac{\mathbf{i}}{m}. \quad (10.4.20)$$

Taking curl of all the terms of this equation we get

$$\frac{\partial w_m}{\partial t} - \frac{\partial w_E}{\partial t} - \nabla \times (\mathbf{u} \times w_m) + \nabla \times (\mathbf{u} \times w_E) + \nabla \times (\mathbf{i} \times w_m) - \nabla \times (\mathbf{i} \times w_E)$$

$$= -\frac{\partial w_B}{\partial t} + \nabla \times (\mathbf{u} \times w_B) - \nabla \times (\mathbf{i} \times w_B) - \frac{Ne^2}{m}\eta \mathbf{w_E}. \quad (10.4.21)$$

Simplification yields the relation

$$\frac{\partial w_m}{\partial t} - \nabla \times (\mathbf{u} \times w_m) - \frac{\partial}{\partial t}w_E + \nabla \times (\mathbf{u} \times (w_E - w_B)) + \nabla \times (\mathbf{i} \times (\mathbf{w_m} - \mathbf{w_E} + \mathbf{w_B}))$$

$$= -\frac{\partial}{\partial t}\mathbf{w}_B - \frac{w_e^2}{4\pi}\eta \mathbf{w_E}. \quad (10.4.22)$$

Subtracting (10.4.22) from (10.4.17), to eliminate

$$\frac{\partial w_m}{\partial t} - \nabla \times (\mathbf{u} \times w_m),$$

and neglecting terms like

$$\frac{m}{M}\frac{\partial \mathbf{a}}{\partial t} \quad \text{and} \quad \frac{m^2}{M^2}\frac{\partial \mathbf{a}}{\partial t},$$

relative to $(\partial a/\partial t)$, for example, we obtain

$$\frac{\partial \mathbf{w}_E}{\partial t} - \nabla \times \{\mathbf{u} \times (w_E - w_B) + \mathbf{i} \times (w_m - w_E + W_B)\} = \frac{\partial w_B}{\partial t} + \frac{1}{4\pi} w_p^2 \eta \mathbf{w}_B \tag{10.4.23}$$

This equation can be rearranged as

$$\frac{\partial}{\partial t}(\mathbf{w}_E - \mathbf{w}_B) - \nabla \times \{(\mathbf{w}_E - \mathbf{w}_B) - i \times (w_E - w_B - w_m)\} = \frac{w^2 \rho}{4\pi} \eta \mathbf{w_E} \tag{10.4.24}$$

and (10.4.17) as

$$\frac{\partial w_m}{\partial t} - \nabla \times (\mathbf{u} \times \mathbf{w}_m) = -\frac{m}{M}\left[\frac{\partial}{\partial t}(\mathbf{w}_E + \mathbf{w}_B) + \nabla \times \left\{\mathbf{u} \times (\mathbf{w}_B - \mathbf{w}_E)\right.\right.$$

$$\left.\left. + \frac{\mathbf{m}}{M} i \times (\mathbf{w}_B - \mathbf{w}_E) - (\mathbf{i} \times \mathbf{w}_m)\right\} + \frac{1}{4\pi} w_\rho^2 \eta e_E\right]. \tag{10.4.25}$$

This is an extension in rotating media, on the Helmholtz theory for vortices, because equation (10.4.24) does not appear in that theory.

The circulations c_m and c_e, and the magnetic flux ϕ, are defined as

$$c_m = \oint_\Gamma (u \cdot dl) = \iint_\Sigma (w_m \cdot d\Sigma),$$

$$c_e = \oint_\Gamma (i \cdot dl) = \iint_\Sigma (w_i \cdot d\Sigma),$$

$$\phi = \oint_\Gamma (A \cdot dl) = \iint_\Sigma (B \cdot d\Sigma) \tag{10.4.26}$$

in which dl is a line element and $d\Sigma$ is an element of the surface bounded by the contour Γ. The Helmholtz relations for the change of flux of a vortex 'a' through a surface moving with the fluid is

$$\frac{d}{dt}\iint_\Sigma (\mathbf{a} \cdot d\Sigma) = \iint_\Sigma \left(\frac{\partial a}{\partial t} - \nabla \times (\mathbf{u} \times \mathbf{a}) + \mathbf{u}(\nabla \cdot \mathbf{a})\right) d\Sigma. \tag{10.4.27}$$

Assuming now that the density N does not vary along the contour Γ, using (10.4.24) and (10.4.25) in (10.4.26) and (10.4.27), we find that the equations to determine the circulations c_e, c_m and ϕ are

$$\frac{d}{dt}c_m = \frac{d}{dt}\iint_\Sigma (w_m \cdot d\Sigma) = \iint_\Sigma \left\{\left(\frac{\partial w_m}{\partial t} - \nabla \times (\mathbf{u} \times w_m) + \mathbf{u}(\nabla \cdot \mathbf{w}_m)\right) \cdot d\mathbf{\Sigma}\right\}$$

$$= \iint_\Sigma \left(-\frac{m}{M}\right)\left[\left\{\frac{\partial}{\partial t}(\mathbf{w}_E + \mathbf{w}_B)\right\} \cdot d\mathbf{\Sigma}\right] - \frac{m}{M}\oint_\Gamma \{\mathbf{u} \times (\mathbf{w}_B - \mathbf{w}_E) \cdot \mathbf{dl}\}$$

$$+\frac{w}{M}\oint_\Gamma \{(\mathbf{i} \times \mathbf{w}_m) \cdot \mathbf{dl}\} - \frac{m^2}{M^2}\oint_\Gamma \{\mathbf{i} \times (\mathbf{w}_B - \mathbf{w}_E) \cdot \mathbf{dl}\} - \frac{w_e^2}{4\pi}\frac{\eta m}{M}\iint_\Sigma (\mathbf{w_E} \cdot d\mathbf{\Sigma}) \quad (10.4.28)$$

$$\frac{dc_e}{dt} - \frac{e}{mc}\frac{d\phi}{dt} = \oint_\Gamma \{\mathbf{i} \times (\mathbf{w}_m + \mathbf{w}_B - \mathbf{w}_E) \cdot \mathbf{dl}\} + \nu_{ic}c_e, \quad (10.4.29)$$

where $\nu_{ie} = \eta w_\rho^2/4\pi$. Since **u** and $d\mathbf{l}$ are parallel, equations (10.4.28) and (10.4.29) reduce to

$$\frac{dc_m}{dt} = \frac{m}{M}\oint_\Gamma \{(\mathbf{i} \times \mathbf{w}_m) \cdot \mathbf{dl}\} - \frac{m}{M}\iint_\Sigma \left(\left\{\frac{\partial}{\partial t}(\mathbf{w}_E + \mathbf{w}_B)\right\} \cdot d\mathbf{\Sigma}\right)$$

$$-\frac{m^2}{M^2}\oint_\Gamma \{\mathbf{i} \times (\mathbf{w}_B - \mathbf{w}_E) \cdot \mathbf{dl}\} - \frac{m}{M}\nu_{ie}\iint_\Sigma (\mathbf{w}_i \cdot d\mathbf{\Sigma}), \quad (10.4.30)$$

$$\frac{dc_e}{dt} - \frac{e}{mc}\frac{d\phi}{dt} = -\oint_\Gamma \{\mathbf{i} \times (\mathbf{w}_m + \mathbf{w}_B - \mathbf{w}_i) \cdot \mathbf{dl}\} + \nu_{ie}c_e. \quad (10.4.31)$$

10.5 Electric Currents of Non-uniform Rotation

The accelerated motion of electrically conducting bodies is equivalent to the action of additional forces of inertia on the conduction electrons, Landau and Lifshitz (1960, section 50). Therefore, if $\dot{\mathbf{v}}$ is the acceleration of the conductor, then the force $-m\dot{v}$ on an electron affects it in the same way as an electric field $m\dot{v}/e$, $-e$ being the charge on the electron. Thus the effective electric field on the conduction electrons in an accelerated conducting material is

$$E' = E + \frac{m}{e}\dot{v}. \quad (10.5.1)$$

The corresponding current density is given by Ohm's law

$$j = E' = \sigma\left(E + \frac{m}{e}\dot{v}\right). \quad (10.5.2)$$

Then, taking curl of (10.5.1) we obtain

$$\text{curl}\, E' = -\frac{1}{c}\frac{\partial B}{\partial t} + \frac{m}{e}\,\text{curl}\,\dot{v}. \quad (10.5.3)$$

In the rotating frame

$$\mathbf{v} = \mathbf{u} + (\Omega \times \mathbf{r}), \quad (10.5.4)$$

where **u** is the translation velocity. Then

$$\dot{v} = \dot{u} + (\Omega \times \mathbf{v}) + (\dot{\Omega} \times \mathbf{r}) = \dot{u} + (\Omega \times u) + \{\Omega \times (\Omega \times r)\} + (\dot{\Omega} \times r)$$

$$= \dot{u} + (\Omega \times \mathbf{u}) - \frac{1}{2}\nabla(\Omega \times \mathbf{r})^2 + (\dot{\Omega} \times \mathbf{r}) \quad (10.5.5)$$

and

$$\nabla \times (\dot{\Omega} \times \mathbf{r}) = 2\dot{\Omega}. \quad (10.5.6)$$

The first two terms in the right hand side of (10.5.5) being independent of the position vector r, their curls vanish; so, using (10.5.5) in (10.5.3), we find that

$$\nabla \times E' = -\frac{1}{c}\frac{\partial B}{\partial t} + \frac{2\dot{\Omega} m}{e} = -\frac{1}{c}\frac{\partial B}{\partial t}, \tag{10.5.7}$$

where

$$B' = B - \frac{2m}{e} c\Omega. \tag{10.5.8}$$

Taking curl of (10.5.8) we obtain

$$\nabla \times B' = \nabla \times B \tag{10.5.9}$$

because Ω is independent of the position co-ordinates. Since $B = \mu H$, for electrical conductor, this relation can be written as

$$\nabla \times B' = \mu \nabla \times H = \mu \frac{4\pi}{c} j. \tag{10.5.10}$$

It becomes

$$\nabla \times B' = \frac{4\pi}{c} \mu\sigma E' \tag{10.5.11}$$

if the Ohm's law relation (10.5.2) is used. Taking curl of (10.5.11) and using the solenoidal condition for B' we obtain

$$\nabla \times (\nabla \times B') = -\nabla^2 B' = \frac{4\pi}{c}\mu\sigma \left[-\frac{1}{c}\frac{\partial B}{\partial t} + \frac{m}{c} 2\dot{\Omega} \right].$$

With the help of (10.5.8) this relation can be written as

$$\nabla^2 B' = \frac{4\pi}{c}\mu\sigma \frac{\partial B'}{\partial t}, \tag{10.5.12}$$

which is the same as that for B in conductors at rest. Outside the conductor both B and B' satisfy the equation

$$\nabla^2 B = 0. \tag{10.5.13}$$

On the surface of the conductor both B and B' are continuous. However, at infinity B tends to zero and B' tends to the limit $-2mc\Omega/e$. So, the problem of determining the field B' near a non-uniformly rotating conductor is equivalent to that of determining the field B' near a body at rest in a uniform applied magnetic field of induction

$$\mathbf{B} = -\frac{2m}{e} c\boldsymbol{\Omega}. \tag{10.5.14}$$

The required induction field B_0 outside the conductor is obtained by subtracting this B from the solution B' of this latter problem. The non-uniform rotation field B of (10.5.14) is not related to the gyromagnetic effect which appears even when the rotation is at a uniform rate and is negligibly small for our purpose here. the magnetic field of induction B of (10.5.14), like any variable field, induces electric currents in the conductor itself. In a simply connected body, these currents appear in the form of a magnetic moment.

10.6 Effects of Rotation from Analogy with EM Fields

The equation of motion of a point mass m, having charge q, neglecting sophisticated higher order effects, is

$$m\frac{d\mathbf{u}}{dt} = q\mathcal{E}_0 + \frac{1}{c}q(\mathbf{u} \times B_0) - m\mathbf{\Omega} \times (\mathbf{\Omega} \times \mathbf{r}) + 2m(\mathbf{u} \times \mathbf{\Omega}) - m(\dot{\mathbf{\Omega}} \times \mathbf{r}), \tag{10.6.1}$$

where $\mathcal{E}_0$ is the electric field and $\mathbf{B}_0$ is the magnetic field. Here, in addition to the terms for the electrostatic and the Lorentz forces, three terms of the forces of rotation appear. These are

$$F_1 = -m[\Omega \times (\Omega \times r)], \tag{10.6.2}$$

$$F_2 = 2m(u \times \Omega), \tag{10.6.3}$$

$$F_3 = -m(\dot{\Omega} \times r), \tag{10.6.4}$$

where F_1 is the centrifugal force, F_2 is the Coriolis force, and F_3 is the force due to change of Ω in time. An observer, in the rotating frame, will experience these three inertial forces. Taking curl and divergence of (10.6.2)–(10.6.4) we find that

$$(\nabla \times F_1) = 0, \quad (\nabla.F_1) = 2m\Omega^2, \tag{10.6.5}$$

$$(\nabla \times F_2) = 0, \quad (\nabla.F_2) = 0, \tag{10.6.6}$$

$$(\nabla \times F_3) = -2\dot{\Omega}m, \quad (\nabla.F_3) = 0. \tag{10.6.7}$$

So F_2 is both curl free and divergence free, F_1 is curl free and F_3 is divergence free. For finding the equivalent electric field E and the magnetic field B, we write

$$\mathbf{F}_1 + \mathbf{F}_2 = q\mathcal{E}, \tag{10.6.8}$$

$$\mathbf{F}_3 = \frac{q}{c}(\mathbf{u} \times \mathbf{B}) \tag{10.6.9}$$

and obtain the relations

$$\mathcal{E} = -\frac{m}{q}[\mathbf{\Omega} \times (\mathbf{\Omega} \times \mathbf{r})] - \frac{m}{q}(\dot{\mathbf{\Omega}} \times \mathbf{r}), \tag{10.6.10}$$

$$\mathbf{B} = \frac{2m}{q}\mathbf{\Omega}. \tag{10.6.11}$$

These show that

$$E_{\parallel} = 0, \quad B_{\parallel} = 2mc\frac{\mathbf{\Omega}}{q}, \tag{10.6.12}$$

$$E_{\perp} = \frac{m}{q}\mathbf{\Omega}^2\mathbf{r}_{\perp} - \frac{m}{q}(\dot{\mathbf{\Omega}} \times \mathbf{r}), \quad B_{\perp} = 0, \tag{10.6.13}$$

where the subscript $\parallel$ stands for field components parallel to Ω, and $\perp$ stands for the perpendicular components. Next, we put

$$E = \nabla\phi - \frac{1}{c}\frac{\partial A}{\partial t} \tag{10.6.14}$$

in (10.6.3), and obtain

$$\phi = \frac{m}{2q}V^2 = \frac{m}{2q}(\Omega \times r)^2 = \frac{m\Omega^2}{2q}(x^2 + y^2). \tag{10.6.15}$$

$$\frac{1}{c}\frac{\partial A}{\partial t} = -\frac{m}{q}(\dot{\Omega} \times r). \tag{10.6.16}$$

Integration of curl of (10.6.10), with respect to time t, gives

$$\nabla \times A = \frac{2m}{q} c\Omega = B, \tag{10.6.17}$$

where (10.6.11) has been used. These A and ϕ are, respectively, the equivalent vector potential and scalar potential in the Lorentz gauge of the forces of rotation. Taking curl of (10.6.10) and using (10.6.11) we find that

$$(\nabla \times E) = -\frac{\partial}{\partial t}\left(\frac{m}{q}2\Omega\right) = -\nabla \times \left(\frac{1}{c}\frac{\partial A}{\partial t}\right). \tag{10.6.18}$$

Divergence of (10.6.10) gives

$$(\nabla . E) = \frac{2m}{q}\Omega^2. \tag{10.6.19}$$

Equating this divergence with $4\pi\rho^c$ we obtain the equivalent net charge density

$$\rho^c = \frac{m\Omega^2}{2\pi q}. \tag{10.6.20}$$

Since curl $B = 0$, the inhomogeneous Maxwell equation

$$\nabla \times B = \frac{1}{c}\frac{\partial E}{\partial t} + \frac{4\pi}{c} j \tag{10.6.21}$$

determines the equivalent current

$$\mathbf{j} = -\frac{1}{4\pi}\frac{\partial E}{\partial t} = \frac{m}{4\pi q}\frac{\partial}{\partial t}[\{\Omega \times (\Omega \times \mathbf{r}) + (\dot{\Omega} \times \mathbf{r})\}]. \tag{10.6.22}$$

Divergence of this relation gives the condition of continuity of charge

$$(\nabla \cdot j) = -\frac{m}{\pi q}\Omega\dot{\Omega} = -\dot{\rho}. \tag{10.6.23}$$

The LT, for the components of displacement, can be written as

$$\mathbf{r}_\perp = r^0_\perp, \quad \mathbf{r}_\parallel = \gamma(r^0_\parallel - \beta c dt^0), \quad dt = \gamma\left(dt^0 - \frac{\beta}{c} r^0_\parallel\right), \tag{10.6.24}$$

where the relativistic factor is $\gamma (= 1/\sqrt{1-\beta^2})$, $\beta = v/c$, v is the constant velocity of translation in the inertial frame K, relative to the laboratory frame K^0), $r_\perp$ is the displacement perpendicular to v, and $r_\parallel$ is its parallel component. The relations of transformation of the EM field components are

$$E_\perp = \gamma[E^0_\perp + (\beta \times B^0), \quad B_\perp = \gamma[B^0_\perp - (\beta \times E^0)] \tag{10.6.25}$$

$$E_\parallel = E^0_\parallel, \quad B_\parallel = B^0_\parallel. \tag{10.6.26}$$

Hence the parallel field components remain unaffected, and only the perpendicular components are affected, relative to the displacement vector. This is the rule of induction of modification in perpendicular fields relative to the direction of a dynamism, which in STR is the direction of the relativistic distortion of displacement in time. This rule follows Faraday's law for EM induction (when a conducting loop of wire is moved in one direction, the magnetic field perpendicular to that direction is affected and the parallel field remains unchanged). It leads to trivial identities when applied to the equivalent fields of rotation of equations (10.6.12) and (10.6.13), because (1) rotation does not have a component parallel to its axis, and because (2) this rule is the transformation for the field of rotation from one inertial frame to another, about the fixed direction of the velocity v of translation. However, results of application of this rule are not known to us for rotation about an axis which is changing direction with time.

10.6.1 Faraday's law of induction for rotation of a particle

The work done W, on a test mass, carried around a closed path Γ, by the three forces, F_1, F_2, F_3, of rotation, is given by

$$W = \oint_\Gamma ((F_1 + F_2 + F_3)\cdot dl), \tag{10.6.27}$$

where dl is the line element at a point in the loop Γ. Since F_1 is conservative, its integral around Γ is zero. The Coriolis force F_2 is always perpendicular to the path; so it does not contribute to the integral. Only the contribution of F_3 remains in equation (10.6.27). Using the equation (10.6.4) and the identity $((\dot{\Omega} \times r)\cdot dl) = (\dot{\Omega}\cdot (\mathbf{r} \times dl))$, this relation becomes

$$\begin{aligned} W &= -m \oint_\Gamma [(\dot{\Omega} \times r)\cdot dl] = -m \oint_\Gamma (\dot{\Omega}\cdot (\mathbf{r} \times dl)) \\ &= -m \left[\dot{\Omega} \oint_\Gamma |r \times dl|\right] = -2m\dot{\Omega}A, \end{aligned} \tag{10.6.28}$$

where A is the area of the loop Γ, since $\dot{\Omega}$ is uniform and parallel to $(\mathbf{r} \times dl)$, we find that

$$2A = \oint_\Gamma |\mathbf{r} \times dl|. \tag{10.6.29}$$

For $m = 1$, $2\dot{\Omega}A$ is the rate the change of flux of 2Ω, per unit mass, through the circuit Γ, and equations (10.6.27)–(10.6.29) show that the rate of work done by a particle of unit mass in moving along the circuit Γ, due to the forces of rotation, equals the rate of loss of flux of 2Ω through the surface enclosed by Γ. This is Faraday' law of induction for rotation of a particle. The equivalent magnetic field B is supposed to vary in time and increase in the positive z-direction. Then the induced electric field E_i, of the Maxwell equation

$$\nabla \times E_i = -\frac{1}{c}\frac{\partial B}{\partial t} \tag{10.6.30}$$

is tangential to any circle in the plane perpendicular to the direction of Ω, with its center at the origin, and is non-conservative. Only this field supplies energy for the work done W on a charged particle, for motion around a closed path Γ in that plane. Since the magnetic field B, of (10.6.11), is always perpendicular to this path,

$$W = q\oint_{\Gamma}(E_i \cdot dl) = q\iiint_{\Sigma}[(\nabla \times E_i)\cdot d\Sigma]$$

$$= -\frac{q}{c}\iint_{\Sigma}\left(\frac{\partial B}{\partial t}\cdot d\Sigma\right) = -\frac{q}{c}\dot{B}A = -2m\dot{\Omega}A. \quad (10.6.31)$$

where Σ is a surface containing the circuit of Γ, and $d\Sigma$ is a surface element on Σ, and Stokes' theorem has been applied to change a line integral into a surface integral. Thus, (10.6.28) is recovered from the analogous EM consideration

10.6.2 Clarifications and Comments

When Ω increases with time, a rotating observer sees a uniform field of Ω, that is increasing with time, and can measure the forces : F_3 of relation (10.6.4), F_1, of (10.6.2), and F_2 of (10.6.3). The transverse force is tangential to circles in the XY plane with center in OZ, when Ω is parallel to OZ. The non-conservative force F_3 of (10.6.4) and the force of the induced emf have the same direction, and give analogous expressions for the work, equations (10.6.28) and (10.6.31). Equations (10.6.10) and (10.6.11), similarly, show that a positive charge q, in the field of E and B, behaves like a particle of mass m in a rotating frame.

A charge q in a plane, bound elastically to a fixed point, starting from rest in presence of an applied field, experiences the Larmor precession, like the precession of the Foucault pendulum, due to gravitational field. Hence, the field-like effects of rotation-dependent forces, in a rotating frame, follow the effects on charged particles of electric and magnetic forces in Newtonian frames. Like the determination of the magnetic induction B, from estimation of the torque on a closed current loop, or from precession of a rotating charged body, the Ω field can be estimated by measuring the precessional velocity of a top in a rotating frame of reference, either with no applied torque, or with the torque necessary to keep its axis fixed.

10.6.3 Basic assumptions

The following assumptions, among others, are implied in the derivation of the EM analogy : (1) The mass densities are normal, so that there is no trace of very high density matter as found in white dwarfs and neutron stars; (2) The motions are much lower than the speed of light (often STR effects are found to hide the GTR effects); (3) The kinetic or potential energy of the bodies is much smaller than their energy of the rest mass; (4) The fields are always weak enough to permit superposition; (5) The distance between objects is not so large that it becomes necessary to take into account the effects of retarded time of radiation from the sources.

10.7 Consequences of the Analogy

The equivalent $E \times B$ drift, denoted by v_E is given by

$$V_E = \frac{C}{B^2}(E \times B) = -\frac{1}{2\Omega}\left[\Omega^2(k \times (ix + jy)) + \dot{\Omega}(ix + jy)\right], \quad (10.7.1)$$

where i, j, k are the unit vectors along the positive directions of the co-ordinates axes. This drift is independent of the mass of the particle, and is perpendicular to the direction of the axis of rotation. The equivalent densities of electric and magnetic energies, the Poynting flux and its divergence are

$$W_E = \frac{E^2}{8\pi} = \frac{m^2}{8\pi q^2}(x^2+y^2)(\Omega^4+\dot{\Omega}^2), \tag{10.7.2}$$

$$W_H = \frac{B^2}{8\pi} = \frac{4m^2\Omega^2 c^2}{8\pi q^2}, \tag{10.7.3}$$

$$S = \frac{c}{4\pi}(E\times B) = -\frac{m^2c^2}{2\pi q^2}\left[-\Omega^2(k\times(ix+jy)+\dot{\Omega}(ix+jy)\right], \tag{10.7.4}$$

$$(\nabla\cdot S) = -\frac{m^2}{8\pi q^2}\frac{\partial}{\partial t}(4c^2\Omega^2). \tag{10.7.5}$$

Also, since

$$(E\cdot j) = -\frac{m^2}{4\pi q^2}\frac{\partial}{\partial t}\left[(x^2+y^2)(\Omega^4+\dot{\Omega}^2)\right], \tag{10.7.6}$$

the Poynting theorem breaks up into the following two energy conservation relations

$$(E\cdot j) = -\frac{\partial}{\partial t}\left(\frac{E^2}{8\pi}\right), \tag{10.7.7}$$

$$(\nabla\cdot S) = -\frac{\partial}{\partial t}\left(\frac{B^2}{8\pi}\right). \tag{10.7.8}$$

Equation (10.7.7) shows that the energy acquired by the medium, through the action of the centrifugal force, and the force due to variation of the angular velocity with time, is lost through dissipation inside the medium. And (10.7.8) shows that the energy, acquired by the action of the Coriolis force, is lost through the boundary surface of the medium. Since

$$\frac{W_E}{W_H} = \frac{(x^2+y^2)}{4c^2}\left(\Omega^2+\frac{\dot{\Omega}^2}{\Omega^2}\right),$$

the sum of the energies of the centrifugal force and that of the force of the time derivative of Ω is much greater than that of the Coriolis force. The elements of the corresponding Maxwell stress tensor are given by

$$T_{ij} = \frac{m^2}{4\pi q^2}\begin{bmatrix} \begin{matrix}\frac{1}{2}(\Omega^4-\dot{\Omega}^2xx^2+y^2)\\ +4\Omega^2\dot{\Omega}xy-2c^2\Omega^2\end{matrix} & (\Omega^4-\dot{\Omega}^2)xy-\Omega^2\dot{\Omega}(x^2-y^2) & 0 \\ (\Omega^4-\dot{\Omega}^2)xy-\Omega^2\dot{\Omega}(x^2+y^2) & \begin{matrix}\frac{1}{2}\{(\Omega^4-\dot{\Omega}^2)(x^2-y^2)\\ +4\Omega^2\dot{\Omega}xy-2c^2\Omega^2\}\end{matrix} & 0 \\ 0 & 0 & 2c^2\Omega^2 \end{bmatrix}. \tag{10.7.9}$$

The phase velocity of Alfven wave-like waves, propagating parallel to the direction of the axis of rotation, and acting on particles, each of mass m and charge q, in a medium of mass density ρ^m, is

$$c_A = \frac{mc\Omega}{q(\pi\rho^m)^{1/2}} = \left(\frac{2mc^2}{q}\right)^{1/2}. \tag{10.7.10}$$

It is a universal constant; for an electron its values is about 60 m/sec. The equivalent Alfven wave frequency W_A must be smaller than the equivalent ion gyration frequency $\Omega_i(= qH/MC) = (2m/M)\Omega = \Omega/920$, because ion mass $M = 1840$ m), so

$$W_A \ll (2m/M)\Omega.$$

As the rotation frequency Ω is itself a small angular velocity, for the validity of the basic field equations, W_A must be a still smaller quantity.

Since the steady flow of electrically conducting fluids is similar to the steady vortex motion in incompressible, non-viscous fluids, following Shafranov (1958) and Yoler (1961), we write

Equations for the kinetics of vortex motion	Magnetohydrostatic equations
$(\nabla \cdot u) = 0$	$(\nabla . E) = 0$
$(\nabla \times u) = W$	$(\nabla \times H) = \frac{4\pi}{c} j$
$(W \times u) = \nabla \left(\frac{p}{\rho} + \frac{u^2}{2} \right)$	$\frac{1}{2}(j \times B) = \nabla p$

where W is the vorticity. Taking curl of the Navier-Stoke's relation

$$\frac{\partial u}{\partial t} = \nabla \times (u \times W) + \nu \nabla^2 u,$$

we find that

$$\frac{\partial W}{\partial t} = \nabla \times (u \times W) + \nu \nabla^2 W,$$

where ν is the fluid viscosity. The MHD equation for the magnetic field, in a fluid of electrical conductivity σ, is

$$\frac{\partial H}{\partial t} = \nabla \times (u \times H) + \frac{\varepsilon^2}{4\pi\sigma} \nabla^2 H.$$

Comparison between these equations, for W and H, gives the relation

$$H = \alpha W,$$

where α is a constant. Then, since $\mathbf{V} = (\Omega \times r)$, we obtain

$$\nabla \times V = 2\Omega = \nabla \times \mathbf{u} = W$$

if $V = u$.

A vortex sheet, therefore, is analogous to a current sheet, and a current ring, in a perfect conductor, to a vortex ring. In steady flow, moreover, the fluid vorticity corresponds to the magnetic field. Conclusions on the stability of one system, cannot, however, be transferred to its analogue in the other system, because, for stability study, one must consider flows in presence of perturbation fields, which make the flow unsteady.

10.8 Examples having no EM Analogy

The spatial tensor components of the gravitational field have no EM analogy. The space is curved by the mass of an object, and the motion in the flat space, due to the gravitational force tensor, can be replaced by the forceless motion in curved space. The gravitational precession of the perihelion of the planet mercury, or the bending of light rays, also do not have any EM analogy. The precession of a spinning satellite, which follows from the EM analogy, is smaller than that from the space curvature effect.

10.9 Lagrangian of Rotation from this Analogy

A Lagrangian L, for motion of charged particles, EM fields and their interaction, Chakraborty (1987), is

$$\mathcal{L} = -\sum\left(mc^2\sqrt{1-\frac{u^2}{c^2}}\right) + \rho^c\phi + \frac{1}{c}(A\cdot j) + \frac{1}{8\pi}(E^2 - B^2). \tag{10.9.1}$$

We take ρ^c form (10.6.20), j from (10.6.22), ϕ from (10.6.15), A from (10.6.16), E from (10.6.10) and B from (10.6.11); also we replace U by $U^0 + v$, taking $\mathbf{v} = \Omega \times \mathbf{r}$, and write

$$\rho^c = \frac{m\Omega^2}{2\pi q}, \tag{10.9.2}$$

$$\begin{aligned}\mathbf{j} &= \frac{m}{4\pi q} - \frac{\partial}{\partial t}[(\Omega \times (\Omega \times \mathbf{r})) + (\dot{\Omega} \times \mathbf{r})] \\ &= \frac{m}{4\pi q}[(\dot{\Omega} \times (\Omega \times \mathbf{r})) + (\Omega \times (\dot{\Omega} \times \mathbf{r})) + (\Omega \times (\Omega \times \dot{\mathbf{r}})) + (\ddot{\Omega} \times \mathbf{r}) + (\dot{\Omega} \times \dot{\mathbf{r}})],\end{aligned} \tag{10.9.3}$$

$$\phi = \frac{m}{2q}V^2, \tag{10.9.4}$$

$$A = -\frac{mc}{q}(\Omega \times r), \tag{10.9.5}$$

$$E = -\frac{m}{q}[(\Omega \times (\Omega \times r)) + (\dot{\Omega} \times r)] \tag{10.9.6}$$

$$B = \frac{2m}{q}c\Omega. \tag{10.9.7}$$

Thus (10.9.1) becomes

$$\begin{aligned}\mathcal{L} = &-\sum\left(mc^2\sqrt{\left\{1 - \frac{1}{c^2}(u^0 + V)^2\right\}}\right) + \frac{m\Omega^2}{2\pi q}\frac{m}{2q}V^2 - \frac{1}{c}\frac{mc}{q}(\Omega \times r)\left(\frac{m}{4\pi q}\right) \\ &\times[(\dot{\Omega} \times (\Omega \times r) + (\Omega \times (\dot{\Omega} \times r)) + (\Omega \times (\Omega \times \dot{r})) + (\ddot{\Omega} \times r) + (\dot{\Omega} \times \dot{r}) \\ &+\frac{m^2}{8\pi q^2}[\{(\Omega \times (\Omega \times r)) - (\dot{\Omega} \times r)\}^2 - 4c^2\Omega] \\ = &-\sum\left(mc^2\sqrt{1 - \frac{1}{c^2}(u^2 + V)^2}\right) + \frac{m^2\Omega^2}{4\pi q^2}\left(V^2 - \frac{c^2}{2}\right) \\ &-\frac{m^2}{4\pi q^2}[(\Omega \times r)\cdot\{(\Omega \times (\Omega \times \dot{r})) + (\ddot{\Omega} \times r) + (\dot{\Omega} \times \dot{r})\}] \\ &+\frac{m^2}{8\pi q^2}[(\Omega \times (\Omega \times r))^2 + (\dot{\Omega} \times r)^2]\end{aligned}$$

$$= -\sum\left(mc^2\sqrt{1-\frac{1}{c^2}(u^0+V)^2}\right) - \frac{m^2\Omega^2}{4\pi q^2}\left(\frac{c^2}{2}-v^2\right)$$

$$-\frac{m^2}{4\pi q^2}[\Omega(-iy+jx)\cdot\{\Omega^2(-ix-jy)+\dot{\Omega}(-iy+jx)+\ddot{\Omega}(-iy+jx)\}]$$

$$+\frac{m^2}{8\pi q^2}[\Omega^4(x^2+y^2)+\dot{\Omega}^2(x^2+y^2)]$$

$$= -\sum\left(mc^2\sqrt{1-\frac{1}{c^2}(u^0+V)^2}\right) - \frac{m^2\Omega^2}{4\pi q^2}\left(\frac{c^2}{2}-V^2\right)$$

$$-\frac{m^2}{4\pi q^2}[\Omega^3(\dot{x}y-x\dot{y})+\Omega\dot{\Omega}(x\dot{x}+y\dot{y})] - \frac{m^2}{8\pi q^2}(x^2+y^2)(2\Omega\ddot{\Omega}-\dot{\Omega}^2-\Omega^4). \quad (10.9.8)$$

10.10 Equation of Motion

The equation of motion, in the linearized approximation, has one term, $-\nabla\phi$, for the static gravitation, one term, $-\partial A/\partial t$, for the gravitational induction effect, and $V\times(\nabla\times A)$, for the gravitational equivalent of the Lorentz force proportional to $(\mathbf{v}\times B)$. Therefore, the force equation is

$$F = -m\nabla\phi - \frac{m}{c}\frac{\partial A}{\partial t} + \frac{m}{c}V\times(\nabla\times A) = mG + \frac{m}{c}(V\times P), \quad (10.10.1)$$

where G is analogous to E, and equals $-\nabla\phi, P$ is analogous to the solenoidal magnetic field B and equals $\nabla\times A$. If the test body is spinning, and has an angular momentum L, then the gravitational torque on it, by analogy, is

$$N = \frac{1}{2}(L\times P). \quad (10.10.2)$$

10.11 Relativistic Motion of a Spinning Charged Particle

The 4-co-ordinate, 4-velocity and 4-momentum are

$$x_\mu = (r, ict). \quad (10.11.1)$$

$$\frac{dx_\mu}{dT} = V_\mu = \gamma(u_x, u_y, u_z, ic), \quad (10.11.2)$$

$$p_\mu = (p_x, p_y, p_z, i\varepsilon/c), \quad (10.11.3)$$

where γ is the relativistic factor $(1-\beta^2)^{-1/2}$, $\beta = v/c$,

$$\mathcal{E} = m\gamma c^2, \quad (10.11.4)$$

$$c^2dT^2 = -ds^2 = c^2dt^2 - dx^2 - dy^2 - dz^2, \quad (10.11.5)$$

$$\mathbf{u} = (u_x, u_y, u_z) = \frac{(dx, dy, dz)}{dt}, \quad p = mru \quad (10.11.6)$$

$\mathcal{E}$ is the kinetic energy plus the rest energy of the particle, m is its rest mass, T is the proper time and V_μ is the Minkowshi 4-velocity, which is tangential to the world line of the particle. The 4-potential is

$$A_\mu = (A, i\phi), \tag{10.11.7}$$

where the vector potential is A and the scalar potential is ϕ. Also we may write

$$(P, iW/c) = P_\mu = p_\mu + \frac{e}{c}A_\mu, \tag{10.11.8}$$

where P_μ is the generalized or the cannonical 4-momentum, or the canonical energy-momentum tensor; W has the dimensions of energy. Then

$$P = p + \frac{e}{c}A, \tag{10.11.9}$$

$$W = \mathcal{E} + e\phi, \tag{10.11.10}$$

$$(p_\mu V_\mu) = \gamma(p\cdot u) - \gamma\mathcal{E} = -m_0c^2. \tag{10.11.11}$$

Hence the invariant parameter m, having the dimensions of mass, is defined by

$$m_0 = -\frac{1}{c^2}(p_\mu V_\mu) = -\frac{1}{c^2}\left[\left(\left(P - \frac{e}{c}A\right)\cdot\gamma u\right) + iW\cdot i\gamma\right]. \tag{10.11.12}$$

This relation can be written as

$$W = \xi + e\phi = \frac{1}{c^2}\left(\left(\rho - \frac{e}{c}A\right)\cdot u\right) + \frac{m_0}{\gamma}c^2. \tag{10.11.13}$$

The relation (5.11.3), for a spinning particle, can be written in the covariant form

$$\dot{S}_{\mu\nu} + \frac{1}{c}(V_\mu p_\mu - V_\nu p_\mu) = N_{\mu\nu}, \tag{10.11.14}$$

where dot denotes differentiation with respect to the proper time T of relation (10.11.7). Hence, S and N are axial vectors because $(\mathbf{v}\times\mathbf{p})$ is an axial vector.

Next, we introduce two polar vectors G and K in terms of elements of the axial vectors $\mathbf{S}$ and $\mathbf{N}$, in the following manner

$$N_{\mu\nu} = \begin{bmatrix} 0 & \frac{\gamma}{c}N_z & -\frac{\gamma}{c}N_y & \frac{i\gamma}{c}G_x \\ -\frac{\gamma}{c}N_z & 0 & \frac{\gamma}{c}N_x & \frac{i\gamma}{c}G_y \\ \frac{\gamma}{c}N_y & -\frac{\gamma}{c}N_x & 0 & \frac{i\gamma}{c}G_z \\ -\frac{i\gamma}{c}G_x & -\frac{i\gamma}{c}G_y & -\frac{i\gamma}{c}G_z & 0 \end{bmatrix}, \quad S_{\mu\nu} = \begin{bmatrix} 0 & S_z & -S_y & ik_x \\ -S_z & 0 & S_x & ik_y \\ S_y & -S_x & 0 & ik_z \\ -ik_x & -ik_y & -ik_z & 0 \end{bmatrix}. \tag{10.11.15}$$

We can, therefore, write

$$ic(N_{41}, N_{42}, N_{43}) = \gamma G, \tag{10.11.16}$$

$$i(S_{41}, S_{42}, S_{43}) = K, \tag{10.11.17}$$

$$c(N_{23}, S_{31}, S_{12}) = \gamma N, \tag{10.11.18}$$

$$(S_{23}, S_{31}, S_{12}) = S. \tag{10.11.19}$$

The axial vector components of (10.11.14) yield (10.9.5) and the polar vector components follow the equation

$$\frac{d\mathbf{k}}{dt} + \frac{1}{c}\mathcal{E}\mathbf{v} - c\mathbf{p} = \mathbf{G}. \tag{10.11.20}$$

We now multiply (10.11.14) by V_ν, use the Einstein summation convention and obtain

$$V_\nu \dot{S}_{\mu\nu} + V_\nu V_\mu p_\nu - V_\nu V_\nu p_\nu = N_{\mu\nu} N_\nu. \tag{10.11.21}$$

Then using (10.11.12) in the second term of the left hand side and (10.11.12) in the third term it follows that

$$p_\mu = mV_\mu - \frac{(\dot{S}_{\mu\nu} V_\nu + N_{\mu\nu} V_\nu)}{c}. \tag{10.11.22}$$

Hence using (10.11.2) and (10.11.15) we find that

$$p = m\gamma v - \frac{\gamma^2}{c^2}\left(v \times \frac{dS}{dt}\right) + \frac{\gamma^2}{c}\frac{dk}{dt} - \frac{\gamma^2}{c}\left(G - \frac{1}{c}(v \times N)\right), \tag{10.11.23}$$

$$\mathcal{E} = m\gamma c^2 + \frac{\gamma^2}{c}\left(v \cdot \left(\frac{dk}{dt} - G\right)\right). \tag{10.11.24}$$

Elimination of p_μ between (10.11.14) and (10.11.22) gives

$$\dot{S}_{\mu\nu} = \frac{1}{c}(\dot{S}_{\mu\sigma} V_\mu - S_{\mu\sigma} V_\nu) V_\sigma + N_{\mu\nu} - (N_{\mu\sigma} V_\nu - S_{\mu\sigma} V_\nu)\frac{V_\sigma}{c^2}. \tag{10.11.25}$$

Differentiating (10.11.14) with respect to t gives

$$\dot{N}_{\mu\nu} = \dot{S}_{\mu\nu} + (\dot{V}_\mu p_\nu + V_\mu \dot{p}_\nu - \dot{V}_\nu p_\mu - V_\nu \dot{p}_\mu)/c. \tag{10.11.26}$$

Multiplying both sides of this relation by V_ν and using the Einstein convention to denote summation over the repeated index we obtain

$$\frac{1}{c}(\dot{N}_{\mu\nu} V_\nu - \ddot{S}_{\mu\nu} V_\nu) = \frac{1}{c^2}(\dot{V}_\mu V_\nu p_\nu + V_\mu V_\nu \dot{p}_\nu - V_\nu \dot{V}_\nu p_\mu) + \dot{p}_\mu, \tag{10.11.27}$$

where, with the help of (10.11.2), we have used the relations

$$V_\nu V_\nu = -c^2, \quad V_\nu \dot{V}_\nu = 0, \quad p_\mu V_\mu = -mc^2. \tag{10.11.28}$$

After some readjustments relation (10.11.27) gives

$$m V_\mu - \ddot{S}_{\mu\nu} V_\nu = \dot{p}_\mu + \frac{1}{c}(V_\nu \dot{p}_\nu) V_\nu - \dot{N}_{\mu\nu} V_\nu. \tag{10.11.29}$$

Next, multiplying both sides of (10.11.14) with V_ν and summing over, we obtain

$$\dot{S}_{\mu\nu} \dot{V}_\nu + \frac{1}{c}\dot{V}_\nu p_\nu - \frac{1}{c} V_\nu \dot{V}_\nu p_{\mu\nu} = N_{\mu\nu} \dot{V}_\nu. \tag{10.11.30}$$

Multiplying (10.11.14) by the antisymmetric tensor $F_{\mu\nu}$ and summing over both μ and ν we obtain

$$\frac{1}{2}\dot{S}_{\mu\nu} F_{\mu\nu} = \frac{1}{2} N_{\mu\nu} F_{\mu\nu} - \frac{1}{c} V_\nu F_{\mu\nu} p_\nu, \tag{10.11.31}$$

where we have used the fact that $-V_\nu p_\mu F_{\mu\nu} = p_\mu V_\mu F_{\mu\nu}$. Now using, in this relation (10.11.22) to eliminate p_μ, we obtain

$$\frac{1}{2}\dot{S}_{\mu\nu}F_{\mu\nu} = \frac{1}{2}N_{\mu\nu}F_{\mu\nu} + \frac{1}{c}V_\mu F_{\mu\nu}S_{\nu\sigma}V_\sigma - \frac{1}{c}V_\mu F_{\nu\nu}N_{\mu\sigma}V_\sigma, \tag{10.11.32}$$

where we have used the fact that $\sum_{\mu,\nu}(V_\mu V_\nu F_{\mu\nu}) = 0$, because $F_{\mu\nu}$ is antisymmetric. The relation between $F_{\mu\nu}$ and the four potential A_μ is

$$F_{\mu\nu} = \begin{bmatrix} 0 & H_z & -H_y & -iE_x \\ -H_x & 0 & H_x & -iE_y \\ H_y & -H_x & 0 & -iE_z \\ iE_x & iE_y & iE_z & 0 \end{bmatrix} \tag{10.11.33}$$

(with $\rightarrow \nu$ labelling the columns and $\downarrow \mu$ the rows)

$$= \partial_\mu A_\mu - \partial_\mu A_\mu. \tag{10.11.34}$$

The elements of the antisymmetric magnetic moment tensor are

$$\mu_{pq} = \begin{bmatrix} 0 & \mu_z & -\mu_y & i\nu_x \\ -\mu_z & 0 & \mu_x & i\nu_y \\ \mu_y & -\mu_x & 0 & i\nu_z \\ -i\nu_x & -i\nu_y & -i\nu_z & 0 \end{bmatrix}. \tag{10.11.35}$$

The torque N for a particle at rest being

$$\mathbf{N} = (\mu \times H) + (\nu \times E). \tag{10.11.36}$$

We can write the tensor relation

$$T_{\mu\nu} = \frac{1}{c}(F_{\mu\sigma}\mu_{\sigma\nu} - \mu_{\mu\sigma}F_{\sigma\nu}). \tag{10.11.37}$$

Then, for a particle at rest, the axial vector components of the antisymmetric tensor $N_{\mu\nu}$, with the help of (10.11.18), reduce to (10.11.36). The polar components of (10.11.37) then give the vector relation

$$\gamma G = (\nu \times H) - (\mu \times E). \tag{10.11.38}$$

For a moving particle the general form of (10.11.36) is

$$\gamma N = (\mu \times H) + (\nu \times E). \tag{10.11.39}$$

Since the intrinsic electric moment $\boldsymbol{\nu}$ vanishes in the rest system of the particle (actually ν arises only from the motion of the magnetic moment $\boldsymbol{\mu}$),

$$\boldsymbol{\nu} = \frac{1}{c}(\mathbf{u} \times \boldsymbol{\mu}). \tag{10.11.40}$$

In the tensor form this relation reads

$$\mu_{\lambda\sigma}V_\sigma = 0. \tag{10.11.41}$$

Then using (10.11.37), equation (10.11.32) simplifies to

$$\frac{1}{2}S_{\rho\sigma}F_{\rho\sigma} = \frac{1}{c^2}V_\rho F_{\rho\sigma}\dot{S}_{\sigma\alpha}V_\alpha = -\frac{1}{c}V_\rho V_{\rho\sigma}p_\sigma \tag{10.11.42}$$

and equation (10.11.22) reduces to

$$p_\mu = mV_\mu - \frac{1}{c}\dot{S}_{\mu\nu}V_\nu - \frac{1}{c^2}\mu_{\mu\sigma}F_{\sigma\nu}V_\nu. \tag{10.11.43}$$

Hence we find that

$$\dot{m}c = -\frac{1}{c}V_\rho \dot{f}_p + \frac{1}{c^2}\dot{V}_\rho\dot{S}_{\rho\sigma}V_\sigma = \frac{1}{c^3}V_\rho\dot{\mu}_{\rho\sigma}F_{\sigma T}V_T. \tag{10.11.44}$$

The 4-force, which acts on the charge and on the electric and magnetic dipoles, is given by

$$\dot{p}_\mu = \frac{e}{c^2}F_{\mu\nu}V_\nu + \frac{1}{2}\mu_{\rho\sigma}\frac{\partial}{\partial x_\mu}F_{\rho\sigma}. \tag{10.11.45}$$

Putting this value of $\dot{p}_\mu$ in (10.11.44) and remembering the fact that $V_\mu V_\nu F_{\mu\nu} = 0$ (because $F_{\mu\nu}$ is antisymmetric), we find that

$$\dot{m}c = -\frac{1}{2}\frac{d}{ds}(\mu_{\rho\sigma}F_{\rho\sigma}) + \frac{1}{c^2}\dot{V}_\sigma\dot{S}_{\sigma\rho}V_\rho - \frac{1}{c^3}V_\rho\dot{\mu}_{\rho\sigma}F_{\sigma\nu}V_\nu + \frac{1}{2c}\dot{\mu}_{\rho\sigma}F_{\rho\sigma}. \tag{10.11.46}$$

10.12 Electron Theory in a Rotating Frame

The electron theory of Lorentz for small rotational velocities, maintaining the form invariance, has been developed, Shiozawa (1973). According to this theory the material medium is assumed to consist of heavy ions in the nuclei, each with the net charge $+e$ and light mobile electrons, each with the charge $-e$. Also, each electron in the medium is electrically bound to the corresponding nucleus and is displaced with respect to the nucleus by EM fields, but the heavy nuclei are assumed to remain unaffected by them. The EM fields interact with the aggregate of the charged particles, which are positively charged nuclei and electrons.

The Coriolis force and the centrifugal force of rotation influences the EM field, but the change in the macroscopic properties of the medium, due to acceleration acting upon it is neglected. So the constitutive material parameters ε, μ, σ are assumed to be constants, and independent of the acceleration of the material body. In the absence of EM fields the average number of electrons and ions per unit volume are equal and equal to N, say; also their velocities are equal. But, in presence of the EM field, the number of electrons per unit volume is changed to $N + N_1$ say, and the velocities of electrons and ions differ. So

$$\rho^c = -N_1 e, \tag{10.12.1}$$

$$j = Neu_i - (N + N_1)eU_e, \tag{10.12.2}$$

where u_i and u_e are, respectively, the average velocity of the ions and electrons at an instant of time at a point.

Let R be the average position vector of an electron in the absence of the field. In the presence of the field let R_p be its position vector. So, if s is the small displacement of the electrons, then

$$R_p = R + S. \tag{10.12.3}$$

The electron velocity $\mathbf{u}_e$ at the position R_p is the vector sum of the velocity $\Omega \times R$ of the nucleus at the position R and the time derivative of the displacement S. So,

$$u_e = (\Omega \times r) + \frac{dS}{dt}. \tag{10.12.4}$$

The displacement S is not only influenced by the EM field, but also by the rotation with the angular velocity Ω relative to the inertial frame K^0. Moreover, the velocity of the nucleus at the point R_p is given by

$$\mathbf{u}_i = (\Omega \times R_p) = (\Omega \times (R + S)). \tag{10.12.5}$$

Hence the current vector j of (10.12.2) can be written as

$$\begin{aligned} J &= Ne\Omega \times (R+S) - (N+N_1)e\left(\Omega \times R + \frac{dS}{dt}\right) \\ &= -Ne\left(\frac{dS}{dt} - \Omega \times S\right) - N_1 ev - N_1 e\frac{dS}{dt}, \end{aligned} \tag{10.12.6}$$

where we have put

$$v = (\Omega \times R) \tag{10.12.7}$$

and evaluated the displacement S at R neglecting the second order quantity $N_1 e(dS/dt)$. We obtain for the first order current density

$$j = -Ne\left(\frac{dS}{dt} - \Omega \times S\right) - N_1 ev = -Ne\left[\frac{\partial S}{\partial t} + (v\cdot\nabla)S - (\Omega \times S)\right] - \rho^c v. \tag{10.12.8}$$

In terms of S, correct up to the first order, the continuity condition gives

$$\rho^c = \operatorname{div}(N_0 eS) \tag{10.12.9}$$

for the charge density. Then since

$$\begin{aligned} \nabla \times [\mathbf{v} \times (NeS)] &= Ne(S\cdot\nabla)\mathbf{v} - (v\cdot\nabla)(NeS) + \mathbf{v}\cdot\nabla\cdot(NeS) - NeS(\nabla\cdot\mathbf{v}) \\ &= -Ne(v\cdot\nabla)S + \Omega \times (NeS) + \rho^c v + 0 \end{aligned} \tag{10.12.10}$$

equation (10.12.8) can be written as

$$j = -We\frac{\partial S}{\partial t} + \nabla \times [\mathbf{v} \times (NeS)]. \tag{10.12.11}$$

The polarization vector P and the magnetization M are defined as

$$P = -NeS, \quad M = -(v \times P)/c. \tag{10.12.12}$$

When a polarized dielectric slab is moving, a surface current flows in the direction of motion on one surface of the slab and another in the opposite direction on the other surface, forming a net current loop which gives rise to magnetic moment. The polarized dielectric slab consists of the aggregate of electric dipoles. So P and M obtain from the electron theory of Lorentz. In terms of ρ^c and j these are

$$\rho^c = -\nabla.P, \quad j = \frac{\partial P}{\partial t} + (\nabla \times M). \tag{10.12.13}$$

Also B and D can be defined as

$$B = H + 4\pi M, \quad D = E + 4\pi\rho^c. \tag{10.12.14}$$

► **Example 10.12.1** *Find the Lagrangian of this field and motion, the Poynting flux and* $(E \cdot j)$.

10.13 Sun, Earth and Other Space Bodies

In plasmas of space bodies thermal energy is larger than the magnetic energy. The magnetic Reynolds umber is small (≤ 30). So, the action of motion is weaker than the action of Ohmic dissipation. Small speed magnetic fields arise naturally in a plasma from the action of thermal motion or from plasma instabilities. A problem of the MHD dynamo theory for these bodies is to construct a velocity field, depending on the parameters specifying the body and its motion, which is capable of indefinite maintenance and amplification of an initially given seed field.

The huge planet Jupiter has radius $= 7 \times 10^9$ cm and surface angular velocity of rotation $= 10^{-4}$/sec. It has a liquid metal core of radius $= 5 \times 10^9$ cm and electrical conductivity $= 10^4$ cm^2/sec. The time of Ohmic dissipation of fields in the core is about 10^7 years. Pulsar magnetic fields of order of 10^{12} gauss are produced by the compression of magnetic fields of stars from which pulsars emerge. On large spatial scales the Universe is homogeneous and isotropic. Cosmic or global magnetic fields violate the isotropy of the Universe. However, these fields do not violate the homogeneity of the Universe.

10.13.1 Sun

The sun is a spheroidal body of radius about 7×10^{10} cm, mass 10^{33} gms, mean density 142 gm/cm^3, mean surface gravitational acceleration 2.73×10^4 cm/sec^2, and the period of rotation about axis 2.2×10^6 sec. At the center of the sun the gravitational pressure is greater than 10^{11} atmosphere and the temperature is 1 to 2 KeV. Hydrogen is belive to exist there at more than one thousand times liquid density.

The magnetic field at the surface of the sun is 1 to 2 gauss and sound speed is 10^6 cm sec^{-1}. The photosphere is a layer of thickness 1000 km, and inner radius 7×10^5 km, temperature 6000°K, mass density 10^{-7} gm cm^{-3}, the Alfven velocity in photosphere is about 10^3 cm/sec. In the inner part of the sun, the temperature is about 16 million degrees; so thermonuclear fusion action occurs there. Most of the solar energy is radiated from the photosphere. The solar corona is a low density high temperature ($= 10^6$ °K) region in solar surface. It radiates UV waves. The solar wind is the flux of ions at a steady rate from the solar corona.

Energy of radiation from the sun (and also from some stars) is produced in thermonuclear fusion reactions involving the hydrogen nuclei or ions. This reaction is less efficient than the reaction with $D-$ and $T-$ions; but the reaction volume in stars is very large. So the temperature is of the order of ten million degrees, which is considerably lower than the temperature of the laboratory thermonuclear plasma for a more favourable reaction.

The polar strength of the solar magnetic field is about 25 gauss, corresponding to a dipole moment 4.2×10^{33} gauss cm^3. Sunspot magnetic fields are about 4000 gauss. The solar magnetic field decreases rapidly further towards the solar corona from the photosphere. The greater part of the solar energy is contained in the range of radiation wavelength from 200 to 1200 nanometers (one nanometer is $10A^0 = 10^{-7}$ cm), the average spectral density is of the order of 1.2 Wm^{-2}/nm, and the total energy flux is about 1.2×10^3 W/m^2.

From the core to its surface, the sun is in a plasma state. Solar energy reaching the entire surface of the earth is about one by a billion of the energy radiated by sun. The solar energy is produced in the core of sun, core diameter is about one-third of the solar diameter (= 1.4 million km). Solar surface temperature is about 5,500°C, and solar core temperature is comparatively very high (of the order of 14,000,000°C). For this reason the core does not collapse inspite of the huge pressure exerted by a billion tonnes of overlying matter on every square meter. This causes collision between atoms at great speeds. The speed brings the ions close, for their fusion, inspite of the force of mutual electrostatic repulsion. Orbital electrons of the atoms are thus knocked out, causing ionization of matter and creation of positive ions (or, the bare nuclei). The solar plasma consists mainly of free electrons and ions of hydrogen nuclei. Usually, four protons fuse together in such fusion reactions, creating the α particle, which is helium nucleus. Though the probability of protons fusing together in sun is very small, the number of fusion reactions per second is sufficient to keep the sun burning.

The two processes causing fusion reaction are : (1) Proton-proton cycle, (2) Carbon cycle. In (1) two protons fuse together and the intermediate products are deuterons (deuterium nuclei) and helium 3 nuclei. In (2) the carbon acts as a catalyst. In both (1) and (2) a considerable amount of energy is liberated, because the mass of an alpha particle is about 99.3 per cent of the sum total of the masses of four protons; the missing mass is converted into energy, according to the Einstein formula $\xi = mc^2$. In the sun, thus 657 million tonnes of hydrogen are converted to 652.5 million tonnes of helium every second; the missing 4.5 million tonnes are transformed into that enormous amount of energy.

10.13.2 Earth

The earth is approximately a sphere of radius 6370 km, and mass 6×10^{24} kgm. It has a solid crust of thickness 35 km, a mantle of thickness 2900 Km and a core of radius 3500 Km. It's mean density is 5.5 gm/cm^2, mean distance from sun is 150×10^6 Km, period of rotation about sun is 1.000 year, mean surface gravitational acceleration is 981 cm/sec^2, escape velocity is 11.2 Km/sec, period of rotation about axis is 23.93 hours and orbital eccentricity is 0.017.

The velocity of earth round the sun is 30 Km/sec. The radiation pressure of sun light is 0.45×10^{-4} dynes cm^{-2}; the force of this pressure on the whole earth is 5.7×10^7 Kg wt; the gravitational attraction of sun on earth is 3.6×10^{21} Kg wt. cm^{-3}; so, comparatively, the radiation pressure is small for the earth. The radiation pressure being a surface force and gravity being a volume force, the ratio of the two becomes greater if the volume density is smaller. The deflection of a comet, when it comes near the sun, can be explained in this way.

The core, up to a radius of about 1200 km consists of solid alloys of iron and nickel at 2 to 3 million atmosphere, and temperature 5000°K. The electrical conductivity of the core is 3×10^{15} sec^{-1} and of the mantle is 3×10^{11} sec^{-1}. The magnetic viscosity, there, is 3×10^4 cm^2/s. The magnetic field in the inner core is 10 gauge, at the core surface it is 4 gauss, on the earth surface at the equator it is 0.3 gauss, at the north and south poles it is 0.6 gauss and at other surface regions it is 0.5 gauss. This field is equivalent to that of a dipole of moment 8.1×10^{25} gauss cm^3, at the center of the earth, if measured at distances greater than 10^5 Km. The axis of this field intersects the earth's surface in antipodal points situated at the latitude 78.5°S longitude 111°E, and at the latitude 78.5°N, longitude 69°W. The field strength decreases by about 0.1 per cent per year. The magnetic field has existed for not less than 10^9 years. Time of Ohmic decay of the field in the outer core layer, between 0.19 to 0.53 times the earth radius, is about 10^4 years. An intermediate liquid core, between 1200 and 3500 km, is of molten iron mainly, and some amount of sulphur and silicon. The outer solid mantle, covering from the radial distance of 3500 km up to the outer surface, is of solid ferromagnesium silicate. The magnetic field on earth's surface is about 1 gauss.

The acceleration from Sun's gravitational field on the orbit of the earth is about 0.5/sec^2 and equals the centrifugal acceleration due to earth's orbital motion. The effect of moon on the earth is an acceleration approximately equal to 4×10^{-3} cm/sec^2. The angular velocity of a point in the earth, about an axis pointed towards the polaris, is

$$\frac{2\pi}{24 \times 60 \times 60} \text{ rad/sec} = 7.3 \times 10^{-5} \text{ rad/sec.}$$

The angular velocity Ω of the solid mantle undergoes slow changes, both in direction and magnitude. The axis of rotation has a very slow precession Ω_p, about the normal to the plane of rotation about the sun; the time period is 2.6×10^4 years, or about 7.7×10^{-12} rad/sec and the corresponding cone has the semivertical angle 23.5°. An object at rest, on the surface of the earth, is acted upon by centrifugal forces, resulting from the diurnal rotation of the earth about its axis, the annual rotation of the earth about the sum, and the monthly rotation of the center of the earth about the center of gravity of the earth-moon system. This precession is due to the net torque exerted by sun and moon on the equatorial protrusion of the earth, and possibly leads to non-rigid rotation of the intermediates liquid core. Since the ellipticities of the two regions differ, the mean density of the core is much greater than that of the mantle. So the cores and the mantle can have very different directions of precessional rotation after several thousand years, due to the uncoupling of their motion, and larger relative velocities between the core and the mantle are possible. But the core and the mantle are strongly coupled together by viscous and EM means, keeping the precessional velocities of the two regions same, both in direction and magnitude. This precessional torque is a strong source of the core motion, leading to the magnetic field of earth (cf. Moffat, 1978, page 9, section 4.4). Other sources are the convection due to buoyancy forces of thermal or sedimentary origin. A source of the precessionally driven dynamo is a turbulent flow in the core due to instability of the boundary layer, where angular momentum is transferred from mantle to core and consequently secondary flows are developed.

In the liquid metallic core of the rotating earth the torque of the Coriolis force possibly balances the torque of the Lorentz force, generating thereby, slowly oscillating hydromagnetic waves. These slow oscillations are probably important in the production of the geomagnetic field by the hydromagnetic dynamo forces, Achesion and Hide (1973).

The high energy electrons and ions, due to radiation from the sun produces a high degree of ionization in the ionosphere in the day time. The ionization decreases at night. Moreover, other effects like the polar glow also occur.

The plasma ejected from the solar surface is captured by the terrestrial magnetic field to generate some radiation belts around the earth. The farthest such belt is situated between

20,000 km to 30,000 km and the nearest one is between 70 km and few thousand km. In the inner belt the radiation intensity is about 10^4 times greater than due to cosmic radiation; and the ion energies are 40 MeV and above. In the outer belt ion energies are about 0.1 MeV to a few MeV, and electron energies about 40 KeV.

10.14 Rotation, Magnetic Field and Turbulence

10.14.1 Hydromagnetic turbulence

In the turbulent motion whirls (eddies) appear, which are distinct fluid elements in a state of rotation. The larger eddies (rotations at large radii) are formed by the applied field. The smaller eddies have considerable viscous stress. The theory of turbulence deals with the mechanisms of transfer of energy of the larger eddies to smaller and smaller eddies, and with the modes of distribution of energy amongst the eddies of various size, in a stationary state. Though the large eddies make a medium anisotropic, it gradually loses its correlation with the structures of these eddies, due to the continuous transfer of their energy to the small eddies. Thus, turbulence becomes homogeneous, isotropic, and independent of the linear dimensions of a system, as well as of the manner in which energy is supplied; and the fluctuations acquire a certain order. Turbulence much depends on the rates of dissipation of energy per unit mass, and on the kinematic coefficient of viscosity, ν. As the viscosity decreases, the size of energy dissipating eddies also decreases. Two initially close elements of a moving fluid eventually diffuse some distance apart, tending, thereby, to lengthen the moving lines of vorticity in turbulent motion, like the magnetic field lines, and to draw out the magnetic lines, increasing thereby, the field intensity in conducting media. Diffuse motion of fluids draw out the magnetic lines of force, and dissipate energy, due to viscosity and electrical resistance, magnetic moment field generation and some other mechanisms are responsible for exchange of kinetic energy and magnetic energy. For large values of Reynolds number, the viscous damping is overcome and mean square vorticity maintains a steady value, though the magnetic energy is amplified (cf. equation (6.20) of Ferraro and Plumpton, 1966). Equipartition of energy between small-scale components of the motion and the corresponding small-scale field components, is more probable than that between the former and the large-scale field components.

A weak seed magnetic field is magnified to a large turbulent field in MHD turbulence. A large applied magnetic field inhibits the onset of turbulence; when the magnetic energy is much greater than the average kinetic energy of motion, the onset of convective turbulence in a fluid is delayed. The tension of the field lines excites Alfven waves (Exercise 4.4.3) in electrically conducting fluids, and prevents stretching by turbulence in a steady state.

Electromagnetic and hydrodynamic forces interact even in the absence of applied electric or magnetic field and electric charges. The average effects of turbulence amplify or suppress the small disturbing electric and magnetic fields produced by factors extraneous to the fluid motion. When turbulence is homogeneous, time is available for the distribution of turbulent energy of conducting fluids uniformly (on the average) over position space.

Ionized media are seldom completely quiet and stable; so, these are largely found in the turbulent state. Generally, the non-linear problems in plasma are studied like fluid turbulence problems, and with the help of the role which stochastic processes play in each case. The theory of non-linear effects of waves of large amplitude, causing mode coupling, is also identified with the theory of turbulence. Then the plasma is made strongly unstable by convective or turbulent means of transport. A turbulent plasma is different from the hypothetical plasma which appears coherent on a macroscopic scale, but which is stochastic on a microscopic level.

A state of turbulence of plasma is described statistically, like the turbulence of ordinary neutral fluids. Then energy is continuously exchanged amongst the different modes, and ordered energy of beams of charged particles and the energy of discrete oscillations are transferred into energy of random motion, and ultimately into heat due to turbulence. The anomalous resistivity of turbulence slows down the electrons by collisions with random electric field fluctuations, rather than with ions. Heat and mass transport in turbulent plasma depend on the energy of electromagnetic and kinetic fluctuations, after the turbulent instability is fully developed. The total energy is the sum of this energy and the free potential energy.

10.14.2 Turbulent fluctuations

The equation for the vorticity

$$\frac{\partial \Omega}{\partial t} = \nabla \times (v \times \Omega) + \nu \nabla^2 \Omega \tag{10.14.1}$$

is determined from the Navier-Stokes equation

$$\frac{\partial v}{\partial t} + (v\cdot\nabla)v = -\nabla\left(\frac{1}{\rho}p\right) + \nu\nabla^2 v \tag{10.14.2}$$

with the help of the vector formula

$$(\mathbf{v}\cdot\nabla)\mathbf{v} = \nabla\frac{v^2}{2} - \mathbf{v}\times(\nabla\times\mathbf{v}). \tag{10.14.3}$$

If Ω_0 is the constant angular velocity of rotation, then

$$\mathbf{v} = (\Omega_0 \times \mathbf{r}), \quad (\nabla\times\mathbf{v}) = 2\Omega_0 = \Omega. \tag{10.14.4}$$

Comparing with the equation for the magnetic field $\mathbf{H}$,

$$\frac{\partial \mathbf{H}}{\partial t} = \nabla\times(\mathbf{v}\times H) + \frac{c^2}{4\pi\sigma}\nabla^2 H, \tag{10.14.5}$$

we can write

$$H = \alpha\Omega, \tag{10.14.6}$$

$$\nu = c^2/4\pi\sigma, \tag{10.14.7}$$

where α is a constant. Hence the lines of magnetic force move with the fluid and H rotates at the rate Ω.

Local properties of turbulent flow (fluctuation spectrum in turbulent fluid) are studied for internal scale lengths λ_0 (Characteristic lengths of the smallest eddies) much smaller than the largest characteristic scale length l, which gives the size of the region in which turbulent flow occurs. In terms of l (the principal scale or the external scale), the Reynolds number R of the turbulent flow is given by

$$R = \frac{ul}{\nu},$$

where u is a typical speed of flow and is of the order of the change in mean velocity over the distance l.

The energy of a mode of scale length λ is transferred only to modes of scale nearly equal to λ, because energy is transferred between modes through resonant or parametric interactions. Thus, energy of modes of larger scale lengths is converted into energy of modes of smaller scale lengths, and the entire range of scales of almost up to the length λ, is obtained. For each scale λ the value of ε (the energy dissipated in unit mas of the fluid, in unit time, which determines the local properties of turbulence) is determined only by the fluctuation level at this scale. Over a wide range of scales of turbulent motion, the fluid viscosity is negligible when R is large, and energy diffuses as waves into the short wave length regions ($\lambda < \lambda_0$), where dissipation (negligible) occurs.

A steady turbulent velocity is reached over times which are of the order of the periods of the corresponding turbulent fluctuations, but are small compared with the total time during which the system is observed. If, at any two points of a turbulent flow, the instantaneous values of Ω are uncorrelated, the magnetic field H at these points will rotate in an uncorrelated manner, and their relative direction will vary randomly with time.

The correlation between the velocities v_1 and v_2, at two points 1 and 2, at a distance λ apart, is determined mainly by the eddies. We write $\Delta v = v_2 - v_1$; then the law of Kolmogorov and Obukhov for finding this correlation, Landau and Lifshitz (1960, section 55, pages 234–236), is

$$\overline{\Delta v_1 \Delta v_k} = \lambda^{2/3} \text{ for } \lambda \gg \lambda_0. \tag{10.14.8}$$

$$\overline{\Delta v_1 \Delta v_k} = \lambda^{2} \text{ for } \lambda \ll \lambda_0. \tag{10.14.9}$$

These relations also determine the correlation of angular velocities Ω_1 and Ω_2 of (10.14.6). In the derivatives of $\overline{\Delta v_i \Delta v_k}$ with respect to the co-ordinates of the point 1 once, and once of 2, putting $2\Omega_1 = \nabla \times v_1$, $2\Omega_2 = \nabla \times v_2$, we find that

$$\overline{\Omega_{1i}\Omega_{2k}} = \begin{cases} \left(\dfrac{1}{\lambda^2}\right)\lambda^{2/3} = \lambda^{-4/3} \text{for } \lambda \gg \lambda_0 \\[2ex] \left(\dfrac{1}{\lambda^2}\right)\lambda^{2} = \text{ constant for } \lambda \ll \lambda_0. \end{cases}$$

Hence, appreciable correlation between the angular velocities exists only at distances up to the order λ_0, falling rapidly at greater distances.

The energy of a mode of scale length λ is transferred only to modes of scale nearly equal to [illegible]

[illegible]

A [illegible]

[illegible]

[illegible]

[illegible]

[illegible]

Chapter 11

Tensor Results

11.1 Introduction

"If a transformation exists", writes Fock (1964), p. 54, on covariance, "such that the new functions, expressed in terms of the new variables, satisfy equations of the same form as do the old functions in the old variables, equations are termed covariant." In other words, covariant form means tensor description, be it in inertial frames, or in non-inertial frames. Pseudo cartesian tensors are necessary for classical electrodynamics, and Riemann tensors for electrodynamics of non-inertial frames. So, for convenience of reading, we here give some relevant results.

Relativistically, all rigid co-ordinate frames of reference, moving with uniform velocity relative to the fixed stars, are equivalent for study of dynamics. Thus, under Lorentz transformation, the dynamical equations of physics are covariant.

According to Einstein, both inertial systems and non-inertial systems are equivalent for the dynamical behaviour of material bodies. In accelerated systems, some force terms appear in the equation of motion, which are designated as fictitious forces, because these have no connection with the physical behaviour of systems. For example, in rotating frames the Coriolis force, the centrifugal force, etc. are the fictitious forces. In accelerated systems of references the fictitious forces are the expressions of difference between the basic dynamical equations in uniformly moving media and in accelerated media. Both these systems of references are to be regarded as completely equivalent for the field equations. The fictitious forces are to be regarded as real forces. In accelerated systems these forces occur because the distant masses of the fixed stars are accelerated relative to those systems. So, the fictitious forces are a type of gravitational force, and the acceleration of the distant masses produces a gravitational field which does not appear in inertial systems. The systems of inertia are special in the sense that it is not necessary to include the gravitation of the distant masses in these. The general principle of relativity is that all systems of reference are equivalent with respect to the fundamental laws of physics.

Like the gravitational forces (of the earth or sun as well as of those of the distant stars) the fictitious forces give all free bodies the same acceleration whatever be their masses. So, a gravitational field has a gravitational acceleration which is independent of the mass of a body. All types of gravitational fields are of the same nature and follow the same fundamental physical laws of nature. This is the principle of equivalence. The gravitational fields of the earth, sun and other nearby masses cannot be made to disappear by any choice of the reference systems. These fields are regarded as permanent gravitational fields.

11.2 Tensors

11.2.1 Flat space tensors

A scalar field variable $\phi(x, y, z)$ has only one component, or element. So $\phi(x, y, z)$ is a tensor of zero order. A vector F has 3-components or elements in the position space. So, F is a first order tensor in position space. More generally, in the pseudo cartesian 3 + 1 space (a 4-space) of the space time continuum the vector $\mathbf{F}$ has 4-components F_μ, where $\mu = 1, 2, 3, 4$ and so this F is a first order tensor. A second order tensor $\mathbf{u}$ has nine ($= 3^2$) components in 3-space and sixteen ($= 4^2$) components in 4-space. So,

$$F_{ij} = \begin{bmatrix} F_{11} & F_{12} & F_{13} \\ F_{21} & F_{22} & F_{23} \\ F_{31} & F_{32} & F_{33} \end{bmatrix} \quad \text{for } (i, j) = (1, 2, 3). \tag{11.2.1}$$

$$F_{\mu\nu} = \begin{bmatrix} F_{11} & F_{12} & F_{13} & F_{14} \\ F_{21} & F_{22} & F_{23} & F_{24} \\ F_{31} & F_{32} & F_{33} & F_{34} \\ F_{41} & F_{42} & F_{43} & F_{44} \end{bmatrix} \quad \text{for } (\mu, \nu) = (1, 2, 3, 4). \tag{11.2.2}$$

The rule of transformation from one linear space (unprimed) to another linear space (primed) for $F_{\mu\gamma}$ is

(11.2.3)

	F'_{11}	F'_{12}	F'_{13}	F'_{14}
F_{11}	α_{11}	α_{12}	α_{13}	α_{14}
F_{12}	α_{21}	α_{22}	α_{23}	α_{24}
F_{13}	α_{31}	α_{32}	α_{33}	α_{34}
F_{14}	α_{41}	α_{42}	α_{43}	α_{44}

This gives for example the algebraic linear relations

$$F_{12} = \alpha_{21}F'_{11} + \alpha_{22}F'_{12} + \alpha_{23}F'_{13} + \alpha_{24}F'_{14},$$

$$F'_{12} = \alpha_{12}F_{11} + \alpha_{22}F_{12} + \alpha_{32}F_{13} + \alpha_{42}F_{14},$$

The coefficients $\alpha_{\mu\gamma}$ are given by

$$\frac{\partial x_\nu}{\partial x'_\mu} = \alpha_{\mu\nu}. \tag{11.2.4}$$

For two 4-vectors **A** and **B**, the product elements

$$(\mathbf{AB}) = \begin{bmatrix} A_1B_1 & A_1B_2 & A_1B_3 & A_1B_4 \\ A_2B_1 & A_2B_2 & A_2B_3 & A_2B_4 \\ A_3B_1 & A_3B_1 & A_3B_3 & A_3B_4 \\ A_4B_1 & A_2B_2 & A_4B_3 & A_4B_4 \end{bmatrix} \tag{11.2.5}$$

have the transformation law,

		B_1	B_2	B_3	B_4
		A_1	A_2	A_3	A_4
B'_1	A'_1	α_{11}	α_{12}	α_{13}	α_{14}
B'_2	A'_2	α_{21}	α_{22}	α_{23}	α_{24}
B'_3	A'_3	α_{31}	α_{32}	α_{33}	α_{34}
B'_4	A'_4	α_{41}	α_{42}	α_{43}	α_{44}

(11.2.6)

Then if AB is a tensor, the elements $A_\mu B_\nu$ are its components. An example of transformation of $A_\mu B_\gamma$ to $A'_\mu B'_\gamma$ is

$$\begin{aligned} A'_2B'_3 &= (\alpha_{21}A_1 + \alpha_{22}A_2 + \alpha_{23}A_3 + \alpha_{24}A_4)(\alpha_{31}B_1 + \alpha_{32}B_2 + \alpha_{33}B_3 + \alpha_{34}B_4) \\ &= \alpha_{21}\alpha_{31}A_1B_1 + \alpha_{21}\alpha_{32}A_1B_2 + \alpha_{21}\alpha_{33}A_1B_3 + \alpha_{21}\alpha_{34}A_1B_4 \\ &\quad +\alpha_{22}\alpha_{31}A_2B_1 + \alpha_{22}\alpha_{32}A_2B_2 + \alpha_{22}\alpha_{33}A_2B_3 + \alpha_{22}\alpha_{34}A_2B_4 \\ &\quad +\alpha_{23}\alpha_{31}A_3B_1 + \alpha_{23}\alpha_{32}A_3B_2 + \alpha_{23}\alpha_{33}A_3B_3 + \alpha_{23}\alpha_{34}A_3B_4 \\ &\quad +\alpha_{24}\alpha_{31}A_4B_1 + \alpha_{24}\alpha_{32}A_4B_2 + \alpha_{24}\alpha_{33}A_4B_3 + \alpha_{24}\alpha_{34}A_4B_4 \\ &= \sum_{\mu=1}^{4}\sum_{\nu=1}^{4} \alpha_{2\mu}\alpha_{3\nu}A_\mu B_\nu. \end{aligned} \tag{11.2.7}$$

The general scheme

$$A_{\mu\nu} = \begin{bmatrix} A_{11} & A_{12} & A_{13} & A_{14} \\ A_{21} & A_{22} & A_{23} & A_{24} \\ A_{31} & A_{32} & A_{33} & A_{34} \\ A_{41} & A_{42} & A_{43} & A_{44} \end{bmatrix} \tag{11.2.8}$$

is a tensor if its components transform according to the rule

$$A'_{\mu\nu} = \sum_{\alpha=1}^{4}\sum_{\beta=1}^{4} \alpha_{\mu\alpha}\alpha_{\nu\beta}A_{\alpha\beta}. \tag{11.2.9}$$

The radius vector OP drawn in the 4-dimensional Minkowski world from the origin O to the point $P(x_1, x_2, x_3, x_4)$, is a 4-vector. Its components follow the transformation rules of (2.16.1), and satisfy the invariant relation (2.15.1). The displacement vector P_1P_2 of the two world points P_1 and P_2 in 4-space is a 4-vector. If the co-ordinates of P_1 and P_2 are (x_1, y_1, z_1, t_1) and (x_2, y_2, z_2, t_2) respectively, then

$$(x_2 - x_1)^2 + (y_2 - y_1)^2 + (z_2 - z_1)^2 - c^2(t_2 - t_1)^2$$
$$= (x_2^0 - x_1^0)^2 + (y_2^0 - y_1^0)^2 + (z_2^0 - z_1^0)^2 - c^2(t_2^0 - t_1^0)^2 \tag{11.2.10}$$

and $x_2 - x_1, y_2 - y_1, z_2 - z_1, t_2 - t_1$ transform according to the rules of (11.2.3).

The 4th component of the displacement vector $\mathbf{A}$ being $A_4 = ic(t_2 - t_1) = icu$ (say), we obtain

$$A^2 = A_1^2 + A_2^2 + A_3^2 - A_4^2 \tag{11.2.11}$$

A^2 may be positive, negative or zero. If $A^2 > 0$, A is a space-like or a spatial vector. If $A^2 < 0, A$ is a time-like or a temporal vector. If $\mathbf{A} = 0$ without being identically equal to zero, it is a singular vector. By rotation of axes, a space-like vector A coincides with the space co-ordinate axes OX_1, OX_2, OX_3 but not with time axis OX_4 which is imaginary. Thus, a time-like vector t can coincide with time axis only but not with any of the space axes since t is imaginary. The scalar product

$$(\mathbf{A.B}) = A_1B_1 + A_2B_2 + A_3B_3 + A_4B_4 = A_\mu B_\mu \tag{11.2.12}$$

of the two 4-vectors A and B is an invariant, where $\mu = 1, 2, 3, 4$. The space-like and time-like vectors have the character of the time axis and the space axis. Transformation for any real rotation can never make a space-axis coincide with the time axis, because the length measuring rod can never be used as a clock. So, we cannot describe this space-time continuum by a 4-dimensional system of co-ordinates. We describe this 4-space by a 3 + 1 dimensional system, which is called *the 4-space of pseudo cartesian co-ordinates.*
In this 4-space, the contravariant 4-vector of a point p is

$$x^\mu \equiv (r, ct), \tag{11.2.13}$$

where r is the position vector (a 3-vector) of the point p in the position space. The corresponding covariant 4-vector is

$$x_\mu \equiv (r, -ct). \tag{11.2.14}$$

The square of the 4-space subinterval ds is given by

$$ds^2 = dx^2 + dy^2 + dz^2 - (cdt)^2 = \eta_{\mu\nu} dx^\mu dx^\nu, \tag{11.2.15}$$

where

$$\eta_{\mu\nu} = \begin{bmatrix} 1 & 0 & 0 & 0 \\ 0 & 1 & 0 & 0 \\ 0 & 0 & 1 & 0 \\ 0 & 0 & 0 & -1 \end{bmatrix} \tag{11.2.16}$$

$\eta_{\mu\nu}$ is called the *fundamental metric tensor.* The square ds^2 is not a positive definite. When $ds^2 < 0$, the 4-vector ds_μ is time-like; when $ds^2 > 0$, ds_μ is space-like; and when $ds^2 = 0$, ds_μ is a null-vector. Since ds^2 can change sign, the increments ds_μ follow the rules for transformation of pseudo cartesian tensors.

A tensor $A_{\mu\nu}$ is symmetric if

$$A_{\mu\nu} = A_{\nu\mu}, \tag{11.2.17}$$

anti-symmetric if

$$A_{\mu\nu} = -A_{\nu\mu} \tag{11.2.18}$$

and unsymmetrical if for all μ and ν

$$A_{\mu\nu} \neq \pm A_{\mu\nu} \tag{11.2.19}$$

A tensor is resolved into a symmetric and an anti-symmetric tensor (the addition rule being the usual addition of terms), because

$$A_{\mu\nu} = \frac{1}{2}(A_{\mu\nu} - A_{\nu\mu}) + \frac{1}{2}(A_{\mu\nu} + A_{\nu\mu}). \tag{11.2.20}$$

The part $A_{\mu\nu} - A_{\nu\mu}$ is antisymmetric (since it changes sign if μ and ν are interchanged) and $A_{\mu\nu} + A_{\nu\mu}$ is symmetric (since it does not change sign if μ and ν are interchanged).

The antisymmetric tensor

$$T_{\mu\nu} = A_\mu B_\nu - A_\nu B_\mu \tag{11.2.21}$$

has six independent components T_{12}, T_{13}, T_{14}, T_{23}, T_{24}, T_{34} and the 4-diagonal components vanish. It is a 6-vector, and is an antisymmetric tensor of rank (or order) two. Such six vectors are represented as

$$T_{\mu\nu} = \begin{bmatrix} 0 & T_{12} & T_{13} & T_{14} \\ -T_{12} & 0 & T_{23} & T_{24} \\ -T_{13} & -T_{23} & 0 & T_{34} \\ -T_{14} & -T_{24} & -T_{34} & 0 \end{bmatrix}, \tag{11.2.22}$$

where $T_{\mu\nu} = -T_{\nu\mu}$ with $\mu \neq \nu$. All its terms, excepting those containing the index four only once, are real. 6-vectors are the 4-dimensional generalizations of vector products in 3-dimensions, because $(A \times B)(= A_y B_z - A_z B_y, A_z B_x - A_x B_z, A_x B_y - A_y B_z)$ is the vector product of two vectors A and B in 3-dimensions.

11.2.2 Transformation relations

Consider the passage from a frame of reference $0^0X^0Y^0Z^0t^0$, denoted by K^0, another frame of reference $OXYZt$, denoted by K; K moving relative to K^0 with velocity v. The co-ordinates of a point P with respect to K^0 are denoted by x^{μ^0} and with respect to K by x^μ with $\mu = 1, 2, 3, 4$. For transformation of the co-ordinates of P from K^0 to K write

$$x^\mu = x^\mu(x^{1^0}, x^{2^0}, x^{3^0}, x^{4^0}). \tag{11.2.23}$$

The right hand side being a well behaved function of the four co-ordinates x^{μ^0}, the rule of transformation of the differential of x^μ is represented by

$$dx^\mu = \alpha^\mu_{\nu^0} dx^{\nu^0}, \tag{11.2.24}$$

where the Einstein rule of use of repeated index to denote summation over that index is followed, with

$$\alpha^{\mu}_{\nu^0} = \frac{\partial x^{\mu}}{\partial x^{\nu^0}}. \tag{11.2.25}$$

The reverse rule of transformation is

$$dx^{\mu^0} = \alpha^{\mu^0}_{\nu}\, dx^{\nu} \tag{11.2.26}$$

with

$$\alpha^{\mu^0}_{\nu} = \frac{\partial x^{\mu^0}}{\partial x^{\nu}}. \tag{11.2.27}$$

The contravariant and convariant 4-differentials dx^{μ} and dx_{μ}, respectively, are given by

$$dx^{\mu} \equiv (dx, dy, dz, cdt), \quad U^{\mu} = \frac{dx^{\mu}}{dT}. \tag{11.2.28}$$

$$dx_{\mu} \equiv (dx, dy, dz, -cdt), \quad U_{\mu} = \frac{dx_{\mu}}{dT}. \tag{11.2.29}$$

Differentiation with respect to the components of a covariant 4-vector, Jackson (1975, section 1.16), gives a contravariant 4-differential operator, and analogously, differentiation with respect to the components of a contravariant 4-vector gives a covariant 4-differential operator; so

$$\partial^{\mu} \equiv \frac{\partial}{\partial x_{\mu}} \equiv \left(\nabla, \frac{\partial}{\partial x_4}\right) \equiv \left(\nabla, -\frac{\partial}{\partial x^4}\right), \quad \frac{\partial}{\partial x_4} = -\frac{1}{c}\frac{\partial}{\partial t}, \tag{11.2.30}$$

$$\partial_{\mu} \equiv \frac{\partial}{\partial x^{\mu}} \equiv \left(\nabla, \frac{\partial}{\partial x^4}\right) \equiv \left(\nabla, -\frac{\partial}{\partial x_4}\right), \quad \frac{\partial}{\partial x^4} = \frac{1}{c}\frac{\partial}{\partial t}, \tag{11.2.31}$$

Hence the invariant 4-divergence of a 4-vector A^{μ} is

$$\partial_{\mu} A^{\mu} = \partial^{\mu} A_{\mu} = \frac{\partial A^{\mu}}{\partial x^{\mu}} = \frac{\partial A_{\mu}}{\partial x_{\mu}}. \tag{11.2.32}$$

If ds is the invariant space-time interval, then

$$ds^2 = g_{\mu\nu} dx^{\mu} dx^{\nu} = g_{\mu^0\nu^0} dx^{\mu^0} dx^{\nu^0}. \tag{11.2.33}$$

The metric elements of $g_{\mu\gamma}$ being symmetric,

$$g_{\mu\gamma} = g_{\gamma\mu}. \tag{11.2.34}$$

During motion matter is concentrated in a somewhat finite 3-dimensional region V in position space; and this matter in motion induces the metric tensor $g_{\mu\gamma}$. In an arbitrary system of reference, $g_{\mu\gamma}$ represents the gravitational field of an isolated system. The deviations of the quantities $g_{\mu\gamma}$ from $\eta_{\mu\gamma}$ of (11.2.16),

$$\delta g_{\mu\nu} = g_{\mu\nu} - \eta_{\mu\nu}, \tag{11.2.35}$$

should vanish together with their derivatives at infinite distances in position space at least inversely to the first power of that distance.

The rules of transformation of the first order contravariant and covariant tensors, respectively, are

$$A^{\mu} = \alpha^{\mu}_{\nu^0} A^{\nu^0}. \tag{11.2.36}$$

$$A_{\mu} = \alpha^{\nu^0}_{\mu} A_{\nu^0}. \tag{11.2.37}$$

These, of the second order tensors, $A^{\mu\gamma}$ and $A_{\mu\gamma}$, are, respectively,

$$A^{\mu\nu} = \alpha^{\mu}_{\eta_0} \alpha^{\nu}_{\zeta_0} A^{\eta^0 \zeta^0}. \tag{11.2.38}$$

$$A_{\mu\nu} = \alpha^{\eta^0}_{\mu} \alpha^{\zeta^0}_{\nu} A_{\eta^0 \zeta^0}. \tag{11.2.39}$$

The proper time T is defined by the relations

$$ds^2 = g_{\mu\gamma} dx^{\mu} dx^{\gamma} = -c^2 dT^2, \tag{11.2.40}$$

where (11.2.33) has been used. The relativistic factor Γ is given by the relation

$$\left(\frac{dT}{dt}\right)^2 = \frac{1}{\Gamma^2}. \tag{11.2.41}$$

Another relativistic factor γ and the two 4-velocities u^{μ} and U^{μ} are defined as

$$\gamma^2 = \frac{1}{1-\beta^2}, \quad \beta = \frac{v}{c} \tag{11.2.42}$$

$$u^{\mu} = \frac{dx^{\mu}}{dt}, \tag{11.2.43}$$

$$U^{\mu} = \frac{dx^{\mu}}{dT}. \tag{11.2.44}$$

Evidently,

$$u^2 = u^{\mu} u_{\mu} = g_{\mu\nu} u^{\mu} u^{\nu} = g_{\mu\nu} \frac{dx^{\mu}}{dt}\frac{dx^{\nu}}{dt} = \left(\frac{ds}{dt}\right)^2 \tag{11.2.45}$$

and

$$U^2 = U^{\mu} U_{\mu} = g_{\mu\nu} U^{\mu} U^{\nu} = \left(\frac{ds}{dt}\right)^2 \left(\frac{dt}{dT}\right)^2 = c^2 = \Gamma^2 g_{\mu\nu} \frac{dx^{\mu}}{dt}\frac{d\,\cdot}{dt} \tag{11.2.46}$$

u^{μ} may be called the *ordinary 4-velocity* and U^{μ} the Minkowski 4-velocity tensor as in STR. The introduction of the infinitesimal time element dt, called the *ordinary time interval* for two events which occur with the same spatial co-ordinates x^1, x^2, x^3, along with the proper time sub-interval dT define these two 4-velocities. For stationary points in K, $U^i = 0$ and so

$$\Gamma = \gamma. \tag{11.2.47}$$

An important quantity in tensor theory is the Christoffel symbol $\Gamma^{\delta}_{\beta\gamma}$, defined by

$$\Gamma^{\delta}_{\beta\gamma} = \frac{1}{2g_{\alpha\delta}} \left[\frac{\partial g_{\alpha\beta}}{\partial x^{\gamma}} + \frac{\partial g_{\alpha\gamma}}{\partial x^{\beta}} - \frac{\partial g_{\beta\gamma}}{\partial x^{\alpha}}\right]. \tag{11.2.48}$$

The Schouten square bracket convention, Schouten (1954a,b), is used to denote the alternate part of a tensor

$$u^{[\mu}{}_{v}{}^{\nu]} = \frac{1}{2!}(u^{\mu} v^{\nu} - u^{\nu} v^{\nu}). \tag{11.2.49}$$

$$u^{[\lambda}{}_v{}^{\mu}{}_w{}^{\nu]} = \frac{1}{3!}(u^\lambda v^\mu w^\nu + u^\mu v^\nu w^\lambda + u^\nu v^\lambda w^\mu - u^\lambda v^\nu w^\mu - u^\mu v^\lambda w^\nu - u^\nu v^\mu w^\nu), \tag{11.2.50}$$

$$\partial_{[\mu} A_{\nu]} = \frac{1}{2!}(\partial_\mu A_\nu - \partial_\nu A_\mu), \tag{11.2.51}$$

$$P_{[\alpha\beta\gamma]} = \frac{1}{3!}(P_{\alpha\beta\gamma} + P_{\beta\gamma\alpha} + P_{\gamma\alpha\beta} - P_{\alpha\gamma\beta} - P_{\beta\alpha\gamma} - P_{\gamma\beta\alpha}), \tag{11.2.52}$$

$$P_{(\alpha\beta\gamma)} = \frac{1}{3!}(P_{\alpha\beta\gamma} + P_{\beta\gamma\alpha} + P_{\gamma\alpha\beta} + P_{\alpha\gamma\beta} + P_{\beta\alpha\gamma} + P_{\gamma\beta\alpha}) \tag{11.2.53}$$

The transformation of the 4-gradient vector $\partial\phi/\partial x_\mu$ is

$$\frac{\partial\phi}{\partial x'_1} = \frac{\partial\phi}{\partial x_1}\frac{\partial x_1}{\partial x'_1} + \frac{\partial\phi}{\partial x_2}\frac{\partial x_2}{\partial x'_1} + \frac{\partial\phi}{\partial x_3}\frac{\partial x_3}{\partial x'_1} + \frac{\partial\phi}{\partial x_4}\frac{\partial x_4}{\partial x'_1}, \tag{11.2.54}$$

where the 4-vector **F** now stands for x_μ of a point P in the 4-space of position co-ordinates and time. We find that

$$\frac{\partial\phi}{\partial x'_1} = \alpha_{11}\frac{\partial\phi}{\partial x_1} + \alpha_{12}\frac{\partial\phi}{\partial x_2} + \alpha_{13}\frac{\partial\phi}{\partial x_3} + \alpha_{14}\frac{\partial\phi}{\partial x_4} = \alpha_{1\mu}\frac{\partial\phi}{\partial x_\mu}. \tag{11.2.55}$$

Hence the transformation rule is

$$\frac{\partial\phi}{\partial x'_\mu} = \alpha_{\mu\nu}\frac{\partial\phi}{\partial x_\nu}. \tag{11.2.56}$$

The 4-divergence

$$\operatorname{div}\mathbf{A}\left(= \frac{\partial A_1}{\partial x_1} + \frac{\partial A_2}{\partial x_2} + \frac{\partial A_3}{\partial x_3} + \frac{\partial A_4}{\partial x_4}\right) \tag{11.2.57}$$

is an invariant, because

$$\frac{\partial A'_\mu}{\partial x'_\mu} = \left(\alpha_{\mu\nu}\frac{\partial}{\partial x_\nu}\right)(\alpha_{\mu\eta}A_\eta) = \frac{\partial A_\mu}{\partial x_\mu}. \tag{11.2.58}$$

If $A_\mu = \partial\phi/\partial x_\mu$, then

$$\operatorname{div}\operatorname{grad}\phi = \frac{\partial}{\partial x_\mu}\frac{\partial}{\partial x_\mu}\phi = \Box^2\phi \tag{11.2.59}$$

is an invariant, where $\Box^2$ is the D'Alembertian. The curl of a 4-vector,

$$(\operatorname{curl} A)_{\mu\nu} = \frac{\partial A_\nu}{\partial x_\mu} - \frac{\partial A_\mu}{\partial x_\nu} \tag{11.2.60}$$

is an antisymmetric tensor of the second order.

11.3 The Spatial Metric Tensor γ_{ij} and Some 3-Vectors

The relation (11.2.16) can be written as

$$ds^2 = dl^2 - \left(a_i dx^i - \sqrt{-g_{44}}dx^4\right)^2 \tag{11.3.1}$$

with

$$dl^2 = \gamma_{ij}dx^i dx^j, \tag{11.3.2}$$

$$\gamma_{ij} = g_{ij} - \frac{1}{g_{44}} g_{4i} g_{4j} = g_{ij} + a_i a_j, \tag{11.3.3}$$

$$a_i = \frac{1}{\sqrt{-g_{44}}} g_{4i} \tag{11.3.4}$$

γ_{ij} is the ijth component of the spatial metric tensor given by (11.3.2). Then from the relation (11.2.42) we find that

$$\frac{1}{\Gamma} = \frac{dT}{dt} = \frac{1}{c}\left[\left(a_i u^i - c\sqrt{-g_{44}}\right)^2 - i^2\right]^{1/2}, \tag{11.3.5}$$

where

$$u^i = \frac{dx^i}{dt}, \quad i^2 = \left(\frac{dl}{dt}\right)^2 = \gamma_{ij} u^i u^j. \tag{11.3.6}$$

Since both metrices $g_{\mu\gamma}$ and $g^{\mu\gamma}$ are non-singular, we have

$$g^{i\mu} g_{\mu k} = g^{ij} g_{jk} + g^{i4} g_{4k} = \delta^i_k, \tag{11.3.7}$$

$$g^{i\mu} g_{\mu\alpha} = g^{ij} g_{j4} + g^{i4} g_{44} = \delta^i_4 = 0, \tag{11.3.8}$$

$$g^{4\mu} g_{\mu 4} = g^{41} g_{41} + g^{44} g_{44} = \delta^4_4 = 1. \tag{11.3.9}$$

The relation (11.3.8) gives

$$g^{i4} = -\frac{1}{g_{44}} g^{ij} g_{j4}. \tag{11.3.10}$$

This relation together with (11.3.7) yields

$$g^{i4} = -\frac{1}{g_{44}} g^{ij} g_{j4}. \tag{11.3.11}$$

Using, therefore, (11.3.3) we can now write

$$g^{ij} \gamma_{ij} = \delta^i_k. \tag{11.3.12}$$

Then, since

$$\gamma^{ij} \gamma_{jk} = \delta^i_k, \tag{11.3.13}$$

it is evident that

$$\gamma^{ij} = g^{ij}. \tag{11.3.14}$$

Equations (11.3.4), (11.3.8) and (11.3.10) give

$$\gamma^{ij} a_j \sqrt{-g_{44}} + g^{i4} g_{44} = 0. \tag{11.3.15}$$

Then writing

$$a^i = \sqrt{-g_{44}} g^{i4} \tag{11.3.16}$$

equation (11.3.15) gives

$$a^i = \gamma^{ij} a_j. \tag{11.3.17}$$

So, a^i may be regarded as the contravariant 3-vector, corresponding to the covariant 3-vector a_i. Equations (11.3.9) show that

$$g^{44} = \frac{1}{g_{44}}(1 - g^{4j} g_{j4}) = \frac{1}{g_{44}}(1 - a^j a_j). \tag{11.3.18}$$

Next, denoting the determinant of the spatial matrix γ_{ij} by the symbol $\hat{\gamma}$ we write

$$\det \gamma_{ij} = |\gamma_{ij}| = \hat{\gamma}, \tag{11.3.19}$$

where

$$\hat{\gamma} = \det \gamma_{ij} = \begin{vmatrix} \gamma_{11} & \gamma_{12} & \gamma_{13} \\ \gamma_{21} & \gamma_{22} & \gamma_{23} \\ \gamma_{31} & \gamma_{32} & \gamma_{33} \end{vmatrix} = \frac{1}{g_{44}} \begin{vmatrix} \gamma_{11} & \gamma_{12} & \gamma_{13} & \gamma_{14} \\ \gamma_{21} & \gamma_{22} & \gamma_{23} & \gamma_{24} \\ \gamma_{31} & \gamma_{32} & \gamma_{33} & \gamma_{34} \\ \gamma_{41} & \gamma_{42} & \gamma_{43} & \gamma_{44} \end{vmatrix}. \tag{11.3.20}$$

A property of determinants is that they are unchanged if the elements of one column are multiplied by a common factor, and after that, these are added to the corresponding elements of another column. So, adding the 4-quantities $(g_{\mu 4} g_{14})/g_{44}$ with the elements of the first column of (11.3.20), we obtain

$$\hat{\gamma} g_{44} = \begin{vmatrix} \gamma_{11} + \dfrac{1}{g_{44}} g_{14} g_{14} & \gamma_{12} & \gamma_{13} & g_{14} \\ \gamma_{21} + \dfrac{1}{g_{44}} g_{24} g_{14} & \gamma_{22} & \gamma_{23} & g_{24} \\ \gamma_{31} + \dfrac{1}{g_{44}} g_{34} g_{14} & \gamma_{32} & \gamma_{33} & g_{34} \\ 0 + g_{44} g_{14}/g_{44} & 0 & 0 & g_{44} \end{vmatrix}. \tag{11.3.21}$$

Then using (11.3.3) this relation simplifies to

$$\hat{\gamma} g_{44} = \begin{vmatrix} g_{11} & \gamma_{12} & \gamma_{13} & g_{14} \\ g_{21} & \gamma_{22} & \gamma_{23} & g_{24} \\ g_{31} & \gamma_{32} & \gamma_{33} & g_{34} \\ g_{41} & 0 & 0 & g_{44} \end{vmatrix}. \tag{11.3.22}$$

Treating, similarly, the second and third columns of (11.3.20) we obtain the useful relations

$$\hat{\gamma} g_{44} = \begin{vmatrix} g_{11} & g_{12} & g_{13} & g_{14} \\ g_{21} & g_{22} & g_{23} & g_{24} \\ g_{31} & g_{32} & g_{33} & g_{34} \\ g_{41} & g_{42} & g_{43} & g_{44} \end{vmatrix}, \tag{11.3.23}$$

$$\hat{\gamma} = \det \gamma_{ij} = \frac{g}{g_{44}}. \tag{11.3.24}$$

Equations (11.3.13) gives the following nine algebraic relations

$$1 = \gamma^{11}\gamma_{11} + \gamma^{12}\gamma_{12} + \gamma^{13}\gamma_{13}, \tag{11.3.25}$$

$$0 = \gamma^{11}\gamma_{12} + \gamma^{12}\gamma_{22} + \gamma^{13}\gamma_{23}, \tag{11.3.26}$$

$$0 = \gamma^{11}\gamma_{13} + \gamma^{12}\gamma_{23} + \gamma^{13}\gamma_{33}, \tag{11.3.27}$$

$$0 = \gamma^{21}\gamma_{11} + \gamma^{22}\gamma_{21} + \gamma^{23}\gamma_{31}, \tag{11.3.28}$$

$$1 = \gamma^{21}\gamma_{12} + \gamma^{22}\gamma_{22} + \gamma^{23}\gamma_{32}, \tag{11.3.29}$$

$$0 = \gamma^{21}\gamma_{13} + \gamma^{22}\gamma_{23} + \gamma^{23}\gamma_{33}, \tag{11.3.30}$$

$$0 = \gamma^{31}\gamma_{11} + \gamma^{32}\gamma_{21} + \gamma^{33}\gamma_{31}, \tag{11.3.31}$$

$$0 = \gamma^{31}\gamma_{12} + \gamma^{32}\gamma_{22} + \gamma^{33}\gamma_{32}, \tag{11.3.32}$$

$$1 = \gamma^{31}\gamma_{13} + \gamma^{32}\gamma_{23} + \gamma^{33}\gamma_{33}. \tag{11.3.33}$$

Solving for $\gamma_{11}, \gamma_{12}, \gamma_{13}$ from the relations (11.3.25), (11.3.28) and (11.3.31) we obtain

$$\gamma_{11} = \frac{1}{\hat{\gamma}}(\gamma^{22}\gamma^{33} - \gamma^{23}\gamma^{32}), \tag{11.3.34}$$

$$\gamma_{12} = \frac{1}{\hat{\gamma}}(\gamma^{23}\gamma^{31} - \gamma^{21}\gamma^{33}), \tag{11.3.35}$$

$$\gamma_{13} = \frac{1}{\hat{\gamma}}(\gamma^{21}\gamma^{32} - \gamma^{31}\gamma^{23}), \tag{11.3.36}$$

because

$$\det \gamma^{ij} = \hat{\gamma} = \det \gamma_{ij}. \tag{11.3.37}$$

The vector

$$C = (A \times B) \tag{11.3.38}$$

is dual to the antisymmetric tensor whose ijth component is c_{ij}, where

$$C_{ij} = A_j B_i - A_i B_j. \tag{11.3.39}$$

The relations of transformation of C are

$$C_i = \frac{1}{2}\sqrt{\hat{\gamma}}\, e_{ijk} c^{jk} = \sqrt{\hat{\gamma}}\, e_{ijk} A^j B^k = (A \times B)^i. \tag{11.3.40}$$

$$C^i = \frac{1}{2\sqrt{\hat{\gamma}}}\, e^{ijk} c_{ijk} = \frac{1}{\sqrt{\hat{\gamma}}}\, e^{ijk} A_j B_k = (A \times B)_i. \tag{11.3.41}$$

11.4 Covariant Derivatives

The covariant derivatives of some contravariant and covariant tensors in terms of the Christoffel symbols are

$$A^{\mu}_{;\nu} = \frac{\partial A^{\mu}}{\partial x^{\nu}} + \Gamma^{\mu}_{\lambda\nu} A^{\lambda}$$

$$A_{\mu;\nu} = \frac{\partial A_{\mu}}{\partial x_{\nu}} - \Gamma^{\lambda}_{\mu\nu} A^{\lambda}$$

$$A^{\mu\nu}_{;\lambda} = \frac{\partial A^{\mu\nu}}{\partial x^{\lambda}} + \Gamma^{\lambda}_{\alpha\lambda} A^{\alpha\nu} + \Gamma^{\nu}_{\alpha\lambda} A^{\mu\alpha}$$

$$A^{\lambda}_{\mu;\nu} = \frac{\partial A^{\lambda}_{\mu}}{\partial x^{\lambda}} - \Gamma^{\alpha}_{\mu\nu} A^{\lambda}_{\alpha} + \Gamma^{\lambda}_{\alpha\nu} A^{\alpha}_{\mu},$$

$$A_{\lambda\mu;\nu} = \frac{\partial A_{\lambda\mu}}{\partial x^{\nu}} - \Gamma^{\alpha}_{\lambda\nu} A_{\alpha\mu} - \Gamma^{\alpha}_{\mu\nu} A_{\lambda\alpha}.$$

Here semi-colon as a subscript denotes the covariant derivative with respect to the independent variable which appears after it and a coma denotes the ordinary derivative. The generalized divergence of a 4-vector is

$$A^{\mu}_{;\mu} = \partial(\sqrt{g}A^{\nu})\sqrt{g}\partial x^{\mu}.$$

Transformation relation of the 4-integral of a 4-divergence over a 4-volume R into the 3-integral over a 3-dimensional hyper surface S is

$$\iiiint_R A^{\mu}_{;\mu} d\mathbf{R} = \iiint_S A^{\mu}\sqrt{g}ds_{\mu}.$$

Some other useful covariant derivatives are

$$\text{curl}\, A_{\lambda\mu} = A_{\mu;\lambda} - A_{\lambda;\mu} = \frac{\partial A_{\mu}}{\partial x_{\nu}} - \frac{\partial A_{\nu}}{\partial x_{\mu}}.$$

$$\Box^2\phi = \text{div grad}\,\phi = \frac{1}{\sqrt{g}}\frac{\partial}{\partial x^{\mu}}\left(\sqrt{g}g^{\mu\nu}\frac{\partial g}{\partial x^{\nu}}\right)$$

$$\text{curl}_{\lambda}\mu_{\nu}A = A_{\lambda\mu;\nu} + A_{\mu\nu;\lambda} + A_{\nu\lambda;\mu}$$

where e_{ijk} and e^{ijk} are the antisymmetric unit tensors of third order, Appendix 4, with

$$e_{123} = e^{123} = 1,$$

etc., and the sign changes under the transformation of indices. The converse relations, with respect to (11.3.39), are

$$c_{ij} = \sqrt{\hat{\gamma}}e_{ijk}c^{i}, \quad c^{ij} = \frac{1}{\hat{\gamma}}e^{ijk}c_k. \tag{11.4.1}$$

Chapter 12

Relevant Electrodynamics in Inertial Frames

12.1 Introduction

This chapter is supplementary to the coverage of classical electrodynamics in the earlier chapters. For convenience of subsequent presentation of the non-inertial electrodynamics, in a gradual manner, the fundamentals of the classical theory particularly the uncovered general and basic aspects are considered. Both vector form and tensor form (pseudo cartesian) descriptions and the transformation properties of the field quantities have been discussed.

12.2 Covariant Formulation of the Maxwell Equations in Inertial Frames

In vacuum, the Lorentz gauge condition, in tensor form, reads

$$\partial_\mu A^\mu = 0, \quad \text{or,} \quad \frac{\partial A^\mu}{\partial x^\mu} = 0, \quad \text{or,} \quad \frac{\partial A_\mu}{\partial x_\mu} = 0, \quad \text{or,} \quad \partial^\mu A_\mu = 0, \tag{12.2.1}$$

where the 4-vector A^μ is given by

$$A^\mu \equiv (A, -\phi) \quad \text{or,} \quad A_\mu \equiv (A, \phi), \tag{12.2.2}$$

where A is the vector potential and ϕ is the scalar potential. The X components of the equations (1.9.1) and (1.9.3) are

$$B_x = \frac{\partial A_z}{\partial y} - \frac{\partial A_y}{\partial z} = \frac{\partial A^3}{\partial x_2} - \frac{\partial A^2}{\partial x_3} = \partial^2 A^3 - \partial^3 A^2, \tag{12.2.3}$$

$$E_x = -\frac{1}{c}\frac{\partial A_x}{\partial t} + \frac{\partial \phi}{\partial x} = \frac{\partial A'}{\partial x_4} - \frac{\partial A^4}{\partial x_1} = \partial^4 A' - \partial' A^4. \tag{12.2.4}$$

These lead to the tensor relations

$$B_i = \partial^j A^k - \partial^k A^j, \quad E_i = \partial^4 A^i - \partial^i A^4, \tag{12.2.5}$$

where $(i, j, k) = (1, 2, 3)$. Then, writing

$$B^i = F^{jk} = \partial^j A^k - \partial^k A^j, \tag{12.2.6}$$

$$E_i = F^{4i} = \partial^4 A^i - \partial^i A^4, \tag{12.2.7}$$

we note that the 3-vectors B and E become the elements of a second order antisymmetric contravariant tensor $F_{\mu\gamma}$, with

$$F^{\mu\nu} = \partial^\mu A^\nu - \partial^\nu A^\mu = \frac{\partial A^\nu}{\partial x_\mu} - \frac{\partial A^\mu}{\partial x_\nu}, \tag{12.2.8}$$

where $(\lambda, \mu, \nu) = (1, 2, 3, 4)$. Similarly, we have

$$F_{\mu\nu} = \partial_\mu A_\nu - \partial_\nu A_\mu = \frac{\partial A_\nu}{\partial x^\mu} - \frac{\partial A_\mu}{\partial x^\nu}. \tag{12.2.9}$$

The matrix of $F^{\mu\gamma}$ is then found to be

$$F^{\mu\nu} = \begin{matrix} & \to \nu \\ \downarrow \\ \mu \end{matrix} \begin{bmatrix} 0 & B_z & -B_y & -E_x \\ -B_z & 0 & B_x & -E_y \\ B_y & -B_x & 0 & -E_z \\ E_x & E_y & E_z & 0 \end{bmatrix} \tag{12.2.10}$$

The x component of (1.4.3),

$$\frac{\partial E_z}{\partial y} - \frac{\partial E_y}{\partial z} = -\frac{1}{c}\frac{\partial B_x}{\partial t} \tag{12.2.11}$$

with the help of (12.2.10), takes the form

$$\frac{\partial F^{4k}}{\partial x_j} + \frac{\partial F^{4j}}{\partial x_k} = \frac{\partial F^{jk}}{\partial x_4}. \tag{12.2.12}$$

For the covariant tensor $F_{\mu\gamma}$, consider the relations

$$B_i = F^{jk} = F_{jk}. \tag{12.2.13}$$

$$E_i = F^{4i} = F_{4i}. \tag{12.2.14}$$

These, or (12.2.10), give

$$F_{\mu\nu} = \begin{matrix} & \to \nu \\ \downarrow \\ \mu \end{matrix} \begin{bmatrix} 0 & B_z & -B_y & -E_x \\ -B_z & 0 & B_x & -E_y \\ B_y & -B_x & 0 & -E_z \\ -E_x & -E_y & -E_z & 0 \end{bmatrix}. \tag{12.2.15}$$

Equations (12.2.11), (12.2.13), and (12.2.14) yield the tensor relation

$$\frac{\partial F_{k4}}{\partial x^j}+\frac{\partial F_{4j}}{\partial x^k}+\frac{\partial F_{jk}}{\partial x^4}=0. \tag{12.2.16}$$

With the help of B_i of (12.2.5), equation (1.4.2) takes the form

$$\frac{\partial F^{jk}}{\partial x_i}+\frac{\partial F^{ki}}{\partial x_j}+\frac{\partial F^{ij}}{\partial x_k}=0 \tag{12.2.17}$$

$$\text{or,}\ \frac{\partial F_{jk}}{\partial x^i}+\frac{\partial F_{ki}}{\partial x^j}+\frac{\partial F_{ij}}{\partial x^k}=0. \tag{12.2.18}$$

Equations (12.2.12) and (12.2.17) combine to give

$$\frac{\partial F^{\mu\nu}}{\partial x_\lambda}+\frac{\partial F^{\nu\lambda}}{\partial x_\mu}+\frac{\partial F^{\lambda\mu}}{\partial x_\nu}=0 \quad \text{or,} \quad \partial^{[\lambda}F^{\mu\nu]}=0; \tag{12.2.19}$$

and equations (12.2.16) and (12.2.18) give

$$\frac{\partial F_{\mu\nu}}{\partial x^\lambda}+\frac{\partial F_{\nu\lambda}}{\partial x^\mu}+\frac{\partial F_{\lambda\mu}}{\partial x^\nu}=0 \quad \text{or,} \quad \partial_{[\lambda}F_{\mu\nu]}=0, \tag{12.2.20}$$

where the square bracket convention of (11.2.53) has been used in the last relation of both (12.2.19) and (12.2.20). These first order partial differential equations are also relations of third order in tensor indices. Tensorially, this is the only choice for expressing the homogeneous Maxwell equations (1.4.2) and (1.4.3) in a covariant manner.

The inhomogeneous pair, (1.4.1) and (1.4.4), has, on the right hand side, the sources j and ρ^c. Since $(j, c\rho^c)$ form a 4-vector, we write

$$j^\mu \equiv (j, c\rho^c), \quad j_\mu \equiv (j, -c\rho^c). \tag{12.2.21}$$

To express the terms of (1.4.1) and (1.4.4) as contravariant first order tensors, we note that the field vectors H and D of these equations are not directly identified with B and E. To identify H and D with another tensor, we write the x-component of (1.4.4) as

$$\frac{\partial H_z}{\partial y}-\frac{\partial H_y}{\partial z}=\frac{1}{c}\frac{\partial D_x}{\partial t}+\frac{4\pi}{c}j_x, \tag{12.2.22}$$

or as

$$\frac{\partial H_k}{\partial x^j}-\frac{\partial H_j}{\partial x^k}=\frac{\partial D_i}{\partial x^4}+\frac{4\pi}{c}j^i. \tag{12.2.23}$$

Also, (1.4.1) can be written as

$$\frac{\partial D_i}{\partial x^i}=\frac{4\pi}{c}j^4. \tag{12.2.24}$$

Let us now put

$$D_i=\mathcal{H}^{4i} \tag{12.2.25}$$

in (12.2.24) and obtain the relation

$$\frac{\partial \mathcal{H}^{4i}}{\partial x^i}=\frac{4\pi}{c}j^4, \tag{12.2.26}$$

which is of first order in tensor indices; then equation (12.2.23) becomes

$$\frac{\partial H_k}{\partial x^j} - \frac{\partial H_j}{\partial x^k} + \frac{\partial \mathcal{H}^{i4}}{\partial x^4} = \frac{4\pi}{c} j^i. \tag{12.2.27}$$

Therefore, setting

$$H_i = \mathcal{H}^{jk}, \tag{12.2.28}$$

we write (12.2.27) as

$$\frac{\partial \mathcal{H}^{ij}}{\partial x^j} + \frac{\partial \mathcal{H}^{ik}}{\partial x^k} + \frac{\partial \mathcal{H}^{i4}}{\partial x^4} = \frac{4\pi}{c} j^i. \tag{12.2.29}$$

This relation, when combined with (11.2.26), gives the 4-space tensor relation

$$\frac{\partial \mathcal{H}^{\mu\nu}}{\partial x^\nu} = \frac{4\pi}{c} j^\mu. \tag{12.2.30}$$

The corresponding result for $\mathcal{H}_{\mu\gamma}$ is

$$\frac{\partial \mathcal{H}_{\mu\nu}}{\partial x_\nu} = \frac{4\pi}{c} j_\mu. \tag{12.2.31}$$

Equations (12.2.25) and (12.2.28) give the matrix of $\mathcal{H}^{\mu\gamma}$;

$$\mathcal{H}^{\mu\gamma} = \begin{bmatrix} 0 & H_z & -H_y & -D_x \\ -H_z & 0 & H_x & -D_y \\ H_y & -H_x & 0 & -D_z \\ D_x & D_y & D_z & 0 \end{bmatrix} \quad (\downarrow \mu, \rightarrow \nu). \tag{12.2.32}$$

Comparison with (12.2.10) shows that the $\mathcal{H}^{\mu\gamma}$ elements of (12.2.32) are obtained from there by replacing B by H and E by D. Next we note that

$$D_i = \mathcal{H}^{4i} = -\mathcal{H}_{4i}, \quad H_i = \mathcal{H}^{ik} = \mathcal{H}_{jk}, \tag{12.2.33}$$

and so obtain, for $\mathcal{H}_{\mu\gamma}$, the matrix :

$$\mathcal{H}_{\mu\gamma} = \begin{bmatrix} 0 & H_z & -H_y & D_x \\ -H_z & 0 & H_x & D_y \\ H_y & -H_x & 0 & D_z \\ -D_x & -D_y & -D_z & 0 \end{bmatrix} \quad (\downarrow \mu, \rightarrow \nu). \tag{12.2.34}$$

In the left hand side of (12.2.5)–(12.2.7), (12.2.13), (12.2.14), (12.2.25) and (12.2.28), the subscripts i, j, k, etc., are not those of any covariant 4-vector. For instance, D_i is simply the component of the 3-vector D along the ith direction in position space. These were temporarily used for finding the field tensors from the vector and scalar field variables correctly for the covariant formulation of the field equations.

The dual tensor $*F_{\mu\gamma}$ to $F_{\mu\gamma}$ is defined as

$$*F^{\alpha\beta} = \frac{1}{2} e^{\alpha\beta\gamma\delta} F_{\gamma\delta}, \tag{12.2.35}$$

where $e^{\alpha\beta\gamma\delta}$ is the fourth order Levi Civita symbol. The matrices of $*F^{\alpha\beta}$ and $*F_{\alpha\beta}$ are, respectively, given by

$$*F^{\alpha\beta} = \begin{bmatrix} 0 & E_z & -E_y & B_x \\ -E_z & 0 & E_x & B_y \\ E_y & -E_x & 0 & B_z \\ -B_x & -B_y & -B_z & 0 \end{bmatrix}. \tag{12.2.36}$$

(rows $\downarrow \alpha$, columns $\rightarrow \beta$)

$$*F_{\alpha\beta} = \begin{bmatrix} 0 & E_z & -E_y & -B_x \\ -E_z & 0 & -E_x & -B_y \\ E_y & E_x & 0 & -B_z \\ B_x & B_y & B_z & 0 \end{bmatrix}. \tag{12.2.37}$$

(rows $\downarrow \alpha$, columns $\rightarrow \beta$)

In terms of $*F_{\alpha\beta}$, the Maxwell equation (12.2.19) reads

$$\frac{\partial * F_{\alpha\beta}}{\partial x^\beta} = 0. \tag{12.2.38}$$

The dual $*\mathcal{H}_{\mu\gamma}$ to $\mathcal{H}_{\mu\gamma}$ is similarly defined. Then, for $\mu = 1$, in terms of $*\mathcal{H}_{\mu\gamma}$, equation (12.2.30) reads

$$\frac{\partial}{\partial x^2} * \mathcal{H}_{34} + \frac{\partial}{\partial x^3} * \mathcal{H}_{42} + \frac{\partial}{\partial x^4} * \mathcal{H}_{23} = \frac{4\pi}{c} j^i. \tag{12.2.39}$$

The general form, therefore, is

$$\frac{\partial}{\partial x^\beta} * \mathcal{H}_{\gamma\delta} + \frac{\partial}{\partial x^\gamma} * \mathcal{H}_{\delta\beta} + \frac{\partial}{\partial x^\delta} * \mathcal{H}_{\beta\gamma} = \frac{4\pi}{c} j^\alpha. \tag{12.2.40}$$

In empty space (ether), the constitutive relations being $B = H$, $D = E$, evidently

$$\mathcal{H}^{\mu\nu} = F^{\mu\nu} = \eta^{\mu\alpha}\eta^{\nu\beta} F_{\alpha\beta}, \tag{12.2.41}$$

where the metric tensor of (11.2.16) has been used. Then equation (12.2.30) gives

$$\partial_\nu F^{\mu\nu} = \frac{4\pi}{c} j^\mu, \tag{12.2.42}$$

$$B^2 - E^2 = \frac{1}{2} F^{\mu\nu} F_{\mu\nu}, \quad (E \cdot B) = \frac{1}{4} F_{\mu\nu} * F^{\mu\nu}. \tag{12.2.43}$$

So $B^2 - E^2$ and $(E \cdot B)$ are invariants.

Relations (12.2.41) show that the Gaussian field identification $B = H, D = E$ due to the complete absence of material polarization mechanism in free space, is a Lorentz invariant identification. Since the relation (12.2.4) in free 4-space involves the invariant Minkowskian metric $\eta_{\mu\nu}$, the invariance of that identification can be used to define the Lorentz group, and to show that the vacuum Maxwell equations (1.4.13)–(1.4.16) have a built-in Lorentz invariance, and these equations are equivalent to the invariance of the metric $\eta_{\mu\nu}$. This invariance defines the Lorentz group as a symmetry property of that 4-space continuum for uniform translation.

12.3 Transformation behaviour of Field Vectors

Since $\mathcal{H}^{\mu\gamma}$, a tensor density of weight + 1, is given by (12.2.32), and $F_{\mu\gamma}$, a tensor density of weight −1, is given by (12.2.15), the transformation relations for these two tensors are

$$F_{\mu\nu} = \alpha_\mu^{\xi^0} \alpha_\nu^{\eta^0} F_{\xi^0\eta^0}, \qquad (12.3.1)$$

$$\mathcal{H}^{\mu\nu} = |\Delta|^{-1} \alpha^\mu_{\xi^0} \alpha^\nu_{\eta^0} \mathcal{H}^{\xi^0\eta^0} \qquad (12.3.2)$$

with

$$\Delta = \det \alpha^\mu_{\xi^0}. \qquad (12.3.3)$$

12.3.1 Results for the infinitesimal LT

For $\mu = 1$ and $\nu = 2$, relation (12.3.1) gives

$$\begin{aligned} F_{12} = B_3 = \alpha_1^{\xi^0} \alpha_2^{\eta^0} F_{\xi^0\eta^0} &= \alpha_1^{1^0} \alpha_2^{2^0} F_{1^02^0} + \alpha_1^{4^0} \alpha_2^{2^0} F_{4^02^0} + \alpha_1^{1^0} \alpha_2^{4^0} F_{1^04^0} \\ &= B_{z^0} + \beta_x(-E_{y_0}) + \beta_y E_{x^0}. \end{aligned} \qquad (12.3.4)$$

Hence, evidently,

$$B = B^0 - (\beta \times E^0). \qquad (12.3.5)$$

Next we put $\mu = 1$ and $\nu = 4$ in (12.3.1) and obtain

$$\begin{aligned} F_{14} = E_1 = \alpha_1^{1^0} \alpha_4^{2^0} F_{1^02^0} &+ \alpha_1^{1^0} \alpha_4^{3^0} F_{1^03^0} + \alpha_1^{1^0} \alpha_4^{4^0} F_{1^04^0} \\ &+ \alpha_1^{4^0} \alpha_4^{1^0} F_{4^01^0} + \alpha_1^{4^0} \alpha_4^{2^0} F_{4^02^0} + \alpha_1^{4^0} \alpha_4^{3^0} F_{4^03^0}. \end{aligned} \qquad (12.3.6)$$

The corresponding vector relation is

$$E = E^0 + (\beta + B^0) - (\beta \cdot E^0)\beta. \qquad (12.3.7)$$

For the first order β it reduces to

$$E = E^0 + (\beta \times B^0). \qquad (12.3.8)$$

Using (7.9.1), the transformation relation

$$\mathcal{H}^{12} = H_3 = |\Delta|^{-1} \alpha^1_{\xi^0} \alpha^2_{\eta^0} \mathcal{H}^{\xi^0\eta^0} \qquad (12.3.9)$$

gives

$$H_z = H_{z0} - \beta_y(-D_x) - \beta_x D_y. \qquad (12.3.10)$$

The vector relation, obtained from (12.3.9), is

$$H = H^0 - (\beta \times D^0). \tag{12.3.11}$$

Similarly,

$$\mathcal{H}^{14} = -D_x = \alpha'_{\xi^0}\alpha^4_{\eta^0}\mathcal{H}^{\xi^0\eta^0} = -D_{X^0} + (\beta_z H_{Y^0} - \beta_y H_{z^0}) - \beta_x(\beta\cdot D^0). \tag{12.3.12}$$

Hence, up to first order of β,

$$D = D^0 + (\beta + H^0). \tag{12.3.13}$$

12.3.2 General transformation results

The general results of transformation are evaluated with the help of (12.3.1)–(12.3.3) and (7.6.5)–(7.6.8). But, it is simpler to generalize the results, deduced from (12.3.1) and (12.3.2), with the help of (7.2.10) and (7.2.11), taking $F_{\mu\gamma}$ from (12.2.15). These simpler relations give

$$F_{12} = B_z = F_{1^0 2^0} = B^0_z, \quad F_{13} = -B_y = -\gamma(B_{y^0} - \beta E_{x^0}) = -\gamma\left[B^0 - (\beta \times E^0)\right]_y \tag{12.3.14}$$

$$F_{14} = E_x = \gamma\beta(-\beta_{y_0}) + \gamma E_{x^0} = \gamma\left[E^0 + (\beta \times B^0)\right]_x, \quad \text{etc.} \tag{12.3.15}$$

The vector forms of these relations are

$$E_\perp = \gamma\left[E^0_\perp + (\beta \times B^0)\right], \tag{12.3.16}$$

$$B_\perp = \gamma\left[B^0_\perp - (\beta \times E^0)\right], \tag{12.3.17}$$

$$E_\parallel = E^0_\parallel, \quad B_\parallel = B^0_\parallel. \tag{12.3.18}$$

Hence the subscripts $\parallel$ and $\perp$ denote parallel and perpendicular components of the field vectors relative to **v**. These equations combine to give

$$E = \gamma\left[E^0 + (\beta \times B^0)\right] - \frac{\gamma^2}{\gamma+1}\beta(\beta\cdot E^0), \tag{12.3.19}$$

$$B = \gamma\left[B^0 - (\beta \times E^0)\right] - \frac{\gamma^2}{\gamma+1}\beta(\beta\cdot B^0). \tag{12.3.20}$$

Hence a purely electric or magnetic field, in one inertial system, gives rise to a mixture of these in another such system, and these fields have no independent existence in an inertial frame. Being thus completely inter related, the fields are referred to, appropriately, as the elements of $F^{\mu\gamma}$ or $F_{\mu\gamma}$, in place of the apparently distinguishable field vectors E and B, which, as such, do not give the correct picture of the situation.

The vectors D and H are similarly evaluated from (12.3.2), with the help of (7.2.11). We find that

$$\mathcal{H}^{12} = H_3 = \alpha'_{\xi^0}\alpha^2_{\eta^0}\mathcal{H}^{\xi^0\eta^0} = H_{3^0}, \tag{12.3.21}$$

$$\mathcal{H}^{13} = -H_2 = -\gamma\left[H^0 - (\beta \times D^0)\right]_\perp, \tag{12.3.22}$$

$$\mathcal{H}^{14} = -D_1 = -\gamma(D_{1^0} - \beta H_{2^0}) = -\gamma\left[D^0 + (\beta \times H^0)\right]_\perp, \tag{12.3.23}$$

$$\mathcal{H}^{34} = -D_3 = -D_{3^0}. \tag{12.3.24}$$

Hence it is evident that

$$H_\perp = \gamma\left[H^0 - (\beta \times D^0)\right]_\perp, \quad H_\parallel = H^0_\parallel, \tag{12.3.25}$$

$$D_\perp = \gamma\left[D^0 + (\beta \times H^0)\right]_\perp, \quad D_\parallel = D^0_\parallel. \tag{12.3.26}$$

More compactly these are

$$H = \gamma\left[H^0 - (\beta \times D^0)\right] - \frac{\gamma^2}{\gamma+1}\beta\,(\beta\cdot H^0), \tag{12.3.27}$$

$$D = \gamma\left[D^0 + (\beta \times H^0)\right] - \frac{\gamma^2}{\gamma+1}\beta\,(\beta\cdot D^0). \tag{12.3.28}$$

For the LT, as may be seen from (7.2.10), the perpendicular components of the displacement are unaffected, and only the parallel components change. But the relations (12.3.25) and (12.3.26) of field transformation show that the perpendicular components of fields are affected, and the parallel components remain invariant. This conclusion is in line with the basic EM features of Faraday's law of induction, which are that when a conducting wire is moved in one direction, the magnetic field perpendicular to that is affected, and the parallel magnetic field remains completely unchanged. For some analogy between the effects of rotation and EM induction is discussed in Chapter 8.

The results for the reverse transformation of the field vectors, which follow from the relations (12.3.19), (12.3.20), (12.3.27) and (12.3.28), are

$$E^0 = \gamma\left[E - (\beta \times B)\right] - \frac{\gamma^2}{\gamma+1}\beta\,(\beta\cdot E), \tag{12.3.29}$$

$$B^0 = \gamma\left[B + (\beta \times E)\right] - \frac{\gamma^2}{\gamma+1}\beta\,(\beta\cdot B), \tag{12.3.30}$$

$$H^0 = \gamma\left[H + (\beta \times D)\right] - \frac{\gamma^2}{\gamma+1}\beta\,(\beta\cdot H), \tag{12.3.31}$$

$$D^0 = \gamma\left[D - (\beta \times H)\right] - \frac{\gamma^2}{\gamma+1}\beta\,(\beta\cdot D). \tag{12.3.32}$$

The matrix form of the relations (12.3.27) and (12.3.28), for the transformation of D and H, is

$$\left[\mathcal{DH}\right] = \mathbf{Q}\cdot\begin{bmatrix} D^0 \\ H^0 \end{bmatrix}, \tag{12.3.33}$$

where $\mathbf{Q}$ is the 6×6 transformation matrix

$$\mathbf{Q} = \begin{bmatrix}
\gamma - \frac{\gamma^2\beta_x^2}{\gamma+1} & -\gamma^2\frac{\beta_x\beta_y}{\gamma+1} & -\gamma^2\frac{\beta_x\beta_z}{\gamma+1} & 0 & -\gamma\beta_z & \gamma\beta_y \\
-\gamma^2\frac{\beta_y\beta_x}{\gamma+1} & \gamma - \frac{\gamma^2\beta_y^2}{\gamma+1} & -\gamma^2\frac{\beta_y\beta_z}{\gamma+1} & \gamma\beta_z & 0 & -\gamma\beta_x \\
-\gamma^2\frac{\beta_z\beta_x}{\gamma+1} & -\gamma^2\frac{\beta_z\beta_y}{\gamma+1} & \gamma - \frac{\gamma^2\beta_z^2}{\gamma+1} & -\gamma\beta_y & \gamma\beta_x & 0 \\
0 & \gamma\beta_z & -\gamma\beta_y & \gamma - \frac{\gamma^2\beta_x^2}{\gamma+1} & -\gamma^2\frac{\beta_x\beta_y}{\gamma+1} & -\gamma^2\frac{\beta_x\beta_z}{\gamma+1} \\
-\gamma\beta_z & 0 & \gamma\beta_x & -\gamma^2\frac{\beta_y\beta_x}{\gamma+1} & \gamma - \frac{\gamma^2\beta_y^2}{\gamma+1} & -\gamma^2\frac{\beta_y\beta_z}{\gamma+1} \\
-\gamma\beta_y & \gamma\beta_x & 0 & -\gamma^2\frac{\beta_z\beta_x}{\gamma+1} & -\gamma^2\frac{\beta_z\beta_y}{\gamma+1} & \gamma - \frac{\gamma^2\beta_z^2}{\gamma+1}
\end{bmatrix} \tag{12.3.34}$$

When the velocity $\mathbf{v}$ of the frame of reference K relative to the frame K^0 is parallel to the Z axis, this transformation matrix $\mathbf{Q}$ simplifies to

$$\mathbf{Q} = \begin{bmatrix} \gamma & 0 & 0 & 0 & -\beta\gamma & 0 \\ 0 & \gamma & 0 & \beta\gamma & 0 & 0 \\ 0 & 0 & 1 & 0 & 0 & 0 \\ 0 & \beta\gamma & 0 & \gamma & 0 & 0 \\ -\beta\gamma & 0 & 0 & 0 & \gamma & 0 \\ 0 & 0 & 0 & 0 & 0 & 1 \end{bmatrix}. \tag{12.3.35}$$

12.4 Covariant Formulation of the Lorentz Force and the Equation of Motion

12.4.1 Lorentz force Law

For a formulation of the Lorentz 4-force, acting on a particle of charge q, stationary or moving, the rules of transformation of the field tensors $F_{\mu\gamma}$ and $\mathcal{H}^{\mu\gamma}$ do not throw any light. The 4-force F_μ can be defined as a first order tensor, either contravariant or covariant, Ise (1957). Specifically, if force is defined as the product of mass and acceleration, it is contravariant. Force can also be defined from the concept of work, which is the product of force and displacement, and is accepted as an invariant scalar quantity. Then displacement, being a contravartiant 4-vector, force must be a covariant 4-vector. Here, following the first definition, we regard force as a contravariant 4-vector :

$$F^\mu = \frac{1}{c} F^{\mu\nu} j_\nu, \tag{12.4.1}$$

where $F^{\mu\gamma}$ is given by (12.2.10) and j^μ by (12.2.21). We find that

$$F^i = \frac{1}{c} F^{i\nu} j_\nu = \frac{1}{c}\rho c E^i + \frac{1}{c}(j \times B)_i, \tag{12.4.2}$$

and

$$F^4 = \frac{1}{c}(E \cdot j). \tag{12.4.3}$$

Since F^i represents the 3-vector Lorentz force F, F^μ is the corresponding 4-vector. The invariant relation

$$F^\mu j_\mu = 0 \tag{12.4.4}$$

obtains easily for all inertial systems. The tensor form of the 4-Lorentz force, per unit charge, is defined as

$$\bar{F}^\mu = \frac{1}{c} F^{\mu\nu} U_\nu, \tag{12.4.5}$$

where the first order tensor U_μ is given by (11.2.45). We find that

$$\bar{F}^\mu = \gamma \left[E + \frac{1}{c}(U \times B),\ \frac{1}{c}(E \cdot U) \right]. \tag{12.4.6}$$

We can define another 4-vector K^μ in the following manner :

$$\bar{K}^\mu = \frac{1}{c}\mathcal{H}^{\mu\nu}U_\nu. \tag{12.4.7}$$

The pseudo 4-vectors $*F^\mu$ and $*K^\mu$ are also similarly defined :

$$*\bar{F}^\mu = \frac{1}{c} * F^{\mu\nu}U_\nu, \quad *\bar{K}^\mu = \frac{1}{c} * \mathcal{H}^{\mu\nu}U_\nu, \tag{12.4.8}$$

where $*F^{\mu\gamma}$ is given in (12.2.36) and $*\mathcal{H}^{\mu\gamma}$ can be similarly obtained from (12.2.32). These give in section 12.6 the covariant formulation of the constitutive relations.

12.4.2 Equation of motion

Equation for motion of a charge, can be expressed as

$$\frac{dp}{dt} = \frac{q}{c}(v \times B) + \frac{q}{c}u^4 E, \quad u^4 = c. \tag{12.4.9}$$

The corresponding relation in 4-space is

$$\frac{dp^\mu}{dt} = \frac{q}{c}F^{\mu\nu}u_\nu = \frac{q}{c}\left(\frac{\partial A^\nu}{\partial x_\mu} - \frac{\partial A^\mu}{\partial x_\nu}\right). \tag{12.4.10}$$

Since A^μ is a function of x^μ, we find that

$$\frac{dA^\mu}{dt} = \frac{\partial A^\mu}{\partial x_\nu}\frac{dx_\nu}{dt} = \frac{\partial A^\mu}{\partial x_\nu}u_\nu. \tag{12.4.11}$$

Hence (12.4.10) can be expressed as

$$\frac{d}{dt}\left(p^\mu + \frac{q}{c}A^\mu\right) = \frac{q}{c}\frac{\partial A^\nu}{\partial x_\mu}u_\nu. \tag{12.4.12}$$

Now writing

$$P^\mu = p^\mu + \frac{q}{c}A^\mu, \tag{12.4.13}$$

the equation of motion obtains in the form

$$\frac{dP^\mu}{dt} = \frac{q}{c}\frac{\partial A^\mu}{\partial x_\mu}u_\nu \tag{12.4.14}$$

P^μ is the canonical (or the generalized or conjugate) 4-momentum.

In terms of the proper time T, equation (12.4.10) for motion reads

$$\frac{dp^\mu}{dT} = \frac{q}{c}F^\mu_{\ ,\nu}U^\nu \tag{12.4.15}$$

with

$$F^\mu_{\cdot\nu} = \eta_{\nu\xi}F^{\mu\xi} \tag{12.4.16}$$

and $\eta_{\mu\nu}$ given by (11.2.16). Similarly, equation (12.4.14) reads

$$\frac{dp^\mu}{dT} = \frac{q}{c}\frac{\partial A^\nu}{\partial x_\mu}U_\nu. \tag{12.4.17}$$

For extension to non-inertial frames, the more appropriate form of the equation of motion, involving the invariant 4-interval, is

$$mc\frac{dw^\mu}{\sqrt{-ds^2}} = -\frac{q}{c}F^\mu_{\cdot\nu}W^\nu \tag{12.4.18}$$

with

$$W^\mu = \frac{dx^\mu}{\sqrt{-ds^2}}. \tag{12.4.19}$$

The minus sign in the right hand side of (12.4.18) is necessary for identification with (12.4.15).

12.4.3 The energy momentum tensor

With the help of (12.2.30) equation (12.4.1) becomes

$$\frac{1}{c}F_{\mu\nu}j^\nu = F_\mu = \frac{1}{4\pi}F_{\mu\nu}\mathcal{H}^{\mu\xi}_{\ ,\xi} = \frac{1}{4\pi}\left[\frac{\partial}{\partial x^\xi}(F_{\mu\nu}\mathcal{H}^{\nu\xi}) - \mathcal{H}^{\nu\xi}\frac{\partial F_{\mu\nu}}{\partial x^\xi}\right]. \tag{12.4.20}$$

In the second term, on the right hand side, interchanging the running co-ordinates ξ and ν, we obtain

$$\mathcal{H}^{\nu\xi}\frac{\partial F_{\mu\nu}}{\partial x^\xi} = \mathcal{H}^{\xi\nu}\frac{\partial F_{\mu\xi}}{\partial x^\nu} = \mathcal{H}^{\nu\xi}\frac{\partial F_{\xi\mu}}{\partial x^\nu} = \frac{1}{2}\mathcal{H}^{\nu\xi}\left(\frac{\partial F_{\mu\nu}}{\partial x^\xi} + \frac{\partial F_{\xi\mu}}{\partial x^\nu}\right) = -\frac{1}{2}\mathcal{H}^{\nu\xi}\frac{\partial F_{\nu\xi}}{\partial x^\mu}, \tag{12.4.21}$$

where the relation (12.2.20) has been used. Hence, equations (12.4.20) become

$$F_{\mu\nu}j^\nu = F_\mu = -\frac{1}{4\pi}\frac{\partial}{\partial x^\xi}(F_{\mu\nu}\mathcal{H}^{\xi\nu}) + \frac{1}{8\pi}\mathcal{H}^{\nu\xi}\frac{\partial F_{\nu\xi}}{\partial x^\mu}. \tag{12.4.22}$$

These relations can be written as

$$\begin{aligned}\frac{1}{c}F_{\mu\nu}j^\nu = F_\mu = &-\frac{1}{4\pi}\frac{\partial}{\partial x^\xi}(F_{\mu\nu}\mathcal{H}^{\xi\nu}) + \frac{1}{16\pi}\frac{\partial}{\partial x^\mu}(\mathcal{H}^{\nu\xi}F_{\nu\xi})\\ &+\frac{1}{16\pi}\left[\mathcal{H}^{\nu\eta}\frac{\partial F_{\nu\eta}}{\partial x^\mu} - F_{\nu\xi}\frac{\partial \mathcal{H}^{\nu\eta}}{\partial x^\mu}\right].\end{aligned} \tag{12.4.23}$$

Further manipulations, using (12.2.30), for j^μ, enable us to write

$$\frac{1}{c}F_{\mu\nu}j^\nu + \frac{1}{16\pi}\left[F_{\nu\eta}\frac{\partial \mathcal{H}^{\nu\eta}}{\partial x^\mu} - \mathcal{H}^{\nu\eta}\frac{\partial F_{\nu\eta}}{\partial x^\mu}\right] = -\frac{\partial}{\partial x^\xi}S^\xi_\mu = \operatorname{div}_\xi(S^\xi_\mu) \tag{12.4.24}$$

with

$$S^\xi_\mu = -\frac{1}{4\pi}(F_{\mu\nu}\mathcal{H}^{\xi\nu}) + \frac{1}{16\pi}\delta^\xi_\mu\mathcal{H}^{\nu\eta}F_{\nu\eta}. \tag{12.4.25}$$

The quantity S^ξ_μ is the energy momentum tensor, also called the *Minkowski energy momentum tensor.* In vacuum (12.4.24) becomes

$$\frac{1}{2}F_{\mu\nu}j^\nu = F_\mu = \operatorname{div}_\xi(S^\xi_\mu). \tag{12.4.26}$$

Taking summation over the repeated index μ, in the left hand side of (12.4.25), we obtain

$$S^\mu_\mu = -\frac{1}{4\pi}(F_{\mu\nu}\mathcal{H}^{\mu\nu}) + \frac{1}{16\pi}\mathcal{H}^{\nu\eta}F_{\nu\eta} = 0. \tag{12.4.27}$$

$$S_2^1 = -\frac{1}{4\pi} F_{2\nu} \mathcal{H}^{1\nu} = \frac{1}{4\pi}(B_x H_y + D_x E_y), \tag{12.4.28}$$

$$S_1^2 = -\frac{1}{4\pi} F_{1\nu} \mathcal{H}^{1\nu} = \frac{1}{4\pi}(B_y H_x + D_y E_x), \tag{12.4.29}$$

$$S_1^1 = -\frac{1}{4\pi} F_{1\nu} \mathcal{H}^{1\nu} + \frac{1}{16\pi} \mathcal{H}^{\nu\eta} F_{\nu\eta} = \frac{1}{4\pi}(B_x H_x + D_x E_x) - \frac{1}{8\pi}[(B\cdot H) + (D\cdot E)], \tag{12.4.30}$$

$$S_4^1 = -\frac{1}{4\pi} F_{4\nu} \mathcal{H}^{1\nu} = \frac{1}{4\pi}(E_y H_z - E_z H_y) = \frac{1}{c} S_x, \tag{12.4.31}$$

$$S_1^4 = -\frac{1}{4\pi} F_{1\nu} \mathcal{H}^{4\nu} = \frac{1}{4\pi}(B \times D)_x, \tag{12.4.32}$$

$$S_4^4 = -\frac{1}{4\pi} F_{4\nu} \mathcal{H}^{4\nu} + \frac{1}{16\pi} \mathcal{H}^{\nu\eta} F_{\nu\eta} = \frac{1}{8\pi}[(E\cdot D) + (B\cdot H)], \tag{12.4.33}$$

where the Poynting flux

$$S = \frac{c}{4\pi}(E \times H). \tag{12.4.34}$$

The quantities S_4^i, S_4^4, S_i^4, in every system of co-ordinates, should be interpreted as the densities of EM energy flux, energy and momentum respectively.

In the rest system K^0, the second term in the left hand side of (12.4.24)

$$= \frac{1}{16\pi}\left[F_{\nu^0\eta^0}\frac{\partial \mathcal{H}^{\nu^0\eta^0}}{\partial x^{\mu^0}} - H^{\nu^0\eta^0}\frac{\partial F_{\nu^0\eta^0}}{\partial x^{\mu^0}}\right]$$

$$= \frac{1}{8\pi}\left[\left(B^0\cdot\frac{\partial H^0}{\partial x^{\mu^0}}\right) - \left(E^0\cdot\frac{\partial D^0}{\partial x^{\mu^0}}\right) - \left(H^0\cdot\frac{\partial B^0}{\partial x^{\mu^0}}\right) + \left(D^0\cdot\frac{\partial E^0}{\partial x^{\mu^0}}\right)\right]$$

$$= -\frac{1}{8\pi}\left[|H^0|^2\frac{\partial \mu}{\partial x^{\mu^0}} + |E^0|^2\frac{\partial \varepsilon}{\partial x^{\mu^0}}\right]. \tag{12.4.35}$$

For homogeneous isotropic bodies ε and μ are constants; so, this term vanishes. This being a tensor should be valid in all inertial frames; hence (12.4.23) reduces to

$$F_\mu = \frac{1}{c} F_{\mu\nu} j^\nu = -\frac{1}{4\pi}\frac{\partial}{\partial x^\xi}\left(F_{\mu\nu}\mathcal{H}^{\xi\nu}\right) + \frac{1}{16\pi}\frac{\partial}{\partial x^\mu}\left(\mathcal{H}^{\nu\xi}F_{\nu\xi}\right). \tag{12.4.36}$$

Relations (12.4.28)–(12.4.30) yield the single relation

$$S_j^i = \frac{1}{4\pi}(B_i H_j + D_i E_j) - \frac{1}{8\pi}[(B\cdot H) + (D\cdot E)]\delta_j^i. \tag{12.4.37}$$

So S_j^i is the ijth component of the Maxwell stresses. Moreover, equation (12.4.31) shows that S_i^4 is the ith component of the Poynting flux vector S of (12.4.34). From (12.4.31) and (12.4.32) it is evident that in a material medium

$$S_i^4 - S_4^i = \frac{1}{4\pi}(\varepsilon\mu - 1)(E \times H)_i \neq 0. \tag{12.4.38}$$

The reason is that S_4^i is the ith component of the energy flux and S_i^4 is the ith component of the EM momentum density, and these are not necessarily identical physical entities. The non-symmetry of the Minkowski energy momentum tensor has given rise to some discussion in the literature, Moller (1972, section 7.7).

12.5 The 4-current Density of Charges, the Continuity Condition and Ohm's Law

Charges being discrete quantities, residing in mathematical points, the charge density ρ^c is such that the integral

$$\iiint_V \rho^c dV \tag{12.5.1}$$

over a certain volume V, equals the sum of the charges contained in it. ρ^c can be expressed with the help of the delta function of Dirac :

$$\rho^c = \sum_\alpha q_\alpha \delta(r - r_\alpha), \tag{12.5.2}$$

where the sum is taken over all the charges and $\mathbf{r}_\alpha$ is the position vector of the charge q_α.

On a classical particle, the charge q is basically an invariant quantity. So, it does not change from one co-ordinate system to another whatever be the relative motion between them, uniform or non-uniform. The charge density ρ^c is not, however, an invariant quantity, necessarily.

To find the transformation behaviour of ρ^c, in conformity with the relation (1.4.19), consider the scalar quantity

$$dq = \rho^c dV \tag{12.5.3}$$

and the identity

$$dq dx^\mu = \rho^c dV \cdot c dt \frac{dx^\mu}{dt} = \rho^c dV \cdot dx^4 \cdot \frac{dx^\mu}{dx^4}. \tag{12.5.4}$$

Since

$$dV dx^4 (= d\mathbf{x} = dx^1 dx^2 dx^3 dx^4)$$

is a scalar quantity, as may be seen from (7.2.15), the left hand side of (12.5.4) is a 4-vector; so, in the right hand side, $\rho^c dx^\mu/dt$ (or $\rho^c dx^\mu/dx^4$) as well, is a 4-vector.

The 4-current j^μ is defined as

$$j^\mu = \rho^c \frac{dx^\mu}{dt} \tag{12.5.5}$$

j^μ can be written as

$$j^\mu = \rho \frac{dx^\mu}{dT} \frac{dT}{dt}. \tag{12.5.6}$$

Since the particle number density N remains invariant,

$$dN = dN^0 \tag{12.5.7}$$

and the particle charge density is

$$\rho^c = q\frac{dN}{dV} = q\frac{dN^0}{dV^0}\frac{dV^0}{dV} = \frac{1}{V}\rho^0, \tag{12.5.8}$$

where the transformation of the volume element dV follows from (7.2.15). Hence, the transformation of the charge element dq, of (12.5.3), is

$$dq = \rho^c dV = \frac{1}{\gamma}\rho^{co}\gamma dV^0 = \rho^{co} dV^0 = dq^0. \tag{12.5.9}$$

So the conclusion, also drawn from (1.4.18), Jackson ((1975), p. 548), is that the net charge in a volume element remains invariant. For electric charges in vacuum, only the convection current of these is present. The current densities j in the two inertial systems K^0 and K are respectively $\rho^{co}u^0$ and $\rho^c u$, where u^0 and u are the velocities with which the charges move in K^0 and K. In the K^0 frame, the conservation of electric charges, following from the continuity equation, is given by

$$\frac{\partial \rho^{co}}{\partial t^0} + \frac{\partial}{\partial x_i}(\rho^{co} u_i^0) = 0. \tag{12.5.10}$$

The tensor form of this equation is

$$\frac{\partial j^\mu}{\partial x^\mu} = 0, \quad \text{or,} \quad j^\mu_{,\mu} = 0. \tag{12.5.11}$$

Assume that j^μ is transformed to $j^{\mu 0}$, Moller (1972, Appendix 2) as

$$j^{\mu 0} = f^\mu(j^1, j^2, j^3, j^4), \tag{12.5.12}$$

where the functions f^μ must satisfy the continuity relation

$$\frac{\partial j^\mu}{\partial x^\mu} = \partial_\mu j^\mu = 0. \tag{12.5.13}$$

Equation (12.5.12) gives

$$\frac{\partial f^\mu}{\partial x^{\mu^0}} = 0 = \frac{\partial f^\mu}{\partial j^\nu}\frac{\partial j^\nu}{\partial x^\eta}\frac{\partial x^\eta}{\partial x^{\mu_0}} = \left(\frac{df^\mu}{\partial j^\nu}\alpha^\eta_{\mu^0}\right)\frac{\partial j^\nu}{\partial x^\eta} \tag{12.5.14}$$

for all values of the twenty variables j^μ and $\partial j^\mu/\partial x^\eta (= j^\mu_{,\eta})$ that are subject to the restriction (12.5.13); the transformation coefficients are defined in (11.2.19).

We now multiply equation (12.5.13) by the Lagrangian multiplier λ and, using the Kronecker delta notation, write it as

$$\lambda j^\nu_{,\eta}\delta^\eta_\nu = 0 \tag{12.5.15}$$

λ may be a function of the variables j^μ. Adding minus times (12.5.15) with (12.5.14) we obtain

$$\left[\frac{\partial f^\mu}{\partial j^\nu}\alpha^\eta_{\mu^0} - \lambda\delta^\eta_\nu\right]\frac{\partial j^\nu}{\partial x^\eta} = 0. \tag{12.5.16}$$

Since this relation must hold for arbitrary independent variations of j^μ and $j^\mu_{,\eta}$, the variation of $j^\mu_{,\eta}$ gives the equation

$$\frac{\partial f^\mu}{\partial j^\nu}\alpha^\eta_{\mu^0} = \lambda(j^\xi)\delta^\eta_\nu \tag{12.5.17}$$

where the orthogonality condition connecting the transformation coefficients is

$$\alpha^\mu_{\nu^0}\alpha^{\nu_0}_\lambda = \delta^\mu_\lambda \tag{12.5.18}$$

and we find that (12.5.17) becomes

$$\frac{\partial f^\mu}{\partial j^\nu}\alpha^\eta_{\mu_0} = \lambda(j^\xi)\alpha^\eta_{\mu^0}\alpha^{\mu^0}_\nu \tag{12.5.19}$$

This relation gives the equation

$$\frac{\partial f^\mu}{\partial j^\nu} = \lambda(j^\xi)\alpha^{\mu^0}_\nu \tag{12.5.20}$$

Differentiating next (12.5.17), with respect to j^ξ, we obtain

$$\frac{\partial^2 f^\mu}{\partial j^\nu \partial j^\xi} = \frac{\partial \lambda}{\partial j^\xi}\delta^\eta_\nu. \tag{12.5.21}$$

Interchanging ξ and ν in the left hand side, we find that

$$\frac{\partial^2 f^\mu}{\partial j^\nu \partial j^\xi} = \frac{\partial^2 f^\mu}{\partial j^\xi \partial j^\nu} = \frac{\partial \lambda}{\partial j^\xi}\delta^\eta_\nu = \frac{\partial \lambda}{\partial j^\nu}\delta^\eta_\xi. \tag{12.5.22}$$

Taking now $\eta = \xi \neq \nu$, we obtain

$$\frac{\partial \lambda}{\partial j^\nu} = 0. \tag{12.5.23}$$

Hence λ must be independent of the variables j^ν. If, therefore, λ is constant in (12.5.20), then integrating it with respect to j^ν, we find that

$$f^\mu(j^\nu) = \lambda \alpha^{\mu 0}_\nu j^\nu + \beta^\mu, \tag{12.5.24}$$

where β^μ is a constant. If the electric charge density is zero in the K system, then $j^\mu = 0$ and, therefore, $j^{\mu 0} = 0$; consequently $\beta^\mu = 0$ if (12.5.12) is used. Then (12.5.12) or (12.5.24) becomes

$$j^{\mu^0} j_{\mu^0} = \lambda^2 \alpha^{\mu^0}_\nu \alpha^\nu_{\mu^0} j^\nu j_\nu. \tag{12.5.25}$$

Relations (12.5.16) and (12.2.21) give

$$j^{\mu^0} j_{\mu_0} = \lambda^2 \alpha^{\mu^0}_\nu \alpha^\nu_{\mu^0} j^\nu j_\nu. \tag{12.5.26}$$

It can be simplified to

$$\rho^{c_0^2}(u^{0^2} - c^2) = \lambda^2 \rho^{c^2}(u^2 - c^2). \tag{12.5.27}$$

When the two inertial frames K and K^0 interchange their role, K^0 moves relative to K with velocity $\mathbf{v}$ in the direction of OZ; then changing the signs on OX and OZ, we can write

$$\rho^2(u^2 - c^2) = \lambda^2 \rho^{0^2}(u^{0^2} - c^2), \tag{12.5.28}$$

$$\lambda^4 = 1. \tag{12.5.29}$$

Since the positive directions of OY and OY^0 are the same, $\rho^2(u^2 - c^2)$ and $\rho^{0^2}(u^{0^2} - c^2)$ must have the same sign; so

$$\lambda^2 = 1. \tag{12.5.30}$$

Moreover, since the charge density remains a scalar, under the reflections at the origin, we can take

$$\lambda = 1. \tag{12.5.31}$$

Then (12.5.25) becomes

$$j^{\mu^0} = \alpha^{\mu^0}_\nu j^\nu, \tag{12.5.32}$$

and (12.5.28) gives

$$\rho^{c^2}\left(1 - \frac{u^2}{c^2}\right) = \rho^{c_0^2}\left(1 - \frac{u^{0^2}}{c^2}\right). \tag{12.5.33}$$

This invariant relation can be written as

$$\rho\frac{dT}{dt} = \rho^{c^0}\frac{dT}{dt^0}, \tag{12.5.34}$$

where T is the proper time. So, j^μ, given by (12.5.5), behaves like a 4-vector. Its space component is the current density vector

$$j = \rho^c \mathbf{u}, \tag{12.5.35}$$

where $u(= dr/dt)$ is the velocity of the charge at the given point; consistent with the relations (12.2.2), the time component is $\rho^c c$.

Since j^μ transforms like a 4-vector,

$$j_{x_0} = j_x, \quad j_{y_0} = j_y, \quad j_{z_0} = \gamma(j_z + \beta\rho c), \quad c\rho^0 = \gamma(c\rho + \beta j_z). \tag{12.5.36}$$

In other words,

$$j_{\perp^0} = (j + \rho^c u)_\perp, \quad j_{\parallel^0} = \gamma(j_\parallel + \beta\rho^c c). \tag{12.5.37}$$

So, using vector notation, we can write

$$j^0 = j + \frac{\gamma v}{v^2}\left\{\rho^c v^2 + (j\cdot u)\left(1 - \frac{1}{\gamma}\right)\right\}, \tag{12.5.38}$$

$$\rho^{c^0} = \gamma\left\{\rho^c + \frac{1}{c}(j\cdot u)\right\}. \tag{12.5.39}$$

The reverse relations of transformation are

$$j = j^0 - \frac{\gamma}{v^2}u\left\{\rho^{c^0}u^2 - (j^0 u)\left(1 - \frac{1}{\gamma}\right)\right\}, \tag{12.5.40}$$

$$\rho^c = \gamma\left\{\rho^{c^0} - \frac{1}{c^2}(j^0\cdot u)\right\}. \tag{12.5.41}$$

When u is small, these relations simplify to

$$j = j^0 - \rho^{c^0} u, \tag{12.5.42}$$

$$\rho^c = \rho^{c^0} - (j^0\cdot u)/c^2. \tag{12.5.43}$$

In the K^0 system, Ohm's law reads

$$j^0 = \sigma E^0. \tag{12.5.44}$$

In tensor form in 4-space, it is

$$j^\nu = \frac{\sigma}{c}F^{\mu\nu}U_\nu. \tag{12.5.45}$$

However, in moving media, the more general vector form Ohm's law is given in (1.12.2); the corresponding covariant tensor form is

$$j^\mu + \frac{1}{c^2}(j^\nu U_\nu)U^\mu = \frac{\sigma}{c}F^{\mu\nu}U_\nu. \tag{12.5.46}$$

For $\mu = 1, 2, 3$ the results, when combined vectorially, give

$$j + \gamma^2(\beta\cdot j)\beta - \gamma^2\rho u = \sigma\gamma\left\{E + \frac{1}{c}(u\times B)\right\}, \tag{12.5.47}$$

where u is the velocity of the material in the K systems.

12.6 Constitutive Relations and their Covariant Formulation

The constitutive relations in vacuum,

$$D^0 = E^0, \tag{12.6.1}$$

$$B^0 = H^0, \tag{12.6.2}$$

valid in the rest frame, K^0, are used in the transformation relations (12.3.29) and (12.3.32), yielding the constitutive relation

$$D + (\beta \times H) - \frac{\gamma}{\gamma+1}\beta(\beta\cdot D) = \varepsilon\{E + (\beta \times B)\} - \frac{\varepsilon\gamma}{\gamma+1}\beta\{\beta.(D - \varepsilon E)\} \tag{12.6.3}$$

in the moving frame K. It can be simplified to

$$D - \varepsilon E = -\beta \times (H - \varepsilon B). \tag{12.6.4}$$

Now, since the scalar product of (12.6.4) with B gives

$$\{\beta.(D - \varepsilon E)\} = 0, \tag{12.6.5}$$

this relation further simplifies to

$$D + (\beta \times H) = \varepsilon E + \varepsilon(\beta \times B). \tag{12.6.6}$$

Similarly, equations (12.3.30) and (12.3.31) show that

$$(\beta\cdot B) - \mu(\beta\cdot H) = 0, \tag{12.6.7}$$

$$B - (\beta \times E) = \mu H - \mu(\beta \times D). \tag{12.6.8}$$

If B is eliminated from (12.6.6), with the help of (12.6.8), then the resulting equation takes the form of (1.6.5). By similarly eliminating D from (12.6.8) with the help of (12.6.6), the resulting equation takes the form of (1.6.6). So the formulation of the constitutive relations, given in section 1.6, is relativistically correct, Kong (1975, Chapters 2 and 7).

Equations (12.6.6) and (12.6.8) can also be expressed as

$$\bar{D} = \varepsilon\bar{E}, \tag{12.6.9}$$

$$\bar{B} = \mu\bar{H}, \tag{12.6.10}$$

where

$$\bar{D} = D + (\beta \times H), \tag{12.6.11}$$

$$\bar{E} = E + (\beta \times B), \tag{12.6.12}$$

$$\bar{B} = B - (\beta \times E), \tag{12.6.13}$$

$$\bar{H} = H - (\beta \times D). \tag{12.6.14}$$

With help of (12.6.12), equations (12.4.6) and (12.4.7) become

$$\bar{F}^\mu = \gamma\left\{\bar{E}, \frac{1}{c}(\bar{E}\cdot u)\right\}. \tag{12.6.15}$$

$$\bar{K}^{\mu} = \gamma\left\{\bar{D}, \frac{1}{c}(\bar{D}, u)\right\}. \tag{12.6.16}$$

Then, in view of (12.4.5) and (12.5.7), the tensor form of equation (12.6.9) is

$$\mathcal{H}^{\mu\nu}U_{\nu} = \varepsilon F^{\mu\nu}U_{\nu}. \tag{12.6.17}$$

Also, equations (12.4.8) give

$$*\bar{F}^{\mu} = \frac{1}{c} * F^{\mu\nu}U_{\nu} = \gamma B - \gamma(\beta \times E), \quad \frac{1}{c}(u \cdot B) = -\gamma\left[\bar{E}, \frac{1}{c}(\bar{E} \cdot u)\right]. \tag{12.6.18}$$

$$*\bar{k}^{\mu} = \frac{1}{c} * \mathcal{H}^{\mu\nu}U_{\nu} = -\gamma H + \gamma(\beta \times D), \quad \frac{1}{c}(H \cdot U) = -\gamma\left[\bar{H}, \frac{1}{c}(\bar{H} \cdot u)\right]. \tag{12.6.19}$$

Then, corresponding to (12.6.17), the other constitutive relation in tensor form follows :

$$*F^{\lambda\mu}U_{\nu} = \mu^{*}\mathcal{H}^{\lambda\nu}U_{\nu}. \tag{12.6.20}$$

12.7 Electromagnetic Lagrangian

12.7.1 Different parts of the EM Lagrangian

The Lagrangian of a system of particles depends only on their co-ordinates and velocities at one and the same time only for infinite speed of propagation of interaction. However, in the EM theory, this velocity being finite, the field is an independent system with its own "degrees of freedom". For a system of interacting charged particles, this system and the field of interaction, are therefore, considered simultaneously. The EM Lagrangian is further discussed in Chapters 2, 4, 6, 8, 10 and 14. The total EM Lagrangian L is resolved in the following manner

$$\mathcal{L} = \mathcal{L}_m + \mathcal{L}_f + \mathcal{L}_I, \tag{12.7.1}$$

where $\mathcal{L}_m$ is the matter Lagrangian, $\mathcal{L}_f$ is that for the EM field and $\mathcal{L}_I$ is that for the interaction. Similarly, for the Lagrangian densities, we write

$$L = L_m + L_f + L_I. \tag{12.7.2}$$

Then, following Kalman (1961), the action S is written as

$$S = S_m + S_f + S_I = \iiiint\limits_R (L_m + L_f + L_I)dx, \tag{12.7.3}$$

where $d\mathbf{x} = dx^1\ dx^2\ dx^3\ dx^4$. The variations of S_m and S_f yield the equations of motion of a free particle, and of a free field respectively, Rund (1962). The variation of S_I with respect to particle co-ordinates gives the equation of motion in the field, and with respect to the field variables gives the inhomogeneous terms in the field equations. Through this procedure conservation laws are defined unambiguously, and the transitions from classical to quantum mechanics is studied, Rund (1962).

Since mass and energy are related, and all fields are carriers of energy, the separation of L into the parts L_m, L_f and L_I, as in (12.7.1), is not very rigid. In fact, L_m, L_f and L_I contain common terms; in the sum Lagrangian L these common terms are taken only once. The same rule applies to the division of L, as in (12.7.10).

For fields in vacuum,

$$\mathcal{L}_f = \iiint\limits_V \frac{1}{8\pi}(E^2 - H^2)dV, \tag{12.7.4}$$

where for E and H, the long range electric and magnetic field intensity vectors, respectively, should be used. For dielectrics, H^2 is simply replaced by the scalar product $(B\cdot H)$, and E^2 by $(E\cdot D)$. Since the quantum mechanical effects are not considered here, the electrostatic potential, depending on the set of position co-ordinates of all the particles, is that of all the short range forces only of charges residing in vacuum.

Actually the electric and magnetic forces on the particles, due to long range electric and magnetic fields (having wave lengths much greater than the mean path), existing in the vacuum around the particles, arise from the interaction Lagrangian

$$\mathcal{L}_I = \sum_i q_i[\{\dot{r}_i(t)\cdot A(r_i,t)\} + \phi(r,t)]. \tag{12.7.5}$$

It produces the EL equations for discrete particles, where the potentials A and ϕ are evaluated at the position of each particle. For continuous distributions of charges, the corresponding density L_I should be used, where

$$\mathcal{L}_I = \iiint\limits_V L_I dV \tag{12.7.6}$$

with

$$L_I = (A\cdot j) + \rho^c\phi. \tag{12.7.7}$$

$$\rho^c(r,t) = \frac{1}{V}\sum_i [V_i\{\delta(r - r_i(t)\}]. \tag{12.7.8}$$

$$j(r,t) = \frac{1}{V}\sum_i [q_i\dot{r}_i(t)\delta\{r - r_i(t)\}] \tag{12.7.9}$$

$\rho^c(r,t)$ is the charge density, and $j(r,t)$ is the current density at the 4-point (r,t), $\delta(r - r_i)$ is the delta function. The Lagrangian is usually expanded in series of multipole moments; in dielectrics, the monopole or free charge density vanishes, but not in plasmas. Also, in many cases, the electric quadrupole, the magnetic dipole and higher moments are neglected.

The Lorentz force F, for a particle of charge q and mass m, is written as, Chakraborty (1987, section 17),

$$F_i = -\frac{d}{dt}\left(\frac{\partial L_I}{\partial \dot{x}_i}\right) + \frac{\partial L_I}{\partial x_i} \tag{12.7.10}$$

with

$$L_I = \frac{q}{c}(U\cdot A) + q\phi. \tag{12.7.11}$$

Hence the Lagrangian of a charge in an applied field is

$$L_1 = L_m + L_I = -\frac{m}{\gamma}c^2 + \frac{q}{c}(U\cdot A) + q\phi \tag{12.7.12}$$

because, Chakraborty (1987, section 6),

$$L_m = -\frac{m}{\gamma}c^2. \tag{12.7.13}$$

Differentiation of L_1 with respect to u_i and x_i give

$$\frac{\partial L_1}{\partial u_i} = m\gamma u_i + \frac{q}{c}A_i, \tag{12.7.14}$$

$$\frac{\partial L_1}{\partial x_i} = q\frac{\partial \phi}{\partial x_i} + \frac{q}{c}\frac{\partial A_j}{\partial x_i}u_j, \tag{12.7.15}$$

where the summation convention is implied over the repeated index j. The EL equation

$$\frac{dL}{dt}\left(\frac{\partial L}{\partial u_i}\right) - \frac{\partial L}{\partial x_i} = 0, \tag{12.7.16}$$

together with the gauge relations (1.9.1) and (1.9.3), yields the equation of motion (1.8.5). The canonical momentum being

$$P^c = \frac{\partial L_1}{\partial U}, \tag{12.7.17}$$

for the EM field, we find that

$$P^c = m\gamma\mathbf{u} + \frac{q}{c}A. \tag{12.7.18}$$

Solving for U from this equation we obtain

$$u = \frac{(cP - qA)}{\sqrt{\{P - (qA/c)\}^2 + m^2c^2}}. \tag{12.7.19}$$

12.7.2 EM Hamiltonian

The total Hamiltonian is given by

$$\mathcal{H} = (P^c \cdot u) - L_1 \tag{12.7.20}$$

the other relation for $\mathcal{H}$ is

$$\nabla\mathcal{H} = -\frac{dP^c}{dt} = -\frac{d}{dt}\left(\frac{\partial L_1}{\partial u}\right). \tag{12.7.21}$$

It is easy to find that

$$\mathcal{H} = c\sqrt{(mc^2 + p^2)} - qQ = \sqrt{(c\rho^c - qA)^2 + m^2c^4} - qQ. \tag{12.7.22}$$

Using (12.7.19) we write (12.7.21) as

$$\nabla\mathcal{H} = \frac{1}{\sqrt{(m^2c^2 + p^2)}}\left[(p\cdot\nabla)\left(\frac{q}{c}A\right) + (p\times\nabla)\times\left(-\frac{q}{c}A\right)\right] + q\nabla\phi$$

$$= -\frac{q}{c}(u\cdot\nabla)A - \frac{q}{c}(u\times(\nabla\times A)) + q\nabla\phi. \tag{12.7.23}$$

The total energy $\mathcal{E}$ of the particle is the sum of the free particle energy

$$(m^2c^2r^2u^2 + m^2c^4)^{1/2} \tag{12.7.24}$$

and the potential energy $-q\phi$. So $\mathcal{E} = -\mathcal{H}$, and

$$(\mathcal{E} + q\phi)^2 - (cP^c - qA)^2 = m^2c^4. \tag{12.7.25}$$

Hamilton's equations of motion, therefore, are

$$\dot{x}_i = \frac{\partial \mathcal{H}}{\partial P_i}, \tag{12.7.26}$$

$$P_i = -\frac{\partial \mathcal{H}}{\partial x_i}. \tag{12.7.27}$$

12.7.3 Formulation for free charges in vacuum

Equation (12.4.18) shows that the Lagrangian L_1 of motion can be expressed as

$$L_1 = \left(p^\mu + \frac{q}{c}A^\mu\right) U_\mu = \frac{mcu^\mu u_\mu}{\sqrt{-u_\mu u^\mu}} + \frac{1}{c}A^\mu j_\mu. \tag{12.7.28}$$

The partial derivative of L_1 with respect to u_μ is

$$\begin{aligned}\frac{\partial L_1}{\partial u_\mu} &= \frac{mc\eta^{\mu\nu}(u_\nu + u_\xi \delta^\xi_\nu)}{2\sqrt{-u_\alpha u_\beta \eta^{\alpha\beta}}} + \frac{q}{c}A^\mu \\ &= \frac{mc\eta^{\mu\nu}u_\nu}{\sqrt{-u_\alpha u^\alpha}} + \frac{q}{c}A^\mu = m\gamma u^\mu + \frac{q}{c}A^\mu = P^\mu,\end{aligned} \tag{12.7.29}$$

where (12.4.13) has been used. Differentiating L_1 with respect to x_μ, and using (12.7.29), we obtain

$$\frac{\partial L_1}{\partial x_\mu} = \frac{q}{c}\frac{\partial A^\mu}{\partial x^\mu}u_\mu. \tag{12.7.30}$$

Since x^μ and u^μ (or p^μ) are independent of each other,

$$\frac{\partial x^\mu}{\partial p^\mu} = 0, \quad \frac{\partial p^\mu}{\partial x^\mu} = 0 \tag{12.7.31}$$

and consequently, the EL equations

$$\frac{dP^\mu}{dt} = \frac{d}{dt}\left(\frac{\partial L_1}{\partial u_\mu}\right) = \frac{\partial L_1}{\partial x_\mu} \tag{12.7.32}$$

recover the equation (12.4.18) for motion.

For application of Hamilton's principle of least action to derive the equation of motion (12.4.19), the expression for the Lagrangian L_1 of (12.7.28) suggests the adoption of the following action integral

$$S = \int_a^b \left[-mc^2\sqrt{-ds^2} + \frac{q}{c}A_\mu dx^\mu\right], \tag{12.7.33}$$

which suggests the Lagrangian

$$L = -m\sqrt{-g_{\mu\nu}\frac{dx^\mu}{\sqrt{-ds^2}}\frac{dx^\nu}{\sqrt{-ds^2}}} + \frac{q}{c}A_\mu\frac{dx^\mu}{\sqrt{-ds^2}}. \tag{12.7.34}$$

It yields the EL equation

$$\frac{d}{\sqrt{-ds^2}}\left[\frac{\partial L}{\partial\left(dx^\mu/\sqrt{-ds^2}\right)}\right]-\frac{\partial L}{\partial x^\mu}=0 \tag{12.7.35}$$

and the conjugate momentum

$$\bar{P}^\mu=-\frac{\partial L}{\partial\left(dx^\mu/\sqrt{-ds^2}\right)}=-\frac{\partial L}{\partial w^\mu}=mw_\mu+\frac{q}{c}A_\mu, \tag{12.7.36}$$

where

$$w^\mu=\frac{dx^\mu}{\sqrt{-ds^2}}. \tag{12.7.37}$$

Since A^μ satisfies a second order wave equation, argued Rohrlich (1965, section 4–9), L_f, the Lagrangian for fields, should contain only the first derivatives $\partial A^\mu/\partial x^\nu$. So, one possible choice for L_f is

$$L_f=-\frac{1}{8\pi}(\partial^\mu A^\nu)(\partial_\mu A_\nu)+\frac{1}{c}A^\mu j_\mu, \tag{12.7.38}$$

where

$$\partial^\mu=\frac{\partial}{\partial x_\mu},\quad \partial_\mu=\frac{\partial}{\partial x^\mu}. \tag{12.7.39}$$

Chakraborty (1987, section 5). Variation of the action integral for L_f gives

$$\frac{\partial L_f}{\partial A^\mu}=\frac{d}{dx_\nu}\left[\frac{\partial L_f}{\partial(\partial^\mu A^\nu)}\right]. \tag{12.7.40}$$

From (12.7.38) we obtain

$$\frac{\partial L_f}{\partial A^\mu}=\frac{1}{c}j^\mu, \tag{12.7.41}$$

$$\frac{\partial L_f}{\partial(\partial^\mu A^\nu)}=\frac{1}{4\pi}F_{\mu\nu}. \tag{12.7.42}$$

Equation (12.7.42) suggests for L_f the expression

$$L_f=-\frac{1}{16\pi}F_{\mu\nu}F^{\mu\nu}+\frac{1}{c}A^\mu j_\mu. \tag{12.7.43}$$

The first term is denoted by L_{em}; so

$$L_{em}=-\frac{1}{16\pi}F_{\mu\nu}F^{\mu\nu}. \tag{12.7.44}$$

Both this L_f and L_1 of (12.7.28) contain the common term $A^\mu j_\mu$. Taking this term once only in the sum, and denoting by Σ summation over all the species of particles per unit volume, we write

$$\begin{aligned}L&=\sum(p^\mu u_\mu)+\frac{1}{c}A^\mu j_\mu-\frac{1}{8\pi}(\partial^\mu A^\nu)(\partial_\mu A_\nu)\\&=\sum(p_\mu u^\mu)+\frac{1}{c}A_\mu j^\mu-\frac{1}{16\pi}F_{\mu\nu}F^{\mu\nu}.\end{aligned} \tag{12.7.45}$$

In terms of vector and scalar fields, this formula appears to be

$$L = -\sum\left(mc^2\sqrt{1-\frac{u^2}{c^2}}\right) + \rho\phi + \frac{1}{c}(A\cdot j) + \frac{1}{8\pi}(E^2 - B^2). \qquad (12.7.46)$$

The last term of (12.7.46) depends only on the properties of the EM field itself, and should exist even in the absence of charges.

For all sources of charges in a volume V, the total Lagrangian is

$$\mathcal{L}_T = -\sum_i\left[m_i c^2\sqrt{1-\frac{u_i^2}{c^2}}\right] + \iiint\limits_V \left[\frac{1}{8\pi}(E^2 - B^2) + \rho\phi + \frac{1}{c}(A\cdot j)\right] dV. \qquad (12.7.47)$$

Using (1.9.1), (1.9.3) and (1.9.5) and the vacuum conditions $\varepsilon = 1$, $\mu = 1$ in the second term of the right side, it becomes

$$\iiint\limits_V \left[E\cdot\left(-\frac{1}{c}\dot{A} + \nabla\phi\right) - B.(\nabla\times A)\right] dV$$

$$= -\frac{1}{c}\frac{d}{dt}\iiint\limits_V (E\cdot B)dV + \frac{1}{c}\iiint\limits_V (A\cdot E)dV + \iiint\limits_V \nabla.(A\phi)dV$$

$$- \iiint\limits_V \phi(\nabla\cdot E)dV + \iiint\limits_V [\nabla.(B\times A)]dV - \iiint\limits_V (A.(\nabla\times E))dV$$

$$= -\frac{1}{c}\frac{d}{dt}\iiint\limits_V (E\cdot A)dV + \iiint\limits_V (A.(\nabla\times H))dV - \frac{4\pi}{c}\iiint\limits_V (A\cdot j)dV$$

$$+ \iint\limits_\Sigma \phi(A\cdot d\Sigma) - 4\pi\iiint\limits_V \rho\phi dV + \iint\limits_\Sigma \{(B\times A)\cdot d\Sigma\} - \iiint\limits_V (A.(\nabla\times B))dV$$

(12.7.48)

in vacuum. For a system which does not send dipole radiation field to outer space, the two surface integrals of (12.7.46) would vanish at the infinitely far-off surface surrounding the system; and the second term cancels with the last term in vacuum, leaving

$$\iiint\limits_V (E^2 - B^2)dV = -4\pi\left[\iiint\limits_V \left\{\frac{1}{c}(A\cdot j) + \rho^c\phi\right\} dV - \frac{1}{c}\frac{d}{dt}\iiint\limits_V (E\cdot A)dV.\right. \qquad (12.7.49)$$

So, equation (12.7.47) reduces to

$$L_T = -\sum_i\left(m_i c^2\sqrt{1-\frac{u_i^2}{c^2}}\right) + \frac{1}{2}\iiint\limits_V \left\{\rho^2\phi + \frac{1}{c}(A\cdot j)\right\} dV - \frac{1}{c}\frac{d}{dt}\iiint\limits_V (E\cdot A)dV. \quad (12.7.50)$$

Since L (or L_T) contains terms of both mutual interaction between particles and field, and the contribution for a particle in the absence of the field, it is not necessarily associated with a definite mechanical system; as such, it does not have to be regarded or given as the difference of a kinetic and a potential energy. The difference between the first term of (12.7.46) and (12.7.47) is that in the former summation is carried over particles of unit volume at the field point **r**, and in the later the summation is over the volume V of charges.

12.7.4 Hamiltonian formalism

The Hamiltonian density H for particles and fields produced by them, expressed in terms of the field potentials A_μ, Moisewitsch (1966 p. 82), is

$$\mathcal{H} = P^\mu A_\mu - L_f, \tag{12.7.51}$$

where the canonical momentum

$$P_\mu = \frac{\partial L_f}{\partial \dot{A}^\mu} \quad \text{or,} \quad P^\mu = \frac{\partial L_f}{\partial \dot{A}_\mu}. \tag{12.7.52}$$

Then

$$P_\mu = -\frac{\partial \mathcal{H}}{\partial A^\mu} + \frac{\partial}{\partial x^\nu}\left[\frac{\partial \mathcal{H}}{\partial(\partial A^\mu/\partial x^\nu)}\right], \tag{12.7.53}$$

$$\frac{\partial A_\mu}{\partial t} = \frac{\partial \mathcal{H}}{\partial P^\mu}. \tag{12.7.54}$$

Employing vector and scalar fields, (12.7.51) reads

$$\mathcal{H} = \frac{1}{8\pi}(E^2 - B^2) - \frac{1}{4\pi c}(\dot{A}\cdot E) + \frac{1}{8\pi}\left((\nabla\cdot A) - \frac{1}{c}\dot{\phi}\right)^2 + \frac{1}{4\pi c}\dot{\phi}\left((\nabla\cdot A) - \frac{1}{c}\dot{\phi}\right) - \rho\phi - \frac{1}{c}(A\cdot j). \tag{12.7.55}$$

Taking the value of E from (1.9.3), it reduces to

$$\mathcal{H} = \frac{1}{8\pi}(E^2 + B^2) - \frac{1}{4\pi}(E\cdot \nabla\phi) + \frac{1}{8\pi}\left\{(\nabla\cdot A)^2 - \frac{1}{2}\dot{\phi}^2\right\} - \rho\phi - \frac{1}{c}(A\cdot j). \tag{12.7.56}$$

Apply the Lorentz gauge condition (1.9.5), and obtain

$$\mathcal{H} = \frac{1}{8\pi}(E^2 + B^2) - \frac{1}{4\pi}(E\cdot \nabla\phi) - \rho\phi - \frac{1}{c}(A\cdot j). \tag{12.7.57}$$

Assuming that the surface integral

$$\iint_\Sigma \phi(E\cdot d\Sigma)$$

vanishes at infinity, with respect to the system, the expression for the corresponding total Hamiltonian is

$$\mathcal{H}^T = \iiint_V \mathcal{H}dV = \iiint_V \left[\frac{1}{8\pi}(E^2 + B^2) - \frac{1}{c}(A\cdot j)\right] dV, \tag{12.7.58}$$

where (1.4.15) is used to eliminate the term following from the third term in the right hand side of (12.7.57).

12.7.5 Lagrangian for Dielectrics

For dielectrics equation (12.7.44) is replaced by

$$L_{em} = -\frac{1}{16\pi}F_{\mu\nu}\mathcal{H}^{\mu\nu}, \tag{12.7.59}$$

where $\mathcal{H}^{\mu\gamma}$ is given in (12.2.32). Then the Lagrangian L of (12.7.45) is replaced by

$$L = \sum(p^\mu u^\mu) + \frac{1}{c}(A_\mu j^\mu) - \frac{1}{16\pi}F_{\mu\nu}F^{\mu\nu} \tag{12.7.60}$$

and the Lagrangian L of (12.7.46) by

$$L = -\sum\left(mc^2\sqrt{1-\frac{u^2}{c^2}}\right) + \rho\phi + \frac{1}{c}(A\cdot j) + \frac{1}{8\pi}\{(E\cdot D) - (B\cdot H)\}. \tag{12.7.61}$$

Since

$$\frac{\partial L}{\partial(\partial_\mu A_\nu)} = -\frac{\mathcal{H}^{\mu\nu}}{4\pi}, \tag{12.7.62}$$

and

$$\frac{\partial L}{\partial A_\mu} = \frac{j^\mu}{c}, \tag{12.7.63}$$

equation (12.2.30) or (12.2.31) obtains from the EL equation

$$\frac{\partial L}{\partial A_\mu} - \frac{d}{dx^\nu}\left\{\frac{\partial L}{\partial(\partial_\mu A_\nu)}\right\} = 0, \tag{12.7.64}$$

which is an alternate form of (12.7.40).

Corresponding to the Lagrangian $\mathcal{L}$ of (12.7.61), the total Lagrangian

$$\mathcal{L} = -\sum_i\left[m_i c^2\sqrt{1-\frac{u_i^2}{c^2}}\right] - \frac{1}{8\pi c}\frac{d}{dt}\iiint\limits_V (D\cdot A)dV + \frac{1}{2}\iiint\limits_V\left\{\rho\phi + \frac{1}{c}(A\cdot j)\right\}dV$$

$$+\frac{1}{8\pi}\iint\limits_\Sigma [\{\phi D + (H\times A)\}\cdot d\Sigma], \tag{12.7.65}$$

where the transformation leading to the relation (12.7.50), but not the approximation adopted there, have been used, $d\Sigma$ is a surface element of the surface Σ of volume V.

12.7.6 EM energy momentum tensor from the variational principle

Consider the action integral, Landau and Lifshitz (1975, section 32)

$$S = \iiiint\limits_R L\left(A_{\mu j}\frac{\partial A_\mu}{\partial x^\nu}\right) d\mathbf{x}. \tag{12.7.66}$$

This L does not depend explicitely on the co-ordinates χ^μ; so the system is assumed closed here; L actually depends on the potentials A^μ and their partial derivatives $\partial A_\mu/\partial x^\mu$ only. Then Hamilton's principle of least action gives the EL equation

$$\frac{\partial}{\partial x^\mu}\left[\frac{\partial L}{\partial(\partial A_\nu/\partial x^\mu)}\right] - \frac{\partial L}{\partial A_\nu} = 0. \tag{12.7.67}$$

Since

$$\frac{\partial L}{\partial x^\mu} = \frac{\partial L}{\partial A_\nu}\frac{\partial A_\nu}{\partial x^\mu} + \frac{\partial L}{\partial(\partial A_\nu/\partial x^\lambda)}\frac{\partial}{\partial x^\mu}\left(\frac{\partial A_\nu}{\partial x^\lambda}\right), \tag{12.7.68}$$

interchanging the indices λ and μ in the second factor of the last term and in the first term, taking $\partial\mathcal{L}/\partial A_\nu$ from (12.7.69), we obtain.

$$\frac{\partial L}{\partial x^\mu} = \frac{\partial}{\partial x^\lambda}\left[\frac{\partial L}{\partial(\partial A_\nu/\partial x^\lambda)}\right]\frac{\partial A_\nu}{\partial x^\mu} + \frac{\partial L}{\partial(\partial A_\nu/\partial x^\lambda)}\frac{\partial}{\partial x^\lambda}\left(\frac{\partial A_\nu}{\partial x^\mu}\right) = \frac{\partial}{\partial x^\lambda}\left[\frac{\partial L}{\partial(\partial A_\nu/\partial x^\lambda)}\frac{\partial A_\nu}{\partial x^\mu}\right]. \tag{12.7.69}$$

Consequently

$$\frac{\partial}{\partial x^\lambda}\left[\frac{\partial L}{\partial(\partial A_\nu/\partial x^\lambda)}\frac{\partial A_\nu}{\partial x^\mu} - L\delta_\mu^\lambda\right] = 0, \tag{12.7.70}$$

where δ_μ^λ is the Kronecket delta. Putting, therefore,

$$T_\mu^\lambda = \frac{\partial L}{\partial(\partial A_\nu/\partial x^\lambda)}\frac{\partial A_\nu}{\partial x^\mu} - L\delta_\mu^\lambda, \tag{12.7.71}$$

in this relation, we obtain the conservation condition

$$\frac{\partial T_\mu^\lambda}{\partial x^\lambda} = 0. \tag{12.7.72}$$

The particular component

$$T_4^4 = \frac{\partial L}{\partial \dot{A}_\nu}A_\nu - L \tag{12.7.73}$$

is the energy density of the system occupying the volume V. Integration of (12.7.72), over the 4-volume R, and appliation of the Gauss theorem to reduce that to a triple integral yield

$$\iiint\limits_V T_\mu^\nu dS_\nu = \text{ constant.} \tag{12.7.74}$$

The component integral for the total energy of the system is

$$\iiint\limits_V T_4^4 dV. \tag{12.7.75}$$

For anisotropic dielectrics the tensor relation

$$\mathcal{H}^{\lambda\mu} = \frac{1}{2}\chi^{\lambda\mu\nu\eta}F_{\nu\eta} \tag{12.7.76}$$

gives the constitutive relation (the constitutive tensor is discussed in Chapter 14. Then (12.7.57) becomes

$$L_{em} = -\frac{1}{32\pi}\chi^{\mu\nu\xi\eta}F_{\mu\nu}F_{\xi\eta} = -\frac{1}{32\pi}\chi^{\mu\nu\xi\eta}(\partial_\mu A_\nu - \partial_\nu A_\mu)(\partial_\xi A_\eta - \partial_\eta A_\xi). \tag{12.7.77}$$

The variation of (12.7.44) gives

$$\delta L = -\frac{1}{8\pi}F^{\mu\nu}\delta F_{\mu\nu} = -\frac{1}{8\pi}F^{\mu\nu}\left[\delta\left(\frac{\partial A_\nu}{\partial x^\mu}\right)\right] = -\frac{1}{4\pi}F^{\mu\nu}\delta\left(\frac{\partial A_\nu}{\partial x^\mu}\right), \tag{12.7.78}$$

where the last result follows from the antisymmetric property of $F_{\mu\gamma}$. Hence

$$\frac{\partial L}{\partial(\partial A_\nu/\partial x^\mu)} = -\frac{1}{4\pi}F^{\mu\nu} \tag{12.7.79}$$

and, evidently,

$$T_\mu^\nu = -\frac{1}{4\pi}\frac{\partial A_\nu}{\partial x^\mu}F^{\nu\lambda} + \frac{1}{16\pi}\delta_\mu^\nu F_{\lambda\xi}F^{\lambda\xi}. \tag{12.7.80}$$

Chapter 13

Electrodynamics for Small Rotational Velocities

13.1 Introduction

In this chapter we consider the EM field equations, involving the concept of absolute time, for small rotational velócities. To make electrodynamics for small rotational motion look easy for handling for application, development of the vector form of these equations is given some attention. Handling, however, will show how far these are from the equations of the classical theory, and how much difficult are these without drastic approximation for application. The equations of electrodynamics, including the constitutive relations, follow from their covariant tensor formulation. These have been discussed.

13.2 Transformation of the Field Equations

13.2.1 Curl and divergence of 3-vectors in curvilinear co-ordinates

After briefly discussing the curvilinear co-ordinate system we give the expressions for $\nabla \times F$ and $\nabla \cdot F$ in this system and then write the formula for $\nabla \times F$ in a cylindrical polar system following the general results. If u_1, u_2, u_3 are the three independent co-ordinates of a point P in a curvilinear co-ordinate system and x, y, z are its co-ordinates in the stationary orthogonal rectangular co-ordinate system $OXYZ$, then its position vector can be expressed as

$$\begin{aligned}\mathbf{r} &= i_0 x(u_1, u_2, u_3) + j_0 y(u_1, u_2, u_3) + k_0 z(u_1, u_2, u_3)\\ &= i u_1(x, y, z) + j u_2(x, y, z) + k u_3(x, y, z),\end{aligned}$$

where (i, j, k) are the unit vectors along the normal, respectively, to the surfaces $u_1 =$ constant, $u_2 =$ constant, $u_3 =$ constant, and (i_0, j_0, k_0) are the same parallel to OX, OY, OZ respectively. The scaling factors h_1, h_2, h_3 are defined as

$$h_p = \left|\frac{\partial r}{\partial u_p}\right|, \quad p = 1, 2, 3,$$

so that $ih_1 = \partial r/\partial u_1$, etc. Then the formula for $\nabla \times F$ and $\nabla \cdot F$, for the variable vector function F, are

$$\nabla \times F = \begin{vmatrix} h_1 i & h_2 j & h_3 k \\ \dfrac{\partial}{\partial u_1} & \dfrac{\partial}{\partial u_2} & \dfrac{\partial}{\partial u_3} \\ h_1 F_1 & h_2 F_2 & h_3 F_3 \end{vmatrix},$$

$$\nabla \cdot F = \frac{1}{h_1 h_2 h_3}\left[\frac{\partial}{\partial u_1}(h_2 h_3 F_1) + \frac{\partial}{\partial u_2}(h_3 h_1 F_2) + \frac{\partial}{\partial u_3}(h_1 h_2 F_3)\right].$$

Equating coefficient of i from both sides of the formula for $\nabla \times F$ gives

$$(\nabla \times F)_1 = \frac{1}{h_2 h_3}\left[\frac{\partial}{\partial u_2}(h_3 F_3) - \frac{\partial}{\partial u_3}(h_3 F_3)\right].$$

Similarly, from the coefficients of j and k we obtain the expression for $(\nabla \times F)_2$ and $(\nabla \times F)_3$ respectively. In cylindrical polar co-ordinates, the components of curl of the vector E become

$$(\nabla \times E)_r = \frac{1}{r}\frac{\partial E_z}{\partial \theta} - \frac{1}{r}\frac{\partial}{\partial z}(rE_\theta), \tag{13.2.1}$$

$$(\nabla \times E)_\theta = \frac{\partial E_\theta}{\partial z} - \frac{\partial E_z}{\partial r}, \tag{13.2.2}$$

$$(\nabla \times E)_z = \frac{1}{r}\frac{\partial}{\partial r}(rE_\theta) - \frac{1}{r}\frac{\partial}{\partial \theta}(E_r). \tag{13.2.3}$$

13.2.2 Transformation of the EM laws

In the inertial laboratory rest frame K^0, transforming from the rectangular cartesian to cylindrical polar co-ordinates[1], the components of equation (1.4.3) (the Faraday law) parallel to directions of OX_1, OX_2, OX_3 are

$$\frac{\partial}{\partial x^2}\left(\frac{1}{r}E_3\right) + \frac{\partial}{\partial x^3}(-E_2) + \frac{\partial B_1}{\partial x^4} = 0, \tag{13.2.4}$$

$$\frac{\partial E_1}{\partial x^3} + \frac{\partial B_2}{\partial x^4} + \frac{\partial}{\partial x^2}(-E_3) = 0, \tag{13.2.5}$$

$$\frac{1}{r}\frac{\partial}{\partial x^1}(rE_2) + \frac{\partial}{\partial x^2}\left(-\frac{1}{r}E_1\right) + \frac{\partial}{\partial x^4}(B_3) = 0, \tag{13.2.6}$$

where

$$dx^1 = dr, \quad dx^2 = d\theta, \quad dx^3 = dz, \quad dx^4 = cdt. \tag{13.2.7}$$

Multiplication of all the terms of (13.2.6) by r gives

$$\frac{\partial}{\partial x^1}(rE_2) + \frac{\partial}{\partial x^2}(-E_1) + \frac{\partial}{\partial x^4}(rB_3) = 0. \tag{13.2.8}$$

[1] In this section, for convenience, we do not use the subscript 0 on the variables of the K^0 frame.

Similarly (13.2.4) and (13.2.5) become

$$\frac{\partial}{\partial x^2}(E_3) + \frac{\partial}{\partial x^3}(-rE_3) + \frac{\partial}{\partial x^4}(rB_1) = 0, \tag{13.2.9}$$

$$\frac{\partial}{\partial x^3}(-E_1) + \frac{\partial}{\partial x^4}(-B_2) + \frac{\partial}{\partial x^1}(E_3) = 0. \tag{13.2.10}$$

Moreover, the equation (1.4.2) for Gauss' law for the magnetic induction reads

$$\frac{\partial}{\partial x^1}(rB_1) + \frac{\partial}{\partial x^2}(B_2) + \frac{\partial}{\partial x^3}(rB_3) = 0. \tag{13.2.11}$$

The tensor form of the equations (13.2.8)–(13.2.11) is (12.2.19) where

$$F_{\mu\nu} = \begin{vmatrix} 0 & rB_3 & -B_2 & E_1 \\ -rB_3 & 0 & rB_1 & rE_2 \\ B_2 & -rB_1 & 0 & E_3 \\ -E_1 & -rE_2 & -E_3 & 0 \end{vmatrix} \tag{13.2.12}$$

$F_{\mu\nu}$ is called the *field intensity tensor*.

The components of the vector equation (1.4.4) can similarly be written as

$$\frac{\partial}{\partial x^2}\left(\frac{1}{r}H_3\right) + \frac{\partial}{\partial x^3}(-H_2) - \frac{\partial D_1}{\partial x^4} = \frac{4\pi}{c}j^1, \tag{13.2.13}$$

$$\frac{\partial H_1}{\partial x^3} + \frac{\partial}{\partial x^1}(-H_3) - \frac{\partial D_2}{\partial x^4} = \frac{4\pi}{c}j^2, \tag{13.2.14}$$

$$\frac{1}{r}\frac{\partial}{\partial x^1}(rH_2) + \frac{\partial}{\partial x_2}\left(-\frac{1}{r}H_1\right) - \frac{\partial D_3}{\partial x^4} = \frac{4\pi}{c}r\rho c. \tag{13.2.15}$$

And, equation (1.4.1) reads

$$\frac{\partial}{\partial x^1}(rD_1) + \frac{\partial}{\partial x^2}(D_2) - \frac{\partial}{\partial x^4}(rD_3) = \frac{4\pi}{c}r\rho c. \tag{13.2.16}$$

Multiplication of all the terms of (13.2.15) by the factor r reduces it to

$$\frac{\partial}{\partial x^1}(rH_2) + \frac{\partial}{\partial x^2}(-H_1) - \frac{\partial}{\partial x^4}(rD_3) = \frac{4\pi}{c}rj^3. \tag{13.2.17}$$

Similarly, (13.2.13) and (13.2.15) are written as

$$\frac{\partial H_3}{\partial x^2} + \frac{\partial}{\partial x^3}(-rH_2) - \frac{\partial}{\partial x^4}(rD_1) = \frac{4\pi}{c}rj^1, \tag{13.2.18}$$

$$\frac{\partial}{\partial x^3}(H_1) + \frac{\partial}{\partial x^1}(-H_3) + \frac{\partial}{\partial x^4}(-D_2) = \frac{4\pi}{c}j^2. \tag{13.2.19}$$

Relations (13.2.28) and (13.2.29) show that equation (13.2.19) can be readjusted to give

$$\frac{\partial}{\partial x^1}(-H_3) + \frac{\partial}{\partial x^3}(rH_1) + \frac{\partial}{\partial x^4}(-D_2) = \frac{4\pi}{c}rj^2. \tag{13.2.20}$$

Appearance of the factor r in the right side is explained after equation (14.5.4). Adopting tensor notation, equations (13.2.16)–(13.2.18) and (13.2.20) are combined together to give

$$\frac{1}{r}\frac{\partial}{\partial x^\nu}(r\mathcal{H}^{\mu\nu}) = \frac{4\pi}{c}j^\mu, \tag{13.2.21}$$

$$\text{or,}\quad \frac{\partial}{\partial x^\nu}(r\mathcal{H}^{\mu\nu}) = \frac{4\pi}{c}rj^\mu \tag{13.2.22}$$

with

$$\mathcal{H}^{\mu\nu} = \begin{bmatrix} 0 & \frac{1}{r}H_3 & -H_2 & -D_1 \\ -\frac{1}{r}H_3 & 0 & \frac{1}{r}H_1 & -\frac{1}{r}D_2 \\ H_2 & -\frac{1}{r}H_1 & 0 & -D_3 \\ D_1 & \frac{1}{r}D_2 & D_3 & 0 \end{bmatrix}. \tag{13.2.23}$$

With the help of (8.13.24), equation (13.2.22) reads

$$\frac{\partial}{\partial x^\nu}\left(\sqrt{-g}\mathcal{H}^{\mu\nu}\right) = \frac{4\pi}{c}\sqrt{-g}j^\mu \tag{13.2.24}$$

$\mathcal{H}^{\mu\nu}$ is called the *electromagnetic tensor.*

To explain the elements of second row of (13.2.23), which are obtained from (13.2.20), consider the θ-component of (1.4.14)

$$(\nabla \times H)_\theta = \frac{1}{c}\frac{\partial D_\theta}{\partial t} + \frac{4\pi}{c}rj_\theta. \tag{13.2.25}$$

It can be written as

$$\frac{\partial H_r}{\partial \theta} - \frac{\partial H_z}{\partial r} = \frac{1}{c}\frac{\partial D_\theta}{\partial t} + \frac{4\pi}{c}rj_\theta, \tag{13.2.26}$$

or as

$$\frac{1}{r}\left[\frac{\partial}{\partial x'}(-H_z) + \frac{\partial H_r}{\partial x^3} + \frac{\partial}{\partial x^4}(-D_\theta)\right] = \frac{4\pi}{c}j_\theta = \frac{4\pi}{c}j^2. \tag{13.2.27}$$

Comparison with (13.2.21) suggests the form

$$\frac{1}{r}\left[\frac{\partial}{\partial x'}\left(\frac{1}{r}H^{21}\right) + \frac{\partial}{\partial x^3}(r\mathcal{H}^{23}) + \frac{\partial}{\partial x^4}(r\mathcal{H}^{24})\right] = \frac{4\pi}{c}j^2, \tag{13.2.28}$$

where obviously

$$\mathcal{H}^{21} = -\frac{1}{r}H_z,\quad \mathcal{H}^{23} = \frac{1}{r}H_r,\quad \mathcal{H}^{24} = -\frac{1}{r}D_\theta. \tag{13.2.29}$$

13.3 Electrodynamic Equations for Small Rotational Velocities

We start from the relevant equations in the laboratory frame K^0, and obtain the field equations in the rotating frame K, using for small rotational velocities, a vector method, Modesitt (1970). The equations of electrodynamics, including the Lorentz force law and the equations of conservation of charge and momentum are identical with those of Schiff (1939) and others derived with the help of the 4-spaee metric for GN rotation (section 8.13).

13.3.1 Charge and momentum conservation

The momentum transfer equation, in the non-relativistic limit, in the rotating frame K, becomes

$$m\frac{du}{dt} = qE + \frac{q}{c}(u \times B) - 2m(\Omega \times \dot{r}) + m\{(\Omega \times r) \times \Omega\}, \tag{13.3.1}$$

where $\mathbf{u}$ is the particle velocity in the K frame, and m is the rest mass. The results of differentiation of the unit vectors in the K frame, taken from section 8.13, are the last two terms in the right side of (13.3.11); the first of these two, $2m(\Omega \times r)$, is the Coriolis force; the other, $m\{(\Omega \times r) \times \Omega\}$, is the centrifugal force of rotation. The condition (1.4.17), for charge conservation, can be written as

$$\frac{\partial N}{\partial t} + \nabla\cdot(N\mathbf{u}) = 0, \tag{13.3.2}$$

where N is the number density of charges.

13.3.2 Transformation of current and charge

Transformation of the 4-current j^μ is given by

$$j^\mu = \rho^c\frac{dx^\mu}{dt} = \rho^c\frac{dx^\mu}{\partial x^{\nu 0}}\frac{dx^{\nu 0}}{dt^0}\frac{dt^0}{dt}. \tag{13.3.3}$$

Relation (8.13.6) then shows that

$$j^r = \rho^{c_0} r^0 = j^{r^0}, \quad j^\theta = j^{\theta^0} - \rho^c \mathbf{v}. \tag{13.3.4}$$

Hence, transformation of the current 3-vector j, as in (12.5.41), is

$$j = j^0 - \rho^{c_0}\mathbf{v}, \tag{13.3.5}$$

the last result of (13.3.4) gives the condition

$$\rho^c = \rho^{co} \tag{13.3.6}$$

for charge conservation which also follows from the general invariance of $\rho^c dV$, given in (1.4.18) and the invariance condition (7.2.15) for the volume element dV. Consistent with (13.3.5) and (13.3.6), equation (13.6.8) can be written as

$$j' = j - \rho^c v.$$

13.3.3 The 4-potential A^μ

Consider the transformation relations

$$A = A^0, \quad \phi = \phi^0 - (\beta\cdot A) \tag{13.3.7}$$

for the components of A^μ where

$$\beta = \frac{1}{c}v = \frac{1}{c}(\Omega \times r). \tag{13.3.8}$$

These relations are used in (1.9.3) (which we assume here to be valid in the K frame as well) to find the electric intensity vector E.

$$E = -\frac{1}{c}\frac{\partial A}{\partial t} + \nabla\phi$$

$$= -\frac{1}{c}\frac{\partial A^0}{\partial t} + \nabla\phi^0 + \frac{1}{c}\{(v\cdot\nabla)A^0 + (A^0\cdot\nabla)v + \{(v\times(\nabla\times A^0)\} + \{A^0\times(\nabla\times v)\}, \quad (13.3.9)$$

where, for $\nabla(\beta\cdot A^0)$, the vector relation (A3.10) has been used, the first term in the right side is simplified with the help of (10.2.17), the last term using (10.2.10) and the fourth term using (10.2.11). Finally, we obtain

$$E = -\frac{1}{c}\left[\frac{\partial A^0}{\partial t^0} + (v\cdot\nabla)A^0 - (\Omega\times A^0) + (\nabla\phi^0) + \frac{1}{c}(v\cdot\nabla)A^0 + \frac{1}{c}(\Omega\times A^0) + \frac{1}{c}(v\times B^0) + \frac{1}{c}(A^0\times\Omega)\right]$$

$$= -\frac{1}{c}\frac{\partial A^0}{\partial t^0} + (\nabla^0\phi) + \frac{1}{c}(v\times B^0) = E^0 + \frac{1}{c}(v\times B^0). \quad (13.3.10)$$

13.3.4 The homogeneous pair of Maxwell equations

The transformation of the Lorentz force density F, developed with the help of (13.3.5), (13.3.6) and (13.3.10), is given by

$$F = \rho^c E + \frac{1}{c}(j\times B) = \rho^{co}E + \frac{1}{c}\rho^{co}(v\times B^0) + \frac{1}{c}(j^0\times B^0) - \frac{1}{c}\rho^{co}(v\times B^0). \quad (13.3.11)$$

Next writing

$$B = B^0 \quad (13.3.12)$$

we find that

$$F = \rho^{co}E^0 + \frac{1}{c}(j^0\times B^0) = F^0. \quad (13.3.13)$$

Thus the conservation of the Lorentz force F has been established for GN rotation. Also equation (13.3.12) for the transformation of the magnetic field intensity vector, shows that

$$H = H^0 \quad (13.3.14)$$

Equations (10.2.7) and (13.3.12) give the invariance of the solenoidal property of magnetic induction :

$$(\nabla\cdot B) = (\nabla^0\cdot B^0) = 0. \quad (13.3.15)$$

For transformation of the time derivative of B, from (13.3.11), (10.2.17) and (1.4.4), we obtain

$$\frac{\partial B^0}{\partial t^0} = \frac{\partial B^0}{\partial t} + (\nabla\times(v\times B^0)) = \frac{\partial B}{\partial t} + \nabla\times(v\times B). \quad (13.3.16)$$

Then equations (1.4.3) and (13.3.12) give

$$(\nabla\times E) + \frac{1}{c}\frac{\partial B}{\partial t} = \nabla^0\times\left\{E^0 + \frac{1}{c}(v\times B^0)\right\} + \frac{1}{c}\left\{\frac{\partial B^0}{\partial t^0} - \nabla^0\times(v\times A^0)\right\}$$

$$= (\nabla^0\times E^0) + \frac{1}{c}\frac{\partial B^0}{\partial t^0} = 0. \quad (13.3.17)$$

Equations (13.3.15) and (13.3.17) establish the form invariance of the homogeneous pair of Maxwell equations (1.4.3) and (1.4.2) for transformation from an inertial frame K^0 to the rotating frame K.

13.3.5 The inhomogeneous pair of Maxwell equations

For D^0 equations (10.2.17), (10.2.20) and (1.4.1) give

$$\frac{\partial D^0}{\partial t^0} = \frac{\partial D^0}{\partial t} + \nabla \times (\mathbf{v} \times B^0) - 4\pi\rho\mathbf{v}. \tag{13.3.18}$$

Since polarization of the medium is not changed by rotation, using the constitutive relation (1.6.2), this equation becomes

$$\frac{\partial E^0}{\partial t^0} + \frac{4\pi}{\varepsilon c}\rho v = \frac{\partial E^0}{\partial t} + \nabla \times (\mathbf{v} \times E^0). \tag{13.3.19}$$

Equations (13.3.6) and (13.3.10) show that (13.3.19) can be written as

$$\frac{\partial E^0}{\partial t^0} + \frac{4\pi}{\varepsilon c}\rho^{co} v = \frac{\partial}{\partial t}\{E - (\beta \times B)\} + \nabla \times \{E - (\beta \times B)\} \tag{13.3.20}$$

and finally as

$$\frac{\partial E^0}{\partial t^0} + \frac{4\pi}{\varepsilon c}\rho^{co} v = \frac{\partial E}{\partial t} + 4\pi j_s, \tag{13.3.21}$$

where

$$4\pi j_s = -\frac{\partial}{\partial t}(\boldsymbol{\beta} \times B) - \nabla \times [(\mathbf{v} \times E) - \beta \times (\mathbf{v} \times B)]. \tag{13.3.22}$$

Divergence of (13.3.10) gives

$$(\nabla \cdot E) = (\nabla \cdot E^0) + \nabla \cdot (\beta \times B) = (\nabla^0 \cdot E^0) + \nabla^0 \cdot (\beta \times B^0) = \frac{4\pi}{E}\rho^c + 4\pi\rho_s^c, \tag{13.3.23}$$

where

$$4\pi\rho_s^c c = \nabla^0 \cdot (v \times B^0) = B^0 \cdot (\nabla^0 \times v) - v \cdot (\nabla^0 \times B^0) \tag{13.3.24}$$

and

$$(\nabla^0 \cdot D^0) = \varepsilon(\nabla^0 \cdot E^0) = 4\pi\rho^{co} = 4\pi\rho^c \tag{13.3.25}$$

ρ_s^c is called the *Schiff charge of rotation* and j_s is the Schiff current because these were first determined by Schiff (1939). These are due to rotation of the observer and vanish for non-rotating observers. So, these do not have natural origin and therefore are spurious effects. Equation (13.3.23), with the help of (10.2.10) gives

$$(\nabla \cdot D) = 4\pi\rho^c + \frac{\varepsilon\mu}{c}\nabla \cdot (v \times H^0), \tag{13.3.26}$$

where the relations (1.6.3) and

$$D = \varepsilon E = \varepsilon E^0 + \frac{\varepsilon}{c}(v \times B^0) = D^0 + \frac{\varepsilon}{c}(v \times B^0) \tag{13.3.27}$$

for the transformation of D, have been assumed. Equations (1.4.4), (1.6.3) and (10.2.8) give

$$\nabla^0 \times H^0 = \nabla \times H^0 = \frac{\varepsilon}{c}\frac{\partial E^0}{\partial t^0} + \frac{4\pi}{c}j + \frac{4\pi}{c}\rho^c v$$

$$= \frac{E}{c}\left[-\frac{4\pi}{\varepsilon c}\rho^{c_0}v + \frac{\partial E}{\partial t} + 4\pi j_s\right] + \frac{4\pi}{c}j + \frac{4\pi}{c}\rho^c v \tag{13.3.28}$$

$$= \frac{1}{c}\frac{\partial D}{\partial t} + \frac{4\pi}{c}j + \frac{4\pi}{c}j_s, \tag{13.3.29}$$

where equations (13.3.5) and (13.2.20) have also been used.

Equation (13.3.23), with the help of (13.3.24), can be written as

$$\nabla\cdot\left\{E - \frac{1}{c}(\mathbf{v}\times B)\right\} = \frac{4\pi}{\varepsilon}\rho^c. \tag{13.3.30}$$

Also equation (13.3.28), with the help of (13.3.22), becomes

$$\nabla\times\left[H^0 + \frac{\varepsilon}{c}\{(v\times E) - \beta\times(v\times B)\}\right] = \frac{\varepsilon}{c}\frac{\partial}{\partial t}\{E - (\beta\times B)\} + \frac{4\pi}{c}j. \tag{13.3.31}$$

These two equations are written as

$$\nabla\cdot D = 4\pi\rho^c,$$

$$\nabla\times[H^0 + (\beta\times D)] = \frac{1}{c}\frac{\partial D}{\partial t} + \frac{4\pi}{c}j, \tag{13.3.32}$$

with D given by

$$D = \varepsilon E - \varepsilon(\beta\times B) = \varepsilon E^0 = D^0. \tag{13.3.33}$$

This relation of transformation of D is different from that of (13.3.27). The significance of this difference will be clarified now.

13.4 Schiff Charge Effect

In place of (13.3.14) we put

$$H = H^0 + (\beta\times D) \tag{13.4.1}$$

and write (13.3.32) as

$$\nabla\times H = \frac{1}{c}\frac{\partial D}{\partial t} + \frac{4\pi}{c}j. \tag{13.4.2}$$

Equations (13.3.25), together with the first relation of (13.3.24), become

$$\nabla^0\cdot D^0 = \varepsilon(\nabla\cdot E) - 4\pi\varepsilon\rho_s^c = \varepsilon\nabla\cdot E - \varepsilon\nabla\cdot(\beta\times B), \tag{13.4.3}$$

so that (13.3.33) is valid, and consequently

$$\nabla\cdot D = \varepsilon\nabla\cdot E - \varepsilon\nabla\cdot(\beta\times B) = 4\pi\rho^c + 4\pi\varepsilon\rho_s^c - \varepsilon\nabla\cdot(\beta\times B) = 4\pi\rho^c. \tag{13.4.4}$$

This value of $(\nabla\cdot D)$ does not tally with $(\nabla\cdot D)$ of (13.3.26), which depends on the Schiff charge ρ_s^c of rotation. Using (13.3.33) equation (13.4.1) becomes

$$H = H^0 + (\beta\times D^0). \tag{13.4.5}$$

The original derivation of the Maxwell equations, in rotating frames, were complicated by the appearance of the Schiff charge and current of rotation. These remained as unexplained paradoxes. The manipulations, leading to the relation (13.3.33) and the consequent elimination of the Schiff charge, follow from GTR, Schiozawa (1973, 1974, 1978).

Equations (13.3.26) and (13.3.29) become

$$(\nabla \times H) = \frac{1}{c}\frac{\partial D}{\partial t} + \frac{4\pi}{c}(j + \varepsilon j_s). \tag{13.4.6}$$

$$(\nabla \cdot D) = 4\pi(\rho^c + \varepsilon \rho_s^c). \tag{13.4.7}$$

The charge density ρ_s^c, and the corresponding current j_s are consequences of the state relation $D = eE + (e/c)(v \times H)$, where $\mathbf{v} = (\Omega \times \mathbf{r})/c$, of Minkowski theory of electrodynamics of moving media. These spurious effects vanish as rotation ceases to exist.

Taking divergence of (13.3.22), and using (10.2.10), we obtain

$$\begin{aligned}(\nabla \cdot j_s) &= -\frac{1}{4\pi}\frac{\partial}{\partial t}\{\nabla \cdot (\beta \times B^0)\} \\ &= -\frac{1}{4\pi c}\frac{\partial}{\partial t}\left[(B^0 \cdot (\nabla \times v) - (v \cdot (\nabla^0 \times B^0))\right] \\ &= -\frac{1}{2\pi c}\left(\frac{\partial B^0}{\partial t} \cdot \Omega\right) + \frac{1}{4\pi c}\left\{v \cdot \left(\nabla^0 \times \frac{\partial B^0}{\partial t}\right)\right\}.\end{aligned} \tag{13.4.8}$$

The continuity condition for the Schiff charge ρ_s^c is derived with the help of (13.3.23)

$$(\nabla \cdot j_s) + \frac{\partial \rho_s^c}{\partial t} = 0. \tag{13.4.9}$$

Taking divergence of (13.4.6), the general continuity condition for charges, with the help of (13.4.7), obtains easily

$$\nabla(j + \varepsilon j_s) + \frac{\partial}{\partial t}(\rho^c + \varepsilon \rho_s^c) = 0. \tag{13.4.10}$$

Hence the continuity condition (1.4.5) remains valid for $\mathbf{j}$ and ρ^c in the K frame as well.

13.4.1 Generalization of the Poynting theorem

Take the scalar product of the terms of (13.4.6) with E and the same of (13.3.17) with H. Then

$$\begin{aligned}(E \cdot \mathbf{j}) &= \frac{c}{4\pi}\left[E \cdot (\nabla \times H) - \frac{1}{c}\left(E \cdot \frac{\partial D}{\partial t}\right) - \varepsilon (E \cdot j)\right] \\ &= -(\nabla \cdot S) + \frac{c}{4\pi}(H \cdot (\nabla \times E)) - \frac{1}{4\pi}\left(E \cdot \frac{\partial D}{\partial t}\right) - \varepsilon (E \cdot j_s),\end{aligned}$$

where the Poynting flux is $\mathbf{S}(= c(E \times H)/4\pi)$. Hence, a standard form of this theorem is

$$(E \cdot j) + (\nabla \cdot \mathfrak{s}) + \frac{\partial W_E}{\partial t} + \frac{\partial W_H}{\partial t} = -\varepsilon (E \cdot j_s), \tag{13.4.11}$$

where

$$W_E = \frac{\varepsilon E^2}{8\pi}, \quad W_H = \frac{\mu H^2}{8\pi} \tag{13.4.12}$$

W_E and W_H are the densities of energy in the electric and magnetic field forms respectively.

13.4.2 Transformation to cylindrical co-ordinates

In cylindrical co-ordinates (r, θ, z), equations (13.3.25) and (13.3.28) become

$$(\nabla\cdot D) = 4\pi\rho^c + \frac{2}{c}\varepsilon\mu H_z - \frac{\varepsilon\mu}{c}v(\nabla\times H)_\theta, \tag{13.4.13}$$

$$\begin{aligned}(\nabla\times H)_r &= \frac{1}{r}\left(\frac{\partial H_z}{\partial r} - r\frac{\partial H_\theta}{\partial z}\right)\\ &= \frac{1}{c}\frac{\partial D_r}{\partial t} + \frac{4\pi}{c}j_r - \frac{\varepsilon\mu}{c}\beta\frac{\partial H_z}{\partial t} + \frac{\varepsilon B}{r}\frac{\partial E_r}{\partial \theta} - \frac{\varepsilon\mu}{c}\beta^2\frac{\partial H_z}{\partial\theta},\end{aligned} \tag{13.4.14}$$

$$\begin{aligned}(\nabla\times H)_\theta &= \frac{\partial H_r}{\partial z} - \frac{\partial H_z}{\partial r}\\ &= \frac{1}{c}\frac{\partial D_\theta}{\partial t} + \frac{4\pi}{c}rj_\theta - \varepsilon\beta\frac{\partial E_z}{\partial z} - \varepsilon\frac{\partial}{\partial r}(\beta E_r) - \varepsilon\mu\beta^2\frac{\partial H_r}{\partial r} + \varepsilon\mu\frac{\partial}{\partial r}(\beta^2 H_z),\end{aligned} \tag{13.4.15}$$

$$\begin{aligned}(\nabla\times H)_z &= \frac{1}{r}\frac{\partial}{\partial r}(rE_\theta) - \frac{1}{r}\frac{\partial E_r}{\partial\theta}\\ &= \frac{1}{c}\frac{\partial D_z}{\partial t} + \frac{4\pi}{c}j_z - \frac{\varepsilon\mu}{c}\beta H_r + \frac{\varepsilon\beta}{r}\frac{\partial E_z}{\partial\theta} + \frac{\varepsilon\mu}{r}\beta^2\frac{\partial H_r}{\partial\theta}.\end{aligned} \tag{13.4.16}$$

Equation (13.4.15) is written as

$$\begin{aligned}(\nabla\times H)_\theta &= \frac{1}{c}\frac{\partial D_\theta}{\partial t} + \frac{4\pi}{c}rj_\theta - \varepsilon\beta\frac{\partial E_z}{\partial z} - \varepsilon\frac{\partial}{\partial r}(\beta E_r) - \varepsilon\mu\beta^2(\nabla\times H)_\theta + 2\varepsilon\mu\frac{\Omega^2}{c^2}rH_z\\ &= \frac{1}{c}\frac{\partial D_\theta}{\partial t} + \frac{4\pi}{c}rj_\theta - \frac{\Omega}{c}\left[r(\nabla\cdot D) - \frac{\partial D_\theta}{\partial\theta}\right].\end{aligned} \tag{13.4.17}$$

Solving for $(\nabla\cdot D)$ and $(\nabla\times H)$ from (13.4.13) and (13.4.17) we obtain

$$(\nabla\cdot D) = 4\pi(1+\varepsilon\mu\beta^2)\rho^c + \frac{2\varepsilon\mu}{c}\Omega H_z - \frac{\varepsilon\mu}{c}\beta\left(\frac{\partial D_\theta}{\partial t} + 4\pi rj_\theta + \Omega\frac{\partial D_\theta}{\partial\theta}\right),$$

$$(\nabla\times H)_\theta = \frac{1}{c}\frac{\partial D_\theta}{\partial t} + \frac{4\pi}{c}rj_\theta + \frac{\Omega}{c}\frac{\partial D_\theta}{\partial\theta} - 4\pi\beta\rho^c. \tag{13.4.18}$$

13.4.3 The case of axially symmetric steady motion

In this case $\partial/\partial t = 0$ and $\partial/\partial\theta = 0$; then (13.4.14), (13.4.17) and (13.4.18) give

$$(\nabla\times H) = \frac{4\pi}{c}j - 4\pi\beta\rho^c\hat{\theta}, \tag{13.4.19}$$

$$(\nabla\cdot D) = 4\pi(1+\varepsilon\mu\beta^2)\rho^c + \frac{2}{c}\varepsilon\mu(\Omega\cdot H), \tag{13.4.20}$$

where $\hat{\theta}$ is the unit vector along the direction of $(r\times z)$. If, moreover, both true charge density ρ^c and current density j are negligible, then these equations reduce to

$$(\nabla\times H) = 0, \tag{13.4.21}$$

$$4\pi\rho_s^c = (\nabla\cdot D) = \frac{2\varepsilon\mu}{c}(\Omega\cdot H). \tag{13.4.22}$$

Hence, under static conditions, the spurious Schiff current vanishes, and the non-zero Schiff charge per unit volume, is

$$(\Omega\cdot H)/2\Omega c. \tag{13.4.23}$$

13.4.4 Critical comments

A spherical conductor, in rotation about a diameter, generates a magnetic field outside and in the absence of rotation no field is observed outside the conductor, Schiff (1939). But, an observer, rotating about a condenser at rest, cannot detect any field because if there is no field for one observer, there is no field for any observer. So a rotating observer detects the same magnetic field as the observer at rest when he sees the condenser rotating. There is no field for the observer rotating about the condenser because it is possible that the laws of electrodynamics in rotating systems are different from these in inertial frames, and these different laws are likely to be such as to allow no field in this case. Since a part of the electric field in a moving frame is $(\mathbf{v}\times B)/C$, as may be seen from (13.3.9), its divergence gives

$$\nabla\cdot(v\times B) = 2(\Omega\cdot B) \tag{13.4.24}$$

because $\mathbf{v} = (\Omega\times\mathbf{r})$. Hence, observations, made from a rotating frame, would record a space charge when none existed in the beginning, if the magnetic induction field B is Curl free.

In fluids rotation induces a component of vorticity in the direction of the angular velocity vector Ω, and so the stream lines become closely wound spirals with motions mainly confined to the family of parallel planes which are perpendicular to the rotation axis.

The Schiff charge and current are obtained by a GN transformation for rotation. If these quantities are physically true, then an observer rotating round an electrical conductor would detect these, whereas, these will be absent if the observer is non-rotating and the conductor is rotated. Appearance of the Schiff charge in a rotating solid body, in the presence of a magnetic field parallel to the axis of rotation, is a consequence of the theory of Minkowski for electrodynamics of moving media, which gives the state relation $D = \varepsilon E + \varepsilon C(\Omega\times r)\times B)/c$. Hence, remarked Goldstein (1961), this theory is not directly applicable to a rotating system, because there will be a space-charge density whenever there is vorticity and a magnetic field along the axis of the vorticity in any conducting medium. Numerically, this effect is very small in any terrestrial experiment. So, the question of its existence cannot be experimentally settled.

If there is vorticity in a dielectric medium whose refractive index is not unity, then due to an applied electric field along the axis of vorticity, a magnetic field must appear, and due to an applied magnetic field along that axis an electric field must appear. Moreover, for a solid body in rotation, the induced magnetic field is of the same strength as that produced by a true surface charge equal to the "bound" charge on the faces of the disc, and rotating with the disc. With a dielectric fluid in shearing rotational motion between counter-rotating discs, if the fluid motion is kept laminar and nearly in circles about the axis, the resulting effect will be different. Experiments employing interaction of a magnetic or electric field with vorticity (rotation) may decide the question of the validity of the theory of Minkowski in these interactions.

The rigorous theory of ERF is a covariant tensor formulation of it (chapter 14). A set of familiar vector form equations of electrodynamics, including the constitutive relations for small rotational velocities, obtain following the GN rules of transformation (section 13.6), Shiozawa

(1974). This theory, and particularly, the relations (13.6.1)–(13.6.10) for transformation are discussed in Chapters 11 and 15. These relations do not create the problem of the spurious Schiff charge and current of rotation.

13.5 EM Lagrangian for GN Rotation

In the rotating frame K, the expression for the Lagrangian L_F of motion of a non-relativistic particle is

$$L_F = -mc^2\left[1 - \frac{1}{2c^2}(u+v)^2\right] \tag{13.5.1}$$

The total differential dL_F of this Lagrangian, Landau and Lifshitz (1978, p. 126–128), is

$$\begin{aligned} dL_F &= m[(u\cdot du) + (du\cdot v) + \{(u\cdot(\Omega\times dr)) + (\Omega\times r)\cdot(\Omega\times dr)\}] \\ &= m[\{(u+v)\cdot du\} + \{dr\cdot((u\times\Omega)\} + \{dr\cdot(\Omega\times r)\times\Omega)\}] \\ &= m(u+v)\cdot dr + \Omega\cdot(dr\times(u+v)). \end{aligned} \tag{13.5.2}$$

So,

$$\frac{\partial L_F}{\partial u} = m(u+v), \tag{13.5.3}$$

$$\frac{\partial L_F}{\partial r} = -m[(u\times\Omega) + (v\times\Omega)]. \tag{13.5.4}$$

Consequently,

$$\frac{d}{dt}\left(\frac{\partial L_F}{\partial u}\right) - \frac{\partial L_F}{\partial r} = \frac{d}{dt}(mu) + 2m(\Omega\times u) - m(v\times\Omega) \tag{13.5.5}$$

and the Lagrangian equation

$$\frac{d}{dt}\left(\frac{\partial L}{\partial \dot{x}_i}\right) - \frac{\partial L}{\partial x_i} = F_i^0. \tag{13.5.6}$$

Chakraborty (1987), equation (6.29)), yields (13.3.1), which is the non-relativistic equation of motion of a particle of charge q and mass m, in the presence of the Lorentz force, the Coriolis force and the centrifugal force.

The generalized momentum P, for small rotational velocities, is

$$P = \frac{\partial L_m}{\partial u} = m(u+v) + \frac{q}{c}A, \tag{13.5.7}$$

where,

$$L_m = -mc^2 + \frac{1}{2}m(u+v)^2 + \frac{q}{c}(A\cdot v) + q\phi. \tag{13.5.8}$$

The Hamiltonian function $\mathcal{H}_m$ is defined as

$$\mathcal{H}_m = (P\cdot u) - L_m, \tag{13.5.9}$$

so that

$$\nabla\mathcal{H}_m = \frac{dp}{dt} = -\frac{d}{dt}\left(\frac{\partial L_m}{\partial u}\right). \tag{13.5.10}$$

And, since $\mathcal{H}_m$ is also the energy $\mathcal{E}$ of the particle,

$$\mathcal{E} = \mathcal{H}_m = m\left[c^2 + \frac{1}{2}(u^2 - v^2)\right] - q\phi \tag{13.5.11}$$

$\mathcal{H}_m$ does not contain any term linear in the velocity u. Rotation of the frame adds a term depending on the co-ordinates of the particle, and proportional to the square of its angular velocity. The centrifugal potential energy term, $-mv^2/2$, arises due to the centrifugal force of rotation. We solve for mu, from (13.5.7), and using that value in (13.5.11) find that

$$\begin{aligned}\mathcal{E} = \mathcal{H}_m &= mc^2 - \frac{1}{2}mv^2 + \frac{1}{2m}\left(P - mv - \frac{q}{c}A\right)^2 - q\phi \\ &= mc^2 + \frac{1}{2m}P^2 - \frac{q}{mc}(P\cdot A) - \left\{\Omega\cdot r \times \left(P - \frac{1}{c}qA\right)\right\} + \frac{q^2A^2}{2mc^2} - q\phi.\end{aligned} \tag{13.5.12}$$

If u^0 is the particle velocity in the inertial frame K^0, and P^0 is the corresponding generalized momentum, then

$$\mathbf{u} = \mathbf{u}^0 + \mathbf{v} = \mathbf{u}^0 + (\Omega \times \mathbf{r}), \tag{13.5.13}$$

$$P^0 = \frac{\partial L_m^0}{\partial u^0} = mu^0 + \frac{1}{c}qA^0, \tag{13.5.14}$$

where L_m^0 is the Lagrangian, and A^0 is the vector potential in the K^0 frame. Moreover, if ϕ^0 is the corresponding scalar potential, equation (13.5.14) shows that

$$L_m^0 = \frac{1}{2}mu_0^2 - mc^2 + q\phi^0 + \frac{1}{c}q(A^0\cdot u^0). \tag{13.5.15}$$

From (13.5.7), (13.5.13) and (13.5.15) we obtain

$$\mathbf{P} = \mathbf{P}^0 + \frac{q}{c}(\mathbf{A} - \mathbf{A}^0). \tag{13.5.16}$$

The relation between the generalized angular momentum vectors L and L^0, defined in the two frames by

$$L(\mathbf{r} \times P), \quad L^0 = (\mathbf{r} \times P^0), \tag{13.5.17}$$

is found to be

$$L = L^0 \times \frac{q}{c}\{\mathbf{r} \times (A - A^0)\}. \tag{13.5.18}$$

The energy $\mathcal{E}^0$ in the K^0 frame is given by

$$\mathcal{E}^0 = mc^2 + \frac{1}{2}mu^2 - q\phi^0. \tag{13.5.19}$$

Then evidently

$$\mathcal{E} + \frac{q}{c}(A\cdot v) + q\phi = \mathcal{E}^0 - \frac{1}{2}mv^2 + q\phi^0 + \frac{1}{2}m(u^2 - {u^0}^2) = \mathcal{E}^0 - m(u^0\cdot v) + q\phi^0, \tag{13.5.20}$$

$$(L\cdot\Omega) = (P\cdot v) = m(u\cdot v) + mv^2 + \frac{1}{c}q(A\cdot v), \tag{13.5.21}$$

$$(L^0\cdot\Omega) = (P^0\cdot v) = m(v\cdot v^0) + \frac{1}{c}q(A^0\cdot v). \tag{13.5.22}$$

The Lagrangian L for a system of charged particles, with their currents, self fields and other fields in inertial frames, in several forms, has been considered. For rotational, velocities, the transformation relation of the field quantities, give

$$L = -\rho^2 c^2 \left[1 - \frac{1}{c^2}(u + v)^2\right]^{1/2} + \rho^c\phi - \rho^c(\beta\cdot A) + \frac{1}{c}A\cdot(j + \rho^c v) + \frac{1}{8\pi}\{E - (\beta \times B)\}^2$$
$$- \frac{1}{8\pi}\{B + (\beta \times E)\}^2. \quad (13.5.23)$$

This relation easily simplifies to

$$L = -\rho^2 c^2 \left[1 - \frac{1}{c^2}(u + v)^2\right]^{1/2} + \rho^c\phi + \frac{1}{c}(A\cdot j) + \frac{1}{8\pi}(E^2 - B^2)$$
$$+ \frac{1}{8\pi}[(\beta \times B)^2 - (\beta \times E)^2]. \quad (13.5.24)$$

13.6 An Elimination of the Shiff Charge

Consider the transformation which includes both uniform translation at the rate v_0, and rotation at the uniform angular rate Ω with velocity ($= \Omega \times r$). It is convenient to employ three co-ordinate systems, namely the systems K^0, K and K', where K^0 is the inertial LF, at rest, K is the inertial frame moving relative to K^0 at the uniform rate v_0 parallel to the z axis, and K' is rotating relative to K at the rate Ω about the z axis. Then the motion of K' relative to K^0 is $v_0 + v$.

The useful relations of transformation from the K frame to the frame K' are (10.2.8), (13.3.5), (13.3.6), (13.3.10), (13.3.12), (13.3.15) and (13.3.18). The K frame quantities are free from primes and other superscripts; the field quantities in the K' frame bear prime as superscripts. These relations are derived from the corresponding tensor forms in Riemann spaces, Shiozhwa (1974). So we can write

$$\nabla = \nabla', \quad (13.6.1)$$

$$\rho^c = \rho^{c'}, \quad (13.6.2)$$

$$j = j' + \rho^{c'}\mathbf{v}, \quad (13.6.3)$$

$$\frac{\partial B}{\partial t} = \frac{\partial B}{\partial t} + \nabla' \times (\mathbf{v} \times B'), \quad (13.6.4)$$

$$\frac{\partial D}{\partial t} = \frac{\partial D'}{\partial t} + \nabla' \times (\mathbf{v} \times B'), \quad (13.6.5)$$

$$B = B', \quad (13.6.6)$$

$$E = E' - \frac{1}{c}(\mathbf{v} \times B'), \quad (13.6.7)$$

$$A = A', \quad \phi = \phi' - (B\cdot A'), \quad (13.6.8)$$

$$D = D', \quad (13.6.9)$$

$$H = H' + \frac{1}{c}(v \times D'), \quad (13.6.10)$$

where **v** is given by (13.3.8). It is then easy to see that

$$\nabla' \cdot B' = (\nabla \cdot B) = 0, \tag{13.6.11}$$

$$(\nabla' \cdot D') = (\nabla \cdot D) = 4\pi\rho^c = 4\pi\rho^{c'}, \tag{13.6.12}$$

$$\begin{aligned}\nabla' \times E' &= \nabla \times \{E + (\beta \times D)\} = -\frac{1}{c}\frac{\partial B}{\partial t} + \nabla \times (\beta \times B) \\ &= -\frac{1}{c}\frac{\partial B}{\partial t} - \frac{1}{c}\nabla' \times (\mathbf{v} \times B) + \nabla \times (\beta \times D) = -\frac{1}{c}\frac{\partial B}{\partial t},\end{aligned} \tag{13.6.13}$$

$$\begin{aligned}(\nabla' \times H') &= \nabla \times \{H - cB \times D)\} = \frac{1}{c}\frac{\partial D}{\partial t} + \frac{4\pi}{c}j - \nabla \times (\beta \times D) \\ &= \frac{1}{c}\left\{\frac{\partial D'}{\partial t} + \nabla' \times (\mathbf{v} \times D') - 4\pi\rho' v\right\} + \frac{4\pi}{c}(j' + \rho^c \mathbf{v}') - \nabla \times (\beta \times D') \\ &= \frac{1}{c}\frac{\partial D'}{\partial t} + \frac{4\pi}{c}j'.\end{aligned} \tag{13.6.14}$$

The relations of transformation from the K^0 frame to the K frame are

$$E^0 = E + (\beta_0 \times B), \tag{13.6.15}$$

$$B^0 = B - (\beta_0 \times E), \tag{13.6.16}$$

$$D^0 = D + (\beta_0 \times H), \tag{13.6.17}$$

$$H^0 = H - (\beta_0 \times D), \tag{13.6.18}$$

$$j^0 = j - \rho^c v, \tag{13.6.19}$$

$$\rho^{c^0} = \rho^c + \frac{1}{c}(j \cdot \beta)_0, \tag{13.6.20}$$

where

$$\beta_0 = \frac{v_0}{c}. \tag{13.6.21}$$

Transformation of the 4-vector $A^\mu (= A. - \phi)$ follows from the LT (7.2.10)

$$A^0 = A - \beta_0\phi, \tag{13.6.22}$$

$$\phi^0 = \phi - (\beta_0 \cdot A). \tag{13.6.23}$$

Equations (13.6.1)–(13.6.10), for transformation from the K frame to the frame K', together with the equations (13.6.15)–(13.6.23) for transformation form the K^0 frame to the K frame, give the equations of transformation from the K^0 frame to the frame K'. Since both velocities **v** and $\mathbf{v}_0$ are small, terms of the order of their product $(v \cdot v_0)$ and squares and other higher powers, are neglected. Thus we find the following relations of transformation.

$$\rho^{co} = \rho^{c'} - (j' \cdot \beta_0), \tag{13.6.24}$$

$$j^0 = j' + \rho^c(v - v_0), \tag{13.6.25}$$

$$B^0 = B' - (\beta \times E'), \tag{13.6.26}$$

$$D^0 = D' + (\beta_0 \times H'), \tag{13.6.27}$$

$$H^0 = H' + (\beta - \beta_0) \times D', \tag{13.6.28}$$

$$E^0 = E' - (\beta - \beta_0) \times B', \tag{13.6.29}$$

$$A^0 = A' - \beta_0 \phi', \tag{13.6.30}$$

$$\phi^0 = \phi' - (\beta \cdot A') - (\beta_0 \cdot A'). \tag{13.6.31}$$

The constitutive relations (12.6.1), (12.6.2) and (12.5.43), together with (13.6.24)–(13.6.31), give rise to the following constitutive relations in the rotating frame K' :

$$D' + (\beta \times H') = \varepsilon E' - \varepsilon(\beta - \beta_0) \times B', \tag{13.6.32}$$

$$B' - (\beta \times E') = \mu H' + \mu(\beta - \beta_0) \times D', \tag{13.6.33}$$

$$j' + \rho^c (v + v_0) = \sigma E' - \sigma(\beta - \beta_0) \times B'. \tag{13.6.34}$$

The relations (13.6.32) and (13.6.33) being complicated due to $\boldsymbol{\beta}(= \mathbf{r}\boldsymbol{\Omega}/c)$, the potentials A' and ϕ', together with the Lorentz gauge condition (1.9.5), in a rotating frame, cannot be directly evaluated.

In vacuum, for small rotational velocities only, we put $\varepsilon = 1, \mu = 1$ and $v_0 = 0$, and find that (13.6.32) and (13.6.33) reduce to

$$D' = E' - (\beta \times B'). \tag{13.6.35}$$

$$B' = H' + (\beta \times D'). \tag{13.6.36}$$

These relations show that even vacuum becomes anisotropic in presence of rotation, Heer (1964). The observable phenomena are then independent of uniform translation, and the effects of rotation are independent of the center of rotation.

In a rotating material, we put $v_0 = 0$ in (13.6.32)–(13.6.34) and obtain the consecutive relations

$$D' = \varepsilon E' - \varepsilon(\beta \times D'), \tag{13.6.37}$$

$$B' = \mu H' + \mu(\beta \times D'), \tag{13.6.38}$$

$$j' + \rho^{c'} v = \sigma E' + \sigma(\beta \times B'). \tag{13.6.39}$$

Using (13.6.24)–(13.6.31), the Poynting theorem becomes

$$\begin{aligned}(\nabla' \cdot S') = &-\frac{\partial}{\partial t'}(W_{E'} + W_{H'}) - \sigma E'^2 + \frac{1}{c}\left(B_0 \cdot \frac{\partial S'}{\partial t'}\right) + \rho'(E'.(v + v_0)) \\ &-\frac{1}{4\pi}(\beta - \beta_0) \cdot \left\{\left(\frac{\partial D'}{\partial t'} \times \mu H'\right) + \left(\varepsilon E' \times \frac{\partial B'}{\partial t'}\right)\right\} \\ &+\sigma E' \cdot \{(\beta \cdot \beta_0) \times B'\} + \sigma E' \cdot \{(\beta - \beta_0) \times \beta'\},\end{aligned} \tag{13.6.40}$$

where

$$S' = \frac{c}{4\pi}(E' \times H'), \quad W_{E'} = \frac{\varepsilon}{8\pi} E'^2, \quad W_{H'} = \frac{\mu}{8\pi} H'^2. \tag{13.6.41}$$

Chapter 14

Electrodynamics in Accelerated Frames

14.1 Introduction

This chapter is concerned with tensor formulation, in curved (Riemannian) space, of electrodynamics. So, accelerated frames of reference, and in particular, rotating frames for rotation about a fixed axis at a constant rate are thus covered. It is implied that the rotation rate is small enough so that the rotation velocity at large radius does not exceed the vacuum velocity of light. The tensor formulas have been transformed to the familiar vector form. The covariant formulation of the Lagrangian in the Riemann space and the 4-force, defined both as a covariant and as a contravariance 4-vector, have been discussed. Covariant formulation of the constitutive relations is not unique. Some of these have been reduced to the familiar vector forms to expose the inherent ambiguity. Ring lasers, are modern devices for rotation sensing. These have been considered. However, for this coverage.

Fundamentals of electrodynamics in rotating and other accelerated frames of reference are not considered at all in text books on electrodynamics. A discussion of the basic equations of this electrodynamics requires a considerable amount of Riemannian geometry in the $3+1$ space of position co-ordinates $\mathbf{r}$ and time t. The Riemannian tensors are not otherwise necessary for electrodynamics. So, this chapter is not over burdened with Riemann tensors

14.2 Basic Equations and their Considerations

The covariant Maxwell equations, in an arbitrary system of co-ordinates, following from (2.2.30) and (2.2.20), for inertial frames, are

$$\mathcal{H}^{\mu\nu}_{;\nu} = \frac{4\pi}{c} j^{\mu}, \tag{14.2.1}$$

$$F_{\lambda\mu;\nu} + F_{\mu\nu;\lambda} + F_{\nu\lambda;\mu} = 0, \tag{14.2.2}$$

where $\mathcal{H}^{\mu\gamma}$ and $F_{\mu\gamma}$ are respectively the EM field tensors of (2.2.32) and (2.2.15), j^{μ} is the 4-current, c is the vacuum velocity of light, the comma is replaced by the semicolon and the indices following a semicolon denote the covariant derivative with respect to the corresponding co-ordinates. $\mathcal{H}^{\mu\gamma}$ is called the EM *induction tensor* and $F_{\mu\gamma}$ is called the ***field intensity tensor***. Equations (14.2.1) and (14.2.2) are also written in the form

$$\frac{1}{\sqrt{-g}}\frac{\partial}{\partial x^{\nu}}\left(\sqrt{-g}\,\mathcal{H}^{\mu\nu}\right) = \frac{4\pi}{c}j^{\mu}. \tag{14.2.3}$$

$$\frac{\partial F_{\lambda\mu}}{\partial x^{\nu}} + \frac{\partial F_{\mu\nu}}{\partial x^{\lambda}} + \frac{\partial F_{\nu\lambda}}{\partial x^{\mu}} = 0. \tag{14.2.4}$$

These relations show that $F_{\mu\gamma}$ is a tensor density of zero weight and $\mathcal{H}^{\mu\gamma}$ is a tensor density of weight + 1. Another form of (14.2.4) is

$$\partial_{[\lambda}F_{\mu\nu]} = 0, \tag{14.2.5}$$

where [ ] denotes the alternative symbol explained in equations (11.2.50)–(11.2.54).

The antisymmetric tensor $F_{\mu\gamma}$ satisfies the relations

$$F_{\mu\nu} = \frac{\partial A_{\nu}}{\partial x^{\mu}} - \frac{\partial A_{\mu}}{\partial x^{\nu}} = A_{\nu,\mu} - A_{\mu,\nu} \tag{14.2.6}$$

with $A^{\mu} = (A, -\phi)$. In curved space this relation is replaced by

$$F_{\mu\nu} = A\nu;\mu - A_{\mu;\nu}.$$

However, tensorially, Landau and Lifshitz ((1995), equation 86.12). We have

$$F_{\mu\nu} = A_{\nu;\mu} - A_{\mu;\nu} = A_{\nu,\mu} - A_{\mu,\nu}. \tag{14.2.7}$$

So (14.2.6) does not change in the transition from inertial to non-inertial frames.

In vacuum the connections between the components of $\mathcal{H}^{\mu\gamma}$ and $F_{\mu\gamma}$ are established with the help of the metric tensor $g_{\mu\gamma}$. These, called the *covariant vacuum constitutive relations*, are

$$\sqrt{-g}\,\mathcal{H}^{\mu\nu} = \sqrt{-g}\,F^{\mu\nu} = \sqrt{-g}\,g^{\mu\xi}g^{\nu\eta}F_{\nu\eta}, \tag{14.2.8}$$

$$\text{or,}\quad F_{\mu\nu} = \mathcal{H}_{\mu\nu} = \frac{1}{\sqrt{-g}}g_{\mu\xi}g_{\nu\eta}\sqrt{-g}\,\mathcal{H}^{\xi\eta}. \tag{14.2.9}$$

Equation (14.2.3) is then replaced by

$$\frac{\partial}{\partial x^{\nu}}\left(\sqrt{-g}\,F^{\mu\nu}\right) = \frac{4\pi}{c}\sqrt{-g}\,j^{\mu} \tag{14.2.10}$$

and the Maxwell equations are expressed in terms of only one of the following four possible pairs of vectors (E, H), (D, H), (E, B) and (D, B), that is, in terms of only two field vectors, one of the electric type and the other of the magnetic type. For instance, Schiff (1939) has used the set (E, H), and Gron (1984) the set (E, B).

Equations (14.2.2) and (14.2.3) are evidently independent of the choice of reference frames. So, these covariant Maxwell equations are *super metric* in the sense that the metric tensor $g_{\mu\gamma}$ does not occur in them explicitly. Only the determinant g of $g_{\mu\gamma}$ appears in (14.2.3). Equation (2.2.20) is completely carried over unchanged to (14.2.4) in curved spaces.

Since g is a function of the 4-co-ordinates x^{μ}, the 4-vector A^{μ} does not necessarily satisfy the Lorentz gauge condition (1.9.5) in curved spaces. So, in general, $\mathcal{H}^{\mu\gamma}$ cannot be identified with $F^{\mu\gamma}$ even in vacuum, in a rotating frame, and the relation (14.2.8) or (14.2.9) is not valid beyond the limits of inertial frame theory of electrodynamics.

The covariant vacuum constitutive relations (14.2.8) or (14.2.9) follow from (14.12.1) and (14.12.2), which are those of Tamm (1924) and Mandel' Shtam (1952), and also from (14.13.12) and (14.13.13), which are those of Heer (1964), both under vacuum conditions (1.4.11). Owing to the involvement of a 4th order tensor $\chi^{\alpha\beta\gamma\delta}$ in the covariant constitutive relations adopted by PYT (Post, Yildiz and Tang) (section 14.15) it is not possible to show, at once, that their equation (14.15.1) is equivalent to (14.2.3) in vacuum. But, as mentioned after relation (14.13.45), the constitutive relations (14.15.35) and (14.15.36) are identical to these of Heer (1964) and Khromykh (1966) in vacuum. So, the constitutive relations (14.2.3) and (14.15.1) are basically same under vacuum conditions.

14.2.1 Relations between tensor elements and non-covariant field-variables

Following Plebanski (1960), define the non-covariant 3-dimensional field vectors D, E, B, H, in the following manner :

$$D_i = -\sqrt{-g}\,\mathcal{H}^{i4}, \tag{14.2.11}$$

$$H_i = \frac{1}{2} e_{ijk}\sqrt{-g}\,\mathcal{H}^{jk}, \tag{14.2.12}$$

$$E_i = F_{4i}, \tag{14.2.13}$$

$$B_i = -\frac{1}{2} e_{ijk} F_{jk}, \tag{14.2.14}$$

$$j_i = \sqrt{-g}\, j^i, \tag{14.2.15}$$

and the charge density

$$\rho^c = \frac{1}{c}\sqrt{-g}\, j^4. \tag{14.2.16}$$

Here e_{ijk} is the perfectly antisymmetric unit pseudotensor of third order, known as the *3-dimensional Levi Civita symbol* (Appendix 4),

$$g = \det g_{\mu\gamma}. \tag{14.2.17}$$

These relations yield the Maxwell equations (1.4.1)–(1.4.4) in material media, at rest, in an inertial frame. The quantities introduced in the relations (14.2.11)–(14.2.15) are not independent.

To find the vector form constitutive relations which would be valid for accelerated frames in vacuum, we also make use of the following tensor relations (11.3.3), (11.3.4), (11.3.13), (11.3.14), (11.3.16), (11.3.23), (11.3.36) and the relation

$$E^i = \gamma^{il} E_l. \tag{14.2.18}$$

Thus we find that

$$\begin{aligned} F^{12} &= \mathcal{H}^{12} = g^{1\xi} g^{2\eta} F_{\xi\eta} \\ &= \left(g^{11}g^{22} - g^{12}g^{21}\right) F_{21} + g^{23}\left(g^{11}F_{13} + g^{12}F_{23}\right) - g^{13}\left(g^{21}F_{13} + g^{22}F_{23}\right) \\ &\quad + \left(g^{11}g^{24} - g^{14}g^{21}\right) F_{14} + \left(g^{12}g^{24} - g^{14}g^{22}\right) F_{24} + \left(g^{13}g^{24} - g^{14}g^{23}\right) F_{34}. \end{aligned} \tag{14.2.19}$$

We now use (14.2.12), (14.2.13), (11.3.4), (11.3.14) and (11.3.16) in this relation and obtain

$$\frac{H_3}{\sqrt{-g}} = \left(g^{11}g^{22} - g^{12}g^{21}\right)F_{12} + g^{23}g^{1i}F_{i3} - g^{13}g^{2i}F_{i3}$$

$$-\frac{E_1}{\sqrt{-g_{44}}}\left(\gamma^{11}a^2 - \gamma^{21}a^1\right) - \frac{E_2}{\sqrt{-g_{44}}}\left(\gamma^{12}a^2 - \gamma^{23}a^1\right) + \frac{E}{\sqrt{-g_{44}}}\left(\gamma^{13}a^2 - \gamma^{23}a^1\right)$$

$$= \left(g^{11}g^{22} - g^{12}g^{21}\right)F_{12} + \left(g^{23}g^{1i} - g^{13}g^{2i}\right)F_{i3} - \frac{1}{\sqrt{-g_{44}}}\left(a^2E^1 - a^1E^2\right). \quad (14.2.20)$$

With the help of (11.3.23) this relation becomes

$$H_3 = -\sqrt{-g}\left[\left(g^{11}g^{22} - g^{12}g^{21}\right)B_3 - \left(g^{23}g^{11} - g^{13}g^{21}\right)B_2 + \left(g^{23}g^{12} - g^{13}g^{22}\right)B_1\right]$$
$$+\sqrt{\hat{\gamma}}\,(a\times E)_3. \quad (14.2.21)$$

Similarly for $\mu = 1$ and $\gamma = 3$, equation (14.2.8) gives

$$H_2 = \sqrt{-g}\left[B_3\left(g^{11}g^{22} - g^{13}g^{31}\right) - B_2\left(g^{11}g^{33} - g^{13}g^{31}\right) + B_1\left(g^{12}g^{33} - g^{13}g^{32}\right)\right]$$
$$+\sqrt{\hat{\gamma}}\,(a\times E)_2. \quad (14.2.22)$$

Putting, next, $\mu = 1$ and $\gamma = 4$, in (14.2.8), we obtain

$$F^{14} = \mathcal{H}^{14} = -\frac{D_1}{\sqrt{-g}} = -\left[(\gamma^{11}a^2 - \gamma^{12}a^1)B_3 + (\gamma^{11}a^3 - \gamma^{13}a^1)B_2 + (\gamma^{12}a^3 - \gamma^{13}a^2)B_1\right]$$

$$\times\frac{1}{\sqrt{-g_{44}}} - \left(\gamma^{11}g^{44} + \frac{a^1a^1}{g_{44}}\right)E_1 - \left(\gamma^{12}g^{44} + \frac{a^1a^2}{g_{44}}\right)E_2 - \left(\gamma^{13}g^{44} + \frac{a^1a^3}{g_{44}}\right)E_3$$

$$= -\frac{B_3}{\sqrt{-g_{44}}}(\gamma^{11}a^2 - \gamma^{12}a^1) + \frac{B_2}{\sqrt{-g_{44}}}(\gamma^{11}a^3 - \gamma^{13}a^1) - \frac{B_1}{\sqrt{-g_{44}}}(\gamma^{12}a^3 - \gamma^{13}a^2)$$

$$-g^{44}(\gamma^{11}E_1 + \gamma^{12}E_2 + \gamma^{13}E_3) - \frac{a_1}{g_{44}}(a^1E_1 + a^2E_2 + a^3E_3)$$

$$= -\frac{\gamma^{11}}{\sqrt{-g_{44}}}(a^2B_3 - a^3B_2) - \frac{\gamma^{12}}{\sqrt{-g_{44}}}(a^3B_1 - a^1B_3) - \frac{\gamma^{13}}{\sqrt{-g_{44}}}(a^1B_2 - a^2B_1)$$

$$-g^{44}\gamma^{1i}E_i - \frac{a^1}{g_{44}}(a^iE_i)$$

$$= -\{\gamma^{11}(a\times B)_1 + \gamma^{12}(a\times B)_2 + \gamma^{13}(a\times B)_3\}\frac{1}{\sqrt{-g_{44}}} - g^{44}E' - \frac{(a\cdot E)}{g_{44}}a^1$$

$$= -\frac{1}{\sqrt{-g_{44}}}\gamma^{1i}(a\times B)_i - g^{44}E^1 - (a\cdot E)\frac{a^1}{g_{44}}$$

$$= -\frac{(a\times B)^1}{\sqrt{-g_{44}}} - g^{44}E^1 - (a\cdot E)\frac{a^1}{g_{44}}. \quad (14.2.23)$$

Hence, using (11.3.23) in this relation, we write the vector equation

$$D = \sqrt{\hat{\gamma}}\,(a \times B) + \sqrt{-g}\,g^{44}E + \frac{1}{g_{44}}\sqrt{-g}\,(a\cdot E)a$$

$$= g^{44}\sqrt{-g}\left[E + \frac{(a \times B)}{g_{44}\sqrt{-g_{44}}} + \frac{(a\cdot E)a}{g^{44}g_{44}}\right]. \quad (14.2.24)$$

Putting next $\mu = 1$ and $\gamma = 2$ in (14.2.9) we obtain

$$\mathcal{H}_{12} = F_{12} = g_{1\xi}g_{2\eta}\mathcal{H}^{\xi\eta}$$

$$= (g_{11}g_{22} - g_{12}g_{21})\mathcal{H}^{12} + (g_{11}g_{23} - g_{13}g_{12})\mathcal{H}^{13} + (g_{11}g_{24} - g_{14}g_{21})\mathcal{H}^{14}$$

$$+(g_{12}g_{23} - g_{13}g_{22})\mathcal{H}^{23} + (g_{12}g_{24} - g_{14}g_{22})\mathcal{H}^{24} + (g_{13}g_{24} - g_{14}g_{21})\mathcal{H}^{34}. \quad (14.2.25)$$

It can be written as

$$B_3 = -\frac{H_1}{\sqrt{-g}}(g_{12}g_{23} - g_{13}g_{22}) + \frac{H_2}{\sqrt{-g}}(g_{11}g_{23} - g_{13}g_{21}) + \frac{H_3}{\sqrt{-g}}(g_{11}g_{22} - g_{12}g_{21})$$

$$+\sqrt{\hat{\gamma}}\,[D_1(g_{11}a_2 - g_{21}a_1) + D_2(g_{12}a_2 - g_{22}a_1) + D_3(g_{13}a_2 - g_{23}a_1). \quad (14.2.26)$$

Elimination of g_{ij} from the coefficients of D_1, D_2, D_3, with the help of (11.3.3), gives

$$B_3 = -\frac{H_1}{\sqrt{-g}}(g_{12}g_{23} - g_{13}g_{22}) + \frac{H_2}{\sqrt{-g}}(g_{11}g_{23} - g_{13}g_{21}) + \frac{H_3}{\sqrt{-g}}(g_{11}g_{22} - g_{12}g_{21})$$

$$+\sqrt{\hat{\gamma}}\,[B_1(\gamma_{11}a_2 - \gamma_{12}a_1)]. \quad (14.2.27)$$

Next we apply the tensor relations (11.3.33)–(11.3.35), in the coefficients of D_1, in (14.2.27), and find that it is

$$= \frac{\sqrt{\hat{\gamma}}}{\hat{\gamma}}[(\gamma^{22}\gamma^{33} - \gamma^{23}\gamma^{32})a_2 - (\gamma^{23}\gamma^{31} - \gamma^{21}\gamma^{33})a_1]$$

$$= \frac{1}{\sqrt{\hat{\gamma}}}[\gamma^{33}\gamma^{2i}a_i - \gamma^{23}a^{3i}a_i]$$

$$= \frac{1}{\sqrt{\hat{\gamma}}}(\gamma^{33}a^2 - \gamma^{23}a^3), \quad (14.2.28)$$

where we have added and subtracted the term $\gamma^{23}a_3$ for our convenience. Similarly, the coefficient of D_2 in (14.2.23) is

$$= \frac{1}{\sqrt{\hat{\gamma}}}(\gamma^{31}a^3 - \gamma^{33}a^1) \quad (14.2.29)$$

and the coefficient of D_3

$$= \frac{1}{\sqrt{\hat{\gamma}}}(\gamma^{32}a^1 - \gamma^{31}a^2). \quad (14.2.30)$$

Hence the terms containing D_1, D_2, D_3 in (14.2.27), taken together, are

$$= \frac{1}{\sqrt{\hat{\gamma}}}\left[(\gamma^{33}a^2 - \gamma^{23}a^3)D_1 + (\gamma^{31}a^3 - \gamma^{33}a^1)D_2 + (\gamma^{32}a^1 - \gamma^{31}a^2)D_3\right]$$

$$= \frac{1}{\sqrt{\hat{\gamma}}}\sum_{(i,j,k)}\left[D_i(a^j\gamma^{3k} - a^k\gamma^{3j})\right], \quad (14.2.31)$$

with $(i, j, k) = (1, 2, 3), (2, 3, 1), (3, 1, 2)$. It seems that the equation (14.2.3) or (14.2.4), with the help of (14.2.7)–(14.2.12), cannot be further developed for the field components to obtain familiar types of constitutive relations in vector form, section 14.7.

14.2.2 Polarization and magnetization tensors

The electric polarization vector, per unit volume, P, and the magnetic polarization vector, also called the *magnetization*, per unit volume, M, are related to the field vectors in empty space (section 1.12), in the following manner :

$$D = \varepsilon_0 E + 4\pi P, \tag{14.2.32}$$

$$\frac{1}{\mu_0} B = H + 4\pi M. \tag{14.2.33}$$

Thus if D, H and E, B combine into the bivectors $\mathcal{H}^{\mu\gamma}$ and $F_{\mu\gamma}$, then P and M also constitute a bivector in 4-space, with the following identifications :

$$M^{4i} = 4\pi P_i. \tag{14.2.34}$$

$$M^{ij} = -4\pi M_k. \tag{14.2.35}$$

A covariant rendition of (14.2.32) and (14.2.33) is

$$\mathcal{H}^{\alpha\beta} - M^{\alpha\beta} = \frac{1}{2}\chi_0^{\alpha\beta\gamma\delta} F_{\gamma\delta}, \tag{14.2.36}$$

where $\chi_0^{\alpha\beta\gamma\delta}$ represents the free-space properties of the constitutive tensor on general coordinates; see section 14.15.

To separate the properties of matter from those of vacuum fields, in perfect analogy with the definition for media at rest, in an inertial frame, define the antisymmetric polarization tensor of the medium, $M^{\mu\gamma}$, by the relation

$$\sqrt{-g}\,\mathcal{H}^{\mu\nu} = \sqrt{-g}\,(F^{\mu\nu} + M^{\mu\nu}) \tag{14.2.37}$$

$$\text{or} \quad \sqrt{-g}\,\mathcal{H}^{\mu\nu} = \sqrt{-g}\,g^{\mu\xi}g^{\nu\eta}(F_{\xi\eta} + M_{\xi\eta}). \tag{14.2.38}$$

Then the dielectric polarization 3-vector P and the magnetization 3-vector M of the medium, both noncovariant, are defined by

$$4\pi P_i = M_{4i}, \tag{14.2.39}$$

$$4\pi M_i = \frac{1}{2} e_{ijk} M_{ik}. \tag{14.2.40}$$

14.3 Fields in Non-inertial Frames

In inertial frames, in the absence of fields of gravitation, the state relations (14.2.8) or (14.2.9) reduce to

$$D = E. \tag{14.3.1}$$

$$B = H. \tag{14.3.2}$$

The medium exerts an influence only on the resulting EM fields which are E and B, that is, on the components of $F_{\mu\gamma}$. In presence of specified charges and currents, E is changed by a factor ε, compared with its vacuum value. So, we write

$$E_{\text{material}} = \varepsilon E_{\text{vacuum}}. \tag{14.3.3}$$

The same relation is also written as

$$D = \varepsilon E, \tag{14.3.4}$$

where D stands for E_{material} and E for E_{vacuum}. Similarly, the presence of a magnetic medium leads to a change in the intensity of the magnetic field B by a factor μ, while the induction H remains unchanged. So, we write

$$H = \frac{1}{\mu}B. \tag{14.3.5}$$

Here ε and μ are respectively the dielectric constant and the magnetic permeability of the medium.

In non-inertial frames (or in presence of a gravitational field) there are no such simple relations between the induction vector B and the electric displacement vector D, and the intensity vectors E and H. Then it is seen that a difference appears between the vectors D and E on one hand, and H and B on the other, even for vacuum, such that the quantities E and B tend to their limiting values D and H in the limit from a non-inertial to an inertial frame. Hence the vectors E and B are given the same resultant-field meaning that they have in a dielectric medium that is at rest in an inertial frame; the other two vectors, D and H, are also given the earlier meaning of induction; obviously the vectors E and B then lead to D and H in vacuum on going over to an inertial frame in the absence of gravitation fields.

The Maxwell equations are written in a covariant form in an instantaneous rest frame, and, by the rules of transformation of tensor algebra, remains valid in an arbitrary frame. For macroscopic matter, these are (14.2.1) and (14.2.2). The field tensors $F_{\mu\gamma}$ and $\mathcal{H}^{\mu\gamma}$ have the dimension of $(\text{energy})^{1/2}$ when the metric tensor elements $g_{\mu\gamma}$ are dimensionless quantities. Although $F_{\mu\gamma}$ and $\mathcal{H}^{\mu\gamma}$ can be written in terms of a polar vector and an axial vector, in a covariant manner, a non-covariant notation, given in Plebanski (1960), is found more convenient and consistent with conventional techniques. Non-covariant vectors, defined in (14.2.11)–(14.2.16) lead to the conventional Maxwell equations (1.4.1–1.4.4). Evidently, the operations curl and divergence are defined only for cartesian ordering of the differentiation.

An instantaneous rest frame, Moller (1972, Chapter 8), is regarded as a system of inertial frames K^0, for which the metric tensor is $\eta_{\mu\nu}$. If (x^0, y^0, z^0, t^0) and (x, y, z, t) are regarded as instantaneously at rest with respect to each other, then the transformation coefficients $\alpha^x_{t^0}$ and $\alpha^{x^0}_t$ vanish (where α^μ and α^{μ^0} are given by (11.2.19) and (11.2.21) respectively) :

$$\alpha^x_{t^0} = \frac{\partial x}{\partial t^0} = 0, \quad \alpha^{x^0}_t = \frac{\partial x^0}{\partial t} = 0. \tag{14.3.6}$$

14.4 Covariant Formulation of EM Lagrangian

14.4.1 Remarks on the Lagrangian formalism

The Lagrangian formalism is useful in classical electrodynamics including that of covariant formulation. Einstein is known to have become convinced of the validity of his general theory of space, time and gravitation only after he was able to define the action principle in a covariant manner and obtain from it the geodesic equations. An advantage of this formalism is that thus a single expression, called the *Lagrangian density*, is made available as a function of all the field variables including those of the EM theory, in which, for the Lorentz model, both the numbers of field variables and field equations can be very large. Euler-Lagrange equations, follow from the variational technique on the basis the principle of least action. The Lagrangian function yields all the field equations of a problem. For this reason, covariant development of the variation principle and the expression for the Lagrangian of motion including that for motion of charged particles and for the interaction of the system with the fields acting on them in non-inertial frames have become important topics.

According to the principle of least action, motion of a mechanical system is such that there exists a certain integral S, called the *action*, which has an extreme value for the actual motion. In other words, this principle states that the actual path of motion is the geodesic, or the path having an extreme length or an extreme proper time according as ds is space-like or time-like.

The Lagrangian formalism and the variational aspects of EM theory have been discussed for inertial frames, using both vector method and the method of pseudo cartesian tensors in 4-space, Chapter 6. After the treatment of Lagrangian for Newtonian rotation, in familiar vector notation, in sections 12.7 and 13.7, we develop here, systematically, in Riemann space, the Lagrangian formalism in covariant tensor form, to complete the treatment of this formalism for the EM theory.

The local Lagrangian density L has the dimension of the energy density in the 4-space of space time. L is a density of weight +1; in the manifold of 4-space. The action should therefore be regarded as a scalar integral expression, invariant for general transformation in 4-space. When S is an invariant with respect to a Lorentz transformation, the equation of motion will be covariant. Proper application of the principles of least action on this Lagrangian density yields the field equations, the momentum transfer equation and the continuity condition.

14.4.2 EM Lagrangian

The EM Lagrangian in inertial frames, equation (12.7.43), is

$$L_0 = (p_\mu u^\mu) + \frac{1}{c}(j^\mu A_\mu) - \frac{1}{16\pi} F_{\mu\nu} F^{\mu\nu}. \tag{14.4.1}$$

For generalization to non-inertial frames, we separate it into two parts

$$L_0 = L_f + L_p, \tag{14.4.2}$$

with

$$L_f = -\frac{1}{16\pi} F_{\mu\nu} F^{\mu\nu} + \frac{1}{c}(A^\mu j_\mu), \tag{14.4.3}$$

$$L_p = (p^\mu u_\mu) = -mc\sqrt{-u_\mu u^\mu}. \tag{14.4.4}$$

In accelerated frames, equation (14.4.4) is replaced by

$$L_p = -mc\sqrt{-g_{\mu\nu}\frac{dx^\mu}{dt}\frac{dx^\nu}{dt}}. \tag{14.4.5}$$

The Lagrangian L_f is a scalar function of weight + 1; so, in non-inertial frames, equation (14.4.3) becomes

$$L_f = -\frac{1}{16\pi}\sqrt{-g}\,(F_{\mu\nu} F^{\mu\nu}) + \frac{1}{c}\sqrt{-g}\,(A^\mu j_\mu). \tag{14.4.6}$$

Hence, the Lagrangian density L_0 of (14.4.1), when transformed to accelerated frames, becomes

$$L = -\frac{\sqrt{-g}}{16\pi}(F_{\mu\nu}F^{\mu\nu}) + \frac{1}{c}\sqrt{-g}\,(A^\mu j_\mu) - mc\sqrt{-g_{\mu\nu}\frac{dx^\mu}{dt}\frac{dx^\nu}{dt}}. \tag{14.4.7}$$

As in section 12.7, for dielectrics, in the first term, $F^{\mu\nu}F_{\mu\nu}$ is to be replaced by $F_{\mu\nu}\mathcal{H}^{\mu\nu}$.

14.5 4-current and Ohm's Law

The 4-current j^μ is defined by

$$j^\mu = \frac{\rho^c c}{\sqrt{-g_{44}}}\frac{dx^\mu}{dx^4}. \tag{14.5.1}$$

For $\mu = 4$ this equation, for the charge density, gives

$$\rho^c = \frac{1}{c}\sqrt{-g_{44}}j^4. \tag{14.5.2}$$

The 4-quantities dx^μ/dx^4, being the rates of change of the 4-co-ordinates x^μ, with respect to the *time* co-ordinate x^4, do not form a 4-vector, though the 4-quantities dx^μ/dT form a 4-vector where dT is the invariant proper-time subinterval.

The 3-vector electric current $j(= j^1, j^2, j^3)$, in cylindrical polar co-ordinates, is given by

$$j \equiv (j_r, rj_\theta, j_z). \tag{14.5.3}$$

Here the second component has the coefficient r because $dx^2 = d\theta$, so that

$$u^2 = \frac{dx^2}{dt} = \frac{d\theta}{dt}. \tag{14.5.4}$$

The corresponding velocity component being $rd\theta/dt$, equals ru^2. For this reason, in the right hand side of (13.2.20), the factor r appears.

The general divergence of the 4-vector j^μ is given by

$$j^\mu_{;\mu} \equiv \text{div}\, j^\mu = \frac{1}{\sqrt{-g}}\frac{\partial}{\partial x^\mu}\left(\sqrt{-g}\,j^\mu\right). \tag{14.5.5}$$

Then the continuity condition, for charges and their currents in 4-spaces, is

$$j^\mu_{;\mu} = 0. \tag{14.5.6}$$

This relation can be written as

$$\frac{1}{\sqrt{-g}}\frac{\partial}{\partial x^\mu}\left(\sqrt{-g}\,j^\mu\right) = 0. \tag{14.5.7}$$

If de is the charge located within the 3-space volume V, then, Landau and Lifshitz (1975, section 90)

$$de = \rho^c dV = \rho^c\sqrt{\gamma}\,dV_0, \tag{14.5.8}$$

where the relation (10.237) of Moller (1972) has been used. Now, multiplying both sides of this relation by dx^μ we obtain

$$dedx^\mu = \rho^c \sqrt{\hat{\gamma}}\, dx^\mu dV_0 = \rho^c \frac{\sqrt{-g}}{\sqrt{-g_{44}}} d\mathbf{x} \frac{dx^\mu}{dx^4}, \tag{14.5.9}$$

where the 4-volume element

$$d\mathbf{x} = dx^1 dx^2 dx^3 dx^4.$$

In the theory of Lorentz, ρ^c and j^μ, at time t, are defined in terms of the Delta functions

$$\rho^c = \sum_n \frac{1}{\sqrt{\hat{\gamma}}} e_n \delta(r - r_n), \tag{14.5.10}$$

$$j^\mu = \sum_n \frac{c}{\sqrt{-g}} e_n \delta(r - r_n) \frac{dx^\mu}{dx^4}, \tag{14.5.11}$$

where r_n is the position vector of the charge e_n and r is the position vector of the running point; both r_n and r are 3-vectors.

The 4-volume element $\sqrt{-g}\, d\mathbf{x}$ is an invariant, Moller (1972, equation (9.63)), where, for our convenience, we now write

$$d\mathbf{x} = dx_1 dx_2 dx_3 dx_4. \tag{14.5.12}$$

Then, in 4-space, the Dirac delta function is defined as

$$\iiiint_R f(x^\mu)\delta^4(x^\mu - x_n^\mu) dx, \tag{14.5.13}$$

where $f(x^\mu) \equiv f(x^1, x^2, x^3, x^4)$,

$$\delta^4(x^\mu - x_n^\mu) = \delta(x^1 - x_n^1)\delta(x^2 - x_n^2)\delta(x^3 - x_n^3)\delta(x^4 - x_n^4) \tag{14.5.14}$$

and

$$\frac{1}{\sqrt{-g}} \delta^4(x^\mu - x_n^\mu) \tag{14.5.15}$$

is a scalar quantity. So the 4-current j^μ can be expressed as

$$\int_{x^4} \delta^4(x^\mu - x_n^\mu) \frac{dx^\mu}{dx^4} dx^4. \tag{14.5.16}$$

Consider the trajectory of the nth particle whose velocity is

$$v_n = \frac{dx_n^i}{dt} = c\frac{dx^i}{dx^4}, \quad (i = 1, 2, 3) \tag{14.5.17}$$

and, for which in place of (14.5.13), there is the 3-dimensional integral

$$\iiint_V f(x^i)\delta^3(x^i - x_n^i) dV = f(x_n^i). \tag{14.5.18}$$

The 4-current j^μ can then be written as

$$j^\mu = \sum \frac{e_n}{\sqrt{-g}} \delta^3(x^i - x_n^i) v_n^\mu, \tag{14.5.19}$$

where

$$v_n^\mu = c\frac{dx_n^\mu}{dx^4}. \tag{14.5.20}$$

Consider a system that, at the time $t(= x^4/c)$, lies entirely inside a finite region of a 3-volume V. Multiply both sides of (14.5.7) by $-g$ and integrate the resulting relation over the co-ordinates x^i $(i = 1, 2, 3)$ of the position space. Thus we obtain

$$\frac{d}{dx^4}\iiint\limits_V \sqrt{-g}j^4 dx^1 dx^2 dx^3 = 0, \tag{14.5.21}$$

because the other three terms of integration in the summation over μ, being partial different coefficients with respect to x^i, vanish. Hence the quantity

$$e = \frac{1}{c}\iiint\limits_V \sqrt{-g_{44}}\sqrt{\tilde{\gamma}}\, j^4 dV \tag{14.5.22}$$

is a constant, where the relation (11.3.23) has been used. The quantity e is interpreted as the total charge of the system.

The tensor form of Ohm's law, given in (12.5.45) for inertial frames, remains valid for non-inertial frames as well. So

$$j^\mu + \frac{1}{c^2}(j^\nu U_\nu)U^\mu = \frac{\sigma}{c}F^{\mu\nu}U_\nu \tag{14.5.23}$$

where j^μ is taken from (14.5.1). If the field point P, with co-ordinates x^μ, in the rotating frame K, is stationary, then this relation gives

$$j_r = \sigma_\gamma(E_r - \beta B_z), \tag{14.5.24}$$

$$rj_\theta = \frac{\sigma}{\gamma}E_\theta, \tag{14.5.25}$$

$$j_z = \sigma\gamma(E_z + \beta B_r), \tag{14.5.26}$$

where $\beta = v/c = (\Omega \times r)/c$, $\gamma = 1/\sqrt{1-\beta^2}$. These relations are combined to give

$$j_\perp = \sigma\gamma\{E - (\beta \times B)\}_\perp, \tag{14.5.27}$$

$$j_\| = \sigma\gamma\{E - \beta(\beta\cdot E)\}_\|. \tag{14.5.28}$$

With the help of relations (11.2.43) and (8.13.28), equation (14.5.23) becomes

$$j^\mu + \frac{\Gamma^2}{c^2}g_{\nu\lambda}j^\nu U^\lambda U^\mu = \frac{\sigma}{c}\Gamma g_{\mu\eta}U^\eta. \tag{14.5.29}$$

Putting $\mu = 1, 2, 3, 4$ in this equation we obtain the following 4-linear simultaneous equations in the components of the 4-current j^μ

$$j^1 + \frac{\Gamma^2}{c^2}\dot{x}\left[(g_{11}\dot{x} + g_{14}c)j^1 + (g_{22}\dot{y} + g_{24}c)j^2 + g_{33}\dot{z}j^3 + (g_{41}\dot{x} + g_{42}\dot{y} + g_{44}c)j^4\right]$$
$$= \frac{\sigma}{c}\Gamma\left[F^{12}(\dot{y} + \beta_y c) + F^{13}\dot{z} + F^{14}\left(\beta_x\dot{x} + \beta_y\dot{y} - \frac{c}{\gamma^2}\right)\right], \tag{14.5.30}$$

$$j^2 + \frac{\Gamma^2}{c^2}\dot{y}\left[(g_{11}\dot{x} + g_{14}c)j^1 + (g_{22}\dot{y} + g_{24}c)j^2 + g_{33}\dot{z}j^3 + (g_{41}\dot{x} + g_{42}\dot{y} + g_{44}c)j^4\right]$$

$$= \frac{\sigma}{c}\Gamma\left[F^{21}(\dot{x} + \beta_x c) + F^{23}\dot{z} + F^{24}\left\{(\beta\cdot u) - \frac{c}{\gamma^2}\right\}\right], \tag{14.5.31}$$

$$j^3 + \frac{\Gamma^2}{c^2}\dot{z}\left[(\dot{x} + \beta_x c)j^1 + (\dot{y} + \beta_y c)j^2 + j^3\dot{z} - \left\{(\beta\cdot u) - \frac{c}{\gamma^2}\right\}j^4\right]$$

$$= \frac{\sigma}{c}\Gamma\left[F^{31}(\dot{x} + \beta_x c) + F^{32}(\dot{y} + \beta_y c) + F^{34}\left\{(\beta\cdot u) - \frac{c}{\gamma^2}\right\}\right], \tag{14.5.32}$$

$$j^4 + \frac{\Gamma^2}{c^2}c\left[(\dot{x} + \beta_x c)j^1 + (\dot{y} + \beta_y c)j^2 + j^3\dot{z} - \left\{(\beta\cdot u) - \frac{c}{\gamma^2}\right\}j^4\right]$$

$$= \frac{\sigma}{c}\Gamma\left[F^{41}(\dot{x} + \beta_x c) + F^{42}(\dot{y} + \beta_y c) + F^{43}\dot{z}\right], \tag{14.5.33}$$

where $\beta_x = -y\Omega/c$, $\beta_y = x\Omega/c$ becomes $\beta = (\Omega \times r)/c$ and only the non-zero elements of $g_{\mu\gamma}$ of (8.13.26) have been retained. These relations can be written as

$$\begin{bmatrix} 1 + \frac{\Gamma^2}{c^2}\dot{x}(\dot{x} + \beta_x c) & \frac{\Gamma^2}{c^2}\dot{x}(\dot{y} + \beta_y c) & \frac{\Gamma^2}{c^2}\dot{x}\dot{z} & -\frac{\Gamma^2}{c^2}\dot{x}\left\{(\beta\cdot u) - \frac{c}{\gamma^2}\right\} \\ \frac{\Gamma^2}{c^2}\dot{y}(\dot{x} + \beta_x c) & 1 + \frac{\Gamma^2}{c^2}\dot{y}(\dot{y} + \beta_y c) & \frac{\Gamma^2}{c^2}\dot{y}\dot{z} & -\frac{\Gamma^2}{c^2}\dot{y}\left\{(\beta\cdot u) - \frac{c}{\gamma^2}\right\} \\ \frac{\Gamma^2}{c^2}\dot{z}(\dot{x} + \beta_x c) & \frac{\Gamma^2}{c^2}\dot{z}(\dot{y} + \beta_y c) & 1 + \frac{\Gamma^2}{c^2}\dot{z}\dot{z} & -\frac{\Gamma^2}{c^2}\dot{z}\left\{(\beta\cdot u) - \frac{c}{\gamma^2}\right\} \\ \frac{\Gamma^2}{c^2}c(\dot{x} + \beta_x c) & \frac{\Gamma^2}{c^2}c(\dot{y} + \beta_y c) & \frac{\Gamma^2}{c^2}c\dot{z} & 1 - \frac{\Gamma^2}{c^2}c\left\{(\beta\cdot u) - \frac{c}{\gamma^2}\right\} \end{bmatrix}\begin{bmatrix} j^1 \\ j^2 \\ j^3 \\ j^4 \end{bmatrix}$$

$$= \frac{\sigma}{c}\Gamma\begin{bmatrix} 0 & F^{12} & F^{13} & F^{14} \\ F^{21} & 0 & F^{23} & F^{24} \\ F^{31} & F^{32} & 0 & F^{34} \\ F^{41} & F^{42} & F^{43} & 0 \end{bmatrix}\begin{bmatrix} \dot{x} + \beta_x c \\ \dot{y} + \beta_y c \\ \dot{z} \\ (\beta\cdot u) - \frac{c}{\gamma^2} \end{bmatrix}$$

$$= \frac{\sigma}{c}\Gamma\begin{bmatrix} 0 & B_z & -B_y & E_x \\ -B_z & 0 & B_x & E_y \\ B_y & -B_x & 0 & E_z \\ -E_x & -E_y & -E_z & 0 \end{bmatrix}\begin{bmatrix} \dot{x} + \beta_x c \\ \dot{y} + \beta_y c \\ \dot{z} \\ (\beta\cdot u) - \frac{c}{\gamma^2} \end{bmatrix}, \tag{14.5.34}$$

where (14.7.14)–(14.7.16) have been used. The non-zero part of the matrix in the pre-factor in the right hand of (14.5.31) is the unit matrix

$$I_4 = \begin{bmatrix} 1 & 0 & 0 & 0 \\ 0 & 1 & 0 & 0 \\ 0 & 0 & 1 & 0 \\ 0 & 0 & 0 & 1 \end{bmatrix}.$$

So (14.5.31) simplifies the matrix equation

$$\begin{bmatrix} j^1 \\ j^2 \\ j^3 \\ j^4 \end{bmatrix} = \frac{\sigma}{c}\Gamma \begin{bmatrix} (u^0 \times B)_x + E_x(\beta\cdot u) - E_x\dfrac{c}{\gamma^2} \\ (u^0 \times B)_y + E_y(\beta\cdot u) - E_y\dfrac{c}{\gamma^2} \\ (u^0 \times B)_z + E_z(\beta\cdot u) - E_z\dfrac{c}{\gamma^2} \\ -(u^0\cdot E) \end{bmatrix}, \tag{14.5.35}$$

where

$$r^0 = u^0 = (\dot{r} + v) = u + v. \tag{14.5.36}$$

Hence

$$j = \frac{\sigma}{c}\Gamma\left[(\dot{r} + v) \times B + E(\beta\cdot u) - E\cdot\frac{c}{\gamma^2}\right], \quad j^4 = -\frac{\sigma}{c}\Gamma E\cdot(\dot{r} + v). \tag{14.5.37}$$

14.6 Lorentz 4-force and Equation of Motion

The rules of transformation of the field tensors $F_{\mu\gamma}$ and $\mathcal{H}^{\mu\gamma}$ do not, however, throw any light on the expression for the force acting on a charge, stationary or moving. The 4-force can be defined as a first order tensor, either contravariant or covariant, Ise (1957). Specifically, if force is defined as the product of mass and acceleration, it is contravariant as in section 12.4. However, force can also be defined from the concept of work, which is that the invariant scalar quantity, known as *work*, is the product of force and displacement. Displacement being a contravariant 4-vector, force, must be, thus, a covariant 4-vector.

According to the first definition, we write

$$F^\mu = \frac{d}{dT}\left(m\frac{dx^\mu}{dT}\right), \tag{14.6.1}$$

where dT is the proper time interval. On the other hand, if dW is the work done, for the infinitesimal displacement dx, then the covariant 4-force F_μ is defined from the invariant relation

$$dW = F_\mu dx^\mu, \tag{14.6.2}$$

Ise (1957) and Moller (1972, p. 368), or the equation (12.4.1); it is evident that

$$F_\mu = \frac{q}{c}F_{\mu\gamma}U^\gamma, \tag{14.6.3}$$

where the 4-velocity U^μ is defined in (11.2.45). The 4-components of F_μ, evaluated from (14.6.3), with the help of (11.2.45), are

$$F_1 = \frac{q}{c}\Gamma(F_{12}\dot{x}^2 + F_{13}\dot{x}^3 + F_{14}\dot{x}^4) = q\Gamma\left[E + \frac{1}{c}(u - v) \times B\right]_r, \tag{14.6.4}$$

$$F_2 = q\Gamma r\left[E + \frac{1}{c}(u \times B)\right]_\theta, \tag{14.6.5}$$

$$F_3 = q\Gamma\left[E + \frac{1}{c}(u - v) \times B\right]_z, \tag{14.6.6}$$

$$F_4 = -\frac{q}{c}\left(u \cdot \left\{E - \frac{1}{c}(v \times B)\right\}\right), \tag{14.6.7}$$

where we write

$$U \equiv (\dot{r}, r\dot{\theta}, \dot{z}) \tag{14.6.8}$$

though

$$\mathrm{U}^{\mu} \equiv (\dot{r}, \dot{\theta}, \dot{z}, c) \tag{14.6.9}$$

and Γ is given in the relation (11.2.42). Since

$$(v \times B)_{\theta} = 0, \tag{14.6.10}$$

equations (14.6.5) and (14.6.7) become

$$F_2 = q\Gamma r\left[E + \frac{1}{c}(u - v) \times B\right]_{\theta} \tag{14.6.11}$$

$$F_4 = -\frac{q}{c}\left[(u \cdot E) + \frac{1}{c}u \cdot (u - v) \times B\right]. \tag{14.6.12}$$

Calculations show that

$$\delta W = F_{\mu}\delta x^{\mu} = F_{\mu}u^{\mu}\delta t = 0.$$

For a charge at rest in the rotating frame K, the components of U vanish, and the relation (11.2.48) gives

$$\Gamma = \gamma. \tag{14.6.13}$$

Hence, in this case, we find that

$$F_1 = q\gamma[E - (\beta \times B)]_r, \tag{14.6.14}$$

$$F_2 = q\gamma r[E - (\beta \times B)]_{\theta}, \tag{14.6.15}$$

$$F_3 = q\gamma[E - (\beta \times B)]_z, \tag{14.6.16}$$

$$F_4 = 0. \tag{14.6.17}$$

The equation of motion in non-inertial frames is

$$mc\left[\frac{dW^{\mu}}{\sqrt{-ds^2}} + \Gamma^{\mu}_{\nu\eta}W^{\nu}W^{\eta}\right] = -\frac{q}{c}g_{\nu\xi}F^{\mu\xi}W^{\nu}, \tag{14.6.18}$$

where $\Gamma^{\mu}_{\nu\eta}$ is the Christofell symbol defined in the relation (11.2.49).

14.7 Maxwell Equations in Vector Form

The elements of the field tensors $F^{\mu\gamma}$ and $\mathcal{H}^{\mu\gamma}$ are some 4-vectors of fields of electrodynamics. Moreover, some properties of covariant derivatives of 4-vectors and anti-symmetric second order tensors are employed to obtain the field equations in forms which resemble the familiar vector form Maxwell equations of classical electrodynamics, Landau and Lifshitz (1975, problem in p. 257) and Moller (1972, section 10.9). We can write

$$\mathcal{H}^{ij}\sqrt{-g_{44}} = H^{ij}, \tag{14.7.1}$$

$$\mathcal{H}^{4\nu} = H^{4\nu}, \tag{14.7.2}$$

$$D^i = \sqrt{-g_{44}}H^{4i}, \tag{14.7.3}$$

$$E_i = F_{i4}, \tag{14.7.4}$$

$$F_{ij} = B_{ij}. \tag{14.7.5}$$

The covariant and contravariant components of the axial vectors H and $\dot{B}$, dual to the corresponding antisymmetric second order tensors H^{ij} and B^{ij}, are given by

$$H_i = \frac{1}{2}\sqrt{\hat{\gamma}}\, e_{ijk}H^{jk}, \tag{14.7.6}$$

$$H^i = \frac{1}{2}\sqrt{\hat{\gamma}}\, e^{ijk}H_{jk}, \tag{14.7.7}$$

$$B_i = \frac{1}{2}\sqrt{\hat{\gamma}}\, e_{ijk}B^{jk}, \tag{14.7.8}$$

$$B^i = -\frac{1}{2}\sqrt{\hat{\gamma}}\, e^{ijk}B_{jk}, \tag{14.7.9}$$

where e^{ijk} is the Levi Civita symbol (a pseudoscalar) and γ_{ij} is the spatial second order metric tensor, defined in equations (11.3.19) and (11.3.20).

The inverse transformation of (14.7.6) is

$$H_{ij} = \frac{1}{\hat{\gamma}}\, e^{ijk}H_k. \tag{14.7.10}$$

Also equations (14.7.6) and (14.7.7) give

$$H^1 = \frac{1}{\sqrt{\hat{\gamma}}}H_{23}, \text{ etc.} \tag{14.7.11}$$

$$H_1 = \sqrt{\hat{\gamma}}\, H^{23}, \text{ etc.} \tag{14.7.12}$$

From these relations we obtain

$$F_{\mu\nu} = \begin{bmatrix} 0 & -\sqrt{\hat{\gamma}}\, B_z & \sqrt{\hat{\gamma}}\, B_y & E_z \\ \sqrt{\hat{\gamma}}\, B_z & 0 & -\sqrt{\hat{\gamma}}\, B_x & E_y \\ -\sqrt{\hat{\gamma}}\, B_y & -\sqrt{\hat{\gamma}}\, B_x & 0 & E_z \\ -E_x & -E_y & -E_z & 0 \end{bmatrix}. \tag{14.7.13}$$

Then, with the help of the tensor relation

$$F^{\mu\nu} = g^{\mu\xi}g^{\nu\eta}F_{\xi\eta},$$

we find that

$$F^{\mu\nu} = \begin{bmatrix} 0 & P_z & -P_y & Q_x \\ -P_z & 0 & P_x & Q_y \\ P_y & -P_x & 0 & Q_z \\ -Q_x & -Q_y & -Q_z & 0 \end{bmatrix}, \tag{14.7.14}$$

where

$$P = -\sqrt{\hat{\gamma}}\,[B + \beta \times (\beta \times B)] - (\beta \times E) = -\sqrt{\hat{\gamma}}\,B + (\beta + Q), \tag{14.7.15}$$

with

$$Q = -\sqrt{\hat{\gamma}}\,(\beta \times B) - E. \tag{14.7.16}$$

Equation (14.2.4) gives

$$\frac{\partial F_{ij}}{\partial x^k} + \frac{\partial F_{jk}}{\partial x^i} + \frac{\partial F_{ki}}{\partial x^j} = 0, \tag{14.7.17}$$

$$\frac{\partial F_{ij}}{\partial x^4} + \frac{\partial F_{j4}}{\partial x^i} + \frac{\partial F_{4i}}{\partial x^j} = 0. \tag{14.7.18}$$

Putting (14.7.4) in the second term of (14.7.14), we obtain

$$\frac{\partial F_{ij}}{\partial x^4} = -\frac{\partial B_{ij}}{\partial x^4} = \frac{1}{c}\frac{\partial}{\partial t}\left(\sqrt{\hat{\gamma}}\,B^k\right) = -\sqrt{\hat{\gamma}}\,(\operatorname{curl} E)^k. \tag{14.7.19}$$

For a 3-vector A, Landau and Lifhtiz (1975, p. 253, foot note), we write

$$\begin{aligned}(\operatorname{curl} A)^i &= \frac{1}{2\sqrt{\hat{\gamma}}}\, e^{ijk}\left(\frac{\partial A_k}{\partial x^j} - \frac{\partial A_j}{\partial x^k}\right) \\ &= \frac{1}{2\sqrt{\hat{\gamma}}}\left[\left(\frac{\partial A_k}{\partial x^j} - \frac{\partial A_j}{\partial x^k}\right) - \left(\frac{\partial A_j}{\partial x^k} - \frac{\partial A_k}{\partial x^j}\right) = \frac{1}{\sqrt{\hat{\gamma}}}\left(\frac{\partial A_k}{\partial x^j} - \frac{\partial A_j}{\partial x^k}\right)\right].\end{aligned} \tag{14.7.20}$$

Using this relation, in the right hand side of (14.7.19), and in its left side using (14.7.5) first and then (14.7.9), we obtain

$$\frac{\partial F_{ij}}{\partial x^4} = -\frac{\partial B_{ij}}{\partial x^4} = \frac{1}{c}\frac{\partial}{\partial t}\left(\sqrt{\hat{\gamma}}\,B^k\right) = -\sqrt{\hat{\gamma}}\,(\operatorname{curl} E)^k. \tag{14.7.21}$$

Hence, in vector notation, (14.7.21) reads

$$\operatorname{curl} E = -\frac{1}{c\sqrt{\hat{\gamma}}}\frac{\partial}{\partial t}\left(\sqrt{\hat{\gamma}}\,B\right). \tag{14.7.22}$$

In (14.7.13), applying (14.7.5) and (14.7.9), we obtain the solenoidal condition

$$\frac{1}{\sqrt{\hat{\gamma}}}\frac{\partial}{\partial x^i}\left(\sqrt{\hat{\gamma}} B^i\right) = \operatorname{div} B = 0 \tag{14.7.23}$$

for the magnetic induction field, where the 3-dimensional divergence of A, Landau and Lifshitz (1975, p. 253, foot note), is given by

$$\operatorname{div} A = \frac{1}{\sqrt{\hat{\gamma}}}\frac{\partial}{\partial x^i}\left(\sqrt{\hat{\gamma}}\,A^i\right). \tag{14.7.24}$$

For generalization of the inhomogeneous pair of Maxwell equations (1.4.1) and (1.4.4), consider the tensor relation (14.2.10)

$$\frac{1}{\sqrt{-g}}\frac{\partial}{\partial x^\nu}\left(\sqrt{-g}\,\mathcal{H}^{\mu\nu}\right) = \frac{4\pi}{c} j^\mu. \tag{14.7.25}$$

Since $\mathcal{H}^{44} = 0$, this equation yields the following two relations

$$\frac{4\pi}{c} j^i = \frac{1}{\sqrt{-g}}\left[\frac{\partial}{\partial x^j}\left(\sqrt{-g}\mathcal{H}^{ij}\right) + \frac{\partial}{\partial x^4}\left(\sqrt{-g}\mathcal{H}^{i4}\right)\right], \tag{14.7.26}$$

$$\frac{4\pi}{c} j^4 = \frac{1}{\sqrt{-g}}\frac{\partial}{\partial x^j}\left(\sqrt{-g}\mathcal{H}^{4i}\right). \tag{14.7.27}$$

In view of (14.7.1)–(14.7.3), these two equations become

$$\frac{4\pi}{c} j^i = \frac{1}{\sqrt{-g}}\left[\frac{\partial}{\partial x^j}\left(\frac{\sqrt{-g}}{\sqrt{-g_{44}}}\mathcal{H}^{ij}\right) - \frac{1}{c}\frac{\partial}{\partial t}\left(\frac{\sqrt{-g}}{\sqrt{-g_{44}}}D^i\right)\right]. \tag{14.7.28}$$

$$\frac{4\pi}{c} j^i = \frac{1}{\sqrt{-g}}\frac{\partial}{\partial x^i}\left(\sqrt{-g}\mathcal{H}^{4i}\right). \tag{14.7.29}$$

Using equation (11.3.23), equation (14.7.28) transforms to

$$\frac{4\pi}{c} j^4 = \frac{1}{\sqrt{-g}}\left[\frac{\partial}{\partial x^j}\left(\sqrt{\tilde{\gamma}}\mathcal{H}^{ij}\right) - \frac{1}{c}\frac{\partial}{\partial t}\left(\sqrt{\tilde{\gamma}}D^i\right)\right] \tag{14.7.30}$$

On using (14.7.30) in the first term in the right side of this equations, we get

$$\frac{4\pi}{c} j^i = \frac{1}{\sqrt{-g_{44}}}\left[\frac{1}{\sqrt{\tilde{\gamma}}}\left(\frac{\partial H_k}{\partial x^j} - \frac{\partial H_i}{\partial x^k}\right) - \frac{1}{c\sqrt{\tilde{\gamma}}}\frac{\partial}{\partial t}\left(\sqrt{\tilde{\gamma}}D^i\right)\right], \tag{14.7.31}$$

where again (11.3.23) has been used. Now, with the help of (14.7.16), in the first term in the right side of (14.7.27), we obtain

$$\frac{4\pi}{c}\sqrt{-g_{44}}\; j^i = (\operatorname{curl} H)^i - \frac{1}{c\sqrt{\tilde{\gamma}}}\frac{\partial}{\partial t}\left(\sqrt{\tilde{\gamma}}D^i\right). \tag{14.7.32}$$

The vector form of (14.7.28) is

$$\operatorname{curl} H = \frac{1}{c\sqrt{\tilde{\gamma}}}\frac{\partial}{\partial t}\left(\sqrt{\tilde{\gamma}}D\right) + \frac{4\pi}{c}\sqrt{-g_{44}}\; j. \tag{14.7.33}$$

With the help of (14.7.28) and (11.3.23), equation (14.7.25) becomes

$$\frac{4\pi}{c} j^4 = \frac{1}{\sqrt{-g_{44}}}\frac{1}{\sqrt{\tilde{\gamma}}}\frac{\partial}{\partial x^j}\left(\frac{\sqrt{-g}}{\sqrt{-g_{44}}}D^j\right). \tag{14.7.34}$$

Further using (1.3.23), it reads

$$\frac{4\pi}{c}\sqrt{-g_{44}}\; j^4 = \frac{1}{\sqrt{\tilde{\gamma}}}\frac{\partial}{\partial x^j}\left(\sqrt{\tilde{\gamma}}D^j\right). \tag{14.7.35}$$

The tensor relation (14.7.20), therefore, takes the form

$$\frac{4\pi}{c}\sqrt{-g_{44}}j^4 = \operatorname{div} D. \tag{14.7.36}$$

Equation of continuity, equation (14.5.6) or (14.5.7), is now

$$j^\mu_{;\mu} = 0, \quad \text{or,} \quad \frac{1}{\sqrt{-g}}\frac{\partial}{\partial x^\mu}\left(\sqrt{-g}j^\mu\right) = 0. \tag{14.7.37}$$

In view of equation (14.5.2), equation (14.7.32) reads

$$\operatorname{div} D = \frac{1}{\sqrt{\hat{\gamma}}}\frac{\partial}{\partial x^j}\left(\sqrt{\hat{\gamma}}D^j\right) = 4\pi\rho^c; \tag{14.7.38}$$

and equation (14.5.1) shows that (14.7.29) can be written as

$$\operatorname{curl} H = \frac{1}{c\sqrt{\hat{\gamma}}}\frac{\partial}{\partial t}\left(\sqrt{\hat{\gamma}}D\right) + \frac{4\pi}{c}\rho^c u, \tag{14.7.39}$$

where $\mathbf{u} \equiv (dx^1, dx^2, dx^3)/dt$.

14.8 Constitutive Relations in Vacuum

The constitutive relations, whether in a vacuum or in dielectrics, determine both D and B in terms of E and H. In vacuum these are obtained from the tensor relations

$$\mathcal{H}^{\mu\nu} = F^{\mu\nu} \equiv g^{\mu\alpha}g^{\nu\beta}F_{\alpha\beta}, \tag{14.8.1}$$

$$F_{4i} = g_{4\alpha}g_{i\beta}F^{\alpha\beta} = g_{4\alpha}g_{i\beta}\mathcal{H}^{\alpha\beta} \tag{14.8.2}$$

which follow from (12.2.41). The relations (14.7.1) and (14.7.2) should be modified in vacuum to

$$\mathcal{H}^{ij}\sqrt{-g_{44}} = \mathcal{H}^{ij} = \sqrt{-g_{44}}F^{ij}, \tag{14.8.3}$$

$$\mathcal{H}^{i4} = H^{i4} = F^{i4}. \tag{14.8.4}$$

Hence (14.7.3) is replaced by

$$D^i = \sqrt{-g_{44}}F^{4i}. \tag{14.8.5}$$

The other necessary tensor relations are (11.3.3), (11.3.4), (11.3.14) and (11.3.16). Moreover, the following transformation relations, for the field elements, in 3-space of position co-ordinates, are necessary

$$D_i = \gamma_{ij}D^j, \tag{14.8.6}$$

$$H_{ij} = \gamma_{il}\gamma_{jk}H^{lk}, \tag{14.8.7}$$

$$E^i = \gamma^{il}E_l, \text{ etc.} \tag{14.8.8}$$

Equations (14.7.4), (11.3.4) and (14.8.7) give

$$\frac{1}{\sqrt{-g_{44}}}(E_i - a^j H_{ij}) = \frac{1}{\sqrt{-g_{44}}}\left[F_{i4} - \sqrt{-g_{44}}g^{j4}\gamma_{il}\gamma_{jk}H^{lk}\right]. \tag{14.8.9}$$

With the help of (14.8.2), (14.8.3) and (11.3.3) this relation becomes

$$\frac{1}{\sqrt{-g_{44}}}(E_i - a^j H_{ij}) = \frac{g_{i\mu} g_{\mu 4}}{\sqrt{-g_{44}}} F^{\mu\nu} - g^{i4}\left(g_{il} - g_{4i}\frac{g_{4l}}{g_{44}}\right)\left(g_{jk} - g_{4j}\frac{g_{4k}}{g_{44}}\right)\sqrt{-g_{44}} F^{lk}. \quad (14.8.10)$$

Readjusting the factors in the last term in the right hand side of this relation, we find that

$$\frac{1}{\sqrt{-g_{44}}}(E_i - a^j H_{ij}) = g_{i\mu}\frac{g_{\nu 4}}{\sqrt{-g_{44}}} F^{\mu\nu} - \sqrt{-g_{44}}\left(g_{il} - g_{4i}\frac{g_{4l}}{g_{44}}\right)\left(g^{j4} g_{jk} - g^{j4}\frac{g_{4j}}{g_{44}} g_{4k}\right) F^{lk}$$

$$= g_{i\mu}\frac{g_{\nu 4}}{\sqrt{-g_{44}}} F^{\mu\nu} - \sqrt{-g_{44}}\left(g_{il} - \frac{g_{4i}}{g_{44}} 4l\right)$$

$$\times \left[\left(\delta_k^4 - g^{44} g_{4k}\right) - \frac{1}{g_{44}}\left(\delta_4^4 - g^{44} g_{44}\right) g_{4k}\right] F^{lk}. \quad (14.8.11)$$

Since the index K takes values only from 1, 2, 3,

$$\delta_k^4 = 0 = \delta_4^k$$

and since $\delta_4^4 = 1$, the two terms within the square brackets, in the right side of this equation, cancel out, leaving

$$\frac{1}{\sqrt{-g_{44}}}(E_i - a^j H_{ij}) = \frac{1}{\sqrt{-g_{44}}}(g_{i\mu} + g_{\nu 4} F^{\mu\nu} - g_{4k} g_{il} F^{lk}) + \frac{g_{4i} g_{4k}}{g_{44}\sqrt{-g_{44}}} g_{4l} F^{lk}. \quad (14.8.12)$$

Since the numerator of the last term of this relation equals $g_{41} g_{44}$, it vanishes, leaving

$$\frac{1}{\sqrt{-g_{44}}}(E_i - a^j H_{ij}) = \frac{1}{\sqrt{-g_{44}}}\left[g_{il} g_{k4} F^{lk} + g_{i4} g_{k4} F^{4k} + g_{i\mu} g_{44} F^{\mu 4} - g_{4k} g_{il} F^{lk}\right]$$

$$= \frac{1}{\sqrt{g_{44}}}(g_{i4} g_{k4} F^{4k} + g_{il} g_{44} F^{k4}). \quad (14.8.13)$$

Using now (14.8.5), this relation becomes

$$\frac{1}{\sqrt{-g_{44}}}(E_i - a^j H_{ij}) = \left(g_{ik} - \frac{g_{i4}}{g_{44}} g_{k4}\right) D^k = \gamma_{ik} D^k = D_i. \quad (14.8.14)$$

Since H_{ij} is antisymmetric, we can write

$$a^j H_{ij} = a^j H_{ij} + a^k H_{ik}, \quad (14.8.15)$$

where summation is not indicated in the right side for repeated index; then (14.8.14) reads

$$D_i = \frac{1}{\sqrt{-g_{44}}}\left[E_i - \sqrt{\tilde{\gamma}}(a^j H^k - a^k H^j)\right], \quad (14.8.16)$$

where (14.7.11) has been used. The vector and tensor relations (11.3.37)–(11.3.41) show that the equation (14.8.16) can be written as

$$D_i = \frac{1}{\sqrt{-g_{44}}}[E_i - (a \times H)_i]. \quad (14.8.17)$$

The quantities C_i and C^j are obviously the covariant and contravariant components of the axial vector C dual to the antisymmetric tensors C^{ik} and C_{jk}, respectively. Therefore, the vector form of the constitutive relation (14.8.17), for the electric displacement vector D (3-vector), in an accelerated frame of reference in vacuum, is

$$D = \frac{1}{\sqrt{-g_{44}}}[E - (a \times H)]. \tag{14.8.18}$$

To find the other constitutive relation, consider the tensor relation

$$\begin{aligned}\frac{1}{\sqrt{-g_{44}}}[H^{ij} - (a^j E^i - a^i E^j)] &= \frac{1}{\sqrt{-g_{44}}}\left[\sqrt{-g_{44}}F^{ij} - (a^j \gamma^{il} - a^i \gamma^{jl})E_l\right] \\ &= F^{il} - (G^{j4} g^{il} - g^{i4} g^{jl})E_l. \end{aligned} \tag{14.8.19}$$

where (11.3.14) and (11.3.16) have been utilized. Now, putting the value of E_l from (14.7.4) in (14.8.19), we obtain

$$\begin{aligned}\frac{1}{\sqrt{-g_{44}}}[H^{ij} - (a^j E^l - a^i E^j)] &= g^{i\mu} g^{j\nu} F_{\mu\nu} - (g^{j4} g^{il} - g^{i4} g^{jl})\, F_{l4} \\ &= g^{il} g^{j\nu} F_{l\nu} + g^{i4} g^{i\nu} F_{4\nu} - g^{j4} g^{il} F_{l4} + g^{i4} g^{4l} F_{l4} \\ &= g^{il} g^{jk} F_{lk} + g^{il} F^{j4} F_{l4} - g^{j4} g^{il} F_{l4} = \gamma^{il} \gamma^{jk} F_{lk} \\ &= -\gamma^{il} \gamma^{jk} B_{lk} = -B^{ij}. \end{aligned} \tag{14.8.20}$$

Hence, with the help of (14.7.6) and (14.7.9), this relation becomes

$$\frac{1}{\sqrt{\gamma}} B_k = \frac{1}{\sqrt{-g_{44}}}\left[\frac{1}{\sqrt{\gamma}} H_k + (a^i E^j - a^j E^i)\right]. \tag{14.8.21}$$

The corresponding vector relation is

$$B = \frac{1}{\sqrt{-g_{44}}}[H + (a \times E)]. \tag{14.8.22}$$

This is the vacuum constitutive relation for the magnetic induction B, in terms of E and H, in an accelerating system of co-ordinates.

14.9 Field Identification in Vacuum

The traditional space-time formulations of the Maxwell equations do not make sufficiently explicit the constitutive properties of free space. While applying the EM theory to rotating and other non-uniformly moving systems, it is not possible to consider solely the mutual motion of observer and object as in uniformly translating systems. A recasting of the Maxwell theory is necessary to delineate the physical and mathematical steps associated with the referred points.

Gaussian field identification is that there is no physical distinction between the electric displacement vector D and the electric field vector E, nor between the magnetic induction B and the magnetic intensity H in vacuum. Actually the Gaussian identifications

$$D = E, \quad B = H \tag{14.9.1}$$

define the EM properties of free space, only as seen from inertial frames because material polarization mechanisms, responsible for non-validity of these relations, are absent in vacuum. This field identification, and the Maxwell equations, produce the free-space D'Alembertian wave equation which is a Lorentz invariant structure.

Since the Minkowski six vectors (section 12.3, 12.4 and 12.6) E, B and D, H in free space, are related to $\eta_{\mu\nu}$ (the Minkowskian fundamental space-time metric, having the signature $(1, 1, 1, -1)$), the Gaussian field identification is equivalent to the invariance of this metric. This invariance defines the Lorentz group as a symmetry property of the space-time continuum, for uniform translation.

The intrinsic properties of a medium are affected by a rotation. Specifically, the constitutive properties, when measured while observer and medium are both at rest in the same rotating frame, are not exactly the same as those observed when observer and medium are both at rest in the same inertial frame. Moreover, the fact that the clockwise (CW) and counter clockwise (CCW) beams of light travel different path lengths before they meet again, as in Sagnac effect or in ring laser, section 14.9 or Post (1967), gives rise to non-reciprocal asymmetric properties between the CW and CCW beams in these experiments. Since a non-reciprocity requires the presence of a mixed space time derivative operator like $\partial^2/\partial t\partial x$, in the wave equation in a rotating frame (where OX is along the direction of wave propagation), such terms cannot be realized by assuming validity of the constitutive relations (14.9.1) in vacuum for rotation. So, either these are not valid in a rotating frame, or the Maxwell equations are affected by rotation. *All existing evidences for the treatment of non-reciprocal phenomena in material media,* wrote Post (1967), *points in the direction of modified constitutive relations,* not modified Maxwell equations. In fact, the Maxwell equations, as the universal laws of macroscopic electrodynamics, are expected to apply to any medium, regardless of symmetry properties of that medium, and regardless of the mutual motion of the frame of reference and the medium". A detectable mutual-motion of medium and reference frame reveals a lack of space time symmetry. Free space has the property that uniform translation is not detectable. Thus space-time symmetry is characterized by the Lorentz group.

14.10 Local Inertial Tetrads

A theory exists of local electrodynamics, which is developed with the help of local inertial tetrads, Mo (1970). The covariant equations of EM fields and a locally valid covariant formalism of the constitutive relations of the macroscopic media are used in this theory. Also used are the formalism of the local inertial tetrads of STR.

Existence of a free space constitutive dependence of D and B of the type considered in equations (13.6.37) and (13.6.38), Post (1967), and Post and Bahulikar (1971) and Post (1972) has been criticized, Mo (1970 section 4). But, constitutive relations like these explain the observations made by Kennard (1917) and Pegram (1917) and others. But, the method of local inertials tetrads cannot be made equivalent to a method of *global* non-inertial references and such an equivalence exists only locally.

14.11 Various Tensor Formulations of the Constitutive Relations

For introducing the constitutive equations in a non-permanent gravitational field the adoption of a procedure similar to that followed by Minkowski, section 1.12, becomes necessary. The constitutive relations are those between the intensity entities E and B, and the quantity entities D and H. So, the elements of $F_{\mu\gamma}$ should be tensorially related with those of $\mathcal{H}^{\mu\gamma}$, for a covariant description of the EM field behaviour in a linear medium.

The deviations in times of equal paths of opposing waves in rotating frames lead to different values of the frequency shift of these waves in ring lasers (section 14.19) and other such rotation sensing optical devices, Volkov and Kiselev (1970); but the differences are not worthy of experimental verification. Some general methods of writing down the covariant constitutive relations have been proposed by different authors. All these give the correct constitutive relations between the field vectors in inertial frames. Up to the first power of $\beta(= r\Omega/c)$, for which the GN rotation rules, under absolute time (section 8.13), are valid, these cannot be considered for acceptance or rejection, because sensitive measuring devices for experimental verification of the conclusions do not exist. Since large angular velocities distort the physical systems, the analysis is restricted to small angular velocities.

Three different covariant formulations of the constitutive relations have been noted by Volkov and Kiselev (1970). These are (1) the formulation by Tamm (1924) and Mandel'shtam (1924), further developed by Volkov and Kiselev (1970); (2) the formulation by Heer (1964) and Khromykh (1966); (3) the formulation by Post (1962), further elucidated by Post and Yildiz (1965), Yildiz and Tang (1966) and Post (1967). These are summarized here, and would be discussed in the following sections in detail.

The constitutive relation of Tamm (1924) and others, section 14.12, is

$$\mathcal{H}^{\mu\nu} = g^{\mu\alpha} g^{\nu\beta} \xi^{\gamma}_{\alpha} \xi^{\delta}_{\beta} F_{\gamma\delta}. \tag{14.11.1}$$

Heer (1964) adopted the relations (see section 14.13)

$$\mu \mathcal{H}^{ij} = F^{ij} = g^{i\lambda} g^{j\mu} F_{\lambda\mu}, \tag{14.11.2}$$

$$\varepsilon F_{i4} = \mathcal{H}_{i4} = g_{i\lambda} g_{4\mu} \mathcal{H}^{\lambda\mu}. \tag{14.11.3}$$

The corresponding generalization for moving points in accelerated frames, section 14.13, are

$$g_{\alpha\xi} g_{\beta\eta} \mathcal{H}^{\xi\eta} u^{\beta} = \varepsilon F_{\alpha\zeta} u^{\zeta}, \tag{14.11.4}$$

$$e_{\alpha\beta\gamma\delta} g^{\beta\xi} g^{\gamma\eta} F_{\xi\eta} u^{\delta} = \mu e_{\alpha\beta\gamma\delta} \mathcal{H}^{\beta\gamma} u^{\delta}. \tag{14.11.5}$$

The theory of Post, Yildiz and Tang (section 14.14) depends on the relation

$$H^{\alpha\beta} = \frac{1}{2} \chi^{\alpha\beta\gamma\delta} F_{\gamma\delta}. \tag{14.11.6}$$

It appears that, (Post (1967)), the covariant constitutive relations of Heer (1964), (14.13.12) and (14.13.13), and equation (14.13.14) of Khromykh (1966) or (14.13.15) of Tamm (1924) and Mandelshtam (1924) do not reveal all the typical experimental ramifications that can occur. However, up to the first order of velocities, qualitatively speaking, the three alternative sets of covariant constitutive relations have the same relative merit; only quantitatively there will be some minor changes in the results.

The problems related to light transmission in accelerated frames of reference could be properly treated within the frame work of GTR through involvement of the constitutive tensor and the covariant constitutive relations. Only the effects of the order of β are observed in a rotating frame. Thus the experiment of Michelson is explained by regarding the trajectory of a light ray as given by the relation $dS^2 = 0$ where ds is the 4-space interval.

14.12 Theory of Tamm (1924) and Mandel'shtam (1924)

14.12.1 Basics of the theory

These authors introduced a constitutive tensor (containing the dielectric constant and magnetic permeability of the medium), $\xi^{\alpha\beta\gamma\delta}$ or $\xi_{\alpha\beta\gamma\delta}$, of 4th order, which transforms according to the rules of transformation of contravariant or covariant tensors of 4thorder on going over from one frame of reference to another. This tensor, together with tensors of the EM field, determines the covariant material equations connecting the elements of the EM field in the medium in any frame of reference, not necessarily inertial. It obviously yields covariant or vector material equations for a medium that is at rest in an inertial frame.

Actually the mixed second order tensor ξ^{μ}_{ν} is introduced, representing the EM properties of the medium. For a medium at rest in the reference frame, specified by the metric tensor $g_{\mu\gamma}$, the vacuum relations (14.2.8) or (14.2.9) are thus replaced by

$$\sqrt{-g}\mathcal{H}^{\mu\nu} = \sqrt{-g}g^{\mu\alpha}g^{\nu\beta}\xi^{\gamma}_{\alpha}\xi^{\delta}_{\beta}F_{\gamma\delta}, \tag{14.12.1}$$

$$\text{or,} \quad \frac{1}{\sqrt{-g}}g_{\alpha\lambda}g_{\beta\nu}\sqrt{-g}\mathcal{H}^{\alpha\beta} = \xi^{\gamma}_{\lambda}\xi^{\delta}_{\nu}F_{\gamma\delta}. \tag{14.12.2}$$

In a linear isotropic medium, at rest in an arbitrary reference frame, the non-zero components of ξ^{λ}_{μ} are

$$\xi^1_1 = \xi^2_2 = \xi^3_3 = \frac{1}{\sqrt{\mu}}, \quad \xi^4_4 = \varepsilon\sqrt{\mu}. \tag{14.12.3}$$

The quantities ε and μ are measured by local observers connected with the medium. For a medium at rest in an inertial frame, these equations generate the relations (14.3.4) and (14.3.5). Relations (14.12.3) show that in vacuum, in an inertial frame,

$$\xi^{\mu}_{\nu} = \delta^{\mu}_{\nu}. \tag{14.12.4}$$

So, (14.12.1) and (14.12.2) become identical to the relations (14.2.8) or (14.2.9).

In place of (14.12.1) and (14.12.2), with the help of a new mixed tensor ξ^{λ}_{μ} for the medium, we could write

$$\bar{\xi}^{\alpha}_{\lambda}\bar{\xi}^{\beta}_{\mu}\sqrt{-g}\mathcal{H}^{\lambda\mu} = \sqrt{-g}g^{\alpha\gamma}g^{\beta\delta}F_{\gamma\delta}, \tag{14.12.5}$$

$$\text{or,} \quad \frac{1}{\sqrt{-g}}g_{\alpha\lambda}g_{\beta\mu}\bar{\xi}^{\beta}_{\delta}\sqrt{-g}\mathcal{H}^{\gamma}\delta = F_{\alpha\beta}. \tag{14.12.6}$$

Then, for a linear isotropic medium, at rest in an arbitrary reference frame, the relations (14.12.3) are replaced by

$$\xi^1_1 = \xi^2_2 = \xi^3_3 = \sqrt{\mu}, \quad \xi^4_4 = \frac{1}{\varepsilon\sqrt{\mu}}. \tag{14.12.7}$$

Equations (14.12.1) and (14.12.2), together with (14.2.1) and (14.2.2), completely describe the EM process in a medium for any arbitrary 4-space metric $g_{\mu\gamma}$. This shows that the tensors ξ^μ_ν (or $\bar{\xi}^\mu_\nu$), $\mathcal{H}^{\mu\gamma}$ and $F_{\mu\gamma}$ are necessary for that purpose. The materiai equations, which relate the components of the tensors of the EM field and depend on the properties of 4-space, are determined completely by $g_{\mu\gamma}$, which characterises all the properties of the geometry of the space in systems of co-ordinates.

Equations (14.12.1)–(14.12.3) and (14.2.4) lead to the tensor relation

$$F_{\mu\nu} + 4\pi M_{\mu\nu} = \xi^\gamma_\mu \xi^\delta_\nu F_{\gamma\delta} \tag{14.12.8}$$

for determination of the magnetization tensor $M_{\mu\gamma}$. Thus the material relations (14.12.1) and (14.12.2) are equivalent to those of (14.2.37) and (14.2.38).

14.12.2 Some results for rotating isotropic media

Introduce the vector fields of rotations (14.2.11)–(14.2.16) to write the covariant Maxwell equations (14.2.1) and (14.2.2) in the form (1.4.1)–(1.4.4), the tensor ξ^μ_ν from (14.12.3) and $g_{\mu\gamma}$ from (8.13.20); then the non-covariant material equations connecting the field vectors in a medium at rest in the rotating frame K, for GN rotation, are

$$D = \varepsilon E - \frac{1}{\mu}(\beta \times B), \tag{14.12.9}$$

$$H = \frac{1}{\mu}B - (\beta \times D), \tag{14.12.10}$$

where $\mathbf{v}$ is given by (8.13.19) and (8.13.1). The EM properties of the medium are described by the equations (1.4.1)–(1.4.4) together with the constitutive relations (14.12.9) and (14.12.1).

14.13 Theory of Heer (1964) and Khromykh (1966)

14.13.1 Constitutive relations for stationary media

For isotropic dielectrics and paramagnetic substances the constitutive equations in an instantaneous rest frame are

$$D_{i^0} = \varepsilon E_{i^0}, \tag{14.13.1}$$

$$B_{i^0} = \mu H_{i^0}, \tag{14.13.2}$$

$$j_{i^0} = \sigma E_{i^0}, \tag{14.13.3}$$

in non-covariant notation. In the instantaneous rest frame, there is no distinction between covariant and contravariant components, and these constitutive equations may be written as

$$\mathcal{H}_{i^0 4^0} = \varepsilon F_{i^0 4^0}, \tag{14.13.4}$$

$$F_{i^0 j^0} = \mu \mathcal{H}_{i^0 j^0}, \tag{14.13.5}$$

$$j_{i^0} = \sigma F_{4^0 i^0}, \tag{14.13.6}$$

$$j_{4^0} = \rho^0 c. \tag{14.13.7}$$

In any frame of reference the components follow from the rules of transformation of second order tensors. Thus, using (14.3.6), we have

$$F_{i4} = \alpha_i^{\xi} \alpha_4^{4^0} F_{\xi^0 r^0}. \tag{14.13.8}$$

Similarly,

$$\mathcal{H}_{i4} = \alpha_i^{\xi^0} \alpha_4^{4^0} F_{\xi^0 \eta^0}. \tag{14.13.9}$$

Hence, using (14.13.4) and the antisymmetric property of $\mathcal{H}_{\mu\gamma}$ and $F_{\mu\gamma}$, we find that

$$\mathcal{H}_{i4} = \alpha_i^{\xi^0} \alpha_4^{4^0} \mathcal{H}_{\xi^0 \eta^0} = \alpha_i^{j^0} \alpha_4^{4^0} \mathcal{H}_{j^0 r^0} = \varepsilon \alpha_i^{\xi^0} \alpha_4^{4^0} F_{\xi^0 4^0}$$

$$= \varepsilon \alpha_i^{j^0} \alpha_4^{4^0} F_{j^0 4^0} = \varepsilon F_{i4}. \tag{14.13.10}$$

Similarly, we can write

$$\mu H^{ij} = F^{ij}. \tag{14.13.11}$$

The corresponding relations between the covariant and contravariant components of the field tensors are

$$\mu \mathcal{H}^{ij} = F^{ij} = g^{i\lambda} g^{j\mu} F_{\lambda\mu}, \tag{14.13.12}$$

$$\varepsilon F_{i4} = \mathcal{H}_{i4} = g_{i\lambda} g_{4\mu} \mathcal{H}^{\lambda\mu}. \tag{14.13.13}$$

For the connection between the components of the covariant and contravariant EM field tensors $F_{\mu\gamma}$ and $\mathcal{H}^{\mu\gamma}$ respectively, Khromykh (1966) has used the following covariant relations in a dielectric medium

$$g_{\alpha\xi} g_{\beta\eta} \mathcal{H}^{\xi\eta} u^{\beta} = \varepsilon F_{\alpha\varsigma} u^{\varsigma}, \tag{14.13.14}$$

$$e_{\alpha\beta\gamma\delta} g^{\beta\xi} g^{\gamma\eta} F_{\xi\eta} u^{\delta} = \mu e_{\alpha\beta\gamma\delta} \mathcal{H}^{\beta\gamma} u^{\delta}, \tag{14.13.15}$$

where $e_{\alpha\beta\gamma\delta}$ is the 4th order Levi Civita symbol (Appendix 4). For media at rest in a frame under GN rotation (Chapter 7) and the non-covariant field vectors of (14.2.11)–(14.2.16), we obtain the following constitutive relations

$$B + \beta \times (\beta \times B) = \mu H + (\beta \times E), \tag{14.13.16}$$

$$D + \beta \times (\beta \times D) = E - (\beta \times H). \tag{14.13.17}$$

Approximately, these relations, correct up to the first order of β, are

$$D = \varepsilon E - (\beta \times H), \tag{14.13.18}$$

$$H = \frac{1}{\mu} B - \frac{1}{\mu} (\beta \times E). \tag{14.13.19}$$

For vacuum put $\varepsilon = 1 = \mu$ in (14.13.16) and (14.13.17), and obtain

$$D = E - \gamma^2 (\beta \times B) - \gamma^2 (\beta \cdot (D - E))\beta, \tag{14.13.20}$$

$$B = H + \gamma^2 (\beta \times D) - \gamma^2 (\beta \cdot (\beta \cdot \mu))\beta. \tag{14.13.21}$$

Correct up to first order of β, we get

$$D = E - (\beta \times B), \tag{14.13.22}$$

$$B = H + (\beta \times D), \tag{14.13.23}$$

violating the rules of Gaussian field identification. Hence, rotation has the effect of making even the vacuum appear anisotropic. This first order effect corresponds to the type of additional terms introduced by Michelson for the description of propagation of light in vacuum in terms of the ether.

14.13.2 Transformation of the 4-current

Transformation of the 4-current j^μ is

$$j_i = \alpha_i^{\xi^0} j_{\xi^0} = \alpha_i^{k^0} j_{k^0} + \alpha_i^{4^0} j_{4^0}, \tag{14.13.24}$$

$$j_4 = \alpha_4^{\xi^0} j_{\xi^0} = \alpha_4^{4^0} j_{4^0} + \alpha_4^{4^0} \rho^{c_0} c, \tag{14.13.25}$$

where, in the final results, the relations (14.3.6) have been used. Relation (14.13.24) can be written as

$$j_i = \sigma \alpha_i^{k^0} F_{4^0 k^0} + \alpha_i^{4^0} j_{4^0}, \tag{14.13.26}$$

$$j_4 = \alpha_4^{\xi^0}. \tag{14.13.27}$$

Then evidently,

$$j_i = \sigma \alpha_i^{k^0} \alpha_{k^0}^{\eta} F_{\xi\eta} + \alpha_i^{4^0} \rho^{c^0} c, \tag{14.13.28}$$

$$j_4 = \sigma \alpha_4^{k^0} F_{4^0 k^0} + \alpha_4^{4^0} \rho^{c^0}. \tag{14.13.29}$$

14.13.3 Energy momentum tensor

As in (12.4.1), define the 4-force F_μ by

$$F_\mu = \frac{1}{c} F_{\mu\nu} j^\nu. \tag{14.13.30}$$

With the help of (12.2.30), this relation becomes

$$F_\mu = \frac{1}{4\pi} F_{\mu\nu} \mathcal{H}^{\nu\xi}_{;\xi}. \tag{14.13.31}$$

In vacuum, corresponding to (12.4.27), for the energy momentum tensor S_μ^ξ, we can write

$$\frac{1}{c} F_{\mu\nu} j^\nu = F_\mu = S^\xi_{\mu;\xi}, \tag{14.13.32}$$

where, as in (12.4.25),

$$S_\mu^\xi = -\frac{1}{4\pi}(F_{\mu\nu}\mathcal{H}^{\nu\xi}) + \frac{1}{16\pi}\delta_\mu^\xi \mathcal{H}^{\nu\eta} F_{\mu\nu}. \tag{14.13.33}$$

Using the covariant differentiation results (see section 11.4 or Moller (1972, equation (9.199))), we have

$$\mathrm{div}_\nu(S_\mu^\nu) = \frac{1}{\sqrt{-g}} \frac{\partial}{\partial x^\nu}(\sqrt{-g} S_\mu^\nu) - \frac{1}{2} \frac{\partial g_{\xi\eta}}{\partial x^\mu} S^{\xi\eta}. \tag{14.13.34}$$

Hence, as a consequence of the field equations (14.2.1), (14.2.2) and (14.13.30), equation (14.13.32) becomes

$$F_\mu = \frac{1}{c} F_{\mu\nu} j^\nu = \frac{1}{\sqrt{-g}} \frac{\partial}{\partial x^\nu}(\sqrt{-g} S_\mu^\nu) - \frac{1}{2} \frac{\partial g_{\xi\eta}}{\partial x^\mu} S^{\xi\eta}. \tag{14.13.35}$$

Equation (12.4.26) then generalizes to

$$\frac{1}{c}F_{\mu\nu}j^{\nu} + \frac{1}{16\pi}[F_{\nu\eta}\mathcal{H}^{\nu\eta}_{;\mu} - \mathcal{H}^{\mu\eta}F_{\nu\eta;\mu}] = S^{\xi}_{\mu;\xi}. \tag{14.13.36}$$

For further manipulations consider the tensor formulas, Landau and Lifshitz (1975, equations (85.12) and (85.41)),

$$A^{\lambda\mu}_{;\nu} = \frac{\partial A^{\lambda\mu}}{\partial x^{\nu}} + \Gamma^{\lambda}_{\xi\nu}A^{\xi\mu} + \Gamma^{\mu}_{\xi\nu}A^{\lambda\xi}, \tag{14.13.37}$$

$$A_{\lambda\mu;\nu} = \frac{\partial A_{\lambda\mu}}{\partial x^{\nu}} - \Gamma^{\xi}_{\lambda\mu}A_{\xi\mu} - \Gamma^{\xi}_{\mu\nu}A_{\lambda\xi}. \tag{14.13.38}$$

Using these relations in the second term in the left hand side of (14.13.36), it transforms to

$$\frac{1}{c}F_{\mu\nu}j^{\nu} + \frac{1}{16\pi}\left[F_{\nu\eta}\left\{\frac{\partial\mathcal{H}^{\nu\eta}}{\partial x^{\mu}} + \Gamma^{\nu}_{\alpha\mu}\mathcal{H}^{\alpha\eta} + \Gamma^{\eta}_{\alpha\mu}\mathcal{H}^{\nu\alpha}\right\} - \mathcal{H}^{\nu\eta}\left\{\frac{\partial F_{\nu\eta}}{\partial x^{\mu}} - \Gamma^{\alpha}_{\nu\mu}F_{\alpha\eta} - \Gamma^{\alpha}_{\eta\mu}F_{\nu\alpha}\right\}\right]$$

$$= \frac{1}{\sqrt{-g}}\frac{\partial}{\partial x^{\xi}}\left(\sqrt{-g}S^{\xi}_{\mu}\right) - \frac{1}{2}\frac{\partial g_{\xi\eta}}{\partial x^{\mu}}S^{\xi\eta}, \tag{14.13.39}$$

where $\Gamma^{\xi}_{\mu\nu}$ is the Christofell symbol, define as

$$\Gamma^{\lambda}_{\mu\nu} = \frac{1}{2}g^{\lambda\mu}\left(\frac{\partial g_{\eta\nu}}{x^{\mu}} + \frac{\partial g_{\eta\mu}}{\partial x^{\nu}} - \frac{\partial g_{\mu\nu}}{\partial x^{\eta}}\right).$$

14.13.4 Constitutive relations for moving media

For particles in motion in the rotating frame K, the constitutive relations (14.13.14) and (14.13.15), formulated by Khromykh (1966), are appropriate in place of the state relations (14.13.12) and (14.13.13) used by Heer (1964).

We here consider equation (14.13.14) together with (8.13.26) for $g_{\mu\gamma}$. For $\alpha = 1$ we obtain

$$g_{1\xi}g_{\beta\eta}\mathcal{H}^{\xi\eta}u^{\beta} = \varepsilon(F_{12}U^2 + F_{13}U^3 + F_{14}U^4). \tag{14.13.40}$$

With the help of (12.2.15), (12.2.31) and (8.14.15), this relation reduces to

$$c\varepsilon\left[E + \frac{1}{c}(U\times B)\right]_x = (U\times H)_X + D_X(U\times B)_Z + C(\beta\times H)_X + \frac{c}{\gamma^2}D_X$$

$$-\beta_Y(U\cdot D) + C[\beta\times(D\times\beta)]_X. \tag{14.13.41}$$

Neglecting higher order terms, this relation leads to the vector relation

$$D = \varepsilon E + \frac{\varepsilon\mu}{c}U\times H - \frac{1}{c}(U\times H) - (\beta\times H). \tag{14.13.42}$$

For $\alpha = 1$ equation (14.13.15) gives

$$e_{1\beta\gamma\delta}g^{\beta\xi}g^{\gamma\eta}F_{\xi\eta}u^{\delta} = \mu e_{1\beta\gamma\delta}\mathcal{H}^{\beta\gamma}U^{\delta}. \tag{14.13.43}$$

The left hand side of this relation is

$$= (e_{1234}g^{2\xi}g^{3\eta}u^4 + e_{1234}g^{2\xi}g^{4\eta}u^3 + e_{1324}g^{3\xi}g^{2\eta}u^4 + e_{1342}g^{3\xi}g^{4\eta}u^2 + e_{1423}g^{4\xi}g^{2\eta}u^3 + e_{1432}g^{4\xi}g^{3\eta}u^2)F_{\xi\eta}$$

$$= u_2(g^{3\xi}g^{4\eta} - g^{3\eta}g^{4\xi})F_{\xi\eta} + u_3(g^{4\xi}g^{2\eta} - g^{4\eta}g^{2\xi})F_{\xi\eta} + C(g^{2\xi}g^{3\eta} - g^{2\eta}g^{3\xi})F_{\xi\eta}$$

$$= 2[CB_x - (u\times E)_x - (v\times E)_x] + \cdots . \qquad (14.13.44)$$

The right hand side

$$= u[e_{1234}\mathcal{H}^{23}u^4 + e_{1243}\mathcal{H}^{24}u^3 + e_{1324}\mathcal{H}^{32}u^4 + e_{1342}\mathcal{H}^{34}u^2 + e_{1423}\mathcal{H}^{42}u^3 + e_{1432}\mathcal{H}^{43}u^2]$$

$$= 2\mu[CH_x - (u\times D)_x]. \qquad (14.13.45)$$

Hence, approximately,

$$B - \frac{1}{c}(u+v)\times B = \mu\left[H - \frac{1}{c}(u\times D)\right]$$

$$\text{or,}\quad B = \mu H - \frac{1}{c}(\varepsilon\mu - 1)(u\times E) + \frac{1}{c}(v\times E). \qquad (14.13.46)$$

When $\varepsilon = \mu = 1$, the relations (14.15.35) and (14.15.36), for linear isotropic media, become identical with (14.13.41) and (14.13.45). Consequently, in this case, the results of Heer (1964) and Khromykh (1966) coincide with those of Post and Yildiz (1965). Yildiz and Tang (1965) and Post (1967).

14.14 Theory of Post, Yildiz and Tang (PYT)

The covariant constitutive relations of PYT resolve various paradoxes and inconsistencies of interpretation that still remain for methodological reasons, in the other covariant formulations, Post and Yildiz (1965) and Yildiz and Tang (1965). The transformation properties of the medium part, in this theory are considered separately from that of the free-space of the constitutive relations. The constitutive tensor for an arbitrarily moving non-conducting material medium, with respect to an accelerated frame of reference, is the result of superposition, of a free space part $\chi_{(0)}$ and a material part $\chi_{(m)}$, which is solely determined by the electric and magnetic polarization properties of the material medium

$$\chi^{\alpha\beta\gamma\delta} = \chi^{\alpha\beta\gamma\delta}_{(0)} + \chi^{\alpha\beta\gamma\delta}_{(m)}. \qquad (14.14.1)$$

We can thus consider the material medium at rest in the rotating frame, so that only $\chi_{(0)}$ is affected by the rotation; and the case when the material medium is at rest in the inertial laboratory system, so $\chi_{(0)}$ and $\chi_{(m)}$ are both affected by the rotation. The free-space part $\chi_{(0)}$ is transformed from its inertial representation to a cylindrical co-ordinate system (r, θ, z, t) rotating with the constant angular velocity Ω around the z axis.

In the covariant formulation of the classical EM theory, this concept of the constitutive relations was proposed by Van Dantzig (1934b–1934d). It was introduced in electrodynamics in a consistent manner by Post (1962), Post and Yildiz (1965), Yildiz and Tang (1966) and Post (1967) (PYT). In this formulation the principle of general covariance, for formulation of the classical EM theory, separates the field equations from the constitutive equations. This formulation of the Maxwell theory, called the *method of natural invariance* (because the invariance comes about independent of the metric tensor), thus bridges the gap between the treatment of the material and the free space medium.

The properties of a general linear medium, including the free space medium, are adequately described by a higher valency constitutive tensor, like that of (14.14.1), which has sufficient parameters to accommodate all the known *linear phenomena.* The Sagnac effect then arises from suitable transformation of this constitutive tensor. Only for the case of the free-space medium may one assume that this tensor is completely determined by the free-space metric.

The 4-space metric $g_{\mu\nu}$ has only 10 independent components, a general anisotropic linear medium can have 6 independent components for the dielectric behaviour, and an equal amount for the magnetic behaviour not counting the parameters needed for describing the Faraday and other optical rotations. Compared with customary presentations by Schiff (1939), Sommerfeld (1952, p. 362, 363), and others, this analysis provides a minor change in the transformation of the field tensor that constitutes the spatial D and H field vectors, which enables a metric independent general covariant formulation of the equations (2.2.20) and (2.2.30), which follow from (14.2.1) and (14.2.2) for inertial frames. This formulation indicates that the derivatives in these equations are ordinary, not covariant derivatives. The field tensor of D and H is normally treated as an ordinary tensor; however, for natural invariance, this tensor should be treated as a tensor density of weight + 1. This method resolves the notorious Schiff (1939) charge and current from rotation effects, Sommerfeld (1952, p. 362–3). In inertial frames $F_{\mu\nu}$ and $\mathcal{H}^{\mu\nu}$ are the skew symmetric tensors of Minkowski, representing (E, B) and (D, H) respectively. The form of (2.2.20) and (2.2.30) is retained for any general, non-linear, but holonomic co-ordinate transformation according to the rules

$$F_{\mu\nu} = \alpha^{\xi^0}_{\mu}\,\alpha^{\eta^0}_{\nu}\,F_{\xi^0\eta^0} \tag{14.14.2}$$

$$\mathcal{H}^{\mu\nu} = |\Delta|^{-1}\alpha^{\mu}_{\xi^0}\alpha^{\nu}_{\eta^0}\mathcal{H}^{\xi^0\eta^0} \tag{14.14.3}$$

with

$$\Delta = \det \alpha^{\mu}_{\xi^0}, \tag{14.14.4}$$

where $\alpha^{\mu}_{\nu^0}$ and $\alpha^{\mu^0}_{\nu}$ are defined in (11.2.19) and (11.2.21). The condition of holonomity, Post (1962, p. 20), is

$$\partial_{[\xi^0}\alpha^{\gamma}_{\eta^0]} = 0 \quad \text{and} \quad \partial_{[\xi}\alpha^{\gamma^0}_{\eta]} = 0 \tag{14.14.5}$$

$$\text{or,}\quad \frac{\partial \alpha^{\gamma}_{\eta^0}}{\partial x^{\xi^0}} = \frac{\partial \alpha^{\gamma}_{\xi^0}}{\partial x^{\eta^0}} \quad \text{and} \quad \frac{\partial \alpha^{\gamma^0}_{\eta}}{\partial x^{\xi}} = \frac{\partial \alpha^{\gamma^0}_{\xi}}{\partial x^{\eta}}. \tag{14.14.6}$$

14.15 Constitutive Tensor of PYT

14.15.1 General theory

For a covariant description of the known EM *field behaviour* in a general linear medium, with instantaneous and local interactions between fields, the elements of $F_{\mu\gamma}$ are tensorially related with those of $\mathcal{H}^{\mu\nu}$.

$$\mathcal{H}^{\alpha\beta} = \frac{1}{2}\chi^{\alpha\beta\gamma\delta}F_{\gamma\delta}, \tag{14.15.1}$$

where $\chi^{\alpha\beta\gamma\delta}$ is the very general contravariant constitutive tensor of 4th order of the EM field. Equation (14.15.1) is the covariant constitutive relation of PYT. The medium may be in arbitrary motion with respect to the frame of reference. The symmetry properties of $\chi^{\alpha\beta\gamma\delta}$ are independent of the symmetry properties of the medium.

The constitutive tensor $\chi^{\alpha\beta\gamma\delta}$ has 256 components. It contains as parts, both the dielectric tensor ε_{ij} and the magnetic permeability tensor μ_{ij}. For simple types of media $\chi^{\alpha\beta\gamma\delta}$ is the covariant analogue of the simplest state relations (1.6.2) and (1.6.3), which are valid in stationary, linear isotropic media in inertial frames. The factor 1/2 in (14.15.1) has been incorporated to compensate for the double summation over γ and δ, and also for identification with the conventional relations of (1.6.2) and (1.6.3).

Since $\mathcal{H}^{\alpha\beta}$ and $F_{\alpha\beta}$ are antisymmetric tensors, $\chi^{\alpha\beta\gamma\delta}$ obeys the following relations

$$\chi^{\alpha\beta\gamma\delta} = -\chi^{\beta\alpha\gamma\delta} = -\chi^{\alpha\beta\delta\gamma}. \tag{14.15.2}$$

Due to this antisymmetric, both the first and second pairs of indices, $\alpha\beta$, and $\gamma\delta$, have 6 independent elements each; thus $\chi^{\alpha\beta\gamma\delta}$ has a total of 36 independent elements. Other conditions for $\chi^{\alpha\beta\gamma\delta}$ also exist in lossless media, for instance, the symmetry condition

$$\chi^{\alpha\beta\gamma\xi} = \chi^{\gamma\delta\alpha\beta}. \tag{14.15.3}$$

These limit the number of independent elements to 21. Another condition, Post (1962 section 6.2, equation 6.18), is

$$\chi^{[\alpha\beta\gamma\delta]} = 0, \tag{14.15.4}$$

which, explicitly, is

$$\chi^{4123} + \chi^{4231} + \chi^{4321} + \chi^{2341} + \chi^{3142} + \chi^{1243} = 0, \tag{14.15.5}$$

reducing the number of independent elements to 20. These elements are independent of the symmetry properties of the medium.

The magnetization of optical materials is negligible for all practical purposes. Also the intrinsic influence of rotation on the medium is neglected here. So, the magnetic permeability μ_r of the medium due to rotation equals unity. Even then, the results of the physical optical analysis carried out, with the help of (14.15.1), will be consistent with the results of the geometrical optical analysis.

Since $\mathcal{H}^{\alpha\beta}$ is a tensor density of weight + 1, and $F_{\alpha\beta}$ is a tensor density of zero weight, $\chi^{\alpha\beta\gamma\delta}$ transforms like a contravariant tensor density of weight + 1. So,

$$\chi^{\alpha\beta\gamma\delta} = |\Delta|^{-1}\alpha^{\alpha}_{\xi^0}\alpha^{\beta}_{\eta^0}\alpha^{\gamma}_{\zeta^0}\alpha^{\delta}_{\lambda^0}\chi^{\xi^0\eta^0\zeta^0\lambda^0}, \quad \Delta = \det\alpha^{\mu}_{\nu^0}. \tag{14.15.6}$$

For arbitrary reference frames, Post (1962), page 132, the table of relations (6.21)), given below is useful

$\chi^{\alpha\beta\gamma\delta}$		41 $-E_1$	42 $-E_2$	43 $-E_3$	23 B_1	31 B_2	12 B_3
41	D_1	$-\varepsilon_{11}$	$-\varepsilon_{12}$	$-\varepsilon_{13}$	γ_{11}	γ_{12}	γ_{13}
42	D_2	$-\varepsilon_{21}$	$-\varepsilon_{22}$	$-\varepsilon_{23}$	γ_{21}	γ_{22}	γ_{23}
43	D_3	$-\varepsilon_{31}$	$-\varepsilon_{32}$	$-\varepsilon_{33}$	γ_{31}	γ_{32}	γ_{33}
23	H_1	$\bar{\gamma}_{11}$	$\bar{\gamma}_{12}$	$\bar{\gamma}_{13}$	χ_{11}	χ_{12}	χ_{13}
31	H_2	$\bar{\gamma}_{21}$	$\bar{\gamma}_{22}$	$\bar{\gamma}_{23}$	χ_{21}	χ_{22}	χ_{23}
12	H_3	$\bar{\gamma}_{31}$	$\bar{\gamma}_{32}$	$\bar{\gamma}_{33}$	χ_{31}	χ_{32}	χ_{33}

(14.15.7)

and also table 2B, page 55 :

$$\begin{bmatrix} 0 & \mathcal{H}^{12} & \mathcal{H}^{13} & \mathcal{H}^{14} \\ \mathcal{H}^{21} & 0 & \mathcal{H}^{23} & \mathcal{H}^{24} \\ \mathcal{H}^{31} & \mathcal{H}^{32} & \mathcal{H}^{33} & \mathcal{H}^{34} \\ \mathcal{H}^{41} & \mathcal{H}^{42} & \mathcal{H}^{43} & 0 \end{bmatrix} = \begin{bmatrix} 0 & H_3 & -H_2 & -D_1 \\ -H_3 & 0 & H_1 & -D_2 \\ H_2 & -H_1 & 0 & -D_3 \\ D_1 & D_2 & D_3 & 0 \end{bmatrix} \tag{14.15.8}$$

which follows from (1.15.6); we find that

$$H^{41} = D_1 = \varepsilon_{1i}E_i + \gamma_{1i}B_i = \varepsilon_{1i}F_{i4} + \gamma_{11}F_{23} + \gamma_{12}F_{31} + \gamma_{13}F_{12} = \frac{1}{2}\chi^{41\gamma\delta}F_{\gamma\delta}. \tag{14.15.9}$$

Comparison of the coefficients of the components of $F_{\mu\nu}$ give

$$\chi^{4112} = \gamma_{13}, \quad \chi^{4113} = -\gamma_{12}, \quad \chi^{4123} = \gamma_{11}, \quad \chi^{414i} = \varepsilon_{1i}; \tag{14.15.10}$$

since

$$\mathcal{H}^{42} = D_2 = \varepsilon_{2i}E_i + \gamma_{2i}B_i = \varepsilon_{2i}F_{i4} + \gamma_{21}F_{23} + \gamma_{22}F_{31} + \gamma_{23}F_{12} = \chi^{42\gamma\delta}F_{\gamma\delta}. \tag{14.15.11}$$

Comparison gives

$$\chi^{4212} = \gamma_{23}, \quad \chi^{4213} = -\gamma_{22}, \quad \chi^{4223} = \gamma_{21}, \quad \chi^{424i} = -\varepsilon_{2i}; \tag{14.15.12}$$

since

$$\mathcal{H}^{43} = D_3 = \varepsilon_{3i}E_i + \gamma_{3i}B_i = \varepsilon_{3i}F_{i4} + \gamma_{31}F_{23} + \gamma_{32}F_{31} + \gamma_{33}F_{12} = \chi^{43\gamma\delta}F_{\gamma\delta}, \tag{14.15.13}$$

Comparison gives

$$\chi^{4312} = \gamma_{33}, \quad \chi^{4313} = \gamma_{32}, \quad \chi^{4323} = \gamma_{31}, \quad \chi^{4341} = \gamma_{3i}, \tag{14.15.14}$$

since

$$\mathcal{H}^{23} = H_1 - \gamma_{1i}E_i + \chi_{1i}B_i = \bar{\gamma}_{1i}F_{4i} + \chi_{11}F_{23} + \chi_{12}F_{31} + \chi_{13}F_{12} = \chi^{23\gamma\delta}F_{\gamma\delta}, \tag{14.15.15}$$

Comparison gives

$$\chi^{2312} = \chi_{13}, \quad \chi^{2313} = -\chi_{12}, \quad \chi^{2323} = \chi_{11}, \quad \chi^{234i} = \bar{\gamma}_{1\nu}; \tag{14.15.16}$$

since

$$\mathcal{H}^{31} = H_2 = -\bar{\gamma}_{2i}E_i + \chi_{2i}B_i = \bar{\gamma}_{2i}F_{4i} + \chi_{21}F_{23} + \chi_{22}F_{31} + \chi_{23}F_{12}, \tag{14.15.17}$$

Comparison gives

$$\chi^{3112} = \chi_{23}, \quad \chi^{3113} = -\chi_{22}, \quad \chi^{3123} = \chi_{21}, \quad \bar{\gamma}^{314i} = \bar{\gamma}_{4i}, \tag{14.15.18}$$

since

$$\mathcal{H}^{12} = H_3 = -\bar{\gamma}_{3i}E_i + \chi_{3i}B_i = \bar{\gamma}_{3i}F_{4i} + \chi_{31}F_{23} + \chi_{32}F_{31} + \chi_{33}F_{12} = \chi^{12\gamma\delta}F_{\gamma\delta}, \tag{14.15.19}$$

Comparison gives

$$\chi^{1212} = \chi_{33}, \quad \chi^{1213} = -\chi_{32}, \quad \chi^{1223} = \chi_{31}, \quad \bar{\gamma}^{124i} = \bar{\gamma}_{3i}. \tag{14.15.20}$$

In empty space we have

$$\varepsilon_{1^0 1^0} = \varepsilon_{2^0 2^0} = \varepsilon_{3^0 3^0} = \varepsilon, \quad \chi_{1^0 1^0} = \chi_{2^0 2^0} = \chi_{3^0 3^0} = \frac{1}{\mu}, \tag{14.15.21}$$

and all the other elements of $\varepsilon_{i^0 j^0}$, $\chi_{i^0 j^0}$, $\gamma_{i^0 j^0}$ vanish. For the corresponding 4th order constitutive tensor $\chi^{\alpha^0\beta^0\gamma^0\delta^0}$ of empty space we find that

$$-\chi^{4^0 1^0 4^0 1^0} = -\chi^{4^0 2^0 4^0 2^0} = -\chi^{4^0 3^0 4^0 3^0} = \varepsilon, \tag{14.15.22}$$

$$\chi^{2^0 3^0 2^0 3^0} = -\chi^{3^0 1^0 1^0 3^0} = \chi^{1^0 2^0 1^0 2^0} = \frac{1}{\mu}. \tag{14.15.23}$$

The other elements vanish. So, evidently,

$$\chi^{\alpha^0\beta^0\alpha^0\beta^0} \neq 0, \tag{14.15.24}$$

and

$$\chi^{\alpha^0\beta^0\gamma^0\delta^0} = 0 \text{ when } \gamma \neq \alpha \text{ or } \delta \neq \beta \text{ or both.} \tag{14.15.25}$$

More specifically

$$\chi^{i^0 j^0 i^0 j^0} = \frac{1}{\mu}, \quad \chi^{4^0 i^0 4^0 i^0} = -\varepsilon, \tag{14.15.26}$$

where $(i, j) = (1, 2, 3), i \neq j$.

The vacuum state relations (1.4.11) are represented as

(14.15.27)

$\chi^{\alpha^0\beta^0\gamma^0\delta^0}$	E^0	B^0
$-D^0$	-1	0
H^0	0	1

For linear isotropic media, on the other hand, we write

(14.15.28)

$\chi^{\alpha^0\beta^0\gamma^0\delta^0}$	E^0	B^0
$-D^0$	$-\varepsilon$	0
H^0	0	$\frac{1}{\mu}$

In an inertial frame, the constitutive relations (1.6.2) and (1.6.3) are expressed in terms of $\chi^{\alpha\beta\gamma\delta}$ as

$\chi^{\alpha^0\beta^0\gamma^0\delta^0}$	E_{x^0}	E_{y^0}	E_{z^0}	B_{x^0}	B_{y^0}	B_{z^0}
D_{x^0}	ε	0	0	0	0	0
D_{y^0}	0	ε	0	0	0	0
D_{z^0}	0	0	ε	0	0	0
H_{x^0}	0	0	0	$\frac{1}{\mu}$	0	0
H_{y^0}	0	0	0	0	$\frac{1}{\mu}$	0
H_{z^0}	0	0	0	0	0	$\frac{1}{\mu}$

(14.15.29)

For the first order LT, this scheme transforms to

$\chi^{\alpha\beta\gamma\delta}$	41 E_x	42 E_y	43 E_z	23 B_x	31 B_y	12 B_z
41 D_x	$\frac{\beta^2}{\mu_0}+\varepsilon_0$	0	0	0	$-\beta\left(\varepsilon_0+\frac{1}{\mu_0}\right)$	0
42 D_y	0	$\frac{\beta^2}{\mu_0}+\varepsilon_0$	0	$\beta\left(\varepsilon_0+\frac{1}{\mu_0}\right)$	0	0
43 D_z	0	0	0	0	0	0
23 H_x	0	$\beta\left(\varepsilon_0+\frac{1}{\mu_0}\right)$	0	$\frac{1}{\mu_0}+\varepsilon_0\beta^2$	0	0
31 H_y	$-\beta\left(\varepsilon_0+\frac{1}{\mu_0}\right)$	0	0	0	$\frac{1}{\mu_0}+\varepsilon_0\beta^2$	0
12 H_z	0	0	0	0	0	$\frac{1}{\mu_0}$

(14.15.30)

14.15.2 Elements of $\chi^{\alpha\beta\gamma\delta}$ for GN rotation

For GN rotation (section 8.8) we first determine the elements of $\chi^{\alpha\beta\gamma\delta}$. Since $\alpha^3_{\mu^0}=0$ when $\mu^0\neq 3$, the elements containing 3 as a super script, only once, vanish. The non-zero elements are

$$\chi^{1212}=\alpha^1_{\xi^0}\alpha^2_{\eta^0}\alpha^1_{\zeta^0}\alpha^2_{\lambda^0},\quad \chi^{\xi^0\eta^0\zeta^0\lambda^0}=\frac{1}{\mu}-\varepsilon\beta^2,\quad \chi^{1214}=-\varepsilon\cdot\frac{x\Omega}{c},$$

$$\chi^{1224} = \varepsilon\cdot\frac{y\Omega}{c},\quad \chi^{1313} = \frac{1}{\mu} - \varepsilon\cdot\frac{y^2\Omega^2}{c^2},\quad \chi^{1323} = \varepsilon\cdot\frac{\Omega^2 xy}{c^2},$$

$$\chi^{1334} = \varepsilon\cdot\frac{\Omega y}{c},\quad \chi^{1412} = \varepsilon\frac{\Omega x}{c},\quad \chi^{1414} = -\varepsilon,\quad \chi^{2313} = \varepsilon\cdot\frac{\Omega^2 xy}{c^2},$$

$$\chi^{2323} = \frac{1}{\mu} - \varepsilon\frac{\Omega^2 x^2}{c^2},\quad \chi^{2343} = \varepsilon\cdot\frac{\Omega x}{c},\quad \chi^{2412} = -\varepsilon\cdot\frac{\Omega y}{c},$$

$$\chi^{2424} = -\varepsilon,\quad \chi^{3413} = \varepsilon\cdot\frac{\Omega y}{c},\quad \chi^{3423} = -\varepsilon\cdot\frac{\Omega x}{c},\quad \chi^{3434} = -\varepsilon, \tag{14.15.31}$$

where

$$\beta = \frac{v}{c} = \frac{1}{c}(\Omega\times r). \tag{14.15.32}$$

In tabular form these read

$\chi^{\alpha\beta\gamma\delta}$	23 B_1	31 B_2	12 B_3	41 $-E_1$	42 $-E_2$	43 $-E_3$
23 H_1	$\frac{1}{\mu} - \varepsilon\cdot\frac{\Omega^2 xy}{c^2}$	$-\varepsilon\frac{\Omega^2 xy}{c^2}$	0	0	0	$\varepsilon\cdot\frac{\Omega x}{c}$
31 H_2	$-\varepsilon\cdot\frac{\Omega^2 xy}{c^2}$	$\frac{1}{\mu} - \varepsilon\cdot\frac{\Omega^2 y^2}{c^2}$	0	0	0	$\varepsilon\cdot\frac{\Omega y}{c}$
12 H_3	0	0	$\frac{1}{\mu} - \varepsilon\beta^2$	$-\varepsilon\cdot\frac{\Omega x}{c}$	$-\varepsilon\cdot\frac{\Omega y}{c}$	0
41 D_1	0	0	$-\varepsilon\cdot\frac{\Omega x}{c}$	$-\varepsilon$	0	0
42 D_2	0	0	$\varepsilon\cdot\frac{\Omega y}{c}$	0	$-\varepsilon$	0
43 D_3	$\varepsilon\cdot\frac{\Omega x}{c}$	$\varepsilon\cdot\frac{\Omega y}{c}$	0	0	0	$-\varepsilon$

(14.15.33)

Hence we find that

$$\mathcal{H}^{12} = H_3 = \frac{1}{2}\chi^{12\nu\eta}F_{\nu\eta} = \frac{1}{\mu}B_z + \varepsilon\{(\beta\times(\beta\times B)\}_z - \varepsilon(\beta\times E)_z. \tag{14.15.34}$$

So

$$H = -D_1 = \frac{1}{2}\chi^{14\nu\eta}F_{\nu\eta} = \varepsilon\cdot\frac{\Omega x}{c}B_3 - \varepsilon E_1, \tag{14.15.35}$$

Again, since

$$\mathcal{H}^{14} = -D_1 = \frac{1}{2}\chi^{14\nu\eta}F_{\nu\eta} = \varepsilon\frac{\Omega x}{c}B_3 - \varepsilon E_1, \tag{14.15.36}$$

we find that

$$D = \varepsilon E - \varepsilon(\beta\times B). \tag{14.15.37}$$

Using (14.15.37), equation (14.15.35) reads

$$H = \frac{1}{\mu}B - (\beta \times D). \tag{14.15.38}$$

The free space part of $\chi^{\alpha\beta\gamma\delta}$, for GN rotation, in cylindrical co-ordinates, is given by

$\chi^{\alpha\beta\gamma\delta}$	E_r	E_θ	E_z	B_r	B_θ	B_z
$-D_r$	$-r$	0	0	0	0	$r\Omega$
$-D_\theta$	0	$-1/r$	0	0	0	0
$-D_z$	0	0	$-r$	$-r\Omega$	0	0
H_r	0	0	$-r\Omega$	$1/r$	0	0
H_θ	0	0	0	0	r	0
H_z	0	0	0	0	0	$1/r$

(14.15.39)

When observed from the rotating frame K, in cylindrical co-ordinates, the table of (14.15.28) is replaced by

$\chi^{\alpha\beta\gamma\delta}$	E_r	E_θ	E_z	B_r	B_θ	B_z
$-D_r$	$-r\varepsilon$	0	0	0	0	$r\Omega\varepsilon$
$-D_\theta$	0	ε/r	0	0	0	0
$-D_z$	0	0	$-r\varepsilon$	$-r\Omega\varepsilon$	0	0
H_r	0	0	$-r\Omega\varepsilon$	$1/\mu r$	0	0.
H_θ	0	0	0	0	r/μ	0
H_z	$r\Omega\varepsilon$	0	0	0	0	$1/\mu r$

(14.15.40)

The factor ε has been introduced in the rows and columns common to E and D, and the factor μ in the rows and columns common to B and H. Moreover,the scheme (14.15.40) describes the case in which the medium is stationary in the inertial frame, and the inertial free space, together with the observer, is observed from the rotating frame.

The constitutive tensor, for the rotating medium, as seen by a stationary interferometer (or observer), is determined by observations made in an inertial frame. This implies that the free space part retains the form (14.15.27) and the medium part, on the rotating frame, has the form $\varepsilon - 1$ and $\mu - 1$ if the intrinsic changes in the material part are considered. When the rotating medium is viewed from the inertial frame, off-diagonal elements appear because the observer of the inertial frame rotates in the opposite direction with respect to the medium that is in absolute motion. Adding the unaffected free-space part with (14.15.40) we obtain

$\chi^{\alpha\beta\gamma\delta}$	E_r	E_θ	E_z	B_r	B_θ	B_z
$-D_r$	$-\varepsilon r$	0	0	0	0	$-r\Omega(\varepsilon-1)$
$-D_\theta$	0	$-\varepsilon/r$	0	0	0	0
$-D_z$	0	0	$-r\varepsilon$	$r\Omega(\varepsilon-1)$	0	0
H_r	0	0	$r\Omega(\varepsilon-1)$	$1/\mu r$	0	0
H_θ	0	0	0	0	r/μ	0
H_z	$-r\Omega/(\varepsilon-1)$	0	0	0	0	$1/\mu r$

(14.15.41)

14.16 Review of ERF by Volkov and Kislev (1970)

In this and the following two section we discuss ERF briefly, following the coverage by Volkov and Kislev (1970), for convenience of some readers. To determine the material equations that connected the EM field vectors in a medium, general method proposed by Tamm (1924) has been used. It is shown that the covariant form of the material equations, which gives correct relations between the field vectors in the inertial system K^0, still does not ensure that these equations hold in a non-inertial system. It is necessary to justify physically the choice of one of the possible methods of writing down the covariant equations. The final formulation of the state equations in Volkov and Kislev (1970) differ from analogous results of Heer (1964) and Khromykh (1966) and also of Post (1967), which have been covered in the preceeding sections.

Tamm (1924) introduced a dielectric permittivity and magnetic permeability tensor of the medium, $S^{\alpha\beta\gamma\delta}$ or $S_{\alpha\beta\gamma\delta}$, which transforms like contravariant and covariant tensors of 4th order on going over from one co-ordinate system to another. This tensor, together with the tensors of the EM field, leads to the covariant material equations connecting the elements of the EM field in the medium in any reference frame. The theory of Tamm (1924) yields covariant or vector material equations for a medium that is at rest in an inertial frame of reference and is considered from an arbitrary frame of reference.

14.16.1 Electrodynamics in arbitrary co-ordinate systems

The covariant Maxwell equations in an arbitrary system of co-ordinates, are

$$\mathcal{H}^{\mu\nu}_{;\nu} = -\frac{4\pi}{c} j^{\mu}, \tag{14.16.1}$$

$$F_{\lambda\mu;\nu} + F_{\mu\nu;\lambda} + F_{\nu\lambda;\mu} = 0, \tag{14.16.2}$$

where $\mathcal{H}^{\mu\gamma}$ and $F_{\mu\gamma}$ are respectively the contravariant and covariant tensors of the EM field, j^{μ} is the current 4-vector and c is the velocity of light in vacuum. The indices following a semi-colon denote the covariant derivatives with respect to the corresponding co-ordinate. $\mathcal{H}^{\mu\gamma}$ is called the EM *induction tensor* and $F_{\mu\gamma}$ is called the *field intensity tensor.*

Equations (14.16.1) and (14.16.2) can also be written in the form

$$\frac{1}{\sqrt{-g}} \frac{\partial}{\partial x^{\nu}} \left(\sqrt{-g}\mathcal{H}^{\mu\nu} \right) = \frac{4\pi}{c} j^{\mu}. \tag{14.16.3}$$

$$\frac{\partial F_{\lambda\mu}}{\partial x^{\nu}} + \frac{\partial F_{\mu\nu}}{\partial x^{\lambda}} + \frac{\partial F_{\nu\eta}}{\partial x^{\mu}} = 0. \tag{14.16.4}$$

Following Plebanski (1960), define the non-covariant 3-dimensional field vectors D, E, B, H in the following manner :

$$D_i = \sqrt{-g}\mathcal{H}^{i4}. \tag{14.16.5}$$

$$H_i = -\frac{1}{2} e_{ijk} \sqrt{-g}\mathcal{H}^{jk}. \tag{14.16.6}$$

$$E_i = F_{4i}. \tag{14.16.7}$$

$$B_i = -\frac{1}{2} e_{ijk} F_{jk}. \tag{14.16.8}$$

$$J_i = \sqrt{-g} j^4. \tag{14.16.9}$$

The charge density is

$$\rho^c = \frac{1}{c}\sqrt{-g}j^4. \tag{14.16.10}$$

Here e_{ijk} is a perfectly antisymmetric unit pseudotensor of third order (also called the *3-dimensional Levi civita symbol*) and

$$g = \det g_{\mu\gamma}. \tag{14.16.11}$$

These relations yield a system of equations for vector fields that coincide with the Maxwell equations in material media at rest in the inertial frame reference :

$$\nabla \times H = \frac{1}{c}\frac{\partial D}{\partial t} + \frac{4\pi}{c}j. \tag{14.16.12}$$

$$\nabla \times E = -\frac{1}{c}\frac{\partial B}{\partial t}. \tag{14.16.13}$$

$$\nabla \cdot D = 4\pi\rho^c. \tag{14.16.14}$$

$$\nabla \cdot B = 0. \tag{14.16.15}$$

The quantities introduced in the relations (14.16.4)–(14.16.9) are not independent. In vacuum, the connection between the components of the tensors $\mathcal{H}^{\mu\gamma}$ and $F_{\mu\gamma}$, estabslished with the aid of the metric tensor $g_{\mu\gamma}$, is

$$\sqrt{-g}\mathcal{H}^{\mu\nu} = \sqrt{-g}g^{\mu\xi}g^{\nu\eta}F_{\xi\eta} \tag{14.16.16}$$

or

$$F_{\mu\nu} = \frac{1}{\sqrt{-g}}g_{\mu\xi}g_{\nu\eta}\sqrt{-g}\mathcal{H}^{\xi\eta}. \tag{14.16.17}$$

In the inertial frame, in the absence of gravitational fields, $g_{\mu\gamma}$ becomes the fundamental unit tensor

$$\eta_{\mu\nu} = \begin{bmatrix} 1 & 0 & 0 & 0 \\ 0 & 1 & 0 & 0 \\ 0 & 0 & 1 & 0 \\ 0 & 0 & 0 & -1 \end{bmatrix}. \tag{14.16.18}$$

So, relations (14.16.16) or (14.16.17) give the vacuum state relations

$$D = E, \quad B = H, \tag{14.16.19}$$

which are also the Gaussian field identifications. In non-inertial frames, or in the presence of a gravitational field, there are no such simple relations between the induction vectors (D and H) and the intensity vectors (E and B).

A difference appears between the vectors D and E on one hand, and H and B on the other, even for vacuum in non-inertial reference frames or in the presence of gravitational fields, such that the quantities E and B tend to their limiting values D and H in the limit from a non-inertial to an inertial frame. Hence the vectors E and B are given the same resultant-field meaning that they have in a dielectric medium that is at rest in an inertial frame; the other two vectors, D and H are also given the earlier meaning of induction; obviously, the vectors E

and B then tend to D and H in vacuum on going over to an inertial frame in the absence of gravitational fields.

The medium exerts an influence only on the resultant EM fields E and B, that is, on the components of $F_{\mu\gamma}$. At specified charges and currents, E is decreased by a factor ε compared with its vacuum value. So we write

$$E_{\text{material}} = \varepsilon E_{\text{vacuum}}.$$

The same relation is also written as

$$D = \varepsilon E, \tag{14.16.20}$$

where D stands for E_{material} and E for E_{vacuum}. Similarly, the presence of a magnetic medium leads to a change in the intensity of the magnetic field B by a factor μ, while the induction H remains unchanged. So we write

$$H = \frac{1}{\mu}B. \tag{14.16.21}$$

Here ε and μ are respectively the dielectric constant and the magnetic permeability of the medium. The electric polarization vector per unit volume P and the magnetic polarization per unit volume M, called *simply the magnetization* are related to the field vectors in empty space by the equations

$$D = \varepsilon_0 E + 4\pi P. \tag{14.16.22}$$

$$\frac{1}{\mu_0}B = H + 4\pi M. \tag{14.16.23}$$

Thus it is evident that if D, H and E, B combine into bivectors (second order tensors), then P and M constitute a bivector in 4-space with the identification

$$M^{4i} = 4\pi P_i, \quad i = 1, 2, 3. \tag{14.16.24}$$

$$M^{23} = -4\pi M_1, \quad M^{31} = -4\pi M_2, \quad M^{12} = -4\pi M_3. \tag{14.16.25}$$

The generally covariant rendition of (14.16.22) and (14.16.23) is

$$\mathcal{H}^{\alpha\beta} - M^{\alpha\beta} = \frac{1}{2}\chi_0^{\alpha\beta\gamma\delta} F_{\gamma\delta}, \tag{14.16.26}$$

where $\chi_0^{\alpha\beta\gamma\delta}$ represents the free-space properties on general co-ordinates.

We now introduce the mixed tensor S^μ_ν, representing the EM properties of the medium. For a medium at rest in the reference frame specified by the metric tensor $g_{\mu\gamma}$, the vacuum relations (14.16.16) or (14.16.17) are replaced by

$$\sqrt{-g}\mathcal{H}^{\mu\nu} = \sqrt{-g}g^{\mu\alpha}g^{\nu\beta}S^\xi_\alpha S^\eta_\beta F_{\xi\eta}, \tag{14.16.27}$$

or

$$\frac{1}{\sqrt{-g}}g_{\alpha\lambda}g_{\beta\nu}\sqrt{-g}\mathcal{H}^{\alpha\beta} = S^\xi_\lambda S^\eta_\nu F_{\xi\eta}. \tag{14.16.28}$$

In a linear isotropic medium at rest in an arbitrary reference frame, the non-zero components of S^λ_μ are

$$S^1_1 = S^2_2 = S^3_3 = \frac{1}{\sqrt{\mu}}, S^4_4 = \varepsilon\sqrt{\mu} \tag{14.16.29}$$

ε and μ are measured by local observers connected with the medium. For a medium at rest in an inertial frame, equations (14.16.27) and (14.16.28) together with (14.16.29), generate the relations (14.16.20) and (14.16.21). Relations (14.16.29) show that in vacuum, in an inertial frame,

$$S^{\mu}_{\gamma} = \delta^{\mu}_{\gamma}. \tag{14.16.30}$$

So, the relations (14.16.27) and (14.16.28) become identical to (14.16.16) and (14.16.17).

In place of (14.16.27) and (14.16.28) we could write

$$\bar{S}^{\alpha}_{\lambda}\bar{S}^{\beta}_{\mu}\sqrt{-g}\mathcal{H}^{\lambda\mu} = \sqrt{-g}g^{\alpha\xi}g^{\beta\eta}F_{\xi\eta}, \tag{14.16.31}$$

$$\frac{1}{\sqrt{-g}}g_{\alpha\lambda}g_{\beta\mu}\bar{S}^{\alpha}_{\xi}\bar{S}^{\beta}_{\eta}\sqrt{-g}\mathcal{H}^{\xi\eta} = F_{\alpha\beta}, \tag{14.16.32}$$

with the help of a new mixed tensor $\bar{S}^{\lambda}_{\mu}$ for the medium. Then, for a linear isotropic medium at rest in an arbitrary reference frame, (14.16.29) should be replaced by

$$\bar{S}^1_1 = \bar{S}^2_2 = \bar{S}^3_3 = \sqrt{-\mu},\ \bar{S}^4_4 = \frac{1}{\varepsilon\sqrt{\mu}}. \tag{14.16.33}$$

Equations (14.16.27) and (14.16.28), together with (14.16.1) and (14.16.2), describe completely the EM process in a medium for any arbitrary 4-space metric $g_{\mu\gamma}$. This shows that the 3-tensors $S^{\mu}_{\gamma}, \mathcal{H}^{\mu\gamma}$ and $F_{\mu\gamma}$ are necessary for that purpose. The material equations, which relate the components of the tensors of the EM field and depend on the properties of 4-space, are determined completely by the metric tensor $g^{\mu\gamma}$, which characterizes all the properties of the geometry of the space in each given system of co-ordinates.

In vacuum

$$\mathcal{H}^{\mu\gamma} = F^{\mu\gamma} = g^{\mu\alpha}g^{\gamma\beta}F_{\alpha\beta}. \tag{14.16.34}$$

It is often desirable to separate the property of matter from the vacuum field. In perfect analogy with the definition for media at rest in an inertial frame, define the antisymmetric polarization tensor of the medium $M^{\mu\gamma}$ by the relation

$$\sqrt{-g}\mathcal{H}^{\mu\nu} = \sqrt{-g}(F^{\mu\nu} + M^{\mu\nu}) \tag{14.16.35}$$

$$\text{or,}\quad \sqrt{-g}\mathcal{H}^{\mu\nu} = \sqrt{-g}g^{\mu\xi}g^{\nu\eta}(F_{\xi\eta} + M_{\xi\eta}). \tag{14.16.36}$$

Then the dielectric polarization 3-vector P and the 3-vector magnetization M of the medium, both non-covariant, are defined

$$P_i = M_{4i}, \tag{14.16.37}$$

$$M_i = \frac{1}{2}e_{ijk}M_{jk}. \tag{14.16.38}$$

With these definitions equations if j is replaced by

$$J = J + \nabla \times M + \frac{\partial P}{\partial t}.$$

Then, in linear isotropic media

$$P_i = \chi E_i = \frac{\varepsilon - 1}{4\pi}E_i, \tag{14.16.39}$$

$$M_i = \frac{\chi}{\mu}B_i = \frac{\mu - 1}{4\pi\mu}B_i, \tag{14.16.40}$$

where K is the coefficient of polarization of the medium (the dielectric susceptibility of the medium), ε is the dielectric constant and χ is the magnetic permeability. The quantities $K, \varepsilon, \chi, \mu$ are the characteristics of the medium measured by a local observer connected with the medium. From (14.16.5)–(14.16.8), (14.16.27)–(14.16.29) and (14.16.40) we see that

$$F_{\mu\nu} + 4\pi M_{\mu\nu} = \zeta^{\xi}_{\nu}\xi^{\eta}_{\nu}F_{\xi\eta}. \tag{14.16.41}$$

Thus, the material relations (14.16.27) and (14.16.28) are equivalent to those of (14.16.35) and (14.16.36).

14.16.2 Some results for rotating isotropic media

Consider rotation of a material medium at the constant angular rate Ω relative to the inertial co-ordinate system. The metric tensor of the rotating frame has been discussed in Chapter 8. Introduce vector fields of (14.16.5)–(14.16.10) to write the covariant Maxwell equations (14.16.1) and (14.16.2) in the form (14.16.12)–(14.16.15). Substitute in the equation (14.16.27) the values of $\mathcal{H}^{\mu\gamma}$ and $F_{\mu\gamma}$ from (14.10.5)–(14.16.10), the tensor S^{μ}_{γ} from (14.16.29) and the metric tensor $g_{\mu\gamma}$ for rotation, from Chapter 8; then, the non-covariant material equations connecting the field vectors in a medium at rest in a rotating frame are

$$D = \varepsilon E - \frac{1}{\mu}(\beta \times B), \tag{14.16.42}$$

$$H = \frac{1}{\mu}\beta - (\beta \times B), \tag{14.16.43}$$

where

$$\beta = \frac{1}{c}(\Omega \times r). \tag{14.16.44}$$

The EM properties of the medium are described by the equations (14.16.12)–(14.16.15) along with (14.16.42)–(14.16.43).

14.16.3 Theory of Heer and Khromykh

Earlier, some authors employed different material equations for the EM field in a medium at rest in the rotating frame. Thus Heer (1964) and Khromykh (1966) used, for the medium at rest, in a rotating frame, vector equations which, in the first order approximation in powers of β, read,

$$D = \varepsilon E - (\beta \times H), \tag{14.16.45}$$

$$H = \frac{1}{\mu}B - \frac{1}{\mu}(\beta \times E) \tag{14.16.46}$$

as may be seen from (14.16.64) and (14.16.65). Of the simple system, for instance, the system with constant linear velocity, linear acceleration, and angular rotation, only angular rotation does not permit the introduction of an instantaneous rest frame for the entire system and the metric cannot be placed in diagonal form. For this reason effects of the order of v/c $(v = r\Omega/c)$ are observed in the rotating frame.

The Maxwell equations are written in a covariant form in an instantaneous rest frame. By the rules of transformation for tensor algebra, these remain valid in an arbitrary frame. For macroscopic matter, the Maxwell equations in covariant form are the equations (14.16.1) and (14.16.2). $F_{\mu\gamma}$ and $\mathcal{H}^{\mu\gamma}$ have the dimensions of $(\text{energy})^{1/2}$. Although $F_{\mu\gamma}$ and $\mathcal{H}^{\mu\gamma}$ can be written in terms of a polar vector and an axial vector in a covariant manner, a non-covariant notation, given in Plebanski (1960), is found more convenient and consistent with conventional techniques. Non-covariant vectors are defined in (14.16.5)–(14.16.10). The Maxwell equations take on their conventional appearance, namely, the equations (14.16.12)–(14.16.15), in this notation. Evidently, the operation curl and divergence are defined only for Cartesian ordering of the differentiation.

An instantaneous rest frame, Moller (1972, Chapter 8), is regarded as a system of inertia K^0, for which the metric tensor is $\eta_{\mu\nu}$, given by (14.16.18). If (x^0, y^0, z^0, t^0) and (x, y, z, t) are regarded as instantaneously at rest with respect to each other, then

$$\alpha^x_{0_t} = 0, \quad \alpha^{0^x}_t = 0. \tag{14.16.47}$$

The components of the tensors $A_{\mu\gamma}$ and $A^{\mu\gamma}$ are related to those in the instantaneous rest frame

$$A_{\mu\nu} = \alpha^{0\xi}_{\mu}\alpha^{0\eta}_{\nu}A^0_{\xi\eta}. \tag{14.16.48}$$

$$A^{\mu\nu} = \alpha^{\mu}_{0\xi}\alpha^{\nu}_{0\eta}A^{0\xi\eta}. \tag{14.16.49}$$

In vacuum

$$\mathcal{H}^{\mu\nu} = F^{\mu\nu}, \tag{14.16.50}$$

which follows from (14.16.16) or (14.16.17), the relationship between the field vectors B, H, E and D follow from (14.16.49) and (14.16.16) or (14.16.17). Macroscopic matter requires further elaboration. Specifically, a procedure similar to that used by Minkowski for introducing the constitutive equations in STR is used to find the constitutive equations in a non-permanent gravitational field. For isotropic dielectrics and paramagnetic substances the constitutive equations in an instantaneous rest frame have their usual form

$$D^0_i = \varepsilon E^0_i, \tag{14.16.51}$$

$$B^0_i = \mu H^0_i, \tag{14.16.52}$$

$$J^0_i = \alpha E^0_i, \tag{14.16.53}$$

in non-covariant notation. In the instantaneous rest frame thee is no distinction between covariant and contravariant components, and these constitutive equations may be written as

$$\mathcal{H}^0_{i4} = \varepsilon F^0_{i4}, \tag{14.16.54}$$

$$F^0_{ij} = \mu \mathcal{H}^0_{ij}, \tag{14.16.55}$$

$$j^0_i = \sigma F^0_{4i}, \tag{14.16.56}$$

$$j^0_4 = \rho^{co} c. \tag{14.16.57}$$

In any other frame of reference the components follow from the rules (14.16.48) and (14.16.49). Thus

$$F_{i4} = \alpha^{0\xi}_i \alpha^{0\eta}_4 F^0_{\xi\eta}, \tag{14.16.58}$$

where the second relation of (14.16.47) has been used. Similarly,

$$\mathcal{H}^{i4} = \alpha^{0\xi}_i \alpha^{04}_4 \mathcal{H}^0_{\xi 4}. \tag{14.16.59}$$

Hence, using (14.16.54) and the antisymmetric property of $\mathcal{H}_{\mu\gamma}$ and $F_{\mu\gamma}$, we find that

$$\mathcal{H}_{i4} = \alpha_i^{0\xi}\alpha_4^{04}\mathcal{H}_{\xi 4}^0 = \alpha_i^{0j}\alpha_4^{04}\mathcal{H}_{j4}^0 = \varepsilon\alpha_i^{0\xi}\alpha_4^{04}F_{\xi 4}^0$$

$$= \varepsilon\alpha_i^{0j}\alpha_4^{04}F_{j4} = \varepsilon F_{i4}. \tag{14.16.60}$$

Similarly we can write

$$\mu\mathcal{H}^{ij} = F^{ij}. \tag{14.16.61}$$

The corresponding relations between the covariant and contravariant components of the field tensors are

$$\mu\mathcal{H}^{ij} = F^{ij} = g^{i\lambda}g^{i\mu}F_{\lambda\mu} \tag{14.16.62}$$

$$\varepsilon F_{i4} = \mathcal{H}_{i4} = g_{i\lambda}g_{4\mu}\mathcal{H}^{\lambda\mu}. \tag{14.16.63}$$

For rotation using the non-covariant field vectors of (14.16.5)–(14.16.10), we obtain

$$B + \beta(\beta \times B) = \mu H + (\beta \times E), \tag{14.16.64}$$

$$D + \beta \times (\beta \times D) = \varepsilon E - (\beta \times H). \tag{14.16.65}$$

Approximately these relations verify (14.16.40) and (14.16.41). In vacuum $\varepsilon = 1 = \mu$, and these relations read

$$D = E - \gamma^2(\beta \times B) - \gamma^2(\beta\cdot(D - E))\beta, \tag{14.16.66}$$

$$B = H + \gamma^2(\beta \times D) - \gamma^2(\beta\cdot(B - H))\beta. \tag{14.16.67}$$

Approximately, in the first order of β, these are

$$D = E - (\beta \times B), \tag{14.16.68}$$

$$B = H + (\beta \times D). \tag{14.16.69}$$

Hence rotation has the effect of making even the vacuum appear anisotropic. This first order effect corresponds to the type of terms introduced by Michelson for the description in terms of the ether.

Transformation of the 4-current j^μ is

$$j_i = \alpha_i^{0\xi}j_\xi^0 = \alpha_i^{0j}j_i^0 + \alpha_i^{04}j_4^0, \quad i = 1, 2, 3; \tag{14.16.70}$$

$$j_4 = \alpha_4^{0\xi}j_\xi^0 = \alpha_4^{04}j_4^0 = \alpha_4^{04}\rho^0 c, \quad \xi = 1, 2, 3, 4; \tag{14.16.71}$$

where, in the final results, the relations (14.16.47) have been used. Relation (14.16.70) can be written as

$$j^i = \sigma\alpha_i^{0j}F_{4j}^0 + \alpha_i^{04}j_4^0. \tag{14.16.72}$$

$$j_4 = \alpha_4^{0\xi}j_\xi^0 = \alpha_4^{0k}j_k^0 + \alpha_4^{04}j_4^0. \tag{14.16.73}$$

Then

$$j^k = g^{k\mu}j_\mu = g^{k\mu}\{(\alpha_\mu^{0j})j_j^0 + \alpha_\mu^{04}j_4^0\} = g^{k\mu}(\alpha_\mu^{0j}\sigma E_j^0 + \alpha_\mu^{04}j_4^0)$$

$$= \sigma g^{k\mu}\alpha_\mu^{0j}F_{4j}^0 + g^{k\mu}\alpha_\mu^{04}j_4^0 = \sigma g^{k\mu}\alpha_\mu^{0j}\alpha_{04}^{\xi}\alpha_{0j}^{\eta}F_{\xi\eta} + g^{k\mu}\alpha_\mu^{04}j_4^0, \tag{14.16.74}$$

$$j^4 = g^{4\mu}j_\mu = \sigma g^{4\mu}\alpha_\mu^{0j}\alpha_{0j}^{\eta}F_{\xi\eta} + g^{4\mu}\alpha_\mu^{04}j_4^0. \tag{14.16.75}$$

These relations show that, Heer (1964),

$$j = \sigma\gamma[E - \beta(\beta\cdot E)], \tag{14.16.76}$$

$$\rho = \gamma\left[\frac{\sigma}{c}(\beta\cdot E) + \rho_0\right], \tag{14.16.77}$$

where

$$\gamma = (1-\beta^2)^{-1/2}. \tag{14.16.78}$$

The energy momentum tensor S_μ^λ is defined as

$$S_\mu^\lambda = F_{\mu\xi}\mathcal{H}^{\lambda\xi} - \frac{1}{4}\delta_\mu^\lambda F_{\xi\eta}\mathcal{H}^{\xi\eta}. \tag{14.16.79}$$

Covariant divergence of S_μ^λ yields

$$\text{div}\,(S_\mu^\lambda) = S_{\mu;\nu}^\lambda = -F_\mu, \tag{14.16.80}$$

where the 4-force F_μ is defined as

$$F_\mu = \frac{1}{c}F_{\mu\nu}j^\nu. \tag{14.16.81}$$

This relation can be written as

$$F_\mu = \frac{1}{4\pi}F_{\mu\nu}\mathcal{H}^{\nu\xi}_{;\xi}. \tag{14.16.82}$$

In inertial frames this relation simplifies to

$$F_\mu = \frac{1}{4\pi}F_{\mu\nu}\mathcal{H}^{\nu\xi}_{,\xi} = \frac{1}{4\pi}\left[\frac{\partial}{\partial x^\xi}(F_{\mu\nu}\mathcal{H}^{\nu\xi}) - \mathcal{H}^{\nu\xi}\frac{\partial F_{\mu\nu}}{\partial x^\xi}\right]. \tag{14.16.83}$$

In the second term on the right hand side, interchanging the running co-ordinates ξ and ν we obtain

$$\begin{aligned}\mathcal{H}^{\nu\xi}\frac{\partial F_{\mu\nu}}{\partial x^\xi} &= \mathcal{H}^{\xi\nu}\frac{\partial F_{\mu\xi}}{\partial x^\nu} = \mathcal{H}^{\mu\xi}\frac{\partial F_{\xi\mu}}{\partial x^\nu}\\ &= \frac{1}{2}\mathcal{H}^{\nu\xi}\left(\frac{\partial F_{\mu\nu}}{\partial x^\xi} + \frac{\partial F_{\xi\mu}}{\partial x^\nu}\right) = -\frac{1}{2}\mathcal{H}^{\nu\xi}\frac{\partial F_{\nu\xi}}{\partial x^\mu},\end{aligned} \tag{14.16.84}$$

where the relation (14.16.2), valid for inertial frames only, has been used. Hence (14.16.83) becomes

$$F_\mu = -\frac{1}{4\pi}\frac{\partial}{\partial x^\xi}(F_{\mu\nu}\mathcal{H}^{\xi\nu}) + \frac{1}{8\pi}\mathcal{H}^{\nu\xi}\frac{\partial F_{\nu\xi}}{\partial x^\mu}. \tag{14.16.85}$$

This relation can be written as

$$F_\mu = -\frac{1}{4\pi}\frac{\partial}{\partial x^\xi}(F_{\mu\nu}\mathcal{H}^{\xi\nu}) + \frac{1}{16\pi}\frac{\partial}{\partial x^\nu}(\mathcal{H}^{\nu\eta}F_{\nu\eta}) + \frac{1}{16\pi}\left[\mathcal{H}^{\nu\eta}\frac{\partial F_{\nu\eta}}{\partial x^\mu} - F_{\nu\xi}\frac{\partial \mathcal{H}^{\nu\eta}}{\partial x^\mu}\right]. \tag{14.16.86}$$

Further manipulations give

$$\frac{1}{c}F_{\mu\nu}j^\nu + \frac{1}{16\pi}\left[F_{\nu\eta}\frac{\partial \mathcal{H}^{\nu\eta}}{\partial x^\mu} - \mathcal{H}^{\nu\eta}\frac{\partial F_{\nu\eta}}{\partial x^\mu}\right] = -\frac{\partial}{\partial x^\xi}S_\mu^\xi = -\text{div}_\xi(S_\mu^\xi) \tag{14.16.87}$$

with

$$S_\mu^\xi = -\frac{1}{4\pi}(F_{\mu\nu}\mathcal{H}^{\xi\nu}) + \frac{1}{16\pi}\delta_\mu^\xi\mathcal{H}^{\nu\eta}F_{\nu\eta}. \tag{14.16.88}$$

S^ξ_μ is the energy momentum tensor. In inertial frames S^μ_ν is called the *Minkowski energy momentum tensor.*

In vacuum (14.16.87) becomes

$$\frac{1}{6}F_{\mu\nu}j^\nu = F_\mu = \text{div}_\xi(S^\xi_\mu). \tag{14.16.89}$$

The corresponding vacuum relation, in an arbitrary frame of reference, is obtained by replacing the ordinary derivative in (14.16.89) by the covariant derivative; so, now,

$$\frac{1}{c}F_{\mu\nu}j^\nu = F_\mu = S^\xi_{\mu;\xi}. \tag{14.16.90}$$

Using the covariant differentiation results, section 11.4, Moller, or [1972, equation (9.199)], we have

$$\text{div}_\nu(S^\mu_\mu) = \frac{1}{\sqrt{-g}}\frac{\partial}{\partial x^\nu}\left(\sqrt{-g}S^\nu_\mu\right) - \frac{1}{2}\frac{\partial g_{\xi\eta}}{\partial x^\nu}S^{\xi\eta}. \tag{14.16.91}$$

Hence (14.16.90) becomes

$$F_\mu = \frac{1}{c}F_{\mu\nu}j^\nu = \frac{1}{\sqrt{-g}}\frac{\partial}{\partial x^\nu}\left(\sqrt{-g}S^\nu_\mu\right) - \frac{1}{2}\frac{\partial g_{\xi\eta}}{\partial x^\mu}S^{\xi\eta}. \tag{14.16.92}$$

In an arbitrary reference frame equation (14.16.88) reads

$$\frac{1}{c}F_{\mu\nu}j^\nu + \frac{1}{16\pi}[F_{\nu\eta}\mathcal{H}^{\nu\eta}_{;\mu} - \mathcal{H}^{\nu\eta}F_{\nu\eta;\mu}] = -S^\xi_{\mu;\xi}. \tag{14.16.93}$$

For the second term in the left hand side, (Landau and Lifshitz (1975) : formulas (85.12) and (85.14)), we obtain

$$A^{\lambda\mu}_{;\nu} = \frac{\partial A^{\lambda\mu}}{\partial x^\mu} + \Gamma^\lambda_{\xi\nu}A^{\xi\mu} + \Gamma^\mu_{\xi\nu}A^{\lambda\xi} \tag{14.16.94}$$

$$= \frac{\partial A_{\lambda\mu}}{\partial x^\nu} - \Gamma^\xi_{\lambda\mu}A_{\xi\mu} - \Gamma^\xi_{\mu\nu}A_{\lambda\xi}. \tag{14.16.95}$$

These relations, together with (14.16.91), transform the energy relation (14.16.93) into

$$\frac{1}{c}F_{\mu\nu}j^\nu + \frac{1}{16\pi}\left[F_{\nu\xi}\left\{\frac{\partial\mathcal{H}^{\nu\eta}}{\partial x^\mu} + \Gamma^\nu_{\alpha\mu}\mathcal{H}^{\alpha\eta} + \Gamma^\eta_{\alpha\mu}\mathcal{H}^{\nu\alpha}\right\} - \mathcal{H}^{\nu\eta}\left\{\frac{\partial F_{\nu\eta}}{\partial x^\mu} - \Gamma^\alpha_{\nu\mu}F_{\alpha\eta} - \Gamma^\alpha_{\eta\mu}F_{\nu\alpha}\right\}\right]$$

$$= -\frac{1}{\sqrt{-g}}\frac{\partial}{\partial x^\xi}\left(\sqrt{-g}S^\xi_\mu\right) + \frac{1}{2}\frac{\partial g_{\xi\eta}}{\partial x^\mu}S^{\xi\eta}. \tag{14.16.96}$$

Relation (14.16.88) shows that

$$S^\mu_\mu = \frac{1}{4\pi}(F_{\mu\nu}\mathcal{H}^{\mu\nu}) + \frac{1}{16\pi}\cdot 4\cdot\mathcal{H}^{\nu\eta}F_{\nu\eta} = 0 \tag{14.16.97}$$

(for summation over the repeated index μ in the left hand side)

$$S^1_2 = -\frac{1}{4\pi}(F_{2\nu}\mathcal{H}^{1\nu}) - 0 = -\frac{1}{4\pi}(F_{23}\mathcal{H}^{13} + F_{24}\mathcal{H}^{14})$$

$$= -\frac{1}{4\pi}[-B_xH_y - E_yD_x] = \frac{1}{4\pi}(B_xH_y + D_xE_y). \tag{14.16.98}$$

$$S^2_1 = -\frac{1}{4\pi}(F_{1\nu}\mathcal{H}^{2\nu}) = -\frac{1}{4\pi}(F_{13}\mathcal{H}^{23} + F_{14}\mathcal{H}^{24})$$

$$= -\frac{1}{4\pi}[-B_yH_x + E_x(-D_y)]. \tag{14.16.99}$$

$$
\begin{aligned}
S_1^1 &= -\frac{1}{4\pi}F_{1\nu}\mathcal{H}^{1\nu} + \frac{1}{16\pi}\mathcal{H}^{\nu\eta}F_{\nu\eta} \\
&= -\frac{1}{4\pi}(F_{12}\mathcal{H}^{12} + F_{13}\mathcal{H}^{13} + F_{14}\mathcal{H}^{14}) + \frac{1}{16\pi}(\mathcal{H}^{12}F_{12} + \mathcal{H}^{13}F_{13} + \mathcal{H}^{21}F_{21} \\
&\quad + \mathcal{H}^{23}F_{23} + \mathcal{H}^{24}F_{24} + \mathcal{H}^{31}F_{31} + \mathcal{H}^{32}F_{32} + \mathcal{H}^{34}F_{34} + \mathcal{H}^{41}F_{41} + H^{42}F_{42} + \mathcal{H}^{43}F_{43}) \\
&= -\frac{1}{4\pi}(F_{12}\mathcal{H}^{12} + F_{13}\mathcal{H}^{13} + F_{14}\mathcal{H}^{14}) \\
&\quad + \frac{1}{8\pi}(\mathcal{H}^{12}F_{12} + \mathcal{H}^{13}F_{13} + \mathcal{H}^{14}F_{14} + \mathcal{H}^{23}F_{23} + \mathcal{H}^{24}F_{24} + \mathcal{H}^{34}F_{34}) \\
&= -\frac{1}{4\pi}[B_zH_z + B_yH_y - E_xD_x] \\
&\quad + \frac{1}{8\pi}[B_zH_z + B_yH_y - E_xD_x + B_xH_x + E_y(-D_y) + E_z(-D_z)] \\
&= -\frac{1}{4\pi}[(B_y \cdot H_y) + B_zH_z - E_xD_x] + \frac{1}{8\pi}[(B\cdot H) - (E\cdot D)] \\
&= -\frac{1}{4\pi}[(B\cdot H) - B_xH_x - E_xD_x) + \frac{1}{8\pi}[(B\cdot H) - (E\cdot D)] \\
&= \frac{1}{4\pi}(B_xH_x + E_xD_x) - \frac{1}{8\pi}[(B\cdot H) + (E\cdot D)] \qquad (14.16.100)
\end{aligned}
$$

$$
S_1^4 = -\frac{1}{4\pi}(F_{1\nu}\mathcal{H}^{4\nu}) + 0 = -\frac{1}{4\pi}(F_{12}\mathcal{H}^{42} + F_{13}\mathcal{H}^{43}) = \frac{1}{4\pi}(B\times D)_x. \qquad (14.16.101)
$$

$$
S_4^1 = -\frac{1}{4\pi}(F_{4\nu}\mathcal{H}^{1\nu}) = -\frac{1}{4\pi}(F_{42}\mathcal{H}^{12} + F_{43}\mathcal{H}^{13}) = \frac{1}{4\pi}(E_yH_z - E_zH_y) = \frac{1}{c}S_x, \qquad (14.16.102)
$$

where the Poynting flux

$$
S = \frac{c}{4\pi}(E\times H). \qquad (14.16.103)
$$

The relations (14.16.98)–(14.16.100) yield the single relation

$$
S_j^i = \frac{1}{4\pi}(B_iH_j + D_iE_j) - \frac{1}{8\pi}\delta_j^i[(B\cdot H) + (D\cdot E)]. \qquad (14.16.104)
$$

So S_j^i is the ijth component of the Maxwell stresses. Equations (14.16.101) and (14.16.104) show that $S_i^4 \neq S_4^i$ in a material medium; S_4^i is the ith component of the energy flux (Poynting flux) and S_i^4 is the ith component of the EM momentum density.

In place of the constitutive relations of Heer (1964), Khromykh (1966) has used the state relations

$$
g_{\alpha\xi}g_{\beta\eta}\mathcal{H}^{\xi\eta}u^{\beta} = \varepsilon F_{\alpha\eta}u^{\eta}, \qquad (14.16.105)
$$

$$
\rho_{\alpha\beta\gamma\delta}g^{\beta\xi}g^{\gamma\eta}F_{\xi\eta}u^{\delta} = \mu\rho_{\alpha\beta\gamma\delta}\mathcal{H}^{\beta\gamma}u^{\delta}. \qquad (14.16.106)
$$

For $\alpha = 1$ equation (14.16.105) gives

$$
g_{1\xi}g_{\beta\eta}\mathcal{H}^{\xi\eta}u^{\beta} = \varepsilon(F_{12}u^2 + F_{13}u^3 + F_{14}u^4).
$$

This relation reduces to

$$g_{11}[u'(g_{14}\mathcal{H}^{14}) + u^2(g_{22}\mathcal{H}^{12} + g_{24}\mathcal{H}^{14}) + g_{33}u^3\mathcal{H}^{13} + u^4(g_{42}\mathcal{H}^{12} + g_{44}\mathcal{H}^{14})]$$
$$+g_{14}[u^1(g_{11}\mathcal{H}^{41}) + u^2(g_{22}\mathcal{H}^{42}) + u^3 g_{33}\mathcal{H}^{43} + u^4(g_{41}\mathcal{H}^{41} + g_{42}\mathcal{H}^{42})]$$
$$= \varepsilon[B_z u_y - B_y u_z + cE_x]. \tag{14.16.107}$$

It further reduces to

$$-u^x\beta_y(-D_x) + u_y\{H_z + \beta_x(-D_x)\} + u_z(-H_y) + c\left[\beta_x H_z - \frac{1}{\gamma^2}(-D_x)\right]$$
$$-B_y[u_x D_x + u_y D_y + u_z D_z + c(-\beta_y D_x + \beta_x D_y)]$$
$$= c\varepsilon\left[E + \frac{1}{c}(u\times B)\right]_x, \tag{14.16.108}$$

where

$$\beta_x = x\frac{\Omega}{c}, \quad \beta_y = y\frac{\Omega}{c}. \tag{14.16.109}$$

This relation can be written as

$$c\varepsilon\left[E + \frac{1}{c}(u\times B)\right]_x$$
$$= u_y H_z - u_z H_y - u_y\beta_x D_x + c\beta_x D_y + \frac{c}{\gamma^2}D_x - \beta_y(u\cdot D) + c[\beta\times(D\times\beta)]_x. \tag{14.16.110}$$

Neglecting terms of higher order, this relation, approximately, gives the vector relation

$$D = \varepsilon\left[E + \frac{1}{c}(u\times H)\right] - \frac{1}{c}(u\times H) - (\beta\times H). \tag{14.16.111}$$

For $\alpha = 1$ equation (14.16.106) gives

$$e_{1\beta\gamma\delta}g^{\beta\xi}g^{\gamma\eta}F_{\xi\eta}u^\delta = \mu e_{1\beta\gamma\delta}\mathcal{H}^{\beta\gamma}u^\delta. \tag{14.16.112}$$

The left hand side of (14.16.112)

$$= e_{1234}g^{2\xi}g^{3\eta}F_{\xi\eta}u^4 + e_{1243}g^{2\xi}g^{4\eta}F_{\xi\eta}u^3 + e_{1324}g^{3\xi}g^{2\eta}F_{\xi\eta}u^4$$
$$+e_{1342}g^{3\xi}g^{4\eta}F_{\xi\eta}u^2 + e_{1423}g^{4\xi}g^{2\eta}u^3 + e_{1432}g^{4\xi}g^{3\eta}F_{\xi\eta}u^2$$
$$= u^2(g^{3\xi}g^{4\eta} - g^{4\xi}g^{3\eta})F_{\xi\eta} + u^3(g^{4\xi}g^{3\eta} - g^{2\xi}g^{4\eta})F_{\xi\eta} + u^4(g^{2\xi}g^{3\eta} - g^{3\xi}g^{2\eta})F_{\xi\eta}$$
$$= u_y(g^{3\xi}g^{4\eta}F_{\xi\eta} + g^{3\eta}g^{4\xi}F_{\eta\xi}) + u_z(g^{4\xi}g^{2\eta}F_{\xi\eta} + g^{4\eta}g^{2\xi}F_{\eta\xi})$$
$$+c(g^{2\xi}g^{3\eta}F_{\xi\eta} + g^{2\eta}g^{3\xi}F_{\eta\xi})$$
$$= 2[-u_z\beta_y F_{12} + u_y\beta_y F_{13} - u_y\beta_x F_{23} - u_x F_{34} + u_z F_{24} + c\beta_x\beta_y F_{13}$$
$$+c(1-\beta_x^2)F_{23} - c\beta_x F_{34}]$$
$$= 2[-u_z\beta_y\beta_z + (u_y + c\beta_x)\beta_y(-\beta_y) + (c - c\beta_x^2 - u_y\beta_x)B_x - (u_y + c\beta_x)E_z + u_z E_y]$$
$$= 2[(c - c\beta_x^2 - u_y\beta_x)B_x - (u_y + c\beta_x)\beta_y B_y - u_z\beta_y B_z - (u_y E_z - u_z E_y) - c\beta_x E_z]$$
$$= -2[c\beta_x - (u\times E)_x - (u\times E)_x]e,$$

where $v = (\Omega \times r)$. The right hand side of this equation is

$$\begin{aligned} &= \mu[e_{1234}\mathcal{H}^{23}u^4 + e_{1243}\mathcal{H}^{24}v^3 + e_{1324}\mathcal{H}^{32}u^4 + e_{1342}\mathcal{H}^{34}u^2 + e_{1423}\mathcal{H}^{42}u^3 + e_{1432}\mathcal{H}^{43}u^2] \\ &= \mu[u_y(\mathcal{H}^{34} - \mathcal{H}^{43}) + u_z(\mathcal{H}^{42} - \mathcal{H}^{24}) + c(\mathcal{H}^{23} - \mathcal{H}^{32})] \\ &= 2\mu[-u_y D_z + u_z D_y + cH_x] = 2\mu[cH_x - (u \times D)_x]. \end{aligned} \tag{14.16.113}$$

Hence, approximately,

$$B - \frac{1}{c}(u + v) \times E = \mu\left[H - \frac{1}{c}(u \times D)\right]$$

$$\text{or,}\quad B = \mu H - \frac{1}{c}(\varepsilon\mu - 1)(u \times E) + \frac{1}{c}(v \times E). \tag{14.16.114}$$

The results of Heer (1964) and Khromykh (1966) coincide when $\varepsilon = \mu = 1$ with those of Post and Yildiz (1965), Yildiz and Tang (1965) and Post (1967).

14.16.4 Theory of Post, Yildiz and Tang (PYT)

For first order effects Galilean rotations under absolute time seems alright. The constitutive tensor for an arbitrarily moving non-conducting material medium, with respect to an accelerated frame of reference, can be regarded as the sum of a free space part $\chi_{(0)}$ and a material part $\chi_{(m)}$ which is solely determined by the electric and magnetic polarization properties of the material medium :

$$\chi^{\alpha\beta\gamma\delta} = \chi_{(0)}^{\alpha\beta\gamma\delta} + \chi_{(m)}^{\alpha\beta\gamma\delta}. \tag{14.16.115}$$

This separation gives rise to the following two possibilities : (1) The material medium is at rest in the rotating frame; only $\chi_{(0)}$ is affected by the rotation. (2) The material medium is at rest in the inertial laboratory system; so $\chi_{(0)}$ and $\chi_{(m)}$ are both affected by the rotation. The free-space part $\chi_{(0)}$ in a cylindrical co-ordinate system (r, θ, z, t) is rotating with an angular velocity Ω around the z axis. For methodological reasons, Heer's (1964) theory does not explain the various experimental results. In the covariant 4-space formulation of the classical EM field theory, the concept of the constitutive relations, proposed first by Van Dantzig (1934), have been discussed much by Post and Yildiz (1965), Yildiz and Tang (1966) and Post (1967). The covariant formulation of the classical EM field, in linear media has been studied by Post (1962). Also the principle of covariance is implemented such that the constitutive relations remain separated from the field equations.

Compared with customary presentations by Schiff (1939) and others [Sommerfeld (1952, p. 362, 363)], the analysis of Post, Yildiz and Tang provides a minor change in the transformation of the field tensor that constitutes the spatial D and H field vectors, which enables a metric independent general covariant formulation of the equations

$$\frac{\partial F_{\lambda\mu}}{\partial x^{\mu}} + \frac{\partial F_{\mu\nu}}{\partial x^{\lambda}} + \frac{\partial F_{\nu\lambda}}{\partial x^{\mu}} = 0. \tag{14.16.116}$$

$$\frac{\partial \mathcal{H}^{\mu\nu}}{\partial x^{\nu}} = -\frac{4\pi}{c}j^{\mu}, \tag{14.16.117}$$

which follow from (14.16.2) and (14.16.1) for inertial frames. This formulation implies that the derivatives in these equations are ordinary, not covariant derivatives. The field tensor of D and

H is normally treated as an ordinary tensor; however, for natural invariance, this tensor should be treated as a tensor density of weight + 1. This method resolves the Schiff (1939) charge and current of rotation, Sommerfeld (1952, p. 362–63). In an inertial frame $F_{\mu\gamma}$ and $\mathcal{H}^{\mu\gamma}$ are the skew symmetric tensors of Minkowski representing (E, B) and (D, H) respectively. The form of (14.16.116) and (14.16.117) is useful for any general, non-linear, but holonomic co-ordinate transformation, according to the rule

$$\mathcal{H}^{\mu'\nu'} = |\Delta|^{-1}\alpha_{\alpha}^{\mu'}\alpha_{\beta}^{\nu'}\mathcal{H}^{\alpha\beta} \tag{14.16.118}$$

because $\mathcal{H}^{\mu\nu}$ is regarded as a tensor density of weight + 1, where

$$\Delta = \det \alpha_{\alpha}^{\mu'}. \tag{14.16.119}$$

The condition of holonomicity is

$$\delta_{[\alpha}\alpha_{\beta]}^{\gamma'} = 0, \tag{14.16.120}$$

which, on expansion, gives

$$\frac{\partial \alpha_{\beta}^{\gamma'}}{\partial x^{\alpha}} + \frac{\partial \alpha_{\alpha}^{\beta'}}{\partial x^{\gamma}} + \frac{\partial \alpha_{\gamma}^{\alpha'}}{\partial x^{\beta}} = 0. \tag{14.16.121}$$

The linear constitutive relation in covariant form is

$$\mathcal{H}^{\lambda\mu} = \frac{1}{2}\chi^{\lambda\mu\nu\eta}F_{\nu\eta}, \tag{14.16.122}$$

where $\chi^{\lambda\mu\nu\eta}$ is the contravariant constitutive tensor of 4th order. This relation explains the EM behaviour of a linear medium for an instantaneous and local interaction between the fields. So, the frame of reference is not necessarily inertial and the medium may be in arbitrary motion with respect to that frame. The non-zero components of $\chi^{\lambda\mu\nu\eta}$, for an isotropic medium at rest in an inertial frame, yield the well-known *vector relations*

$$B = \mu H, \quad D = \varepsilon E. \tag{14.16.123}$$

The constitutive tensor $\chi^{\alpha\beta\gamma\delta}$ obeys the symmetry relations

$$\chi^{\alpha\beta\gamma\delta} = -\chi^{\beta\alpha\gamma\delta} = -\chi^{\alpha\beta\delta\gamma} = \chi^{\gamma\delta\alpha\beta}. \tag{14.16.124}$$

$$\chi^{[\alpha\beta\gamma\delta]} = 0. \tag{14.16.125}$$

It transforms like a contravariant tensor density of weight + 1 :

$$\chi^{\lambda'\mu'\nu'\eta'} = |\Delta|^{-1}\alpha_{\alpha}^{\lambda'}\alpha_{\beta}^{\mu'}\alpha_{\gamma}^{\nu'}\delta_{\delta}^{\eta'}\chi^{\alpha\beta\gamma\delta}. \tag{14.16.126}$$

In an inertial frame in which an isotropic medium is at rest, a comparison with the customary constitutive relations (14.16.123) gives the non-zero elements of $\chi^{\alpha\beta\gamma\delta}$

14.17 Transformation Relations of Post, Yildiz and Tang in Inertial frames

The general Lorentz transformation from the frame K' moving relative to K with velocity $\mathbf{v}$ to the k frame is

$$x' = \left\{1 + (\gamma - 1)\frac{v_x^2}{v^2}\right\}x + (\gamma - 1)\frac{v_x v_y}{v^2}y + (\gamma - 1)\frac{v_x v_z}{v^2}z - \gamma v_x t, \tag{14.17.1}$$

$$y' = (\gamma - 1)\frac{v_y v_x}{v^2}x + \left\{1 + (\gamma - 1)\frac{v_y^2}{v^2}\right\}y + (\gamma - 1)\frac{v_y v_z}{v^2}z - \gamma v_y t \tag{14.17.2}$$

$$z' = (\gamma - 1)\frac{v_z v_x}{v^2}x + (\gamma - 1)\frac{v_z v_y}{v^2}z + \left\{1 + (\gamma - 1)\frac{v_z^2}{v^2}\right\}z - \gamma v_z t \tag{14.17.3}$$

$$t' = -\frac{\gamma v_x}{c^2}x - \frac{\gamma u_y}{c^2}y - \frac{\gamma u_z}{c^2}z + \gamma t. \tag{14.17.4}$$

The corresponding LT from the frame K to the frame K' is obtained by changing the sign of the components of $\mathbf{v}$. The matrix of the transformation coefficients $\alpha_\mu^{\lambda'}$ obtained from (14.17.1)–(14.17.4) is

$$\alpha_\mu^{\lambda'} = \begin{bmatrix} 1 + (\gamma - 1)\frac{v_x^2}{v^2} & (\gamma - 1)\frac{v_x v_y}{v^2} & (\gamma - 1)\frac{v_x v_z}{v^2} & -\gamma v_x \\ (\gamma - 1)\frac{v_y v_x}{v^2} & 1 + (\gamma - 1)\frac{v_y^2}{v^2} & (\gamma - 1)\frac{v_y v_z}{v^2} & -\gamma v_y \\ (\gamma - 1)\frac{v_z v_x}{v^2} & (\gamma - 1)\frac{v_z v_y}{v^2} & 1 + (\gamma - 1)\frac{v_z^2}{v^2} & -\gamma v_z \\ -\gamma\frac{v_x}{c^2} & -\gamma\frac{v_y}{c^2} & -\gamma\frac{v_z}{c^2} & \gamma \end{bmatrix}. \tag{14.17.5}$$

For first order of the components of v, $\gamma = 1$ and this matrix reduces to

$$\alpha_\mu^{\lambda'} = \begin{bmatrix} 1 & 0 & 0 & -\frac{v_x}{c} \\ 0 & 1 & 0 & -\frac{v_y}{c} \\ 0 & 0 & 1 & -\frac{v_z}{c} \\ -\frac{v_x}{c} & -\frac{v_y}{c} & -\frac{v_z}{c} & 1 \end{bmatrix}. \tag{14.17.6}$$

The reverse transformation obviously is

$$\alpha_{\mu'}^{\lambda} = \begin{bmatrix} 1 & 0 & 0 & \frac{v_x}{c} \\ 0 & 1 & 0 & \frac{v_y}{c} \\ 0 & 0 & 1 & \frac{v_z}{c} \\ \frac{v_x}{c} & \frac{v_y}{c} & \frac{v_z}{c} & 1 \end{bmatrix}. \tag{14.17.7}$$

The transformation relations for the field tensors $F_{\mu\nu}$ and $\mathcal{H}^{\mu\nu}$ are

$$F_{\mu'\nu'} = \alpha_{\mu'}^{\xi}\alpha_{\nu'}^{\eta}F_{\xi\eta}, \tag{14.17.8}$$

$$\mathcal{H}^{\mu'\nu'} = |\Delta|^{-1}\alpha_\xi^{\mu'}\alpha_\eta^{\nu'}\mathcal{H}^{\xi\eta}, \tag{14.17.9}$$

where

$$F_{\mu\nu} = \begin{bmatrix} 0 & B_3 & -B_2 & E_1 \\ -B_3 & 0 & B_1 & E_2 \\ B_2 & -B_1 & 0 & E_3 \\ -E_1 & -E_2 & -E_3 & 0 \end{bmatrix}. \tag{14.17.10}$$

$$\mathcal{H}^{\mu\nu} = \begin{bmatrix} 0 & H_3 & -H_2 & -D_1 \\ -H_3 & 0 & H_1 & -D_2 \\ H_2 & -H_1 & 0 & -D_3 \\ D_1 & D_2 & D_3 & 0 \end{bmatrix}. \tag{14.17.11}$$

Therefore,

$$\begin{aligned} F'_{12} &= B'_3 = \alpha^{\xi}_{1'}\alpha^{\eta}_{2'}F_{\xi\eta} = \alpha^{1}_{1'}\alpha^{1}_{2'}F_{12} + \alpha^{4}_{1'}\alpha^{2}_{2'}F_{42} + \alpha^{1}_{1'}\alpha^{4}_{2'}F_{14} \\ &= B_3 + \frac{v_x}{c}(-E_2) + \frac{v_y}{c}E_1. \end{aligned} \tag{14.17.12}$$

Hence, evidently

$$B' = B - \frac{1}{c}(v \times E) = B - (\beta \times E). \tag{14.17.13}$$

$$\begin{aligned} F'_{14} &= \alpha^{1}_{1'}\alpha^{2}_{4'}F_{12} + \alpha^{1}_{1'}\alpha^{3}_{4'}F_{13} + \alpha^{1}_{1'}\alpha^{4}_{1'}F_{14} + \alpha^{4}_{1'}\alpha^{1}_{4'}F_{41} + \alpha^{4}_{1'}\alpha^{2}_{4'}F_{42} + \alpha^{4}_{1'}\alpha^{3}_{4'}F_{43} \\ &= \frac{v_y}{c}B_3 + \frac{v_z}{c}(-B_2) + E_1 + \frac{v_x}{c}\frac{v_x}{c}(-E_1) + \frac{v_x}{c}v_y(-E_2) + \frac{v_x}{c}\frac{v_z}{c}(-E_3) = E'_1, \end{aligned}$$

$$E' = E + (\beta \times B) - \beta(\beta \cdot E),$$

which, for first order of velocities, is

$$E' = E + (\beta \times B), \tag{14.17.14}$$

$$H'^{12} = H'_3 = |\Delta|^{-1}\alpha^{1'}_{\xi}\,\alpha^{2'}_{\eta}\,\mathcal{H}^{\xi\eta},$$

where

$$\Delta = \det \alpha^{\mu'}_{\nu} = \begin{vmatrix} 1 & 0 & 0 & -\dfrac{v_x}{c} \\ 0 & 1 & 0 & -\dfrac{v_y}{c} \\ 0 & 0 & 1 & -\dfrac{v_z}{c} \\ -\dfrac{v_x}{c} & -\dfrac{v_y}{c} & -\dfrac{v_z}{c} & 1 \end{vmatrix} = 1 - \frac{v_z^2}{c^2} - \frac{v_y^2}{c^2} - \frac{v_x^2}{c^2} = 1 - \beta^2 \approx 1.$$

Hence

$$\begin{aligned} H'_3 &= \alpha^{1'}_{1}\,\alpha^{2'}_{2}\,\mathcal{H}^{12} + \alpha^{1'}_{1}\,\alpha^{2'}_{4}\,\mathcal{H}^{14} + \alpha^{1'}_{4}\,\alpha^{2'}_{1}\,\mathcal{H}^{41} + \alpha^{1'}_{4}\,\alpha^{2'}_{2}\,\mathcal{H}^{42} \\ &= \mathcal{H}^{12} - \frac{v_y}{c}\mathcal{H}^{14} - \frac{v_x}{c}\cdot 0 - \frac{v_x}{c}\mathcal{H}^{42} = H_3 - \frac{v_y}{c}(-D_1) - \frac{v_x}{c}D_2. \end{aligned}$$

The corresponding vector relation is

$$H' = H - (\beta \times D). \tag{14.17.15}$$

And since

$$\begin{aligned}
\mathcal{H}'^{14} &= -D_1' = \alpha_\xi^{1'}\alpha_\eta^{4'}\mathcal{H}^{\xi\eta} \\
&= \alpha_1^{1'}\alpha_2^{4'}\mathcal{H}^{12} + \alpha_1^{1'}\alpha_3^{4'}\mathcal{H}^{13} + \alpha_1^{1'}\alpha_4^{4'}\mathcal{H}^{14} + \alpha_4^{1'}\alpha_1^{4'}\mathcal{H}^{41} + \alpha_4^{1'}\alpha_2^{4'}\mathcal{H}^{42} + \alpha_4^{1'}\alpha_3^{4'}\mathcal{H}^{43} \\
&= -\frac{v_y}{c}H_3 + \left(-\frac{v_z}{c}\right)(-H_2) - D_1 + \frac{v_x}{c}\left(-\frac{v_x}{c}\right)D_1 + \frac{v_x}{c}\left(-\frac{v_y}{c}\right)(D_2) + \frac{v_x}{c}\left(-\frac{v_z}{c}\right)D_3 \\
&= -D_x + (\beta_z H_y - \beta_y H_z) - \beta_x(\beta\cdot D),
\end{aligned} \tag{14.17.16}$$

vectorially, up to first order of β,

$$D' = D + (\beta \times H). \tag{14.17.17}$$

14.18 Constitutive Tensor of Post, Yildiz and Tang for Arbitrary Reference Frames

For arbitrary reference frames we use the table of relation (6.21) of Post (1962); then, using, moreover, his table 2B of p. 55, we find that

$$\begin{aligned}
\mathcal{H}^{41} &= D_1 = \varepsilon_{11}E_1 + \varepsilon_{12}E_2 + \varepsilon_{13}E_3 + \gamma_{11}B_1 + \gamma_{12}B_2 + \gamma_{13}B_3 \\
&= -\varepsilon_{11}F_{41} - \varepsilon_{12}F_{42} - \varepsilon_{13}F_{43} + \gamma_{11}F_{23} + \gamma_{12}F_{31} + \gamma_{13}F_{12} = \frac{1}{2}\chi^{41\gamma\delta}F_{\gamma\delta} \\
&= \chi^{4112}F_{12} + \chi^{4113}F_{13} + \chi^{4123}F_{23} + \chi^{4141}F_{41} + \chi^{4142}F_{42} + \chi^{4143}F_{43}.
\end{aligned}$$

Therefore,

$$\begin{aligned}
\chi^{4212} &= \gamma_{23}, \qquad \chi^{4213} = -\gamma_{22}, \qquad \chi^{4223} = \gamma_{21}, \\
\chi^{4241} &= -\varepsilon_{21}, \qquad \chi^{4242} = -\varepsilon_{22}, \qquad \chi^{4243} = -\varepsilon_{23},
\end{aligned}$$

$$\begin{aligned}
\mathcal{H}^{43} &= D_3 = \varepsilon_{31}E_1 + \varepsilon_{32}E_2 + \varepsilon_{33}E_3 + \gamma_{31}B_1 + \gamma_{32}B_2 + \gamma_{33}B_3 \\
&= -\varepsilon_{31}F_{41} - \varepsilon_{32}F_{42} - \varepsilon_{33}F_{43} + \varepsilon_{31}F_{23} + \varepsilon_{32}F_{31} + \varepsilon_{33}F_{12} \\
&= \frac{1}{2}\chi^{43\gamma\delta}F_{\gamma\delta} = \chi^{4312}F_{12} + \chi^{4313}F_{13} + \chi^{4323}F_{23} + \chi^{4341}F_{41} + \chi^{4342}F_{42} + \chi^{4343}F_{43}.
\end{aligned}$$

Therefore,

$$\begin{aligned}
\chi^{4312} &= \gamma_{33}, \qquad \chi^{4313} = -\gamma_{32}, \qquad \chi^{4323} = \gamma_{31}, \\
\chi^{4341} &= -\gamma_{31}, \qquad \chi^{4342} = -\gamma_{32}, \qquad \chi^{4343} = -\gamma_{33},
\end{aligned}$$

$$\begin{aligned}
\mathcal{H}^{23} &= H_1 = -\bar{\gamma}_{11}E_1 - \bar{\gamma}_{12}E_2 - \bar{\gamma}_{13}E_3 + \chi_{11}B_1 + \chi_{12}B_2 + \chi_{13}B_3 \\
&= \bar{\gamma}_{11}F_{41} + \bar{\gamma}_{13}F_{42} + \bar{\gamma}_{13}F_{43} + \chi_{11}F_{23} + \chi_{12}F_{31} + \chi_{13}F_{12} \\
&= \frac{1}{2}\chi^{23\gamma\delta}F_{\gamma\delta} = \chi^{2312}F_{12} + \chi^{2313}F_{13} + \chi^{2323}F_{23} + \chi^{2341}F_{41} + \chi^{2342}F_{42} + \chi^{2343}F_{43}.
\end{aligned}$$

Therefore,

$$\chi^{2312} = \chi_{13}, \quad \chi^{2313} = -\chi_{12}, \quad \chi^{2323} = \chi_{11},$$
$$\chi^{2341} = \bar{\gamma}_{11}, \quad \chi^{2342} = \bar{\gamma}_{12}, \quad \chi^{2343} = \bar{\gamma}_{13},$$

$$\begin{aligned}\mathcal{H}^{31} &= H_2 = -\bar{\gamma}_{21}E_1 - \bar{\gamma}_{22}E_2 - \bar{\gamma}_{23}E_3 + \chi_{21}B_1 + \chi_{22}B_2 + \chi_{23}B_3 \\ &= \frac{1}{2}\chi^{31\gamma\delta}F_{\gamma\delta} = \chi^{3112}F_{12} + \chi^{3113}F_{13} + \chi^{3123}F_{23} + \chi^{3141}F_{41} + \chi^{3142}F_{42} + \chi^{3143}F_{43} \\ &= \bar{\gamma}_{21}F_{41} + \bar{\gamma}_{22}F_{42} + \bar{\gamma}_{23}F_{43} + \chi_{21}F_{23} + \chi_{22}F_{31} + \chi_{23}F_{12}.\end{aligned}$$

Therefore,

$$\chi^{3112} = \chi_{23}, \quad \chi^{3113} = -\chi_{22}, \quad \chi^{3123} = \chi_{21},$$
$$\chi^{3141} = \bar{\gamma}_{21}, \quad \chi^{3142} = \bar{\gamma}_{22}, \quad \chi^{3143} = \bar{\gamma}_{23},$$

$$\begin{aligned}\mathcal{H}^{12} &= H_3 = -\bar{\gamma}_{31}E_1 - \bar{\gamma}_{32}E_2 - \bar{\gamma}_{33}E_3 + \chi_{31}B_1 + \chi_{32}B_2 + \chi_{33}B_3 \\ &= \frac{1}{2}\chi^{12\gamma\delta}F_{\gamma\delta} = \chi^{1212}F_{12} + \chi^{1213}F_{13} + \chi^{1223}F_{23} + \chi^{1241}F_{41} + \chi^{1242}F_{42} + \chi^{1243}F_{43}.\end{aligned}$$

$$\chi^{1212} = \chi_{33}, \quad \chi^{1213} = -\chi_{32}, \quad \chi^{1223} = \chi_{31},$$
$$\chi^{1241} = \bar{\gamma}_{31}, \quad \chi^{1242} = \bar{\gamma}_{32}, \quad \chi^{1243} = \bar{\gamma}_{33}.$$

In empty space we have

$$\varepsilon^0_{11} = \varepsilon^0_{22} = \varepsilon^0_{33} = \varepsilon, \quad \chi^0_{11} = \chi^0_{22} = \chi^0_{33} = \frac{1}{\mu}$$

and all the other elements of ε^0_{ij}, $\chi^0_{ij}\gamma^0_{ij}$, γ_{ij} vanish. For the corresponding 4th order constitutive tensor $\chi^{0\alpha\beta\gamma\delta}$ of empty space, we now find that

$$-\chi^{04141} = -\chi^{04242} = -\chi^{04343} = \varepsilon, \quad \chi^{02323} = -\chi^{03113} = -\chi^{01212} = \frac{1}{\mu}.$$

The other elements vanish. So, evidently we find that

$$\chi^{0\alpha\beta\alpha\beta} \neq 0$$

and $\chi^{0\alpha\beta\gamma\delta} = 0$ when $\gamma \neq \alpha$ or $\delta \neq \beta$ or both. More specifically, $\chi^{0ijij} = \frac{1}{\mu}$, $\chi^{04i4i} = -\varepsilon$, where $(i, j) = (1, 2, 3)$, $i \neq j$.

Since $\chi^{\alpha\beta\gamma\delta}$ is a 4th order tensor density of weight + 1, its transformation from the K^0 frame to the K frame, where the relative motion of K is arbitrary, is given by

$$\chi^{\alpha\beta\gamma\delta} = |\Delta|^{-1}\alpha^{\alpha}_{\xi^0}\alpha^{\beta}_{\eta^0}\alpha^{\gamma}_{\zeta^0}\alpha^{\delta}_{\lambda^0}\chi^{0\xi\eta\gamma\delta},$$

where $\Delta = \det \alpha^{\lambda}_{\mu^0}$.

Consider the transformation relations

$$x^0 = x\cos\Omega t - y\sin\Omega t, \quad y^0 = x\sin\Omega t + y\cos\Omega t, \quad z^0 = z, \quad t^0 = t$$

and the converse relations

$$x = x^0\cos\Omega t + y^0\sin\Omega t^0, \quad y = -x^0\sin\Omega t + y^0\cos\Omega t^0, \quad z = z^0, \quad t = t^0.$$

The treatment here is restricted to only small angular velocities; since, large angular velocities will distort the physical system.

The transformation coefficients $\alpha_{\nu}^{\mu^0}$ and $\alpha_{\nu^0}^{\mu}$ are found to be

$$\alpha^1_{1^0} = \cos\Omega t^0, \quad \alpha^2_{2^0} = \sin\Omega t^0, \quad \alpha^3_{3^0} = 0,$$

$$\alpha^1_{4^0} = \frac{\partial x}{c\partial t^0} = \frac{\Omega}{c}(-x^0\sin\Omega t^0 + y^0\cos\Omega t^0) = \frac{\Omega y}{c},$$

$$\alpha^2_{1^0} = -\sin\Omega t^0, \quad \alpha^2_{2^0} = \cos\Omega t^0, \quad \alpha^2_{3^0} = 0,$$

$$\alpha^2_{4^0} = \frac{\partial y}{c\partial t^0} = -\frac{\Omega}{c}(x^0\cos\Omega t^0 + y^0\sin\Omega t^0) = -\frac{\Omega x}{c},$$

$$\alpha^3_{1^0} = 0, \quad \alpha^3_{2^0} = 0, \quad \alpha^3_{3^0} = 1, \quad \alpha^3_{4^0} = 0,$$

$$\alpha^4_{1^0} = 0, \quad \alpha^4_{2^0} = 0, \quad \alpha^4_{3^0} = 1, \quad \alpha^4_{4^0} = 0,$$

$$\alpha^{1^0}_1 = \cos\Omega t, \quad \alpha^{1^0}_2 = -\sin\Omega t, \quad \alpha^{1^0}_3 = 0,$$

$$\alpha^{1^0}_4 = \frac{\partial x^0}{c\partial t} = -\frac{\Omega}{c}(x\sin\Omega t + y\cos\Omega t) = -\frac{\Omega y^0}{c},$$

$$\alpha^{2^0}_1 = \sin\Omega t, \quad \alpha^{2^0}_2 = \cos\Omega t, \quad \alpha^{2^0}_3 = 0,$$

$$\alpha^{2^0}_4 = \frac{\Omega}{c}(x\cos\Omega t - y\sin\Omega t) = \frac{\Omega x^0}{c}$$

$$\alpha^{3^0}_1 = 0, \quad \alpha^{3^0}_2 = 0, \quad \alpha^{3^0}_3 = 1, \quad \alpha^{3^0}_4 = 0,$$

$$\alpha^{4^0}_1 = 0, \quad \alpha^{4^0}_2 = 0, \quad \alpha^{4^0}_3 = 1, \quad \alpha^{4^0}_4 = 1.$$

$$\Delta = \det\alpha^{\lambda}_{\mu^0} = \begin{vmatrix} \alpha^1_{1^0} & \alpha^1_{2^0} & \alpha^1_{3^0} & \alpha^1_{4^0} \\ \alpha^2_{1^0} & \alpha^2_{2^0} & \alpha^2_{3^0} & \alpha^2_{4^0} \\ \alpha^3_{1^0} & \alpha^3_{2^0} & \alpha^3_{3^0} & \alpha^3_{4^0} \\ \alpha^4_{1^0} & \alpha^4_{2^0} & \alpha^4_{3^0} & \alpha^4_{4^0} \end{vmatrix} = \begin{vmatrix} \cos\Omega t^0 & \sin\Omega t^0 & 0 & \frac{\Omega y}{c} \\ -\sin\Omega t^0 & \cos\Omega t^0 & 0 & -\frac{\Omega x}{c} \\ 0 & 0 & 1 & 0 \\ 0 & 0 & 0 & 1 \end{vmatrix} = 1.$$

$$\begin{aligned}
\chi^{1212} &= \alpha^1_{\xi^0}\alpha^1_{\eta^0}\alpha^1_{\zeta^0}\alpha^1_{\lambda^0}\chi^{0\xi\eta\zeta\lambda} \\
&= \alpha^1_{1^0}\alpha^2_{2^0}\alpha^1_{1^0}\alpha^2_{2^0}\chi^{01212} + \alpha^1_{1^0}\alpha^2_{2^0}\alpha^1_{2^0}\alpha^2_{1^0}\chi^{01221} + \alpha^1_{2^0}\alpha^2_{1^0}\alpha^1_{1^0}\alpha^2_{2^0}\chi^{02112} \\
&\quad +\alpha^1_{2^0}\alpha^2_{1^0}\alpha^1_{2^0}\alpha^2_{1^0}\chi^{02121} + \alpha^1_{1^0}\alpha^2_{3^0}\alpha^1_{1^0}\alpha^2_{3^0}\chi^{01313} + \alpha^1_{1^0}\alpha^2_{3^0}\alpha^1_{3^0}\alpha^2_{1^0}\chi^{01331} \\
&\quad +\alpha^1_{3^0}\alpha^2_{1^0}\alpha^1_{1^0}\alpha^2_{3^0}\chi^{03113} + \alpha^1_{3^0}\alpha^2_{1^0}\alpha^1_{3^0}\alpha^2_{1^0}\chi^{03131} + \alpha^1_{1^0}\alpha^2_{4^0}\alpha^1_{1^0}\alpha^2_{4^0}\chi^{01414} \\
&\quad +\alpha^1_{1^0}\alpha^2_{4^0}\alpha^1_{4^0}\alpha^2_{1^0}\chi^{01441} + \alpha^1_{4^0}\alpha^2_{4^0}\alpha^1_{1^0}\alpha^2_{4^0}\chi^{04114} + \alpha^1_{4^0}\alpha^2_{4^0}\alpha^1_{4^0}\alpha^2_{4^0}\chi^{04141} \\
&\quad +\alpha^1_{2^0}\alpha^2_{3^0}\alpha^1_{2^0}\alpha^2_{3^0}\chi^{02323} + \alpha^1_{2^0}\alpha^2_{3^0}\alpha^1_{3^0}\alpha^2_{2^0}\chi^{02332} + \alpha^1_{3^0}\alpha^2_{2^0}\alpha^1_{2^0}\alpha^2_{3^0}\chi^{03223} \\
&\quad +\alpha^1_{3^0}\alpha^2_{2^0}\alpha^2_{3^0}\alpha^2_{2^0}\chi^{03232} + \alpha^1_{2^0}\alpha^2_{4^0}\alpha^1_{2^0}\alpha^2_{4^0}\chi^{02424} + \alpha^1_{2^0}\alpha^2_{4^0}\alpha^1_{4^0}\alpha^2_{2^0}\chi^{02442} \\
&\quad +\alpha^1_{4^0}\alpha^2_{2^0}\alpha^1_{2^0}\alpha^2_{2^0}\chi^{04224} + \alpha^1_{4^0}\alpha^2_{2^0}\alpha^1_{4^0}\alpha^2_{2^0}\chi^{04242} + \alpha^1_{3^0}\alpha^2_{4^0}\alpha^1_{3^0}\alpha^2_{4^0}\chi^{03434} \\
&\quad +\alpha^1_{3^0}\alpha^2_{4^0}\alpha^1_{4^0}\alpha^2_{3^0}\chi^{03443} + \alpha^1_{4^0}\alpha^2_{3^0}\alpha^1_{3^0}\alpha^2_{4^0}\chi^{04334} + \alpha^1_{4^0}\alpha^2_{3^0}\alpha^1_{4^0}\alpha^2_{3^0}\chi^{04343}
\end{aligned}$$

$$
\begin{aligned}
&= (\alpha^1_{10}\alpha^2_{20}\alpha^1_{10}\alpha^2_{20} - \alpha^1_{10}\alpha^2_{20}\alpha^1_{20}\alpha^2_{10} - \alpha^1_{20}\alpha^2_{10}\alpha^1_{10}\alpha^2_{20} + \alpha^1_{20}\alpha^2_{10}\alpha^1_{20}\alpha^2_{10})\chi^{01212} + 0 \\
&\qquad + (\alpha^1_{10}\alpha^2_{40}\alpha^1_{10}\alpha^2_{40} - \alpha^1_{10}\alpha^2_{40}\alpha^1_{40}\alpha^2_{10} - \alpha^1_{40}\alpha^2_{10}\alpha^1_{10}\alpha^2_{40} + \alpha^1_{40}\alpha^2_{10}\alpha^1_{40}\alpha^2_{10})\chi^{01414} + 0 \\
&\qquad + (\alpha^1_{20}\alpha^1_{40}\alpha^1_{20}\alpha^2_{40} - \alpha^1_{20}\alpha^2_{40}\alpha^1_{40}\alpha^2_{20} - \alpha^1_{40}\alpha^2_{20}\alpha^1_{20}\alpha^2_{40} + \alpha^1_{40}\alpha^2_{20}\alpha^1_{40}\alpha^2_{20})\chi^{02424} + 0 \\
&= (\cos^4\Omega t^0 + 2\cos^2\Omega t^0\sin^2\Omega t^0 + \sin^4\Omega t^0)\chi^{01212} \\
&\qquad + \left[\cos^2\Omega t^0\left(-\frac{\Omega x}{c}\right)^2 - 2\cos\Omega t^0\left(-\frac{\Omega x}{c}\right)\left(\frac{\Omega y}{c}\right)(-\sin\Omega t^0) + \left(\frac{\Omega y}{c}\right)^2(-\sin\Omega t^0)^2\right]\chi^{01414} \\
&\qquad + \left[\sin^2\Omega t^0\left(-\frac{\Omega x}{c}\right)^2 - 2\sin\Omega t^0\left(-\frac{\Omega x}{c}\right)\left(\frac{\Omega y}{c}\right)\cos\Omega t^0 + \left(\frac{\Omega y}{c}\right)^2\cos^2\Omega t^0\right]\chi^{02424} \\
&= \chi^{01212} + \frac{\Omega^2}{c^2}(x^2\cos^2\Omega t^0 - 2\cos\Omega t^0\sin\Omega t^0 xy + y^2\sin^2\Omega t^2)\chi^{01414} \\
&\qquad + \frac{\Omega^2}{c^2}(x^2\sin^2\Omega t^0 + 2\sin\Omega t^0\cos\Omega t^0 xy + y^2\cos^2\Omega t^0)\chi^{02424} \\
&= \chi^{01212} + \frac{\Omega^2}{c^2}(x\cos\Omega t^0 - y\sin\Omega t^0)^2\chi^{01414} + \frac{\Omega^2}{c^2}(x\sin\Omega t^0 + y\cos\Omega t^0)\chi^{02424} \\
&= \chi^{01212} + \frac{x^{0^2}\Omega^2}{c^2}\chi^{01414} + \frac{y^{0^2}\Omega^2}{c^2}\chi^{02424} \\
&= \frac{1}{\mu} - \varepsilon\frac{\Omega^2}{c^2}(x^{0^2} + y^{0^2}) = \frac{1}{\mu} - \varepsilon\left(r^0\frac{\Omega}{c}\right)^2.
\end{aligned}
$$

$$
\begin{aligned}
\chi^{1213} &= \alpha^1_{\xi 0}\alpha^2_{\eta 0}\alpha^1_{\zeta 0}\alpha^3_{\lambda 0}\chi^{0\xi\eta\zeta\lambda} = \alpha^1_{\xi 0}\alpha^2_{\eta 0}\alpha^1_{\zeta 0}\alpha^3_{30}\chi^{0\xi\eta\zeta 3} = 0 \text{ because } \alpha^1_{30} = 0, \alpha^2_{30} = 0 \\
\chi^{1214} &= \alpha^1_{\xi 0}\alpha^2_{\eta 0}\alpha^1_{\zeta 0}\alpha^4_{\lambda 0}\chi^{0\xi\eta\zeta\lambda} = \alpha^1_{\xi 0}\alpha^2_{\eta 0}\alpha^1_{\zeta 0}\alpha^4_{40}\chi^{0\xi\eta\zeta 4} \\
&= \alpha^1_{20}\alpha^2_{40}\alpha^1_{20}\alpha^4_{40}\chi^{02424} + \alpha^1_{40}\alpha^2_{20}\alpha^1_{20}\alpha^4_{40}\chi^{04224} \\
&\qquad + \alpha^1_{10}\alpha^2_{40}\alpha^1_{10}\alpha^4_{40}\chi^{01414} + \alpha^1_{40}\alpha^2_{10}\alpha^1_{10}\alpha^4_{40}\chi^{04114} + 0 \\
&= (\alpha^1_{20}\alpha^2_{40}\alpha^1_{20}\alpha^4_{40} - \alpha^1_{40}\alpha^2_{20}\alpha^1_{20}\alpha^4_{40})\chi^{02424} + (\alpha^1_{10}\alpha^2_{40}\alpha^1_{10}\alpha^4_{40} - \alpha^1_{40}\alpha^2_{10}\alpha^1_{10}\alpha^4_{40})\chi^{01414} \\
&= \left[\sin^2\Omega t^0\left(-\frac{\Omega x}{c}\right) - \left(\frac{\Omega y}{c}\right)\cos\Omega t^0\sin\Omega t^0\right]\chi^{02424} \\
&\qquad + \left[\cos^2\Omega t^0\left(-\frac{\Omega x}{c}\right) - \left(\frac{\Omega y}{c}\right)(-\sin\Omega t^0)\cos\Omega t^0\right]\chi^{01414} \\
&= -\frac{\Omega}{c}\sin\Omega t^0(x\sin\Omega t^0 + y\cos\Omega t^0)\chi^{02424} - \frac{\Omega}{c}\cos\Omega t^0(x\cos\Omega t^0 - y\sin\Omega t^0)\chi^{01414} \\
&= -\frac{y^0\Omega}{c}\sin\Omega t^0\chi^{02424} - \frac{x^0\Omega}{c}\cos\Omega t^0\chi^{01414} \\
&= -\frac{y^0\Omega}{c}\sin\Omega t^0(-\varepsilon) - \frac{x^0\Omega}{c}\cos\Omega t^0(-\varepsilon) \\
&= \varepsilon\frac{\Omega}{c}(x^0\cos\Omega t^0 + y^0\sin\Omega t^0) = \varepsilon\frac{x\Omega}{c}.
\end{aligned}
$$

$$\chi^{1223} = 0$$

$$\begin{aligned}
\chi^{1224} &= \alpha^1_{\xi^0}\alpha^2_{\eta^0}\alpha^2_{\zeta^0}\alpha^4_{\lambda^0}\chi^{0\xi\eta\zeta\lambda} = \alpha^1_{\xi^0}\alpha^2_{\eta^0}\alpha^2_{\zeta^0}\alpha^4_{4^0}\chi^{0\xi\eta\zeta 4}\\
&= (\alpha^1_{2^0}\alpha^2_{4^0}\alpha^2_{2^0}\alpha^4_{4^0} - \alpha^1_{4^0}\alpha^2_{2^0}\alpha^2_{2^0}\alpha^4_{4^0})\chi^{02424} + (\alpha^1_{1^0}\alpha^2_{4^0}\alpha^2_{1^0}\alpha^4_{4^0} - \alpha^1_{4^0}\alpha^2_{1^0}\alpha^2_{1^0}\alpha^4_{4^0})\chi^{01414}\\
&= \left[\sin\Omega t^0\cdot\cos\Omega t^0\left(-\frac{\Omega x}{c}\right) - \frac{\Omega y}{c}\cos^2\Omega t^0\right](-\varepsilon)\\
&\qquad + \left[-\cos\Omega t^0\cdot\sin\Omega t^0\left(-\frac{\Omega x}{c}\right) - \frac{\Omega y}{c}\sin^2\Omega t^0\right](-\varepsilon)\\
&= \varepsilon\frac{\Omega}{c}\cos\Omega t^0(x\sin\Omega t^0 + y\cos\Omega t^0) - \varepsilon\frac{\Omega}{c}\sin\Omega t^0(x\cos\Omega t^0 - y\sin\Omega t^0)\\
&= \varepsilon\frac{y^0\Omega}{c}\cos\Omega t^0 - \varepsilon\frac{x^0\Omega}{c}\sin\Omega t^0 = -\varepsilon\frac{\Omega}{c}(x^0\sin\Omega t^0 - y^0\cos\Omega t^0) = \varepsilon\frac{y\Omega}{c}.
\end{aligned}$$

$$\begin{aligned}
\chi^{1414} &= \alpha^1_{\xi^0}\alpha^1_{\eta^0}\alpha^1_{\zeta^0}\alpha^4_{\lambda^0}\chi^{0\xi\eta\zeta\lambda} = \alpha^1_{\xi^0}\alpha^4_{4^0}\alpha^1_{\zeta^0}\alpha^4_{4^0}\chi^{0\xi 4\zeta 4}\\
&= \alpha^1_{1^0}\alpha^4_{4^0}\alpha^1_{1^0}\alpha^4_{4^0}\chi^{01414} + \alpha^1_{2^0}\alpha^4_{4^0}\alpha^1_{2^0}\alpha^4_{4^0}\chi^{02424} + 0\\
&= -\varepsilon[\cos^2\Omega t^0 + \sin^2\Omega t^0] = -\varepsilon.
\end{aligned}$$

$$\begin{aligned}
\chi^{3423} &= \alpha^3_{\xi^0}\alpha^4_{\eta^0}\alpha^2_{\zeta^0}\alpha^3_{\lambda^0}\chi^{0\xi\eta\zeta\lambda} = \alpha^3_{3^0}\alpha^4_{4^0}\alpha^2_{\zeta^0}\alpha^3_{3^0}\chi^{034\zeta 3}\\
&= \alpha^3_{3^0}\alpha^4_{4^0}\alpha^2_{4^0}\alpha^3_{3^0}\chi^{03443} = -\frac{\Omega x}{c}\varepsilon
\end{aligned}$$

$$\begin{aligned}
\chi^{3413} &= \alpha^3_{\xi^0}\alpha^4_{\eta^0}\alpha^1_{\zeta^0}\alpha^3_{\lambda^0}\chi^{0\xi\eta\zeta\lambda} = \alpha^3_{3^0}\alpha^4_{4^0}\alpha^1_{\zeta^0}\alpha^3_{3^0}\chi^{034\zeta 3}\\
&= \alpha^3_{3^0}\alpha^4_{4^0}\alpha^1_{4^0}\alpha^3_{3^0}\chi^{03443} = \frac{\Omega y}{c}\varepsilon
\end{aligned}$$

$$\chi^{3412} = \alpha^3_{\xi^0}\alpha^4_{\eta^0}\alpha^1_{\zeta^0}\alpha^2_{\lambda^0}\chi^{0\xi\eta\zeta\lambda} = \alpha^3_{3^0}\alpha^4_{4^0}\alpha^1_{\zeta^0}\alpha^2_{\lambda^0}\chi^{034\zeta\lambda} = 0$$

$$\chi^{1324} = 0, \quad \chi^{1413} = 0, \quad \chi^{1423} = 0, \quad \chi^{1434} = 0,$$

$$\chi^{1231} = 0, \quad \chi^{1232} = 0, \quad \chi^{1234} = 0, \quad \chi^{1312} = 0, \quad \chi^{1314} = 0,$$

$$\begin{aligned}
\chi^{1313} &= \alpha^1_{\xi^0}\alpha^3_{\eta^0}\alpha^1_{\zeta^0}\alpha^3_{\lambda^0}\chi^{0\xi\eta\zeta\lambda} = \alpha^1_{\xi^0}\alpha^3_{3^0}\alpha^1_{\zeta^0}\alpha^3_{3^0}\chi^{0\xi 3\zeta 3}\\
&= \alpha^1_{1^0}\alpha^3_{3^0}\alpha^1_{1^0}\alpha^3_{3^0}\chi^{01313} = \alpha^1_{2^0}\alpha^3_{3^0}\alpha^1_{2^0}\alpha^3_{3^0}\chi^{02323} + \alpha^1_{4^0}\alpha^3_{3^0}\alpha^1_{4^0}\alpha^3_{3^0}\chi^{04343}\\
&= \cos^2\Omega t^2\frac{1}{\mu} + \sin^2\Omega t^0\frac{1}{\mu} + \frac{\Omega^2 y^2}{c^2}(-\varepsilon) = \frac{1}{\mu} - \varepsilon\frac{\Omega^2 y^2}{c^2}
\end{aligned}$$

$$\begin{aligned}
\chi^{1323} &= \alpha^1_{\xi^0}\alpha^3_{\eta^0}\alpha^2_{\zeta^0}\alpha^3_{\lambda^0}\chi^{0\xi\eta\zeta\lambda} = \alpha^1_{\xi^0}\alpha^3_{3^0}\alpha^2_{\zeta^0}\alpha^3_{3^0}\chi^{0\xi 3\zeta 3}\\
&= \alpha^1_{1^0}\alpha^3_{3^0}\alpha^2_{1^0}\alpha^3_{3^0}\frac{1}{\mu} + \alpha^1_{2^0}\alpha^3_{3^0}\alpha^2_{2^0}\alpha^3_{3^0}\frac{1}{\mu} + \alpha^1_{4^0}\alpha^3_{3^0}\alpha^2_{4^0}\alpha^3_{3^0}(-\varepsilon)\\
&= \frac{1}{\mu}[\cos\Omega t^0(-\sin\Omega t^0) + \sin\Omega t^0\cdot\cos\Omega t^0] - \varepsilon\frac{\Omega y}{c}\left(-\frac{\Omega x}{c}\right) = 0 + \varepsilon\frac{\Omega^2 xy}{c^2}
\end{aligned}$$

$$\chi^{1334} = \alpha^1_{\xi 0}\alpha^3_{\eta 0}\alpha^3_{\zeta 0}\alpha^4_{\lambda 0}\chi^{0\xi\eta\zeta\lambda} = \alpha^1_{\xi 0}\alpha^3_{30}\alpha^3_{30}\alpha^4_{40}\chi^{0\xi 334}$$

$$= \alpha^1_{40}\alpha^3_{30}\alpha^3_{30}\alpha^4_{40}\chi^{04334} = \varepsilon\frac{\Omega y}{c}$$

$$\chi^{1412} = \alpha^1_{\xi 0}\alpha^4_{\eta 0}\alpha^1_{\zeta 0}\alpha^2_{\lambda 0}\chi^{0\xi\eta\zeta\lambda} = \alpha^1_{\xi 0}\alpha^4_{40}\alpha^1_{\zeta 0}\alpha^2_{\lambda 0}\chi^{0\xi 4\zeta\lambda}$$

$$= \alpha^1_{10}\alpha^4_{40}\alpha^1_{10}\alpha^2_{40}\chi^{01414} + \alpha^1_{10}\alpha^4_{40}\alpha^1_{40}\alpha^2_{10}\chi^{01441}$$

$$+(\alpha^1_{20}\alpha^4_{40}\alpha^1_{20}\alpha^2_{40} - \alpha^1_{20}\alpha^4_{40}\alpha^1_{40}\alpha^2_{20})\chi^{02424} + 0$$

$$= \left[\cos^2\Omega t^0\left(-\frac{\Omega x}{c}\right) - \cos\Omega t^0\frac{\Omega y}{c}(-\sin\Omega t^0)\right](-\varepsilon)$$

$$+\left[\sin^2\Omega t^0\left(-\frac{\Omega x}{c}\right) - \sin\Omega t^0\cdot\frac{\Omega y}{c}\cdot\cos\Omega t^0\right](-\varepsilon)$$

$$= \frac{\varepsilon\Omega}{c}\cos\Omega t^0(x\cos\Omega t^0 - y\sin\Omega t^0) + \frac{\varepsilon\Omega}{c}\sin\Omega t^0(x\sin\Omega t^0 + y\cos\Omega t^0)$$

$$= \frac{\varepsilon\Omega}{c}x^0\cos\Omega t^0 + \frac{\varepsilon\Omega}{c}y^0\sin\Omega t^0 = \varepsilon\frac{\Omega x}{c}$$

$$\chi^{1424} = \alpha^1_{\xi 0}\alpha^4_{\eta 0}\alpha^2_{\zeta 0}\alpha^4_{\lambda 0}\chi^{0\xi\eta\zeta\lambda} = \alpha^1_{\xi 0}\alpha^4_{40}\alpha^2_{\zeta 0}\alpha^4_{40}\chi^{0\xi 4\zeta 4}$$

$$= \alpha^1_{10}\alpha^4_{40}\alpha^2_{10}\alpha^4_{40}\chi^{01414} + \alpha^1_{20}\alpha^4_{40}\alpha^2_{20}\alpha^4_{40}\chi^{01424} + 0$$

$$+\alpha^1_{20}\alpha^4_{40}\alpha^2_{10}\alpha^4_{40}\chi^{02414} + \alpha^1_{20}\alpha^4_{40}\alpha^2_{20}\alpha^4_{40}\chi^{02424}$$

$$= \cos\Omega t^0(-\sin\Omega t^0)(-\varepsilon) + 0 + 0 + \sin\Omega t^0\cos\Omega t^0(-\varepsilon) = 0.$$

$$\chi^{2323} = \alpha^2_{\xi 0}\alpha^3_{\eta 0}\alpha^2_{\zeta 0}\alpha^3_{\lambda 0}\chi^{0\xi\eta\zeta\lambda} = \alpha^2_{\xi 0}\alpha^3_{30}\alpha^2_{\zeta 0}\alpha^3_{30}\chi^{0\xi 3\zeta 3}$$

$$= \alpha^2_{10}\alpha^3_{30}\alpha^2_{10}\alpha^3_{30}\chi^{01313} + \alpha^2_{20}\alpha^3_{30}\alpha^2_{20}\alpha^3_{30}\chi^{02323} + \alpha^2_{40}\alpha^3_{30}\alpha^2_{40}\alpha^3_{30}\chi^{04343}$$

$$= \sin^2\Omega t^0\frac{1}{\mu} + \cos^2\Omega t^0\frac{1}{\mu} + \left(-\frac{\Omega x}{c}\right)^2(-\varepsilon) = \frac{1}{\mu} - \varepsilon\frac{\Omega^2 x^2}{c^2}$$

$$\chi^{2313} = \alpha^2_{\xi 0}\alpha^3_{\eta 0}\alpha^1_{\zeta 0}\alpha^3_{\lambda 0}\chi^{0\xi\eta\zeta\lambda} = \alpha^2_{\xi 0}\alpha^3_{30}\alpha^1_{\zeta 0}\alpha^3_{30}\chi^{0\xi 3\zeta 3}$$

$$= \alpha^2_{10}\alpha^3_{30}\alpha^1_{10}\alpha^3_{30}\chi^{01313} + \alpha^2_{20}\alpha^3_{30}\alpha^1_{20}\alpha^3_{30}\chi^{02323} + \alpha^2_{40}\alpha^3_{30}\alpha^1_{40}\alpha^3_{30}\chi^{04343}$$

$$= -\sin\Omega t^0\cdot\cos\Omega t^0\frac{1}{\mu} + \cos^2\Omega t^0\cdot\sin\Omega t^0\cdot\frac{1}{\mu} + \left(-\frac{\Omega x}{c}\right)\cdot\frac{\Omega y}{c}\cdot(-\varepsilon) = 0 + \varepsilon\frac{\Omega^2 xy}{c^2}$$

$$\chi^{2323} = \alpha^2_{\xi 0}\alpha^3_{\eta 0}\alpha^2_{\zeta 0}\alpha^3_{\lambda 0}\chi^{0\xi\eta\zeta\lambda} = \alpha^2_{\xi 0}\alpha^3_{30}\alpha^2_{\zeta 0}\alpha^3_{30}\chi^{0\xi 3\zeta 3}$$

$$= \alpha^2_{10}\alpha^3_{30}\alpha^2_{10}\alpha^3_{30}\chi^{01313} + \alpha^2_{20}\alpha^3_{30}\alpha^2_{20}\alpha^3_{30}\chi^{02323} + \alpha^2_{40}\alpha^3_{30}\alpha^2_{40}\alpha^3_{30}\chi^{04343}$$

$$= +\sin\Omega t^0\cdot\frac{1}{\mu} + \cos^2\Omega t^0\cdot\frac{1}{\mu} + \left(-\frac{\Omega x}{c}\right)\cdot(-\varepsilon) = \frac{1}{\mu} - \varepsilon\frac{\Omega^2 x^2}{c^2}$$

$$\chi^{2343} = \alpha^2_{\xi 0}\alpha^3_{\eta 0}\alpha^4_{\zeta 0}\alpha^3_{\lambda 0}\chi^{0\xi\eta\zeta\lambda} = \alpha^2_{\xi 0}\alpha^3_{30}\alpha^4_{40}\alpha^3_{30}\chi^{0\xi 343}$$

$$= \alpha^2_{40}\alpha^3_{30}\alpha^4_{40}\alpha^3_{30}\chi^{04343} = -\frac{\Omega x}{c}(-\varepsilon) = \varepsilon\frac{\Omega x}{c}$$

$$\begin{aligned}
\chi^{2412} &= \alpha^2_{\xi 0}\alpha^4_{\eta 0}\alpha^1_{\zeta 0}\alpha^1_{\lambda 0}\chi^{0\xi\eta\zeta\lambda} = \alpha^2_{\xi 0}\alpha^4_{40}\alpha^1_{\zeta 0}\alpha^2_{\lambda 0}\chi^{0\xi 4\zeta\lambda} \\
&= \alpha^2_{10}\alpha^4_{40}\alpha^1_{10}\alpha^2_{40}\chi^{01414} + \alpha^2_{10}\alpha^4_{40}\alpha^1_{40}\alpha^2_{10}\chi^{01414} + \alpha^2_{20}\alpha^4_{40}\alpha^1_{20}\alpha^2_{40}\chi^{02424} \\
&\qquad + \alpha^2_{20}\alpha^4_{40}\alpha^1_{40}\alpha^2_{20}\chi^{02442} + 0 \\
&= \varepsilon[\alpha^2_{10}\alpha^4_{40}\alpha^1_{10}\alpha^2_{40} - \alpha^2_{10}\alpha^4_{40}\alpha^1_{40}\alpha^2_{40}] + \varepsilon[\alpha^2_{20}\alpha^4_{40}\alpha^1_{20}\alpha^2_{40} - \alpha^2_{20}\alpha^4_{40}\alpha^1_{40}\alpha^2_{20}] \\
&= \varepsilon\left[-\sin\Omega t^0\cos\Omega t^0\left(-\frac{\Omega x}{c}\right) - \sin^2\Omega t^0\frac{\Omega y}{c}\right] \\
&\qquad + \varepsilon\left[\cos\Omega t^0\sin\Omega t^0\left(-\frac{\Omega x}{c}\right) - \cos^2\Omega t^0\frac{y\Omega}{c}\right] \\
&= \varepsilon\frac{\Omega}{c}\sin\Omega t^0(x\cos\Omega t^0 - y\sin\Omega t^0) - \varepsilon\frac{\Omega}{c}\cos\Omega t^0(x\sin\Omega t^0 + y\sin\Omega t^0) = -\varepsilon\frac{\Omega}{c}y \\
\chi^{2414} &= \alpha^2_{\xi 0}\alpha^4_{\eta 0}\alpha^1_{\zeta 0}\alpha^4_{\lambda 0}\chi^{0\xi\eta\zeta\lambda} = \alpha^2_{\xi 0}\alpha^4_{40}\alpha^1_{\zeta 0}\alpha^4_{40}\chi^{0\xi 4\zeta 4} \\
&= \alpha^2_{10}\alpha^4_{40}\alpha^1_{10}\alpha^4_{40}\chi^{01414} + \alpha^2_{20}\alpha^4_{40}\alpha^1_{20}\alpha^4_{40}\chi^{02424} \\
&= -\varepsilon[-\sin\Omega t^0\cos\Omega t^0 + \cos\Omega t^0\sin\Omega t^0] = 0 \\
\chi^{2424} &= \alpha^2_{\xi 0}\alpha^4_{\eta 0}\alpha^2_{\zeta 0}\alpha^4_{\lambda 0}\chi^{0\xi\eta\zeta\lambda} = \alpha^2_{\xi 0}\alpha^4_{40}\alpha^2_{\zeta 0}\alpha^4_{40}\chi^{0\xi 4\zeta 4} \\
&= \alpha^2_{10}\alpha^4_{40}\alpha^2_{20}\alpha^4_{40}\chi^{01414} + \alpha^2_{20}\alpha^4_{40}\alpha^2_{20}\alpha^4_{40}\chi^{02424} = -\varepsilon(t\sin^2\Omega t^0 + \cos^2\Omega t^0) = -\varepsilon \\
\chi^{3434} &= \alpha^3_{\xi 0}\alpha^4_{\eta 0}\alpha^3_{\zeta 0}\alpha^4_{\lambda 0}\chi^{0\xi\eta\zeta\lambda} = \alpha^3_{30}\alpha^4_{40}\alpha^3_{30}\alpha^4_{40}\chi^{03434} = -t \\
\mathcal{H}^{12} &= \frac{1}{2}\chi^{12\nu\eta}F_{\nu\eta} = H_3 = \chi^{1212}F_{12} + \chi^{1214}F_{14} + \chi^{1224}F_{24} \\
&= \left(\frac{1}{\mu} - \varepsilon\beta^2\right)B_3 + \varepsilon\frac{x\Omega}{c}E_1 + \varepsilon\frac{y\Omega}{c}E_2 = \frac{B_z}{\mu} + \varepsilon\{(\beta\times(\beta\times B))\}_z + \varepsilon(-\beta_x E_y + \beta_y E_x) \\
&= \frac{1}{\mu}B_z + \varepsilon\{(\beta\times(\beta\times B))\}_z - \varepsilon(\beta\times E)_z.
\end{aligned}$$

Hence

$$H = \frac{B}{\mu} - \varepsilon\beta\times\{E - (\beta\times B)\}$$

$$\begin{aligned}
\mathcal{H}^{14} &= \frac{1}{2}\chi^{14\nu\eta}F_{\nu\eta} = -D_1 = \chi^{1412}F_{12} + \chi^{1414}F_{14} + \chi^{1424}F_{24} \\
&= \varepsilon\frac{\Omega x}{c}B_3 + (-\varepsilon)E_1 + 0, \quad \text{or,} \quad D_x = \varepsilon E_x - \varepsilon\beta_y B_z = \varepsilon E_x - \varepsilon(\beta\times B)_x.
\end{aligned}$$

Hence

$$D = \varepsilon E - \varepsilon(\beta\times B).$$

14.19 Ring Laser

Ring lasers are modern, laser active self-oscillating rotation sensing devices requiring no external references for their operation. The laser source inside a rotating close ring-like cavity emits two stimulated radiations of extreme monochromacity, one propagating in the clockwise (CW) sense, and the other, in the counter-clockwise (CCW) sense. Mode degeneracy is removed by rotation of the cavity containing the laser source. The resulting split in frequency can be measured to high accuracy by using frequency standards.

Ring lasers are generally $H_e - N_e$ gas lasers having stimulated emission of neon lines in a discharge at the wave length 1.15228×10^{-4} cm and frequency near 10^{14} cps, whatever be the cavity configuration. Any closed geometrical path can be used as the ring. Square, rectangular and triangular configurations with perimeters varying from 8 meters to 30 centimeters have been reported by Macek and Davis (1967), Macek, Schneider and Salamon (1964) and others. The dielectric part of the medium and its laser active part can be kept in two different segments of the ring. Then these are separated by one or more reflecting mirrors. Four gas tubes with Brewster angle windows and 4-external corner mirrors are employed. The ring laser has no moving parts, and so noise is not a major problem. The operation of the instrument has considerable mechanical simplicity because only optical, atomic and electronic phenomena are involved.

Fabry-Perot square cavity with flat mirrors has been used, Chao and Heer (1964). The cavity is arranged such that all mirrors are almost flat, aluminized and aligned to form a 92 cm × 92 cm resonant square structure. The beats are observed by a Pb-Se photo conductor. Confocal system of mirrors consisting of one spherical (radius 10 meters) and three flat mirrors were used to observe beat signals. Also 4-plane mirrors were used to observe the same in other experiments.

The cavity containing the laser source is rotated by means of an external drive a bout its input axis or by submitting the apparatus to a torsional vibration. Either mechanical or optical techniques, or a combination of the two, is used to produce beats between the CW and CCW rays of the laser source inside the closed ring.

Sometimes a dielectric moving along the path of the rays is inserted inside the ring. The moving dielectric can be a solid (say a quartz piece), liquid (say CCl_4), or a gaseous medium (say dry air or even plasma). In the experiment of Macek, Schneider and Salamon (1964) with a meter square (perimeter of the ring cavity 408 cm) on ring laser three of the legs are gas tubes and the 4th leg contains the moving dielectric. The solid dielectric is a quartz flat tilted to Brewster's angle of total reflection to minimize the reflection losses. It is rotated about an axis normal to the flat surface in such a manner that the motion has a component along the path of the ring laser. The path of refracted light and the axis of rotation are not coplanar. If the dielectric is a flowing fluid it is placed between two transparent flat surfaces at Brewester's angle such that the flow has a non-zero component parallel to the path of the refracted light. If the dielectric is a flowing gas it is introduced through an axially oriented tube and its flow is monitored by a small bore pitot tube.

Following Macek and McCartney (1966) the triangular ring laser configuration is given in figure 14.19.1. This ring has three reflectors which are either fully reflective prisms or external surface mirrors, or multilayer dielectric coatings with nearly 99.5 per cent reflection. These are fixed such that any ray in either direction retraces the same path during successive round trips and at such angles that a maximum of the circulating light flux is retained within the circuit. The CW and CCW modes are extracted through one of the corner mirrors which is made collinear by external combining optics and mixed on a photo cathode.

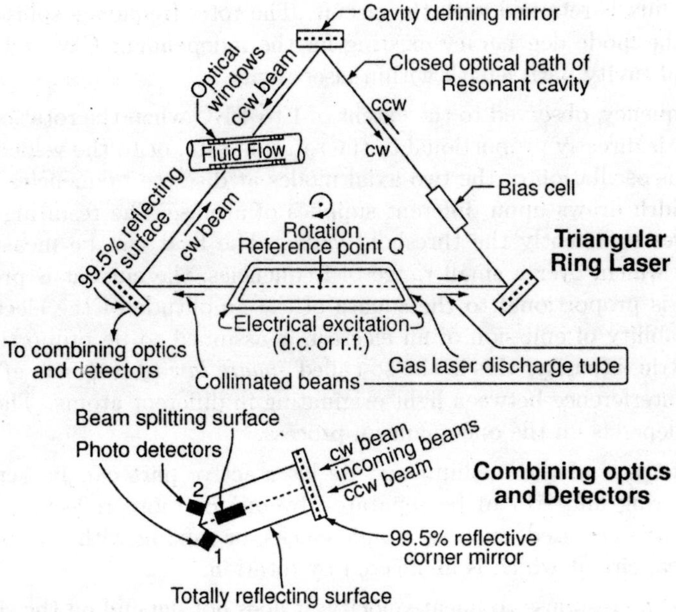

Fig. 14.19.1 Shows a triangular ring and combining optics

Rotation and other accelerated motions affect radiation from the laser source inside the closed optically resonant cavity. When there is no laser source inside, rotation also affects electromagnetic resonance of closed optical cavities, and the coherent CW and CCW beams are derived from a rotating beam splitter. When these are combined the phase differences due to path difference between the beams are manifested as a series of fringes of interference. Since the external light source moving with an interferometer unambiguously determines the light frequency occurring in the interferometer, the mode degeneracy prevails. As a result the frequency difference between the CW and CCW travelling rays must vanish. So

$$\delta\omega = 0. \tag{14.19.1}$$

A Doppler shift affects the CW and CCW beams if the light source is stationary. Then the fringe shift is only affected in higher orders by rotation; equal fringes, that is, the condition (14.19.1), are still necessary for interference.

When a laser source of light exists inside the rotating close ring-line and not simply connected toroidal cavity, two stimulated radiations of extreme monochromacity are emitted; one of these is propagating in the CW sense and the other in the CCW sense. Rotation of the beam splitter adiabatically affects the phase conditions in a laser beam arriving in phase after one circulation because the optical paths of the CW and CCW modes are different in the laser resonant cavity. The phase shift that occurs in the two rays is non-reciprocal and gives rise to a non-degenerate frequency splitting such that the number of nodal points along the closed loop remains invariant. For this reason the ring laser effect is an adiabatic change. The classical Michelson type of rotating mirror interferometers using interference of degenerate modes do not attain the degree of accuracy of the ring laser.

A clockwise oscillation is obtained by directing the waves around the complete optical ring instead of retroreflection them. The reflectors are fixed such that any ray in either direction retrace the same path during successive round trips and at such angles that a maximum of the circulating light flux is retained with the circuit. The total frequency splitting $\Delta\nu$ arises from the removal of the mode degeneracy existing for the independent CW and CCW modes, due to the differential cavity path length within laser cavity.

The beat frequency, observed to the extent of 1 to 4 KC when the rotation rate of the entire system is varied, is directly proportional to the rotation rate, or to the velocity of the dielectric. The simultaneous oscillation of the two axial modes at discrete frequencies within the natural radiative line width draws upon different supplies of atoms. The resulting gain in the active medium surpasses sufficiently the threshold value. The beat can be measured by the photo electric effect in which, over a small range of frequencies, the current is proportional to light intensity, which is proportional to the square of the amplitude of the electric field. For this reason the probability of emission of an electron is assumed to be proportional to the square of the total electric intensity; this is the so-called *square law of response of the detector.* This also implies an interference between light originating in different atoms. The generation of the beat frequency depends on the one-quantum process.

The dielectric part of the medium and its laser active part can be kept in two different segments of the ring and so can be separated by one or more reflecting mirrors. Primary frequency standards are used, such as atomic clocks, or mixing with output of a second laser outside the optical circuit which is unaffected by rotation.

The effect has no response to linear velocity; it does not depend on the shape of the surface area of the ring and is independent of the position of the center of rotation. The beat frequency output to each detector is proportional to the angular rate of rotation. The limits of rotation are set by equipment and technique used and are not the physical limits of the rotation rate-sensing effect. Frequency changes due to the local fluctuations in the active medium, cavity mirror vibrations, thermal effect, etc., will affect both laser modes identically. Also quantum mechanical correlations may exist and reduce the lower limit on the rate sensing.

The beat signal which has been observed when oscillation was at threshold, became, stable and locked at a frequency corresponding to a given angular rotation rate, Cheo and Heer (1964). The amplitude of the beat signal increased in the N_e transition of a $H_e - N_e$ maser. Small dust particles, diffraction edges etc., seem to affect one beam more strongly than the other and cause sufficient phase shift to split the resonant frequency of the two modes to a slight extent. This non-reciprocal effect is perhaps due to non-linear power saturation of the maser.

Mode interlocking of the CW and CCW modes, a disturbing problem, is removed by the application of a bias between them. Rotation of the ring raises or lowers the output frequency from the fixed value due to bias even at zero angular rate. But biasing partly eliminates the initial advantage of no moving parts. For electrical biasing the non-reciprocal Faraday rotation in optimal media in presence of strong magnetic field can be employed, because then the split up CW and CCW rotations travel with different phase and group velocities. The zero point accuracy depends on the constancy of the magnetic field, or on the symmetry of an oscillating bias in the presence of electrically generated magnetic field. In a low loss Faraday bias cell subjected to magnetic field, bi-refringence of glass cause phase delay between the oppositely travelling beams and so can prevent the locking that occurs at low rotation rate. Due to non-inertial energy coupling the CW and CCW beams lock into a single mode. This locking is perfect when the ring laser is in absolute rest; then the frequencies of the two modes approach the same value. Angular rate of the biasing rotation with respect to the inertial platform should be constant and known to *good accuracy* to avoid instantaneous errors in measurement of actual input rate. Application of a bias can prevent the occurrence of absolute inertial rest and so of equal frequencies. Even at very low angular rates the two mode frequencies are separated so much in this way that prevention of interlocking is possible.

Heterodyning (the beating of two frequencies together to get a third frequency) eliminates the energy output at the sum frequency by tuned circuits following it. The two frequencies of the CW and CCW modes heterodyne at the common photo detector[1] Combining optics bring the two modes together at each of the two photo detectors 1 and 2 of figure 14.19.1. At detector 1 the CW beam has undergone two dielectric reflections and the CCW beam one reflection. The inverse is true for detector 2. The resulting phase difference beat frequency output of the two detectors show the direction of rotation.

The superheterodyne detection scheme is employed for the mixing of two laser outputs and for measuring relative frequency shifts between two identical lasers due to velocity, acceleration or gravitational effects. For example, if the relative width of the laser output is given, say $\Delta\nu/\nu = 7 \times 10^{-11}$, then the relative velocities of about 0.2 cm/sec might be observed. Photoelectric mixing occurs between possible signals and the minimum observable signal; then the distance over which two similar lasers may be compared is significant.

Diffraction at an aperture gives rise to an angular spread of the laser output. If the aperture length is 1 cm, the angular spread is given by $\theta = \lambda/l = 3 \times 10^{-4}$ rad. For example, energy of a beam falling on the photomixer of area 1 m^2 spreads over the area 6×10^{14} cm^2 after passing the distance 10^{11} cm. The frequency of the two light sources can be compared up to distance 10^{11} cm. The frequency of the two light sources can be compared up to distances of the order of 10^6 km. This possibility offers scope for space navigation and for the precise checking of gravitational and acceleration effects predicted by the general theory of relativity.

The optical path length inside the cavity of the ring laser must equal an integral number times the wave lengths of the CW and CCW oscillating modes. So the circumferential optical path of the circuit of total length p determines the frequency ν of the preferred mode resonances of the laser field. For the stationary system the paths p of the CW and CCW rays are equal; so the wave lengths are equal and the eigen modes are degenerate in frequency. For the rotating system, since the wave fronts of the CW and CCW modes travel the lengths $p + \Delta p$ and $p - \Delta p$ respectively of figure 14.19.2 to close on themselves, the paths are unequal, and consequently a frequency shift $\Delta\nu$ is obtained between the two modes.

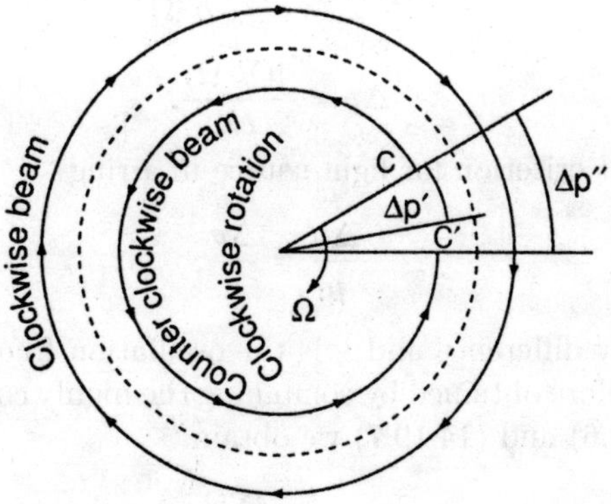

Fig. 14.19.2 It shows the path difference of clockwise and counter-clockwise beams from a source of light in a circular ring rotating in its own plane

[1]Photo electric effect can detect the non-linear mixing of waves. Beats between components of an electromagnetic field are observed as fluctuations in the photoelectric current. Thus the shapes of the spectral lines are measured from the frequency spectrum of the photo current in the superheterodyne detection.

Consider a circular ring of radius R, rotating in the clockwise sense (figure 14.19.2), Post (1967). The wave fronts are constrained to follow the circular path of the ring, and the two beams leave the beam splitter when it is in position c. The CCW circulation is opposite to the sense of rotation of the ring, and meets the beam splitter again in the new position C' which is shifted from c by the amount Δp. For stationary observers, the CW beam travelling in the sense of the rotation of the ring, meets the beam splitter when it is shifted through a distance $\Delta p''$ to C''.

If Ω is the constant rate of rotation of the ring, T' is the time of the CCW motion, T'' is the time of the CW motion and C is the free space velocity of light, then

$$T' = \frac{1}{c}(2\pi R - \Delta p') = \frac{\Delta p'}{R\Omega}, \quad T'' = \frac{1}{c}(2\pi R - \Delta p'') = \frac{\Delta p''}{R\Omega}.$$

In the stationary frame of the observer the time interval between the consecutive positions C' and C'' of the beam-splitter, denoted by $(\Delta T)_s$, is given by

$$(\Delta T)_s = T'' - T' = \frac{4\Sigma R}{c^2 - R^2\Omega^2}, \tag{14.19.2}$$

where Σ is the area enclosed by the optical path of the ring, so $\Sigma = \pi R^2$ and for all practical purposes

$$(\Delta T)_s = \frac{1}{c^2}4\Sigma\Omega. \tag{14.19.3}$$

Hence if Δp is the fringe shift, measured in fringe numbers, then

$$\Delta p = \frac{c}{\lambda}(\Delta T)_s = \frac{4\Sigma\Omega}{c\lambda}, \tag{14.19.4}$$

where λ is the nominal wave length in vacuum for zero rotation. More generally when the axis of rotation is not perpendicular to the plane of the ring, we obtain

$$(\Delta T)_s = \frac{4(\Sigma\cdot\Omega)}{c^2 - R^2\Omega^2}, \tag{14.19.5}$$

$$\Delta p = \frac{4(\Sigma\cdot\Omega)}{c\lambda}. \tag{14.19.6}$$

From the phase agreement criterion for light source in a ring

$$\frac{\Delta p}{p} = \frac{\Delta\nu}{\nu}, \tag{14.19.7}$$

where $\Delta\nu$ is the frequency difference and ν is the oscillation frequency for zero rotation. The beat frequency $\Delta\nu$ is therefore obtained by combining the highly coherent, oppositely circulating beams. Combining (14.19.6) and (14.19.7) we obtain

$$\Delta\nu = \frac{4\Sigma\Omega}{\nu p}. \tag{14.19.8}$$

Thus rotation of the ring in either direction produces a frequency split in each axial mode and the difference between the two corresponding frequencies is proportional to the angular rate. For a square configuration if l is the length of each side, then

$$\Delta\nu = \frac{l\Omega}{\lambda}. \tag{14.19.9}$$

Since the length of a ray path inside the cavity must equal an integral multiple of the wave length, so that the returning wave is in phase with the initiating energy, the round trip phase must be a multiple of 2π radians. Since the length of a ray path inside the cavity (approximately equals to the ring perimeter) is millions of times greater than the optical wave length, this requirement is met by many closely spaced wavelengths. The number of axial modes, which can be sustained concurrently, depends upon (1) the spectral distribution curve of energy emitted by the discharge tube, (2) the threshold energy level required for laser oscillation, and (3) frequency separation between the adjacent axial modes.

The usable energy lies between the cut-off points established by intersection of the threshold level and gain curve. Oscillations occur between these limits at discrete axial modes, with a frequency separation $\Delta\nu'$ (or spectral free range) nearly given by

$$\Delta\nu' = \frac{c}{p}, \tag{14.19.10}$$

where p is the ring perimeter. If $p = 100$ cm; then

$$\Delta\nu' = 300 \text{ MC}.$$

For further discussion of the theory see section 5.5.

The experimental results for air and quartz crystals are consistent with the theoretically predicted values and the increase in sensitivity over the earlier classical fringe shift experiment is expected. But experiments with flowing liquid do not give good correlation between theory and laboratory results; the main cause for that may be the simplified procedure employed to monitor the liquid velocity because, owing to viscosity and turbulence, the recorded mean flow data is different from the effective flow along the cavity.

The influence of rotation of earth on cavity resonance can be studied by adjusting the vertical plane in which the ring laser is placed so that rotational motion of earth is parallel or anti-parallel to the direction of motion of the dielectric. The dielectric is first given a motion parallel to this direction and next, the same speed is given in the anti-parallel direction. The frequency mode split $\Delta\nu_1$, in the two cases, can be measured. It is found that earth's influence is considerable in laser cavity resonance. Theory has been developed of the closed loop line integral of velocity field for an approximate evaluation of the path length changes due to motion of an interferometer as a whole, Post (1972). It is to study ring laser effect by the methods of geometrical optics.

The ring laser effect is considered with the help of cavity resonance method, Heer (1964), and Slater (1950, Ch. IV). The effects in a rotating ring laser based on classical electrodynamics and geometrical optics have been discussed by Menegozzi and Lamp (1973) and Privalov and Fridrikhov (1969), and derived from an adiabatic energy change law by Post (1972a) and Khromykh (1966), with the help of state relations valid in systems rotating with a uniform and constant angular velocity Ω. The roles of ring lasers and gyroscopes in rotation sensing have also been discussed Macek and McCartney (1966). Post (1972a) writes : Our *investigation is prompted by the intrinsic scientific interest of problems concerning non-inertial systems* and by the modern interest *in inertial guidance systems for navigation...* Inertial guidance systems require for their implementation the design and construction of sensitive probes for measuring linear and rotational accelerations. Probes or sensors of this nature give a zero indication when they are at rest in an inertial frame. For an initially assumed inertial frame position rotation sensors are required to have a minimal rate of drift with respect to that position... Inertial sensors acquire their sensitivity from effects related to non-inertial systems... For navigational purposes a measurement of the acceleration associated with rotation is of major importance",... and the *two devices... of extremely sensitive sensors of absolute rotation... are the gyroscope and the ring laser.* Application of the ring laser path interference effect for probing plasmas has been proposed, Khan and Chakraborty (1979). Self-induced precessional rotation of waves in plasmas (cf. Chakraborty (1978), Ch. 12) and in the effect of strong radiation in plasmas, Chakraborty, Khan and Bhattacharyya (1981), may thus be detected.

The high resolution and low threshold provided by a ring laser, suggested application to a radar-type tracking pedestals for measurement of azimuth and elevation. The sensor need not be located at the axis of rotation, as with conventional shaft angle encoders. The only restriction is that the plane of the ring is maintained normal to the rotation axis. So the ring laser is particularly useful for rotation measurement when a physically defined axis of rotation does not exist.

Ring lasers provide for new types of attitude reference systems for target designation and fire control in helicopters and fixed-wing air craft. A strapped down system using three rings, in a rigid orthogonal configuration, mounted directly to the air frame, can measure body angular rates continuously. An incremental digital computer updates a direction cosine matrix at a high iteration rate, transforming co-ordinates and providing attitude data relative to a preselected co-ordinate. Data for rate stabilization of two-axes mounts also can be furnished. Ring lasers in strapped-down systems are cheap, smaller in size and less heavy than the conventional inertial platforms. The sensitivity of a spinning top, precessing due to gravitational field, is about 10^{-9} of an earth rate of rotation. The ring laser has so far not attained the sensitivity of the gyroscope. Its present limit of sensitivity is of the order of 0.1 degree per hour, or approximately 10^{-3} earth rate.

Chapter 15

An Appraisal of Electrodynamics

15.1 Introduction

Critical comments and clarifications on the definition of physical quantities of electrodynamics in rotating frames are considered in this chapter. The basic effects and tendencies are discussed. Topics like metric dependence of electrodynamics in 3-space and 4-space co-ordinasation of events, the topology of fields, local and global aspects of electrodynamics, are considered briefly and non-mathematically.

15.2 Some Comments

The LT from one inertial frame to another frame pointed out Sommerfeld (1952, Section 34), approximately holds for accelerating frames as well when $|\dot{u}|T \ll |u|$, where u is the velocity of a point p with respect to the laboratory frame, $\dot{u}$ is the acceleration, and the time $T = 2a/c$, where $2a$ is the electron diameter. Then the change in velocity in the time taken by a light wave to sweep over the electron cloud is small compared to c.

Rotating objects are usually studied by the Maxwell equations and the constitutive relations of Minkowski, both referred to an inertial frame and the equation of motion, containing only the Coriolis force. For convenience, the centrifugal force is also neglected. Only the homogeneous pair of Maxwell equations, that carry over to rotating co-ordinates without any change of form, can be used easily. For the other equations of Maxwell, if necessary, rotating co-ordinates are avoided, Webster and Whitten (1973) and Schiff (1939). So, a rotating object is practically studied by the Maxwell equations, referred to an inertial frame, Gron (1984). Working rules have been formed (Van Bladel (1973, 1976), MO (1969), Videc and Kooy (1982)) for application to technical problems of rotating objects, which do not follow from the inertial frame theory of electrodynamics.

The Lorentz transformation shows that the perpendicular components of the displacement vector, with respect to the relative velocity of two frames of references, remain unaffected; only the parallel components change. But, the perpendicular components of the EM field vectors are affected, and the parallel components remain unaffected. This distinction between the transformation of the displacement components, and the related transformation of the fields, follows Faraday's law of EM induction (when a conducting wire is moved in one direction, the magnetic field perpendicular to that direction is affected, and the parallel field remains unchanged). This distinction might be generalised for evolution of permanent gravitational fields (like the EM fields) by non-permanent gravitational fields of accelerated motion which includes rotation. Then the LT may be called the LT *for positional displacement* only which is proportional to velocity. A displacement is possible of velocity, which is proportional to acceleration. Non-permanent fields of gravitation, are proportional to their velocity displacement. The generalisation of the referred distinction might be that non-permanent gravitational fields affect permanent gravitational fields of perpendicular directions only. This generalisation leads to a triviality for rotation about a fixed axis at uniform rates, because this rotation implies transformation for rotation from one inertial frame to another.

If a spherical conductor is in rotation about a diameter, a magnetic field is seen to exist outside, Schiff (1939). But, an observer rotating about a conductor at rest, cannot detect any such field. So, a rotating observer does not observe the same magnetic field as the observer at rest when he sees the condenser rotating, because a part of the electric field in a moving frame is $(v \times B)/C$, where $v = \Omega \times r$ (Ω is the angular velocity of rotation and r is the position vector), its divergence gives $\nabla \cdot (v \times B) = 2(\Omega \cdot B)$ when B is constant. Hence, a space charge (the Schiff charge) proportional to the angular velocity of rotation, is possible by this theory, when none existed in the beginning; and an observer, rotating round an electrical conductor, would detect this charge, whereas, this space charge separation will be absent if the observer is non-rotating and the condenser is rotated. It is, therefore, thought that the EM laws are different in rotating and other accelerated frames from those in inertial frames, and these should be changed to allow no field for rotating observers relative to a stationary condenser, when there is a field for stationary observers relative to rotating condensers.

While the LT is a definite improvement over the GN rules of transformation for translation at uniform rates, the general theory of relativity (GTR) is not helpful in furnishing a unique guideline for rotation; for instance, several equally valid covariant formulations are possible for the constitutive relations which give qualitatively, as well as, quantitatively different results even for small rotational velocities.

Van Dantzig (1934a–e) developed the metric-independent form of the field equations of electrodynamics, in arbitrary frames of reference, using tensor calculus of Schouten. However, Birss (1982) and Post (1984) concluded that, only in 3-space of position co-ordinates, a metric-independent description is possible, and, in the space-time continuum, the required set of fundamental differential forms cannot be defined by experiment on a differential manifold without a metric.

An emf was measured in the experiments of Kennard (1912, 1917) by an observer rotating round the axis of symmetry of a magnetic field. Pegram (1917) obtained a surface charge induced in rotating condenser plates. These experiments were explained by some novel ideas, (see section 15.8).

Since magnetic fields are independent of the co-ordinate system, but not the electric field, all celestial magnetised bodies are, on account of their rotation, electrically polarised when seen from a system at rest, though these are electrically neutral, or nearly so on the astronomical scale. On the atomic scale, EM forces far exceed gravitational attractions, so it is the other way round, and, the EM theories and those of gravitation can be studied separately. The rotation dependent spurious Schiff charge and current are avoided by using the covariant tensor form of the ERF for formulation of the vector form EM theory for rotation. The convention that E and B shall appear in the Lorentz force law, and that D and H shall be related to the charge and current, as in the Biot-Savart law, relates the EM field tensors for the covariant equations with the vector and scalar EM field variables.

Theory of electromagnetics is based on two main sets of laws. One set gives the universally valid Maxwell equations which do not contain any material constants. The other set of laws, complementary to these are called the *constitutive relations* (also called the *state relations*), these depend on the nature and intensity of the applied field and on the constitution and motion of the material under examination.

The observations of rotating and stationary observers, relative to rotating conductors, and other difficulties cannot be explained by the traditional renderings of the basic laws of electrodynamics. So, for the valid theory, either the Maxwell equations are to be regarded as valid for accelerated frames, and the constitutive relations are to be changed, or the Maxwell equations are to be modified without changing the constitutive relations or both are to be modified. The current tendency is that the constitutive relations are to be changed without changing the Maxwell equations. Thinking has also been focussed on the nature of field quantities. Two classes of field quantities are found to be involved in experimental measurements and theoretical investigations. One is the observable quantities, called the *primary quantities* or the *domain quantities* :

$$\text{emf} = \oint_{\Gamma} (E \cdot dl)$$

$$\text{field flux} = \iint_{\Sigma} (B \cdot d\Sigma)$$

$$\text{current flux} = \iint_{\Sigma} (j \cdot d\Sigma),$$

where Γ is the closed contour of a curve and Σ is the surface of a material, dl is a line element and $d\Sigma$ is a surface element. Evidently, these are integrals over well defined domains. Faraday's laws of EM induction deal with relations between the observable integrated, primary quantities, like the emf and the rate of change of magnetic flux through a surface. These, being valid over specific domains, are global laws. The other electromagnetic quantities are derived quantities. These are distinct from the domain quantities, but follow from these, obtain their properties from these; and are local point functions of the continuum of space time. Examples are the field vectors E, B, H, D, J. From the generally valid global laws, involving the domain quantities, the local laws (the partial differential equations of electrodynamics of Maxwell), are derived under some assumptions.

So the EM laws of induction and their applications depend on inertial frame observations involving the domain quantities. In the purely spatial domain Faraday originally imagined, for the magnetic flux, the tubes of force which close on themselves, and defined the laws of EM induction. A primary role should, therefore, be given to the observable quantities, with respect to the frame of reference.

Field concepts of EM theory follow from the macroscopic continuum physics. For ERF, both field concepts and the concepts of particle motion in vacuum, in presence of these fields, are simultaneously necessary.

Whether a physical vector is contravariant or covariant is decided by the elementary definition of physical quantities, and not from their tensor characters. The 4-force cannot be defined uniquely, because it can be derived both as a contravariant 4-vector and as a covariant 4-vector. Forces can be created by a change in the frame of reference, so these are non-tensorial quantities. However, the sum of inertial and gravitational forces may be tensorial. The Lorentz force is an example of a tensorial force field if it is regarded as a contravariant vector density.

If force is defined as the product of mass and acceleration, then it is contravariant. Force can also be defined from the concept of work, which is a product of force and displacement. If work is accepted as an invariant scalar quantity, then since displacement is a contravariant vector, force must be a covariant vector.

Several covariant renderings of the constitutive relations are possible, which give non-identical vector form state relations for small rotational velocities, see sections 14.13–14.15. These qualitatively different constitutive relations follow from tensor formulations, Tamm (1924) and Mandel'shtam (1924), Heer (1964) and Khromykh (1969), Post (1967), Post and Yildiz (1965), Yildiz and Tang (1966).

A proper sequential presentation of electrodynamics, is desirable for finding its complete evolution in accelerated frames. Moreover, a systematic presentation of the available information, followed by critical comments and clarifications, would expose the problems of ERF and the ideas working for its emergence from these difficulties.

In Sagnac effect and in ring laser (section 14.19) the difference between the paths of clockwise and anticlockwise rotation of optical beams leads to optical interference, see Anderson and Ryon (1969), Heer (1964), Khromykh (1966), Post (1967, 1972), Post and Yildiz (1965), Volkov and Kisiliev (1970), Yildiz and Tang (1966). However, effects of rotation velocity of order higher than those available from the Galilean-Newtonian ideas of space time, could not be detected thus. So these attempts of application fail to throw any light on the basic problems of ERF.

Application seems possible in pulsar magneto-sphere, where large rotational velocities, high electrical conductivity and very high magnetic field are available. However, so far, only a naive application of electrodynamics, with reference to inertial frames, has been attempted, in which the medium velocity equals the velocity of rotation, at a constant angular rate, which is characteristic of the rotational velocity of the pulsar magnetosphere, Ostriker and Gunn (1969), Gold (1968), Goldreich and Julian (1969) and Michel (1982).

The Maxwell equations in a rotating frame, even for small rotational velocities, do not exist even in the best text books; Sommerfeld (1952), Jackson (1975), Landau and Lifshits (1960, 1975) and Panofsky and Phillips (1964). For study of problems and special features of rotation and for finding answer to the questions : *Why rotation and How rotation* occur in nature. Rotation should be treated as a separate subject of physics. No serious review was found in literature of physical sciences, on the topics of ERF, or of physics in rotating frames (PRF).

Some key expressions of physics of rotation are : Field concepts of rotation; Co-ordinate-free rotation; Metric-independent rotation kinematics; Differential rotation; Lorentz rotation; Holonomic differentials (or frames) of rotation; anholonomic (or non-holomic) differentials (or frames) of rotation; Physics in rotating frames of references; Electrodynamics in rotating frames; Mechanics in rotating frames; kinetics in rotating frames; motion of microbodies and macrobodies in rotating frames.

A rotating observer can assign a vector Ω, equal to the vectorial angular velocity of rotation, to every point of position space, thus defining an Ω field of rotation. When Ω is uniform, this field is uniform. When $\Omega(t)$ increases with time in the rotating frames, a rotating observer sees the field of this increasing Ω, and can measure the corresponding force in addition to the Coriolis force and the centrifugal force. The inertial field of Ω is just as real in the rotating frame as the EM field in the Newtonian frame, since both are defined in the same way.

Problems involving closed loops of tubes, filled with a fluid, which are moved in a rotating system of co-ordinates, can be solved by analogy with an electric current in a magnetic field, with help of Faraday's law of induction for rotation of fluids which is : The change in the flux of 2Ω through a circuit, equals the total impulse, per unit mass, given to the fluid in the tube, see Chapter 8.

The number of papers on EM theory in accelerated frames, (up to about 1984) is not very large; in alphabetical order of the names of the authors, these are: Anderson and Ryon (1969), Atwater (1975), Birss (1982), Bow (1971, 1972), Van Bladel (1973, 1976), Coisson (1973), Corum (1977, 1980), Van Dantzig (1934a–e), Goldstein (1961), Gron (1984), Heer (1964), Irvine (1964), Ise (1957), Ise and Uretskey (1955), Khromykh (1966), Mandel'shtam (1924), Marsh (1982), Modesitt (1970), Plebanski (1960), Post (1967, 1972a, 1972b, 1984), Post and Bahulikar (1971), Post and Yildiz (1965), Schiff (1939), Semon and Schmeig (1981), Shiozawa (1973, 1974, 1978), Tai (1964), Tamm (1924), Trocheris (1949), Volkov and Kisiliev (1970), Webster (1961, 1963), Webster and Whitten (1973), Yildiz and Tang (1966). The Maxwell equations, for accelerated frames, are discussed by Moller (1972, sections 10.9 and 10.10), Landau and Lifshitz (1975, section 90) and Misner et al (1973, section 22.4). Problems of ERF have been briefly considered by Alfven (1953, section 1.3). Ingarden and Jamiolkowski (1985) treated electrodynamics by the method of exterior differential forms, which is relevant to our studies. Some more relevant works have appeared recently.

Experiments with rotating magnets and rotating cylindrical condenser plates, for detection of motion of lines of force and other related effects, are well-known, Barnett (1912), Eichenwald (1903), Kennard (1912, 1917), Pegram (1917), Rontgen (1908), Rowland (1876), Wilson (1905), Wilson and Wilson (1914). Culwick (1959, 1966) has summarised some of these experiments.

15.3 Laws of EM Induction

Faraday's law of EM induction and the laws of Oersted, Biot and Savart, Ampere and Maxwell for induction of time varying EM fields, have been critically considered with the hope of making these basic laws suitable for ERF. Faraday's laws are : (i) If a closed circuit moves across a magnetic field, or if the magnetic field through a circuit changes in time, a current flows though no batteries are present, such that the induced emf is proportional to the rate of change of the magnetic field; (ii) The emf acts in such a way as to oppose the change in the magnetic flux. Faraday's law is that the magnetic fields are generated by a circulating convection current, and it is implied that these fields are measured in inertial frames. Hence, the laws of EM induction and their applications depend entirely on inertial frames; their validity beyond the range of inertial frames should be questioned. For instance, Faraday's law has been treated, almost without exception, as if everything about it was known very thoroughly. But some of the implied notions are not very general. The usual Faraday law is valid only in inertial frames and not beyond these frames. Specifically, Faraday's law deals with the integrated quantities of flux and emf. No restriction is mentioned in this law that the split in the electrically conducting loop for measuring the emf should be stationary in an inertial frame. Therefore, emphasis should be given on primary quantities which are observable quantities, and so are integrals of fields and called *domain* quantities; example emf and flux. They are distinct from the derived quantities, which are the fields themselves, like the electric field vector E or the magnetic induction vector B, in both inertial and non-inertial frames of reference. The derived field quantities obtain their properties from the assumption that the *domain* quantities of flux and emf may be regarded as the integrals of fields; which being observable quantities with respect to reference frames, should be given considerable importance. It seems that the traditional renderings of the Faraday induction law are incomplete and should be generalised.

15.4 Flux Cutting Law

Moving a closed circuit through a magnetic field, an electric field, which is perpendicular to both the magnetic field and the direction of motion, is experienced by a circuit. Therefore, if v is the velocity of the observer relative to the current system, which is due to the magnetic field, B is the magnetic flux density, θ is the angle between them and E is the induced electric field, then the flux cutting law is

$$E = \frac{1}{c} Bv \sin\theta. \tag{15.4.1}$$

Hence, for a conductor placed at right angles to the field, the emf $\mathcal{E}$ is given by

$$\mathcal{E} = \frac{1}{c} Bvlm \tag{15.4.2}$$

where l is the length of the conductor and v is the component of its velocity perpendicular to the field.

Cullwick (1966, p. 100) suggested that the flux cutting law should be used for cases of motion of conductors through constant magnetic field; which are mutually connected through sliding contacts of constant shape, between which there is relative motion. Examples of the use of this principle are the arrangements for unipolar induction, Sommerfeld (1952, p. 288–290) and Trocheris (1949), and homopolar induction, Cullwick (1959, p. 109). But, rigorously speaking, each component of the magnetic field shares the motion of its source, because, relative motion between matter and current circuits, or magnets, should be the fundamental cause of motional inductance. In this way, the emf is induced in each section of the circuit.

15.5 Flux Conservation

The local approach involves the differential equations for the field variables, and the global approach involves the integral relations for the fields valid over given domains. Only with the help of the global approach, Post (1974), the non-inertial aspects of the problem become clear. The law of flux conservation, which is a global aspect of conservation, and is directly related to the Faraday-Maxwell law, because it covers the 4-space integration and not the position space domain integration only, becomes fundamental like the law of conservation of charge. Faraday imagined that the tubes of force of the magnetic flux close into themselves in purely spatial domains due to flux conservation. These tubes of force are oriented in the sense of aligning their circulation directions so that the tubes of force can be oriented and assembled by tying a string around them. Hence, conservation of magnetic flux means that neither the tubes of force can be pulled out of the string, nor can new ones be added. For the given conditions all tubes of force have equal magnitudes $|h/2|$ and similarly all charges inside an enclosure have magnitude $|e|$ where $\hbar$ is the action. This extended Faraday picture covers the case of static magnetic field, and is also compatible for normally conducting solenoids with constant current, provided the complete circuit plus the source is taken as the loop that confines tubes of force, Post (1974).

15.6 Some Experiments

Experiments have been performed on finding influence of rotation on electricity and magnetism. The observation have been explained with the help of interesting theories which ensure elimination of the spurious Schiff charges and current. These theories have been discussed.

15.6.1 Experiments of Pegram and Kennard

A magnetic rotation sensor was searched in vain with respect to inertial space, from measurement of emf by an observer rotating around the axis of symmetry of a magnetic field in the experiments of Kennard (1912, 1917), and Pegram (1917). The relevant earlier experiments on ERF are by Rowland (1876), Rontgen (1888), Eichenwald (1903), Wilson (1905), Wilson and Wilson (1914) and Barnett (1912).

Initially Kennard (1912) used a steel bar, set in rotation inside an insulated metal cylinder connected to earth, and magnetised by a stationary current carrying solenoid. The rotating magnet was enclosed in a second metallic screen of small diameter. The metal cylinder and the screen formed the two plates of a condenser which were connected to the quadrants of an electrometer. His search for charges on the cylinder due to a possible emf in the earthing wire led to a negative result. In the latter arrangement, Kennard (1917) showed the existence of the charge on the inner cylinder, by keeping the solenoid or the magnet stationary and rotating the condenser with radial connection between the cylinders.

Pegram (1917) employed a solenoid, mounted to rotate about a vertical axis; a cylindrical copper condenser was mounted co-axially with the solenoid, with closed ends, except a shielded connection to an electrometer, running through a central hole in top end. When the systems of cylindrical condenser and connections are stationary, and only the solenoid was rotated, Pegram observed no effect. But when the solenoid and condenser were rotated together, the potential difference set up between the condenser armature, when connected together, was simply that calculated from the emf produced in the connecting wire, as a result of its motion in the magnetic field. This potential difference was measured by the electrometer, the armature being separated after the rotation was started, and the system being allowed to come to rest before readings were taken. The same result would be obtained even though the wires connecting the armatures were outside the field.

A conducting ring permits a continuous change of tubes of force it encloses. The condition of conservation is that the change in the number of quantised tubes of force, enclosed by the ring, equals the number of quantised tubes of force travelling through the ring confinement. The rate of increase of the number of tubes of force through the surface Σ is

$$\frac{d}{dt}\iint_{\Sigma}(B\cdot d\Sigma). \tag{15.6.1}$$

It is expressed in terms of the number B of tubes of force per unit area through the surface element $d\Sigma$: The number of tubes of force travelling through the boundary Γ of Σ is

$$\oint_{\Gamma}(B\cdot(\mathbf{u}\times d\mathbf{l})) = \oint_{\Gamma}((B\times\mathbf{u})\cdot d\mathbf{l}), \tag{15.6.2}$$

where $\mathbf{u}$ is the velocity of the tubes of force and $d\mathbf{l}$ is a length element in Γ. Conservation of induction, or the Faraday law, means that

$$\frac{d}{dt}\iint_{\Sigma}(B\cdot d\Sigma) + \oint_{\Gamma}((B\times\mathbf{u})\cdot d\mathbf{l}) = 0. \tag{15.6.3}$$

15.6.2 Experiment proposed by Post

Post (1972) suggested the employment of a disc capacitor with slits in the annular plates to force the electric charges to move with the plates. A simultaneous rotation of the two plates in the same direction generates two convection currents in opposite directions if the capacitor is charged. There will be a corresponding magnetic field which may be measured by a conduction loop in vertical oscillation as shown in fig. 15.6.1. The emf generated in this loop is proportional to the change in magnetic flux through the ring due to the non-uniformity of the field in the vertical direction.

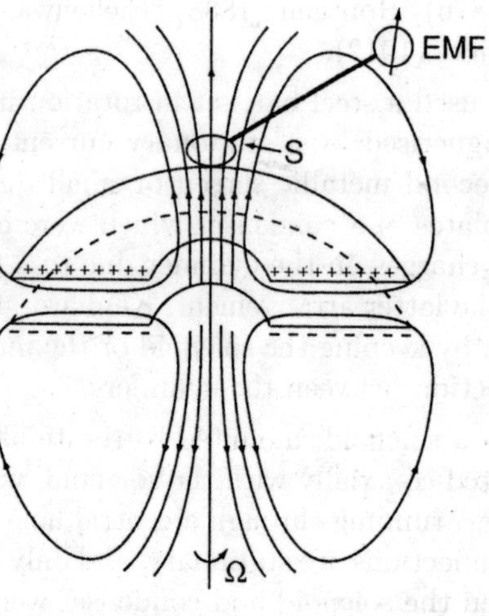

Fig. 15.6.1 Magnetic field due to a circulating convection current generated by rotation of the two plates of a disc capacitor.

In the Faraday law of EM induction no restriction is mentioned that the split S in the loop for measuring the emf should be stationary in an inertial frame. Hence the emf ring can be rotated in itself so that the split S in the loop for measuring the emf rotate at the same rate as two plates of the charged disc capacitor. The same emf and the same B field are expected and a sensor of absolute rotation is obtained, because the entire equipment rotates rigidly with respect to inertial space. Hence the second condition is : The magnetic field for the co-rotating observer is the same as that for the inertial observer, although the convection current vanishes for the co-rotating observer.

15.7 Physics of the Experiments

The experiments of Pegram and Kennard show that the observations are independent of whether the solenoid generating the coaxial field was stationary or rotating, at the same angular velocity as the cylindrical condenser. The interesting point is that the effects still exist even when he solenoid generating the coaxial B field rotates with the same angular velocity as of the cylindrical condenser.

Post and Bahulikar (1971) urged that these effects be considered as observations of a constitutive nature concerning a frame of reference rotating in free space. Then the sensitivity of the Kennard-Pegram sensors depends on the convection current as seen by an inertial observer. If Q is the charge of the convection current and T is the time for one rotation, then the volume current I is given by

$$I = \frac{Q}{T} \text{ stat amperes,} \tag{15.7.1}$$

where Q is in the state Columb and T is the time in second. Since $T = 2\pi/\Omega$, this equation reads

$$I = \frac{\Omega Q}{2\pi}, \tag{15.7.2}$$

where Ω is the uniform angular velocity of rotation in radians per second. I being proportional to Q, the sensitivity increases with increasing Q. A condenser with a high potential, therefore, gives a large Q. The magnetic field should be closed because charges of opposite sign occur on the two sides of the condenser plates. Ignoring the problem of closing the magnetic circuits, for a capacitance of 10^{-8} Faraday with a potential of 10^{-3} volt, the charge would be 10^{-8} Coulomb. The corresponding current I is given by

$$I = \frac{\Omega}{2\pi} \times 3 \times 10^4 \text{ stat amperes.}$$

If the volume current of amount 3×10 stat amperes is measurable, then frequencies as low as $2\pi \times 10^{-3}$ rad/sec are realisable.

In these devices the surface charge induced in the condenser plates produces an additional field such that the total D field will be divergence free; that is, without the space charge. This condition may, therefore, be used to find the expected emf or charge. The sensitivity of these devices depends on the total amount of magnetic energy that can be stored in a condenser. Hence the limitations are determined by space and magnitude of the field intensities.

The divergence of the second term in the right side of (13.6.35) leads to the right side of (13.4.24). Further, using the relation

$$E = -\frac{1}{c}(v \times B), \tag{15.7.3}$$

which is valid inside electrical conductors, it is evident that the right side of (13.4.24) is proportional to the Schiff charge of rotation. Hence an observation, made from a rotating frame, would record a space charge when non-existed in the beginning. In other words, it means that a charge is created merely by having the observer change from an inertial frame to a non-inertial frame, which, physically considering, is an absurd conclusion.

It is possible that the laws of electrodynamics in rotating systems are different from those in the rest system or inertial systems; and these different laws are likely to be such as to allow no field for a rotating observers, when there is no field for inertial observers. For instance, if a spherical condenser is at rest, no field is observed outside it by all observers, if it is set in rotation about a diameter, the field of a magnetic dipole is seen to exist outside. And an observer rotating about a diameter of the condenser at rest, cannot observe any field. The contradiction can be removed if one realises that a measurement of the emf requires the presence of some material object; for example, a set of condenser plates perpendicular to the expected direction of the emf. The surface charges induced in the condenser plates produce del operator an additional field such that the total E field, and so the total D field, will be divergence free, that is, without space charge separation. Discarding the constitutive relations $D = E$, $B = H$ which are valid only for inertial frames in vacuum, the conditions (13.6.35) and $(\nabla \cdot D) = 0$ have been used by Post and Bahulikar (1971) and Post (1972) to find the emf measured by Kennard or the charge expected by Pegram.

Post concluded that (1) The effects observed by these authors are realistic and their validity has to be accepted; (2) Constitutive relations like (13.6.35), which are of the form $D = D(E, B)$, rather than the inertial frame relation $D = E$, explain these observations; (3) A constitutive relation of the form (13.6.35) resolves the difficulty from the appearance of the Schiff charge and the related paradoxes. It has been assumed that (i) Divergence of individual electric field components of E contributing to a total electric displacement D is not physically meaningful; and (ii) Divergence of the *surface* vector D vanishes in the rotating systems when it vanishes in the inertial system. These assumptions can be applied in laboratory problems involving rotation of conductors and circuits. More generally, for a system of charges in vacuum, in the microscopic Lorentz theory, which is suitable, the constitutive relations are equivalent to the vacuum relations

$$D^0 = E^0 \tag{15.7.4}$$

$$D = E + \frac{1}{c}(v \times B) \tag{15.7.5}$$

along with the restriction

$$(\nabla^0 \cdot D^0) = (\nabla \cdot D), \tag{15.7.6}$$

where the field with the subscript 0 are those in the inertial frame K^0, and the fields free from that subscript are those in the rotating frame K.

In the experiment of Post (1972), section 15.6(b), the relevant equations are (1.4.11) and (13.6.36)

$$B^0 = H^0 \tag{15.7.7}$$

$$B = H + (\beta \times D). \tag{15.7.8}$$

Moreover, the inertial observer detects a volume current I of convection enclosed by the loop; so in the inertial frame

$$\iint_{\Sigma} \nabla \times H^0 \cdot d\Sigma = \oint_{\Gamma} (H^0 \cdot dr) = \frac{4\pi}{c} I. \tag{15.7.9}$$

But the rotating observer does not see a convection current; so in the rotating frame

$$\iint_{\Sigma} \nabla \times H \cdot d\Sigma = \oint_{\Gamma} (H \cdot dr) = 0. \tag{15.7.10}$$

Pegram's experiment indicates the possibility of finding the non-diagonal terms in the tensor of electric permeability of the relations (14.15.37)–(14.15.39). Due to the simultaneous rotation of coaxial cylindrical condensers and a solenoid about the common axis, the solenoid is energised and gives a magnetic field in the axial direction between the plates of the condenser. This charge being now trapped on the condenser and a solenoid, about the common axis, can be stopped. The solenoid is thus energised and gives a magnetic field in the axial direction between the plates of the condenser. This charge is observed by breaking the short again. Since the charge is trapped on the condenser, the rotation is stopped. An electrometer measurement then shows that the charge indeed exists, but its magnitude is of the order given by the non-diagonal terms of (14.15.37). The existence of a cross relation between the electric and magnetic fields for rotating frames in vacuum is indicated in the Pegram experiment.

For rotation the Maxwell equations retain their form and acquire extra terms related to the mutual changes in velocity of neighbouring inertial frames attached to the accelerated system. The transformation of constitutive equations is a purely algebraic matter that does not require the complexity of covariant derivatives. In local inertial frames the constitutive relations for free space are simply those of (1.4.11), and in a rotating frame, these are (13.6.35) and (13.6.36).

Since the observational phenomena related to rotation are independent of uniform translation, those are independent of the center of rotation.

Both Kennard and Pegram identified their experiments formally with those on unipolar induction. But, in unipolar induction, Sommerfeld (1952, p. 288–290), Trocheris (1949), Post (1972), each element can be considered to be a complete magnetic field source in linear motion, which is the condition for motional inductance of an electric field. Moreover, unipolar induction measures rotation between two material objects, and here the rotation of a single material object, with respect to inertial space, is measured. Hence the experiments of Kennard and Pegram cannot be identified with those on unipolar induction.

15.8 Explanation of the Experimental Findings

15.8.1 Explanation of findings of Pegram and Kennard

These authors assumed that the divergence of individual electric field components of E, contributing to a total electric displacement D, is not physically meaningful. Divergence of the *surface* vector D is taken and it is insisted that this divergence vanishes also on the rotating system, in which the state relation (13.6.35), $D = E - (\beta \times B)$, is assumed to be valid. In cylindrical co-ordinates, therefore,

$$\frac{1}{r}\frac{\partial}{\partial r}(rD_r) = 0, \tag{15.8.1}$$

and so,

$$D_r = \frac{A}{r}, \tag{15.8.2}$$

where A is a constant.

In Kennard's experiment an open circuit, with no displacement exists. So we put $A = 0$ in (15.8.2), then the Kennard potential V_k is given by

$$V_k = \int_{r_1}^{r_2} E_r dr = \frac{\Omega}{c}\int_{r_1}^{r_2} rB_z dr = -\frac{\Omega}{2c}B_z(r_2^2 - r_1^2). \tag{15.8.3}$$

In Pegram's case A can be obtained from the condition

$$\int_{r_1}^{r_2} E_r dr = 0. \tag{15.8.4}$$

Putting this relation in the radial component of (13.6.35), we find that

$$A = \frac{1}{2}\frac{\Omega B_z(r_2^2 - r_1^2)}{C\log(r_2/r_1)}. \tag{15.8.5}$$

The Pegram charge Q_P, on the condenser, is obtained by integrating D_r over the cylinder surface

$$Q_P = \int_0^{2\pi} D_r r d\theta \cdot l = 2\pi A l. \tag{15.8.6}$$

Hence (15.8.5) shows that

$$Q_P = \frac{\pi l \Omega B_z(r_2^2 - r_1^2)}{C\log(r_2/r_1)}. \tag{15.8.7}$$

The capacitance C_0 of the cylindrical capacitor is therefore given by

$$C_0 = \left|\frac{Q_l}{v_k}\right| = \frac{2\pi l}{\log(r_2/r_1)}. \tag{15.8.8}$$

15.8.2 Theory of Post's experiment

For the rationally symmetric cylindrical device, shown in Fig. 15.6.1, the solution of the solenoidal condition $(\nabla \cdot B) = 0$ for the magnetic induction is

$$B_r = \frac{A}{r}, \tag{15.8.9}$$

where A is a constant. Putting this value of B_r into the radial component of equation (13.6.36), we find that

$$H_r = \frac{A}{r} - \beta D_z. \tag{15.8.10}$$

For the co-moving observer who does not see a convection current, the constant of integration A is determined from the condition

$$\oint (H \cdot dr) = 0. \tag{15.8.11}$$

Neglecting the field contribution outside the condenser plate of Figure 15.6.1, we obtain

$$\oint (H \cdot dr) = \int_{r_1}^{r_2} H_r dr = 0, \tag{15.8.12}$$

where r_1 and r_2 are the inner and outer radii of the annular condenser plates. Hence

$$A = -\frac{\Omega D_z (r_2^2 - r_1^2)}{2c \log(r_2/r_1)}, \tag{15.8.13}$$

$$B_r = -\frac{\Omega D_z (r_2^2 - r_1^2)}{2cr \log(r_2/r_1)}. \tag{15.8.14}$$

The flux Φ of the magnetic induction, emerging from the space between the two plates, is given by

$$\Phi = \int_0^{2\pi} B_r r d\theta l = 2\pi r l B_r = \frac{l\Omega\pi D_z (r_2^2 - r_1^2)}{c \log(r_2/r_1)}, \tag{15.8.15}$$

where l is the distance of the plates. For the inertial observer the constitutive relation $B = H$, with B given by (15.8.9), is used because the Maxwell equations retain their form. However, the loop integral of H is not zero. The integration constant A determined by the relation

$$\oint (H \cdot dr) = \frac{4\pi}{c} I, \tag{15.8.16}$$

where I is the volume current of convection enclosed by the loop; I is given by

$$I = \frac{Q}{T} \text{ stat amperes}, \tag{15.8.17}$$

where Q is in stat Coulombs and $T(=2\pi/\Omega)$ in seconds, is the time for one rotation. Hence,

$$A\log\left(\frac{r_2}{r_1}\right)=\frac{4\pi}{c}I=\frac{4\pi}{c}\frac{\Omega Q}{2\pi}=\frac{2\Omega}{c}\iiint \rho r d\theta dr dz \tag{15.8.18}$$

$$=\frac{2\Omega}{4\pi c}\iiint(\nabla\cdot D)rd\theta drdz=\frac{\Omega}{2\pi c}\iint D_z d\Sigma_z,$$

because

$$(\nabla\cdot D)=4\pi\rho=\frac{\partial D_z}{\partial z}, \tag{15.8.19}$$

where $d\Sigma_z$ is the projection of the surface area element perpendicular to the z-axis; so

$$A=\frac{\Omega D_z(r_2^2-r_1^2)}{2c\log(r_2/r_1)} \tag{15.8.20}$$

because

$$\iint D_z d\Sigma_z=D_z\pi(r_2^2-r_1^2).$$

Since the magnitudes of this A and that of (15.8.13) are identical, the calculations for the flux Φ of B yield the same result. Hence this special case confirms the view that the magnetic field for the rotating observer is the same as that for the observer in the inertial frame.

15.9 Local and Global Aspects of Electrodynamic

Global and topological notions are common to all geometries. So, topology and physics are well related. Actually, the study of topological aspects of physics does not require the specification of the detailed structure of the geometry of the physics, Post (1972, 1972a, 1972b, 1972c, 1974).

The Newtonian equations of motion are local in nature, being differential equations, and are meant for determining the trajectory of mass points or centers of gravity provided initial conditions, constrains and force laws are given. The abstraction known as the *mass point* is an example of a *localizing* of concepts. But Fermats' principle for the minimal length of a light path is a global conceptualisation of a problem. The works of Euler, Lagrange, Hamilton and others on the formalism for the mechanics of a system of particles gave a partial globalisation to the inherently local concepts of point mechanics. The extension and generalisation of the laws of point dynamics, to the dynamics of continuum by Euler, emerge essentially as a local theory. The fundamental law of induction of Faraday, being an integral law, is global. It was made a local differential law by Maxwell. The equations of gravitational field of GTR are local, but the solutions are full of global elements.

Generally an original global law has to be reduced, with the help of the theorems of Gauss and Stokes[1] to a local differential law for application. More generally, an integral with respect to n variables is reduced to integrals with respect to $n-1$ and $n-2$ variables. These theorems, namely, the laws of Gauss, Green and the fundamental theorem of integral calculus, under the name of generalised Stoke's theorem, determine the relations between global and local approaches in mathematical physics. It is not always possible to have meaningful global extensions of all local relations. So, simultaneous development of both the local and global points of view, though desired, is not possible.

[1]These theorems transform volume integrals to surface integrals and surface integrals to line integrals.

The formulation of quantum mechanics by a variation principle by Schroedinfger has a bearing on basic assumptions concerning the geometry of physical space. The mathematical conditions obtaining these stationary states, namely (1) the uniqueness of the wave function and (2) the square integrability of the wave function over the whole space, are global aspects of physics.

The local and global view points are not related to the micro and macro systems in physics. A micro system can have a global character and a macro theory, such as continuum dynamics, may very much lack a proper global formulation.

Local differential geometry deals with a collection of differential expressions relating to local geometry properties in the neighbourhood of points of lines, surfaces or manifolds, in general. Global differential geometry is concerned with the integrals of these local differential relations. When the local discipline of differential geometry fails, the local extension is necessary. Study of local and global aspects of surfaces or manifolds, in general, reflects the distinction between these two aspects of differential geometry.

For the study of curved lines and surfaces, in 3-dimensions, both the local and global results and the concepts of topological configurations in the space of 3-dimensions are required. But the global insights cannot keep up with the local insights in higher dimensional manifolds.

The generalised Stoke's theorem encompassing the theorens of Gauss, Green and Stokes and the fundamental theorem of integral calculus as special cases, referred to already, are concerned with integrals of differentials. The coefficients of these differentials, in a cartesian frame, are the field quantities of physics, which are completely antisymmetric covariant tensors in arbitrary frames of reference. The valence of the tensor is the number of its subscripts and corresponds to the degree of the differential which is the number of integration variables.

The generalised Stoke's theorem relates global and local statements in physics. For instance, the Maxwell equations as relations between field derivatives and local statements are referred from the corresponding global integral statements, like Faraday's induction law, representing experimental observations, through the use of Stoke's theorem. Usually the local statement is inferred through Stoke's law. The multidimensional generalisation of Stoke's theorem in a complete form, with the help of the local and global analyticity of the theory of complex functions, shows the existence of de Rham periods which take the place of the Cauchy residues without necessarily having some of the unwanted features of the residue pole of complex functions.

Analyticity in complex function theory requires that the derivative be independent of the direction along which two points are made to coincide. It follows that the real and imaginary components of the function satisfy the Cauchy-Riemann equations. The condition of local analyticity in the neighbourhood of a point can be extended into a global condition by requiring that the Cauchy-Riemann equations are met in a domain of definition of the function. The function is then called *halomorphic* or *regular* in that domain. Global analyticity can be defined by demanding that certain line integrals along arbitrary contours inside a domain vanish. These contour integrals are defined uniquely when the function is single valued.

For a field that is a differential, a condition corresponding to the Cauchy-Riemann equations, is that the exterior derivative of a field vanishes. The field is called *closed* in the neighbourhood of a point where its exterior derivatives vanishes. Hence the property of a field being closed is a local property. A corresponding global property cannot be obtained by a field which is closed everywhere in a region. Though the familiar notions of Curl and divergence are special cases of the exterior derivative, any curl and divergence operation cannot be translated meaningfully into an exterior derivative operation.

The global counterpart of the local concept of a field being closed can be obtained from global analyticity by requiring that certain contour integrals in a region vanish. For a global extension of the local concept *closed* all *cyclic* integrals of a closed form vanish, or, in other words, the closed form is exact. The term *cyclic* denotes that the integrations are performed over closed domains in the region. A boundary of a region is closed, but any closed domain in a region is not necessarily a boundary. The sum of a number of closed domains can constitute a boundary. For example, an annular region, bounded by two closed cyclic domains, together, form a boundary but not individually.

15.10 Comments on Nature of Tensor Fields

A set of basic fields are prototypes of a specific transformation behaviour. The prototype fields are scalar (example : potential per unit charge), covariant vectors (example : wave vector), densities (example : charge per unit volume) and contravariant vectors (example : particle velocity). The assumed behaviour of transformation of the prototype fields fixes, not necessarily in a unique manner, the behaviour of transformation of the derived fields.

Some convention is necessary for relating the EM field tensors for the covariant equations with the vector and scalar EM field variables. The convention that E and B shall appear in the Lorentz force law, and that D and H shall be related to the charge current density, as in the Biot-Savart law, (1.4.3), turns out to the requirement that the Maxwell equations have the same form in both inertial and rotating frames of reference.

The elementary definition of physical quantities generally refers only to rectangular components, and must be supplemented before deciding on whether the physical vector is contravariance or covariant. An example of the non-uniqueness of the derived fields from the prototype, section 14.6, is the 4-force, which is derived both as a contravariant 4-vector and as a covariant 4-vector. Forces can be created by a change in the frame of reference; so these are non-tensorial, Post (1962, p. 37–38). However, the sum of inertial and gravitational forces may be tensorial. The Lorentz force is another example of a tensorial force field if it is regarded as a contravariant vector density.

The momentum is also defined either as a contravariant 4-vector or as a covariant 4-vector in inertial frames. If, together with the 3-vector particle momentum p, the 4th component is $m\gamma c$ ($m\gamma$ being the effective mass), then in inertial frames, the 4-momentum $(p, m\gamma c)$ is covariant. However, if the 4th component is $-\mathcal{E}/c$, $\mathcal{E}$ being the energy, then the 4-momentum $(p, -\mathcal{E}/c)$ is covariant. The interpretation of the 4-momentum as a contravariant 4-vector, Post (1962, p. 14–15), breaks down for particles without a rest mass. The covariant interpretation, on the other hand, is still meaningful for zero rest mass and can be related to the 4-vector $(K, -\omega/C)$ of wave vector frequency of a wave by the invariant action $\hbar$, thus establishing the intrinsic covariant character of the momentum-energy 4-vector.

Some limitations existing in the use of the metric for associating covariant and contravariant components is a weak argument in favour of an intrinsic transformation behaviour, Post (1962). Actually, the corroborative nature of the different methods of approach encourages the investigation of the meaning of intrinsic transformation behaviour.

15.11 Metric Dependence of Laws of Electrodynamics

The notion of metric, writes Van Dantzig (1934b), is a very complicated one. It requires measurements with clocks and scales, generally with rigid bodies, which are themselves things of extreme complexity. Hence, it seems undesirable to take the notion of a metric as a fundament also of phenomena which are much simpler and independent of it. Formulation of the fundamental laws of physics, both of EM theory and material waves, in a form independent of material geometry, is an important goal. The metric turns out to be a system of some statistical mean values of certain physical quantities. An absolute 4-space, however, remains the mathematical background of the theory, at least, as far as classical mechanics is concerned. Consequently, each special kind of geometry has to become an a posteriori property of the physical quantities. Moreover, metrical geometry has led to the introduction of several kinds of connections.

In the ordinary, covariant tensor theory, quantities characterizing the metric $g_{\mu\nu}$ and quantities like the magnetic permeability μ^{ij} and the dielectric permittivity $\mathcal{E}^{ij}$ are necessary. However, it is possible to avoid $g_{\mu\nu}$ and to make use of quantities only, which characterize matter in the sense of *environment*, including the empty space, the so-called *ether*, but not including the EM field itself.

A connection, Van Dantzig (1934b), is, an arbitrary method of linking up quantities in different points in space time. Choosing a special connection in order to express the equations of physics in an invariant manner is a process of quite the same nature as choosing a special system of co-ordinates, viz., a mathematical (*formal*) act. Hence, all equations which occur should be made *naturally* invariant, under arbitrary transformations of co-ordinates.

The Maxwell equations are sometimes written in the metric-independent form, so that they are formally equivalent to these in a macroscopic medium in the Minkowskian 4-space, Tamm (1920), Plebanski (1960). At several places in their voluminous monumental treatise on gravitation, Misner et al (1973, 12.5, p. 302-3; 13.2, p. 305) have discussed metric independent field equations of classical physics.

Minkowski (1908) showed that the equations of Maxwell for macroscopic electrodynamics are invariant under the affine group of space time, the fundamental metric $\eta_{\mu\nu}$ is adjoined, and this invariance is equivalent to invariance under the Lorentz group. This affine group is extended to the group of all arbitrary, continuous transformations in space-time, if the ordinary derivatives, belonging to the Riemannian connections, is determined by the metric elements $g_{\mu\nu}$.

Cunnigham (1910) and Bateman (1910, 1920) found that the phenomenological Maxwell equations are invariant under a still wider group, namely the group of all conformal transformations in space time. Schouten and Haantjes (1934) obtained these equations in a form, invariant under arbitrary transformations of co-ordinates in space-time by adjoining, instead of the metric tensor $g_{\mu\nu}$, its density

$$\bar{g}_{\mu\nu} = (-g)^{-1/4} g_{\mu\nu}, \quad g = \det g_{\mu\nu}. \tag{15.11.1}$$

The electrodynamical equations of Maxwell can be shown to be altogether independent of metric-projection, Riemannian or conformal, Van Dantzig (1934a).

The topology of fields can be built on a 4-dimensional space-time in terms of a set of fundamental differential forms which are assumed to exist for all physical systems having a field theory. An interpretation of flux, charge and angular momentum quantisation is possible in terms of a set of naturally independent concepts relating to the periods of one-, two- and

3-dimensional cyclic integrals in space-time. Since the topological invariance of the periods of these integrals is much stronger than the metric invariance, the space-time manifold itself is neither equipped with a particular topology, nor indeed, endowed with a metric which would, of course, automatically, induce a topology. But the required set of fundamental differential forms can be defined on a differentiable manifold without a metric, when the physical theory is the 4-dimensional formulation of electrodynamics?

The co-ordinasation of events involves not the raw data received by an observer but his logging of events after allowing for the time taken for signals to reach him. Hence, the required set of fundamental differential forms, chapter 9 (say, the two differential forms F, involving B and E, and G, involving H and D), cannot be defined by experiment on a differential manifold without a metric. The corollary that space-time must be regarded from the outset as being endowed with a semi-definite metric makes it easier to accommodate the fact that all known residues of two forms like G appear to relate to 2-dimensional cyclic domains of a purely spatial nature.

The metric-independent properties of the medium-independent form of the Maxwellian laws for electrodynamics have also been referred to, by Post (1984), as the premetric properties of the precencepts pertaining to position space as well as to space time. The electrodynamical laws in this form were established by Van Dantzig (1934a–c) and others, see Post (1984). In this demonstration the premetric condition was found to be applicable to the preconstitutive Maxwell laws for all conceivable media, including the free-space (ether) medium with its physical properties determined by the metric. Birss (1982) found that the Maxwellian law can be made to be independent of the spatial metric, but not independent of the space-time metric.

For determination of the metric independent field equations of electrodynamics, it is necessary to postulate a metric locally (that is, to define an inner product in the tangent space) for electrodynamics formulated either in the 3-dimensional space of position co-ordinates or in the 4-dimensional space of space-time, Birss (1982), Post (1984). They concluded that in 3-space a metric-independent formulation is possible, but in 4-space the required set of fundamental differential forms cannot be defined by experiment on a differential manifold without a metric.

A comparison between the spatial and the space-time cases of counting quanta illustrate the essence of the metric distinction made by Birss (1982). In the spatial cases quanta can be frequently directly perceived, say, in terms of discrete levels of magnetic flux, which can be recorded regardless of the choice of units or the calibration of the instrument, provided the sensitivity is there. However, in space-time, quantas are being perceived indirectly through the intermediary of metric measurement. The metric specifies, through drop outs in the end, honour the earlier theory in which the metric concepts were used in electrodynamics in space, as well as, in space time.

Appendix A

Delta Functions of Dirac

Properties of the delta function :

1. $\delta(x - x_0) = 0$ if $x \neq x_0$ and $\delta(x - x_0)$ is undefined at $x = x_0$.

2. $\displaystyle\int_{x_0-\varepsilon}^{x_0+\varepsilon} \delta(x - x_0)dx = 1$ if $\varepsilon > 0$.

3. $$\int_a^b f(x)\delta(x - x_0)dx = \begin{cases} f(x_0) \text{ when } a < x < b \\ 0 \text{ when } x < a \text{ or } x > b \text{ if } a < b. \end{cases}$$

4. $\displaystyle\int_a^b F(x)\delta(x)dx = F(0)$ if $a < 0$ and $b > 0$.

5. $\delta(-x) = \delta(x)$.

6. $x\delta x = 0$.

7. $\delta(ax) = \dfrac{\delta(x)}{|a|}$.

8. $f(x)\delta(x - a) = f(a)\delta(x - a)$

9. (a) $\displaystyle\int \delta(a - x)\delta(x - b)dx = \delta(a - b)$; (b) $\displaystyle\int f(x)\delta(x - a)dx = f(a)$;

   (c) $\displaystyle\int f(x)\delta'(x - a)dx = -f'(a)$;

   (d) $\displaystyle\iint f(x, y)\delta'(x - a)\delta'(y - b)dxdy$

$$= \iint \left[\frac{\partial^2 f}{\partial x \partial y}\delta(x - a)\delta(y - b)\frac{\partial^2 f}{\partial x \partial y}\delta(x - a)\delta(y - b) - \frac{\partial f}{\partial x}\delta(x - a)\delta'(y - b) - \frac{\partial f}{\partial y}\delta'(x - a)\delta(y - b)\right] dxdy$$

$$= 0 - f_{xy}(a, b) + f_{xy}(a, b) + f_{xy}(a, b) = f_{xy}(a, b).$$

10. $\delta(x^2 - a^2) = \dfrac{\delta(x-a) + \delta(x+a)}{2|a|}.$

11. $\delta[\phi(x)] = \sum_i \dfrac{\delta(x - x_i)}{\left[\left(\dfrac{d\phi}{dx}\right)_{x=x_i}\right]}.$

So, $\displaystyle\int_{-\infty}^{\infty} \delta[\phi(x)] f(x) dx = \sum_i \frac{f(x_i)}{\left|\dfrac{d\phi}{dx}\right|_{x=x_i}},$

where x_i are the roots of the equation $\phi(x) = 0$.

12. $\nabla^2(1/r) = -4\pi\delta(r)$.

13. In three dimensions $\delta(r - r_1) = \delta(x - x_i)\ \delta(y - y_i)\ \delta(z - z_i)$ where the components of r and r_i in rectangular cartesian coordinates are (x, y, z) and (x_1, y_1, z_1) respectively.

Appendix B

Theorems of Gauss, Green and Stokes

Let A be a vector field and let ϕ and ψ be scalar fields. Also let V be a volume bounded by the surface Σ. If n is the unit vector normal to the surface element $d\Sigma$ then $nd\Sigma = d\Sigma$, and Gauss's theorem reads

$$\iiint\limits_V (\nabla\cdot A)dV = \iint\limits_\Sigma (A\cdot d\Sigma) \quad \text{or} \quad \nabla\cdot A = \lim_{V\to 0}\left(\frac{1}{V}\iint\limits_\Sigma (A\cdot d\Sigma)\right) \tag{A2.1}$$

Green's Theorem :

$$\iint\limits_\Sigma \left(\psi\frac{\partial\phi}{\partial n} - \phi\frac{\partial\psi}{\partial n}\right)d\Sigma = \iiint\limits_V (\psi\nabla^2\phi - \phi\nabla^2\psi)dV \tag{A2.2}$$

where $\frac{\partial\phi}{\partial n} = (n\cdot\nabla)\phi$.

Stoke's Theorem :

$$\iint\limits_\Sigma ((\nabla\times A)\cdot d\Sigma) = \oint_\Gamma (A\cdot dl), \tag{A2.3}$$

where Γ is a closed contour drawn on Σ and dl is a line element of Γ. Other relevant formulas are

$$\iiint\limits_V \nabla\phi dV = \iint\limits_\Sigma \phi n d\Sigma = s\int \phi d\Sigma, \tag{A2.4}$$

$$\iint\limits_\Sigma ((n\times\nabla\phi)d\Sigma) = \oint_\Gamma \phi dl,$$

$$\iint\limits_\Sigma ([B\times(\nabla\times A) - A\times(\nabla\times B)]\cdot d\Sigma) = \iiint\limits_V (A\cdot[\nabla\times(\nabla\times B)] - B\cdot[\nabla\times(\nabla\times A)])dV,$$

$$\iiint_V \nabla \times B dV = \iint_\Sigma (d\Sigma \times B),$$

$$\iiint_V (\nabla \cdot \mathbf{T}) dV = \iint_\Sigma (\mathbf{T} \cdot d\Sigma),$$

where **T** is a tensor.

Appendix C

Vector Formulae

1. $A\cdot(B\times C) = B\cdot(C\times A) = C\cdot(A\times B) = (ABC)$. (A3.1)

2. $A\times(B\times C) = (A\cdot C)B - (A\cdot B)C$. (A3.2)

3. $(A\times B)\times(C\times D) = (ABD)C - (ABC)D = (ACD)B - (BCD)A$. (A3.3)

4. $(A\times B)\cdot(C\times D) = (A\cdot C)(B\cdot D) - (A\cdot D)(B\cdot C)$. (A3.4)

5. $\mathrm{grad}(\phi\psi) = \phi\,\mathrm{grad}\,\psi + \psi\,\mathrm{grad}\,\phi$. (A3.5)

6. $\mathrm{div}(\phi A) = \phi(\nabla\cdot A) + (A\cdot\mathrm{grad}\,\phi)$. (A3.6)

7. $\mathrm{curl}(\phi A) = \phi\,\mathrm{curl}\,A + (\mathrm{grad}\,\phi\times A)$. (A3.7)

8. $\mathrm{div}(A\times B) = (B\cdot\mathrm{curl}\,A) - (A\cdot\mathrm{curl}\,B)$. (A3.8)

9. $\mathrm{curl}(A\times B) = (B\cdot\nabla)A - (A\cdot\nabla)B + A(\nabla\cdot B) - B(\nabla\cdot A)$. (A3.9)

10. $\mathrm{grad}(A\cdot B) = (B\cdot\nabla)A + (A\cdot\nabla)B + (B\times(\nabla\times A)) + (A\times(\nabla\times B))$. (A3.10)

11. $\mathrm{curl}\,\mathrm{curl}\,A = \mathrm{grad}\,\mathrm{div}\,A - \nabla^2 A$. (A3.11)

12. $\mathrm{div}\,\mathbf{r} = 3$ where r is a position vector. (A3.12)

13. $\mathrm{curl}\,\mathbf{r} = 0$ where $\mathbf{r}$ is a position vector. (A3.13)

14. $\mathrm{curl}\,\mathrm{grad}\,\phi = 0$. (A3.14)

15. $\mathrm{div}\,\mathrm{curl}\,A = 0$. (A3.15)

16. $(A\times\nabla)\times B - A\times(\nabla\times B) = (A\cdot\nabla)B - A(\nabla\cdot B)$. (A3.16)

Appendix D

Kronecker Delta and Levi Civita Symbols

The Kronecker delta is defined as

$$\delta_{ij} = \begin{cases} 1 \text{ if } i = j, \\ 0 \text{ if } i \neq j. \end{cases} \tag{A4.1}$$

So

$$\delta_{ij} = \begin{bmatrix} 1 & 0 & 0 \\ 0 & 1 & 0 \\ 0 & 0 & 1 \end{bmatrix}. \tag{A4.2}$$

The alternative unit tensor ε_{ijk} is defined as

$$\varepsilon_{ijk} = \begin{cases} +1 & \text{if } i, j, k \text{ are in cyclic order : } 123,\ 231,\ 312 \\ -1 & \text{if } i, j, k \text{ are in noncyclic order : } 321,\ 132,\ 213 \\ 0 & \text{if any of the subscripts is repeated.} \end{cases} \tag{A4.3}$$

The Levi Civita symbols of fourth order, $\varepsilon_{\alpha\beta\gamma\delta}$ is anti-symmetric in all the four indices $\alpha, \beta, \gamma, \delta$. The nonvanishing components of $\varepsilon_{\alpha\beta\gamma\delta}$ are those for which all four indices are different, and they are equal to $+1$ or -1 according as $(\alpha, \beta, \gamma, \delta)$ is obtained through an even or odd permutation of (1, 2, 3, 4). So, $\varepsilon_{\alpha\beta\gamma\delta}$ vanishes if any two or more of these indices is identical.

Bibliography

1. Alexeff and Jones. *Phys. Fluids*, **12**, 345–346, (1969).
2. Acheson, D.J., R. Hide. *Hydromagnetics of rotating fluids*, Rept. Prog. Phys., **36**, 159–221, (1973).
3. Adler, R.M., M. Bazin and M. Schiffer. *An introduction to general relativity*, McGraw Hill, New York, (1975).
4. Alfven, H. *Cosmical electrodynamics*, Oxford University Press, London, (1953).
5. Anderson, J.L. and J.W. Ryon. *Electromagnetic Radiation in Accelerated Systems*, Phys. Rev. **181**, 1765–1775, (1969).
6. Arons, J. and C.E. Max., *Self precession and frequency shift for electromagnetic waves in homogeneous plasma, Phys. Fluid,* **17**, 1983–1994, (1994).
7. Armstrong, J.A., N. Bloembergen, J. Ducuing and P.S. Pershan. *Interactions between light waves in nonlinear dielectrics, Phys. Rev.* **127**, 1918–1939, (1962).
8. Atwater, B.A. *The electromagnetic field in rotating coordinate frames*, Proc. IEEE (Lett.) **63**, 316–318, (1975).
9. Arzelies, H. *Relativistic kinematics*, Pergamon Press, Oxford, (1966).
10. Arzelies, H. *Relativistic point dynamics*, Pergamon Press, Oxford, (1972).
11. Barnett, S.J. *On Electromagnetic induction and relative motion*, Phys. Rev. XXXV, 323–336, (1912).
12. Bateman, H. *The transformation of the electrodynamical equations*, Proc. London, Math. Soc. **8**, 223–264, (1910).
13. Bateman, H. *On lines of electric induction and the conformal transformation of a space of four dimensions*, Proc. London Math. Soc. **21**, 256–270, (1920).
14. Bateman, H. *Electrical and Optical wave motion*, Cambridge University Press, Cambridge, 1915.
15. Birss, R.R. *The metric dependence of four-dimensional formulation of electromagnetism*, Jour. Math. Phys. (USA), **23**, 1153, (1982).
16. Bekefi, G. *Radiation Processes in plasmas*, Wiley, New York, (1966).

17. Bladel, J. Van. *Electromagnetic field in presence of rotating bodies,* Proc. IEEE, **64,** 30–318, (1976).

18. Bladel, J. Van. *Relativistic theory of rotating disks,* Proc. IEEE, **61,** 260–268, (1973).

19. Balescu, R. *Statistical mechanics of charged particles.* Interscience, New York, (1964).

20. Bellman, R. *Perturbation techniques in mathematics, Physics and Engineering,* Holt, Rinehart and Winston, New York, (1966).

21. Bloembergen, N. and P.S. Pershan. *Light waves at the boundary of nonlinear media,* Phys. Rev. **128,** 606–612, (1962).

22. Bloembergen, N. and Y.R. Shen. *Quantum theoretical comparison of nonlinear susceptibilities in parametric media,* Laser and Raman Lasers, Phys. Rev. **133,** A37–A49, (1964).

23. Bloembergen, N. *Nonlinear Optics,* W.A. Benjamin, New York, (1965).

24. Bow, Y.F. *Note on the Hamiltonians of a charged particle in a rotating frame,* Amer. Jour. Phys. **39,** 452, (1971).

25. Bow, Y.F. *The Hamiltonian of a charged particle in a rotating frame,* Amer, Jour. Phys. **40,** 252–256, (1972).

26. Boyd, T.J.M. and J.J. Sanderson. *Plasma Dynamics,* Nelson, London, (1969).

27. Bobroff, L. *Independent space variables for small signal electron beam analysis,* IEEE Trans. Electrical Devices, ED **6,** 68–78, (1959).

28. Brandstatter, J.J. *An introduction to waves rays and radiation in plasma media,* McGraw Hill, New York, (1963).

29. Bremmer, H. *Propagation of electromagnetic waves,* Handbuch der Physik, Band XVI, Springer Verlag, Berlin, (1958).

30. Brinkman, H.C. *Magnetohydrodynamics of rotating plasmas,* Riso Report No. 18, DAEC, Riso, Denmark, p. 55–60, (1960).

31. Bulletin of American Physical Society. *Physics and Astronomy Classification Scheme – 1977,* **22,** 218, (1977).

32. Corben, H.C. *Classical and Quantum Theories of Spinning Particles;* Holden Day Inc. San Francisco, (1968).

33. Cambel, A.B. *Plasma Physics and Magnetofluid Mechanics,* McGraw Hill, New York, (1963).

34. Caldirola, P. *A relativistic theory of classical electron,* La Rivista del Nuovo Cimento, **2,** 1–49, (1979).

35. Chakraborty, B. *Variational Classical Electrodynamics,* Physics Report, **156,** 1–65, (1987).

36. Chakraborty, B. and S. Sarkar. *Physics in rotating frames I on uniform rotation about a fixed axis in some holonomic and anholonomic frames,* Annals of Phys. (USA), **163,** 167–198, (1985).

37. Chakraborty, B. M. Khan and B. Bhattacharyya. *Impact of strong radiation on neutral atoms in the plasma background,* Phys. Rev. **23A,** 345–351, (1981).

38. Chandrasekhar, S. *Hydrodynamic and Hydromagnetic stability,* Oxford University Press, London, (1961).

39. Chapman, S. and T.G. Cowling. *The mathematical theory of nonuniform gases,* Cambridge University Press, London, (1960).

40. Chakraborty, B. *Principles of Plasma Mechanics,* New Age International, New Delhi, (Third Edition), (1997).

41. Chakraborty, B., S.N. Paul, M. Khan and B. Bhattacharyya. *Nonlinear precessional rotation of strong EM waves in a plasma,* Physics Reports, **114,** 181–317, (1984).

42. Clemmow, P.C. and J.P. Dougherty. *Electrodynamics of particles and plasmas.* Addision Wesley, Reading, (1969).

43. Coisson, R. *On the vector potential of Coriolis forces,* Amer. Jour. Phys., **41,** 585–586, (1973).

44. Corum, J.F. *Relativistic rotation and the anholonomic object,* Jour. Math. Phys. **18,** 770–776, (1977).

45. Corum, J.F. (1980), *Relativistic covariance and rotational electrodynamics,* Jour. Math. Phys. **21,** 2360–2364, (1980).

46. Courant, R. and D. Hilbert. *Methods of mathematical physics,* Vol. I, Interscience, New York, (1953).

47. Cowling, T.G. *Magnetic fields of Sunspots,* Math. Royal Astronism. Soc., **94,** 39–48, (1934).

48. Cullwick, E.G. *The fundamentals of electromagnetism,* University Press, Cambridge, UK, (1966).

49. Cullwick, E.G. *Electromagnetism and Relativity,* (Longmans Green, London), (1959).

50. Cunningham, E. *The Principle of Relativity in Electrodynamics and an Extension thereof,* Proc. London Math. Soc., **8,** 77, (1910).

51. Cheo, P.K. and C.V. Heer. *Beat frequency between two travelling waves in Fabry-Petro square cavity,* Applied optics, **3,** 786–789, (1964).

52. Dantzig, D. Van. *The fundamental equations of electromagnetism independent of metrical geometry,* Proc. Cambridge Phil. Soc., **30,** 421–27, (1934a).

53. Dantzig, D. Van. *Electromagnetism independent of metrical geometry, 2. Variational Principles and further generalization of the theory,* Proc. Royal Acad. Amsterdam, **37,** 521–25, (1934b).

54. Dantzig, D. Van. *Electromagnetism independent of metrical geometry,* 2. *Variational Principles and further generalization of the theory,* Proc. Royal Acad. Amsterdam, **37,** 526–31, (1934c).

55. Dantzig, D. Van. *Electromagnetism independent of metrical geometry,* 3. Mass and Motion, Proc. Royal Acad. Amsterdam, **37,** 643–52, (1934d).

56. Dantzig, D. Van. *Electromagnetism independent of metrical geometry,* 4. Momentum and energy, Proc. Royal Acad. Amsterdam, **37,** 825–36, (1934e).

57. De Groot, S.R. *The Maxwell equation,* North Holland, Amsterdam, (1969).

58. Delcroix, J.L. *Plasma Physics,* Vol. I & II, Wiley, New York, (1965).

59. De Groot, S.R. and L.G. Suttorp. *Foundation of Electrodynamics,* North Holland, Amsterdam, (1972).

60. Deschamps, G.A. Pages 111–161 in *Mathematics, Applied to Physics,* eidted by E. Roubine, Springer Verlag, Berlin, (1970).

61. Dixon, W.G. *Special Relativity,* Cambridge University Press, Cambridge, (1978).

62. Dutta, A.K. *Electromagnetism, electron theory and special theory of relativity,* Macmillan, Delhi, (1977).

63. Einstein, A. *In the Principle of Relativity,* Original in German in Annalender der Physik, **17,** 891, Dover publication, New York, (1905).

64. Eichenwald, A. *Ann. der Physik,* **II,** 421, (1903).

65. Engeles, E. and F. Verheest. *Wave propagation in rotating plasma and influcence of the Coriolis force,* Astrophys. Space Science, **37,** 427–440, (1970).

66. Fano, R.M., L.J. Chu and R.B. Adler. *Electromagnetic fields, energy and forces,* Wiley, New York, (1960).

67. Ferraro, V.C.A. and C. Plumpton. *An introduction to magnetofluid mechanics,* Clarendon Press, Oxford, (1966).

68. Fock, V. *Theory of Space, time and gravitation,* Pergamon Press, Oxford, (1964).

69. Forward, R.L. *General Relativity for the experimentalist,* Proc. IRE **49,** 829–904 (1961).

70. Flanders, H. *Differential forms,* Academic Press, New York, (1963).

71. Ghosh, P.K. *Lectures on Theory of Generalized Functions,* Calcutta University (unpublished), (1957).

72. Gold, T. *Rotating Neutron stars at the origin of the pulsating radio sources,* Nature, **218,** 731–732, (1968).

73. Goldreich, P. and W.H. Julian. *Pulsor Electrodynamics,* Astrophys. Jour. **157,** 869–880, (1969).

74. Goldstein, S. *Concerning a Continuum Theory of Electrodynamics of Gaseous Plasmas,* Polytechnique Institute of Brooklyn, Report, p. 65–80, (April, 1961).

75. Goldstein, H. *Classical Mechanics,* Addision Wesley, Reading, (1980).

76. Gron, O. *Application of Schiff's Rotating Frame,* Electrodynamics, Internat. Jour. Theor. Phys. **23,** 441–448, (1984).

77. Gron, O. *The asynchronous formulation of relativistic statics and thermodynamics,* Nuovo Cimento, **17B,** 141–165, (1973).

78. Heer, C.V. *Resonant Frequencies of an Electromagnetic Cavity in an Acclerated system of Reference,* Phys. Rev. **134,** A799–A803, (1964).

79. Hughes, W.F. and F.J. Young. *The Electromagnetodynamics of Fluids,* John Wiley, New York, (1966).

80. Hide, R. and P.H. Roberts. *Some elementary problems in Magnetohydrodynamics; in Advances in Applied Mechanics,* vol. **VII,** p. 216–236 and p. 274–285, (1962).

81. Ingarden, R.S. and A. Jamiolkowski. *Classical Electrodynamics,* Elsevier, Amsterdam, (1985).

82. Irvine, W.M. *Electrodynamics in a Rotating System of Reference,* Physica, **30,** 1160–1170.

83. Ise, J. and J.I. Uretsky. *Vacuum Electrodynamics on a Merry-go-round,* Amer. Jour. Phys. **21,** 431–435.

84. Ise, J. *Relativistic Electrodynamics of Rotating Media,* UCRL–3636, p. 1–24, (1957).

85. Jackson, J.D. *Classical Electrodynamics,* John Wiley, New York, (1975).

86. Kalman, G. *Note on the Lagrangian Formalism in Relativistic Mechanics,* Phys. Rev. **123,** 384–390, (1961).

87. Kennard, E.H. *Unipolar induction,* Phil. Mag. **23,** 937–941, (1912).

88. Kennard, E.H. *On Unipolar Induction : Another Experiment and its significance as Evident for the Existence of Ether,* Phil. Mag., **33,** 179–190, (1917).

89. Khromykh, A.M. *Ring Generator in a Rotating Reference Frame,* Sov. Phys. JETP, **23,** 185–186, (1966).

90. Kong, J.A. *Theory of Electromagnetic Waves,* John Wiley, New York, (1975).

91. Keller, J.B. and B.R. Levy. *Scattering of short waves,* Pergamon Press, New York, (1963).

92. Kline, M. and I.W. Kay. *Electromagnetic theory and geometrical optics,* Interscience, New York, (1965).

93. Kittel, C. *Introduction to Solid State Physics,* Wiley, New York, (4th Edition), (1976).

94. Kobe, D.H. *Derivation of Maxwell equations from the local gauge invariance of quantum mechanics,* Amer. Jour. Phys., **46,** 342–348, (1978).

95. Klemperer, O. *Electron Physics,* Butterworth, London, (1959).

96. Khan, M.R. and V. Chakraborty. *Precessional rotation and other nonlinear effects due to standing waves and other two-wave resonant interactions in a magnetized plasma,* Applied Phys., **50,** 7977–85, (1979).

97. Lamb, Jr. W.E. *Theory of Optical Maser,* Phys. Rev., **A134,** 1429–1450, **(1964).**

98. Landau, L.D. and E.M. Lifshitz. *Electrodynamics of Continuous Media,* Pergamon Press, Oxford, (1960).

99. Landau, L.D. and E.M. Lifshitz. *The Classical Theory of Fields,* Pergmon Press, Oxford, (1975).

100. Landau, L.D. and E.N. Liftshitz. *Mechanics,* Pergamon Press, Oxford, (1978).

101. Landu, L.D. and E.N. Liftshitz. *Statistical Mechanics,* Pergamon Press, Oxford, (1969).

102. Lehnert, B. *Magnetohydrodynamic waves under the action of Coriolis force,* Astrophys Jour., **119,** 647, (1954).

103. Lehnert, B. *Rotating Plasmas,* Nuclear Fusion, **11,** 485–533, **(1971).**

104. Lorentz, H.A. *Theory of electrons,* (1915), Dover, New York, Reprint, (1952).

105. Low, F.E. *A Lagrangian formulation of the Boltzmann Vlasov equations for plasmas,* Proc. Roy. Soc., **A245,** 282–287, (1958).

106. Luneburg, R.K. *Mathematical theory of optics,* University of California Press, Berkeley, (1965).

107. Mathews, J. *Coordinate-free rotation formalism,* Amer. Jour. Phys., **44,** 1210, (1976).

108. Mandel'shtam, L.I. *Collected Works,* 1. Acad. Nauk, USSR, Original Paper, (1924).

109. Marsh, J.S. *Magnetic and electric fields of rotating charge distribution,* Amer. Jour. Phys. **50(1),** 51–53, (1982).

110. Mazur, P. *On statistical mechanics and electromagnetic properties of matter,* Advances in Chem. Phys., **1,** 309–360, (1958).

111. Mazur, P. and B.R.A. Nijboer. *On the statistical mechanics of matter in an electromagnetic field,* Phyisca, **19,** 971–980, (1953).

112. Maxwell, J.C. *A Treatise on Electricity and Magnetism,* Oxford University Press, London, 1891 (2 volumes reprint by Dover, New York, 1954).

113. Michel, F.C. *Theory of pulser magnetospheres,* Rev. Mod. Phys., **54,** 1–66, (1982).

114. Minkowski, H. *Die Grund gleichunger fur die electromagnetisichen vorgange in Bewegten Korper,* Gottinger, Nachrichten, 53–116 (1908), English translation : Space and Time, p. 75–91 in the Principle of relativity, Dover Publication, New York, (1952).

115. Misner, C.W., K.S. Thorne and J. Wheeler. *Gravitation,* W.H. Freeman, San Francisco, (1973).

116. Minorskey, N. *Nonlinear oscillations,* Van Nostrand, Princeton, (1962).

117. Mo, T.H. *Theory of electrodynamics in media in noninertial frames and applications,* Jour. Math. Phys., **II,** 2589–2610, (1969).

118. Modesitt, G.E. *Maxwell equations in a rotating reference frame,* Amer. Jour. Phys., **38,** 1487–1489, (1970).

119. Moiseiwitsch, B. *1. Variational Principles,* Interscience Publishers, New York, (1966).

120. Moller, C. *The Theory of Relativity,* Oxford University Press, London, (1972).

121. Meierovich, B.E. *Electromagnetic collapse,* Equilibrium of a Dense Pinch, Phys. Reports, **92(3),** 83–133, (1982).

122. Macek, W.M. and D.T.M. Davis. *Rotation Rate Sensing with Travelling wave Ring Lasers,* Appl. Phys. Lett., **2,** 67–68, (1963).

123. Macek, W.M., J.R. Schneider and R.M. Salamon. *Measurement of Fresnel Drag with Ring Laser,* J. Appl. Phys., **35,** 2566–2577, (1964).

124. Macek, W.M. and E.J. McGartney. *The Ring Laser,* Sperry Eng. Rev., **19,** 8–15, (1966).

125. Menegozzi, L.N. and Lamb, W.E. Jr. *Theory of a ring laser,* Phys. Rev., A, **8,** 2103–2125, (1972).

126. Moffat, H.K., *Magnetic field generation in electrically conducting fluids,* University Press, Cambridge, (1978).

127. Montgomery, D.C. and D.A. Tidman, *Plasma Kinetic theory,* McGraw Hill, New York, (1965).

128. Nelson, D.E. *Electric, Optic and acoustic interaction in dielectrics,* Wiley, New York, (1979).

129. Newburgh, R.G. *Relativistic thermodynamics : Termperature transformations, invariance and measurement,* Nuovo Cimento, **52B,** 219–228, (1979).

130. Ostriker, J.P. and J.E. Gunn, *On the nature of Pulsars,* Astrophys. Jour., **157,** 1395–1417, (1969).

131. Parker, E.N. *Cosmic Magnetic Fields,* Clarendon Press, Oxford, (1979).

132. Pedersen, P.O. *The propagation of radio waves,* Denmarks Natus Vidense Kabellge Samfund, Copenhagen, (1927).

133. Panofsky, W.K.H. and M. Phillips, *Classical Electricity and Magnetism* (Addision Wesley, Reading Mass), (1964).

134. Parker, E.N. *Cosmic Magnetic field,* Clarendon Press, Oxford, (1979).

135. Pars, L. A. *A Treaties on Analytical Dynamics,* (1965).

136. Pegram, G.B. *Unipolar induction and electron theory,* Phys. Rev., **10, 591–600, (1917).**

137. Peng, Y.M. and F.W. Crawford. *Microscopic Lagrangian description of warm plasma I,* Formulation of the Lagrangian, Radio Science, **18,** 1355–57, (1983).

138. Piddington, J.H. *Cosmic Electrodynamics,* Wiley, New York, (1969).

139. Pirani, F.A.E. *Bull. Acad. Polona Science,* CCIII **V,** 145, (1957).

140. Penfield, P. Jr. and H.A. Haus. *Electrodynamics of Moving Media,* M.I.T. Press, Cambridge, Mass., USA, (1967).

141. Physics Today, *Physics and Astronomy Classification Scheme* – 1985, PACS 1–31, (after p. 96), December, (1985).

142. Plebanski, J. *Electromagnetic Waves in Gravitational Fields,* Phys. Rev. **118,** 1396–1408, (1960).

143. Post, E.J. *Formal Structure of electromagnetics* (North Holland, Amsterdam), (1962).

144. Privalov, V.E. and E.A. Fridrikov. *The ring gas laser,* Sov Phys. Uspekhi **12,** 153–167, (1969) (original Russian, Uspekhi, **97,** 377, (1969)).

145. Portis, A.M. *Electromagnetic fields,* Wiley, (1978).

146. Post, E.J. Geometry and Physics : *A Global Approach, In Problems in Foundations of Physics,* edited by M. Bunge (Springer Verlag, Berlin), p. 57–78, (1971).

147. Post, E.J. *Sagnac Effect,* Rev. Mod. Phys. **39,** 475–493, (1967).

148. Post E.J. *Non-inertial phenomena and Inertial sensing,* Klein Heubacher Ber. (Fermal detech, Zentralamt, Darmstadt, Germany), (1972a).

149. Post, E.J. *Electromagnetism and the Principle of Equivalence Annals of Phys.* (USA), (1972b).

150. Post, E.J. *The constitutive Map and Some of its Ramifications,* Annals of Physics (USA), **71,** 497–518, (1972c).

151. Post, E.J. *Ramifications of flux Quantization,* Phys. Rev. D, **12,** 3379–3386, (1974).

152. Post, E.J. *The metric dependence of four-dimensional formulation of electromagnetism,* Jour. Math. Phys. (USA), **25,** 612, (1984).

153. Post, E.J. and D.D. Bhaulikar, *Note on the electrodynamics of accelerated systems,* Jour. Math. Phys. **12,** 1098–1099, (1971).

154. Post, E.J. and A. Yildiz, *Cavity Resonance in Accelerated System,* Phys. Rev. Lett. **15,** 177–178, (1965).

155. Post E.J. *Linking and enclosing magnetic flux,* Phys. Lett. **A119,** 47–49, (1986).

156. Post, E.J. *Quantum Hall effect as experimentum crucis,* Phys. Lett. **A125,** 225–227, (1987).

157. Poverlain, M. *Sommerfeld Runge law in three and fours dimensions,* Phys. Rev. **128,** 957–964, (1962).

158. Post E.J. *Interferometric path length changes due to motion,* Jour. Opt. Soc. Amer., **62,** 234–239, (1972d).

159. Ray, B.S. *Lectures on Advanced Theory of Electricity and Magnetism,* Calcutta University (unpublished), (1957).

160. Rontgen, W.C. *Annder Physik and Chem,* **35,** 264, (1988).

161. Robinson, F.N.H. *Macroscopic electromagnetism*. Pergomon Press, Oxford, (1973).

162. Rosenfeld, L. *Theory of electrons*, North Holland, Amsterdam, (1951).

163. Rund, H. *Note on the Lagrangian formalism in relativistic Mechanics*, Nuovo Cimento, **23,** 227–232, (1962).

164. Russakoff, G. *A deviation of the macroscopic Maxwell equations*, Amer. Jour. Phys. **38,** 1188–1195, (1970).

165. Rohrlich, F. *Classical Charged Particles* (Addision Wesley, Reading, Mass), (1965).

166. Rowland, H.A. *Ann. der Physik*, CLVIII, 487, (1976).

167. Schiff, L.I. *A question of general relativity*, Proc. National Acad. Sci., U.S. **25,** 391–395, (1939).

168. Scheliffer, N. *Differential form as a basis for vector analysis with applications to electrodynamics*, Amer Jour. Phys. **51,** (12), 1139–1145, (1983).

169. Schouten, J.A. Ricci Calculus, *An Introduction to Tensor Analysis and its Geometrical Applications*, (Springer Verlag, Berlin), (1954a).

170. Schouten, J.A. *Tensor Analysis for Physicists*. (Oxford University Press, London) (1954b).

171. Schouten, J.A. and J. Hantjes, *Uberdie Konform invariante Gestatt Berungen*, Physica, **I,** 869, (1934).

172. Semon, M.D. and G.M. Schmieg, *Note on the analogy between inertial and electromagnetic forces*, Amer. Jour. Phys. **49,** 689–690, (1981).

173. Shafranov, V.D. *On magnetohydrodynamical Equilibrium configurations*, Jour. Eksp. Teoret. Fiz. (USSR), (in Russian), **33,** 710–722; *English translation* : Sov. Phys. (JETP) (1958), 545–554, (1957).

174. Shiozawa, T. *Phenomenological and electron theoretical study of electrodynamics of rotating systems*, Proc. IEEE **61,** 1694–1702, (1973).

175. Sard, R.D., *Relativistic mechanics*, W.A. Benjamin, New York, (1970).

176. Schiff, L.I., *Motion of a gyroscope according to Einstein theory of gravitation*, Proc. Nat. Acad. Science, USA, **46,** 871–882, (1960).

177. Shercliff, J.A., *A Text book of Magneto Hydrodynamics*, Pergamon Press, Oxford, (1965).

178. Shiozawa, T., *Constitutuve relations for a rotating electron plasma*, Proc., IEEE, **62,** 1283–1284, (1974).

179. Shiozawa, T., *Electrodynamics of rotating bodies, 19th General Assembly of USRI*, Helisinki, (1978).

180. Silverstein L. *Annalender Physik*, **22,** 579, **24,** 784, (1907).

181. Simon, A. and Harris, E.G., *Kinetic equations for plasma and radiation*, Phys. Fluids, **3,** 245–254, (1960).

182. Slater, J.C., *Microwave electronics* (Van Nostrand, Princeton), (1960).

183. Sokolov, A.A. and I.M. Ternov, *Syncotron Radiation,* Pergamon Press, Oxford, (1968).

184. Sommerfeld, A., *Electrodynamics,* (Acad. Press, New York), (1952).

185. Stamper, J.A., *Review on spontaneous magnetic fields in laser produced plasmas : Phenomena and Measurements,* Laser and Particle Beams, **9,** 841–862, (1991).

186. Stedman, G.E. *Ring laser tests of fundamental physics and geophysics,* Rept. Prog. Phys, **60,** 615–688, (1997).

187. Steele, M.C. and B. Vural, *Wave interaction in solid state plasma,* McGraw Hill, New York, (1969).

188. Steiger, A.D. and C.H. Woods, *Intensity dependent propagation characteristics of circularly polarized high power laser radiation in a dense electron plasma,* Phys. Rev. **45,** 1467–74, (1972).

189. Stratton, J.D., *Electromagnetic theory,* McGraw Hill, New York, (1941).

190. Sturrock, P.A., *A variational principle and energy theorem for small, amplitude disturbance of electron beams, and of electron-ion plasma,* Annals of Physics, (New York), **4,** 306–324, (1958).

191. Syng, J.L. and B.A. Griffith, *Principles of Mechanics,* McGraw Hill, New York, (1949).

192. Tai, C.T. *A review study of electrodynamics of moving media,* Proc., IEEE, **52,** 685–689, (1964).

193. Tamm, I.E., *Jour. of Russ. Phys. Chem. Soc.,* Physics Part, **56,** 248, (1924).

194. Tassoul, J.L., *Theory of rotating stars,* Princeton University Press, Princeton, USA, (1978).

195. Ter Haar, D and H. Wergerland, *Thermodynamics and statistical mechanics of the special theory of relativity,* Physica, **IC,** 31–54, (1971).

196. Ter Haar, D. *Introduction to the physics of many body system,* Interscience, New York, (1958).

197. Tolman, R.C. *The principles of statistical mechanics,* Oxford University Press, London, (1938).

198. Trocheris, M.G., *Electrodynamics in a rotating frame of reference,* Phil. Mag. **40,** 1143–1154, (1949).

199. Van Vleck, J.H., *The theory of electric and magnetic susceptibilities,* Oxford University Press, London, (1932).

200. Videc, M.F. and C. Kooy, *On the geometric optic formulation of radiation in a rotating medium,* SIAM Jour. Appl. Maths. **42,** 471–485, (1982).

201. Volkov, A.M. and V.A. Kisilev, *Proper Frequencies of a Rotating Ring Resonator,* Sov. Phys. JETP **30,** 733–736, (1970).

202. Webestenholtz, C. Van, *Differential forms in mathematical physics,* (North Holland, Amsterdam), (1981).

203. Webester, D.L. and R.C. Whitten, *Which electromagnetic equations apply in rotating coordinates.* Astrophys and Space Sci., **24,** 323–333, (1973).

204. Webester, D.L., *Relativity of moving circuits and magnets,* Amer. Jour. Phys. **29,** 262–268, (1961).

205. Webester, D.L., *Schiff's charges and currents in rotating matter.* Amer. Jour, Phys., **31,** 590–597, (1963).

206. Wells, D.A. *Lagrangian dynamics, Schaum's outline series,* McGraw Hill, New York, (1970).

207. Whitham, G.B., *Linear and nonlinear waves,* Wiley, New York, (1974).

208. Wilson, H.A and M. Wilson, *On the electric effect of rotating a magnetic insulator in a magnetic field,* Proc. Roy. Soc. (London), **89,** 99–106, (1974).

209. Wilson, H.A., *On the electric effect of rotating a dielectric in a magnetic field,* Phil. Trans. Roy. Soc. **204,** 121–137, (1905).

210. Yildiz, A. and C.H. *Tang, Electromagnetic cavity resonance in an accelerated systems of reference,* Phys. Rev. **146,** 947–954 (1966).

211. Yoler, Y.A., *A review of magnstohydrodynamics, in plasma physics,* edited by J.E. Drummond, Chapter 6, McGraw Hill, New York, (1961).

212. Yuen, C.K., *Lorentz transformation for thermodynamic quantities.* Amer. Jour. Phys. **38,** 246–252, (1970).

213. Zeldovich, Ya. B. and I.D. Novikov, *Relativistic Astrophysics,* Vol. I, *Stars and Relativity,* University of Chicago, Chicago, (1978).

Subject Index

Author Index